Intelligent Computation and Analytics on Sustainable Energy and Environment

Proceedings of ICICASEE 2023, GKCIET, Malda, West Bengal

Edited by

Dr. Amarjit Roy
Department of Electrical Engineering, Ghani Khan Choudhury Institute of Engineering and Technology (GKCIET), Malda, India

Dr. Chiranjit Sain
Department of Electrical Engineering, Ghani Khan Choudhury Institute of Engineering and Technology (GKCIET) Malda, India

Dr. Raja Ram Kumar
Department of Electrical Engineering, Ghani Khan Choudhury Institute of Engineering and Technology (GKCIET) Malda, India

Dr. Sandip Chanda
Department of Electrical Engineering, Ghani Khan Choudhury Institute of Engineering and Technology (GKCIET) Malda, India

Prof. Valentina Emilia Balas
Aurel Vlaicu University of Arad, Romania

Prof. Saad Mekhilef
Swinburne University of Technology, Australia

CRC Press
Taylor & Francis Group
Boca Raton London New York

CRC Press is an imprint of the
Taylor & Francis Group, an **informa** business

First edition published 2025
by CRC Press
4 Park Square, Milton Park, Abingdon, Oxon, OX14 4RN

and by CRC Press
2385 NW Executive Center Drive, Suite 320, Boca Raton FL 33431

CRC Press is an imprint of Informa UK Limited

British Library Cataloguing-in-Publication Data
A catalogue record for this book is available from the British Library

ISBN: 9781032888903 (pbk)
ISBN: 9781003540199 (ebk)

DOI: 10.1201/9781003540199

Typeset in Sabon LT Std
by HBK Digital

Contents

List of Figures

List of Tables

About the Book

The book of proceedings from the 1st International Conference on Intelligent Computation and Analytics on Sustainable Energy and Environment, GKCIET, Malda, West Bengal, India, is likely to hold significant importance for emerging areas of technology, particularly in the fields of sustainability, energy, and environmental science.

This conference has brought together experts, researchers, and practitioners from various domains. The proceedings document present discussions, and findings shared during the conference, providing a comprehensive repository of knowledge. This exchange of ideas fosters collaboration and innovation in emerging areas of technology.

The featured presentations of cutting-edge research and advancements in technology. The proceedings capture these developments, offering insights into the latest methodologies, algorithms, and applications in intelligent computation and analytics for sustainable energy and environment.

This proceeding will surely serve as valuable reference material for researchers, students, and professionals interested in the intersection of intelligent computation, analytics, and sustainability. They provide a curated collection of papers that cover a wide range of topics, from renewable energy systems and environmental monitoring to predictive modeling and optimization techniques.

Research presented in the proceedings may have implications for policy-making and decision-making processes related to sustainable energy and environmental management. Policymakers and stakeholders can leverage insights from the proceedings to inform strategies, regulations, and investments in emerging technologies.

The proceedings can also serve as educational resources for academic programs and professional development initiatives. They offer case studies, methodologies, and best practices that can enhance learning and training in fields related to sustainable energy, environmental science, and computational intelligence.

In summary, the book of proceedings from the 1st International Conference on Intelligent Computation and Analytics on Sustainable Energy and Environment plays a crucial role in advancing knowledge, fostering collaboration, informing policy, and supporting education in emerging areas of technology aimed at addressing global sustainability challenges.

Preface and acknowledgement

The 1st International Conference on Intelligent Computation and Analytics on Sustainable Energy (ICICASEE 2023) was held at Ghani Khan Choudhury Institute of Engineering & Technology (GKCIET), Malda, West Bengal, India. GKCIET is a premier engineering institute located in Malda, West Bengal, India. Being established in 2010, at present the institute offers B.Tech and Diploma Civil Engineering, Mechanical Engineering, Electrical Engineering, Computer Science and engineering and Food processing technology. The conference was aimed to provide a platform for researchers, academicians, industry professionals, and students to exchange knowledge and ideas on intelligent computation, analytics, and their applications in sustainable energy systems. The Department of Electrical Engineering of the institute hosted the conference from September 21–23, 2023. The conference received overwhelming response throughout the country and about 125 were submitted for possible presentation and publication in the conference. Among these papers after peer review about 80 papers were selected for presentation in the conference. The presentations were divided into five tracks – (i) AI/ML in Power Systems and Smart Grids, (ii) Applications of AI/ML in Power Converters and Industrial Drives, (iii) Intelligent Computation in Renewable Energy & Energy Storage, (iv) Applications of AI/ML in Signal Processing, and (v) Control and Communication and Environmental Sustainability using AI/ML. The conference commenced with an inaugural session on September 21, 2023, featuring esteemed guests Shri. Prashant Pole, Hon'ble Chairman BoG, GKCIET, Malda, Prof. Parameswara Rao Alapati, Hon'ble Director, GKCIET, Malda, Prof. Narendranath S., Hon'ble Director, NERIST, Prof. Sailendra Jain, Former Director SLIET, Longowal, and Professor MNIT, Bhopal, Dr. Marta Zurek-Mortka, Institute for Sustainable Technologies, Radom, Poland, Dr. Souvik Pal, Series Editor, CRC Press, Taylor and Francis (UK), Dr. Susanta Ray, Conference Chair IEEE Kolkata Section and Honorary Secretary IET, Assoc. Prof. Jadavpur University, Deans, Professors, Faculty, Staff, Students of GKCIET, Malda, and authors and participants and participants of the conference.

Apart from thought-provoking presentations from the paper presenters the conference witnessed keynote session from Prof. Sailendra Jain, Former Director SLIET, Longowal, and Professor MNIT, Bhopal, Dr. Marta Zurek-Mortka, Institute for Sustainable Technologies, Radom, Poland. The conference was fortunate enough to host some special invited lecture sessions from Prof. Shaikh Faruque Ali, IIT Madras, Prof. Abhinandan De, IIEST, Shibpur, Prof. R. H. Laskar, NIT Silchar, Prof. M. K. Bhuyan, IIT Guwahati, Prof. R. K. Saket, IIT (BHU) Varanasi, Dr. Santosh K. Singh, IIT (BHU) Varanasi and Dr. Arnab Ghosh, NIT Rourkela. The conference also welcomed the participation of renowned academicians and professionals as session chairs Prof. Kiran Yarrakula, GKCIET, Malda, Prof. Kshirod Kumar Dash, GKCIET, Malda, Dr. Souvik Pal, Series Editor, CRC Press, Taylor and Francis (UK), Prof. Sutanu Samanta, GKCIET, Malda, Dr. Koushik Paul, GKCIET, Malda, Dr. Kundan Kumar, NIT Manipur, Dr. Dipu Sarkar, NIT Nagaland, Dr Prem Prakash Mishra, NIT Nagaland, Dr. Alok Srivastav, JIS College of Engineering, Prof. Dalbir Singh, GKCIET, Malda and Dr. Subhajit Kar, IEM, Kolkata, Dr. Arnab Kumar Maji, North-Eastern Hill University, Dr. Dharmeswar Dash, GKCIET, Malda, Dr. Rinku Rabidas, Assam University, Silchar, Dr. Mohiul Islam, VIT, Vellore, Dr. Shekha Rai, NIT Rourkela, Dr. Devesh Shukla, NIT Calicut and Dr. Snehashis Pal, University of Maribor, Slovenia. The voluntary contribution of many other faculty members in the peer review process was the basis of the success of the conference.

We are sincerely thankful to Prof. Parameswara Rao Alapati, Hon'ble Director, GKCIET for supporting and standing at all times with us, whether it's good or tough times and given ways to concede us. "Sir without you we would have been lost". Starting from the Call for papers till the finalization of chapters, all the organizing committee members given their contributions amicably, we are thankful to all the Deans, HoDs, Officers, Faculty and staff of the institute for being there always. We thank our publication partner CRC press and technical sponsors the Institute of Engineering and Technology (IET, Kolkata section) and Institute of Electrical and Electronics Engineers (IEEE, Kolkata) and their association lifted the conference to pinnacle of success. We thank all the authors of the papers and participants for choosing this conference and sharing your innovative and scholarly ideas with us.

Finally, we would like to wish you to have good success in your presentations and social networking. Your strong supports are critical to the success of this conference. We hope that the participants not only enjoyed the technical program in conference but also found eminent speakers and delegates in the virtual platform.

Thanking you all,

The Editors,
Proceedings of ICICASEE, 2023

Organizing Committee and Key Members

Patron
Shri. Prashant Pole, Chairman BoG, GKCIET, Malda

Honorary Chair
Prof. P. R. Alapati, Director, GKCIET, Malda

Program Advisory Board
Dr. Kshirod Kumar Dash, GKCIET, Malda
Dr. Kiran Yarrakula, GKCIET, Malda
Dr. Koushik Paul, GKCIET, Malda
Mr. Subrata Roy, GKCIET, Malda
Dr. Shib Shankar Chowdhury, GKCIET, Malda
Dr. Dharmeswar Dash, GKCIET, Malda
Dr. Debashish Ghurui, GKCIET, Malda
Dr. Soutick Nandi, GKCIET, Malda
Dr. Rakesh Das, GKCIET, Malda

Program Chair
Dr. Sandip Chanda, GKCIET, Malda

Organizing Chair(s)
Dr. Raja Ram Kumar, GKCIET, Malda
Dr. Chiranjit Sain, GKCIET, Malda
Dr. Amarjit Roy, GKICET, Malda

Organization Committee Members
Dr. Surajit Chattapadhyay, GKCIET, Malda
Dr. Tapash Kumar Das, GKCIET, Malda
Mr. Goutam Kumar Ghorai, GKCIET, Malda
Mr. Amiungshu Karmakar, GKCIET, Malda
Mr. Pranab Mandal, GKCIET, Malda
Mr. Rajeev Kumar, GKCIET, Malda
Mr. Dhaju Mohhamad, GKCIET, Malda
Mr. Sankar Mukherjee, GKCIET, Malda
Ms. Smita Anand, GKCIET, Malda
Mr. Mehbub Alam GKCIET, Malda
Mr. Ayan Banik GKCIET, Malda
Mr. Amit Koley GKCIET, Malda

About the Editor(s)

Dr. Amarjit Roy received the M. Tech degree from NIT Silchar in 2014, Ph.D. degree from NIT Silchar in Electronics and Communication Engineering in the year of 2018. After that, he joined as an Assistant Professor in BML Munjal University in 2017. In July 2020, he joined as an Assistant Professor Sr. Grade 1 in VIT AP University. Now, he is working as an Assistant Professor at Ghani Khan Choudhury Institute of Engineering and Technology, Malda. His research interests include image noise removal, soft computing, biomedical image processing, etc. He has published many papers in journals such as IEEE, Elsevier, Springer, etc.

Chiranjit Sain (IEEE, Senior Member) is working as an Assistant Professor in the Electrical Engineering Department at Ghani Khan Choudhury Institute of Engineering & Technology (GKCIET), Malda. He has about 12 years of academic experience, 8 years of research experience and 1 year of industrial experience. His present research of interest includes senorless and vector control of permanent magnet motor drives, digital control in power electronics converters, electric vehicles, renewable energy systems. Additionally, he is guiding four (04) PhD scholars in joint collaboration with NIT Mizoram. He has published several papers in national/international level journals/conferences/book chapters and two books in the relevant field.

Dr. Raja Ram Kumar received the B. Tech. Degree in Electrical Engineering from the WBUT, Kolkata, India in 2010 and M. Tech. as well as Ph.D. Degree in Electrical Machines & Drives from the IIT (BHU), Varanasi, India in 2012 and 2018, respectively. He is currently working as an Assistant Professor in the Department of Electrical Engineering, Ghani Khan Choudhury Institute of Engineering & Technology, Malda, West Bengal. He has published several research papers in international/national journals and conference proceedings. His current research interests include Renewable Energy, Electrical Machines, Electric Drives, Power Electronic Converter Design and Control.

Dr. Sandip Chanda is a B.E from Jadavpur University in Electrical Engineering. He completed his M.Tech Degree in Electrical Engineering from applied Physics Department of Science college and he was awarded Ph.D. (Engineering) form IIEST, Shibpur (Formerly Shibpur B.E. College) in 2015. Dr. Chanda has 17 years of teaching experience including 13 years of research experience in the field of Electrical Power System. He has worked 2.5 years as Principal and 8 years as head of Electrical Engineering Department of reputed engineering colleges. He has published 46 journals and conference papers in Elsevier, Springer, IEEE and in other reputed publications. He has also published 2 books on Smart Grid Research and 4 book chapters available in IET digital library and other reputed publications. Currently he is working as Dean of Faculty Welfare and Head of Electrical Engineering in Ghani Khan Choudhury Institute of Engineering and Technology, Malda, a CFTI under Ministry of Education, Govt. of India. His research area includes – Power system Optimization, Smart Gird, Renewable Energy Sources and Micro Grid.

Valentina E. Balas is currently Full Professor in the Department of Automatics and Applied Software at the Faculty of Engineering, "Aurel Vlaicu" University of Arad, Romania. She holds a Ph.D. in Applied Electronics and Telecommunications from Polytechnic University of Timisoara. Dr. Balas is author of more than 270 research papers in refereed journals and International Conferences. Her research interests are in Intelligent Systems, Fuzzy Control, Soft Computing, Smart Sensors, Information Fusion, Modeling and Simulation. Dr. Balas is the director of Intelligent Systems Research Centre in Aurel Vlaicu University of Arad and Director of the Department of International Relations, Programs and Projects in the same university.

Prof. Saad Mekhilef is an IEEE and IET Fellow. He is a Distinguished Professor at the School of Science, Computing and Engineering Technologies, Swinburne University of Technology, Australia, and an Honorary Professor at the Department of Electrical Engineering, University of Malaya. He authored and co-authored more than 500 publications in academic journals and proceedings and five books with more than 40,000 citations, and more than 70 Ph.D. students who graduated under his supervision. He serves as an editorial board member for many top journals, such as IEEE Transactions on Power Electronics, IEEE Open Journal of Industrial Electronics, IET Renewable Power Generation, etc.

गनी खान चौधरी अभियांत्रिकी तथा प्रौद्योगिकी संस्थान
Ghani Khan Choudhury Institute of Engineering and Technology
(A Centrally Funded Technical Institute under the Ministry of Education ,Govt. of India.)
Narayanpur, Dist. Malda, PIN- 732141, West Bengal
Website : www.gkciet.ac.in

Date: 16.09.2023

Shri Prashant Pole
Hon'ble Chairman, Board of Governors
GKCIET, Malda

Message

It is with immense pleasure and anticipation that I extend a warm welcome to present to you the "1st International Conference on Intelligent Computation and Analytics on Sustainable Energy and Environment" to be organized from 21st to 23rd September, 2023 at Ghani Khan Choudhury Institute of Engineering and Technology, Malda, West Bengal, India. This conference represents a significant step towards harnessing the power of Artificial Intelligence (AI) and Machine Learning (ML)-based computational intelligence to address some of the most pressing challenges faced by our planet today. As we gather here, we stand at a critical juncture in human history. The need for sustainable development in the domains of energy and environment has never been more urgent. The impact of climate change, resource depletion, and environmental degradation is felt worldwide, underscoring the necessity for innovative solutions. Artificial intelligence and machine learning have emerged as powerful tools with the potential to revolutionize how we approach these challenges. From optimizing energy systems to predicting environmental trends and managing natural resources, the applications of AI and ML in this context are vast and promising. This conference brings together experts, researchers, and stakeholders from diverse backgrounds, all with a shared commitment to explore the possibilities and limitations of AI and ML in promoting sustainable development. Over the course of the conference, we will delve into topics ranging from renewable energy integration and smart grid technologies to climate modeling, conservation, and ecological monitoring. It is crucial that our discussions not only focus on technological advancements but also address the ethical, social, and policy implications of AI and ML in the context of sustainability. As we develop these powerful tools, we must ensure they are deployed in ways that are fair, inclusive, and respectful of our planet's fragile ecosystems. I want to express my deepest appreciation to the organizing team of GKCIET, the sponsors, CRC Press Taylor and Francis, UK, SERB, Govt. of India and participants who have come together to make this conference a reality. Your dedication to the cause of sustainable development is truly commendable, and I have no doubt that the ideas and collaborations that emerge from this gathering will pave the way for a brighter, more sustainable future. In closing, I encourage each of you to actively engage in the presentations, workshops, and discussions throughout this conference. Let us seize this opportunity to collectively advance our understanding of how AI and ML can be harnessed to protect our environment, conserve our resources, and ensure a better world for generations to come.

Thank you for your commitment to this important mission, and I look forward to the innovative solutions and partnerships that will undoubtedly emerge from our time together.

Sd/-
(Shri. Prashant Pole)

गनी खान चौधरी अभियांत्रिकी तथा प्रौद्योगिकी संस्थान
Ghani Khan Choudhury Institute of Engineering and Technology
(A Centrally Funded Technical Institute under the Ministry of Education ,Govt. of India.)
Narayanpur, Dist. Malda, PIN- 732141, West Bengal
Website : www.gkciet.ac.in

Date: 16.09.2023

Prof. Parameswara Rao Alapati
Director, GKCIET, Malda

Message

Computational modeling and simulation play a significant role in addressing sustainability challenges in the energy and environmental sectors. Sustainable energy and environmental systems are inherently complex, involving numerous variables and interactions. Computational models allow researchers to simulate these systems, understand their behavior, and predict outcomes under different scenarios. Artificial intelligence and machine learning-based computational tools can optimize various processes, such as energy production, distribution, and consumption. For example, they can find the most efficient configuration for renewable energy installations, reducing costs and resource use. These simulations can also assist in the design and optimization of energy storage systems, making them more efficient and cost-effective. These methods enable the analysis of vast datasets, including satellite imagery, sensor data, and climate records, to extract insights and trends relevant to sustainability and environmental research. Thus, computational approaches have become essential in recent past in the pursuit of sustainable energy and environmental goals. They enable researchers, policymakers, and engineers to make informed decisions, optimize processes, and develop innovative solutions to address the complex challenges of sustainability and environmental protection. I congratulate, the Department of Electrical Engineering for selecting this topic for organizing 1st International Conference on Intelligent Computation and Analytics on Sustainable Energy and Environment at Ghani Khan Choudhury Institute of Engineering and Technology, Narayanpur, Malda, India. The research works to be presented and discussed in the this conference will be undoubtedly helpful for developing innovative solutions to the present day problems in creating sustainable source of energy adhering to environmental concerns like global warming, pollution and waste management. Under the able leadership of our prime minister, the country is leading the world as "Vishuguru" in G20 summit with the message "One Earth, one family, one future" and this future lies on building sustainable green infrastructure which is not possible without substantial research on energy and environment. I thank CRC Press, Taylor and Francis UK for being the publication partner of this conference. I also thank IEEE, Kolkata section and IET, UK, Kolkata section for technically sponsoring this conference. Special thanks to SERB, Govt. of India for funding this conference and understanding the importance of the topic. I am sure that the outcome of this conference will be contributing and in realizing this dream of Govt. of India and the whole world into reality and also in understanding the meaning of what was written in Maha Upanishad "Vasudhaiva Kutumbakam" – "One earth one family".

Jai Hind

(Prof. Parameswara Rao Alapati)

1 Environmental sustainability using AI

Shib Shankar Chowdhury[a]

Department of Humanities and Social Sciences, Ghani Khan Choudhury Institute of Engineering & Technology, Malda, India

Abstract

Artificial intelligence (AI) has been playing a significant role in developing the areas of environmental issues such as water management, and biodiversity. The study has shown that AI can improve environmental sustainability by monitoring pollution, conserving natural resources, and renewing energy. Thus, AI can easily and innovatively manage environmental consequences. This report has developed a better understanding of the role and significance of environmental sustainability. The application of AI also has been discussed in the report. The impact and application of "The Internet of Things and AI" for environmental sustainability also have been highlighted. AI also has been playing a huge in improving the overall productivity and sustainability of this business. This report has highlighted the environmental sustainability of AI. The impact and role of AI on environmental sustainability has been focused on in this report. This report has highlighted the environmental sustainability of AI. The impact and role of AI on "environmental sustainability" have been focused on in this report. Artificial intelligence (AI) is harnessing the power of the grid and intelligent systems to manage energy demand and supply, such as accurate environmental forecasting, optimizing efficiency rates, reducing costs, and generating negligible pollution carbon.

Keywords: Environmental sustainability, AI, environmental management, robots.

Introduction

This report has been focused on the environmental sustainability of artificial intelligence (AI). The impact and role of AI on environmental sustainability have been discussed in this report. AI is a powerful tool that has capabilities and intelligent systems to manage the demand and supply of energy.

Sustainable technologies

Sustainable technologies focus on innovation to develop a social and economic structure as well as reduce ecological risks. AI has been playing a significant role in developing the areas of environmental issues such as water management, and biodiversity. By using the process of machine learning, biodiversity has improved in predicting the services of the ecosystem. Environmental management can be executed by using AI tools for rehabilitation and conservation purposes [8]. Thus, AI can easily and innovatively manage environmental consequences. For example, AI self-driving vehicles can reduce the emission of pollutants by 50% which can be effective in protecting the environment's sustenance.

Improve the conservation of renewable energy

Reduce pollution

Improve supply chain logistics

In addition, using AI in business organizations can be beneficial to improve the conservation of renewable energy and maintain supply chain management effectively. AI can magnificently improve production units, clean energy, and sanitation, and can provide a healthy working ambiance. Therefore, AI can develop the working culture and can provide a sustainable environment with economic growth.

Quick developments in digital technologies have paved the path of innovation in the fields of agriculture, water management systems, and reducing pollutants from vehicles. AI can improve environmental sustainability by monitoring pollution, conserving natural resources, and renewing energy [9]. Hence, AI has improved the functioning of energy storage, reduced carbon pollution, and improved accuracy in weather forecasting. The study has revealed that most of the population has preferred to use electric cars to improve environmental conditions. Therefore, using AI can reduce air pollution and traffic congestion, and supply chain logistics can be improved efficiently. The study has depicted that AI can detect crop diseases easily which can be effective for the farmers to solve the agricultural issues quickly. AI robots in agricultural work can reduce the use of pesticides [1]. Hence, AI applications have improved the development of the environment with innovative techniques.

AI helps to reduce disaster risk by predicting the effects of disasters like floods, earthquakes, etc. AI helps to increase the sector's resource efficiency and decrease the effect on the environment. As per the

[a]shibshankar@gkciet.ac.in

DOI: 10.1201/9781003540199-1

view of Xiang et al. (2021), the application of AI is important to execute the implication of ethics by using AI for the sustainability of the environment and transparently secure the technology use. The use of AI in environmental sustainability has developed the understanding and ability to manage the ecosystem. The technology of AI is used to gather and analyze a huge amount of biodiversity data. The machines are being used in water science to develop the quality and discharge the forecasting.

Research data

Artificial intelligence is applying powerful capabilities and intelligent systems of the grid to manage the demand and supply of energy such as proper environment forecast, optimize the rate of efficiency, and reduce the cost and unimportant pollution generated the carbon. As per the opinion of Feroz et al. (2021), this helps to develop the storage of energy, efficiency, and management and assists in integration and reliability. This helps to facilitate the dynamic cost and trading outcome in the incentives of the market. The uses of AI optimize the forecasting of environmental outcomes in development by around 30% [4]. The outcome of the management of the plant increases the production of energy and decreases the emission of carbon. The benefits of EVs are defects in the environment and GHG emissions by the car throughout the sources of changing climate. The negative issues of health impart the emissions of GHG in their lifespan. AI integrated the sense of changing the use of land, forest clover, the fallout of disasters, and detecting diseases which is possible with the use of AI-generated agriculture. As per the insights of Van Wynsberghe (2021), this system is involved with automation to ensure the corrective and collection of data and makes decisions. This helps to streamline the input of agriculture based on the demand and supply outcome in developed environments and enhances the efficiency of resources in the industry.

The decreasing rate of use of pesticides, fertilizers, and water devastated the ecosystems by integrating the system of AI [6]. The data on the environment and quality of air adopt the rate of efficiency of filtration. The simulation of the efficiency of AI allows for sources of pollution to be detected before time. This is the way to take the time and place through the LED lighting and compare the form of a tradition of lighting. Smart LED lighting offers the resources to use energy to stay last longer. AI refers to the systems that perform the work that need intelligence and develop them depending on the data to collect the data. As per the thoughts of Nishant et al. (2020), AI plays an important function in tracking the challenging life environment from designing the efficiency of energy to monitoring deforestation for optimizing the deployment of energy. This platform operates the database and is verified by the emissions of methane. AI strategically related to the data with the policy, transparency for informing the decision of data-driven factors. The technology of IMEOs allows the integration of diversity to the emissions of methane stream to make the copy of verified emissions of methane at the rate of accuracy.

Analysis

The advancement of technologies has been playing a key role in the sustainability of the environment. As per the remarks of Rosário and Dias, "Artificial Intelligence" has been considered one of the most effective tools of technology which have been widely adopted by "business organizations" for their "sustainability practices" [7]. In addition to this, the main purpose of using AI is to increase the efficiency and effectiveness of "business activities, operations, and tasks" [2]. As a result, it helps "business organizations" to make minimum use of resources, raw materials, and labor which also helps to save a huge amount of funds.

On the other hand, the advanced tools and software of "Artificial intelligence" are also used mostly for improving the "supply chain management" so that the company can make sure to reduce their waste. The opinions of Musleh Al Sartawi et al. have suggested that AI also has been largely adopted for cost-cutting, increased efficiency, and operational optimization [5]. Moreover, it has been found that there has been a significant increase in the implementation of AI and other technological tools for the sustainability of the environment.

Besides that, AI tools also have the capability and potential to predict the time and date of delivery which helps the suppliers to improve the management of supply and demand without making zero waste. The views of Wu et al. have noted that the primary goal of AI is to provide "accurate forecasting of weather" to prevent any kind of disruption in the supply chain [9]. In contrast, there is a wide range of technological tools and modern software that helps companies to reduce the emission of greenhouse gases, release of carbon, and other harmful substances. However, it has been that the "business organizations" are responsible and accountable for reducing the negative impact of the "business activities, operations, and tasks" on the "society, the environment, and the world" as a whole. In this same context, the application of AI also has been playing huge in improving the overall productivity and sustainability of this business. According to the opinions of Tsolakis et al., the application of AI tools has also helped companies to recycle waste materials and achieve effective management of waste [8].

As a result, business organizations also have been determined to achieve "renewable energy" and make fewer negative impacts on the environment [3]. On the other hand, AI for optimizing their weather forecasting has been considered as the key to achieving effective management of their plants, boosting renewable energy production, and decreasing carbon emissions. Furthermore, "powerful predictive capabilities and intelligent grid systems" also has been considered to be one of the most efficient tools for monitoring and tracking the exact locations of the trucks to manage the supply and demand of renewable energy. As per the ideas of Musleh Al Sartawi et al., "sustainable technologies" have a significant positive impact on the well-being of "society, and the environment" [5].

Conclusions

Thus, it can be concluded that "environment sustainability" has been considered one of the core parts of a business which helps them to lead to a positive impact on the environment. Henceforth, "business organizations" also need to adopt various kinds of strategies and policies to make sure that the supplies are also reducing overall wastes. In this context, the application of AI helps a company monitor and track the delivery time, day, and location of the products. As a result, effective management of inventory and supply chain helps to minimize waste materials.

Acknowledgement

My deepest gratitude goes out to everyone who has assisted me in bringing this research project to fruition. I would first like to express my gratitude to everyone who has assisted me in gathering data for the study. I want to express my sincere gratitude to my professors for their assistance. They have aided in my comprehension of the subject and have guided me in drawing conclusions from my research. I also want to express my gratitude to my friends and fellow teammates who helped me sufficiently to accomplish a specific objective. I declare myself solely responsible for any shortcomings in this research and thank my classmates, supervisors, and professors for their support.

References

1. Bharany, S., Sharma, S., Bhatia, S., Imam Rahmani, Md. K., Shuaib, M., and Lashari, S. A.. (2022). Energy efficient clustering protocol for FANETS using moth flame optimization. *Sustainability,* 14(10), 6159.
2. Dash, B. and Sharma, P. (2022). Role of artificial intelligence in smart cities for information gathering and dissemination (a review). *Acad. J. Res. Sci. Publish.*, 4(39), 58-75.
3. Javaid, M., Haleem, A., Singh, R. P., and Suman, R. (2022). Artificial intelligence applications for industry 4.0: A literature-based study. *J. Indus. Integ. Manag.*, 7(01), 83–111.
4. Kaack, L. H., Donti, P. L., Strubell, E., Kamiya, G., Creutzig, F., and Rolnick, D. (2022). Aligning artificial intelligence with climate change mitigation. *Nat. Climate Change,* 12(6), 518–527.
5. Al-Sartawi, M., Abdalmuttaleb, M. A., Hussainey, K., and Razzaque, A. (2022). The role of artificial intelligence in sustainable finance. *J. Sustain. Fin. Inves.* 1–6.
6. Onyelowe, K. C., Denise-Penelope, N. K., Ebid, A. M., Dabbaghi, F., Soleymani, A., Jahangir, H., and Nehdi, M. L. (2022). Multi-objective optimization of sustainable concrete containing fly ash based on environmental and mechanical considerations. *Buildings,* 12(7), 948.
7. Rosário, A. T. and Dias, J. C. (2022). Sustainability and the digital transition: A literature review. *Sustainability,* 14(7), 4072.
8. Tsolakis, N., Zissis, D., Papaefthimiou, S., and Korfiatis, N. (2022). Towards AI driven environmental sustainability: an application of automated logistics in container port terminals. *Inter. J. Prod. Res.*, 60(14), 4508–4528.
9. Wu, C.-J., Ramya, R., Gupta, U., Acun, B., Ardalani, N., Maeng, K., Chang, G., et al. (2022). Sustainable AI: Environmental implications, challenges and opportunities. *Proc. Mac. Learn. Sys.* 4, 795–813.

2 A demand response algorithm to improve transient stability margin for smart electrical grid

Shubham Anand[1,a], *Gaurang Ashok Humne*[1,b], *Ravi Kumar*[1,c], *Suparna Maity*[2,d], *Papun Biswas*[3,e], *Sandip Chanda*[1,f] *and Abhinandan De*[4,g]

[1]Department of Electrical Engineering, Ghani Khan Choudhury Institute of Engineering and Technology, Malda, India

[2]Department of Electrical Engineering, Global Institute of Management & Technology, Krishnanagar, India

[3]Department of Electrical Engineering, JIS College of Engineering, Kalyani, India

[4]Department of Electrical Engineering, Indian Institute of Engineering Science and Technology (IIEST), Shibpur, Howrah, India

Abstract

This work presents a demand response (DR) algorithm for the improvement of transient stability in smart grid in pre-perturbation condition. The operating point in a smart power grid is not only a balance between supply and demand but also optimally of price of electricity, congestion, transmission loss, voltage profile and use of renewable energy sources (RES) needs to be taken care of. Introduction of dynamic loads and RES may perturb the operating point as they vary throughout the day for techno-economic reasons and weather. This introduces transient stability problem in the grid. The proposed algorithm takes care of rotor angle stability in a way that it minimizes the relative deviation of rotor angles of the generators of the grid by optimally scheduling the generation and the load demand. This assists the machines to stay in synchronism and also helps in reaching a stable operating point quicker is post perturbation condition. The proposed algorithm referred here as transient stability constrained social welfare optimization as it also causes optimal benefit simultaneously to GENCOs, TRANSCOs and DISCOs adhering to transient stability constraint of the grid. This work also validates the results of the optimization (optimal generation and load schedule), demonstrating how the proposed solution provides shorter time for rotor angle restoration and comparatively lesser relative deviation of rotor angles of the generators during a few cases of simulated faults in MATLAB SIMULINK environment. For the complexity of the objective function a differential evolution modified quantum particle swarm optimization (DEQPSO) algorithm has been proposed in this work.

Keywords: Renewable energy sources (RES), smart grid, optimization framework, social welfare, demand response (DR) programme, DEQPSO

Acronyms

C_T = Cost in Rs/hr
N_g = Generator number
V_{min} = Per unit bus voltage (minimum)
P_{ijmax} = Line ij maximum flow limit in MW
P_{lmax} = Maximum active power loss
P_1, P_2, P_3 = Penalties for optimization in Rs/MW
C_n = Generation cost of nth unit
T_{lmaxj} = The minimum limit of active power loss of jth transmission line
γ_i = Coefficient of consumer cost benefit function in Rs
P_{1ij} = The penalty in Rs./MW for excess flow of power
P_{2ij} = The penalty in Rs./MW for excess transmission loss
P_3 = The penalty in Rs./MW for excess deviation of voltage
P_4 = The penalty in Rs./MW for demand imbalance
δ_{max} = Maximum deviation of power angle from center of inertia
δ_i = Power angle of ith generator
P_{gi} = Generation in MW of bus i
P_{di} = Requested or scheduled demand in MW at bus i
Q_{gi} = Reactive power generation in MVAr at ith bus
Q_{di} = Reactive power demand in MVAr at ith bus
V_i = Bus voltage at ith generating unit
G_{ij} = Conductance of line ij
B_{ij} = Susceptance of line ij
θ_{ij} = Impedance angle
P_{gi}^{min}, Q_{gi}^{min} = Minimum active and reactive power at ith generating unit
P_{gi}^{max}, Q_{gi}^{max} = Maximum active and reactive power at ith generating unit
P_{ij}^{max}, P_{ij}^{min} = Minimum and maximum power flow in line ij in MW

[a]Anandshubham813@gmail.com, [b]gauranghumne@gmail.com, [c]Kumarravipatel2001@gmail.com, [d]sm.suparna@gmail.com, [e]head_ee.jisce@jisgroup.org, [f]sandipee1978@gmail.com, [g]abhinandan.de@gmail.com

DOI: 10.1201/9781003540199-2

V_i^{max}, V_i^{min} = Minimum and maximum voltage at ith generating unit

1. Introduction

Smart grids represent a unique generation of power systems that incorporate advanced technologies to improve efficiency and sustainability. The grids use advanced sensors and control systems to monitor, analyze, and control power generation, transmission, and distribution in real-time. Smart grids combine these technologies to maximize energy efficiency, reduce energy waste, and improve power system performance.

Power system stability is a key challenge in smart grid operation when loads, generation sources, and other dynamic factors change. To address this challenge, transient stability constrained optimal power flow (TSCOPF) can be used. To ensure continued system stability, TSCOPF optimizes generation and transmission of electricity based on the dynamic characteristics of the power system, including voltage and frequency fluctuations.

In smart grids, renewable energy sources like wind and solar power pose significant challenges due to their intermittency and unpredictability. By optimizing the distribution and utilization of renewable energy sources and taking into consideration the transient stability constraints of the system. TSCOPF is able to mitigate these challenges. For example, renewable energy sources can be managed in order to match demand, energy storage systems can be utilized to smooth out fluctuations in energy supply, and electric vehicle charging can be controlled to avoid overloading the grid.

Furthermore, TSC-OPF facilitates the efficient operation of demand response, energy storage, and electric vehicle charging, which are essential components of smart grids. Smart grid operators can achieve optimal system performance by integrating these components into the TSC-OPF framework while maintaining stability and reliability. As a result, energy costs can be reduced, system efficiency increases, and carbon emissions can be reduced.

2. Literature review

Transient stability constrained optimal power flow (TSCOPF) is a powerful optimization technique that has gained increasing attention in recent years for its potential to address the challenges associated with the integration of renewable energy sources (RES) in smart grids. In this literature review, we will examine some of the recent studies that have investigated TSC-OPF in the context of smart grids with renewable energy sources.

Francisco Arredondo et al., (2022) suggests a novel approach to addressing the issue of non-synchronous renewable energy generation and its impact on system stability during critical events. The proposed method involves incorporating synthetic inertia into the TSCOPF formulation. By doing so, the optimization problem can determine the optimal operating point of the system while accounting for a significant portion of non-synchronous renewable generation. The study's findings indicate that the use of synthetic inertia in renewable power plants can reduce electromechanical oscillations, lowering the cost of maintaining transient stability during critical events. Furthermore, the system's stability improves when traditional generation methods are decommissioned due to renewable promotion policies. Finally, the proposed model can calculate the parameters of the synthetic inertia control, making it a useful tool in optimizing system performance [1].

Z. Zou et al. (2020) introduces a new model for reactive optimal power flow (ROPF), which incorporates short-term voltage stability constraints. The paper starts by discussing the phenomenon of transient overvoltage and the causes of DC commutation failure. The article then examines the effect of allocating SCs (shunt capacitors) to reduce transient overvoltage. To prevent photovoltaic (PV) plants from tripping due to transient overvoltage, the paper introduces short-term voltage stability constraints based on the high-voltage ride-through (HVRT) requirements of PV plants. These constraints are linearized using trajectory sensitivity and integrated into the ROPF model. Finally, the paper presents simulation results based on the Qinghai power grid, demonstrating the effectiveness of the proposed model [2].

R. de Girardier focuses on optimal operation of microgrids, which face challenges due to the high penetration of renewable generation and increasing interconnections with the main grid. To address these challenges, the paper presents a TSCOPF model that considers system dynamic constraints governed by relevant differential equations. The proposed model is then tested on a generic, islanded microgrid consisting of both conventional and renewable generators to demonstrate the benefits for its transient performance. The study shows that the inclusion of transient stability constraints enables the determination of a dispatch plan that avoids violating any stability limits, thereby enhancing the system's overall stability.

D. M. Barbosa de Siqueira et al (2022) presents a TSCOPF formulation to determine the optimal operating point of synchronous generators in distribution networks concerning the dispatch of their active and reactive powers. However, the complexity of the problem arises due to the nonlinear constraints and a high number of variables. To overcome these challenges, the paper proposes an optimization problem that employs a set of mathematical approximations in the constraints of the active and reactive power balance equations. Moreover, the swing equation of synchronous generators belonging to a group of coherent machines is approximated concerning the active power injection from the generators. The proposed TSCOPF model is then tested on a 31-bus radial distribution system with two and four generators, showing that the running time to solve the optimization problem with approximations is significantly reduced compared to the time required without them.

B. Zaker et al. describes an approach to increase the system's critical clearing time (CCT) in order to improve transient stability. The formulation of optimal power flow with transient stability constraints and the search for a practical solution to this problem have recently received a lot of attention. Because it alters the power flow of the line and the output powers of the generators, this improvement has an impact on the prices of electricity in the electricity market as well as local marginal prices (LMP). DIgSILENT software is utilized to evaluate the test system's transient stability, and MATLAB software is utilized to calculate LMPs. The proposed method for improving transient stability and its effects on LMPs are examined using the New England 39-bus test system.

G. Geng proposes GPU acceleration as a coprocessor for TSCOPF. GPUs can be used as plug-and-play coprocessors for time-consuming linear algebra operations in TSCOPF solving thanks to the revealed two-level decomposition parallelism in the reduced-space interior point method. The efficiency of solving TSCOPF is greatly enhanced by the use of multi-GPU processing and the mixed-precision iterative refinement technique without having to redesign and implement the existing algorithm framework. The proposed GPU-based approach is effective for large-scale TSCOPF problems, according to numerical studies based on a series of test cases with up to 1,2, 9,5,1 buses.

In view of this literature survey this work compares all the available algorithms of generation and demand rescheduling and selects the most suitable algorithm for power system optimization adhering to transient stability constraints. The benchmark IEEE 30 bus system was tested with the results of optimization algorithm used in MATLAB SIMULINK platform. The results and outcome looked promising.

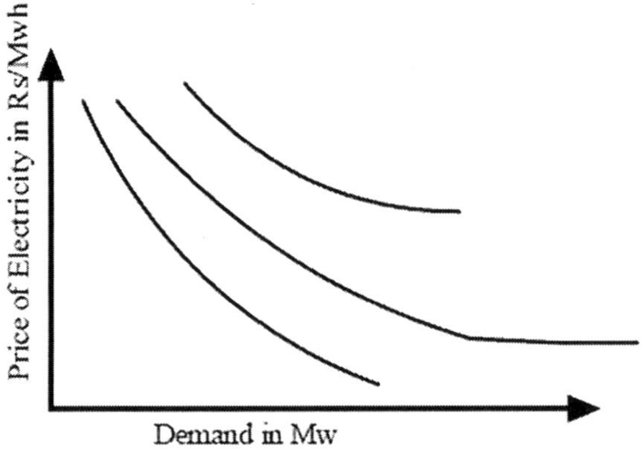

Figure 2.1 Price elasticity of demand

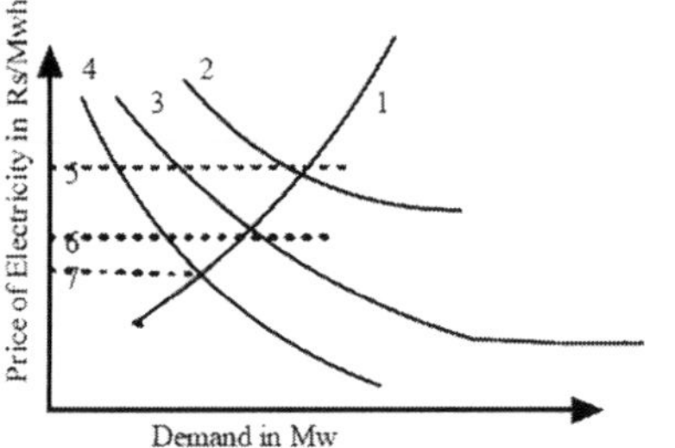

Figure 2.2 Nodal pricing with price elasticity of demand

2.1 Social welfare optimization with transient stability constraint

The price dependent characteristics as per willingness to pay of the consumers are depicted in Figures 2.1 and 2.2.

The objective of social welfare maximization algorithm (SWM) is to optimize electricity delivery to the consumers and simultaneously profiting GENCOs, TRNSCOs, DISCOs. Equation 1 shows the objective function of the same ("SWMWP").

$$\text{Maximize } f(C_n, P_{lmaxj}, T_{Lmaxj}, V_{min}) = \sum_{i=1}^{i=n_d} \alpha_i P_{di}^2 + \beta_i P_{di} + \gamma_i - \Big(\sum_{n=1}^{n_g} C_n + \sum_{j=1}^{nlns} P_{lmaxij} \cdot p_{1ij} + \sum_{ij=1}^{nlns} T_{Lmaxij} \cdot p_{2ij} + V_{min} \cdot p_3 + P_4 . Sqrt\left(\frac{\sum_{k=1}^{ng} (\text{Coi} - \delta \text{i})^2}{ng}\right)\Big) \tag{1}$$

The first part of this objective function is the consumer cost benefit function and the second part is the constrained generation cost function. The second part as visible, includes the transient stability constrain which tries to minimize the overall deviation of power angles of the generators from center of inertia CoI. If all the generators continue to operate in coherence, the re-dispatching results can maintain adequate transient stability margin for the power system.

Equation 2 depicts CCurt which is the limit of load curtailment as developed in ref [24]. As the demand is variable here and as the generation can also be variable owing to intermittent nature of RES, this equation calculates the permissible limit of load curtailment that can ensure minimum requested demand of the consumers.

$$\text{CCurt}=\frac{\sum_{i=1}^{ng} P_{gimax}-\sum_{i=1}^{ng} P_{gi}}{\sum_{i=1}^{ng} P_{gimax}-\sum_{i=1}^{ng} P_{gimin}} x\ (Pdimax - Pdimin) \quad (2)$$

2.2. The general operation of all the algorithms tested

Differential evolution modified quantum particle swarm algorithm (DEQPSO) has been used for solving all the constrained optimization problems proposed and tested in this work. The flow chart in Figure 2.3 represents the steps of DEQPSO algorithm.

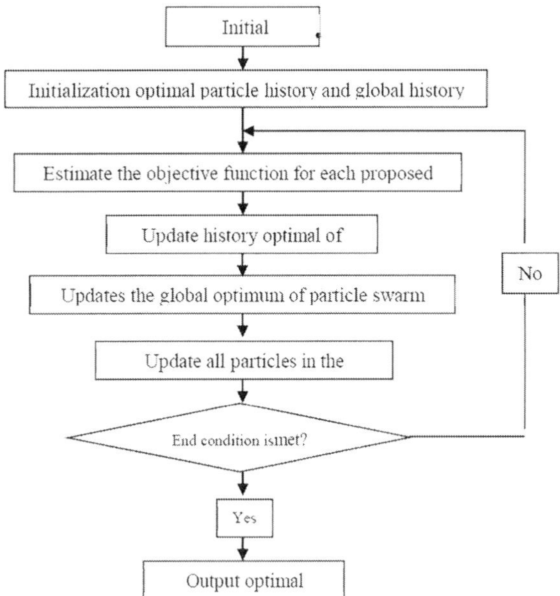

Figure 2.3 The flowchart of quantum particle swarm algorithm

Result

The description of the modified IEEE 30 bus test system has been presented in Table 2.1 and Figure 2.3.

Table 2.1 IEEE 30 bus systems

System parameters	Description
No. of buses	15
Generator	6
Synchronous condensers	3
Load buses	24
Transmission lines of which 4 branches (6–9), (6–10), (4–12), and (28–27) are with the tap setting transformer	41
Total demand(MW)	554.6

The generator cost coefficients of 6 generators of the network and price responsive consumer characteristics of 27 load centers/consumers is given in Tables 2.2 and 2.3.

The algorithms proposed and tested have been referred here as "GenCostMinWOP", "GenCostMinWP", "LossMinWP", "VolMaxWOP", "DeltaSDO" and "SWMWP". The perturbation of the power angles of the generators of the healthy system has been depicted in Figure 2.4. In case of the rotor angle standard deviation minimization (DeltaSDO) least nonconformity of rotor angles has been observed while the voltage maximization algorithms (VolMaxWOP) nearly fails to maintain the coherence in the generators.

The social welfare maximization with transient constraint penalty (SWMWP) provides second best solution and successfully minimizes both deviation and relative deviation of rotor angles.

Figure 2.5 depicts the comparison of these proposed algorithms with respect to their conservative load curtailment solution. It can be observed that the load curtailment is less than the limit of load curtailment (CCurt) in case of generation cost minimization algorithm (GenCostMinWOP) and social welfare maximization algorithm (SWMWP).

To validate further the schedule of these algorithms were forced into a SIMULINK model of IEEE 30 bus system. Use of SIMULINK model was compelling as we wanted to verify the transient response of these schedules. Near bus no. 2, generator (a three phase) to ground fault was simulated between 0.02 s and 0.1 s as shown in Figures 2.6 and 2.7. After withdrawal of the fault, the social welfare maximization algorithm depicts lesser deviation of rotor angle and also comparatively lesser recovery time.

Conclusions

This work tests some traditional algorithms and one proposed algorithm used for power system generation and load scheduling in presence of demand response in smart grid. The primary objective was to minimize the deviations of the rotor angles/

Table 2.2 Cost coefficients of generators

Bus No.	Output of the generator in MW		Generator cost coefficients		
	Min	Max	a_i (Rs/MW2)	b_i (Rs/MW)	c_i (Rs)
1 (Thermal)	100	200	0.0071	7.3335	1881
2 (Thermal)	10	35	0.0083	8.22	2020
5 (Thermal)	15	50	0.092	10.11	2350
8 (Wind)	50	115	0	1.1682	2886.1
11 (Solar)	50	80	0	1.308	2320
13 (Thermal)	10	30	0.01	11.2	2500

Table 2.3 Cost coefficients of consumer/LDC cost benefit function

Bus No.	Load dispatch		Co-efficient of benefit function		
	Min	Max	α_i (Rs / MW2)	β_i (Rs / MW)	γ_i (Rs)
2	11.7	21.2	-0.2	67.1	0
3	3.2	5.5	-0.10	70.5	0
4	8	10.6	-0.3	65.5	0
5	114.2	215	-0.2	87	0
7	12.7	40.2	-0.035	60	0
8	20	55	-0.21	70	0
10	7.8	10.6	-0.35	78	0
12	10.5	15.8	-0.16	60	0
14	5	17.5	-0.125	60.6	0
15	7.2	10.5	-0.25	50	0
16	2.5	15	-0.25	60	0
17	5.5	14.5	-0.005	55.5	0
18	2.5	6.2	-0.17	90.2	0
19	8.25	10.1	-0.255	56.2	0
20	3.15	5	-0.15	80	0
21	27.5	60	-0.152	63.8	0
23	2.5	5.2	-0.52	75	0
24	3.8	8.5	-0.12	70	0
26	2.5	5.7	-0.13	70.2	0
29	1.2	5.5	-0.15	82	0
30	11.2	17	-0.12	72	0

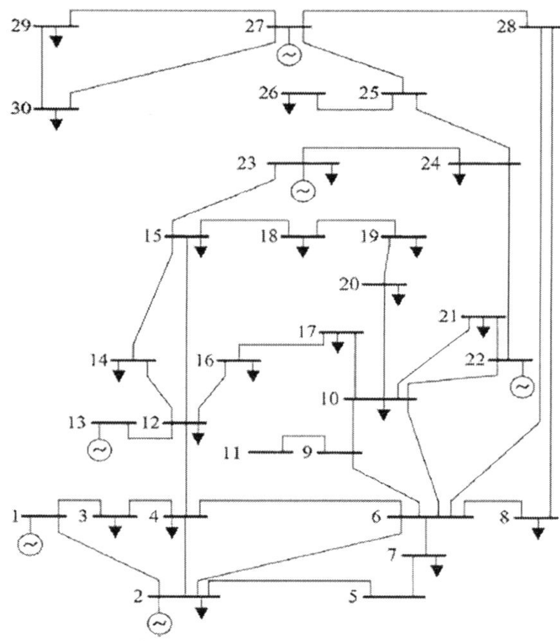

Figure 2.4 The IEEE 30 bus system

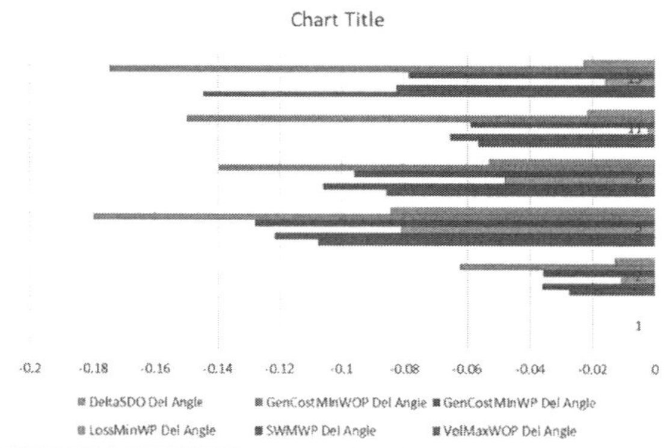

Figure 2.5 Deviation of rotor angles of the generators

power of the generators, so that the coherence in the operation of the generators may get improved, thus, improving transient stability of the system. Among the six algorithms tested, the social welfare maximization algorithm not only ensures load curtailment and other economic and technical parameters within permissible limit, but also minimizes the deviation of the rotor angles of the system. For validation of the work the results of the scheduling

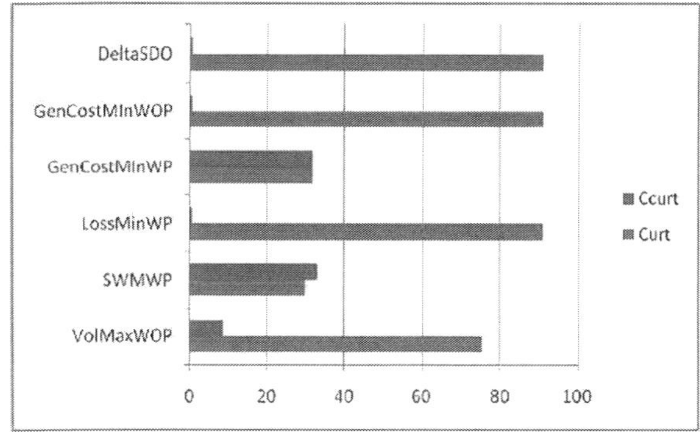

Figure 2.6 Load curtailment in different algorithms

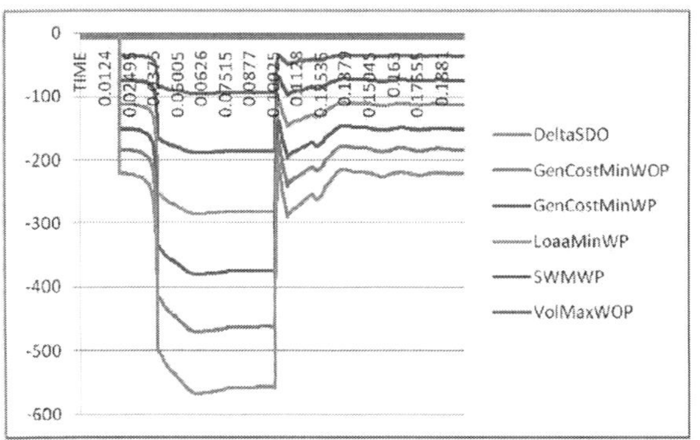

Figure 2.7 Comparison of recovery time of different algorithms

were imposed in SIMULINK model of IEEE 30 bus system in presence of a fault and it showed that the system recovery is much faster in case of the proposed social welfare optimization algorithm.

Acknowledgement

The authors gratefully acknowledge the students, staff, and authority of Electrical Engineering Department for their cooperation in the research.

References

1. Al-Omar, B., Al-Ali, A. R., Ahmed, R., and Landolsi, T. (2012). Role of information and communication technologies in the smart grid. 3(5), 707–716. ISSN 2079-8407
2. Zhaou, W. and Jianhui, W. (2015). Self-healing resilient distribution systems based on sectionalization into microgrids. *Power Sys. IEEE Trans.*, 30(6), 1–11. DOI:10.1109/TPWRS.2015.2389753.
3. Ruofei, M., Chen, H.-H., Huang, Y.-R., and Meng, X. (2013). Smart grid communication: Its challenges and opportunities. 4(1), 36–46. DOI: 10.1109/TSG.2012.2225851, *INSPEC Accession* Number: 13355369.
4. Phebe, A. O. and Samuel, A.-S. (2016). A review of renewable energy sources, sustainability issues and climate change mitigation. 3(1), 2016. https://doi.org/10.1080/23311916.2016.1167990.
5. Hirotaka, T., Asuka, K., Hisao, T., and Atsumi, O. (2018). A basic study on incentive pricing for demand response programs based onsocial welfare maximization. 8(1), 136–144. https://doi.org/10.1080/22348972.2018.1477092.
6. Behzad, M., Abbas, K., Seyed Ali, N. N., and Seyed, Y. D. (2022). A novel approach for transient stability constrained optimal power flow using energy function sensitivity method. 13(3), 13–24. http://dx.doi.org/10.22108/isee.2021.128735.1473.
7. Jianqiang, L., Fei, T., and Siqi, B. (2020). Stability-constrained power system scheduling: A review. 8, 219331–219343.
8. Al-Bahrani, L. T. (2020). Transient stability improvement based on optimal power flow using particle swarm optimization. 870, 012118. doi:10.1088/1757-899X/870/1/012118.
9. Deqiang, G., Robert, J. T., and Ray, D. Z. (). A transient stability constrained optimal power flow. 129–134, August 22-27, 2004.
10. Guangchao, G., Shrirang, A., Xiaoyu, W., and Dinavahi, V. (2017). Slution technique fortransient stability-constrained optimal power flow. 11(12), 3186–3193, https://doi.org/10.1049/iet-gtd.2017.0346.
11. Amel, Z., Louis, A. D., Ridha, H., Robert, T. F. Ah, K. and Innocent, K. (2014). Statistical approach for transient stability constrained optimal power flow. 9(14), 1856–1864. ISSN 1751-8687, doi: 10.1049/iet-gtd.2014.0689www.ietdl.org.
12. Deqiang, G., Robert, J. T., and Ray, D. Z. (2000). Stability-constrained optimal power flow. *IEEE Trans. Power Sys.*, 15(2), 535–540.
13. Maniraj, P. and Yuvaraj, M. (2018). Transient stability analysis for 6 bus system using E-TAP. 118 (20), 2277–2285. ISSN: 1314-3395 (on-line version), url: http://www.ijpam.eu.
14. Anulekha, S., Prasenjit, D., Aniruddha, B., and Boonruang, M. (2019). SSA - A new meta-heuristic algorithm for solving transient stability constrained optimal power flow. 1(02), 1–11. http://paper.ieti.net/SCEE/.
15. Yousef, O., Salem, A., and Mohammad, A. A. (2016). Improved PSO applied to the optimal power flow with transient stability constraints. *JES 2016 On-line*. 672–686. journal/esrgroups.org/jes12-4.
16. Yogesh, P. K. (2015). A review on transient stability constrained optimal power flow. 4(3), 1298–1303. ISSN (Print): 2320–3765, ISSN (Online): 2278–8875. (An ISO 3297: 2007 Certified Organization).
17. Tran, T. T. and Vo, D. N. (2016). Transient stability constrained optimal power flow using improved particle swarm optimization. *GMSARN Inter. J.*, 10, 87–94.
18. Alireza, B., Damien, E., Yves, V., Quentin, G., Camille, P. and Patrick, P. (2021). Extended equal area criterion revisited: A direct method for fast transient stability analysis. *Energies*, 14, 7259. https://doi.org/10.3390/en14217259,

19. Gaikwad, S. R. and Harikrishnan, R. (2021). Social welfare maximization in smart grid: Review. *IOP Conf. Series Mater. Sci. Engg.*, 1099, 012023. doi:10.1088/1757-899X/1099/1/012023.
20. Ignacio, A. C. (2015). Transient stability constrained optimal power flow: Improved models and practical applications. *Leganés (Madrid).*
21. Renuka, K., Manoj, K. and Ganga, A. (2014). Transient stability analysis and enhancement of IEEE - 9 bus system. *Elec. Comp. Engg. Inter. J. (ECIJ)*, 3(2), June 2014, 41–51.
22. Sandip, C., Suparna, M., and Abhinandan, De. (2022). A differential evolution modified quantum PSO algorithm for social welfare maximisation in smart grids considering demand response and renewable generation. 13 nov, 2022. DOI 10.1007/s00542-022-05399-1.
23. Sandip, C. and De, A. (2014). A multi-objective solution algorithm for optimum utilization of smart grid infrastructure towards social welfare. *Inter. J. Elec. Power Energy Sys.*, 58, 307–318. ISSN: 0142-0615.

3 Design of a maximum power point tracking controller for a three-phase grid connected photovoltaic system using fuzzy logic-based non-linear backstepping control approach

Chiranjeeb Kumar Das[a] *and Mrinal Buragohain*[b]

Department of Electrical Engineering, Jorhat Engineering College, Jorhat, India

Abstract

Specifically, this paper is concerned with achieving maximum power point tracking in an on-grid photovoltaic configuration under varying irradiation. For this, a fuzzy-logic based hybrid non-linear back-stepping controller is designed which regulates the boost converter's duty cycle. The fuzzy-logic controller which is incorporated with the back-stepping controller delivers a sequence of predicted voltage points that serve as a reference for the controller to monitor and achieve optimal efficiency. Moreover, in the second stage, a synchronous reference-frame control strategy is employed to regulate the current of the voltage source inverter to satisfy the grid characteristics. To achieve the above objectives in the system, an elaborated description of all the controllers is presented along with the MATLAB Simulink simulation results. Finally, a comparison is made between the proposed controller with a traditional MPPT control method.

Keywords: DC/DC boost converter, perturb & observe MPPT method, voltage source inverter, PWM generator, total harmonics distortion

Introduction

A photovoltaic system consisting of solar cells connected together into arrays are used to capture solar energy. However, a photovoltaic system which consists of DC/DC converters, inverters, batteries and voltage regulators has a disadvantage of inferior power conversion efficiency. Efficiency of a PV system can be as low as 16% [1]. The output of a PV system fluctuates throughout the day depending on the changes in metrological factors like solar irradiation and air temperature. Thus, unable to transmit the maximum power to the grid. For this reason, maximum power point tracking (MPPT) methods are being developed. The proposed fuzzy-logic based back-stepping controller is a hybrid MPPT approach. To keep a track on the point of maximum power in a photovoltaic system, a back-stepping controller requires a reference voltage. Therefore, the designer uses perturb and observe or P&O method to generate the reference voltage [2], whereas, the study [3] uses a regression plane technique.

This paper proposes an on-grid photovoltaic system as shown in Figure 3.1. The system can be divided into two stages. The first stage includes a DC\DC boost converter with the proposed fuzzy-logic based non-linear backstepping MPPT controller and the second stage consist of a voltage source inverter with a synchronous reference frame-based controller which satisfies the grid characteristics. In order to increase efficiency and avoid complexities, the proposed PV system debars from including an energy storage system. The performance of the system is observed in varying irradiation while keeping the temperature constant at 25°C. Finally, the system is compared with a traditional system by replacing the novel MPPT controller with a perturb & observe (P&O) controller in the same system and under same conditions.

Mathematical interpretation of the PV array

The photovoltaic module employed in this work is Sunpower SPR-305E-WHT-U. The module's current output "I_{pv}" can be represented by:

$$I_{pv} = N_p \times I_p \left[\exp\left(\frac{\frac{V}{N_S}+I\times\frac{R_S}{N_P}}{n\times V_t}\right) - 1\right] - I_{sh} \quad (1)$$

where, I_p, I_o and I_{sh} represents the photo current, saturating current and current through the shunt resistor, respectively. Table 3.1 displays the characteristics of the PV array used in the system.

The current and power characteristics against voltage of the photovoltaic array under varying solar irradiance, G = (1000, 700, 500, 300 and 100) W/m^2 at temperature, T= 25°C is depicted in Figure

[a]chiranjeebkumardas07@gmail.com, [b]mrinalburagohain@gmail.com

DOI: 10.1201/9781003540199-3

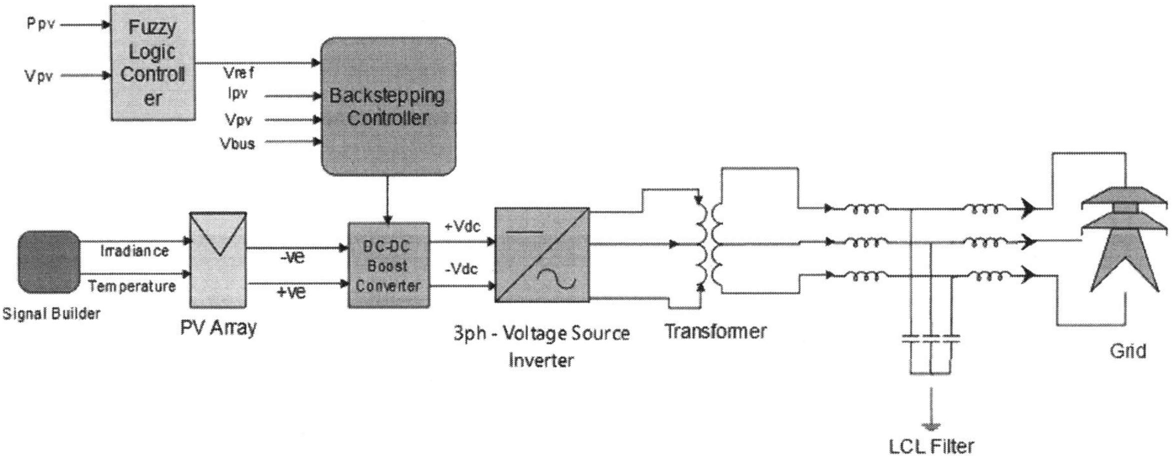

Figure 3.1 Block diagram of the on-grid photovoltaic system

Table 3.1 PV array characteristics

Parameter	Symbol	Values
Number of parallel strings	N_p	60
Number of modules connected per string	N_s	10
Cells per module	N_{cell}	96
Output power (maximum)	W	305.23 W
Maximum power point current	I_{mp}	5.6 A
Maximum power point voltage	V_{mp}	54.7 V
Short circuit current	I_{SC}	5.959 A
Open circuit voltage	V_{OC}	64.19 V
Light generated current	I_L	5.965 A
Shunt resistance	R_{sh}	393.21 Ω
Diode saturation current	I_o	6.308 × 10^{-12} A
Series resistance	R_s	0.3743 Ω

3.2. The simulation study is performed under constant temperature of 25°C at different irradiation level ranging from G = (0–1000) W/m^2 portrayed by the irradiation (W/m^2)-time(s) graph in Figure 3.3.

Maximum power point tracking control strategy

The primary emphasis is on establishing a control strategy that vouches global stability while manipulating the DC bus voltage to deliver robust tracking of the highest power point. Consequently, a back-stepping controller regulates the DC/DC boost converter's duty cycle.

Equations (2) and (3) are realized by applying Kirchhoff's principles to the converter circuit illustrated in Figure 3.4.

$$\dot{V_{pv}} = \frac{I_{pv}}{C} - \frac{I_L}{C} \tag{2}$$

$$\dot{I_L} = \frac{V_{pv}}{L} - \frac{V_{bus}}{L}.u \tag{3}$$

where, $u = 1 - \theta$ and $0 \leq \theta \leq 1$.

Back-stepping controller

The backstepping controller obtains a master control law "θ" which stabilizes the system using Lyapunov function at each sub-system.

Let, $x_1 = V_{pv}$, $x_2 = I_L$ and $u = \mu$. Therefore, from the Equations (2) and (3) it can be written as:

$$\dot{x_1} = \frac{I_{pv}}{C} - \frac{x_2}{C} \tag{4}$$

$$\dot{x_2} = \frac{x_1}{L} - \frac{V_{bus}}{L}.(1 - \theta) \tag{5}$$

where θ is the master control law and it can be expressed as:

$$\theta = 1 - \left[\frac{\mathrm{L.C}}{\mathrm{V_{bus}}}(-\mathrm{k_2 e_2} + \mathrm{e_1} - \dot{\varphi}) + \frac{\mathrm{V_{pv}}}{\mathrm{V_{bus}}}\right] \tag{6}$$

Here, is the error signal and since, hence

$$\dot{e_1} = \dot{V_{ref}} - \frac{I_{pv}}{C} + \frac{x_2}{C} \tag{7}$$

Again, a Lyapunov function $V = \frac{1}{2}e_1^2$ is taken into consideration, and stability is assured by turning its derivative into a negative definite. Therefore, virtual command (φ) of x_2/C is:

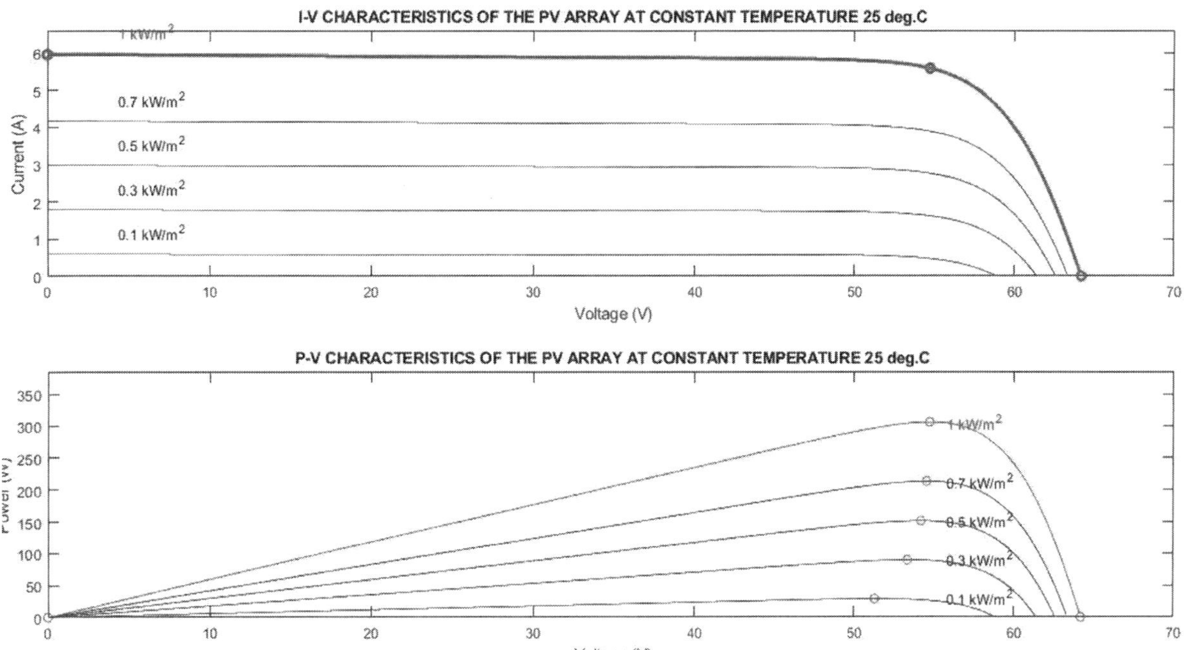

Figure 3.2 Current and power versus voltage characteristics under different irradiation (G) at 25°C constant temperature

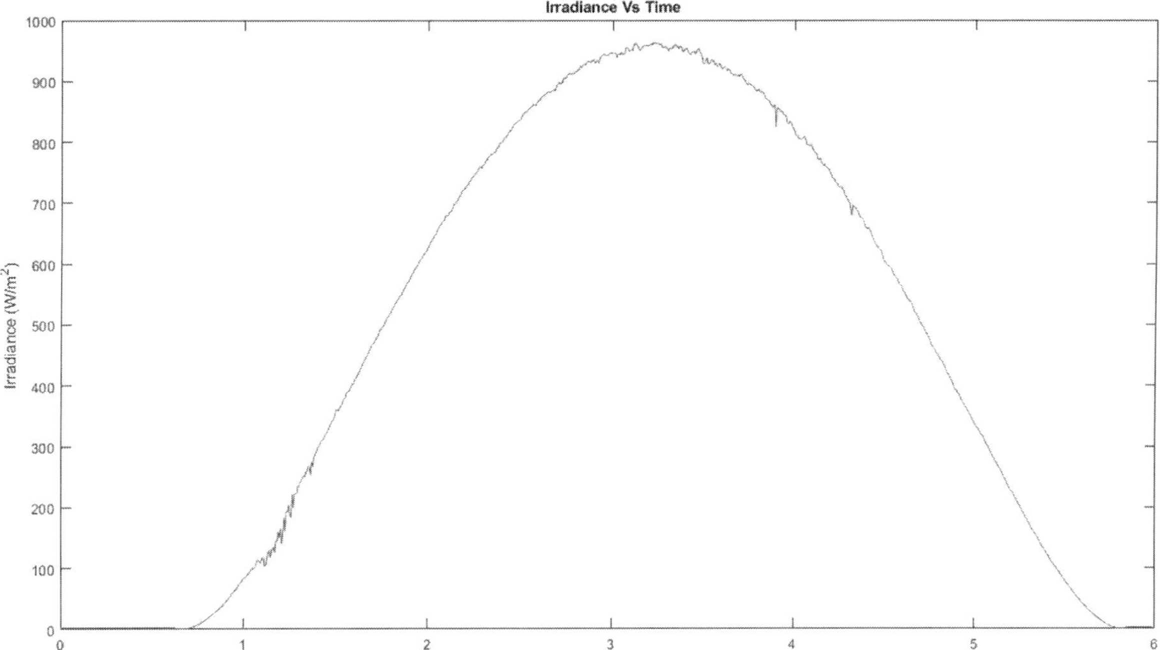

Figure 3.3 Solar irradiation profile at 25°C constant temperature

$$\varphi = -k_1 e_1 - \dot{V}_{ref} + \frac{I_{pv}}{C} \tag{8}$$

Taking the derivative –

$$\dot{\varphi} = \dot{e}_2 + \frac{V_{pv}}{L.C} - \frac{V_{bus}}{L.C}(1-\theta) \tag{9}$$

where, the derivative of the second error e_2 is $e_2 = e_1 - k_2 e_2$.

The equations are realised in MATLAB Simulink. The designed controller has V_{pv}, I_{pv}, V_{bus} and V_{ref} as inputs and "θ" as the controller output. Where, "θ" is the DC/DC boost converter's control signal.

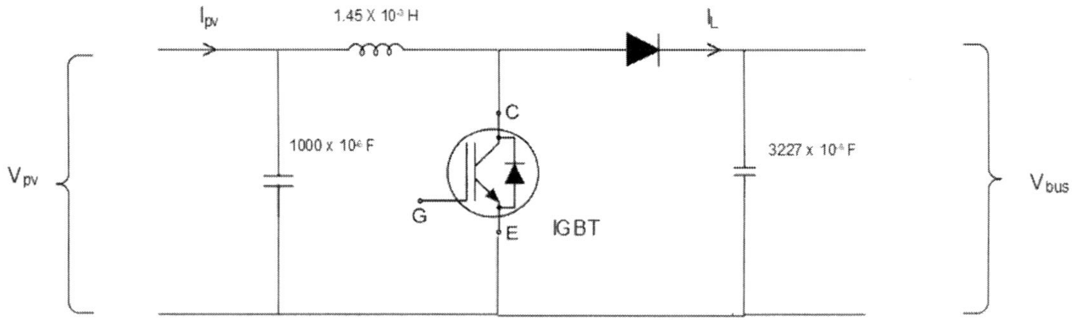

Figure 3.4 DC/DC boost converter circuit diagram

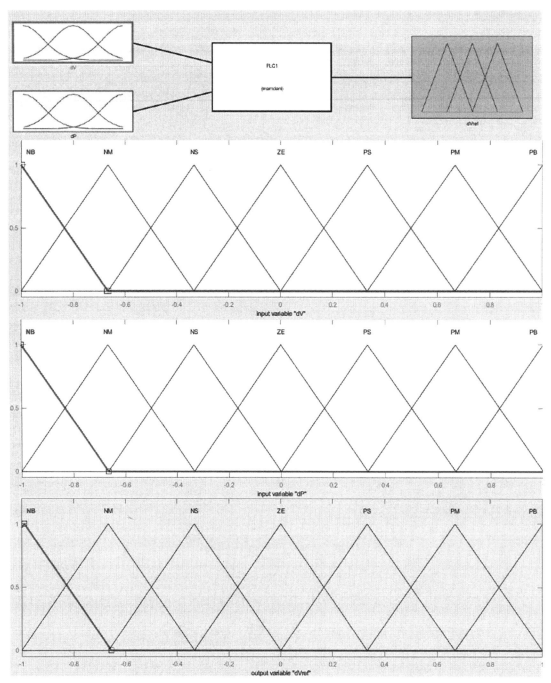

Figure 3.5 Triangular membership functions of δV, δP and δV_{ref}

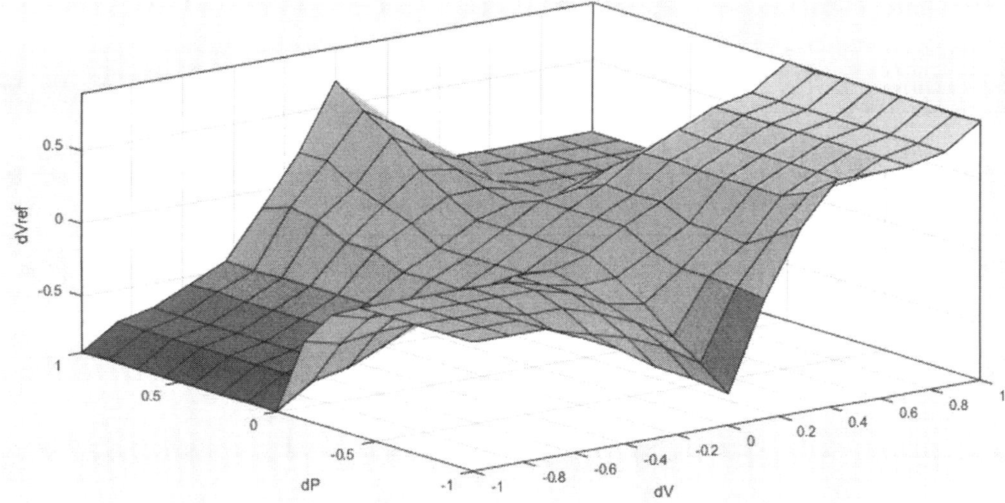

Figure 3.6 Graphical representation of δV_{ref} with respect to δV and δP

The photovoltaic module provides the V_{pv} and I_{pv}. Whereas, the converter output is V_{bus} and the fuzzy-logic based controller forecasts the reference voltage V_{ref}.

Fuzzy-logic based reference voltage estimating controller

With an eye towards forecasting the reference for the proposed backstepping controller to monitor maximum power point, an appropriate fuzzy-logic based controller (FLC) is being constructed [4]. The FLC scrutinizes the changes that occur in the output voltages (δV) and power (δP) of the PV module due to the changes in solar irradiation levels in a specific duration of time and provides a calculated reference voltage V_{ref} using a set of 49 control rules. The controller uses a Mamdani fuzzy inference system along with center of gravity method to transfer fuzzy inference results into crisp values. The triangular memberships functions of the input and output variables are depicted in Figure 3.5. Whereas, Figure 3.6 displays the three-dimensional graphical representation of the predicted reference voltage.

DC/AC three-phase voltage source inverter system with synchronous reference frame-based control approach

A three-phase DC/AC voltage source inverter-system is employed in stage two. To regulate the inverter current and satisfy the grid quality, a synchronous reference-frame theory-based controller is used along with a three-phase step-up transformer having star-delta configuration and a LCL filter to filter out any harmonics present before feeding the current to the grid without compromising power quality. Synchronous reference frame theory-based control also known as direct-quadrant-zero (dq0) controller is a method to simplify analysis. It is an amalgamation of Clarke and Park's transformation [5]. Clarke transformation converts the three-phase AC current and voltage I_a, I_b, I_c/V_a, V_b, V_c into I_α, I_β/V_α, V_β by alternating its reference frame [6]. Whereas, Park's transformation converts the elements in I_α, I_β/V_α, V_β to I_d, I_q/V_d, V_q domain [7]. This, conversion of AC waveform to DC signals by rotating its reference frame favors regulation through simple PI controllers [8]. The schematic representation of the control scheme is shown in Figure 3.7.

Results and discussion

This paper succeeds in meeting the set objectives of designing a fuzzy-logic based hybrid non-linear backstepping controller to monitor maximum power-point and integration of the system with the grid via a voltage-source inverter managed by a dq0 controller. The proposed system is realised in the MATLAB/SIMULINK environment.

The controller is studied for a total time-span of 6 s. During this time, the temperature is maintained at 25°C while the sun irradiation levels are gradually and uniformly adjusted between (0-1000-0) W/m2 to analyze the controller's performance in varying irradiance.

level is zero until 0.7 s and gradually rises to almost 1000 W/m^2 at 3.2 s and again gradually decreases until 5.8 s where it is again zero.

Figure 3.8 demonstrates that the boost converter's output voltage V_{bus}, coincides along the reference

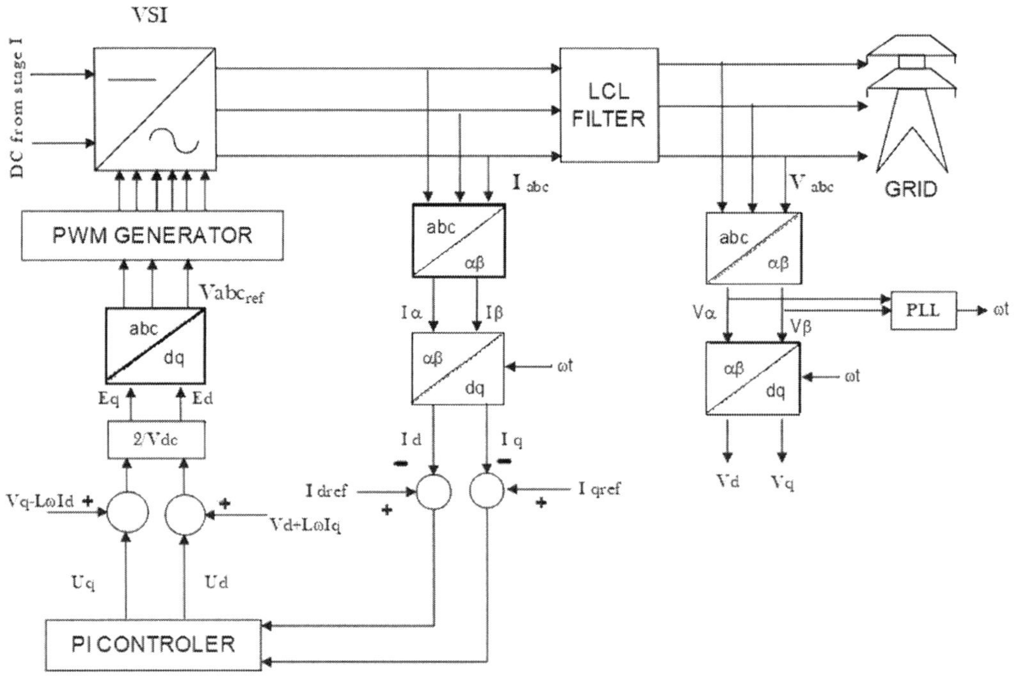

Figure 3.7 dq-controller block diagram

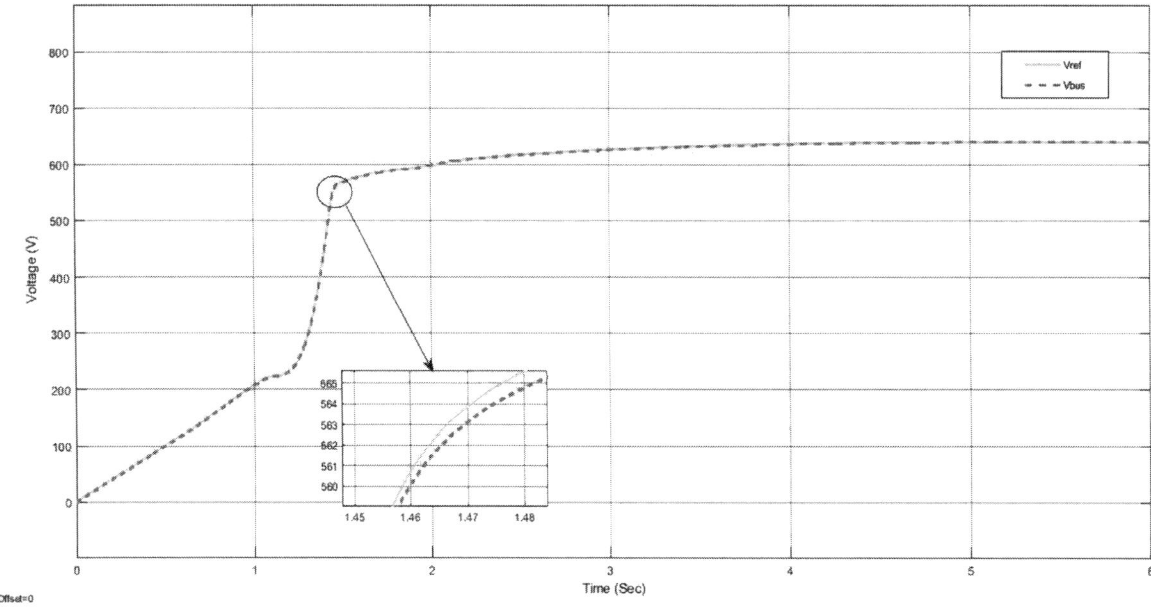

Figure 3.8 V_{bus} through fuzzy-logic based back-stepping controller

voltage curve V_{ref} estimated by the fuzzy-logic based controller. Since the proposed hybrid back-stepping controller controls the boost converter, this action guarantees the controller's ability to track maximum power point. Hence, proves the effectiveness of the controller's design. Figure 3.9 displays the three-phase AC output voltage (V_{abc}) and output current (I_{abc}) of the system.

Moreover, output current (I_{abc})'s total harmonics distortion (THD) percentage was also recorded. The THD recorded for 150 cycles starting from 1 s at 50 Hz frequency is 1.67%, shown in Figure 3.10.

Finally, the proposed controller being compared with a traditional method to track maximum power in a PV system by keeping every component of the system constant and replacing the designed

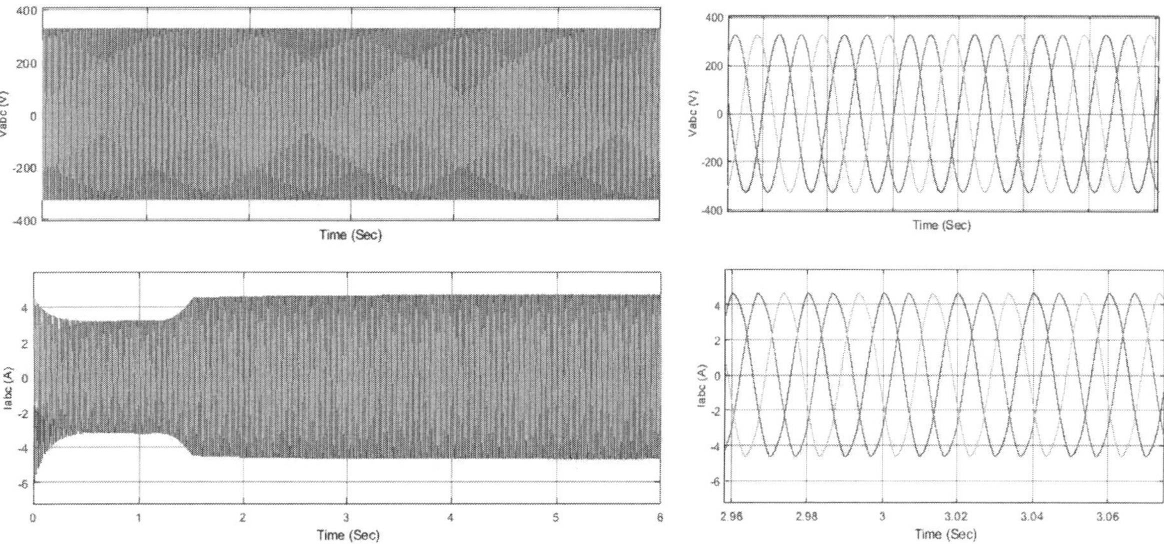

Figure 3.9 V_{abc} and I_{abc}

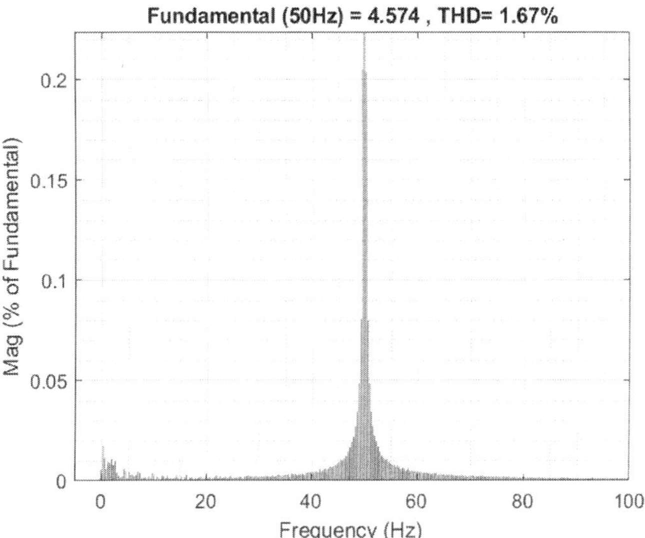

Figure 3.10 THD% of FLC based BSC

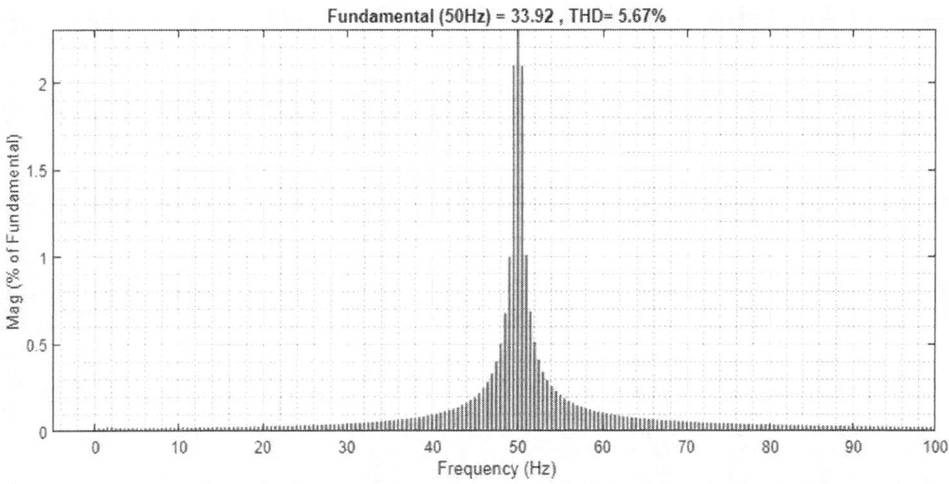

Figure 3.11 THD% of classical P&O based MPPT controller

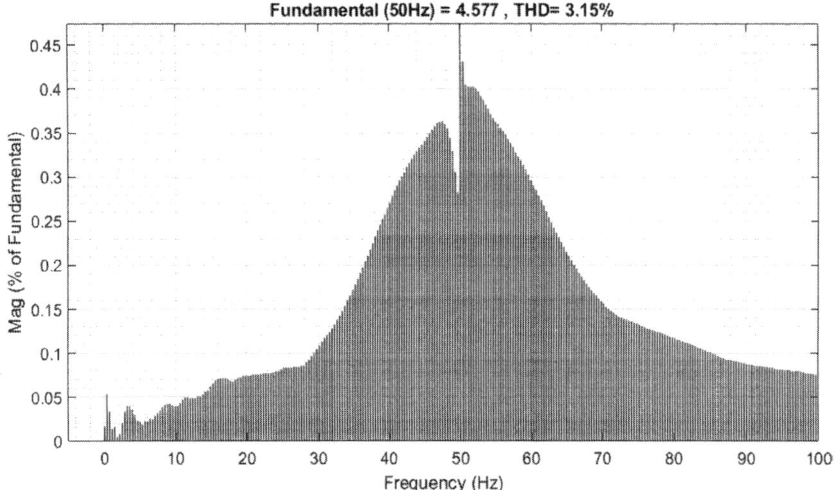

Figure 3.12 THD% of modified P&O based MPPT controller

controller with a P&O and modified P&O based tracker and analysing the outcome. The three-phase current output of both the scenarios were compared by computing their THD percentage at similar condition of 150 cycles at 50 Hz frequency and starting from 1 s which is 5.67% and 3.15%, respectively as shown in Figures 3.11 and 3.12.

Conclusions

The boost converter's output voltage's ability to follow the reference voltage curve generated by the fuzzy-logic based predictor in a constantly changing irradiation condition implies the controller's creditable and robust performance. The three-phase AC current generated had a THD% of 1.67% which is fairly within 5% as set by IEEE 512–1992. On comparing the proposed controller to classically used P&O MPPT method and modified P&O methods, the FLC based BSC showed far lower THD percentage. Thus, it is possible to infer that the proposed fuzzy-logic-based non-linear hybrid back-stepping MPPT controller is a stable, robust, accurate and a better alternative in tracking maximum power point in a constantly changing solar irradiation condition. The synchronous reference frame theory-based controller is successful in controlling the VSI and hence produced a stable three-phase AC that satisfies the grid characteristics. The active power (P) obtained is also positive, which confirms the system's performance.

Acknowledgement

The authors gratefully acknowledge the staff and authority of the Electrical department for their cooperation in the research.

References

1. Borekci, S., Ekrem, K., and Ali, K. (2015). A simpler single-phase single-stage grid-connected PV system with maximum power point tracking controller. *Elektronika ir elektrotechnika*, 21(4), 44–49. doi: 10.5755/j01.eee.21.4.12782.
2. Abbou, A., Salahddine, R., and Mahdi, S. (2017). Maximum power point tracking of photovoltaic systems using backstepping controller. *2017 Inter. Conf. Engg. Technol. (ICET)*, 1–6. doi: 10.1109/ICEngTechnol.2017.8308147.
3. Zaghar, F., Hekss, Z., Rafi, M., and Ridah, A. (2022). Backstepping approach-based control for three phase grid connected photovoltaic system. *IFAC-Papers OnLine*, 55(12), 520–525. doi: 10.1016/j.ifacol.2022.07.364.
4. Marhraoui, S., Ahmed, A., Zineb, C., Salah, E. R., and El Hichami, N. (2020). Fuzzy logic-integral backstepping control for PV grid-connected system with energy storage management. *Inter. J. Intell. Engg. Sys.*, 13(3), 359–372. doi: 10.22266/ijies2020.0630.33.
5. Park, R. H. (1929). Two-reaction theory of synchronous machines generalized method of analysis-part I. *Trans. Am. Inst. Elec. Engr.*, 48(3), 716–727. doi: 10.1109/T-AIEE.1929.5055275.
6. O'Rourke, C. J., Mohammad, M. Q., Matthew, R. O., and James, L. K. (2019). A geometric interpretation of reference frames and transformations: dq0, clarke, and park. *IEEE Trans. Energy Conver.*, 34(4), 2070–2083. doi: 10.1109/TEC.2019.2941175.
7. Debasish, M., Abhijit, C., and Aparajita, S. (2020). Chapter 2 - Fundamental Models of Synchronous Machine. Power System Small Signal Stability Analysis and Control (Second Edition). Academic Press. 15–40. doi: 10.1016/B978-0-12-817768-6.00002-0.
8. Dragicˇ evicˇ, T. and Li, Y. (2018). AC and DC microgrid control. *Control Power Elec. Conver. Sys.*, 167–200. doi: 10.1016/B978-0-12-816136-4.00018-X.

4 Performance analysis of H-bridge multilevel inverter for application in renewable energy

Md Asif Ahmad[a] *and Soumitra Kumar Mandal*[b]

Department of Electrical Engineering NITTTR, Kolkata, Kolkata, India

Abstract

This paper suggested a design of cascade H-bridge multilevel inverters for green energy production and its integration. Green energy has certain distinct electrical features therefore, injecting straight to grid is very challenging. Traditional inverter suffers with complication and need many devices. In order to get rid from such difficulties, cascade H-bridge multilevel inverters are designed. In this proposed inverter, the performance gets enhanced with increase in levels as harmonics decline and supply high standard voltage. In this paper, performance of three different cascade H-bridge multilevel inverters (five level, seven level and nine level) is presented.

Keywords: Cascade H-bridge multilevel inverter, distortion, green energy production and integration

1. Introduction

Demand for multilevel inverter constantly hiked in several fields. Multilevel inverter formed a massive enhancement in output voltage hence its role becomes extremely high for upgrading voltage quality. Output voltage of such inverter works at low switching frequency. Due to its capability to generate waveform with improved harmonic it is made for purpose of high voltage. While changing direct current to alternating current most significant thing that required to emphasis on total harmonic distortion which must kept small. When levels of inverter upgraded its performance gets boosted. Voltage generated by multilevel inverter is stairway form and appears identical to sinusoidal form also carries reduced number of harmonics as compare to traditional inverter. If inverter output rises up to N levels THD reduce to zero. For proposed inverter a number of H-bridges are required along with a isolated DC source. This inverter has capacity to control voltage for such feature. Maximum power point tracking control is gained from individual string. Multilevel inverter is economical and turns out to be excellent choice. Voltages given by five level inverter are 0, +V, +2V, –V, –2V. Each separate bridge excites switches only at basic frequency and produce modified square wave formed by phase change. Quality of power gets improved when voltage levels get increased. When switching elements increases voltage levels at output also increases this makes more expensive, complex and its efficiency gets minimized. During fault state such inverter maintains their action, voltage at output remains unchanged but levels fall into declined. Maximum voltage provided by planned inverter is 13.8 kilo volt and power is 30 mega volt amp, also its modular structure offers more consistency. Applying appropriate pulse width modulation method energy lost during transient period can be decreased. So many novel technology and methods are established for quality enhancement of voltage. Generally, multilevel inverter has three different types – diode clamped inverter, flying capacitor inverter and cascade H-bridge multilevel inverter. Among them cascade H-bridge multilevel inverter are highly acceptable and desirable, quite basic, follows uncomplicated technology. The foremost drawback of diode clamped inverter is the requirement of diodes which are more to upraise levels of voltage thus making whole device and system ineffectual also deliver half of applied voltage. The flying capacitor inverter has lots of attractive functional quality because its work is independent from transformer also have uniformly distributed switching stress between switches but need many capacitors therefore, its rate and complication become high. Unlike other two inverter which requires additional diode or capacitor, planned inverter is free from additional components. Specific harmonic deduction method of PWM provides appropriate control to harmonic band. This feature is extremely beneficial for converters and performs at low switching frequency in order to lower down harmonics. This paper primarily emphasis on the design and execution of different level cascade H-bridge multilevel inverters. And analogies are made for the THD evaluation.

[a]asifsbg5@gmail.com, [b]skmandal@ nitttrkol.ac.in

DOI: 10.1201/9781003540199-4

II. H-bridge cell

H-bridge also known as basic building blocks of cascade H-bridge inverter. It has several cells that are coupled into series where each and every cell carries DC link which is isolated. Primary cell generates three distinct levels of voltage 0, V and –V at its output after primary cell each additional cell generate two different levels. Therefore, when two bridge is coupled into series output will be the summation of the two bridges as both H-bridge function autonomously. Thus, by joining cells of two voltage level at output will be five for N number of cells voltage levels be (2n+1) where n represents numerical quantity of DC sources. Methodical figure of cell is shown in Figure 4.1.

III. Cascade H-bridge inverter

3.1 Five level cascade H-bridge inverter
This inverter comprise of coupling of two H-bridge which are connected into series. These two separate H-bridges have two DC sources and eight power electronic switches. Figure 4.2 shows the basic structure of five level cascade H-bridge multilevel inverter.

Switching mechanism
To acquire five different levels of voltage in Simulink model of inverter, the switching functions are discussed below:
First half period function

- For zero voltage levels
  To acquire zero voltage level in Simulink model of inverter as shown in Figure 4.2 here are two opportunities such as (1) S_1 and S_2 are activated and made open circuit and (b) S_1 and S_3 are activated and made short circuited. At both stages, zero level is provided by the model as shown in Figure 4.2. Therefore, S_1, S_2, S_5, S_6 are turn on to get zero level.

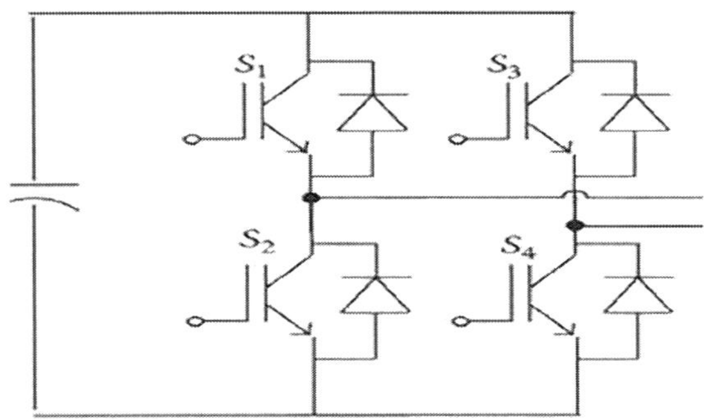

Figure 4.1 H-bridge cell

- For second level (V)
  Opposite switches of first H-bridge are activated. This opposite switch will provide positive V. So S_1, S_4, S_5, S_6 are activated. S_1 and S_4 will give V and S_5 and S_6 will provide zero all together second level V are achieved by this configuration.
- For third level (2V)
  Opposite switches of both H-bridge are activated. These opposite switches will provide V. So S_1, S_4, S_5, S_8 are turn on. S_1 and S_4 will give V level and S_5 and S_6 will also provide V level with all together 2V levels are achieved.
- For fourth level (V)
  Same action will take place as in case of second level.
- To get five level (0)
  Same action will take place as in case of 0 level.

Second half cycle function

- For negative V level (–V)
  S_2, S_3 from first H-bridge and S_5 and S_6 from other H-bridge are activated. S_2, S_3 will provide –V and S_5 and S_6 will give zero, combination of all switches gives –V.
- For negative 2V level (–2V)
  S_1, S_3 from first H-bridge and S_6 and S_7 from other H-bridge are activated. S_1, S_3 will provide –V and S_6 and S_7 will give –V, combination of all switches gives –2V.
- For negative V level (–V)

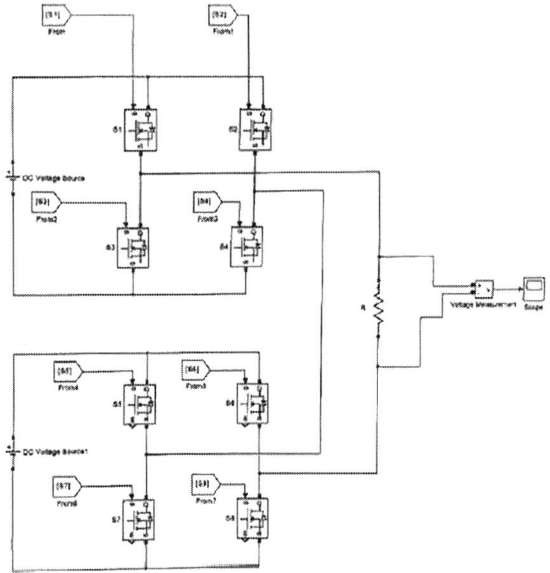

Figure 4.2 Simulink model five level cascade H-bridge inverter

Similar action will take place as mentioned in first case.

Switching scheme of cascade H-bridge five level inverter and pulse width modulation and delay values are shown in Tables 4.1 and 4.2.

3.2 Seven level cascade H-bridge inverter

For seven level inverter three H-bridge, having three separate DC sources and twelve switches are coupled into series. This is controlled by pulse generating through pulse width modulation method. Like five level cascade H-bridge inverter same method and senses are applied while simulating of seven and nine level inverters. Figure 4.3 shows Simulink model of seven level cascade H-bridge inverter. Switching technique of cascade H-bridge seven level multilevel inverter, pulse width modulation with delay values are shown in Tables 4.3 and 4.4.

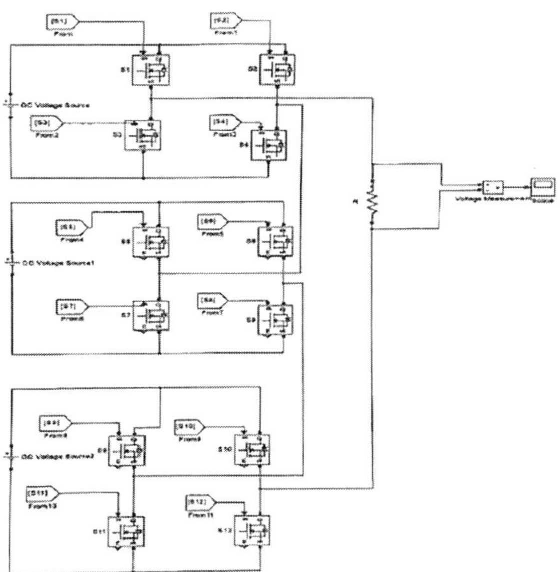

Figure 4.3 Simulink model seven level cascade H-bridge inverter

Table 4.1 Switching scheme of cascade H-bridge five level inverter

Mode of operation	M0	M1	M2	M3	M4	M5	M6	M7
Level of voltage	0	V	2V	V	0	-V	-2V	-V
Switches								
S_1	1	1	1	1	1	0	0	0
S_2	1	0	0	0	1	1	1	1
S_3	0	0	0	0	0	1	1	1
S_4	0	1	1	1	0	0	0	0
S_5	1	1	1	1	1	1	0	1
S_6	1	1	0	1	1	1	1	1
S_7	0	0	0	0	0	0	1	0
S_8	0	0	1	0	0	0	0	0

Table 4.3 Switching scheme of cascade H-bridge seven level multilevel inverter

Mode of action	M0	M1	M2	M3	M4	M5	M6	M7	M8	M9	M10	M11
Voltage level	0	V	2V	3V	2V	V	0	V	-2V	-3V	2V	-V
Switches												
S_1	1	1	1	1	1	1	1	0	0	0	0	0
S_2	1	0	0	0	0	0	1	1	1	1	1	1
S_3	0	0	0	0	0	0	0	1	1	1	1	1
S_4	0	1	1	1	1	1	0	0	0	0	0	0
S_5	1	1	1	1	1	1	1	1	0	0	0	1
S_6	1	1	0	0	0	1	1	1	1	1	1	1
S_7	0	0	0	0	0	0	0	0	1	1	1	0
S_8	0	0	1	1	1	0	0	0	0	0	0	0
S_9	1	1	1	1	1	1	1	1	1	0	1	1
S_{10}	1	1	1	0	1	1	1	1	1	1	1	1
S_{11}	0	0	0	0	0	0	0	0	0	1	0	0
S_{12}	0	0	0	1	0	0	0	0	0	0	0	0

Table 4.2 Pulse width modulation and delay with values of cascade H-bridge five level inverter

Switches	Pulse width modulation	Delay (s)
S_1	5/8×100	0.02×8×0
S_2	1/8×100, 4/8×100	0, 0.02/8×4
S_3	3/8×100	0.02/8×5
S_4	3/8×100	0.02/8×1
S_5	6/8×100, 1/8×100	0, 0.02/8×7
S_6	2/8×100, 5/8×100	0, 0.02/8×3
S_7	1/8×100	0.02/8×6
S_8	1/8×100	0.02/8×2

Table 4.4 Pulse width modulation and delay with values of cascade H-bridge seven level inverter.

Switches	Pulse Width Modulation	Delay
1	7/12 x100	0.02/12 x 0
2	1/12 x 100, 6/12 x 100	0.02/12 x 0, 0.02/12 x6
3	5/12 x 100	0.02/12 x 7
4	5/12 x100	0.02/12 x 1
5	8/12 x100	0.02/12 x 0
6	2/12 x 100, 6/12 x 100	0.02/12 x 0, 0.02/12 x 5
7	3/12 x 100	0.02/12 x 8
8	3/12 x 100	0.02/12 x 2
9	9/12 x 100, 2/12 x 100	0.02/12 x 0, 0.02/12 x 10
10	3/12 x 100, 8/12 x100	0.02/12 x 0, 0.02/12 x4
11	1/12 x 100	0.02/12 x 9
12	1/12 x 100	0.02/12 x 3

3.3 Nine level cascade H-bridge inverter

Nine level cascade H-bridge multilevel inverter comprise of four H-bridge connected into series. It has altogether sixteen switches and four separate DC sources. Simulation of nine level cascade H-bridge multilevel inverter is shown in Figure 4.4. Switching method of nine level cascade H-bridge inverter, pulse width modulation with delayed values are shown in Tables 4.5 and 4.6.

IV. Results

The Simulink results of different models are demonstrated in this paper. The proposed model has a R load of 100 Ohms, L of 30 mH and 220V. Wave form and THD of output voltage with R and RL load from proposed model are shown in Figure [5–13b]. Out voltage, current and THD with R and RL load of single phase five level inverter are shown in Figures 4.5–4.7. Out voltage, current and THD with R and RL load of single phase seven level inverter are shown in Figures 4.8–4.10. Out voltage, current and THD with R and RL load of single phase nine level inverter are shown in Figures 4.11–4.13.

Total harmonic distortion analysis

Cascade H-bridge multilevel inverter can generate sinusoidal output voltage. On ascending the levels of voltage quality of voltage gets better. This can be understood through three different levels of inverter. Table 4.7 shows THD performance.

Table 4.5 Switching scheme of cascade H-bridge nine level

Mode of operation	M0	M1	M2	M3	M4	M5	M6	M7	M8	M9	M10	M11	M12	M13	M14	M15
Level of voltage	0	V	2V	3V	4V	3V	2V	V	0	-V	-2V	-3V	-4V	-3V	-2V	-V
Switches																
S_1	1	1	1	1	1	1	1	1	1	0	0	0	0	0	0	0
S_2	1	0	0	0	0	0	0	0	1	1	1	1	1	1	1	1
S_3	0	0	0	0	0	0	0	0	0	1	1	1	1	1	1	1
S_4	0	1	1	1	1	1	1	1	0	0	0	0	0	0	0	0
S_5	1	1	1	1	1	1	1	1	1	1	0	0	0	0	0	1
S_6	1	1	0	0	0	0	0	1	1	1	1	1	1	1	1	1
S_7	0	0	0	0	0	0	0	0	0	0	1	1	1	1	1	0
S_8	0	0	1	1	1	1	1	0	0	0	0	0	0	0	0	0
S_9	1	1	1	1	1	1	1	1	1	1	1	0	0	0	1	1
S_{10}	1	1	1	0	0	0	1	1	1	1	1	1	1	1	1	1
S_{11}	0	0	0	0	0	0	0	0	0	0	0	1	1	1	0	0
S_{12}	0	0	0	1	1	1	0	0	0	0	0	0	0	0	0	0
S_{13}	1	1	1	1	1	1	1	1	1	1	1	1	0	1	1	1
S_{14}	1	1	1	1	0	1	1	1	1	1	1	1	1	1	1	1
S_{15}	0	0	0	0	0	0	0	0	0	0	0	0	1	0	0	0
S_{16}	0	0	0	0	1	0	0	0	0	0	0	0	0	0	0	0

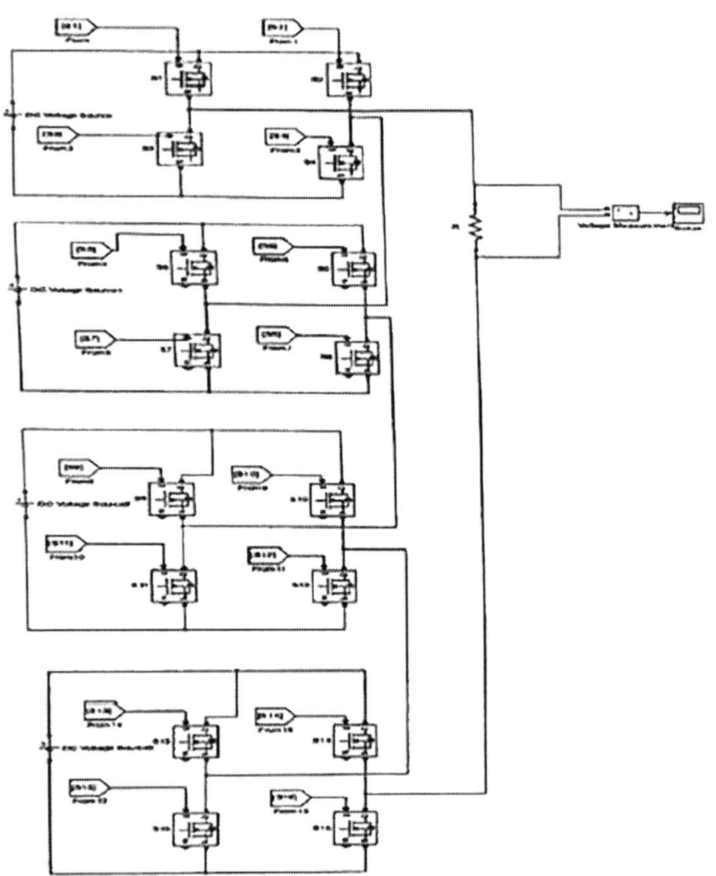

Figure 4.4 Simulink model nine level cascade H-bridge

Table 4.6 Pulse width modulation and delay with values of cascade H-bridge nine level inverter

Switches	Pulse Width Modulation	Delay
S_1	9/16 x100	0.02/16 x 0
S_2	1/16 x100, 8/16 x100	0.02/16 x 0, 0.02/16 x 8
S_3	7/16 x100	0.02/16 x 9
S_4	7/16 x100	0.02/16 x 1
S_5	10/16 x100, 1/16 x100	0.02/16 x 0, 0.02/16 x 15
S_6	2/16 x100, 9/16 x100	0.02/16 x 0, 0.02/16 x 7
S_7	5/16 x100	0.02/16 x 10
S_8	5/16 x100	0.02/16 x 2
S_9	11/16 x100, 2/16 x100	0.02/16 x 0, 0.02/16 x 14
S_{10}	3/16 x100, 10/16 x100	0.02/16 x 0, 0.02/16 x 6
S_{11}	3/16 x100	0.02/16 x 11
S_{12}	3/16 x100	0.02/16 x 3
S_{13}	3/16 x100, 12/16 x100	0.02/16 x 0, 0.02/16 x 13
S_{14}	4/16 x100, 11/16 x100	0.02/16 x 0, 0.02/16 x 5
S_{15}	1/16 x100	0.02/16 x 12
S_{16}	1/16 x100	0.02/16 x 4

V. Conclusions

The performance investigation of proposed inverter for grid coupled PV system has been demonstrated in this paper. Different levels of cascade H-bridge multilevel inverter are simulated in MATLAB output results are compared and THD observation are also demonstrated. This study of different phase shifted pulse width modulation techniques applied on five, seven, and nine level cascade H-bridge multilevel inverter and simulation results shows that THD decline with accelerating number of levels of H-bridge inverter. As per analyzed results the above nine level cascade

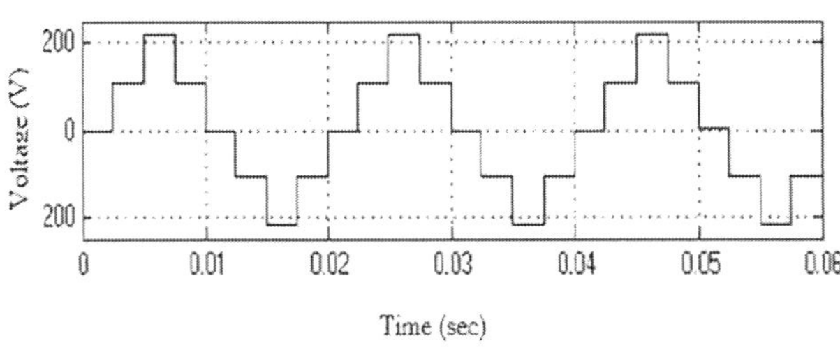

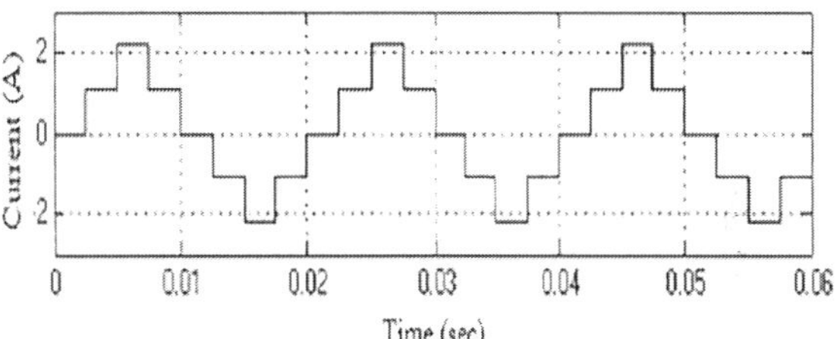

Figure 4.5 Output voltage and current with R load

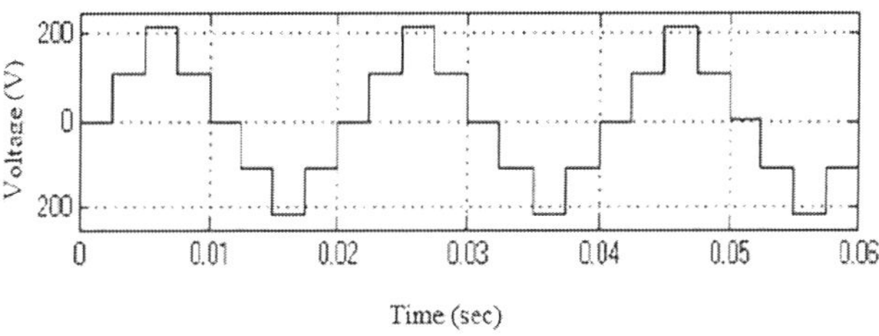

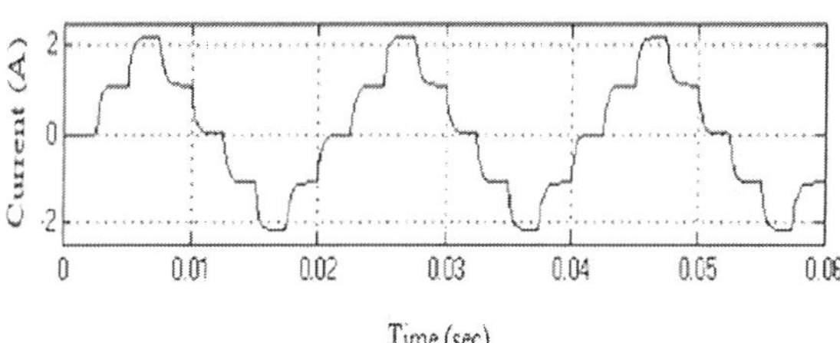

Figure 4.6 Output voltage and current level with R-L load

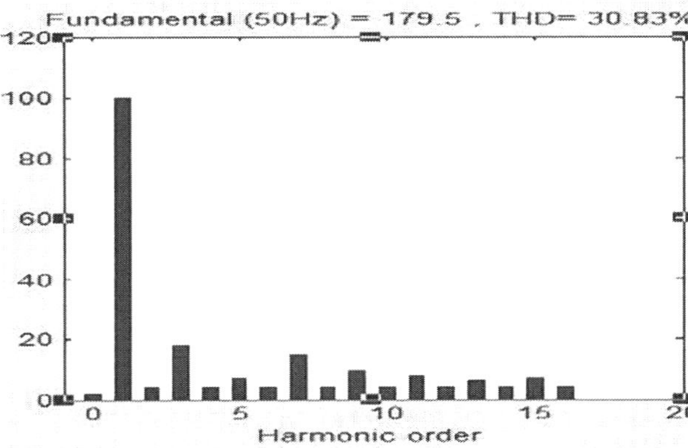

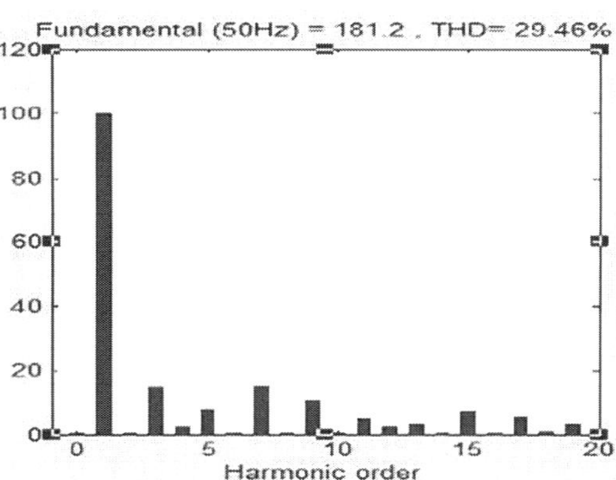

Figure 4.7 (a) THD on R load (b) THD on R-L

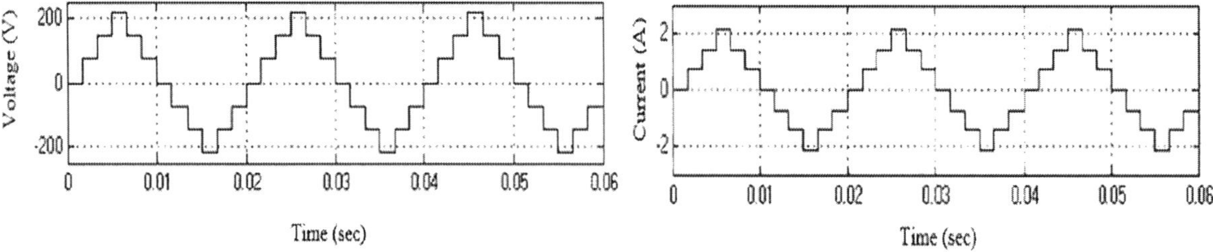

Figure 4.8 Output voltage and current with R load

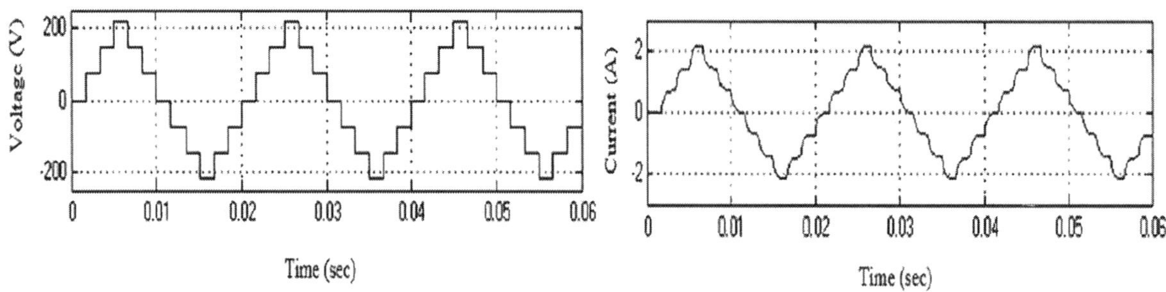

Figure 4.9 Output voltage and current with R-L load

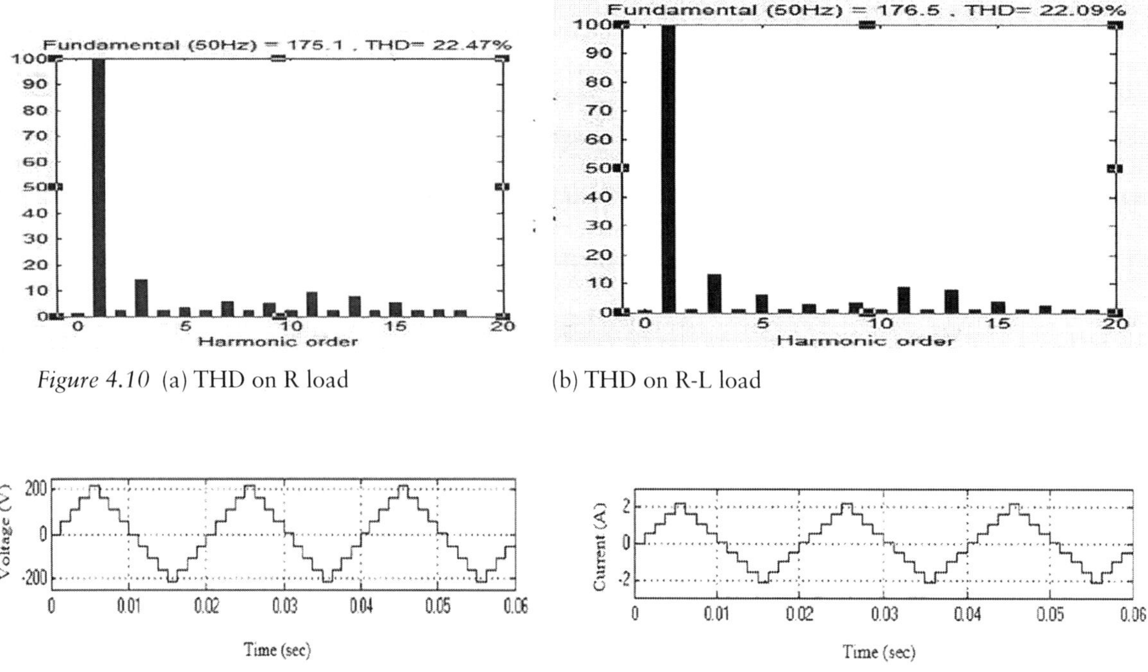

Figure 4.10 (a) THD on R load (b) THD on R-L load

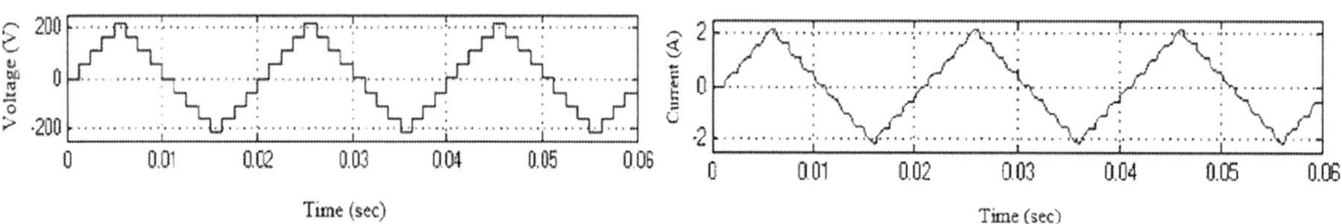

Figure 4.11 Output voltage and current with R load

Figure 4.12 Output voltage and current with R-L load

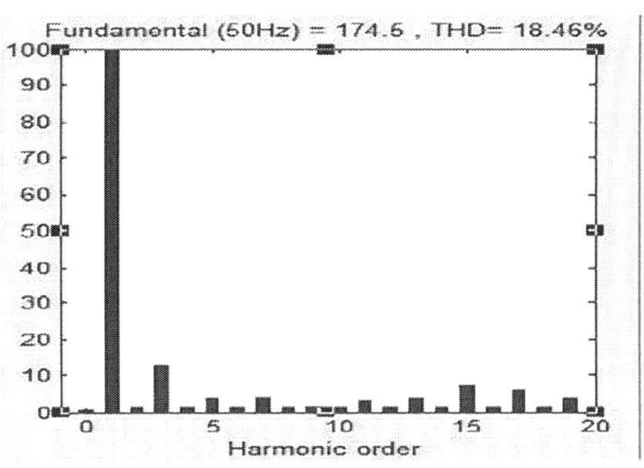

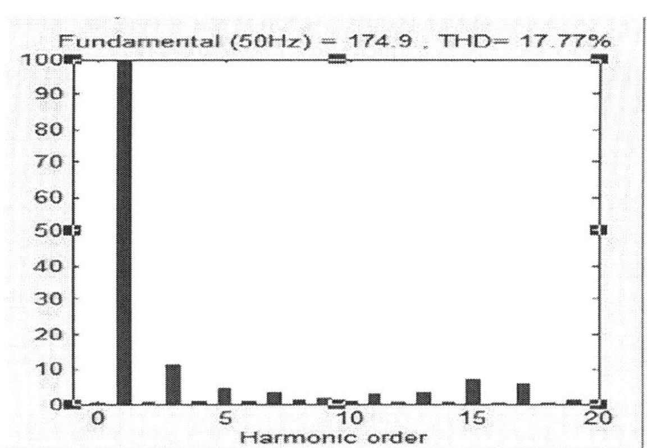

Figure 4.13 (a) THD on R load (b) THD on R-L load

Table 4.7 Cascade H-bridge inverter/THD

Level of inverter	THD (R load)	THD (R-L load)
Five level inverter	30.83%	29.46%
Seven level inverter	22.47%	22.9%
Nine level inverter	18.46%	17.77%

H-bridge multilevel inverter topology is efficient to minimize total harmonic, generate better quality output voltage and can be used to grid for connecting PV array.

VI. References

1. Jayaram, N. and Asharaf, A. (2016). Improved performance of multilevel Inverter. *IEEE 2016 Inter. Conf. Microelec. Comput. Comm. (MicroCom)*. DOI: 10.1109/MicroCom.2016.7522514, 978-1-4673-6621
2. Rodriguez, J., Lai, J. S., and Peng, F. Z. (2002). Multilevel inverters: A survey of topologies, controls, and applications. *IEEE Trans. Ind. Electrons*, 49(4), 724–738.
3. Bailu, X., Ke, S., Jun, M., Faete, F., and Leon, M. (2012). Tolbert energy conversion congress exposition. 15-20 September 2012. 10.1109/ECCE.2012.6342474.
4. Mcgrath, B. P. and Holmes, D. G. (2000). Multi carrier PWM strategies for multilevel inverter. *IEEE Trans. Indus. Elec.*, 49(4), 858–867.
5. Lezana, P., Pou, J., Meynard, T. A., Rodriguez, J., Ceballos, S., and Richardeau, F. (2010). Survey on fault operation on multilevel inverters. *IEEE Trans. Ind. Electron*, 57(7), 2207–2218.
6. Wen, J. and Smedley, K. M. (2008). Synthesis of multilevel converters based on single- and/or three-phase converter building blocks. *IEEE Trans. Power Electron.*, 23(3), 1247–1256.
7. Zhang, J. Z. J., Zou, Y. Z. Y., Zhang, X. Z. X., and Ding, K. D. K. (2001). Study on a modified multilevel cascade inverter with hybrid modulation. *4th IEEE Int. Con. Power Electron. Drive Syst. IEEE PEDS 200i -Indones. Proc. (Cat. No. 01 TH8594)*, 379-383.
8. Rodriguez, J., Franquelo, L. G., Kouro, S., Leon, J. I., Portillo, R. C., Prats, M. A. M., and Perez, M. A. (2009). Multilevel converters: an enabling technology for high-power applications. *Proc IEEE*, 97(11), 1786–1817.
9. Asha, G. and Pallavi, A. A. (2016). Study of cascade H- Bridge Multilevel Inverter. *2016 Inter. Conf. Autom. Con. Dyn. Optim. Techniq. IEEE*. DOI: 10.1109/ICACDOT.2016.7877574, 179-182.
10. Lai, J. S. and Peng, F. Z. (1996). Multilevel converters–A new breed of power converters. *IEEE Trans. Ind. Applicant.*, 32, 509–517.
11. Lu and Corzine, K. A. (2007). Advanced control and analysis of cascaded multilevel converters based on P-Q compensation. *IEEE Trans. Power Electron.*, 22(4), 1242–1252.
12. Hemant, G., Arvind, Y., and Sanjay, M. (2016). Multi carrier PWM and selective harmonic elimination technique for cascade multilevel inverter. *IEEE Trans. Adv. Elec. Electron. Inform. Comm. Bio-Inform.*, 98–102. DOI 101109 AEEI/2016. 7538405.
13. Carrara, G., Marchesoni, M., Gardella, S., Sciutto, G., and Salutari, R. (1992). A new multilevel PWM method: A theoretical analysis. *IEEE Trans. Power Elec.*, 7(3), 497–505.
14. Holmes, D. G. and Lipo, T. A. (2003). Pulse width modulation for power converters: Principles and practice. John Wiley & Sons, 18,1-744.
15. Lhami, C., Ersan, K., and Ramzan, B. (2011). Review of multilevel voltage source inverter topologies and control schemes. *Energy Conver. Manag.*, 52(2), 1114–1128.
16. José, R., Bin, W., Jorge, O. P., and Samir, K. (2007). Multilevel voltage-source- converter topologies for

industrial medium-voltage drives. *IEEE Trans. Indus. Elec.*, 54(6), 2930–2945.

17. Mostafa, B. K. and Murtaza, F. (2014). A new switched capacitor boost inverter for renewable energy sources and POD control modulation. *ICEE*, 636–641.
18. Roozbeh, N. and Abdolreza, R. (2008). Phase-shifted carrier PWM technique for general cascaded inverters. *IEEE Trans. Power Electron.*, 23(3), 1257–1269.
19. Somanatham, R., Venkatakrishna, A. and Sandeep Reddy, M. (2014). Phase shifted and level based cascaded multilevel inverter fed induction motor drive. *Inter. J. Curr. Engg. Technol*,350-354.
20. Villarruel-Parra, I., Araujo-Vargas, N., Mondragon, E., and Forsyth, A. J. (2010). Control of a hybrid seven-level PWM inverter. *Power Electron. Congr. - CiEP, no. 1000*, 191–196.

5 ANN and ANFIS controller for load frequency control and automatic voltage regulation

Ashim Sonowal[a] *and Tilok Boruah*[b]

Department of Electrical Engineering, Jorhat Engineering College, Assam, India

Abstract

In today's world due to advancement in electrical technologies, load variation is unpredictable. Therefore, a proper control of voltage and frequency is required. This paper studies artificial neural network (ANN) and adaptive neuro fuzzy inference system (ANFIS)-based controller to perform load frequency management and automatic voltage control in an isolated power system and results are compared to that of a standard proportional integral and derivative controller (PID controller). A step load change is applied to the generator to observe dynamic frequency and per unit voltage change. The primary goal of this paper is to show the superiority of the ANN and ANFIS controller over a PID controller. Simulation is performed at the MATLAB software.

Keywords: Load frequency control (LFC), artificial neural network (ANN), automatic voltage regulation (AVR), adaptive neuro fuzzy inference system (ANFIS), transfer function (TF), proportional integral and derivative controller (PID)

Introduction

The control of voltage and frequency management is required to maintain the power system stability. During load fluctuation, real and reactive power fluctuates. The load frequency control (LFC) is in charge of the real power and frequency. The automatic voltage regulation (AVR), on the other hand, controls the reactive power and voltage magnitude. This may be accomplished by comparing and sending the minimized error to the governor and amplifier, respectively [1]. Because the time constant of the excitation system is considerably lower than that of the prime mover, a combined model for LFC and AVR is achievable. As a result, the excitation system's transient decays significantly faster and have no effect on the LFC dynamics [2].

The proportional integral and derivative controller (PID controller) is one of the most extensively used controllers for controlling system stability. Several studies are being conducted in order to optimize a PID controller employing meta-heuristics optimization techniques such as Particle Swarm Optimization, Firefly Algorithm and Genetic Algorithm [3–5]. Some researchers have also tried nonlinear sliding mode control for LFC regulation [6].

This paper aims at developing a LFC and AVR control using artificial neural network (ANN) and adaptive neuro fuzzy inference system (ANFIS) controller. The simulation result is compared with PID controller in result section. This paper is organized in sections, where, section 2 is about the modeling of the AVR and LFC model, section 3 describes application of ANN and ANFIS controller, section 4 shows the simulation result and section 5 gives the conclusion.

Modeling of the LFC and AVR model

Load frequency controller

The primary goal of the LFC controller is to keep the frequency stable. When the load changes, the frequency changes, which is amplified and turned into a command signal that controls the torque of the primary mover. The LFC block as shown in Figure 5.1 consist of a governor, turbine, inertia constant, speed regulator and a controller to control the governor speed. The values of the various quantities are given in Table 5.1.

Automatic voltage regulation

The magnitude of the voltage changes when the generator reactive loads changes. A potential transformer senses the voltage variation, which is rectified and compared to a dc set point. The amplified signal is utilized to change the exciter terminal voltage by regulating the exciter field. As a result, the generator field current increases or decreases, resulting in a higher or lower produced electromotive force. By adjusting the terminal voltage to the required value, this creates a new equilibrium. The AVR block shown in the Figure 5.1 consists of a generator, sensor, amplifier, exciter, and a controller to control the error.

[a]ashimsonowal849@gmail.com, [b]tilokb@rediffmail.com

DOI: 10.1201/9781003540199-5

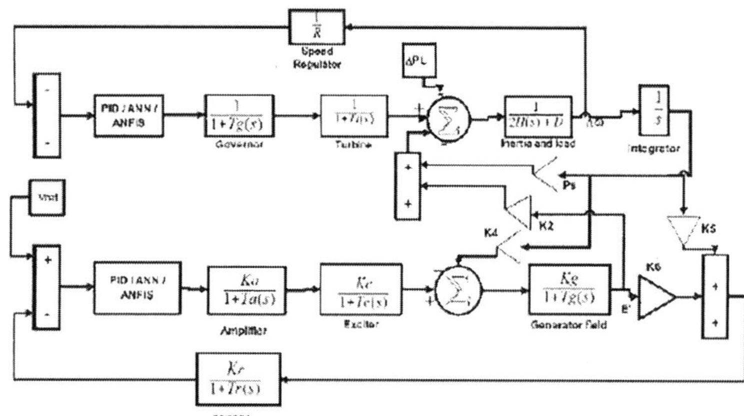

Figure 5.1 T.F. model of integrated AVR and LFC in MATLAB using PID/ANN/ANFIS

Table 5.1 Parameters for designing combined LFC and AVR model

Parameter	Gains	Time constant
Turbine	$K_t = 1$	$T_t = 0.5$
Governor	$K_g = 1$	$T_g = 0.2$
Amplifier	$K_a = 10$	$T_a = 0.1$
Exciter	$K_e = 1$	$T_e = 0.4$
Sensor	$K_r = 1$	$T_r = 0.05$
Inertia	$H = 5$	
Regulation	$R = 0.05$	
Generator	$K_g = 0.8$	$T_G = 1.4$

Combined LFC and AVR model

An isolated power system of thermal power plant is taken as a reference. Where a 0.2% load change is considered for 1% change in frequency. The synchronizing coefficient is taken as $P_s = 1.5$ and voltage coefficient is taken as $K_6 = 1.5$, and the coupling constants are $K_2 = 0.2$, $K_4 = 1.4$ and $K_5 = 0.1$. All the other parameters are given in the Table 5.1 [1]. From the concept of automatic generation control, the equation obtained is:

$$\Delta P_e = P_s * \Delta\delta + K_2 E \quad (1)$$

where, ΔP_e = Small variation in real power

P_s = Coefficient of synchronizing power

$\Delta\delta$ = Deviation of system frequency from its nominal value

K_2 = Gain constant

E = Error in voltage

Equation for impact of rotor angle on terminal voltage is:

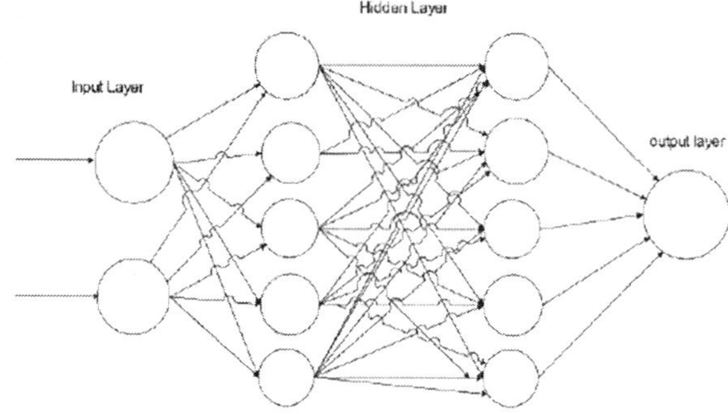

Figure 5.2 ANN architecture

$$\Delta V_t = K_5 * \Delta\delta + K_6 E \quad (2)$$

where, K_5 = variation in the final voltage for a minor change in rotor angle at constant stator emf.

K_6 = Variation in final voltage for a minor change in stator emf at constant rotor angle.

Equation for the impact of rotor angle:

$$E = \frac{K_g}{1+T_g} \times (V_f - K_4 \times \Delta\delta) \quad (3)$$

Where, $\frac{K_g}{1+T_g}$ = T.F of the generator field

V_f = Reference voltage

Application of ANN and ANFIS in the model

ANN

ANN is a machine learning model which is inspired by neurons of human brain, where the process involves training of data using certain algorithms to achieve a faster response than tradition methods. It can be classified into layers: input, hidden and output layer. Figure 5.2 is the basic model of ANN. It is an easy way for solving non-linear problems. But right and accurate training of the data is required to avoid inadequate results.

The input and output data of the PID controller is exported to the workspace after tuning, and the data is trained using the flow chart given in Figure 5.3, after successful training of the data, the model is made using the code: "gensim(net,-1)" where, net is the name of the training data. The trained model is then used in place of the PID controller [7, 8].**ANFIS**

ANFIS is the combination of ANN and fuzzy logic controller. Since, a fuzzy logic controller shows complexities in defining accurate rules for the system. The combined ANN and fuzzy logic helps in solving various nonlinear problems by manipulating the data of the system [9, 11]. But, accurate training is required for obtaining the correct output. The diagram in Figure 5.4 represents ANFIS architecture.

Graphical user interface (GUI) is used for training the ANFIS data. At first, the tuned input and output parameters of the PID controller are exported to the MATLAB workspace and then loaded to GUI interface. By default Takagi Sugeno fuzzy system is present in GUI tool. To generate Fuzzy inference file, select the input value i.e., number of membership function and type, and output value as constant or linear. In this study, output is taken as constant and triangular membership is taken as input because of its simplicity to adapt various complex problems.

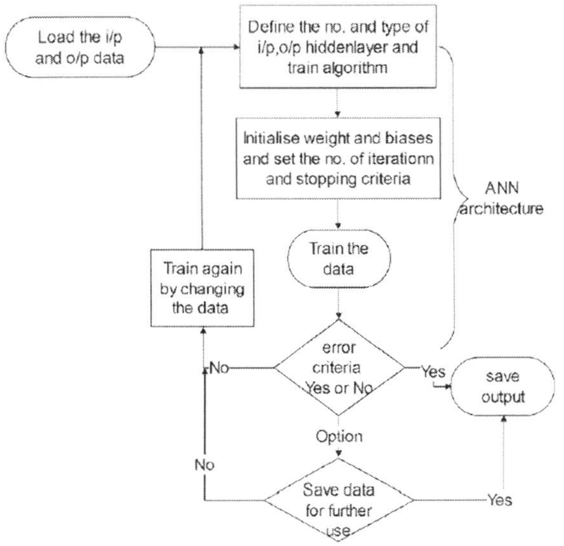

Figure 5.3 ANN training flow chart

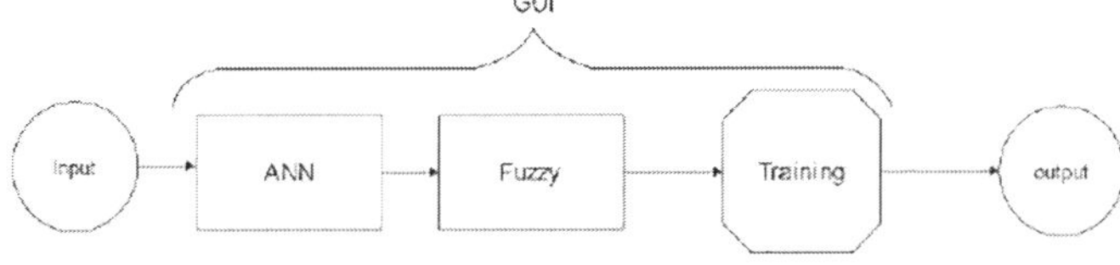

Figure 5.4 Block diagram of ANFIS controller

Simulation and results

The simulation is performed in MATLAB software as shown in Figure 5.5. The simulation time is set to 20 s and solver is set as auto. The data of the various parameters are taken as per Table 5.1. The PID controller is replaced with the ANN and ANFIS controller and simulation graphs (Figures 5.5 and 5.6).

Figure 5.6 is frequency deviation graph using ANN. blue line is for the PID controller and red line is for the ANN controller. The maximum undershoot obtained for PID controller is 0.510%, and the response time is obtained as 1.25 ms (minimum). The maximum undershoot obtained for ANN controller is 0.505%, and the response time is obtained as 1.18 ms. The response and undershoot is improved in the ANN controller.

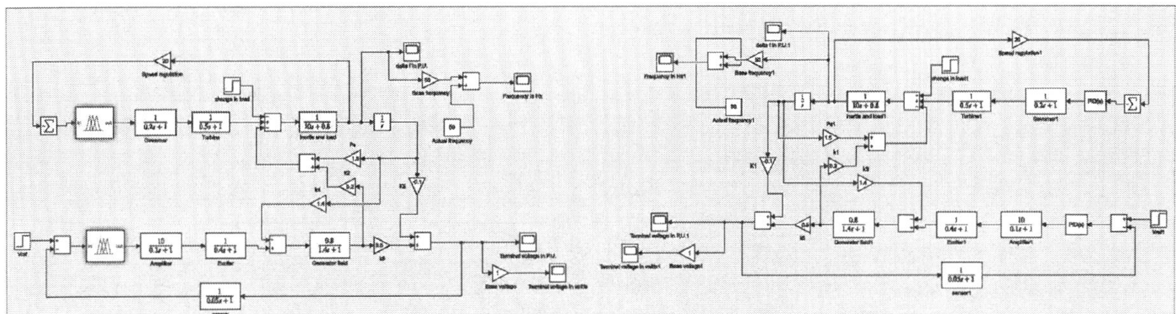

Figure 5.5 Combined MATLAB Simulink model

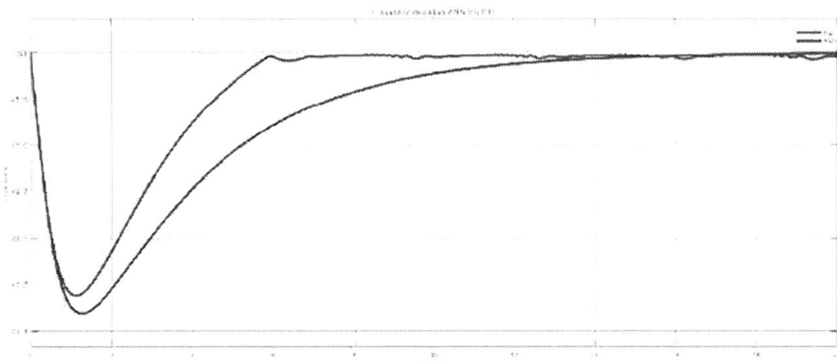

Figure 5.6 Frequency deviation graph using ANN

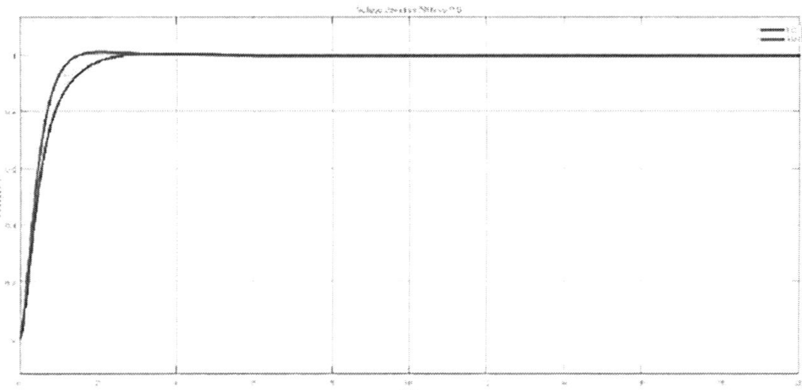

Figure 5.7 Voltage deviation graph using ANN

Figure 5.7 is voltage deviation graph using ANN. red line is for the ANN controller and blue line is for the PID controller. The peak undershoots obtained for PID controller is 1.879%, and the response time is obtained as 3.615 ms (maximum). The peak undershoots obtained for ANN controller is 1.690%, and the response time is obtained as 2.092 ms (maximum). The response is improved in the ANN controller.

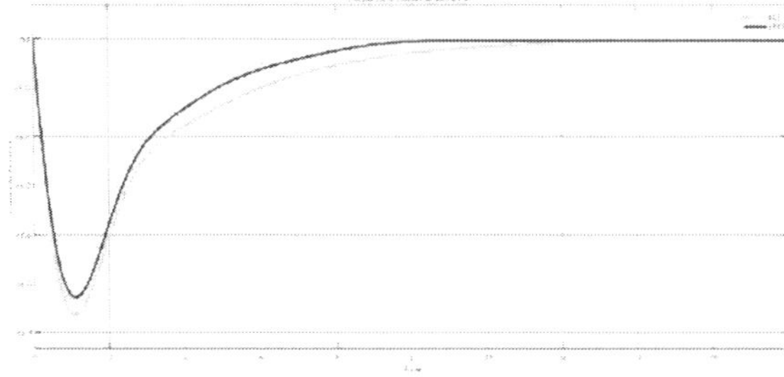

Figure 5.8 Frequency deviation graph using ANFIS

Figure 5.8 is frequency deviation graph while using ANFIS. Red line is for the ANFIS controller and yellow line is for PID controller. For PID controller the maximum overshoot obtained is 1.903%, and the response time is obtained as 1.802 ms (minimum). For ANFIS controller the maximum overshoot obtained is 1.217%, and the response time is obtained as 1.082 ms (minimum). Undershoot and response is improved in the ANFIS controller.

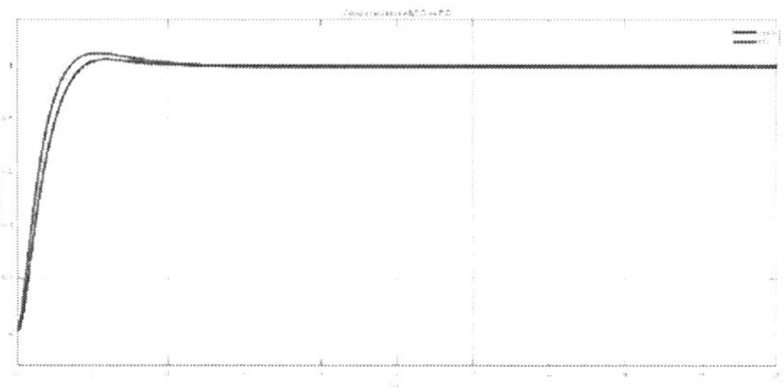

Figure 5.9 Voltage deviation graph using ANFIS

Figure 5.9 is voltage deviation graph using ANFIS. Red line is for the ANFIS controller and blue line is for the PID controller. For PID controller the peak overshoot obtained is 4.737%, and the response time is obtained as 2.175 ms (maximum). For ANFIS controller the peak overshoot obtained is 2.577%, and the response time

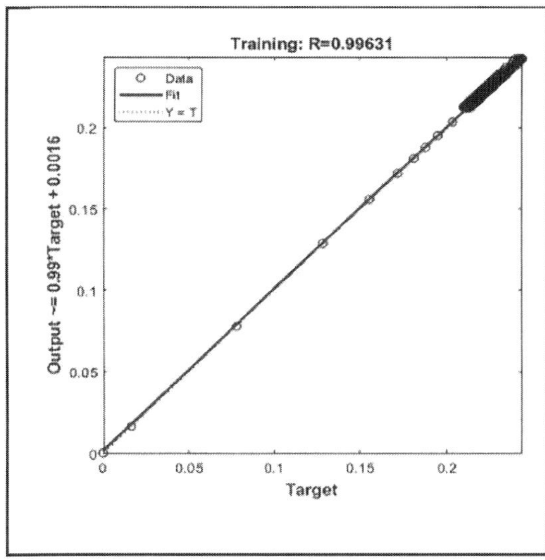

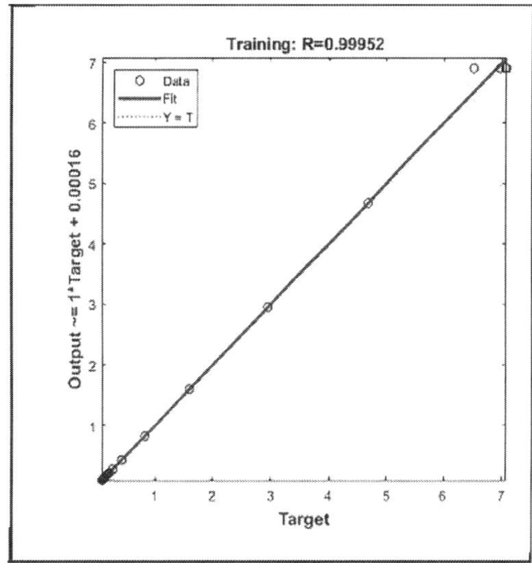

Figure 5.10 (a) LFC regression (b) AVR regression

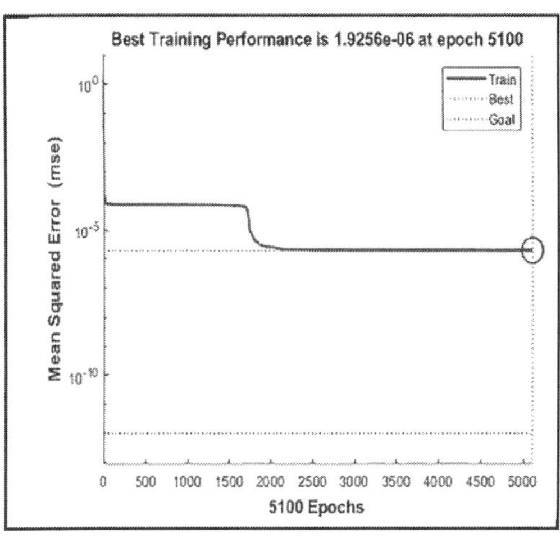

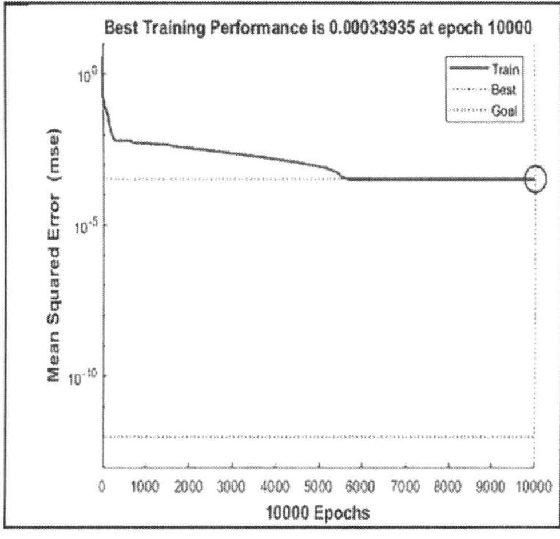

Figure 5.11 (a) For LFC (b) For AVR

is obtained as 2.373 ms (maximum). Response is improved in the ANFIS controller.

The regression graph after ANN training is shown in Figure 5.10. The value of regression for LFC is 0.99631 and for AVR is 0.99952, which is close to 1. Hence the training is successful.

Figure 5.11 is the mean square vs. iteration graph while training the data using ANN. From the graph we can see that mean squared error for LFC is 1.92×10^{-6} at iteration of 5100, and the mean squared error for AVR is 0.33935×10^{-4} at iteration of 10,000.

Conclusions

From the graphs it is observed that ANN and ANFIS give better responses than PID controller. PID controller is most widely used for its simple construction but optimizing the gain parameter is difficult. So ANN and ANFIS can be used as a pre-trained model. The responses can be further improved by accurately manipulating the data of the system. Since in present power scenario smart grid is evolving with increase in technology, so ANN and ANFIS will play a great role in maintaining stability of the power system.

Acknowledgement

The authors gratefully acknowledge the staff and authority of Electrical department for their cooperation in the research.

References

1. Saadat, H. (1999). Power system analysis. McGraw Hill, 12, 527–585.

2. Goswami, M., Goswami, K., and Mishra, L. (2017). Load frequency and voltage control of two area interconnected power system using PID controller. *IJERT*, 4, 301–315. www.researchtrend.net.
3. Salman, G. A., Jafar, A. S., and Ismael, A. I. (2019). Application of artificial intelligence techniques for LFC and AVR systems using PID controller. *Inter. J. Power Elec. Drive Sys.*, 10(3), 1694–1704. http://doi.org/10.11591/ijpeds.v10.i3.
4. Grover, H., Verma, A., and Bhatti, T. S. (2020). Load frequency control & automatic voltage regulation for a single area power system. *2020 IEEE 9th Power India Inter. Conf. (PIICON)*, 1–5. https://doi.org/10.1109/PIICON49524.2020.9112902.
5. Ghosh, A., Ray, A. K., Nurujjaman, M., and Jamshidi, M. (2021). Voltage and frequency control in conventional and PV integrated power systems by a particle swarm optimized Ziegler–Nichols based PID controller. *SN Appl. Sci.*, 3, 1–13. https://doi.org/10.1007/s42452-021-04327-8.
6. Prasad, S., Purwar, S., and Kishor, N. (2019). Load frequency regulation using observer based non-linear sliding mode control. *Inter. J. Elec. Power Energy Sys.*, 104, 178–193. https://doi.org/10.1016/j.ijepes.2018.06.035.
7. Kumari, K., Shankar, G., Kumari, S., and Gupta, S. (2016). Load frequency control using ANN-PID controller. *2016 IEEE 1st Inter. Conf. Power Elec. Intell. Con. Energy Sys. (ICPEICES)*, 1–6. http://doi.org/10.1109/ICPEICES.2016.7853516.
8. Salih, A. M., Humod, A. T., and Hasan, F. A. (2019). Optimum design for PID-ANN controller for automatic voltage regulator of synchronous generator. *2019 4th Sci. Inter. Conf. Najaf (SICN)*, 74–79. http://doi.org/10.1109/SICN47020.2019.9019367.
9. Mosaad, M. I. and Salem, F. (2014). LFC based adaptive PID controller using ANN and ANFIS techniques. *J. Elec. Sys. Inform. Technol.*, 1(3), 212–222. http://dx.doi.org/10.1016/j.jesit.2014.12.004.
10. Ramoji, S. K., Saikia, L. C., Dekaraja, B., Behera, M. K., Bhagat, S. K., Babu, N. R., and Saha, A. (2022). Conflated voltage–frequency control of multi-area multi-source system using fuzzy TID controller and its real-time validation. *Adv. Smart Energy Sys.* 277–294. Singapore: Springer Nature Singapore. http://doi.org/10.1007/978-981-19-2412-5_16.
11. Saadat, S. A., Ghamari, S. M., Mollaee, H., and Khavari, F. (2021). Adaptive neuro–fuzzy inference systems (ANFIS) controller design on single–phase full–bridge inverter with a cascade fractional–order PID voltage controller. *IET Power Elec.*, 14(11), 1960–1972. https://doi.org/10.1049/pel2.12162.

6 Artificial gorilla troops optimizer – Assisted AGC study for improving frequency stability of isolated power systems

Shruthi Nookala[a], *Chandan Kumar Shiva and B. Vedik*[b]

Department of Electrical & Electronics Engineering, SR University, Warangal, India

Abstract

In the study, frequency regulation in an isolated hybrid power system is studied. This system uses wind turbines, solar panels, aqua-electrolysers, fuel cells, and diesel engine generators. Flywheel and battery energy storage systems are also used in a hybrid system. An artificial gorilla troops optimizer technique is suggested for reducing the frequency deviation. The designed controller performance is tested under a wind speed, random load demand, and non-linear solar radiation change. The simulation results showed that the controller with FESS and BESS reduced frequency deviation. Thereby, the suggested controller improves the hybrid system frequency regulation.

Keywords: Frequency stability, isolated power system, artificial gorilla troops optimizer algorithm

I. Introduction

An isolated power system (IPS) or standalone power system runs independently in the main grid. Remote populations like rural or island villages employ IPSs because connecting to the main grid is impractical or expensive. IPSs may use diesel generators, wind turbines, solar panels, hydropower generators, and energy storage systems like batteries or flywheels. The system components work together to meet power demand without external power or grid support. Due to their small size, significant power supply and demand unpredictability, and restricted ability to respond to system variables, IPSs present unique power system stability and control difficulties. Therefore, frequency stability is crucial to isolated power system performance.

1.1 Literature survey

IPS is a system which is not connected with any national/regional grid. To power or electrify this power system there are many renewable energy sources like solar PV (Photo Voltaic), WTG (Wind Turbine Generator), FC (Fuel Cell) using hydrogen as a fuel by an aqua electrolyzer, etc., to be connected with this power system. This IPS is very advantageous as there is less transmission and distribution losses, no maintenance required, less cost of fuel, clean neat and abundant source of energy and a small micro grid can be formed easily from these components for forming a hybrid power system. And to avail this renewable power always, a BESS and FESS, ultra capacitor may be used for storing the energy, when in excess and it can be reused whenever needed [1]. The stability and dependability of the power system have become more important issues for power system operators and decision-makers in recent years. The necessity for stable and effective isolated power systems becomes more critical as the need for dependable and sustainable power supply rises. In isolated power systems, LFC is a vital control mechanism for preserving the equilibrium between power generation and load demand [2, 3]. To improve LFC performance, evolutionary optimization approaches have been extensively researched in [4–12]. This review of the literature offers a thorough summary of studies that have used GA, PSO, and other evolutionary optimization approaches in LFC of isolated power systems.

For LFC optimization in isolated power systems, GA had been widely used. For a hybrid energy generation/storage system using solar thermal, diesel, and wind power, Das et al. [4] presented a GA-based frequency controller. In this article [5], the authors suggested an H-infinite controller for hybrid DG systems to regulate frequency. The controller is built using a combination of the PSO and HSA optimization techniques. Using robust optimization, Pan and Das [6–8] initiated a fractional-order AGC strategy for distributed energy resources. PSO has also been used in this work to optimize LFC in isolated power systems. Insights into energy resource planning and control in hybrid micro-grids were provided by the work of Barik et al. [9]. A peak shaving method for an islanded micro-grid utilizing a battery energy storage

[a]nookala.shruthi@gmail.com, [b]chandankumarshiva@gmail.com

DOI: 10.1201/9781003540199-6

device was proposed by Uddin et al. [12]. Other meta-heuristic algorithms have been investigated for LFC optimization in isolated power systems in addition to GA and PSO [10, 11]. The authors of [10] talk about energy management in a micro-grid with an emphasis on tackling the problems brought on by the maximum penetrations of renewable energy resources and a surplus power output. In a grid of AC/DC micro-grids, Alghamdi and Caizares [11] looked into the coordinated regulation of frequency and voltage. The authors of [12] suggest a peak shaving algorithm for an island micro-grid that makes use of a battery energy storage technology. The study offers a method for managing peak demands in a microgrid by maximizing battery system charging and discharging. In isolated power systems, system characteristics can be controlled and frequency stability can be improved using optimization methods such the AGTO [13–16]. The social interactions and collaboration of wild gorilla troops, which display high levels of coordination, communication, and adaptation, served as the basis for AGTO. In order to react to changes in the system dynamics, such as load variations and disturbances, AGTO is built to optimize the system parameters.

1.2 Motivation

This work addresses the issue of frequency stability in isolated hybrid power systems that incorporate multiple RES. Frequency stability is an essential component of power systems because it directly impacts the quality and dependability of the electrical power supply. In isolated power systems, the absence of interconnections with other power systems can lead to significant frequency deviations, which can cause power disruptions and other issues. AGC is used to regulate the power output of autonomous generating units to match load demand and sustain system frequency within acceptable limits in order to address this issue. However, conventional AGC methods may not be adequate for addressing frequency stability issues in isolated power systems because they neglect the nonlinear and time-varying characteristics of the system. Therefore, this paper proposes the use of an AGTO-assisted AGC strategy for enhancing frequency stability in isolated power systems. The main aim of this paper is to demonstrate that the proposed AGTO-assisted AGC method improves the frequency stability of an isolated power system. The simulation results are demonstrated to show the effectiveness of the proposed methodology.

1.3 Contribution of this article

Utilizing a potent optimization technique, this paper has analyzed the AGC response of the investigated test system configuration. The following is the primary contribution of this work.

(a) The suggested method utilizes an AGTO to enhance frequency stability.
(b) The suggested method optimizes AGC based PID parameters.
(c) The suggested algorithm's convergence plot, dynamic performance analysis, and stability analysis are analyzed.

Studied system

The LFC analysis in this paper uses an isolated HPS model as a test system. The model used in this study relies on solar thermal and photovoltaic (STPG) as its primary energy sources [17]. The combined output powers of the STPG, PV, and ESDs are the power supplied to the load. Figure 6.1 [17] depicts

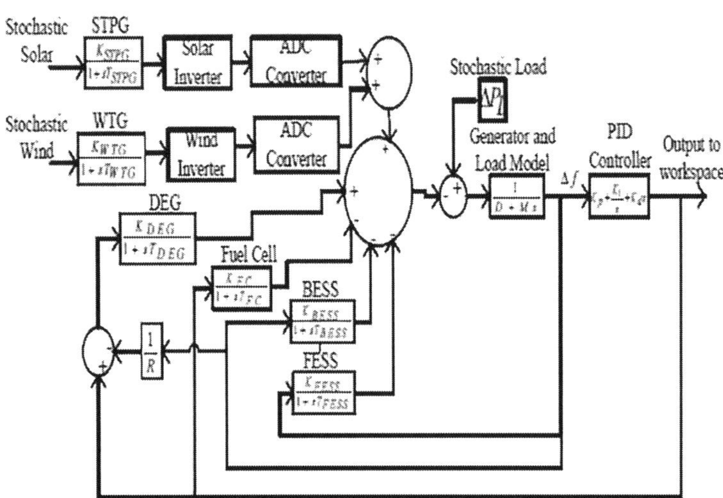

Figure 6.1 Studied system

a block diagram of the proposed test systems, which is dependent on the transfer function.

Control structure

A PID controller is a common control structure for control system research. The PID controller adjusts generator power output depends on the deviation between nominal frequency and the measured frequency. LFC investigations in isolated power systems can benefit from PID controllers' quick response times and low steady-state errors. PID controller performance depends on tuning parameters, which must be modified based on system dynamics and performance needs [16].

$$K_i(s) = K_{pi} + \frac{K_{ii}}{s} + s\,K_{di} \tag{1}$$

Problem formulation

The issue formulation entails the utilization of a PID controller for the purpose of regulating the frequency. The utilized objective function in this particular scenario is the integral of time multiplied absolute error (ITAE), which serves as a metric for quantifying the system's reaction to external perturbation. The objective function aids in the optimization of the controller's performance through the minimization of the ITAE with respect to time. The ITAE objective function is given as: [16]

$$ITAE = \int_{0}^{t_s} |\Delta f_i|\, t\, dt \tag{2}$$

whereas, Δf_i is the deviation in the frequency of the system and t_s is the time horizon.

The gains of the PID controller serve as the optimization task's restrictions. They can be shaped between the minimum and maximum numbers. Constraint (3) demonstrates the bounds of the restrictions [16].

$$\left.\begin{array}{l} K_p^{min} \le K_p \le K_p^{max} \\ K_i^{min} \le K_i \le K_i^{max} \\ K_d^{min} \le K_d \le K_d^{max} \\ N^{min} \le N \le N^{max} \end{array}\right\} \tag{3}$$

Proposed algorithm

The GTO algorithm is a type of swarm intelligence-based optimization technique which got inspired by the behavior of gorilla troops in nature. GTO is a meta-heuristic algorithm that mimics the social behavior of the gorilla troops, in which the members of the troop collaborate communicate with each other to achieve common goals. This GTO is achieved by means of intelligence methods of troops of gorillas for finding the best position for its food and other gorilla troops for their occurrence. These troops have silver back gorilla as a leader [13]. In GTO, a population of solutions is generated and evaluated, and the best solutions are selected for further iterations. The solutions are represented as individuals or agents, and each agent is characterized by its position in the search space. The search space is explored through a series of iterations, during which the agents communicate with each other to share information and improve their individual solutions [13].

Simulation results

The main objective of the paper is to examine the effective of the AGTO algorithm in regulating the frequency of an isolated hybrid DG system. The system comprised various energy sources, including WTG, PV panels, a FC, and a DEG. The system also incorporated FESS and BESS to mitigate frequency deviations. To evaluate the performance of the proposed AGTO-based controller, simulations were conducted considering various sources of uncertainty, such as speed of the wind, randomly varying load demand, and non-linearly varying

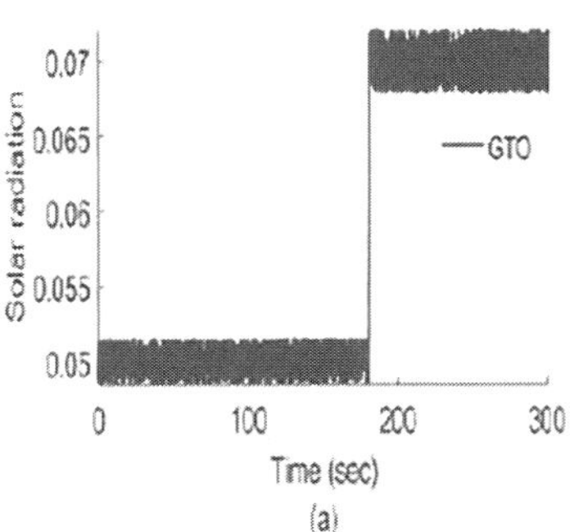

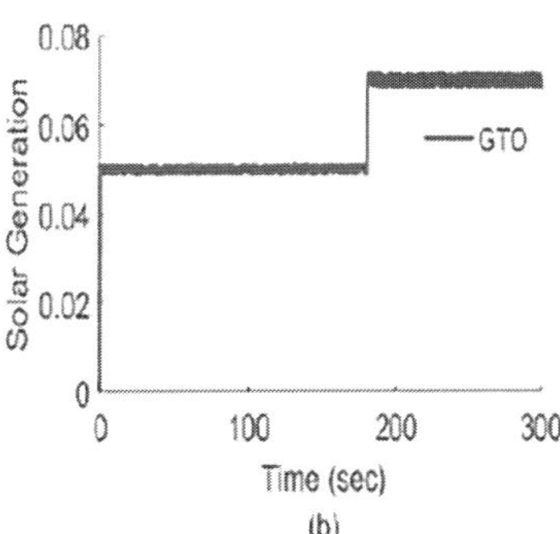

Figure 6.2 Distributed generation details: (a) solar radiation, (b) solar generated generation

solar radiation. The frequency deviation was used as the metric to assess the efficiency of the controller. Figure 6.2 represents the distributed generation details in the solar radiation. Figure 6.2(a) displays the solar radiation data, indicating the intensity of solar energy received by the photovoltaic panels. Figure 6.2(b) shows the power generation from the photovoltaic panels based on the solar radiation.

Figure 6.3 presents the generated power by auxiliary components in the hybrid DG system. Figure 6.3(a) shows the power generated by the DEG, Figure 6.3(b) represents the power generated by the FC.

Figure 6.4 shows the LFC response profile. Figure 6.4(a) represents the system frequency deviation, indicating the variation of the system frequency from its nominal value. It provides insights into the variations in the system frequency caused by load demand changes or disturbances. Figure 6.4(b) displays the load demand.

Figure 6.5 is a convergence plot showing the optimization algorithm's performance over 30 iterations. It demonstrates the progression of the objective function or fitness value during the optimization process. The simulation results obtained from the study indicate that the load following capability of the AGTO algorithm is satisfactory. This means that the algorithm effectively regulates the power output in response to load perturbations, ensuring that the system maintains a stable and balanced operation.

The controller employed in the AGTO algorithm exhibits excellent performance in terms of

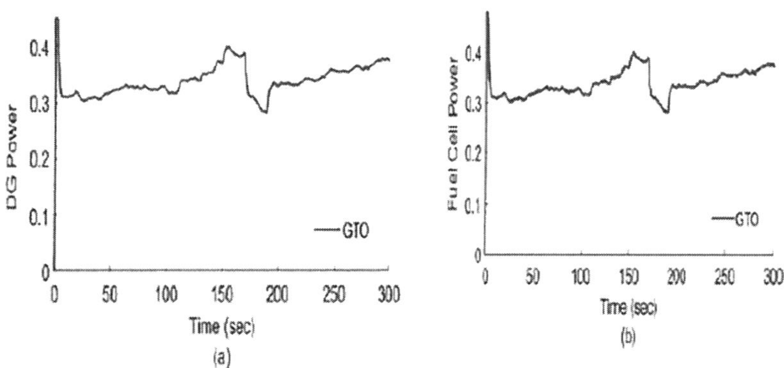

Figure 6.3 Power generated by auxiliary components: (a) DEG, (b) FC

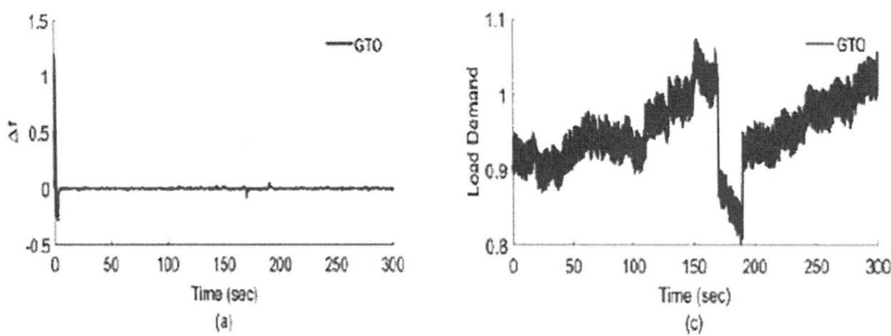

Figure 6.4 LFC response: (a) system frequency deviation, (b) load demand

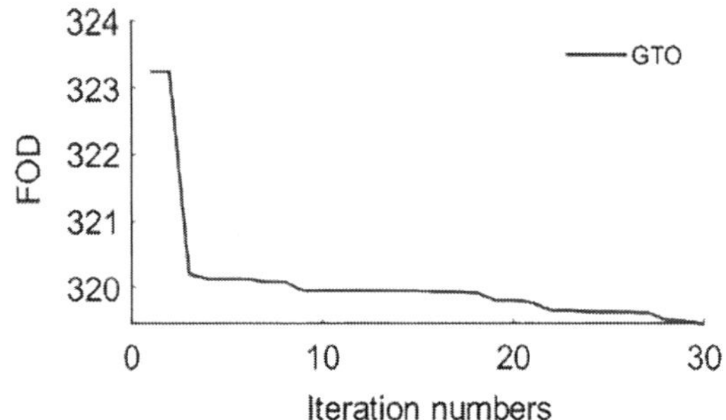

Figure 6.5 Convergence plot for 30 number of iterations

maintaining system frequency within acceptable limits. It effectively regulates the control signals based on feedback information, ensuring that the system frequency remains stable despite varying load conditions or disturbances.

Conclusion

The AGTO technique is used to examine frequency deviation. The study evaluates the AGTO algorithm's frequency deviation reduction with a PID controller. The proposed controller's analyses show a considerable improvement in system performance. FESS and BESS combined with the AGTO algorithm increased the system's frequency deviation in the transient characteristics and the stability, reducing the requirement for expensive and environmentally hazardous fossil fuel power plants.

References

1. Senjyu, T., Nakaji, T., Uezato, K., and Funabashi, T. (2005). A hybrid power system using alternative energy facilities in isolated island. *IEEE Trans. Energy Conver.*, 20(2), 406–414.
2. Maged, N. A., Hasanien, H. M., Ebrahim, E. A., Tostado-Véliz, M., and Jurado, F. (2022). Real-time implementation and evaluation of gorilla troops optimization-based control strategy for autonomous microgrids. *IET Renew. Power Gen.*, 16(14), 3071–3091.
3. Ray, P. K., Mohanty, S. R., and Kishor, N. (2011). Proportional–integral controller based small-signal analysis of hybrid distributed generation systems. *Energy Conv. Manag.*, 52(4), 1943–1954.
4. Das, D. C., Roy, A. K., and Sinha, N. (2012). GA based frequency controller for solar thermal–diesel–wind hybrid energy generation/energy storage system. *Inter. J. Elec. Power Energy Sys.*, 43(1), 262–279.
5. Mohanty, S. R., Kishor, N., and Ray, P. K. (2014). Robust H-infinite loop shaping controller based on hybrid PSO and harmonic search for frequency regulation in hybrid distributed generation system. *Inter. J. Elec. Power Energy Sys.*, 60, 302–316.
6. Pan, I. and Das, S. (2014). Kriging based surrogate modeling for fractional order control of microgrids. *IEEE Trans. Smart Grid*, 6(1), 36–44.
7. Pan, I. and Das, S. (2016). Fractional order fuzzy control of hybrid power system with renewable generation using chaotic PSO. *ISA Trans.*, 62, 19–29.
8. Pan, I. and Das, S. (2015). Fractional order AGC for distributed energy resources using robust optimization. *IEEE Trans. Smart Grid*, 7(5), 2175–2186.
9. Barik, A. K., Jaiswal, S., Das, D. C. (2022). Recent trends and development in hybrid microgrid: A review on energy resource planning and control. *Int. J. Sustain. Energy*, 41(4), 308–322.
10. Tabar, V. S. and Abbasi, V. (2019). Energy management in microgrid with considering high penetration of renewable resources and surplus power generation problem. *Energy*, 189, 116264. https://doi.org/10.1016/j.energy. 2019.116264.
11. Alghamdi, B. and Cañizares, C. (2022). Frequency and voltage coordinated control of a grid of AC/DC microgrids. *Appl. Energy*, 310, 118427. https:// doi.org/10.1016/j.apenergy.2021.118427.
12. Uddin, M., Romlie, M. F., Abdullah, M. F., Tan, C., Shafiullah, G. M., and Bakar, A. H. (2020). A novel peak shaving algorithm for islanded microgrid using battery energy storage system. *Energy*, 196, 117084. https://doi.org/10.1016/j.energy.2020.117084.
13. Abdollahzadeh, B., Soleimanian, G. F., and Mirjalili, S. (2021). Artificial gorilla troops optimizer: A new nature-inspired metaheuristic algorithm for global optimization problems. *Int. J. Intell. Syst.*, 36, 5887–5958. https://doi.org/10.1002/int.22535.
14. Xiao, Y., Sun, X., Guo, Y., Li, S., Zhang, Y., and Wang, Y. (2022). An improved gorilla troops optimizer based on lens opposition-based learning and adaptive β-hill climbing for global optimization. *Comput. Model. Eng. Sci.*, 131(2), 815–850. https://doi.org/10.32604/cmes.2022.019198.
15. Ginidi, A., Ghoneim, S. M., Elsayed, A., El-Sehiemy, R., Shaheen, A., and El-Fergany, A. Gorilla troops optimizer for electrically based single and double-diode models of solar photovoltaic systems. *Sustainability*, 13, 9459–9469. https://doi.org/10.3390/su13169459.
16. Ali, M., Kotb, H., Aboras, K. M., and Abbasy, N. H. (2021). Design of cascaded PI fractional order PID controller for improving the frequency response of hybrid microgrid system using gorilla troops optimizer. *IEEE Acc.*, 9, 150715–150732. https://doi.org/10.1109/ACCESS.2021.3125317.
18. Balaji, K. (2023). Load frequency control in stochastic (dynamic) micro grid (https://www.mathworks.com/matlabcentral/fileexchange/83908-load-frequency-control-in-the-stochastic-dynamic-microgrid), MATLAB Central File Exchange. Retrieved April 29, 2023.

7 Improving power system stability with artificial gorilla troops optimizer-assisted automatic generation control study

Shruthi Nookala[a], *Chandan Kumar Shiva and B. Vedik*[b]

Department of Electrical & Electronics Engineering, SR University, Warangal, India

Abstract

The automatic generation control (AGC) is a control technique in a power system that maintains the balance between power generation and the load demand. A new strategy to improving the AGC in the power system for increasing frequency and regulating tie-line power is proposed in this work. The proposed method optimizes the proportional-integral-derivative controller gain in the AGC system using the artificial gorilla troops optimizer (AGTO) algorithm. The AGTO method is a meta-heuristic optimization algorithm inspired by gorilla troop behavior in the wild. The integral time absolute error was employed as the goal function in this investigation. According to simulation results, the suggested method perform better in terms of preserving system frequency, minimizing overshoot, and settling time, and improving power system stability.

Keywords: Frequency enhancement, artificial gorilla troops optimizer algorithm, optimization

1. Introduction

1.1 Essentials

Because of the greater integration of renewable energy sources, environmental concerns, and rising electricity consumption, power system stability has become a critical issue in recent years. The automatic generation control (AGC) of power plants, which is responsible for maintaining the balance between power generation and load demand, is one of the key issues of power system stability [1, 2]. The AGC idea is critical for managing power production for various locations based on load variations. When the load in any area changes due to a small disturbance, the frequency response changes accordingly, as does the steady state error and tie-line power, and all of these things must be maintained in the best possible ways, which is accomplished through controlling and optimization techniques. To solve this issue, academics have proposed a number of strategies for improving AGC performance, including traditional PID control, fuzzy logic control, and neural network control. Traditional procedures, on the other hand, may not always yield optimal solutions and may suffer from sluggish convergence, oscillations, and overshoot.

1.2 Literature survey

AGC is an essential component of power systems because it maintains a healthy equilibrium between load demand and generation. When power systems are interconnected, the controlling of tie-line power flow becomes an important consideration [3]. Various control techniques and optimization methods have been explored to enhance AGC performance [4–6]. Meta-heuristic techniques, which are optimization algorithms suitable for complex and nonlinear problems like AGC, have gained significant attention in recent years. This literature review provides an overview of the applications of meta-heuristic techniques in AGC. The implementation of intelligent AGC in isolated power systems, such as the one in Japan, has been successful [7]. Control structures for different meta-heuristic algorithms have been discussed in refs [8, 9]. One of the earliest meta-heuristic techniques applied to AGC is the genetic algorithm (GA) [10, 11]. GA has been used to optimize PID controllers for AGC, demonstrating its effectiveness in the maintenance of system frequency and tie-line power. Particle swarm optimization (PSO) is another important meta-heuristic method used in AGC [12]. PSO-based PID controllers have shown success in maintaining system frequency and reducing settling time. Ant Colony Optimization (ACO) is inspired by the foraging behavior of ants and has also been applied to AGC [13]. ACO-based PID controllers have proven effective in maintaining system frequency while reducing overshoot and settling time. GWO, a relatively new meta-heuristic technique, has shown promise in AGC. GWO has been used to optimize the parameters of a fuzzy-PID controller, resulting in improved frequency control and

[a]nookala.shruthi@gmail.com, [b]chandankumarshiva@gmail.com

DOI: 10.1201/9781003540199-7

reduced overshoot and settling time [14]. Another meta-heuristic technique, the artificial bee colony (ABC) algorithm, has been employed for AGC [15]. By optimizing PID controller parameters using ABC, the system frequency can be well-maintained, and the settling time can be reduced. Considering the previous research in this field, a new technique called artificial gorilla troops algorithm [17] is proposed in this paper. The effectiveness of this algorithm in AGC is explored, providing new insights into improving frequency control and reducing settling time.

1.3 Motivation

Heuristic and meta-heuristic algorithms are the approximate methods for solving any optimization problem and provide the optimal solutions in less time compared to other techniques. Meta-heuristics mainly have many applications these days, due to its advantages like, no complex calculations of derivations, easy to implement and simple in finding optimization values using local search regions. This is the motivation for meta-heuristic technique for solving the optimization problem and obtaining optimal solution. Studies have shown that the majority of proposed meta-heuristic algorithms are based on the natural hunting and prey behavior of animals. However, there hasn't yet been any work done to build and create a meta-heuristic approach that simulates the way of life of gorilla groups. This inspired us to develop the GTO, a mathematical representation of gorilla behavior in their search for food, water, and shelter. This paper includes research into the use of AGTO-assisted AGC to raise the system frequency of the power grid. Integral gain, proportional gain and differential gain are just few of the AGC control parameters that the proposed method intends to optimize.

1.4 Contribution of this article

Utilizing a potent optimization technique, this paper has analyzed the AGC response of the investigated test system configuration. In summary, the following are examples of the work performed:

(a) An optimization method for the GTO technique, based on their capability and our aim, has been devised for the aforementioned AGC problem.
(b) The studied technique is applied to AGC issue in the context of standard, four-area power systems.
(c) The convergence plot, dynamic performance analysis, and stability analysis of the suggested algorithm are provided.

2. Studied system and control structure

For investigating the AGC behavior in a power system, an equivalent transfer function-based model is used, which consists of the governor, turbine, power system (load and machine) as the important components. This section shows how the power system models are developed in the analysis of AGC. Figure 7.1 illustrates a simple diagram of an interconnected four areas test system, where areas 1, 2, and 3 consist of identical capacity reheat thermal power systems, and area 4 is a hydro turbine power system. Related and important parameters of the system are listed in Section A.1 of the Appendix, which can be found in work did by Tan [16].

3. Control structure

AGC systems employ PID controllers extensively. The PID controller compares the measured frequency to the set point frequency to the AGC system's reference signal. PID controllers calculate error signals that are used to change system control inputs like governor valves to return frequency to set point. The proportional component part of the PID controller generates control signal proportional to the measured-set point frequency difference. Integrating historical error signals minimizes steady-state errors by adding a control signal [8, 9]. The derivative component prevents overshoot and oscillations by providing a control signal proportionate to the error signal's rate of change [16].

$$K_i(s) = K_{pi} + \frac{K_{ii}}{s} + s\,K_{di} \quad (1)$$

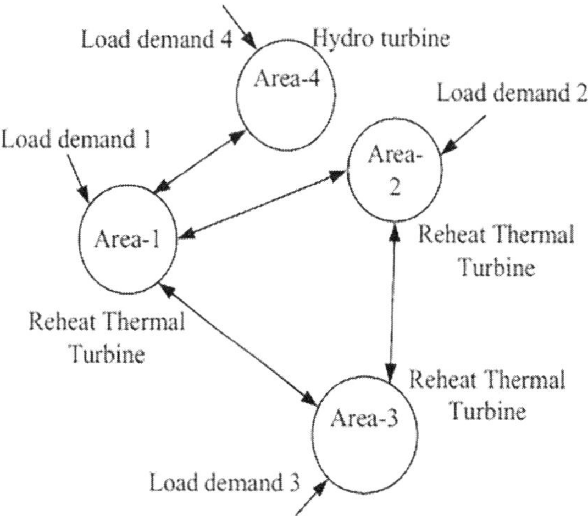

Figure 7.1 Interconnected four area test power system [16]

4. Problem formulation

AGC problem aims for maintaining and regulating the power system frequency within a limited range by controlling the power output of generators. The objective is to minimize the frequency deviation and for restoring the system frequency to its limited set point after any disturbances. In this analysis, the objective function ITAE is used to evaluate performance of the proposed AGTO-assisted AGC method. The ITAE objective function represents the area between the frequency deviation and the set point over time, giving more weight to larger deviations that persist for a longer duration. The ITAE objective function is given as [16]

$$ITAE = \int_{0}^{t_s} \left(|\Delta f_i| + |\Delta P_{tie\,ij}| \right) t \, dt \tag{2}$$

where, Δf_i is the frequency of a system, $\Delta P_{tie\,ij}$ is the set point frequency, and t_s is the time horizon.

5. Proposed algorithm

In this paper, we use meta-heuristic optimization techniques for controlling the different areas, power generation capability according to load changes and this meta-heuristic technique uses the natural intelligence methods of various birds and animals. This paper employs the gorilla troops optimization (GTO) techniques which updates the values based on exploration and exploitation methods of GTO technique [17]. This GTO is achieved by means of intelligence methods of troops of gorillas for finding the best position for its food and other gorilla troops for their occurrence. These troops have silver back gorilla as a leader for them.

6. Simulation results

Analyzing dynamic performance for a four area power system serves to highlight the suggested algorithm's capacity to be tuned. The scenario taken into consideration, in this case is the concurrent application of SLP at an instant of 0.1 p.u.MW in areas 1 and 3. Figure 7.2 displays the study test system's frequency deviations. The LFC response can offer important ideas about the performance and stability of the connected power system under analysis. Here, a flat or gradual change in LFC responses shows that the connected power system is stable and capable of adjusting to changes in load without experiencing substantial frequency deviations. It indicates that the connected power system is stable or completely under control with respect to its control systems.

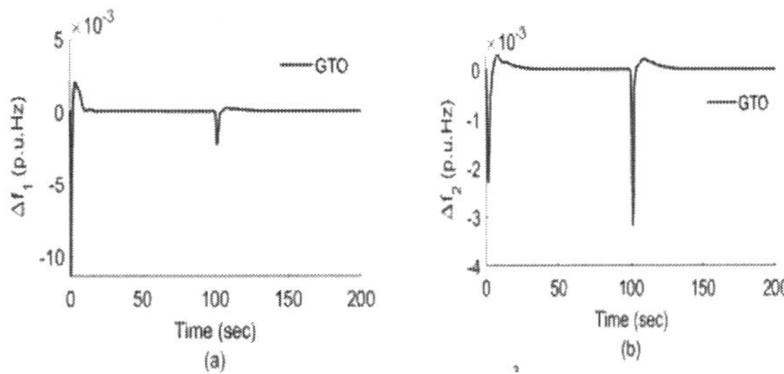

Figure 7.2 Frequency response of a given 4 area test system with GTO technique

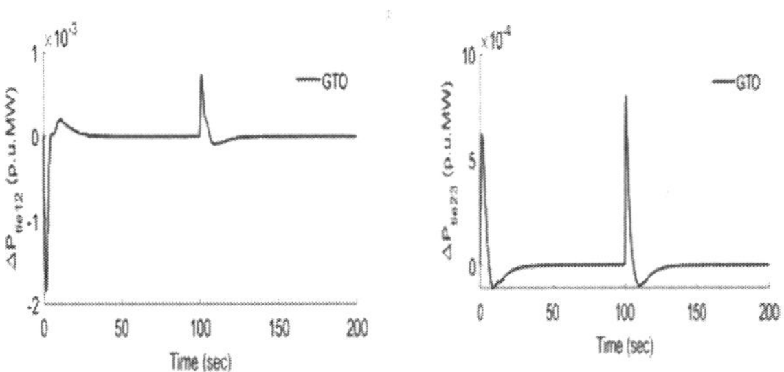

Figure 7.3 Tie-line powers of a given 4 area test system with GTO technique

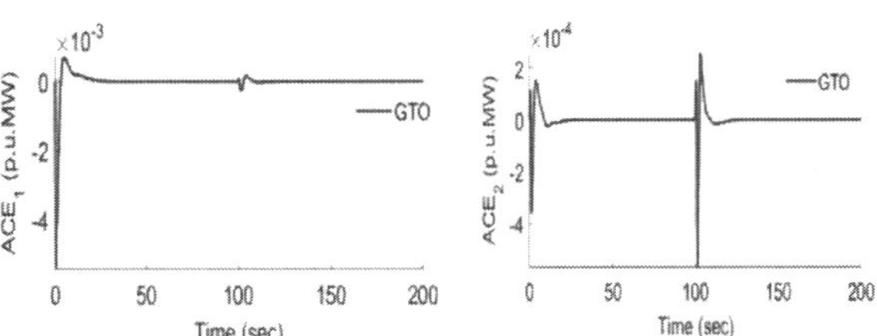

Figure 7.4 Errors of 4 areas of a given test power system using GTO technique

The changes in tie-line power, time domain responses for the proposed test system are given in Figure 7.3. The tie-line power deviation curve shows how power flow on the tie-line affects power system frequency deviation. This curve shows key LFC system performance and related power system stability information. This figure indicates that the power system is steady and insensitive to tie-line power flow variations. The curve indicates that the LFC controls the frequency changes of the interconnected power system and limits it to an acceptable level with its tie-line power flow. Tie-line power flow has a sharp slope and few fluctuations, according to the curve.

Figure 7.4 shows area control error (ACE) time domain simulations. The ACE response curve shows how the ACE affects the interconnected power system's frequency deviation. The ACE is the discrepancy between the scheduled and actual control area interchange. The interconnected power system can

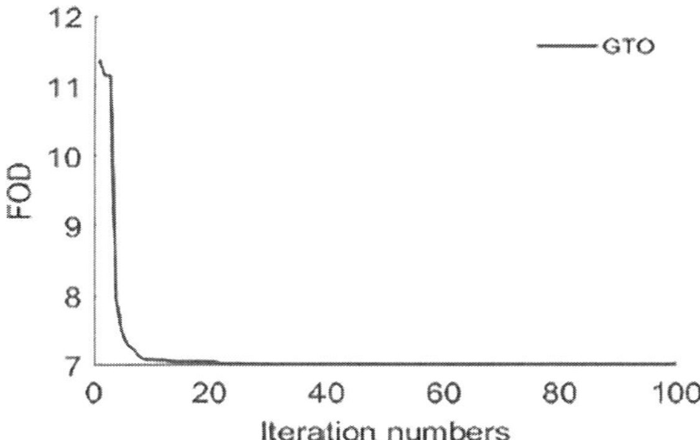

Figure 7.5 Convergence plot for 100 no of iterations

be evaluated under different load conditions using the ACE response curve. The suggested algorithm's rapid ACE response indicates its capability.

In LFC optimization, the convergence curve shows the objective function value's behavior. The convergence curve is important in LFC investigations because it shows the convergence rate and optimization solution quality. The convergence curve in LFC research helps evaluate the optimization method's efficacy and efficiency. A high convergence rate and a small final objective function value (Figure 7.5) indicates that the GTO algorithm found a satisfactory solution and that the control settings are suitable for the LFC problem. The convergence curve also demonstrated that optimization algorithm settings improved LFC problem performance.

7. Conclusions

This work improves power system frequency stability with the AGTO algorithm for AGC. The AGTO algorithm optimizes PID controller gain in the AGTO based AGC approach. The result of simulation studies suggests that the proposed strategy preserves system frequency and lowers overshoot and settling time, improving power system stability.

References

1. Kundur, P., Balu, N. J. and Lauby, M. G. (1994). Power system stability and control. 7, New York: McGraw-Hill (Pages-1-2000).
2. Kothari, D. P. and Dhillon J. S. (2004). Power System Optimisation. India: Prentice Hall (Pages-1-700)
3. Cohn, N. (1957). Some aspects of tie-line bias control on interconnected power systems. *Amer. Inst. Electr. Eng. Trans.*, 75, 1415–1436.
4. Kumar, I. P. and Kothari, D. P. (2005). Recent philosophies of automatic generation control strategies in power systems. *IEEE Trans. Power Syst.*, 20(1), 346–357.
5. Chaturvedi, D. K., Satsangi, P. S., and Kalra, P. K. (1999). Load frequency control: A generalized neural network approachInt. *J. Elect. Power Energy Syst.*, 21(6), 405–415.
6. Wood, A. J. and Wollenberg, B. F. (1996). Power Generation Operation and Control. New York: John Wiley & Sons. (Pages: 1-630)
7. Bevrani, H. (2009). Intelligent Automatic Generation Control. New York: Springer (Pages: 1- 308)
8. Ogata, K. (1995). Modern Control Engineering, Second Ed., India: Printice Hall International (Pages: 1-812).
9. Nise, N. S. (2006). Control System Engineering. Sixth ed. Pomana: John Wiley & Sons (Pages: 1-984).
10. Bhongade, S., Tyagi, B., and Gupta, H. (2011). Genetic algorithm based PID controller design for a multi-area AGC scheme in a restructured power system. *Inter. J. Engg. Sci. Technol. Afr. J. Online (AJOL)*, 3. 10.4314/ijest.v3i1.67649.
11. Golpîra, H. and Bevrani, H. (2011). Application of GA optimization for automatic generation control design in an interconnected power system. *Energy Conv. Manag.*, 52(5), 2247–2255.
12. Ghoshal, S. P. (2004). Optimizations of PID gains by particle swarm optimizations in fuzzy based automatic generation control. *Electric Power Sys. Res.*, 72(3), 203–212.
13. Gad, A. G. (2022). Particle swarm optimization algorithm and its applications: A systematic review. *Arch. Comp. Methods Engg.*, 29, 2531–2561. https://doi.org/10.1007/s11831-021-09694-4.
14. Dorigo, M., Birattari, M., and Stützle, T. (2006). Ant colony optimization. *IEEE Comput. Intell. Mag.*, 1(4), 28–39. https://doi.org/10.1109/mci.2006.329691.
15. Mirjalili, S., Mirjalili, S., and Lewis, A. L. (2014). Grey wolf optimizer. *Adv. Engg. Softw.*, 69, 46–61. https://doi.org/10.1016/j.advengsoft.2013.12.007.
16. Tan W. (2008). Unified tuning of PID load frequency controller for power systems via IMC. *IEEE Trans. Power Sys.*, 25(1), 341–350.
17. Abdollahzadeh, B., Gharehchopogh, F. S., and Mirjalili, S. (2021). Artificial gorilla troops optimizer: A new nature-inspired metaheuristic algorithm for global optimization problems. *Inter. J. Intell. Sys.*, 36(10), 5887–5958. https://doi.org/10.1002/int.22535.

8 A real-time object detection and warning system for Kaziranga National Park elephant corridors with YOLOv8

Himangshu Nath[a] *and Pranabjyoti Haloi*[b]

Department of Electrical Engineering, Jorhat Engineering College, Assam, India

Abstract

The detection and identification of individual objects in real-time, which was considered a difficult task, can now be accomplished with advanced computer vision and deep learning techniques. An object identification system called YOLO (you only look once) is able to recognize objects in real time with incredible speed and precision. Proposed system aims to detect elephant movements in real-time by deploying cameras along the roadside and an alarm system is integrated to alert approaching vehicles about potential elephant crossings. The proposed system addresses the critical issue of human-elephant conflict (HEC) by providing timely warnings to motorists, thereby reducing the risk of accidents and promoting coexistence between elephants and humans. The camera captures live footage, and YOLOv8 is utilized to detect every object within a desired trapezoid area of the video frames in PyCharm. If the movement of elephants is detected, alarms placed at the beginning and end of the elephant corridor are triggered to alert approaching vehicles.

Keywords: Computer visions, deep learning, YOLOv8, object detection, elephant corridors

Introduction

Human-elephant conflict (HEC) has peaked in the last two decades in India. The elephant population in Assam is currently over 5700. Kaziranga National Park has five elephant corridors: Panbari (Kaziranga – Karbi Anglong), Kanchanjuri (Kaziranga – Karbi Anglong) and Haldibari (Kaziranga – Karbi Anglong), Deosur (Kaziranga – East Karbi Anglong), Amguri (Kukurakata – Bagser) [1]. The elephant habitat is under constant strain due to National Highway 36 passing through the national park. Assam ranks high in terms of the number of lives lost due to HEC in India, making it necessary to develop a system that provides timely warnings to vehicles passing through elephant corridors.

YOLO uses a single neural network consisting of multiple deep layers to predict bounding boxes and class probabilities for every object in an image [2]. YOLOv8 [3] is much faster than the previous YOLOv2 [4], YOLOv3 [5], and YOLOv5 models [6], with better accuracy. YOLOv8 also demonstrates excellent performance and strong robustness for aerial image detection [7, 8]. Tracking and continuous monitoring of outdoor animals with tinyYOLOv3 and time-lapse cameras are discussed in ref [9]. Using an NVIDIA Tesla C2075 GPU (a professional graphics card by NVIDIA used in large scale calculations, simulations and for high-end image generation for professional and scientific fields), the detection process took an average of 0.28 s per image [9]. However, this method is not suitable for wild animals as they continuously move without any restrictions, and the time-lapse method may cause delays in the warning system.

An image processing-based method is developed for detecting elephants, where cameras mounted on trees or towers are used to capture images every 5 s [10]. Forest officials are informed by an SMS when elephants are detected near forest border areas. The detection algorithm is developed using smaller dataset of 114 elephant images. The system proposed in this paper utilizes YOLOv8 to process video with a much larger COCO dataset. YOLOv8 is highly flexible, as YOLOv8 supports a range of hardware architectures (CPU-GPU). The comparison of different YOLO detection algorithm models for different-sized objects is described in ref [11]. The computing setup running on the NVIDIA Jetson AGX Xavier embedded system (used for AI workloads and solve optical inspection problems) registers frame rate of ~10 FPS for YOLOv3, as stated in ref [12]. With YOLOv8, NVIDIA GTX1050Ti GPU (mainstream GPU based on the Pascal architecture), speed=1–2 ms pre-process, 80 ms inference, 2 ms post process per image at shape (1,3,640,640) proposed system achieved ~12 FPS.

[a]himangshunath111@gmail.com, [b]pranabjyoti2003@gmail.com

DOI: 10.1201/9781003540199-8

Table 8.1 Kaziranga National Park Elephant Corridors [1]

Corridor name	Length (km)	Width (km)	Frequency of usage by elephants
Panbari (Kaziranga – Karbi Anglong)	1	0.75	Regular
Kanchanjuri (Kaziranga – Karbi Anglong)	0.9-2.6	3	Regular
Haldibari (Kaziranga – Karbi Anglong)	0.1	2.2	Occasional
Deosur (Kaziranga – East Karbi Anglong)	0.8	1.6	Regular
Amguri (Kukurakata – Bagser)	0.5-2		Low

Source: Menon, V., Tiwari, S. K., Ramkumar, K., Kyarong, S., Ganguly, U., and Sukumar, R. (2017). Elephant corridors of India right of passage. Conservation Reference Series No. 3. *Wildlife Trust of India, New Delhi.*

The aim of proposed paper is to develop a real time object detection system for detecting elephants in Kaziranga National Park elephant corridors with YOLOv8 and mainstream GPU to prevent HEC (Table 8.1).

System design

Proposed system is develop using YOLOv8 object detection algorithm in PyCharm.

Python libraries used are Cvzone, Requests, ultralytics, hydra-core, numpy, Torch, Torchvision, matplotlib, PyYAML, scipy, filterpy, opencv-python, scikit-image, Pillow, Tqdm.

For GPU processing cuda toolkit 12.1, NVIDIA cuDNN, Visual Studio is installed. GPU used is NVIDIA GTX1050Ti (Mainstream GPUs are economical than professional GPUs, power consumption is low, support any external display).

Dataset used is COCO (Common Object in Context) dataset.

Continuous live video is obtained from tree or tower-mounted cameras in the elephant corridor and processed on computers equipped with any higher mainstream GPU based on the Pascal architecture. The YOLOv8 object detection algorithm with its excellent accuracy and speed, is used to detect any elephants within the trapezoidal area marked along the road in the computer screen. If any elephant is detected within this area, alarms placed at the beginning and end of the elephant corridor are triggered.

Mathematical model

The proposed system can be described as S = {I, O, F}

I = The input set (which contains real time video frames, trapezoidal coordinate values, alarm audio)

O = Output set (processed video frames by YOLOv8 which switch ON/OFF the alarm by detecting presence of elephant in trapezoidal are)

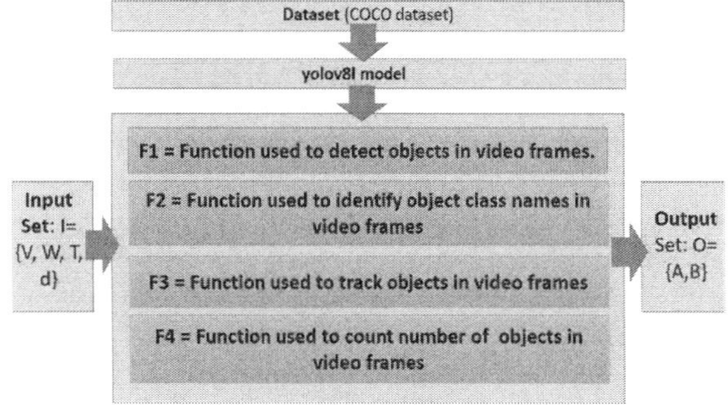

Figure 8.1 Proposed system block diagram

F = Set of functions (YOLOv8 detecting objects, objects class name, object movement and number of object present in video frames)

Set of input values: I= {V, W, T, d}

Where, V= {x1, x2, x3, . . .} (Set of video frames containing elephant inside trapezoid area),

W= {y1, y2, y3, ...} (Set of video frames not contain any elephant inside trapezoid area),

T = {t1,t2,t3,t4,t5} (Set of trapezoid top width value, trapezoid bottom width value, trapezoid height value, trapezoid x offset value, trapezoid y offset value),

d is destination of audio alarm.

Output set: O = {A, B}

Where, A = Audio alarm switch on by V,

B = Audio alarm switch-off by W.

F = {F1, F2, F3, F4} is a set of functions.

F1 stands for the object detection function in video frames.

F2 is the function that locates the names of object classes in video frames,

F3 is a function that tracks moving objects in video frames and

F4 counts the number of moving items in a frame (Figure 8.1).

The system diagram

YOLOv8

The YOLO algorithm's most recent version is YOLOv8. It introduces various improvements that considerably increase its capabilities including feature fusion, context aggregation modules and spatial attention. As a result, YOLOv8 provides faster and more precise object recognition [6, 13].

YOLOv8 advantages

1. Improved speed: YOLOv8 achieves faster inference speeds than existing object detection models while keeping good accuracy.
2. Enhanced accuracy: By adding new approaches and optimizations, YOLOv8 offers higher object detection accuracy than its predecessors YOLOv3, YOLOv5, etc.
3. Customizable architecture: YOLOv8 has highly customizable architecture, for easy modification of the model's structure and parameters.
4. Numerous backbones: YOLOv8 is compatible with a number of backbones, such as EfficientNet, ResNet, and CSPDarknet.
5. Adaptive training: YOLOv8 achieves improved model performance by using adaptive training to balance the loss function and optimize the learning rate during training.
6. Advanced-data augmentation: YOLOv8 makes use of complex data augmentation methods like MixUp and CutMix.
7. Models that have already been trained are available in YOLOv8.

YOLOv8 architecture

1. YOLOv8 improves on earlier iterations of the YOLO object identification models by adding an updated fully convolutional neural network with the backbone and the head as its two primary parts.
2. The foundation of YOLOv8 is a 53-layer convolutional adaptation of the CSPDarknet53 architecture. The architecture uses cross-stage partial connections to improve information flow between levels, ensuring efficient network connectivity
3. The convolutional layers are followed by fully linked layers in the YOLOv8 head. In order to accurately anticipate object bounding boxes, objectness ratings, and class probabilities for recognized objects inside an image, these layers are essential.
4. YOLOv8 has a self-attention mechanism built into its network head that enables it to focus on multiple image regions and dynamically change the significance of particular elements depending on how important they are to the task at hand.
5. YOLOv8 also performs exceptionally well in multi-scale object detection. It makes use of a feature pyramid network to recognize items in an image that are different sizes and scales. This network is made up of several layers, each of which is effective at seeing objects of various sizes. With this skill, the model is able to accurately identify both small and large items in the input image.

Results

The warning alarm is switch on for V, set of video frames containing elephants inside trapezoidal area and switched off for W, set of video frames not contain any elephants inside trapezoid area (Figure 8.2–8.4).

Figure 8.2 Object detection for single video frame

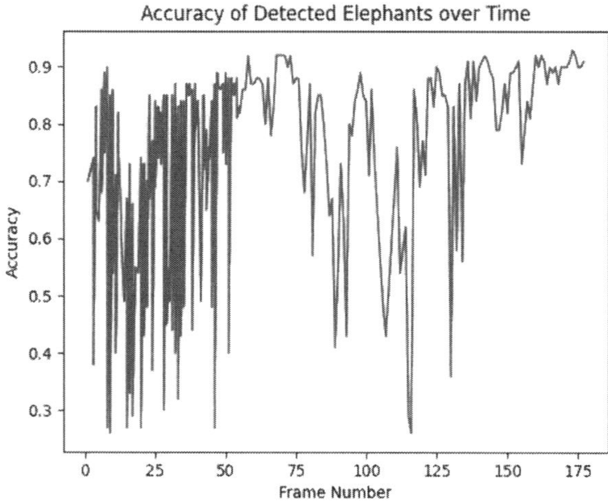

Figure 8.3 Accuracy of detected elephants over 175 video frames

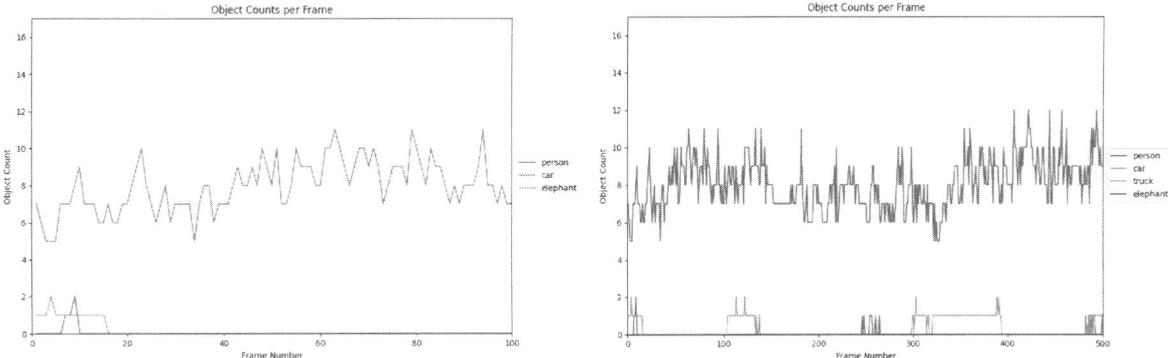

8. (a) Object counts for 100 video frames 8. (b) Object count for 500 video frames

Figure 8.4 Object counts per frame for 100 and 500 video frames

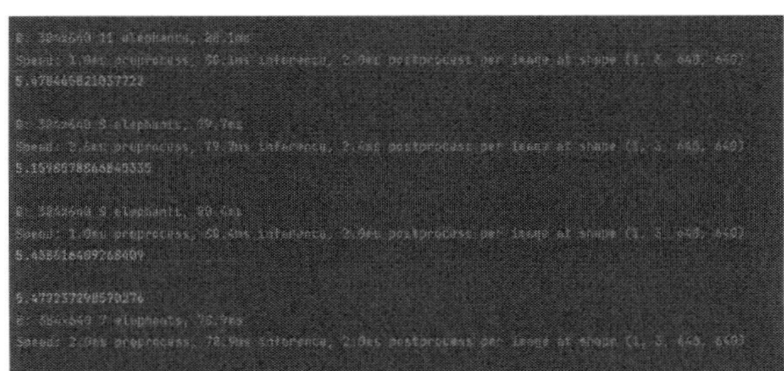

Figure 8.5 System pre-processing, inference, post-process time data for video frames

Accuracy and object counts

System performance

Pre-process: 1–2 ms, Inference: 80 ms, Post-process: 2 ms

Total time per image = Pre-process + Inference + Post-process = 1 ms + 80 ms + 2 ms = 83 ms

Total time per image in seconds = 83 ms / 1000 = 0.083 s

Frames per second (fps) = 1 / Total time per image in seconds = 1 / 0.083 =12.05 fps

So, the proposed system achieves approximately 12 frames per second.

The 80 ms inference time can be reduce by using a high end mainstream NVIDIA GPU of high model number (example – NVIDIA RTX 3090, NVIDIA GeForce RTX 4070) [14] (Figure 8.5).

Conclusions

The detection of elephant in elephant corridor with high real time accuracy is obtained from the system. The pre- and post-processing time is in the range of 1–2 ms for single video frames. Inference time of 80 ms can be reduce by using a high model number NVIDIA GPU. Precise object detection and high speed of YOLOv8 helps the system to run in real time basis.

Acknowledgement

The authors gratefully acknowledge the staff and authority of Electrical engineering department, Jorhat engineering college, Assam for their cooperation in the project.

References

1. Menon, V., Tiwari, S. K., Ramkumar, K., Kyarong, S., Ganguly, U., and Sukumar, R. (2017). Elephant corridors of India right of passage. Conservation Reference Series No. 3. *Wildlife Trust of India, New Delhi*. 508–532.
2. Redmon, J., Santosh, D., Ross, G., and Ali, F. (2016). You only look once: Unified, real-time object detection. *Proc. IEEE Conf. Comp. Vis. Patt. Recogn.*, 779–788.https://doi.org/10.1109/CVPR.2016.91.
3. Solawetz, J. (2023). What is YOLOv8? The Ultimate Guide. [Web log post]. Retrieved from https://blog.roboflow.com/whats-new-in-yolov8/.
4. Redmon, J. and Ali, F. (2017). YOLO9000: better, faster, stronger. *Proc. IEEE Conf. Comp. Vis. Patt. Recogn.*, 7263–7271. https://doi.org/10.1109/CVPR.2017.690.
5. Redmon, J. and Ali, F. (2018). Yolov3: An incremental improvement. *arXiv preprint arXiv:1804.02767*. https://doi.org/10.48550/arXiv.1804.02767.
6. Terven, J. and Diana, C.-E. (2023). A comprehensive review of YOLO: From YOLOv1 to YOLOv8 and beyond. arXiv preprint arXiv:2304.00501. https://doi.org/10.48550/arXiv.2304.00501.
7. Reis, D., Jordan, K., Jacqueline, H., and Ahmad, D. (2023). Real-time flying object detection with YOLOv8. *arXiv preprint arXiv:2305.09972*. https://doi.org/ 10.48550/arXiv.2305.09972.
8. Li, Y., Qingsong, F., Haisong, H., Zhenggong, H., and Qiang, G. (2023). A modified YOLOv8 detection network for UAV aerial image recognition. *Drones*, 7(5), 304. https://doi.org/10.3390/drones7050304.
9. Bonneau, M., Jehan-Antoine, V., Willy, T., and Rémy, A. (2020). Outdoor animal tracking combining neural network and time-lapse cameras.

Comp. Electron. Agricul., 168, 105150. https://doi.org/10.1016/j.compag.2019.105150.

10. Sugumar, S. J. and Jayaparvathy, R. (2014). An improved real time image detection system for elephant intrusion along the forest border areas. *Sci. World J.*, 2014. https://doi.org/10.1155/2014/393958.
11. Lou, H., Xuehu, D., Junmei, G., Haiying, L., Jason, G., Lingyun, B., and Haonan, C. (2023). DC-YOLOv8: Small-size object detection algorithm based on camera sensor. *Electronics*, 12(10), 2323. https://doi.org/10.3390/electronics12102323.
12. Gunasekara, S., Maleen, J., Nalin, H., Lilantha, S., and Gamini, D. (2021). A convolutional neural network based early warning system to prevent elephant-train collisions. *2021 IEEE 16th Inter. Conf. Indus. Inform. Sys. (ICIIS)*, 271–276. https://doi.org/10.1109/ICIIS53135.2021.9660651.
13. Hussain, M. (2023). YOLO-v1 to YOLO-v8, the rise of YOLO and its complementary nature toward digital manufacturing and industrial defect detection. *Machines*, 11(7), 677. https://doi.org/10.3390/machines11070677.
14. Glawion, A. (2023). Nvidia graphics cards list in order of performance [Web log post]. Retrieved fromhttps://www.cgdirector.com/nvidia-graphics-cards-order-performance/.

APPENDIX

PyCharm – It is an integrated development environment used for programming in Python.
Cvzone – A computer vision package that help run image processing and AI functions.
Ultralytics – Establish a uniform framework for training models that can recognize objects, segment instances, and classify images.
Cudatoolkit 12.1 – To create, refine, and deploy applications for embedded GPU-accelerated devices.
cuDNN – Library optimized for CUDA containing GPU implementations.
EfficientNet – EfficientNet is a convolutional neural network design and scaling technique that employs a compound coefficient to consistently scale all depth, breadth, and resolution dimensions.
ResNet – ResNet is a deep learning network in which the layer inputs are used to help the weight layers learn residual functions.
CSPDarknet – A convolutional neural network that employs darknet 53 as its foundation is called CSPDarknet-CSPDarknet53. The base layer's feature map is divided into two parts using a CSPNet technique, and these two parts are then combined using a cross-stage hierarchy. A split-and-merge technique promotes greater gradient flow throughout the network.
MixUp – Data augmentation technique, linearly interpolates input examples and the labels assigned to them.
CutMix – CutMix is a technique for enhancing visual data. It uses a patch from another image to replace the deleted regions rather than merely eliminating pixels. In accordance to the total number of pixels in the merged images, the ground truth labels are also mixed.

9 Simulation of symmetrical conventional cascaded H bridge seven level inverter using LC filter with different PWM techniques for minimization of total harmonic distortion

Niharika Devi[a] *and Pranabjyoti Haloi*[b]

Department of Electrical Engineering, Jorhat Engineering College, Assam, India

Abstract

Now-a-days, multilevel inverters (MLI) are most widely used power electronic devices for both higher and moderate range of voltage workloads due to reduction of voltage stresses, total harmonic distortion (THD). Based on their configuration, multilevel inverters can be cascaded H bridge, diode clamped, flying capacitor. Due to its flexibility, better output waveform and simpler control operations, cascaded type of MLI is taken into consideration for the fabrication of seven level multilevel inverter. This paper's primary goal is to use a passive LC filter to decrease THD. The techniques approached are alternate phase opposition disposition (APOD), phase disposition (PD) as well as phase opposition disposition (POD). MATLAB/SIMULINK has been used to conduct an in-depth investigation of the model that was suggested.

Keywords: Cascaded H bridge (CHB), APOD-PWM, PD-PWM, POD-PWM, total harmonic distortion

Introduction

In recent times, electronic devices have played a significant role in converting and controlling electric power, especially when harnessing renewable energy sources namely tidal, wind and solar energy. DC to AC power conversion encounters traditionally employed conventional bipolar inverters, but they often produce waveforms with a large number of harmonics, resulting in square or rectangular staircase waveforms. To address this issue, multilevel inverters have emerged as a solution [1]. They are widely employed in industrial as well as home appliances viz., solar photovoltaic applications, where electrical energy is generated by utilizing solar energy from solar panels and by the use multilevel inverters converts the variable direct current to alternating current [2]. In electric vehicles, permanent magnet synchronous motors are mostly used, to overcome harmonics, ripple content and to protect the motors from over current multiphase inverter is used [3, 4]. Due to their ability to improve power quality, reduce total harmonic distortion, minimize dv/dt stress in the appliances, lower switching losses, and generate output voltages close to sinusoidal waveforms compared to conventional inverters [5, 6]. Multiple research works have focused on different types of MLIs, including cascaded H bridge, diode clamped as well as flying capacitor [7–9].

Neutral-point clamped inverters, additionally referred to as diode clamped multilevel inverters, are commonly used for moderate-voltage settings due to their outstanding efficiency. These inverters generate voltages with "r" levels using diodes and typically have (r-1) capacitors across the bus for DC power, (r-1) (r-2) diodes that are clamped, and 2(r-1) switches. However, managing active power flow becomes challenging as the assortment of clamping diodes climbs with every successive stage of the inverter [7, 8].

While holding level of the voltage equivalent to diode-clamped MLI the flying capacitor type of inverters address the clamping diode issue and permit managing of both reactive as well as active power. These inverters need (r-1) switches, (r-1) (r-2)/2 flying capacitors as well as (r-1) DC bus capacitors, and to achieve a voltage level of "r". However, it becomes challenging to maintain the voltage across each and every capacitor, and increased in assortment of capacitors guide the way to higher costs and switching losses [7, 8]. In order to advance beyond these issues, cascaded H bridge type of inverters is employed. Capacitors and diodes for balancing are not necessary, offering solution with lower total harmonic distortion [9].

[a]niharikadevi97@gmail.com, [b]pranabjyoti2003@gmail.com

DOI: 10.1201/9781003540199-9

Motivation

The primary goal of the paper is to significantly mitigate the harmonics of the output voltage within the conventional cascaded H bridge MLI of seven level. Harmonics in inverter have always been a major drawbacks. To get intended output of the inverter that is more nearer to sinusoidal curve LC filter is used. Different control schemes are taken forwarded in the paper, based on the simulation findings, a comparative analysis is developed to investigate which modulation technique gives less THD.

Cascaded H bridge inverter
Design of CHB MLI is accomplished by series-connecting H bridges in order to elevate voltage level associated with the output. A conventional H bridge inverter shown in Figure 9.1 is comprised of just one H bridge which consists 4 switches complemented to a dc source. As indicated in Figure 9.2, the output has the knack of yielding to 3 distinct levels $+V_{dc}$, 0 and V_{dc}. Consequently, when switch H1 and H4 are activated $+V_{dc}$ is ensued, on the contrary when H2 and H3 are turned on $-V_{dc}$ is realized. Furthermore, the output voltage created is zero if either switches H1 and H3 or H2 and H4 are switched on [10, 11].

For "n" number of voltage level and "q" dc sources in a cascaded MLI, the relation obtained is

$$n = 2q + 1 \tag{1}$$

Cascaded H bridge MLI is at a greater extend categorized as symmetrical and unsymmetrical multilevel inverter.

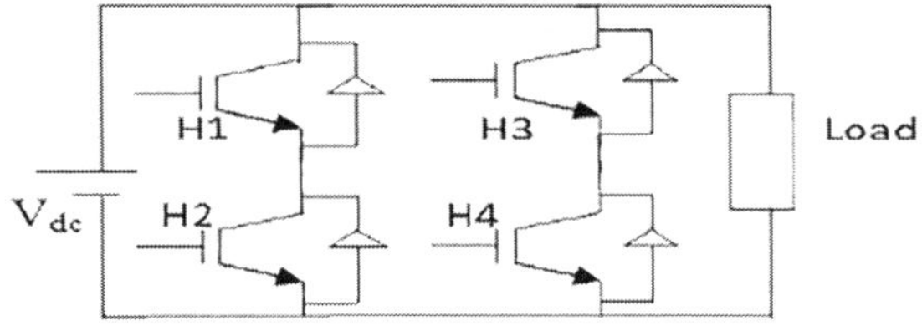

Figure 9.1 Design of single phase CHB MLI

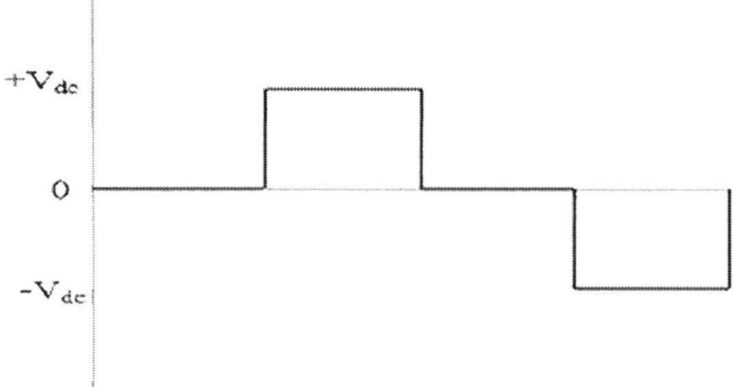

Figure 9.2 Output voltage waveform

When the H bridges are connected by equal dc voltages then it is symmetrical but when the unequal dc voltages are supplied to the H bridges it is unsymmetrical MLI structure [12].

Symmetrical configuration of cascaded MLI
Three H bridges are needed to produce a seven level CHB multilevel inverter, and each H bridge must be coupled to a different dc inputs. Single H bridge consists of four switches, so three H bridges consists of 12 switches as demonstrated in Figure 9.3. Each bridge is connected in cascaded manner. In symmetrical configuration, the dc voltage sources are V_1, V_2 and V_3 equal to V_{dc}. Figure 9.4 depicts the CHB MLI of seven level type's construction.

The output voltage magnitude is given by

$$\text{Vout} = q * \text{Vdc} \tag{2}$$

Where "q" is total amount of dc inputs. For a seven level inverter, q = 3 [11, 13–15].

Control strategies
For the control and performance of inverters, PWM techniques help in eliminating the harmonics and thus increasing the efficiency. The most comprehensively used methods are, Space vector modulation, selective harmonic PWM and phase shift PWM modulation schemes [1, 9, 14].

In PD-PWM, each and every carrier wave as illustrated at Figure 9.5 are in the same phase with

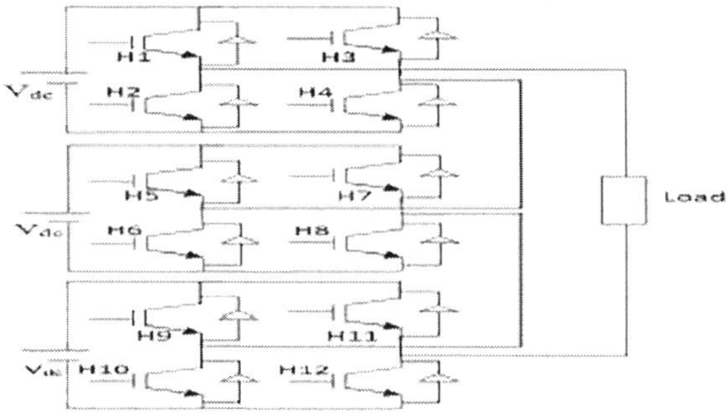

Figure 9.3 Design of seven level CHB MLI

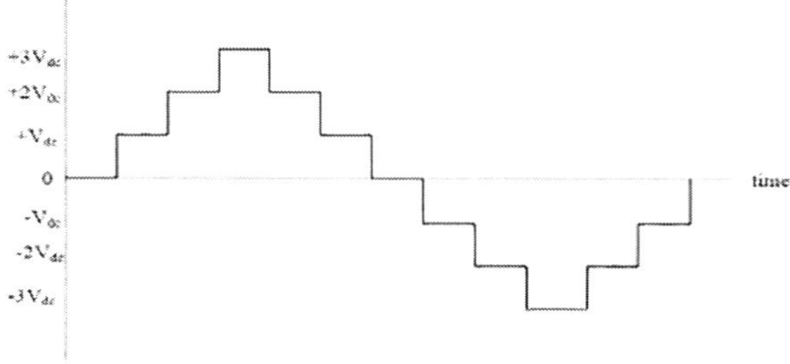

Figure 9.4 Output voltage waveform

Table 9.1 Switching arrangement for a symmetrical seven level cascaded H bridge MLI

Voutput	Activated switches
$3V_{dc}$	H1, H4, H5, H8, H9, H12
$2V_{dc}$	H1, H4, H5, H8, H10, H12
V_{dc}	H1, H4, H6, H8, H10, H12
0	H2, H4, H6, H8, H10, H12
$-V_{dc}$	H2, H3, H6, H8, H10, H12
$-2V_{dc}$	H2, H3, H6, H7, H10, H12
$-3V_{dc}$	H2, H3, H6, H7, H10, H11

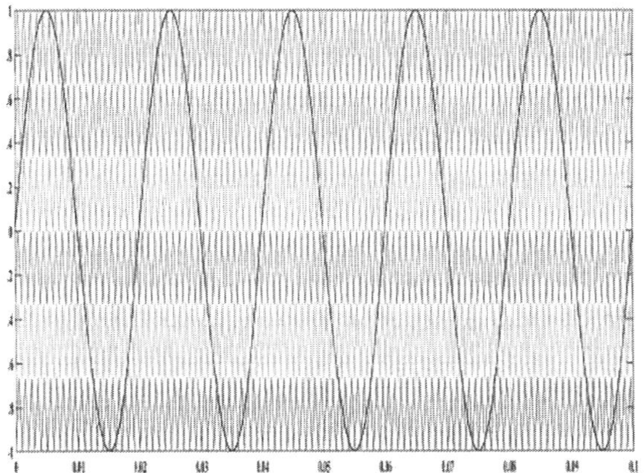

Figure 9.5 Control scheme PD-PWM

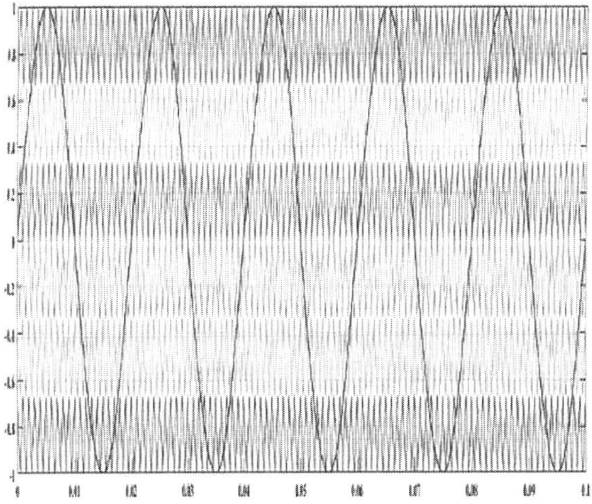

Figure 9.6 Control scheme POD-PWM

regard to the reference line, both above and below [1, 9]. In POD-PWM control strategy, the carrier wave above modulating signal are 180° phase difference in relative to the carrier wave below reference line shown in Figure 9.6 [1, 9]. As demonstrated in Figure 9.7, in APOD-PWM control scheme, each carrier wave is 180° phase shifted

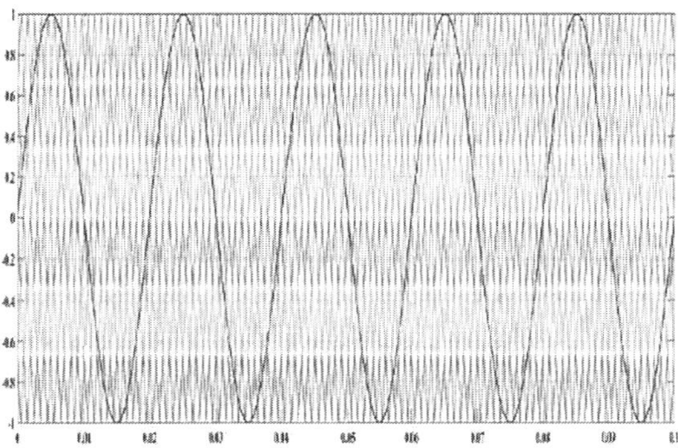

Figure 9.7 Control scheme APOD-PWM

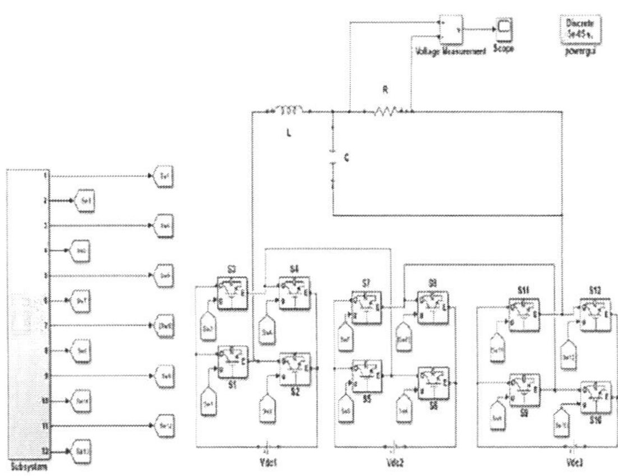

Figure 9.8 Seven level CHB MLI simulation model

from its neighboring one [1, 9]. For the quantity of levels r =7, the carrier waves required is r-1 that 7-1= 6 for all the three strategies.

Simulation and its results

A seven level symmetrical CHB inverter, designed with modulation techniques APOD, PD as well as POD separately to lower content of harmonic present in output voltage. With the aid of Matlab/ Simulink, a passive LC filter is linked to the load as revealed in Figure 9.8. Three 100 V dc sources are supplied to each of the H bridge taking resistance load R = 8Ω. The total harmonic distortion with different modulation schemes are analyzed without filter. Likewise, with the help of LC filter, taking inductance L=15 mH and capacitance C=180 μF, the THD in resulting voltage in the output is monitored. Figures 9.9–9.9.11 shows the THD of output voltage generated in CHB MLI without filter. Moreover, the Figures 9.12–9.9.14 shows the THD of output voltage of CHB inverter with LC filter. According to various control schemes, the THD analysis associated with output voltage under filter and filter-free situations is depicted in Table 9.2.

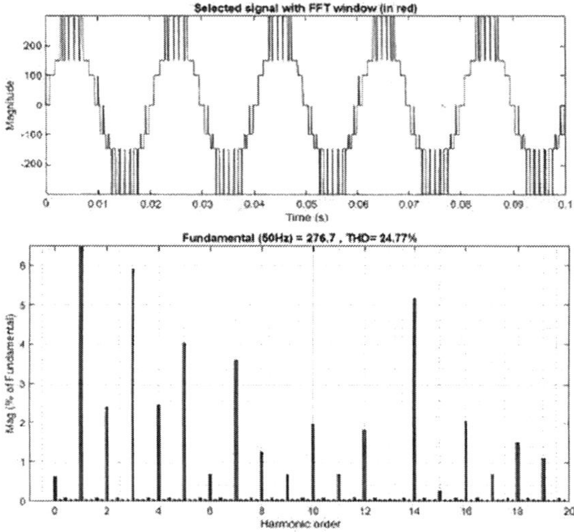

Figure 9.9 PD-PWM-based THD result without a filter

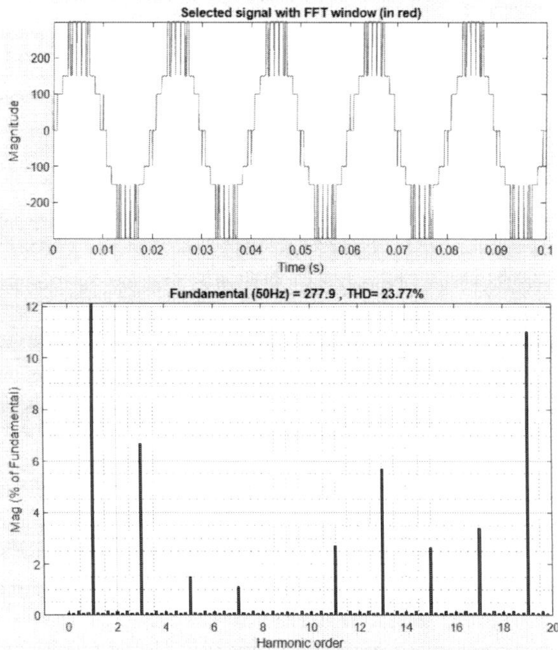

Figure 9.10 POD-PWM-based THD result without a filter

Conclusions

In this paperwork, different control strategies are implemented for the harmonic analysis of the resultant voltage response of seven level CHB MLI using the MATLAB Software. As shown in Table 9.2, POD-PWM has a lower THD% of the output voltage without using a filter. In contrast to the other two modulation approaches, APOD-PWM has a low THD% of the resulting voltage when the model is being simulated with an LC filter. Utilizing a passive LC filter enhances the resultant voltage's

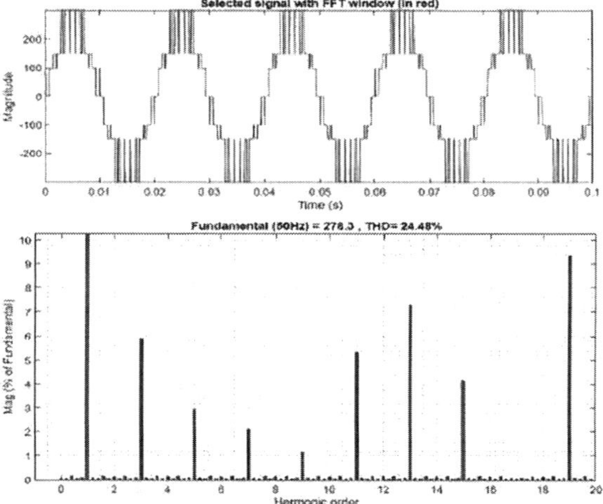

Figure 9.11 APOD-PWM-based THD result without a filter

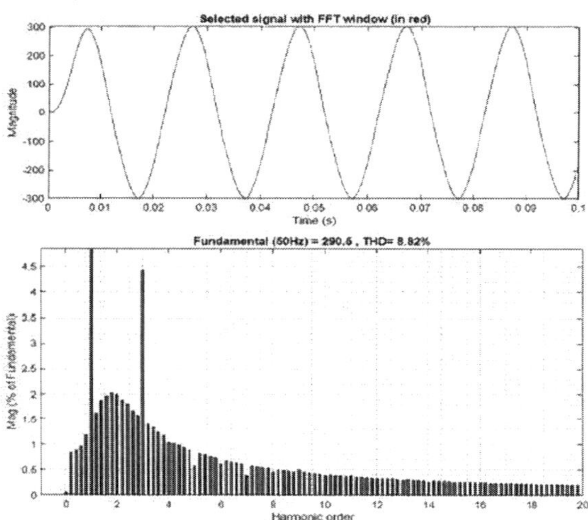

Figure 9.12 PD-PWM-based THD result with LC filter

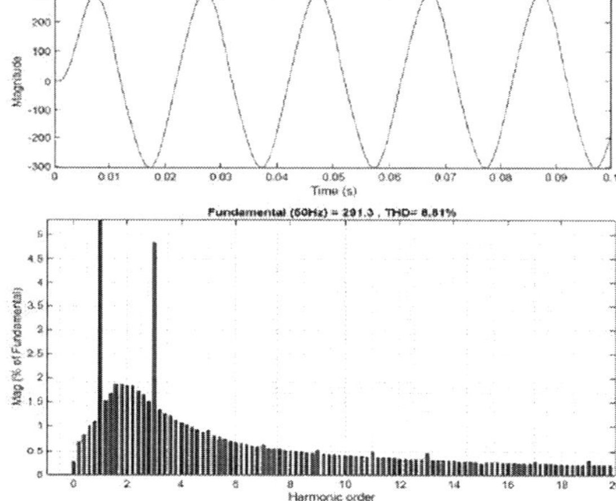

Figure 9.13 POD-PWM-based THD result with LC filter

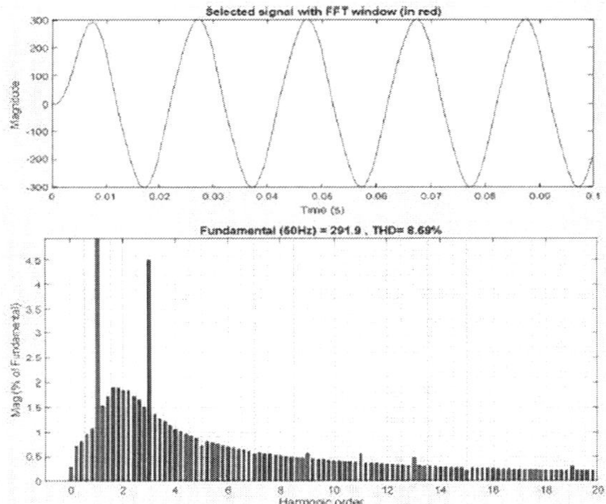

Figure 9.14 APOD-PWM-based THD result with LC filter

Table 9.2 THD analysis with and without filter by PWM schemes

Classification	PD-PWM	POD-PWM	APOD-PWM
Without filter	24.77	23.77	24.48
With filter	8.82	8.81	8.69

quality i.e., the waveform approaches closer to sinusoidal wave. The harmonics are eliminated to a greater extent that it is reduced to about 63% in all the three control strategies.

Acknowledgement

The authors gratefully acknowledge the staff, and authority of Electrical department for their cooperation in the research.

References

1. Shet, G. U. and Shanmukha Sundar, K. (2018). THD comparison using different PWM techniques for single phase eleven level inverter. *Int. J. Adv. Res. Sci. Engg.*, 07(7), 1194–1195. http://www.ijarse.com/images/fullpdf/1525503511_Reva711ijarse.pdf
2. Haq, I., Ajmal, F., Mian Muhammad, A. A., and Muhammad, A. (2023). Design of hybrid multi-level inverter for photovoltaic (PV) application. *Inter. Conf. Appl. Engg. Nat. Sci.*, 1(1), 385–94.https://doi.org/10.59287/icaens.1027.
3. Sain, C., Atanu, B., Pabitra Kumar, B., Ahmad, T. A., and Thanikanti Sudhakar, B. (2022). Design and optimisation of a fuzzy-PI controlled modified inverter-based PMSM drive employed in a light weight electric vehicle. *Inter. J. Autom. Con.*, 16(3–4), 459–488. 10.1504/IJAAC.2022.122603
4. Sain, C., Atanu, B., Pabitra Kumar, B., Thanikanti Sudhakar, B., and Tomislav, D. (2020). Updated PSO optimised fuzzy-PI controlled buck type multi-phase inverter-based PMSM drive with an over-current protection scheme. *IET Elec. Power Appl.*, 14(12), 2331–2339. https://doi.org/10.1049/iet-epa.2020.0165.
5. Siddique, M. D., Saad, M., Noraisyah, M. S., Adil, S., Atif, I., and Mudasir Ahmed, M. (2019). A new multilevel inverter topology with reduce switch count. *IEEE Acc.*, 7, 58584–58594. 10.1109/ACCESS.2019.2914430.
6. Mohan, T. M. and Shakeera, Sk. (2021). Transformer based cascaded multilevel inverter with reduced number of switches. https://www.ijraset.com/best-journal/transformer-based-cascaded-multilevel-inverter-with-reduced-number.
7. El-Hosainy, A., Hany, A. H., Haitham, Z. A., and El-Kholy, E. E. (2017). A review of multilevel inverter topologies, control techniques, and applications. *2017 Nineteenth Inter. Middle East Power Sys. Conf. (MEPCON)*, 1265–1275. 10.1109/MEPCON.2017.8301344.
8. Balal, A., Saleh, D., Farzad, S., Miguel, H., and Yao Lung, C. A review on multilevel inverter topologies. *Emerg. Sci. J.*, 6(1), 185–200. 10.28991/ESJ-2022-06-01-014.
9. Subsingha, W. (2016). A comparative study of sinusoidal PWM and third harmonic injected PWM reference signal on five level diode clamp inverter. *Energy Proc.*, 89, 137–148. https://doi.org/10.1016/j.egypro.2016.05.020.
10. Pawar, S. V. and Abhijeet, R. M. (2023) . Harmonic elimination in single phase multilevel inverter with non equal DC sources using differential evolution algorithm. *Inter. J. Innov. Sci. Engg. Technol.*, 10(01), 50–51. https://ijiset.com/vol10/v10s1/IJISET_V10_I1_06.pdf
11. Barah, S. S. and Sasmita, B. (2021). An optimize configuration of H-bridge multilevel inverter. *2021 1st Inter. Conf. Power Elec. Energy (ICPEE)*, 1–4. 10.1109/ICPEE50452.2021.9358533.
12. Mehta, S. and Vinod, P. (2021). 7 level new modified cascade H bridge multilevel inverter with modified PWM controlled technique. *2021 11th IEEE Inter. Conf. Intell. Data Acquis. Adv. Comput. Sys. Technol. Appl. (IDAACS)*, 1, 560–565. 10.1109/IDAACS53288.2021.9660954.
13. Prayag, A., Sanjay, B., and Raisoni, G. H. (2018). A comparative study of symmetrical and asymmetrical cascaded H-bridge multilevel inverter topology for industrial drive. *Inter. Res. J. Engg. Technol.*, 5(2), 1931–1932. https://irjet.net/archives/V5/i2/IRJET-V5I2407.pdf.
14. Qureshi, M. R., Mukhtiar Ahmed, M., and Abdul Sattar, L. (2020). Harmonic analysis and design of LC filter for a seven-level asymmetric cascaded half bridge multilevel inverter. *Inter. J. Elec. Engg. Emerg. Technol.*, 3(2), 5258. https://ijeeet.com/index.php/ijeeet/article/view/46.
15. Sanoop, P. and Vinita, C. (2016). Seven level inverter topologies: A comparative study. *Inter. J. Innov. Res. Elec. Electron. Instrum. Con. Engg.*, 3(1), 148–163. https://ijireeice.com/wpcontent/uploads/2016/02/nCORETech-30.pdf.

10 Renewable energy integration: Its associated challenges

Chetan Srivastava[a] *and Manoj Tripathy*[b]

Department of Electrical Engineering, Indian Institute of Technology Roorkee, Haridwar, India

Abstract

Integrating renewables with high penetration will fundamentally change the power system configuration. Power system topology, its protection, and operational challenges are entirely changed with renewable energy integration (REI). This work illustrates the Indian and global scenario for the increasing trend of renewable energy and its associated challenges. It depicts the standards for REI and its associated power quality concerns. The approach to effectively balance power generation and consumption significantly impact its operation and planning in the short- and long-terms. The intermittent nature of renewable energy resources (RESs) will challenge power system stability, protection, security, economics, and ancillary services. Hence, high REI will severely affect future power system's development and operation. If failed to effectively address the challenges posed by renewable integration, may lead to the curbing of the large renewables, adversely impacting power system economics and operation. Wind and solar power are the two primary sources of renewable energy.

Keywords: Renewable energy integration, worldwide renewable scenario, renewable standards, Integration challenges

Introduction

The energy guidelines of several countries envisage that renewable accounts for a noteworthy energy share. United Nations Sustainable Development Goal (SDG-7) and Paris Agreement target to ramp up available renewable resources to 8000 GW by 2030 from 2800 GW at present [1]. India has an installed power capacity of 482.232 GW as of 30 April 2023 [2]. India ranks third major generator and energy consumer worldwide. Figure 10.1, sector-wise power generation in India and Figure 10.2 depict the leading countries in renewable energy (RE) capacity as per the International Renewable Energy Agency (IRENA).

The Indian government has fixed a goal to achieve an installed energy capacity by 2030, which consists of the installation of 280 GW solar and 140 GW of wind power [3]. In the last 8.5 years, India's non-conventional power generation capacity has been extended by 396%, which constitutes above 178.9 GW, including big hydropower plants. It reaches nearly 43% of the nation's power capability as of May 2023. India stands at the fourth spot worldwide as per total installed renewable power generation, including large hydropower plants. It also stands fourth in terms of wind and solar capacity, as per REN21-Global Status Report [2]. As per the International Energy Agency (IEA) report, modern renewable power generation will account for 18% of total final consumption by 2030. This is far less than the 32% share required in 2030 for the globe to meet the net zero emissions by 2050 scenario [3]. In such a scenario of high penetration of renewables, its associated challenges are required to be best addressed to meet the desired objectives.

High renewable penetration challenges

Several countries are injecting more than 40% of renewable energy resources. Some countries are generating more power from renewables than their requirement (Srivastava and Tripathy, 2021). The technical issues with high REI include (i) power quality problems, which constitute voltage and frequency oscillations and harmonics injection in the system with renewable penetration; (ii) short- and long-term power fluctuations; (iii) large energy storage issues, and (iv) ideal location for REI.

Table 10.1 represents standards beneficial for REI and power quality concerns [5–7]. High REI brings various challenges, which can be divided into the following sections.

Stochastic dependencies modeling

High renewable power injection into the traditional power system faces several operational and scheduling uncertainties. Uncertain load and renewable generation variation will complicate power system operation and control. Other uncertainties include electric vehicles' high mobility and energy storage systems' buffering nature [8]. The exact modeling of uncertainties is a big challenge as it depends on

[a]csrivastava@ee.iitr.ac.in, [b]manoj.tripathy@ee.iitr.ac.in

DOI: 10.1201/9781003540199-10

various factors. Microgrids can use batteries, heat buffers, and plug-in electric vehicles with vehicle-to-grid systems to more efficiently adapt to intermittent and weather dependency. Stochastic modeling is generally done for microgrids. In order to find out the severity of uncertainties and rank them global

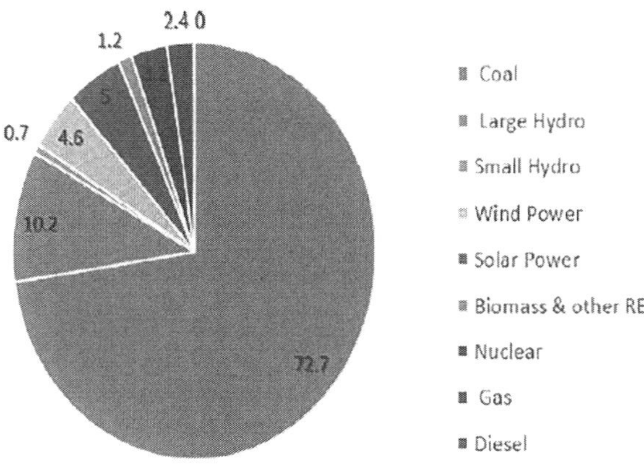

Figure 10.1 Electricity generation by sources in percent India 2021–22

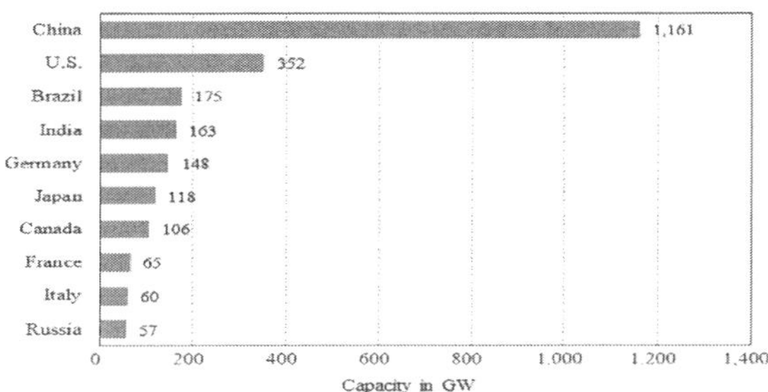

Figure 10.2 Leading countries in installed RE capacity FY worldwide in 2022

sensitivity analysis (GSA) was applied, severe uncertainties affect the damping of oscillations [9].

Power system anticipation

Integration of renewable, whether done in a centralized or distributed manner, will increase the variables and non-linear relation of renewables. Power system anticipation is useful to control different processes and decisions like unit commitment problems, fuel allocation, and power system analysis. Generally, weather prediction is done accurately for larger areas from hours to days. Prediction in a short time is difficult [10]. Accurate wind and solar power prediction substantially affect power systems' operation, control, and economy [4]. Data based probabilistic net load anticipating method explicitly for high PV generation integration was developed [11]. Preferably, a warning should be issued at first before any deviations in the renewable output take place to take some corrective measures. Figure 10.3 represents the weather forecast on a time scale [10].

Flexibility in power system

The flexibility of the power system explains how readily it can accommodate RESs without violating its operational boundaries. A less flexible power system is a major factor in curbing renewable energy and jeopardizing operational economics. The probabilistic flexibility evaluation method explains different factors due to a shortage of flexibility like direction, amount, and frequency [12]. It gives linear relation with the curtailment of renewable energy, so this method can be effectively used for high integration of renewable studies. Dui presents a two-stage process to estimate the ideal

Table 10.1 Standards useful for renewable integration and power quality issues

Standards	Applications
IEEE 1159	Inspecting power quality
IEC-61000-4-6	High-frequency disturbance compatibility levels
IEEE SCC-22	Coordinating committee for standards of power quality
IEEE 519	Harmonic control standard in power system
IEEE 519A	Guidelines to present harmonic limits
IEEE P1453	States voltage flicker permissible limit
IEEE P1564	Represents limits for voltage sag
IEEE P1547	Interconnection standards for distributed resources
IEEE 2800	Interconnection and interoperability of inverter-based sources with power systems
IEEE 2030	Guidelines for smart grid interoperability and recommended practices

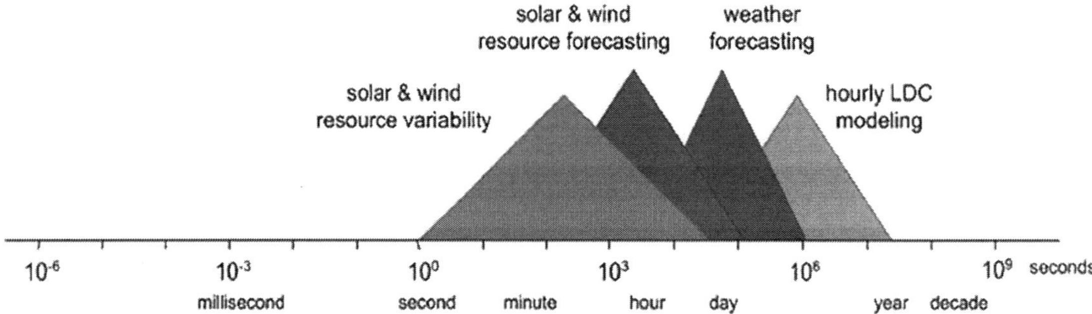

Figure 10.3 Forecasting on a time scale basis

power and battery energy storage system (BESS) [13]. It explained that if BESS is optimally utilized, it can decrease the renewable energy curtailment and improve the operational cost.

Transmission system-associated concerns

The randomness of renewable power generation alters the power flow direction and magnitude of power flowing [4]. The transmission system operation becomes complex due to more direct current (DC) lines. The operation and planning are challenging tasks to lodge such deviations in the transmission system. Different controllable elements like transmission switching, DC system control, and transmission side controllable elements must be considered for high REI. Transmission line switching is beneficial to achieve optimal power flow during line congestion [14]. It also reduces unit commitment costs when the system is unperturbed. The high load variation and fickle renewable power generation require a more operating reserve.

Distribution system and microgrid operation

Integration of renewable changes power system topology and power injection statistics. The power flow pattern becomes bi-directional, contrary to the uni-directional nature of traditional power systems, which poses challenges to microgrids and distributed power system operation. Wang suggested high PV integration at a weak distribution network while maintaining the voltage boundaries with coordinated control of the battery and inverter [15]. The control strategies for microgrid operation include droop control and active load sharing. In the droop control method, the converter output is controlled using droop. The magnitude of droop is obtained from the difference between the reference and converter output current/voltage signal. For the islanded microgrid virtual multi-slack (VMS) type droop control is proposed [16]. In the suggested control, one slack bus physically controls its bus's voltage and phase angle only while using the VMS approach. The generators regulate the bus voltage magnitudes and phase angles indirectly.

Frequency control in power system

Renewable energy resources suffer from a low inertia problem, which causes frequency variation, and this problem deteriorates with high REI. Therefore, solar and wind frequency responses should be synchronized with conventional power systems. The rate at which frequency varies is considered an essential factor in the selection of protection schemes. The double-fed induction generator (DFIG) can improve system's frequency response and can also stabilize frequency fluctuations. Even for high wind penetration, the proposed scheme gives an improved response. Considering the following measures, smoothing power frequency control may be achieved [10].

i. High reserve generation capacity
ii. Generation with better regulation ability
iii. Dispatchable generation with high ramp rates
iv. Better supply-demand response
v. Dispatchable electric storage

Power system stability

Renewable energy integration will lead voltage, frequency, and power to oscillate if not done with great care. Temporal coordination in power systems explicitly addresses the time-varying nature of renewables and explains the effects of intermittent nature on different system components. The rate of power system parameters change and various components' response time is very important to analyze power system operation. The way renewables are electrically interconnected and associated with the load through the transmission and distribution system is termed spatial coordination.

With the increasing trend of renewable integration, converters, and AC/DC grids have been used. The renewables' dynamic characteristics, lower

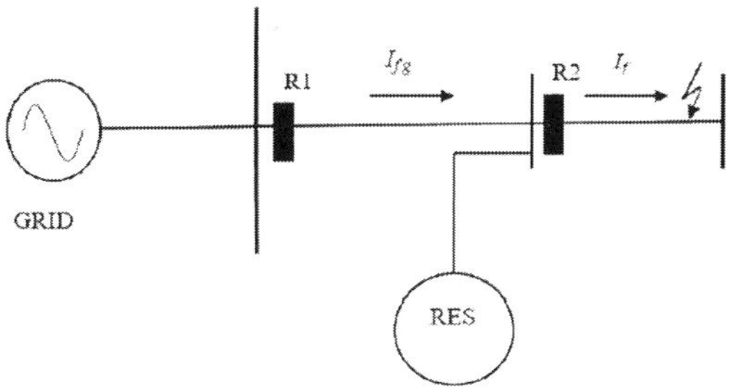

Figure 10.4 Relay R1 blinding operation

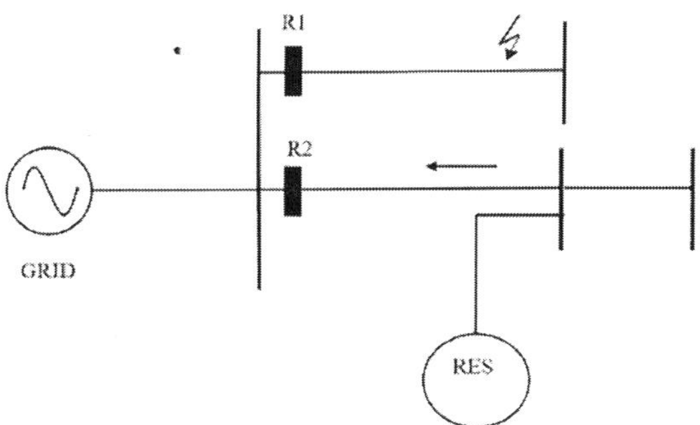

Figure 10.5 Sympathetic tripping of relay R2

inertia, and poor grid-converter control make stability assessment difficult. Ying explained the impact of inertia on oscillation damping at DFIG-based wind turbines [17].

Power system protection

Generally, renewable sources are applied at the distribution side where the reach of conventional power is not technically or economically feasible. The level of RE penetration and its location have a substantial impact on protection requirements. The fault levels and power flow statics are entirely changed with renewable energy penetration into the power system. High REI can lead to false tripping of overcurrent-based relays at the power distribution level, and its impression may further be traversed to transmission relays. The local or differential relaying schemes may provide fast and reliable protection [18]. Maintaining the power balance in an islanded grid mode is challenging. Power flow in traditional power systems is considered uni-directional from generation to the load, and the protection philosophy is designed on the same assumption. The other relaying issues with REI are illustrated below [19].

Protection blinding

The relay's blinding is shown in Figure 10.4 [19]. With the addition of RES the current sensed by the relay R1 is reduced, restricting it to operate termed as relay's blinding operation.

False tripping or sympathetic tripping

The directional over current relays are required due to bi-directional power flow due to renewables, as non-direction relays can't provide adequate protection. Figure 10.5 represents that relay R2 mal-operate for fault at connected line due to its non-direction feature. It may trip prior to the dedicated relay, resulting in large network isolation. This mal-operation is termed as false or sympathetic tripping.

Auto recloser problem and loss of coordination

While moderately isolating the fault, the recloser bus feeds the fault through renewables. It energizes an arc over recloser points and converts a momentary fault to an enduring fault [19]. Due to the aforementioned factors relay may mal-operate, which in turn will lead to successive relays' false operations. A relay's wrong tripping in a cascade fashion is known as coordination loss.

The protection coordination index (PCI) is evaluated as the ratio of change in power injection and variation in coordination time interval (Δ CTI) [20]. This index identifies safe areas for RESs placement where the protective margins are least affected, as represented by Equation 1 [20].

$$PCI = -\frac{\Delta P}{\Delta CTI} \tag{1}$$

Conclusions

This chapter depicts the Indian and global scenario of renewable energy integration and presents different challenges imposed due to the integration of renewable. It presents standard practice for REI and power quality issues. Due to the low inertia of renewables, maintaining constant frequency is a big task, so different frequency control measures are mentioned. Power flow topology and injection statistics are altered on REI, leading to several protection issues. It focuses on the need for weather anticipation tools to convert weather conditions to proportionate power. The relaying aspects with REI are required to be investigated thoroughly to ensure fast, reliable, and accurate fault isolation and classification.

References

1. Gupta, A. and Chopra, A. (2022). Microgrid market share report 2022–2030. Report ID: GMI1187, July 2022.

2. Central Electricity Authority (CEA). (2023). Total installed capacity, April, 2023. https://powermin.gov.in/en/content/ power-sector-glance-all-india. (accessed May 15, 2023).
3. Ministry of New and Renewable Energy (MNRE). (2023). Creating a sustainable world. https://www.investindi. gov.in/sector/renewable-energy. (accessed April 10, 2023).
4. Srivastava, C. and Tripathy, M. (2021). DC microgrid protection issues and schemes: A critical review. *Renew. Sustain. Energy Rev.*, 151, 111546.
5. IEEE Power Quality Standards. (2021). Power standard lab IEEE power quality standards. https://powerquality.blog /2021/04/08/power-quality-standards/. (accessed March 1, 2023).
6. IEEE DERs Standards. Distributed Energy Resources. https://standards.ieee.org/beyond-standards/ieee standardsfor- the- evolving-distributed-energy- resources-der-ecosystem/. (accessed March 1, 2023).
7. IEC 61000-4-6. (2023). The electromagnetic compatibility - Part 4-6: Measurement and Testing techniques - Immunity for disturbances, introduced by radio frequency fields. (accessed March 1, 2023)
8. Yuxin, L., Fan, M., Lin, D., and Li, H. (2016). Integrating high-penetration renewable energy into power system - A case study. *China Int. Conf. Elec. Distrib. CICED*, 1–5.
9. Xu, X., Yan, Z., Shahidehpour, M., Wang, H., and Chen, S. (2018). Power system voltage stability evaluation considering renewable energy with correlated variabilities. *IEEE Trans. Power Syst.*, 33(3), 3236–3245.
10. Von Meier, A. (2014). Challenges to the integration of renewable resources at high system penetration. *California Energy Commission Report (CEC)*, 52.
11. Wang, Y., Zhang, N., Chen, Q., Kirschen, D., Li, P., and Xia, Q. (2018). Data-driven probabilistic net load forecasting with high penetration of behind-the-meter PV. *IEEE Trans. Power Syst.*, 33(3), 3255–3264.
12. Lu, Z., Li, H., and Qiao, Y. (2018). Probabilistic flexibility evaluation for power system planning considering its association with renewable power curtailment. *IEEE Trans. Power. Syst.*, 33(3), 3285–3295.
13. Dui, X., Zhu, G., and Yao, L. (2018). Two-stage optimization of battery energy storage capacity to decrease wind power curtailment in grid-connected wind farms. *IEEE Trans. Power Syst.*, (33)3, 3296–3305.
14. Shi, J. and Oren, S. (2018). Stochastic unit commitment with topology control recourse for power systems with large-scale renewable integration. *IEEE Trans. Power Syst.*, 33(3), 3315–3324.
15. Wang, L., Bai, F., Yan, R., and Saha, T. K. (2018). Real-time coordinated voltage control of PV inverters and energy storage for weak networks with high PV penetration. *IEEE Trans. Power Syst.*, 33(3), 3383–3395.
16. Choi, D., Park, J., and Lee, S. H. (2018). Virtual multi-slack droop control of stand-alone microgrid with high renewable penetration based on power sensitivity analysis. *IEEE Trans. Power Syst.*, 33(3), 3408–3417.
17. Ying, J., Yuan, X., Hu, J., and He, W. (2018). Impact of inertia control of DFIG-based WT on electromechanical oscillation damping of SG. *IEEE Trans. Power Syst.*, 33(3), 3450–3459.
18. Srivastava, C. and Tripathy, M. (2023). Novel adaptive fault detection strategy in DC microgrid utilizing statistical-based method. *IEEE Trans. Ind. Informatics.*, 19(5), 6917–6929.
19. Telukunta, V., Pradhan, J., Agrawal, A., Singh, M., and Srivani, S. G. (2018). Protection challenges under bulk penetration of renewable energy resources in power systems: A review. *CSEE J. Power Energy Syst.*, 3(4), 365–379.
20. Zeineldin, H. H., Mohamed, Y. A. R., Khadkikar, V., and Ravikumar Pandi, V. (2013). A protection coordination index for evaluating distributed generation impacts on protection for meshed distribution systems. *IEEE Trans. on Smart Grid.*, 4(3), 1523–1532.

11 Strength and reliability analysis of IoT-enabled especially designed smartstool cum walking stick for blinds using FEA

Ankur[1], Arjun Gupta[1], Awani Bhushan[1,a], Abhishek Kumar Singh[2], S. C. Ram[3], S. Suman[4] and Raja RamKumar[5]

[1]Schoolof Mechanical Engineering, Vellore Institute of Technology, Chennai, Tamil Nadu, India

[2]School of Advanced Sciences,Vellore Institute of Technology, Chennai, Tamil Nadu, India

[3]Department of Mechanical Engineering, Tula's Institute, Dehradun, Uttarakh and 248011, India

[4]Department of Mechanical Engineering, MIET, Meerut, Uttarakh and 248011, India

[5]Department of Electrical Engineering, GhaniKhan Choudhury Institute of Engineering & Technology, India

Abstract

In this paper, a smart stick guidance model to aid both the visually compromised and senior persons by combining the advantages of two products, a stoolandawalkingstickthatcomesinhandyduringmovement. The product concerning the model is grounded on the designs behind a walking stick and a stool with other multiple additions to make it helpful in movement, embedding, and espousing smart features like an electronic circuit conforming to Global Positioning System (GPS), Global System for Mobile Communications (GSM), an Secure Digital (SD) cardmodule along with ultrasonic sensor, here there are three buttons present on the handle of the stick circle, a square, anda triangle each has a significant function of its own. In accordance to a study conducted it was found that most visualassistive s technologies are to beused by just the disabled, whereas, the system that we have developed is not only forthe visually impaired user but also for the general public/streetwalker who navigates around the user to avoidanyaccidents/mishaps.

Keywords: IoT, machine learning, machine design, switch walking stick

1. Introduction

Vision is one of the species' most significant senses. About 18 million people in India are eyeless or visually impaired. And around 2.2 billion people in the world have near or distant vision impairment. Accordingtothereportproducedbythe International Agency for the Prevention of Blindness (IAPB), 1.1 billion people suffer from loss of vision worldwide. Blindpeoplehavetorelyuponothersforthe completionofbasictasksastheyare unable to deal with daily life obstacles such as doors, other people, steps/stairs, walls, etc. Now to innovate further from the previous generation of walking sticks for three visually impaired, which presented them with a lot of difficulties while navigating new territories. Having seen ideas of an intelligentguidancesystem would in some ways act as a third eye for the user. For example, illustration the smart walking stick with an IR sensor or a white canedevicethatcouldalerttheuserofobstacles in his/her path but also alert on lookers that the user is visually disabled/differently abled. Another notable innovation in the same field for peoplewith blindness and deafness involved vibration and sound processing functionalities, a navigation systembaseduponGIS&RAFIDallowing theusertonavigate using a cellular device with GPS availability. Furtherdevelopment into the same led to the attachment of micro-controlled devices such as GPS-boxed devices, ultrasonicsensors, etc. [1–3].

2. Hardware/circuit diagram

Our invention consists of a compact circuit designed for flawless integration into a smart walking switch stick and a foldable chair specifically created for individuals with visual impairments.This circuit board consists of several remarkable features [5, 6]. especially, it can transmit the stick's position in the absence of Wi-Fi connectivity, thanks to the utilization of GPS technology, ensuring accurate positioning information. Moreover, the attained position data goes beyond being transferred to a dedicated mobile operation. It is also shared via text messages with a designated family member of thevisually challenged person (VCP).This functionality allows the family member to conveniently access the stick's position by simply transferring a text message containing the keyword "POS" to the SIM card number inserted intothe module[7] (Figure 11.1).

[a]awani.bhushan@vit.ac.in

DOI: 10.1201/9781003540199-11

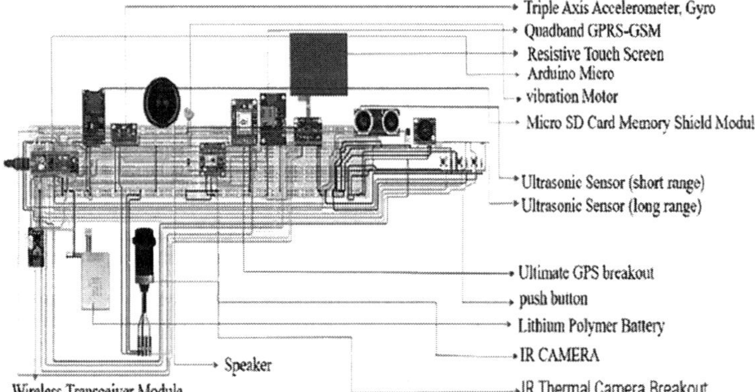

Figure 11.1 Circuit diagram for a smart walking stick

The circuit includes three distinct buttons, each serving a specific purpose. The circular button enables communication through a cloud-based application, facilitating flawless interaction between the VCP and their family members. The square button allows the VCP to send text messages directly to their family members' mobile phones, ensuring swift and reliable communication. Additionally, the triangular button provides voice alerts to family members, delivered to their mobile phones, thus enhancing real-time notifications. To enablecomprehensive shadowing, the circuit continuously monitors the stick's position, resulting in a detailed path show casing the VCP's movements. This inestimable information, along with other applicable data, is stored in the SD card for unborn analysis and reference. Likewise, auser-friendlydisplayisincorporatedtofacilitatethedeletion of unwanted data and the management of stored information.

Operating on a connected battery, the circuit offers an impressive 40-hour runtime on a single charge. Thisextended battery life ensures reliable functionality and reduces the need for frequent recharging, there by enhancing the overall convenience and usability ofthesmartwalking switchstick.

Our compact circuit, designed for the smart walking switch stick and foldable chair, incorporates advanced GPS technology, flawless communication options,nonstop shadowing, and extended battery life.This innovation aims to enhance the safety and independence of visually challenged individuals while furnishingtheir family members with peace of mind through real-time position updates and effective communicationchannels.Thecompactandfeature-richcircuit, with its advanced GPS technology [7], seamless communication options, continuous tracking, and extended battery life, seeks to enhance the safety andindependence of visually challenged individuals. By providing real-time position updates and effective communication channels, our innovation aims to offer peace of mind to family members while empowering VCPs to navigate their surroundings with confidence.

Our invention utilizes a flow chart to outline its operation. Originally, the microcontroller is initialized toestablish the system [8, 9]. Once powered on, the ultrasonic sensor begins collecting nonstop data from the surrounding environment. During any cycle, the measured distance falls below a pre-defined threshold, boththevibration motor and speakerareactivateduntilthedistancesurpassesthethres hold once again. Following the micro controller initialization [9], the3-axisaccelerationsensorgathers data from the real-life environment. If the collected values deviate from or exceed the set threshold, the GPS coordinates are acquired from the GPS moduleand transmitted through GSM for furtheruse [10].

3. Proposedmodel

The proposed model as shown in Figure 10.2 closed and opened view of the product made of aluminum alloy. Hence it is light, portable, and durable. It is feather light owing to the aluminum material, coming up tolower than 3 pounds. The seat is of durable canvas that can hold or support up to a minimal weight of 250 lbs. The seat is supported by steel legs that can fluently be folded. The stool feature easily slides into alocking position. When ready to start walking again, simply unleash the stool and slide back into awalking cane. Locking and unlocking are all accomplished in one simple motion of raising or lowering theseat. Handle has a soft nonslip grip due to the rubber material. There can be multiple uses for stabilitywhile walking or a quick rest by sitting. The height can be adjusted. It can be stored easily due tofoldability.

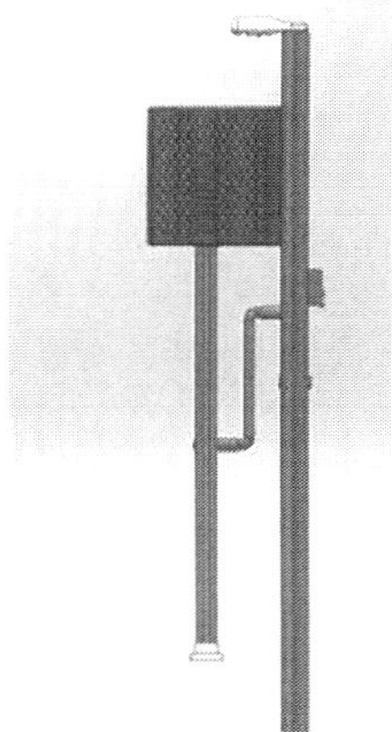

Figure 11.2 (a) Smart switch stick (closed view)

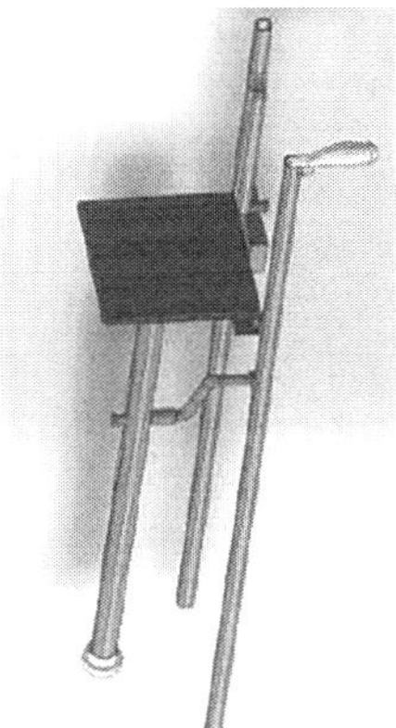

Figure 11.2 (b) Smart switch stick (opened view)

4. Finiteelementanalysis

Finite element analysis model is shownin Figure 10.2. FEAsoftware (Ansysworkbench) for 200kg (2000 N) is done. Themeshing of modelis presented in Figure 10.3.

4.1 Materialproperties

Selecting material properties is very significant because it's necessitated to minimize the weight-ofthestickas well as give sufficient strength. Grade 5154 aluminum alloy is of the aluminum magnesium family(wrought, 5000 or 5xxxx series). It combines features like moderate to high strength and excellent weldability depending upon the magnesium percentage andfurtherheat & surface treatments done, which are commonly used in places like ships andpressure vessels. Forming of this alloy is achievable by rolling,extrusionandforgingsinceit'sawroughtalloy. Moreover, it can be cold worked to produce tempers with an advanced strength but lower ductility [11].

Thealloy composition of 5154 aluminum is:

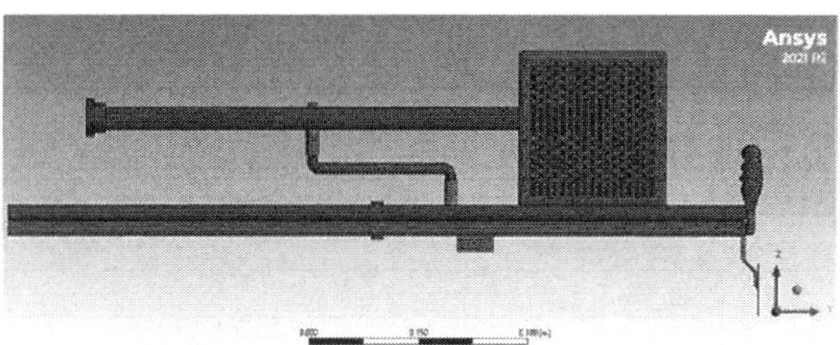

Figure 11.3 Meshing of the smart switch stick (closed view)

Aluminum: 94.4–96.8%
Chromium: 0.15–0.35%
Copper: 0.1%max
Iron: 0.4%max
Magnesium: 3.1–3.9%
Manganese: 0.1%
Silicon: 0.25%max
Titanium: 0.2% max
Zinc: 0.2%max
Residuals: 0.15%max

For the handle of the stick, natural rubber is applied significantly in multitudinous operations andproducts,either alone or in combination with other accouterments. In most of its useful forms, it has a large stretchratio and high adaptability and is extremely waterproof. The physical properties of rubber: Specific gravity, graze resistance, tear resistance, compression set, resilience, elongation, tensile modulus, and tensilepotency.

4.2 Factors of safety calculation and stress analysis

The von-Misesstress distribution for the model is shown in the Figure 10.4. From this analysis, thefactorofsafetyis analyzedforopened andclosedconditionsofsmartswitch stick.

(a) The factor of safety for opened smart switch stick

Applied load open: 2000N

Location: Top face of the foldable seat

Distribution of load: Uniformly distributed over the seat

The factor of safety calculation for 2000N: 1.3654

Fracture points: None at 2000N (seat joint or seat support leg if overloaded by 700N)

Max deformation @2000N: 0.35934mm (in comparison to length of the stick is negligible)

Life: Design can undergo minimum 106 cycles of loading and unloading.

(b) Factor of safety for closed smart switch stick

Applied load closed: 2000N

Location: Top surface of handle

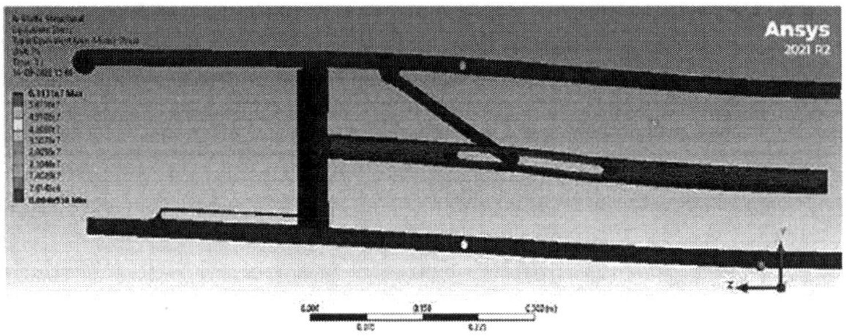

Figure 10.4 Equivalent von-Misesstress

Distribution of load: Equally distributed on handle top surface.

Factor of safety calculation for 2000N: 1.5528

Fracture points: None at 2000N (main support tubing if overloaded by 1000N)

Max deformation @2000N: 0.18596mm (in comparison to length of the stick is negligible)

Life: Design can undergo 106 cycles of loading and unloading.

Hence, for factor of safety calculation, it is analyzed that product life is estimated for infinite life (106 cycles). Although the maximum load assumed for a person is 120 kg (1200N), but it has been taken 200kg weight for safety purpose, that overall increase the reliability of the product.

5. Conclusions

This paper presents a smart stick guidance model that combines the functionalities of a walking stick and a stool to assist visually impaired and elderly individuals in their daily movements. The model incorporatesvarious smart features, including GPS, GSM, an SD card module, and ultrasonic sensors, along with three buttons on the handle for communication and alerts. Unlike mostassistivesystemsthatfocussolelyonthevisuallyimpaireduser,thismodelalso considers the safety of peoplenavigatingaroundthem. By integrating advanced technology and providing real-time location tracking and communication options, thesmart stick enhances the safety, independence, and peace of mind for visually impaired individuals and their family members. The hardware and circuit design of the model ensure accurate positioning, continuoustracking, and extended battery life. The mechanical structure, made of lightweight and durable materials,offers portability and stability, with the added functionality of a foldable stool. Finite element analysisconfirms the strength and reliability of the design under applied loads. Overall, this innovative smart stickguidance model presents a comprehensive solution to address the mobility challenges faced by visuallyimpairedandelderlyindividuals,promotingtheirindependenceandsafetyinnavigatingtheirsurroundings.

6. References

1. Chen, L.-B., Su, J.-P., Chen, M.-C., Chang, W.-J., Yang, C.-H., and Sie, C.-Y. (2019), An Implementation of an intelligent assistance system for visually impaired/blind people; An implementation of an intelligent assistance system for visually impaired/blind people. *2019 IEEE Inter. Conf. Cons. Elec. (ICCE)*. 1–2. http://www.who.int/mediacentre/factsheets/fs282/en/.
2. Chaurasia, S. and Kavitha, K.V.N. (2015). An electronic walking stick for blinds. *2014 Int. Conf. Inform. Comm. Embed. Sys. ICICES 2014*, pp.1–5. https://doi.org/10.1109/ICICES.2014.7033988.
3. Debnath, N., Hailani, Z. A., Jamaludin, S., Syed, I., and Aljunid, A. K. (2001). An electronically guided-walking stick for the blind. *Ann. Inter. Conf. IEEE Engg. Med. Biol. Proc.*, 2, 1377–1379. https://doi.org/10.1109/iembs.2001.1020456.
4. Arunbalaji, T. E., Roshan, R., Prasanth, B., Praveen, S., and Ruthrapathi, R. (2021). Smart blind walking stick. *Inter. J. Creat. Res. Thoughts*, 9(7) pp.b416–b421. www.ijcrt.org.
5. Jadhav, A., Sarkar, J., Patil, R., and Pardeshi, J. (2019). Designs of an effective smartwalking stick for visually disabled. *Proc. 2019 3rd IEEE Inter. Conf. Elec. Comp. Comm. Technol.*, pp.1–5. https://doi.org/10.1109/ICECCT.2019.8869412.
6. Hu, L., Lou, W. Z., Song, R., Gao, C., and Li, X. (2009). A novel design of micro- magnetic sensor-guidance system for the blind. *4th IEEE Inter. Conf. Nano/Micro Eng. Mol. Sys. NEMS2009*, 235–237. https://doi.org/10.1109/NEMS.2009.5068567.
7. Innet, S. and Ritnoom, N. (2009). An application of infrared sensors for electronic white stick. *2008 Inter. Symp.Intell. Sig. Proc.Comm. Sys.*

ISPACS2008, pp. 1-4. https://doi.org/10.1109/ISPACS.2009.4806716.

8. Prashik, C., Kartikesh, A., Siddhesh, B., Rohan, C., and Roshani, R. (2022). Smart blind stick. *IEEE 6th Inter. Conf. Comput. Comm. Con. Autom. (ICCUBEA). pp. 1-4.*
9. Emerson Solomon, F., Prasath, S., Manoj Prasath, T., and Vasuki, R. (2019). Smart walking canefor-blind. *Inter. J. Recent Technol. Engg.*, 8(2Special-Issue3), 791–792. https://doi.org/10.35940/ijrte.B1146.0782S319.
10. Gupta, S., Sharma, I., Tiwari, A., and Chitranshi, G. (2016). Advanced guide cane for the visually impaired people. *Proc. 2015 1st Inter. Conf. Next Gen. Comput. Technol. NGCT2015*, 452–455. https://doi.org/10.1109/NGCT.2015.7375159.
11. ANSYS.Inc(materialslibrary) version 22.

12 Resonance frequency prediction in rectangular microstrip antenna: A novel AI-driven approach

Kundan Kumar[1,a], *S. Chakraborty*[2] *and D. Pal*[2]

[1]Central University of Rajasthan, India

[2]CSIR-CEERI, Pilani, India

Abstract

This work presents the application of artificial neural networks (ANNs) in predicting the resonant frequency for rectangular microstrip antenna design and working out the error between the proposed approach and theoretical approach. ANNs, a subset of artificial intelligence (AI), have the capability to learn from data and execute actions without explicit programming. Microstrip antennas, designed using microstrip technology, consist of metal patches placed on a dielectric substrate. The metal patch functions as the radiating element, while the substrate provides support. The increasing utilization of microstrip antennas in wireless communication systems is attributed to their advantages of low profile, lightweight construction, and ease of manufacturing.

Keywords: Microstrip antennas, feedforward neural networks (FNNs)

1. Introduction

The rapid advancement of artificial intelligence (AI) has opened new possibilities in various fields, including antenna design. In this manuscript, we explore the use of artificial neural networks (ANNs) to predict the resonant frequency in microstrip antenna design. Additionally, we not only investigate the diverse applications of ANNs in this context but also aim to enhance accuracy by leveraging the capabilities of feedforward neural networks (FNNs). It is also known as multilayer perceptron (MLP) [1]. Furthermore, we delve into the methodology to further reduce the associated error obtained from this process. Rectangular microstrip antennas, a popular type of antennas in wireless communication systems, are designed using the microstrip technology. This technology involves placing metal patches on a dielectric substrate, with the patch serving as the radiating element and the substrate providing structural support. Microstrip antennas are valued for their low profile, lightweight nature, and ease of manufacturing. To achieve precise and efficient microstrip antenna design, FNN algorithms come into play. These algorithms, such as Levenberg–Marquardt (LM) with feed-forward backpropagation (FFBPN), offer the ability to learn from data and make accurate predictions without explicit programming. They provide a cost-effective solution for achieving optimal microstrip antenna performance [2, 3].

In this work, we focus on the basic geometry of a conventional rectangular microstrip antenna as shown in Figure 12.1. The parameters considered for variation include patch width (W) and length (L), resonant frequency (fr), and substrate thickness (h) and substrate dielectric constant (εr). By employing FNN techniques, we aim to calculate the resonant frequency of rectangular microstrip antennas with high precision (minimum percentage error between theoretically calculated resonant frequency and predicted resonant frequency) but, at low computational cost. The obtained results will demonstrate the effectiveness of ANN-based approaches in facilitating efficient and accurate microstrip antenna design. Microstrip antennas can be easily printed on a circuit board, so they are becoming increasingly useful in wireless communications, mobile, radio and satellite applications, as well as in spacecraft, missiles, airships, etc. [4].

2. Recent works on application of AI in microstrip antenna

Recently, there have been many studies that explore the application of neural networks for resonance frequency prediction in rectangular microstrip antenna. This radiating patch may be square, rectangular, circular, elliptical, triangular, and any other configuration. The purpose of that work was to predict the resonance frequency in rectangular microstrip antenna using improved algorithm of ANN. In the training process of frequencies for microstrip patch antenna design, various algorithms such as TrainLM, TrainRP and NuRB were used in FBFPN, RPROP and RBF. The performance of different ANN algorithms

[a]kundankumar291@gmail.com

DOI: 10.1201/9781003540199-12

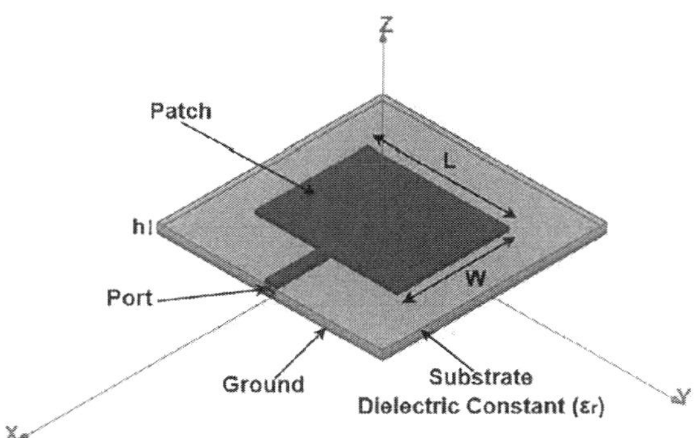

Figure 12.1 Schematic of rectangular microstrip patch

was compared using 41 out of 160 test samples. Some errors were found during this process which can be seen in the work did by Singh [4]. ANNs demonstrated their effectiveness in designing and analyzing microstrip patch antennas, offering precise results with minimal computational expenses. The forward back-propagation algorithm and the trainlm were used in the study did by Kushwah and Tomar [5]. The supported transfer functions include purelin (linear transfer function) and tansig (hyperbolic tangent sigmoid transfer function). During the training of the ANN, it took approximately 100 epochs to achieve a significant reduction in the error from 10^2 to nearly 10^-2 [5]. The multilayer perceptron (MLP) network employed activation functions such as sigmoid and logsig (logarithmic sigmoid transfer function) in the hidden layers, while the output layer utilized a linear activation function known as poslin (positive linear transfer function). The training algorithm of choice was backpropagation, specifically trainrp, a network training function that updates weight and bias values based on the resilient backpropagation algorithm. To train the network, a stopping criterion of either 20,000 epochs or a training error equal to 1.5795% [6]. Both MLP and radial basis function (RBF) networks have been employed in ANN models for micro-strip antennas, with MLP networks being trained with various learning algorithms accuracy is 97.76% [7].

The present work is the best predictive network with an overall accuracy rate of 99% built on only one FNN given the canonical and labeled data, while the above-mentioned previous works mostly used sequential data. Along with this, the number of data has also been increased by more than 50% in the previous works.

3. Basic microstrip antenna geometry

Here, we have considered a conventional rectangular microstrip antenna (Figure 12.1) designed on a dielectric substrate (εr = 2.33, and height h = 1.575 mm). The substrate as well as ground plane dimensions are taken as 65 × 65 mm^2. The patch length (L) and width (W) have undergone variations, adhering to the methodology outlined in refs [8, 9], enabling the theoretical calculation of the antenna's resonant frequency. The schematic of the antenna is shown in Figure 12.1.

4. Utilizing ANN model for work related applications

To predict the resonant frequency for rectangular microstrip antennas and minimize the associated error, a feedforward backpropagation neural network model was employed. The model utilized the training function TRAINLM, adaptation learning function LEARNGD, and performance function MSE. It was implemented in MATLAB [10]. The input layer of the network comprised four parameters (L, W, εr, h), while the target variable (frequency) was represented by a single neuron in the output layer. The model incorporated a PURELIN activation function and included one hidden layer, with consisting of 21 hidden neurons. Also added TANSIG activation function with output layer. Figure 12.2 illustrates the architecture of the network, showcasing the weighted connections and biases.

5. Results and discussions

As discussed earlier, first resonant frequencies of several rectangular microstrip antennas have been theoretically calculated [8, 9]. Now, we utilized various algorithms and techniques within ANNs, to successfully calculate the resonant frequency for a wide range of antenna dimensions. The resonant frequency prediction process was accomplished in MATLAB [10]. The data transformation process was accomplished using PURELIN function, combined with a configuration consisting of four hidden layers. To validate the accuracy of the predictions, data from 11 samples, comprising more than 230 resonant frequency data points, were compared against their corresponding theoretical values. The validation process revealed a maximum error less than 0.05%. A few predicted resonance frequencies for rectangular microstrip antennas for different patch sizes have been shown in Table 12.1. Very close agreement between theoretically computed resonance frequency and predicted resonance frequency has been observed in all the cases.

A. Performance of resonant frequency prediction

The performance of resonant frequency prediction tests such as train, validation and test are depicted in the performance graph of Figure 12.3.

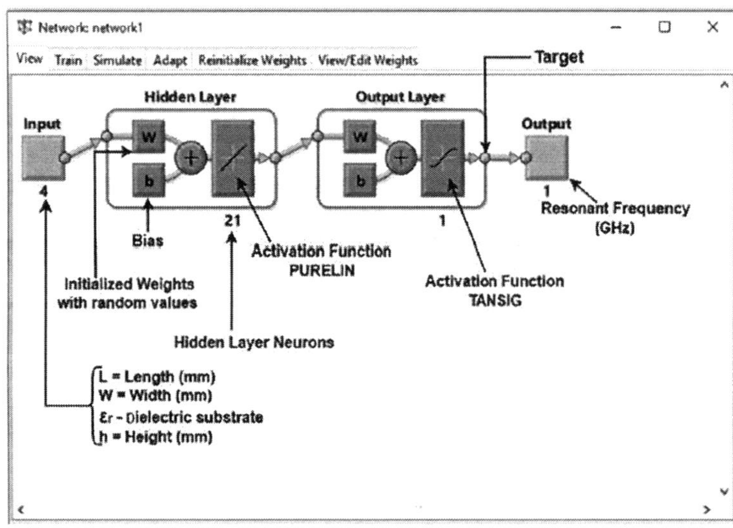

Figure 12.2 Feed forward back propagation algorithm of ANN for predicts the resonant frequency

Table 12.1 Prediction of the resonant frequency of rectangular microstrip patch antenna through ANN

W (mm)	L (mm)	Resonant Freq. in GHz (Theoretical)	Resonant Freq. in GHz (Predicted Through ANN)	Percentage Error (%Error)
9	12	9.3178	8.972426	0.03
8	14	9.8462	9.285277	0.05
8	16	9.5614	9.211883	0.03
22	24	4.2342	4.047580	0.01
21	28	4.3574	4.150537	0.02
25	28	3.7455	3.634985	0.01
23	30	4.0091	3.825199	0.01
20	16	4.7237	4.600558	0.01
22	11	4.4175	4.270059	0.01
24	20	3.9627	3.821224	0.01
24	23	3.9313	3.786652	0.01

Source: Sources must be provided for both figures and tables if they are reproduced/adapted/modified, etc., and permissions may be required.

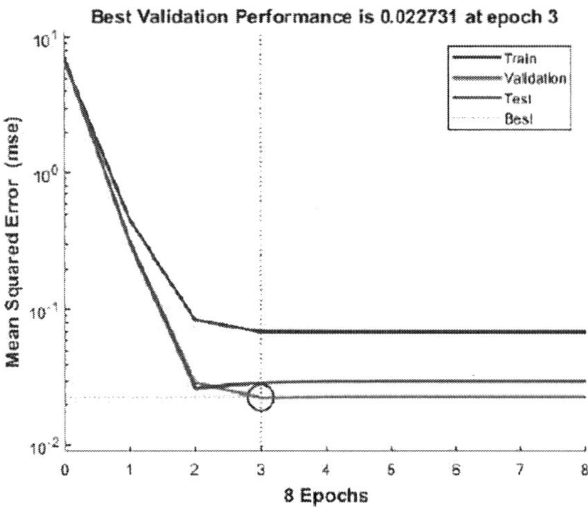

Figure 12.3 Performance of feed forward back propagation algorithm with purelin & tansig function.

The statement "best validation performance is 0.022731 at epoch 3" indicates that during the training of a neural network, the model achieved a validation performance of 0.022731 at the third out of thousand epochs. The validation performance is typically measured using a performance metric, such as mean squared error (MSE), accuracy, or any other relevant measure, depending on the task. In this context, an MSE of 0.022731 implies that the average squared difference between the predicted values and the actual values (targets) on the validation dataset is 0.022731. A lower MSE indicates better performance, as it signifies that the model's predictions are closer to the actual values. The blue line represents the training performance, showing how the network's performance metric changes during training iterations or epochs. The

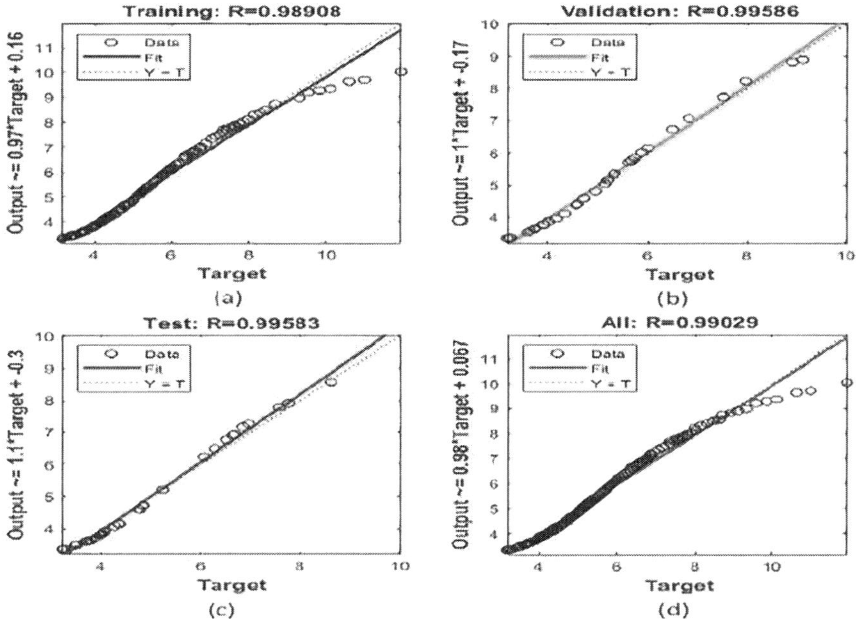

Figure 12.4 Regression graphs using FNN: (a) train, (b) validation, (c) test and (d) error analysis

goal is to minimize this metric, indicating learning and improved predictions. The green line represents the validation performance, measured on a separate dataset not used during training. It helps assess generalization and prevent overfitting, evaluating the network's performance on unseen data. The red line represents the test performance, measured on an independent dataset not used during training or validation. It indicates how well the network is likely to perform in real-world scenarios, on unseen and unknown data. The dotted line represents various elements depending on the visualization. It often denotes a threshold or target value, such as a minimum error rate. It allows comparison against the target and helps assess the network's performance.

B. Regression of resonant frequency prediction

A regression graph in nntool in MATLAB [10] typically includes multiple lines representing different aspects of the model's predictions and actual values, which can be understood through the graphs seen in Figure 12.4. Figure 12.4 (a) represents the predicted values by the neural network model. It visualizes how well the model fits the data and captures patterns or trends in the target variable. Figure 12.4 (b) represents the actual or target values in the regression graph. It serves as a reference to evaluate the model's accuracy and performance. Figure 12.4 (c) can have different interpretations depending on the context. In some cases, it represents the residual or error values, showing the difference between the predicted and actual values. It helps visualize the model's performance in terms of the deviation from the true values. Figure 12.4(d) represents various elements depending on the visualization. It is often used as a baseline or reference level, such as the mean or average value of the target variable. It provides a comparison point to assess the model's performance relative to a simple baseline.

These distinct lines help to fit the data and provide insight into the deviation from its true values with the predictive performance model. The expression "output ≈ 0.98*target + 0.067" represents a linear regression model. It indicates that the output is obtained by multiplying the target by 0.98 and adding a constant term of 0.067. This equation suggests a linear relationship between the output and the target, where the coefficient 0.98 represents the slope or weight of the relationship, and the constant term 0.067 represents the baseline value of the output. In summary, this equation defines a simple linear regression model with a slope of 0.98 and a bias of 0.067, allowing you to predict the output based on a given target value.

6. Conclusions

In conclusion, this study delves into the exploration of ANNs and their application in the estimation of resonant frequency of rectangular microstrip antennas. Much effort has been made in order to minimize the error between theoretically computed resonant frequency and predicted resonant frequency of rectangular microstrip antennas. In future works, the results obtained through present approach will also be validated against measurements.

Acknowledgement

Authors are thankful for the support and guidance provided by the Director of CSIR-CEERI and the Vice-Chancellor of Central University of Rajasthan during the research work.

References

1. Russell, S. and Norvig, P. (2011). A modern approach. *Artif. Intel.*, 175(3rd Edition).
2. Haykin, S. (2009). Neural networks and learning machines. *Neural networks (Computer science).* Pearson Education, Inc.: Prentice Hall.
3. Rumelhart, D. E., Hinton, G. E., and Williams, R. J. (1986). Learning representations by back-propagating errors. *Nature*, 323(6088), 533–536.
4. Singh, B. K. (2015). Design of rectangular microstrip patch antenna based on artificial neural network algorithm. *Conference: 2nd Internat. Conf. Sig. Proc. Integr. Netw. (SPIN), 6-9.*
5. Kushwah, V. S. and Tomar, G. S. (2017). Design and analysis of microstrip patch antennas using artificial neural network. *Trends Res. Microstr. Anten, Chapter 3, 55-75.*
6. Vitaly, F., Esquerrel, R., dos Santos, F. N., Nascimento, S. S., and Fabricio, G. S. F. (2011). Analysis and design of microstrip antennas by artificial neural networks. *Conference: Microwave Optoelec. Conf. (IMOC), SBMO/IEEE MTT-S International, 226-230.*
7. Turker, T. Y. N. and Gunes, F. (2006). Artificial neural networks applied to the design of microstrip antennas. *Turkish J. Elec. Engg. Comp. Sci.*, 14(3), 445–453.
8. Chattopadhyay, S., Biswas, M., Siddiqui, J. Y., and Guha, D. (2009). Rectangular microstrips with variable air gap and varying aspect ratio: Improved formulations and experiments. *Microw. Optic. Technol. Lett:51, 169-173.*
9. Balanis, C. A. (2005). Antenna Theory. Antennas & Propagation. United States: John Wiley & Sons, Inc, (3rd Edition).
10. MATLAB Version 2019a.

13 Coordinated hybrid AC/DC microgrid system with optimized CSA-TLBO-tuned robust 2-DOF-FOPID controller

Indrajit Koley[1,a], *Pralay Roy*[2,b], *Chiranjit Sain*[3,c], *Asim Datta*[4,d], *Goutam Kumar Panda*[5,e] *and Pabitra Kumar Biswas*[1,f]

[1]Department of Electrical Engineering, Siliguri Institute of Technology, Siliguri, India

[2]Department of Electrical Engineering, National Institute of Technology, Mizoram, Mizoram, India

[3]Department of Electrical Engineering, Ghani Khan Choudhury Institute of Engineering & Technology, Malda, India

[4]Department of Electrical Engineering, Tezpur University, Tezpur Assam, India

[5]Department of Electrical Engineering, Jalpaiguri Government Engineering College, Jalpaiguri, India

Abstract

Frequency regulation has become more difficult as the introduction of sustainable energy sources (RES) has grown. The load frequency control (LFC) mechanism is a critical function in an electrical power network for scheduling a balance between power generation and the load in order to avoid frequency deviation (FD). The outcome of this article is to establish a practical LFC topology for a hybrid AC/DC microgrid (MG) system that includes a wind turbine generator (WTG) and a battery energy storage system (BESS). The LFC system is implemented using two degree of freedom fractional order proportional integral derivative controllers (2-DOF-FOPID). Further, a combination of Cuckoo search algorithm (CSA) and TLBO approach is blended to optimize the controller parameters. The results of 2-DOF-FOPID controllers are compared to those of proportional integral double derivative (PIDD) and proportional integral derivative (PID) controllers. While contrasted to the PIDD and PID controllers, the 2-DOF-FOPID controller exhibits superior characteristics in terms of settling time and magnitude of oscillations. Finally, the study reveals that suggested 2-DOF-FOPID -based LFC scheme's robustness is tested under various loading disturbances.

Keywords: LFC, MG system, CSA-TLBO, 2-DOF-FOPID controller, hybrid AC/DC MG

Introduction

Microgrid (MG) integration with traditional power systems aims to address economic and environmental concerns while also increasing the reliability of traditional power systems. A distributed generating unit (DGU), such as a wind turbine, solar panel, or other sources, is combined with a number of additional DGUs to create a hybrid power or storage system called MG. Sustainable energy sources are gaining popularity as a result of the current dilemma with fossil fuels and environmental worries. On the other hand, conventional fossil fuel-based generating continues to be a feasible choice because of its great reliability. An idea that strikes a balance between generation dependability, cost, and environmental issues is combining sustainable and conventional energy sources. There are several difficulties that distributed generation (DGs) based on RES must overcome, including controllability, islanding operation, system stability, and others [1]. The grid controls the voltage and frequency at the DG interconnection points when the system is connected to the grid. The fundamental problem, however, is the stability of a DG running on renewable energy in an island mode. An MG's power storage system is capable of supporting power balancing in non-islanded mode [2], however, to maintain system frequency stability, an effective LFC technique is required.

LFC seeks to decrease frequency excursions in the system by lowering the area control error (ACE), which is brought on by fluctuating load and the inconsistent output of distributed energy resources (DER) units. The LFC system monitors tie-line power flows and system frequency, adjusting regional generation as necessary to preserve the temporal average of the ACE. As a regulation metric, ACE is often employed in LFC. To reduce the ACE, frequency and tie-line power errors should be nearly nonexistent. The high penetration of uncertain RESs, which regularly misaligns supply

[a]indrajit.koley@gmail.com, [b]pralay07@gmail.com, [c]sain.aec@gmail.com, [d]asimdatta2012@gmail.com, [e]gpandaee@gmail.com, [f]pabitra.eee@nitm.ac.in

DOI: 10.1201/9781003540199-13

and demand, causes variations in power generation that lead to frequency volatility in the microgrid. Frequency stability is critical for AC systems. The study's major focus earlier, in keeping with typical AC use, was on AC microgrids [3].

Demand in hybrid AC/DC MGs has grown as a result of the expanding availability of DC power sources comprising of fuel cells, PV, and BESS as well as a broader integration of power electronic converters and DC loads. Hybrid DC/AC MGs offer extra advantages over solely DC or AC MGs, including enhanced power quality, localized energy supply, fewer conversion steps, and reduced power processing costs. Bidirectional power converters (BPCs) and hybrid AC/DC MGs collaborate to link the AC and DC sectors [4]. Although many LFC solutions for AC and DC MGs have been published, because to the coexistence of AC and DC sections, which calls for sophisticated control and optimization processes, these topologies are not particularly relevant to hybrid AC/DC MGs systems.

Many researchers are now concentrating on various issues of power management criteria for AC/DC hybrid MGs. A nonlinear control method based on state-dynamic feedback linearization theory for parallel BPC in grid linked mode was proposed in reference [5]. For sharing the power and voltage support from DC for BPCs, an effective hybrid MG control approach was described in reference [6] [7], which could reduce the circulating current in the parallel BPC working technique. The hybrid AC/DC MG control system developed by Ge et al. [8] for local distributed generation units is based on distributed control and decentralized load management. In comparison to AC or DC MGs, hybrid MGs experience additional problems with energy management because of the fluctuation of both DC and AC inputs and demands. Moreover, the energy storage system of an MG is frequently used to offer long-term power assistance. The capacity and load types of both AC and DC sub grids (SGs) have not been taken into account in earlier studies, which could result in power imbalance. The aforementioned limitations make power regulation in a hybrid AC/DC MG very difficult. An effective control strategy is needed to maintain bus voltage in the DC SGs, balance the power demands in both SGs, and keep the AC SGs operating at their operational frequency.

The literature [9] states that because to their simplicity in design and implementation, the majority of classical controllers have been widely employed in automated generation and control (AGC) of multi-area power systems. These typical controllers, however, exhibit substantially longer dynamic properties in terms of their settling times and amplitude oscillations. The integral-double-derivative (IDD) controller and the PIDD controller are two newly developed classical controllers for AGC [10, 11]. Faster dynamic responses are made possible by the double derivative function, which reduces the settling time and boosts system stability. However, tracking the set point of the power system and eliminating disturbances are not possible with conventional integral order (IO) controllers. The advent of fractional order (FO) calculus has resulted in a significant improvement in the domain of control system design. Computational processes are required for tuning the controller settings in AGC [12]. Furthermore, the majority of recent research uses evolutionary methods due to the nonlinear character of constraint optimization problems and the difficulty in locating the best solution [13].

Many optimization techniques have been used by researchers in recent years to deal with computing control parameters. It is challenging to build controllers in an LFC with the best possible settings. The fruit fly algorithm (FFA), cuckoo search algorithm (CSA), firefly algorithm (FA), bacterial foraging (BFO) optimization, and artificial bees' colony (ABC) algorithm are only a few of the optimization techniques that have been proposed [13].

The two SGs in this study are coupled via a BPC to allow for power interaction in a hybrid AC/DC MG employing a distributed synchronized power control approach. The recommended control strategy enables the two SGs to coordinate their assistance through a power interlinkage that works in the opposite direction. Following are the study's main contributions: (a) When creating a coordinated droop control system, consideration is given to the load sizes in the AC and DC SGs, allowing for power transmission between them. (2) Using 2-DOF-FOPID controllers as controllers, the system's response is contrasted with that of traditional PIDD and PID controllers. The controller design parameters are estimated using a minimized CSA known as an integral-of-time-multiplied-absolute-error (ITAE). The suggested technique's resilience is demonstrated in DC and AC SGs with a 20% load variation.

Proposed system

This study describes a hybrid MG system that integrates battery and wind turbine technologies. The MG is equipped with both AC and DC SGs. For interaction with power service dependability, the BPC connects the two SGs. The discussed mechanism is depicted in Figure 13.1. While the AC SG consists just of wind-based devices, the DC SG is made up of wind turbines and batteries. Only the

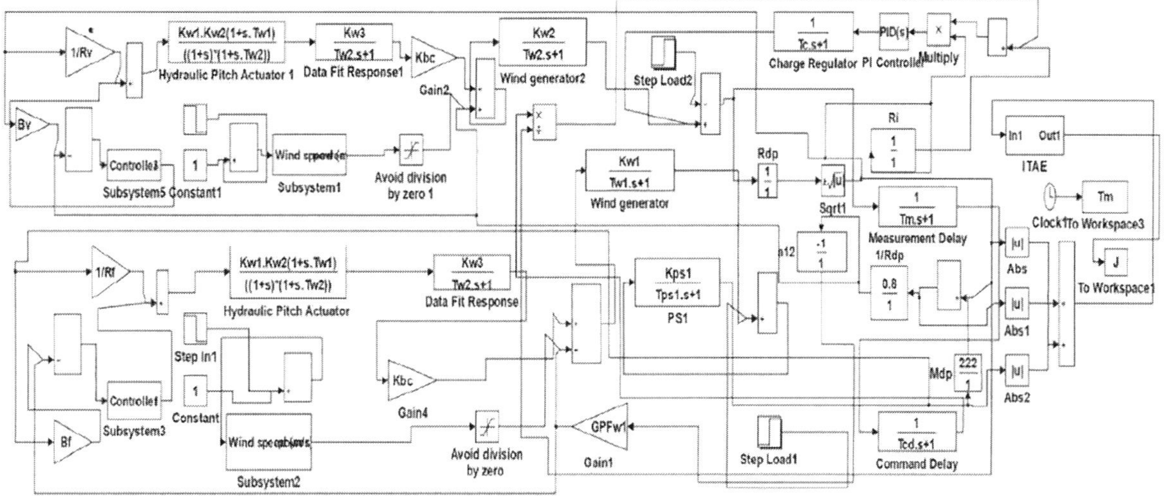

Figure 13.1 System diagram [16]

DC SG has access to the energy storage component, which is used to supply the SG with an efficient power supply. A hybrid AC/DC MG is developed and simulated in order to assess the efficacy of the recommended coordinated control method.

WTG

Wind speed (V_w), pitch angle (β), blade radius (R), air density (ρ), and power co-efficient (C_p) all affect how much power is produced by WTG. The formula used to compute a wind turbine's output is shown below [13]:

$$P_m = \frac{1}{2}\pi\rho C_p(\lambda, \beta) R^2 V_w^3 \tag{1}$$

The wind turbine power unit coefficient is shown in ref [13]. The first order mathematical model of a WTG (G_p) can be expressed as [14]

$$G_p(s) = \frac{\Delta F_W}{\Delta P_{WP}} = \frac{K_{WW}}{1+sT_W} \tag{2}$$

Where, ΔF_W is the deviation in speed of WTG and ΔP_{WG} is the output power variation from the WTG. Moreover, the change of speed of a WTG can be expressed in terms of power control [15].

The hydraulic pitch actuator's TFs and data fit pitch response can be represented as [13]

$$G_{hpa}(s) = \frac{K_{W1}K_{W2}(1+sT_{W1})}{(1+s)(1+sT_{W2})} \tag{3}$$

$$G_{dfr}(s) = \frac{K_{W3}}{sT_{W3}+1} \tag{4}$$

BESS

In MGs, BESS is frequently used to stabilize the power supply when there is intermittent renewable energy. A BESS system is a good example of a black box with an output that can be determined analytically based on inputs and set points. While the output current varies depending on the load, the output voltage is constant. A thorough charge control strategy (CRS) is necessary to guarantee trustworthy process and safeguard the system from harmful situations like overcharging, deep-discharging, etc. The parameters for charge management, measurement delay, and command latency are taken into account in the system description, as shown in Figure 13.1. In this study, a standard PID controller (K_p = 0.222, K_i = 0.513, K_d = 0.513) is also employed to regulate the BESS model under ambiguous circumstances.

The TFs of different components are denoted by G_{md}, G_{cd}, and G_{cc} as [10]:

$$G_{md}(s) = \frac{1}{1+sT_m} \tag{5}$$

$$G_{cd}(s) = \frac{1}{1+sT_{cd}} \tag{6}$$

$$G_{cc}(s) = \frac{1}{1+sT_c} \tag{7}$$

Control strategy

The coordinated control system controls RES generation and manages power exchange via BPC to achieve power stability in each of the SGs. A power imbalance in the relevant area is shown by the FD in AC SGs and the change in voltage in DC SGs. [14]. f tends to be higher (or lower) when the AC SG's power is excessive. The DC droop approach controls energy storage, so when the power of the DC SG is in excess (or deficit) (or lower), the DC bus voltage (V_{dc}) tends to be higher. As a result, we analyze power in DC SG $P_{dc} - V_{dc}^2$ droop in DC SG power regulation. Further, he dynamic linkage between voltage and power is defined as

$$\Delta V_{dc}^2 = R_{dp} \Delta P_{dc} \tag{8}$$

R_{dp} is the droop coefficient, and V_{DC} and P_{dc} are the voltage and power variants in the DC SGs, independently.

While the AC SG raises voltage in the DC SG, the DC SG increases frequency in the AC SG. As a result, changes in DC voltage or AC frequency can happen even when power is being applied to one or more SGs. The power interaction in these instances is coupled to a droop control method.

$$\Delta V_{dc} = M_{dp} \left(f^* - f\right) \tag{9}$$

where f^* and f are the rated and normal frequencies of the system, respectively. According to Equation (9), both the DC voltage and the AC frequency must increase or decrease simultaneously in order for the power fluctuation to be distributed fairly. Since both SGs must interchange power during synchronization, the droop coefficient M_{dp} has a significant impact on the exchange of power. As a result, it is important to take into account the two SGs' capacities and load kinds while choosing the M_{dp} number. The equations of M_{dp} are shown in ref [16]. The ITAE is used to create the objective function of a heuristic optimization method [10].

$$J = ITAE = \int_0^{t_{sim}} \left(|\Delta F_{AC}| + |\Delta V_{DC}| + |\Delta P_{ic}|\right).t.dt \tag{10}$$

2-DOF-FOPID controller

2-DOF protects the behavior of closed-loop control systems by managing the command input and the independent disturbance rejection resources. The transfer functions of a closed-loop control system are individually tunable. Control engineers must take smooth set-point variable tracking and disturbance suppression into consideration when selecting a 2-DOF-FOPID controller. The 2-DOF-FOPID controller, as opposed to the single DOF FOPID controller, produces an output signal when comparing the measured signal to the reference signal [17]. The output of the controller is comprise of the proportional, derivative, and integral actions of the corresponding various signals, and the pressure of the gain parameters determines the actions. The used controller offers quick reaction and increases stability compared to conventional PIDD and PID controllers. In the case of PID, the derivative operator significantly intensifies turbulence, leading to system instability in the presence of nonlinearities.

Proposed algorithm

Deb and Yang created CSA, a meta-heuristic optimization method inspired by nature, in 2009. CSA is triggered by nature and depends on cuckoo birds' brood reproduction strategy to increase their population. It is influenced by the cuckoo species' breeding tactics in relation to some birds' Lévy flight behavior. Lévy flights are arbitrary walks with step lengths derived from the Lévy distribution and random directions. Numerous studies have shown that many animals' flight patterns exhibit the characteristic behavior of Lévy flights. The bird's method of reproduction and sound has made it popular. In the past, cuckoo birds would lay their eggs in other birds' nests. Some species are exceptionally adept at mimicking the color and outline of a select few swarm species' eggs, such as the brood-parasitic tapera. Cuckoos have developed in this way to increase their ability to reproduce.

Due to their significance in locating global optimal values in the presence of multiple uncertainties, hybrid optimization algorithms have recently drawn attention from the research communities. The major goal of the suggested hybrid optimization method is to combine TLBO's quick convergence rate with CSA's strong search capabilities. The suggested technique primarily executes two steps, wherein the CSA will do a Lévy flight to produce new solutions for solutions that have been abandoned. Additionally, the TLBO is employed to improve CS's local search functionality for other solutions. As a result, without sacrificing any of the enticing features of the original CSA and TLBO, the implemented algorithm is more beneficial for a widespread range of applications. Strong global search capabilities and a quick convergence rate make the recommended CSA-TLBO fit for a range of problem areas. Lévy flight and the teaching-learning process are two crucial methods for updating the solutions in the population, much like in the framework. The following sections provide a detailed description of the two primary techniques. Teaching-learning process, Lévy flight, constraint-handing methodology and putting the suggested algorithm into practice. The flowchart of the implemented CSA-TLBO is shown in Figure 13.2.

Results and analysis

As shown in Figure 13.1, the tested MG model is run on the MATLAB/SIMULINK platform with step load perturbations (SLPs) in each of the generating units. While frequency bias (B_i) is maintained constant at the area frequency response attributes, the responses of various controllers, such as PI, PID, and PIDD, are evaluated individually. The best controller settings are chosen using the CSA-TLBO approach. The optimization is carried out by minimizing the performance index, per Equation (1). For each controller, the proportional, integral, and derivative gain factors (K_p, K_i, K_d) as well as the integration and differentiation orders {λ, μ}, are computed using the CSA technique. The coordinated control model (as seen in Figure 13.1) is developed in Simulink (.mdl), and the CSA method is coded in MATLAB. In order to find the least value of J, the Simulink model (.mdl) is used by the Matlab program's CSA method to generate an updated value for J. The outcomes that correspond to J's lowest value are regarded as the best ones. The conceptual flow-chart for calculating the least and a best result is shown in Figure 13.2.

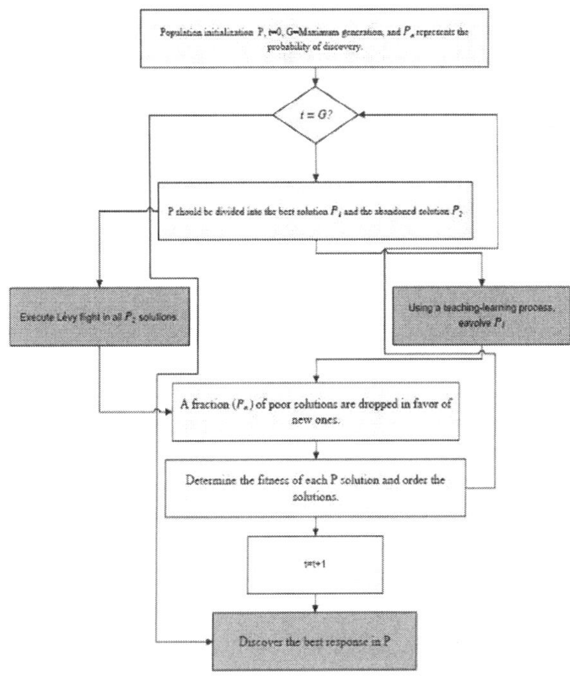

Figure 13.2 Flowchart of the CSA-TLBO method [18]

Table 13.1 displays the top values of many controllers for hypothetical loading scenarios. The optimum controller parameter settings are used to assess the system's dynamic responses. Investigated are the changing interchange power and the dynamic behaviors in the particular area. Results from 2-DOF-FOPID controllers are contrasted with those from traditional proportional-integral (PIDD) and PID controllers.

The time responses (TS) of ΔF_{AC}, ΔV_{DC}, and ΔP_{IC} in both DC and AC SGs owing to SLP of 0.1 p.u and wind power change are shown in Figure 13.3. Figure 13.4 shows the TS of ΔF_{AC}, ΔV_{DC}, and ΔP_{IC} in both DC and AC SGs with a 0.1 p.u load disturbance. The findings show that 2-DOF-FOPID controllers offer fewer fluctuations in tie-line powers and area frequencies. The settling time, peak overshoot, peak undershoot of the CSA tuned implemented controller is far superior than pre-existing PIDD and PID controller as observed from Figures 13.3–13.5. With less iteration, the CSA-TLBO approach can produce excellent outcomes. The DC and AC SG loading is increased from nominal to 20% in order to evaluate the designed controller's resistance to failure. As demonstrated in Figure 13.6(a, b), the dynamic responses found with the loading variations are almost identical to the results obtained with the nominal loading.

Table 13.1 Optimal data attained with the ITAE optimized hybrid CSA-TLBO technique

Area	Generating source	PIDD			PID			2-DOF-FOPID				
		ITAE=1140			ITAE=892			ITAE=525				
		K_p	K_i	K_{dd}	K_p	K_i	K_d	K_p	K_i	K_{dd}	λ_i	μ_i
AC SG	Wind	2.20	0.27	0.12	3.72	0.48	0.14	1.64	0.34	0.15	0.89	0.93
	Battery	2.13	0.02	0.38	1.22	0.05	0.48	1.19	0.008	0.46	0.98	0.69
DC SG	Wind	4.37	2.35	0.48	4.44	2.18	0.042	1.27	0.36	0.09	0.89	0.17

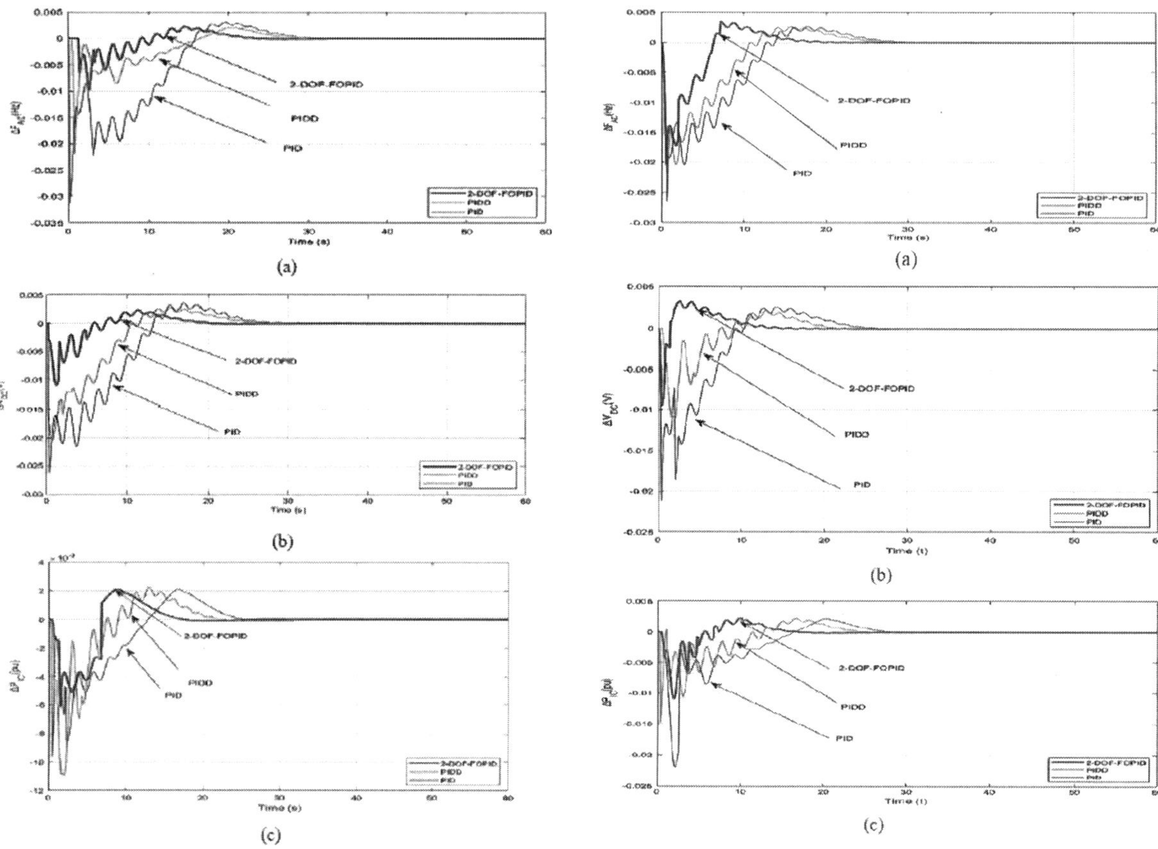

Figure 13.3 TS of (a) ΔF_{AC}, (b) $\Delta V_{DC,}$ and (c) ΔP_{IC} with SLP of 0.1 p.u, and wind turbine output power variations simultaneously in both DC and DC SGs

Figure 13.4 TS in both AC and DC SGs for (a) ΔF_{AC}, (b) $\Delta V_{DC,}$ and (c) ΔP_{IC} with 0.1 pu SLP

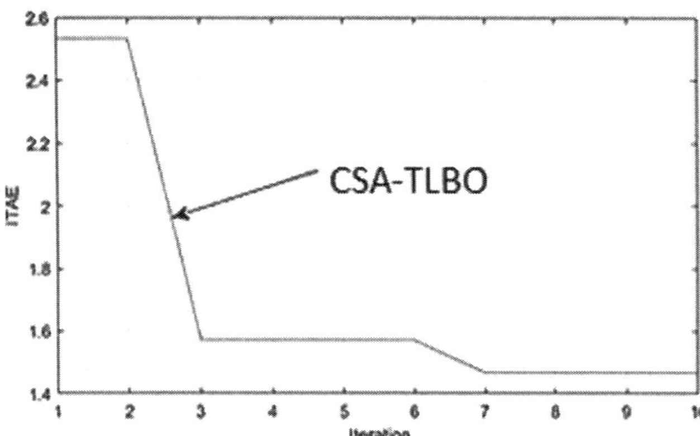

Figure 13.5 Convergence response of CSA-TLBO algorithm

Conclusions

In this study, a 2-DOF-FOPID controller is used to control the frequency of a hybrid AC/DC microgrid system. The dynamic responses produced by the CSA-TLBO hybrid approach are compared in view of the parameters of the PIDD and PID controllers are adjusted. The 2-DOF-FOPID controller's performance is evaluated under various load circumstances to guarantee about the reliability. The results show that in MG systems driven by sustainable energy sources, the suggested CSA-TLBO hybrid adjusted 2-DOF-FOPID controller is very effective and has a better dynamic behavior and stability under various operating points.

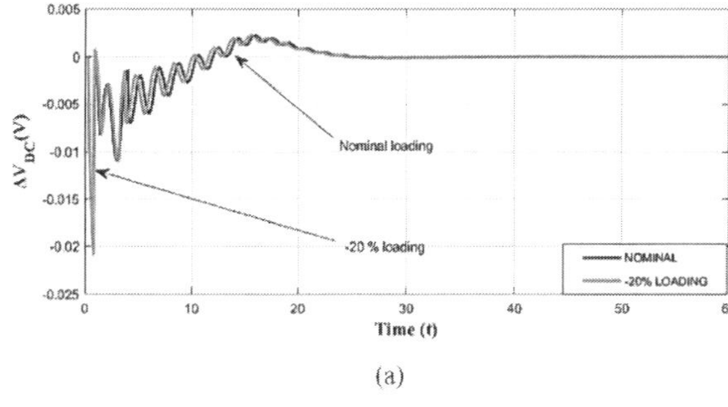

(a)

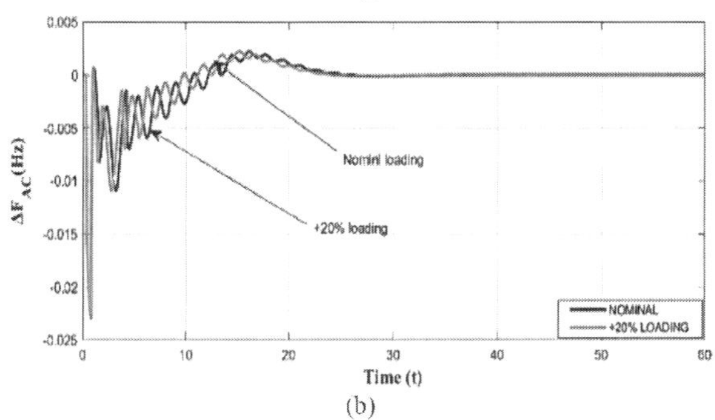

(b)

Figure 13.6 Sensitivity study for both AC and DC SGs with a nominal loading of ±20%

References

1. Pavković, D., Lobrović, M., Hrgetić, M., and Komljenović, A. (2016). A design of cascade control system and adaptive load compensator for battery/ultracapacitor hybrid energy storage-based direct current microgrid. *Energy Conver. Manag.*, 114, 154–167. doi:10.1016/j.enconman.2016.02.005
2. Wei, G., Darvishan, A., Toghani, M., Mohammadi, M., Abedinia, O., and Ghadimi, N. (2018). Different states of multi-block based forecast engine for price and load prediction. *Inter. J. Elec. Power Energy Sys.*, 104, 423–435. doi:10.1016/j.ijepes.2018.07.014 .
3. Nikam, V. and Kalkhambkar, V. (2020). A review on control strategies for microgrids with distributed energy resources, energy storage systems, and electric vehicles. *Inter. Trans. Elec. Energy Sys.*, 1 26. doi:10.1002/2050-7038.12607.
4. Aprilia, E., Meng, K., Hosani, M. A., Zeineldin, H. H., and Dong, Z. Y. (2017). Unified power flow algorithm for standalone AC/DC hybrid microgrids. *IEEE Trans. Smart Grid*, 1–1. doi:10.1109/tsg.2017.2749435 .
5. Parida, D. C. (2018). Stand-alone AC-DC microgrid-based wind-solar hybrid generation scheme with autonomous energy exchange topologies suitable for remote rural area power supply. *Inter. Trans. Elec. Energy Sys.*, 28.
6. Nejad, H. C., Farshad, M., Rahatabad, F. N., and Khayat, O. (2016). Gradient-based back-propagation dynamical iterative learning scheme for the neuro-fuzzy inference system. *Expert Sys.*, 33(1), 70–76. doi:10.1111/exsy.12131 .
7. Xia, Y., Wei, W., Yu, M., Wang, X., and Peng, Y. (2017). Power management for a hybrid AC/DC microgrid with multiple subgrids. *IEEE Trans. Power Elec.*, 1–1. doi:10.1109/TPEL.2017.2705133.
8. Ge, X., Han, H., Xiong, W., Su, M., Liu, Z., and Sun, Y. (2020). Locally-distributed and globally-decentralized control for hybrid series-parallel microgrids. *Inter. J. Elec. Power Energy Sys.*, 116, 105537. doi:10.1016/j.ijepes.2019.105537.
9. Dash, P., Saikia, L. C., and Sinha, N. (2015). Comparison of performances of several FACTS devices using Cuckoo search algorithm optimized 2DOF controllers in multi-area AGC. *Inter. J. Elec. Power Energy Sys.*, 65, 316–324. doi:10.1016/j.ijepes.2014.10.015.
10. Lalit, C. S., Nanda, J., and Mishra, S. (2011). Performance comparison of several classical controllers in AGC for multi-area interconnected thermal system. International Journal of Electrical Power & Energy Systems., 33(3), 394–401. doi:10.1016/j.ijepes.2010.08.036.
11. Fathy, A. and Kassem, A. M. (2018). Antlion optimizer-ANFIS load frequency control for multi-interconnected plants comprising photovoltaic and wind turbine. *ISA Trans.*, S0019057818304774. doi:10.1016/j.isatra.2018.11.035.
12. Datta. G., Bhattacharya, D. M. and Saha, H. (2014). An efficient technique for controlling power flow in a single stage grid-connected photovoltaic system. *Scientia Iranica*, 21, 885–897.
13. Shankar, R., Pradhan, S. R.., Chatterjee, K., and Mandal, R. (2017). A comprehensive state of the art literature survey on LFC mechanism for power system. *Renew. Sustain. Energy Rev.*, 76, 1185–1207. doi:10.1016/J.RSER.2017.02.064.
14. Abazari, A., Monsef, H., and Wu, B. (2019). Coordination strategies of distributed energy resources including FESS, DEG, FC and WTG in load frequency control (LFC) scheme of hybrid isolated micro-grid. *Inter. J. Elec. Power Energy Sys.*, 109, 535–547. doi:10.1016/j.ijepes.2019.02.029.
15. Asim, D., Alejandro, C. A., Indrajit, K., Rishiraj, S., Javier, V. C., Kamalika, D., and Debasree, S. (2021). Coordinated AC frequency vs DC voltage control in a photovoltaic-wind-battery-based hybrid AC/DC microgrid. *Inter. Trans. Elec. Energy Sys.*, 1469–1479. doi:10.1002/2050-7038.13041
16. Alam, M. K., and Khan, F. H. (2013). Transfer function mapping for a grid connected PV system using reverse synthesis technique. *2013 IEEE 14th Workshop Con. Model. Power Elec. (COMPEL)*. doi:10.1109/compel.2013.6626425.
17. Huang, J., Gao, L., and Li, X. (2015). An effective teaching-learning-based cuckoo search algorithm for parameter optimization problems in structure designing and machining processes. *Appl. Soft Comput.*, 36, 349–356. doi:10.1016/j.asoc.2015.07.031.

14 IoT-based weather monitoring station

Subungsha Basumatary, Swapnanil Dekaraja, Prakashyab Poddar, Parmita Saikia and Ganesh Roy[a]

Department of Instrumentation Engineering, Central Institute of Technology Kokrajhar, Kokrajhar, Assam, India

Abstract

A real-time weather monitoring device has been proposed in the present work based on Internet of Thing (IoT) technology. In recent days, IoT-based products are increasing enormously by adding good impact in the society. A number of advantages including power consumption, light weight, access from any point are obtained when a device is deigned in a smart way and connected over the inter network. The proposed system comprises various sensors which can monitor the temperature, humidity, and pressure of a remote region using IoT. At the initial stage, DHT11 and BMP180 sensor senses the input data and send to the ESP32 microcontroller. The microcontroller sends output to the cloud server. Cloud server sends the data to a personal computer (PC) or smartphone for monitoring and recording purpose. Finally, a comparative study has been performed and observed average accuracy of 95.65% and 97.94% in the measurement of temperature and humidity, respectively.

Keywords: IoT, sensor, microcontroller, weather, cloud server

Introduction

A weather station provides necessary information regarding the state of the past, present or future weather conditions of some places. Now-a-days, internet helps us to track the weather situations through any smart devices like mobile phones, laptop, tablet, etc. But the circumstances changes for remote areas where internet facilities are not available. By motivated with these situations, the present research work has been initiated and it is found that the design of weather station is important for remote areas where internet is not accessible. In earlier days, the weather station may be built up with a combination of various physical instruments like thermometer for temperature measurement, barometer for atmospheric pressure measurement, wind vanes for measuring air direction, etc. After recording the measurements, all the reports are circulated to the newspapers, news channels or any other media for necessary information to the inhabitants. The difficulties for the existing system involve bulkiness, heavy maintenance, manual operated, huge power requirement, etc. To remove those limitations smart systems are in use recently. A smart weather station can be prepared by the use of microcontroller, different sensors, monitoring device and use of software packages [1, 2]. The developed device is very light weight, easily portable and also consumes very low power. In literature, many researchers have made their effort to build different type of weather station that can be used to monitor and forecast weather [3–5]. The applications are also enriched with the combination of IoT [1, 6–8]. Real-time weather stations and weather forecasts differ from one another. The purpose of a weather forecasting device is to make weather predictions for a specific location and time. The user can monitor the sensor measurements always. Even a tablet or smartphone can be used to see the results. The present paper describes how to create a weather station that can provide users with real-time data access from anywhere. In order to allow users to get weather information anywhere in real-time, the proposed IoT weather station is made. The weather station has been constructed using IoT. A comparative performance study has been performed with the help of other standard methodologies.

Methodology

In the block diagram shown in Figure 14.1, two respective sensors have been utilized to measure three different weather parameters. The sensors are temperature, humidity sensor (DHT11), and pressure sensor (BMP180), which are activated by providing power input of 3.3V. All the sensors are directly connected to the ESP32 microcontroller board, which is driven by 5V. The ESP32 board has an inbuilt WIFI module and ADC converter. Also, ESP32 can calculate the data and shows the live weather report either on LCD or cloud server through internet. The process of sending data to the cloud server is repetitive using WIFI for a constant interval of time. The data can be stored in the cloud server for future use. After uploading, the user may visit the web page or

[a]g.roy@cit.ac.in

DOI: 10.1201/9781003540199-14

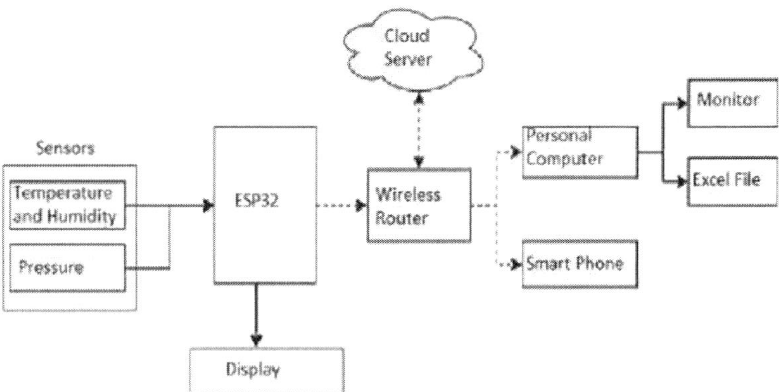

Figure 14.1 Block diagram for the proposed system

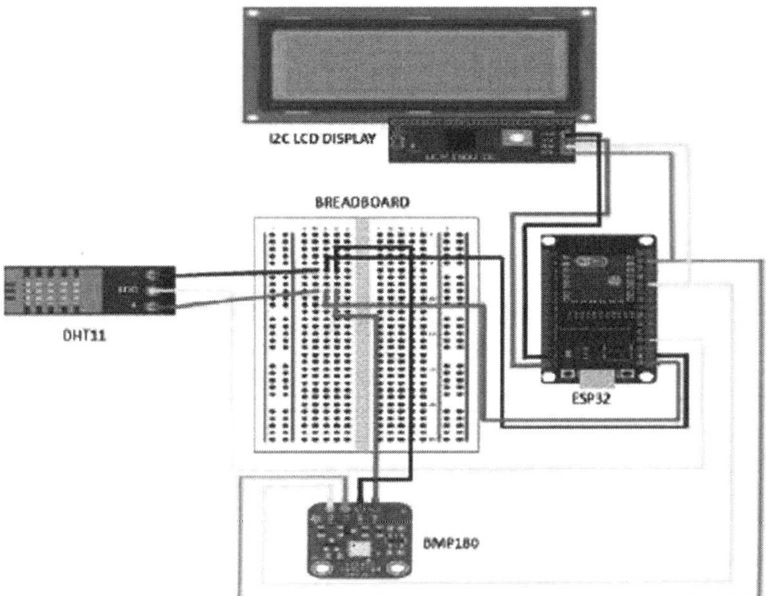

Figure 14.2 Connection diagram of the proposed system

cloud server through PC or smartphone to view or monitor the data. The previously stored data can be accessed as ".csv" format.

Hardware implementation

The major hardware components of the weather monitoring device are the DHT11, BMP180 sensor, and microcontroller. The interfaces in between sensor with the microcontroller along with the display unit have been provided in Figure 14.2. Blynk software IDE is used to programme the microcontroller. After successful programme, the microcontroller is ready to receive signals from the sensors. The system architecture is made up of integrated electronic components are ESP32 microcontroller, DHT11 sensor, BMP180 sensor, and I2C LCD display. The ESP32 is used in the system development as the main microcontroller, and the sensors are connected through digital and analogue ports. A reliable voltage power supply is used for the ESP32 microcontroller. The ESP32 microcontroller collects data from DHT11 and BMP180 sensors, processes it, and then sends the results on the display unit. The measured information is also transmitted to the cloud server through the wireless router. Data on the cloud server can be remotely accessible through PCs and smartphones. The steps for drawing the connections diagram are enlisted as follows and full circuit diagram has been shown in Figure 14.2.

i. DHT11 and BMP180 positive terminal have been connected to 3.3V power supply of the ESP32 board and the negative terminals in GND.

ii. The output terminal of DHT11 is connected to D2 terminal in ESP32 board.
iii. VCC terminal of LCD display is connected to 5V power supply in ESP32 board and GND is in GND terminal.
iv. Both SDA, SCL terminals of BMP180 and LCD display is connected to ESP32 board. D21 and D22 are connected to SDA and SCL, respectively.

Software implementation

A C++ compiler has been used to implement the software design on the ESP32 microcontroller. The programme that implemented in the ESP32 microcontroller is shown as a flowchart in Figure 14.3. The overall working procedure has been provided in Figure 14.1. In Figure 14.3, a working flowchart has been provided from the sensors to the ESP32 device board for measuring temperature, humidity, and pressure data and sends to the Blynk server for every minute and also displays live data in LCD. Temperature, humidity, and pressure data are deposited in the cloud server databases. The minimum and maximum limit of temperature has been set as 0°C and 50°C, respectively. Similarly, for 20% and 50% are the corresponding minimum and maximum humidity. The sensor data has been collected by the ESP32 device board and sent it to the Blynk cloud server wirelessly. Data which is kept in the storage database server may be monitored in PC or smartphone by the user, and the user can also read the data in graph form.

Developed prototype

The final prototype of the developed system has been provided in Figure 14.4. The developed prototype is a practical application of the concept. The system includes an ESP32, sensors, a WIFI module, etc., encased inside the box. Inside the system, the sensors monitor changes in the weather and climate, including pressure, temperature, humidity, and air concentrations. The information obtained on the websites can also be used as a future reference. The system might also include an app that notifies users of significant weather changes, which serves as a sound alarm system. The device is used to predict the conditions of the atmosphere for a given location and time and detect, record, and display various weather parameters.

Results and discussions

The device is capable of measuring current temperature in degree centigrade, atmospheric pressure in Pascal, and humidity in percentage. It can monitor the small environmental changes to determine present and prior weather conditions. The data received from the sensors are in the form of weather parameters like temperature, pressure, and humidity that are required to send into Blynk interface through

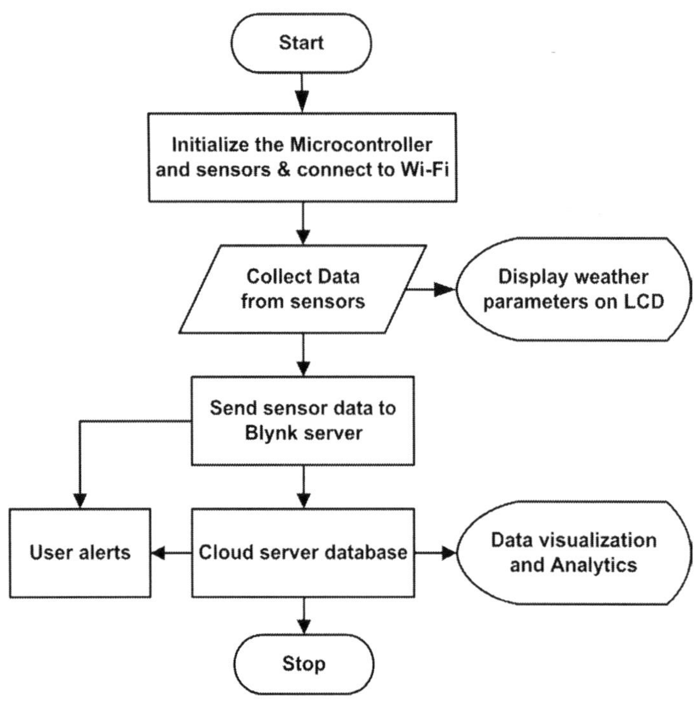

Figure 14.3 Flow chart diagram of operation

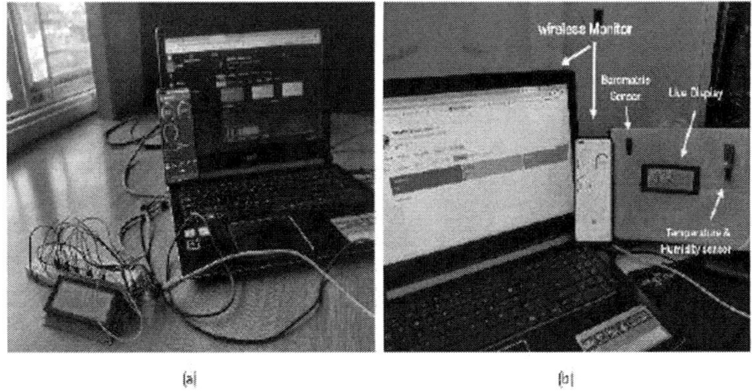

Figure 14.4 Prototype device (a) components without casing, (b) components with proper casing

Timestamp	Date	Time	Temprature(C)	Humidity%	Pressure(Pa)
11/25/2022 14:54:05	11/25/2022	10:00:00 AM	28.75	59	933.3
11/25/2022 14:54:42	11/25/2022	11:00:00 AM	28.8	57.4	931.71
11/25/2022 14:55:42	11/25/2022	12:00:00 PM	29.2	57	929.8
11/25/2022 14:56:42	11/25/2022	12:30:00 PM	28.8	57	929.73
11/25/2022 14:57:50	11/25/2022	2:00:00 PM	29.41	57.59	926.36
11/25/2022 15:06:09	11/25/2022	2:30:00 PM	30	56	925.61
11/25/2022 15:07:48	11/25/2022	3:00:00 PM	30.12	56.28	925.38
11/25/2022 15:56:47	11/25/2022	3:39:00 PM	30.5	56	924.27
11/25/2022 16:32:16	11/25/2022	4:30:00 PM	28.62	58	928.35
11/25/2022 17:00:23	11/25/2022	5:00:00 PM	27.7	64	932.94
11/25/2022 19:09:46	11/25/2022	5:30:00 PM	23.4	74.33	941.03
11/25/2022 19:11:06	11/25/2022	6:00:00 PM	24	70	941.44
11/25/2022 19:13:03	11/25/2022	6:30:00 PM	23.8	72.04	942.82
11/25/2022 19:13:50	11/25/2022	7:00:00 PM	23.4	72	943.68
11/25/2022 20:34:45	11/25/2022	7:30:00 PM	23.12	74	944.98
11/25/2022 21:01:16	11/25/2022	8:00:00 PM	23.13	73.94	945.06
11/25/2022 21:02:13	11/25/2022	8:30:00 PM	21.02	79.08	950.27
11/25/2022 21:03:20	11/25/2022	9:00:00 PM	21.6	77.04	948.98
11/25/2022 23:43:27	11/25/2022	9:30:00 PM	21.2	80	949.62
11/27/2022 0:20:22	11/26/2022	10:00:00 PM	20.82	78.8	931.9

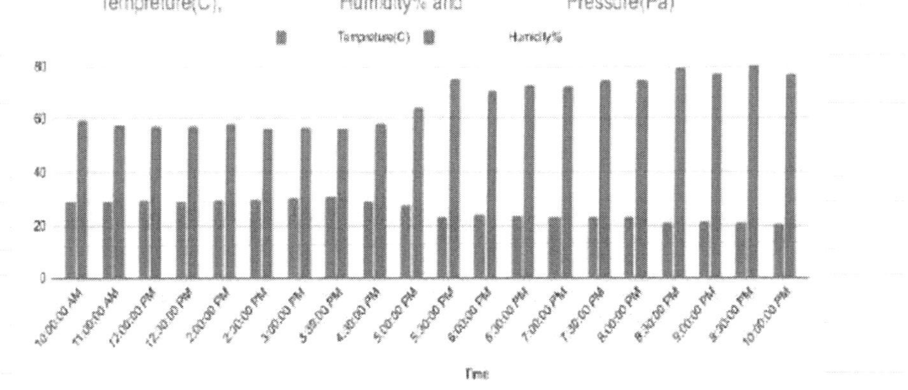

Figure 14.5 Day 1 weather monitoring

microcontroller. The Blynk interface shows the parameters in Google Sheets. As an example one Excel sheets for day one weather reports [9] have been demonstrated in Figure 14.5. A comparative performance has been tested with respect to a standard website reading is shown in the Table 14.1. The overall average accuracy for temperature and humidity reading for consecutive three days in a winter season have been obtained as 95.65% and 97.94%, respectively. The accuracy has been calculated by following Equation 1.

$$\%Acc = 100 - \frac{(Proposed\ dev\ reading - Other\ web\ reading) \times 100}{Proposed\ dev\ reading} \quad (1)$$

Table 14.1 Comparative results

Observations	Temperature and Humidity						Accuracy (%)	
	Proposed device reading			Temperature and Humidity				
Day 1	Time	Temperature	Humidity	Time	Temperature	Humidity	Temperature	Humidity
	10:05am -	23.10°C	66%	10:05am -	22°C	68%	95.24	96.97
	12:30pm -	25.52°C	60%	12:30pm -	24°C	60%	95.05	100
	2:30pm -	25.30°C	60%	2:30pm -	25°C	59%	98.82	98.33
	4:50pm -	23°C	65%	4:50pm -	22°C	66%	95.66	98.47
Day 2	10:05am -	22.30°C	71.10%	10:05am -	21°C	70%	94.17	98.45
	12:30pm -	25.20°C	62%	12:30pm -	23°C	61%	91.30	98.38
	2:30pm -	25.30°C	60%	2:30pm -	24°C	60%	94.86	100
	4:50pm -	23°C	65%	4:50pm -	22°C	62%	95.65	95.38
Day 3	10:05am -	22.24°C	71%	10:05am -	21°C	70%	94.42	98.59
	12:30pm -	25°C	66%	12:30pm -	24°C	64%	96.00	96.96
	2:30pm -	24.50°C	65%	2:30pm -	24°C	63%	97.95	96.92
	4:50pm -	22.30°C	62%	4:50pm -	22°C	60%	98.65	96.77
Average 23							**95.65**	97.94

Conclusions

For people living in the remote area, the proposed device helps them in numerous structures, including well-being, security, and consistent arranging. The IoT likewise helps in better control over asset usage. When levels of a needed parameter fall below or exceed a certain threshold, the connected sensors also aid disaster management by prompting immediate action. This weather station may monitor and predict the weather based on environmental factors. It is an intelligent way to continuously monitor the environmental parameters efficiently and cost-effectively by achieving average accuracy of temperature and humidity are 95.65% and 97.94%, respectively. The present model of IoT-based weather stations can be updated to monitor the climate of a particular area in the near future. The system can be implemented in cities/villages for constant monitoring environment and its conditions. The device can be implemented as a part of the more giant space exploration robot with more sensors like fire, gas, pollution, etc. After all the weather forecasting would be based on predicting tools like machine or deep learning.

References

1. Rahut, Y., Afreen, R., Divya, K., and Sheebarani Gnanamalar, S. (2018). Smart weather monitoring and real time alert system using IoT. *Inter. Res. J. Engg. Technol.*, 5(10), 848–854.
2. Bahga, A. and Vijay, M. (2014). *Internet of Things: A hands-on approach*.
3. Susmitha, P. and Sowya Bala, G. (2014). Design and implementation of weather monitoring and controlling system. *Inter. J. Comp. Appl.*, 97(3), 0975-8887.
4. Krishnamurthi, K., Suraj, T., Lokesh, K., and Arun, P. (2015). Arduino based weather monitoring system. *Inter. J. Engg. Res. Gen. Sci.*, 3(2), 452–458.
5. Munandar, A., Hanif, F., Muhammad Ilham, R., Rian Putra, P., Jony, W. W., and Irfan, A. F. A. (2017). Design of real-time weather monitoring system based on mobile application using automatic weather station. *2017 2nd Inter. Conf. Automat. Cogn. Sci. Optics Micro Elec-Mech. Sys. Inform. Technol. (ICACOMIT)*, 44–47.
6. Mohapatra, D. and Bidyadhar, S. (2022). Development of a cost-effective IoT-based weather monitoring system. *IEEE Cons. Elec. Mag.*, 11(5), 81–86.
7. Kodali, R. K. and Snehashish, M. (2016). IoT based weather station. *2016 Inter. Conf. Con. Instrum. Comm. Comput. Technol. (ICCICCT)*, 680–683.
8. Ladi, K., Manoj, A. V. S. N., and Deepak, G. V. N. (2017). IOT based weather reporting system to find dynamic climatic parameters. *2017 Inter. Conf. Energy Comm. Data Analyt. Soft Comput. (ICECDS)*, 2509–2513.
9. The weather channel, www.weather.com (accessed 27th December, 2022).

15 45-nm CMOS ring oscillator

Rachana Arya[1,a] *and B. K. Singh*[2,b]

[a]Research Scholar, V. M. S. B. Uttrakhand Technical University, Dehradun, India

[b]Director, BIAS, Bhimtal, Uttrakhand, India

Abstract

The 5 MHz frequency band is tremendously used for military and organizational purposes. The I. T. U. regulations necessitate seclusion from anterior radio frequency services. Frequently (5250–5450 KHz) frequency space is used. To transmit and receive the required frequencies, a local oscillator is premeditated. For this, a three-stage current-starved ring oscillator (CSRO) is designed. The cadence tool through 45-nm CMOS technology at a 1V supply has executed the circuit simulation. Initially, the oscillator is aimed at a 1 pF fixed capacitor, the total achieved delay is 40 ps, and the attainable frequency at 1 V is 5 MHz at 27°C. The attainable time-period of this oscillator is 1.6369 µs. The complete delay (38.2 psec) is the least compared to other processes, and the calculated transient time is 252.6 ns. Another circuit is designed with 1–100 pF variable capacitors. For this, the calculated power dissipation is 1.732 µW at 50 Ω output impedance. The overall achieved delay is 4.523 ns with a 10-step size. Through the scaling of the CMOS transistor, it will control the process changeability, increase chip density, and be cost-limited. A transient analysis with variable capacitors is performed.

Keywords: Ring oscillator, impedance, circuit delay, time period

Introduction

The ring oscillator (RO) is inference-based and conclusive for the electronic industry. With the considerable broadening of the VLSI arena, their standards have exceptionally increased. Voltage control oscillators, phase lock loops, and analog to digital circuits are quite adept at handling these. RO comprises a ring of "n" inverters, which have been connected in series to deliver the oscillated output. Most of these oscillators are used for clock generation, microprocessors, and frequency synthesis. VCO is credibly the most substantial block of locked loop [1, 2]. The middle range of frequencies can be achieved by altering the bandwidth of the circuit. The alteration in the center frequency will deviate from the tuning range of the circuit. In the phase lock loop (PLL), the output signal is reverted to the input via a mixer to generate the required output voltage. The essential blocks of PLL are the phase detector (PD), low pass filter (LPF), charge pump (CP), and controlled oscillator.

Experimental

CMOS inverter

The inverter is the main block of RO. Odd delay cells are used in series toward oscillations. For this, three delay stages are connected in series. The length of all transistors is 45 nm and the width is 120 nm, respectively. The advantages of the CMOS configuration are the great noise margin due to the full rail-to-rail swing. In steady-state conditions, the path of current is always V_l or ground, so it always provides low output impedance [12, 13]. Equally, there is no shortest path from current to ground, so no static power dissipation occurs [5]. The delay is dependent on the load resistance (R_L) and capacitance of the transistor. If the input voltage is less than the threshold voltage (V_{th}), the NMOS is cut off, so the drain current is zero [14]. The NMOS to PMOS charge ratio is almost 2.66 times [6, 7], so both have the same current driving capability with equal charging and discharging times. Figure 15.1 signifies the transient behavior of the CMOS inverter. This governs the speed of operation. The circuit's propagation delay (P_d) is correlated with the network's time constant [3, 4]. The time constant is the function of the resistor (R_n) and capacitor (C_L). The high-to-low transition for P_d is t_{PHL}= f (R_n, C_L), and $R_n = R_{eq} = \frac{3V_{dd}}{4I_{dsat}}$. Moreover, $C_L = C_I + C_E + C_W$. So, $C_L = C_{ox} W_p L_p + C_{ox} W_n L_n$, where: C_{ox}= capacitor density of gate oxide = 25 fF/ µm², gate capacitance $C_g = C_{ox}$ W L (F) and trans-conductance $(\beta) = k_n \frac{W}{L}$ (A/V²).

Equations 1 and 2 can describe the nMOS resistance and pMOS resistance.

$$R_n=L_n/(W_n k_n (V_{dd}-V_{th}))=1/(\beta_n (V_{dd}-V_{th})),$$
$$\beta_n=\mu_n C_{ox} (W/L)_n \quad (1)$$

[a]rachna009@gmail.com

DOI: 10.1201/9781003540199-15

$$R_p=L_p/(W_p k_p(V_{dd}-V_{th}))=1/(\beta_p(V_{dd}-V_{th}))$$
$$\beta_p=\mu_p C_{ox}(W/L)_p \quad (2)$$

Furthermore, $R_n=R_p$ if the rise and fall times are equal. As of $t = 0.69\ R_n\ C_L$, we have reached the 50% wave point. Given that the waveform's time at 90% is $t = 2.2\ R_n C_L$, $PD = (t_{PHL}+t_{PLH})/2 = 0.69\ C_L\ (R_n+R_p)$. By reducing the values of R and C through 3 to 7, the delay performance can be improved. The key profits of RO are less designing space, less area, low power density, ease of modification, and being tunable to deliver extensive bandwidth [8, 9]. Nevertheless, it will suffer from pitiable phase noise. The channel length of the transistor and supply voltage will refine the RO performance [10, 11]. As a result, the oscillator attains a high frequency range, minimal power use, and an outstanding rating of merit.

Power utilization in ring oscillator

In the observation, the power consumption is consequently significant for the design of the oscillator. The formation of RO comprises a source, along with the identical gain stages. The power dissipation is separated by two parts [16]. The primary power is disbursed by the source, and additional power is provided via the complete circuit [15]. Figure 15.1, supposing an idyllic input to analyze the energy disbursed by the supply, shows dynamic power at the low-to-high transition (3–5):

$$E_{V_{dd}} = \int_0^{\infty} I_{V_{dd}}(t)V_{dd}\ dt = V_{dd}\int_0^{\infty} C_L \frac{dV_o}{dt}\ dt = C_L V_{dd}\int_0^{V_{dd}} dV_o = C_L\ V_{dd}^2 \quad (3)$$

Now, using the energy stored at C_L, we get:

$$E_{charge} = \int_0^{\infty} I_{V_{dd}}(t)V_o\ dt = \int_0^{\infty} C_L \frac{dV_o}{dt} V_o dt = C_L \int_0^{V_{dd}} V_o\ dV_o = \frac{C_L V_{dd}^2}{2} \quad (4)$$

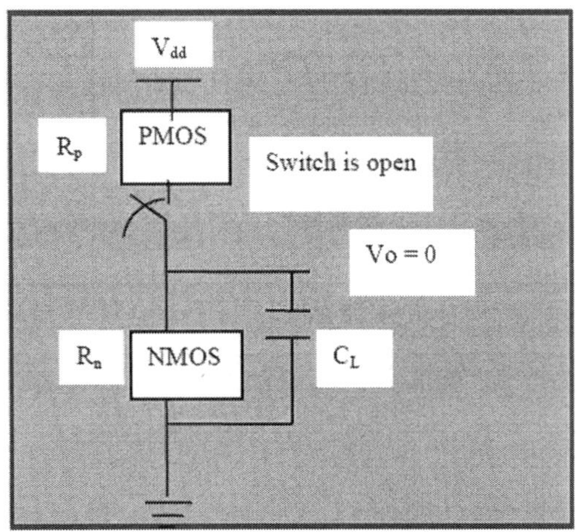

Figure 15.1 Circuit switch with load capacitor (C_L)

$$E_{V_{dd}} = C_L\ V_{dd}^2 \text{ and } E_{charge} = \frac{C_L V_{dd}^2}{2} \quad (5)$$

After analyzing these equations, the requisite energy to charge load capacitance is double the energy kept by the capacitor [17]. It means PMOS resistance loses half its energy. Supposing I/P deviations are 0→1, now for the high-to-low transition, the gate is grounded, and charge flows via the NMOS transistor to ground. The degenerated energy is the whole energy kept by the load capacitance.

$$E_{discharge} = \int_0^{\infty} I_{discharge}(t)V_o\ dt$$
$$= \int_0^{\infty} C_L \frac{dV_o}{dt} V_o dt = C_L \int_{V_{dd}}^{0} V_o\ dV_o = \frac{C_L V_{dd}^2}{2} \quad (6)$$

The total energy is observably equivalent to the supplied energy.

$$E_{charge} + E_{discharge} = E_{V_{dd}} = C_{Load}\ V_{dd}^2 \quad (7)$$

By these equations, at each sequence, the circuit disperses a static extent of energy, free from device size. The charge kept by the capacitor is the operative energy [18]. For low-to-high transition ($f_{0\to 1}$) times/second and vice versa, the dynamic power expenditure is $P_{dynamic} = C_{Load}\ V_{dd}^2\ f_{0\to 1}$. The power consumption of the gate is a feature of transferring frequency. The $P_{dynamic}$ is express as the inverter activity factor (α) to the prospect of the O/P switch: $P_{dynamic} = C_L\ V_{dd}^2 f\alpha = C_E\ V_{dd}^2 f$, where, C_E is the effective capacitance of complex circuits that describes the capacitance switches at each cycle. The energy degenerate through short-circuit power is

$$E_{sc} = V_{dd}\frac{I_p t_{sc}}{2} + V_{dd}\frac{I_p t_{sc}}{2} V_{dd}\ t_{sc} I_p \quad (8)$$

In addition, the typical power consumption is $P_{sc}=V_{dd}\ t_{sc}\ I_{peak}\ f$. Supposing a linear I/P alteration $t_{sc} = \frac{V_{dd}-2V_{th}}{V_{dd}} * \frac{t_{rise\ (fall)}}{0.8}$, to diminish the P_{sc}, a high-load capacitor is required. Therefore, the CMOS inverter never proved any conducting path (in steady state), resultant certainly not any static power dissipation. Meanwhile, static power is frequently disbursed, so the power dissipation can be expressed as $P_{static}=I_{static}V_{dd}$, and the total power dissipation is the sum of all three components of the inverter, $P_{total}=P_{dynamic}+P_{static}+t_{sc}$.

$$P_{total} = C_L\ V_{dd}^2 f\alpha + I_{static}V_{dd} + V_{dd}\ t_{sc} I_p f \quad (9)$$

$$PDP = P_{dynamic}\, t_{PD} = C_L\ V_{dd}^2 f\ t_{PD} = \frac{C_L\ V_{dd}^2}{2} \quad (10)$$

Meanwhile, mutual power and PDP contribute to energy reduction.

Ring oscillator with sweep capacitor

The 3-stage RO is designed with 3-inverters situated in cascading mode. The source of all the NMOS (M_1, M_2 and M_3) is connected to ground, and the drain of all the NMOS (P_1, P_2 and P_3) is wired to the supply. The C_1, C_2 and C_3 are the bypass capacitors to maintain the gain of all the stages, as shown in Figure 15.2. Applying a high signal to the first inverter causes it to send a low signal to the output while the PMOS is off [20]. This signal is sent to the following state and supplied to the following step once more. When a low signal is applied, the PMOS is on, producing a high signal on output. This alteration will switch on and off the both transistors and sustain the oscillations [21]. For the appropriate procedure of RO, an odd number of cells are required. The signal's peak and fall times can be used to predict it. $Rise\ time\ tr = \frac{4C_L}{\mu_p \frac{W_p V_{dd}}{L_p}}$, and $Fall\ time\ tp = \frac{4C_L}{\mu_n \frac{W_n V_{dd}}{L_n}}$, where: μ_p and μ_n = mobility of NMOS and PMOS transistors.

The frequency of oscillation can be determined by $f_{oscillation} = \frac{1}{n\,(t_{rise} + t_{fall})}$ (Hz) and the transfer function for a 3-stage RO circuit, $H\,(s) = -\frac{A_0^3}{\left[1+\frac{s}{\omega_0}\right]^3}$, where A_0 is single stage gain. Each stage of the RO contributes a 60° phase shift to the circuit. $\omega_{oscillation} = \omega_0 tan^{-1}\frac{\pi}{3} = \sqrt{3}\,\omega_0$. Moreover, when $\omega = \omega_{oscillation}$, the loop gain of the circuit is unity. So $A_0 = 2$. Therefore, the oscillation frequency is $(\sqrt{3}\,\omega_0)/\pi$. If all stage gain is two, then every stage will provide a 60° phase shift to the circuit [19]. Once, the RO continuously oscillated until it reached the running point.

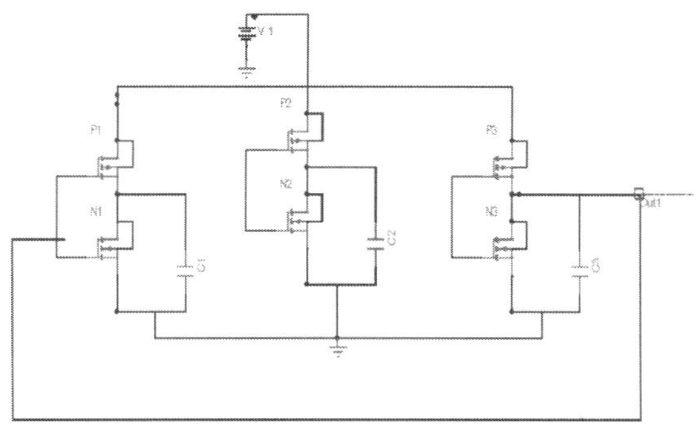

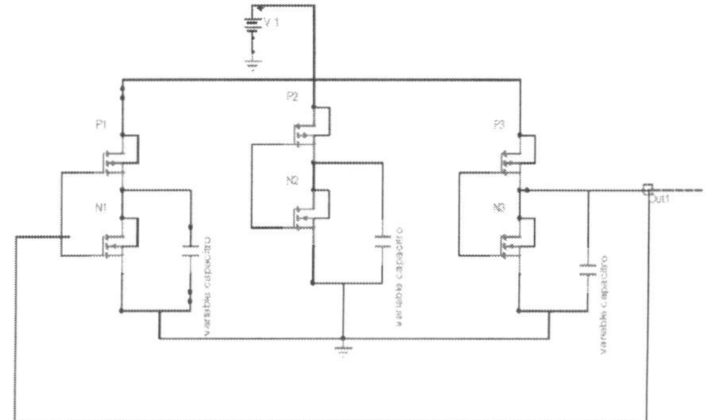

Figure 15.2 (a), (b) Simulated model of CMOS based RO, simulated circuit using variable capacitor

Variable capacitor ring oscillator

The variable capacitor is utilized in conjunction with the circuit schematic (Figure 15.2) of a three-stage RO. The capacitance of the capacitor is 1–100 pF. The output waveforms show that the circuit produces a high frequency since the value of the capacitor is low. From the simulation analysis, it is clear that the capacitor and the frequency are

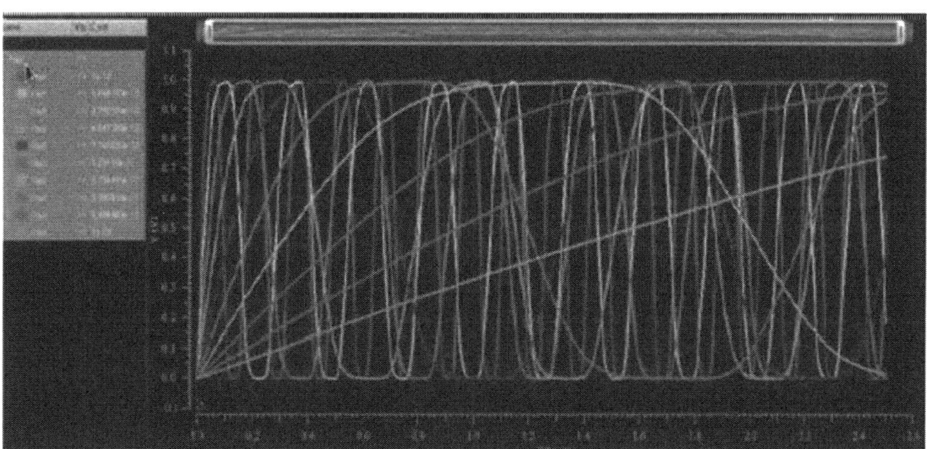

Figure 15.3 The output waveforms for 1–100 pF capacitor

Table 15.1 Analysis of the ring oscillator's results using variable capacitors

Parameter	Values
Technology	45 nm
Width of NMOS (Wn)	120 nm
Width of PMOS (Wp)	120 nm
Voltage supply	1 V
NMOS threshold voltage	0.22 V
PMOS threshold voltage	-0.22 V
Initial capacitor value	1 pF
Final capacitor value	100 pF
Total steps	10
Start time	0
Stop time	105 ns
Step size	10 ns
Maximum step	200 ns
No. of stages	3
Number of components	10

Table 15.2 Frequency cravings on capacitor (values)

Capacitor's values (pF)	Time-period (μs)	Obtained frequency (MHz)
1	0.20	5
17	0.28	4.5
2.8	0.49	2.2
4.75	0.80	1.2
7.8	1.5	0.71
0.154	2.8	0.421

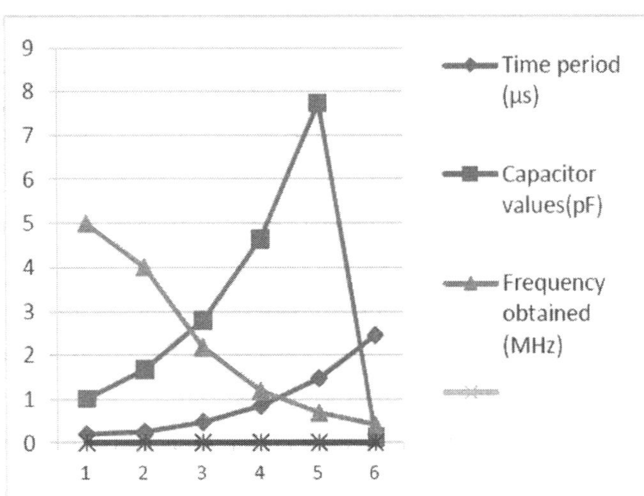

Figure 15.5 Time-period and frequency zone diagram

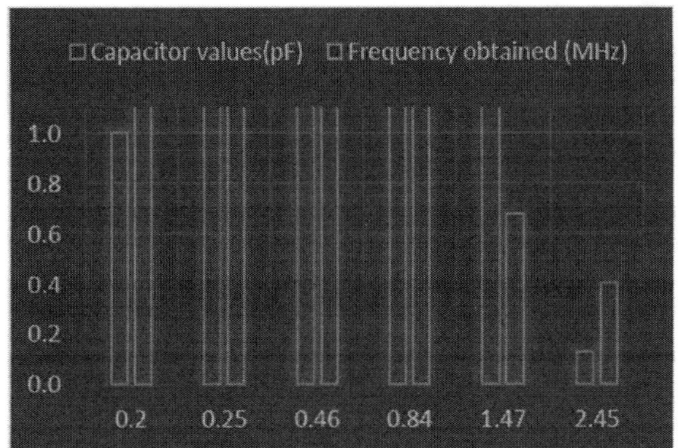

Figure 15.4 Bar graph of the variable capacitor and frequency obtained

reciprocal. Figure 15.3 represents the simulated O/P signal with variable capacitances.

Result and discussion

Tables 15.1 and 15.2 show the dissimilar parameter values acquired by the frequency vs. voltage analysis. The O/P waveforms at different capacitor values, along with the various frequencies, are obtained. As the outcome, via upward capacitor values, the acquired frequency is low. Consequently, mutually, these parameters remain contrariwise, and Figures 15.4 and 15.5 represent it.

Conclusions and future work

The simulated design of 3-stage RO in 45-nm CMOS via cadence tools is completed. The RO contains three delay stages with a fixed capacitor of 1 pF. The considered propagation interval is 38.2

psec at 1V. The simulated frequency is 5 MHz at a 1V supply. This band has unexpected propagation advantages, bonding the chief side for a 40–80 m range. This is also used for stable navigation and satellite communications during the course of dusk and sunset. The operative frequency depends on some leading parameters, such as the capacitor's value, time period, and number of stages of the delay cell. The suggested oscillators use a variable capacitance of 1–100 pF to create a frequency band of 0.41–5 MHz. This document presents a number of curves that illustrate the link between varied frequency and variable values of capacitors. At a 1V supply, the circuit uses 1.732 μW of power overall. The key areas of the proposed RO are data transfer and recapture systems, high-performance transmitters, clock causation, continuous-wave radio frequency communication, and small power applications.

Acknowledgement

The authors gratefully acknowledge the Electronics and Communication Engineering Department, B. T. K. I. T. Dwarahat, Almora, Uttrakhand for their cooperation in the research.

References

1. Reddy, J. N., Tiwari, M., and Reddy, V. P. (2021). Power efficient two transistor exclusive or gate for full adder using GDI in 45nm. *Turk. J. Comp. Math. Educ.*, 12(2), 1342–1347.
2. Faruqe, O., Akhter, R., and Amin, M. T. (2020). A low power wideband varactorless VCO using tunable active inductor. *Telkomnika Telecomm. Comput. Elec. Con.*, 18(1), 264–271.
3. Hathwalia, S. and Grover, N. (2019). A comparative study of ring VCO & LC-VCO based PLL. *Inter. J. Engg. Res. Appl. (IJERA)*, 9(2), 52–54.
4. Arya, R. and Singh, B. K. (2023). Ring oscillator for 60-meter bandwidth. *Comp. Sys. Sci. Engg.*, 46(1), 93–105.
5. Kumar, G. C. K. and Mahalakshmi, K. (2020). Advanced improvement in speed of operation of 3-stage CMOS ring oscillator clock generation using CPG. Inter. *J. Innov. Technol. Explor. Engg. (IJITEE)*, 9(5), 772–777.
6. Dandare, S. N. and Deshmukh, A. H. (2018). Design of DPLL using sub-micron 45 nm CMOS technology and implementation using Microwind 3.1 software. *J. Sci. Technol.*, 3(1), 22–32.
7. Lakshmi, A. and Reddy, P. C. (2019). Design and implementation of conventional D flip-flop for registers. *Inter. J. Comp. Appl.*, 181(39), 24–28.
8. Abbas, W., Mehmood, M., and Seo, M. A. (2020). Band phase-locked loop with a novel phase-frequency detector in 65 nm CMOS. *Electronics*, 9(9), 1–12.
9. Kumar, N. and Kumar, M. (2018). Design of CMOS-based low-power high-frequency differential ring VCO. *Inter. J. Elec. Lett.*, 7 (2), 143–153.
10. Kumari, U. and Yadav, R. (2018). Design and implementation of digital phase lock loop: A review. *Inter. J. Adv. Res. Elec. Comm. Engg.*, 7(3), 197–201.
11. Koithyar, A. and Ramesh, T. K. (2021). Modeling of the submicron CMOS differential ring oscillator for obtaining an equation for the output frequency. *Cir. Sys. Sig. Proc.*, 40(2), 1–8.
12. Hemlatha, M., Urkude, V. R., and Sunil, K. J. (2021). Low power high speed GDI 4 bit RCA circuit design using 45 nm CMOS technology. *Turk. Online J. Qualit. Inq.*, 12(6), 8856–8865.
13. Badal, M. T. I., Alam, M. J., Reaz, M. B. I., Bhuiyan, M. A., and Jahan, N. A. (2019). High-resolution time to digital converter in 0.13 μM CMOS process for RFID phase locked loop. *J. Engg. Sci. Technol.*, 14(4), 1776–1788.
14. Sabu, N. A. and Batri, K. (2018). Review of low power design techniques for flip-flops. *Inter. J. Pure Appl. Math.*, 120(6), 1729–1749.
15. Cuautle, E. T., Castañeda-Aviña, P. R., Trejo-Guerra, R., and Carbajal-Gómez, V. H. (2019). Design of a wide-band voltage-controlled ring oscillator implemented in 180 nm CMOS technology. *Electronics*, 8(1156), 2–17.
16. Zouaq, K., Bouyahyaoui, A., Alami, M., Aitoumeri, A. (2018). A simple and novel scheme of LC-VCO for ultra low power low phase noise applications. *Inter. J. Appl. Engg. Res.*, 13(9), 6760–6765.
17. Bitla, L., Saraswathi, V., and Akano, H. M. (2020). Analysis of NAND gate-based phase frequency detector for phase locked loop. *J. Crit. Rev.*, 7(19), 172–176.
18. Easwaran, M. and Ganapathy, R. (2018). Dark soliton generation using CMOS ring oscillator. *ARPN J. Engg. Appl. Sci.*, 13(3), 852–858.
19. Shakya, M. and Agrawal, S. (2018). Design low power CMOS D-flip flop using modified SVL techniques. *Inter. J. Res. Analyt. Rev. (IJRAR)*, 5(4), 55 65.
20. Arya, R. and Jain, A. (2021). Investigation of D flip-flop designing using INDEP with bi-triggering (Trig01) NAND gate approach. *Inter. J. Engg. Res. Technol.*, 14(7), 674–678.
21. Gunda, S., Ravindran, R. S. E., Hemalatha, B., and Divya, C. (2019). Design and simulation of digitally controlled oscillator of ADPLL. *Inter. J. Adv. Sci. Technol. Engg. Manag. Sci.*, 5(3), 1–4.

16 Fundus image datasets: A valuable resource for deep neural network-based glaucoma research

Vijaya Kumar Velpula[1,a], *Amarjit Roy*[2,b], *Lakhan Dev Sharma*[1,c] *and Diksha Sharma*[3,d]

[1]School of Electronics Engineering, VIT-AP University, Amaravati, 522237, Andhra Pradesh, India

[2]Department of Electrical Engineering, Ghani Khan Choudhury Institute of Engineering Technology, Narayanpur, Maligram, 732141, West Bengal, India

[3]Department of Nanoscience and Technology, Central University of Jharkhand, Ranchi, India

Abstract

Glaucoma is a progressive optic neuropathy that can cause irreversible vision loss. Fundus image, a non-invasive imaging technique, is widely used to diagnose and monitor glaucoma. In recent years, numerous fundus image datasets have been developed for use in glaucoma research. This review provides an overview of the major fundus image datasets used in glaucoma research and their characteristics. In the classification problem where a deep neural network has been used, a valuable resource of the database is much required considering the common utility. The review highlights the details and strengths of each dataset and provides insights into the types of research questions that can be addressed using these datasets. The datasets discussed in this review include 21 publicly available datasets. The review concludes with a discussion of the challenges and opportunities for future research using fundus image datasets in the field of glaucoma, it will be helpful to the research community in selecting the correct image datasets for their studies.

Keywords: Cup-to-disc ratio, fundus image, glaucoma, machine learning, medical imaging, ophthalmology, deep neural network

Introduction

According to the World Health Organization, medical imaging has developed greatly and now plays an important role in medicine [1]. Glaucoma is a collection of eye conditions that result in harm to the optic nerve, which transmits visual data from the eye to the brain. Increasing intraocular pressure (IOP) causes damage to the optic nerve, leading to irreversible blindness, which has already affected around 80 million individuals globally. Timely identification and management of glaucoma are crucial in avoiding vision impairment. Fundus images play a critical role in glaucoma research, as it provide a non-invasive and accurate way to visualize the optic nerve head (ONH), retina, and retinal nerve fiber layer (RNFL), which are important structures for diagnosis and monitoring of the disease. Optical equipment called a fundus camera [2] is used to take images of the retina of the eye. According to the National Eye Institute [3] and the Glaucoma Research Foundation [4], glaucoma has been categorized as primary and secondary. There are a total of eight variants of glaucoma, such as normal-tension, angle-closure (close-angle), open-angle, and congenital comes under primary glaucoma and uveitic, exfoliation, pigmentary, and neovascular comes under secondary glaucoma. The most common glaucoma forms are open-angle and angle-closure, which belong to primary. All these types can affect either one or both eyes. Figure 16.1(a) displays the fundus picture of the eye, illustrating the optical disc (OD), optical cup (OC), neuroretinal rim (NRR), and blood vessels. The region between OD and OC is referred to as NRR. As the cup and disc diameters increase in glaucoma images, there is a simultaneous rise in both the overall color intensity and the complexity of textural characteristics. The variation of cup-to-disc ratio (CDR) changes in sample images are illustrated in Figure 16.1(b–d). Glaucoma severity can be evaluated using these variations in CDR and NRR. Publicly available datasets have the potential to be useful sources of information for health services research and innovation because they simplify the process of data collection and preparation and enable the development and evaluation of more effective methods. In this review, we aim to provide an overview of the major fundus image datasets used in glaucoma research, their strengths and limitations, and opportunities for future research. The purpose of this is to aid researchers in choosing suitable datasets that align with their research inquiries.

[a]vijaya.20phd7043@vitap.ac.in, [b]royamarjit90@gmail.com, [c]lakhan.sharma@vitap.ac.in, [d]diksha.hec@gmail.com

DOI: 10.1201/9781003540199-16

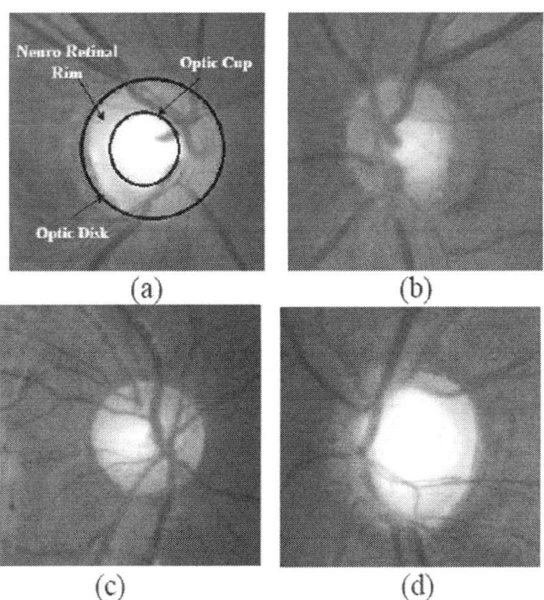

Figure 16.1 (a). Fundus image showing CDR and NRR. (b–d). Fundus images demonstrate the differences in CDR and NRR among normal, early, and advanced glaucoma cases

The remaining part of the article is structured in the following manner: Details about the fundus image datasets, outlining the search approach and dataset descriptions. The methods for examining fundus images. We finally concluded this article with difficulties and future directions.

Available fundus image datasets

Search strategy

To begin the process of finding publicly available fundus image datasets, a general search is typically initiated using a search engine like Google. This search involves using specific keywords, such as "publicly available fundus image datasets," to identify relevant websites and resources that offer access to such datasets. Online repositories such as Kaggle and GitHub are also used to search the datasets by using keywords like "fundus," "retina," "ophthalmology," or "eye images." Academic search engines like Google Scholar and PubMed are also used for finding academic papers that discuss or use fundus image datasets. Each result returned by the search includes direct links, an explanation of the details of datasets, the image format, etc. We are focusing on publicly available retinal fundus image datasets and have excluded other ophthalmology-related datasets. Based on the accessibility the datasets are available as open access, access on request and some are accessible with the registration process by providing personal details like email, name, mobile number, etc. Table 16.1 shows a complete listing of all the retrieved features along with related information.

Description of datasets

Publicly available datasets:

- RIGA dataset: The retinal images for glaucoma analysis (RIGA) dataset is a publicly available fundus image dataset that was specifically developed for glaucoma research [5]. It contains a total of 500 fundus images, including 247 healthy images and 253 glaucoma-positive images. A total of six expert eye doctors manually highlighted and commented on the OC and OD boundaries on these images. The RIGA dataset was developed to tackle some shortcomings, including a small sample size, an absence of image diversity, and variable image quality.
- CHASE-DB: The images in the Child Heart and Health Study in England (CHASE)-DB dataset were taken using a 768×584 pixel Topcon TRV-50 fundus camera [6]. A total of 39 subjects had their left and right eyes examined; 16 of those subjects had glaucoma, while 23 were in good health. This dataset has been criticized for its tiny sample size and probable lack of diversity; it is also very modest compared to other available datasets like ORIGA and RIGA.
- Drishti-GS1: A dataset called Drishti-GS1 was created to validate notching, OD segmentation, and OC detection. Images in the dataset is compiled and characterized by Aravind Eye Hospital in Madurai. To address challenges and pertinent issues in OD and OC segmentation and optic nerve head segmentation, this dataset has been made available [7].
- HRF: Researchers from Brazil's Universidade Federal do Paraná created the dataset [8]. There are a total of 45 images, divided into 15 categories. Every group has pictures of normal eyes, eyes with diabetic retinopathy, and eyes with glaucoma. There were a total of 23 images used for testing, and 22 for training. To take these pictures, we utilized a Canon CR6-45NM retinal camera. Every image in the HRF set comes with an annotated segmentation map of the optic disc, macula, and exudates. A retinal specialist double-checked these annotations after they were completed by professional graders to make sure they were accurate and consistent.
- G1020: This is a publicly available, extensive collection of retinal fundus images used for glaucoma categorization [9]. The dataset was revised by generally recognized ophthalmology standards, and its intended use is as a bench-

Table 16.1 Extracted details of available datasets

Dataset name	Dataset URL	Image format	Number of images			Resolution (in pixels)
			Healthy	Glaucoma	Total	
RIGA	RIGA	JPG and TIFF	NA	NA	500	1440×960 2240×1488 2304×1536
CHASE-DB	CHASE	PNG and JPG	48	20	68	768×584
DRISHTI-GS1	DRISHTI-GS1	PNG	70	31	101	2048×2048
High-resolution fundus (HRF) image database	HRF	JPG	NA	NA	45	1440×960
Akshi IMAGE	Akshi IMAGE	JPEG/PNG	NA	NA	1345	2448×3264 2048×1536 1920×1440
Retinal fundus glaucoma challenge (REFUGE)	REFUGE	JPEG	1080	120	1200	2048×2048
G1450	G1450	PNG	551	899	1450	NA
ORIGA	ORIGA	JPG	482	168	650	3072×2048
G1020	G1020	JPG	724	296	1020	B/W 1944×210 and 2426×3007
ACRIMA	ACRIMA	JPG	309	396	705	2048×1536
LAG	LAG	JPEG	6882	4878	11760	500×500
Harvard dataverse	HVD(3 class)	PNG	788	289 early 467 advance 788 normal	1544	240×240
RFMiD	RFMiD	PNG	NA	NA	3200	2144×1424
STructured analysis of the retina (STARE)	STARE	PPM	NA	NA	400	NoA
RIM-ONE r1	RIM-ONE	JPG	118	51	169	2144×1424
RIM-ONE r2		JPG	255	200	455	2144×1424
RIM-ONE r3		JPG	85	74	159	2144×1424
DR HAGIS dataset	DR HAGIS	JEPG	NA	NA	40	4752×3168
INSPIRE –arteriovenous ratio	INSPIRE-AVR	TIF	NA	NA	40	768×1019
INSPIRE—Stereo	INSPIRE-Stereo	TIF	NA	NA	30	768×1019
Retina	Retina	PNG	300	301	601	2592×1728

mark dataset for glaucoma diagnosis. This set of 1020 high-resolution colored fundus photos comes with ground truth marks including vertical CDR, OD and OC segmentation, nasal and temporal quadrants, inferior and superior NRR sizes, and the location of the OD's bounding box for glaucoma detection, among other things.

- ACRIMA: The ACRIMA database, which has 705 labeled photos, has been made public. It is made up of 396 images of glaucoma and 309 images of healthy eyes [10]. The python scripts used to get the findings shown are also accessible. The Ministerio de Economy Competitividad of Spain produced this dataset for the ACRIMA project, released on March 15, 2019, and written by Andres Diaz-Pinto. Only classification tasks could be performed using the initial version of this database. It is a widely used dataset in glaucoma research in the machine and deep learning areas.
- Harvard Dataverse: According to Kim et al. [11], Ungsoo Kim was the one who uploaded the dataset from Kim's Eye Hospital to Harvard. Images ranging from 467 showing severe glaucoma to 289 showing early glaucoma and

788 showing normal eye health make up this collection. Due to its pre-processing, it is ready to be used in deep learning applications. According to Kim and Ungsoo (2018), the dataset is titled "Machine Learning for Glaucoma" and may be found on the Harvard Dataverse.

- RFMiD: It is a new open-access dataset, and it contains, 3200 retinal images with a total of 45 different diseases [12]. These images are classified into 45 different groups, and the dataset is divided into three subsets as test 20%, validation 20%, and training 60%. The major objective of this dataset is to give information on a number of conditions that are frequently seen in clinical practice, including glaucoma.
- RIMONE: In 2011, the Retinal IMage database for Optic Nerve Evaluation (RIMONE) [13] was released in its first version, referred to as RIMONE v1. Two other versions RIMONE v2 and RIMONE v3 are released in 2014 and 2015, respectively. The RIMONE datasets are widely used for glaucoma evaluation among the public fundus datasets. A medical expert manually segments 200 glaucoma photos and 255 healthy images into RIM-ONE v2. These images were captured at the Miguel Servet University Hospital and the Hospital Universitario de Canarias. Included in the 2015 version of RIM-ONE v3 are 85 normal images and 74 glaucoma images. The segmentation of the OC and OD was done manually as a result of the availability of stereo pictures.
- DR HAGIS: The database DR HAGIS can be downloaded free of cost for research on glaucoma [14]. It was developed to support the creation of vascular extraction algorithms suitable for retinal screening programs. This dataset can be used by researchers' segmentation methods. Included are 39 fundus images obtained from a screening program for diabetic retinopathy in the UK. Here are the four comorbidity subgroups that can be found in the database: Images 1–10 are associated with glaucoma, images 11–20 with hypertension, images 21–30 with diabetic retinopathy, and images 31–40 with age-related macular degeneration. This fundus image is from a patient with diabetic retinopathy and age-related macular degeneration, hence images 24 and 32 are identical.
- INSPIRE-stereo: Iowa Normative Set for Processing Images of the Retina (INSPIRE) can be a freely downloadable dataset just by filling out a form on their website [15]. It contains 30 color fundus images and is from glaucoma patients. This is the only dataset with objective depth and non-telemetry continuous as ground truth.
- INSPIRE-AVR: It is one of the benchmark datasets for glaucoma research, and it has 40 fundus images [16]. The ground truth provides the optic disc and vessels of the image. The reference standard is the average of the ratings given to the photos by two experts.
- Retina dataset: The retina dataset contains four categories: 101 images of glaucoma, 100 images of cataracts, 100 images of retinal diseases, and 300 normal images [17].

Datasets available on request

- AKSHI image: Retinal color fundus photos captured using three different pieces of equipment are included in the database [18]. Ground truth for the OD and OC to evaluate segmentation performance and a binary conclusion was provided by five experienced ophthalmologists. The 1345 fundus images in the dataset are split into 1050 for training and 295 for testing respectively. To calculate important parameters like the vertical, horizontal, and area of CDR, the database provides data on OD and OC width/area/height.
- G1450: The dataset contains a total of 1450 fundus images, including 899 glaucoma images and 551 healthy images provided by Kaohsiung Chang Gung Memorial Hospital, Taiwan [19]. This is available at the IEEE data port website, https://doi.org/10.21227/4bcp-2z21.
- ORIGA-light: A total of 650 fundus images from the Singapore Eye Research Institute are included in the ORIGA-light [20] collection and have been interpreted by qualified experts. Numerous picture indications are important for glaucoma diagnosis, and they are all annotated. More clinical ground-truth photos will be added to the system continuously. On request, it can be accessed online.
- REFUGE: Researcher Huazhu et al. (2019) [21] from the University of Texas at Arlington and the University of Iowa collaborated to produce the dataset. Clinical glaucoma labels and ground truth segmentations for OD and OC are applied to 1200 fundus pictures. There are a total of 1080 healthy-looking images and 120 glaucoma images in the collection. Ground truth labels indicating glaucoma presence or absence have been attached to every image in the dataset.
- LAG: The 11,760 fundus images in the LAG database relate to 4878 suspected and 6882 undiagnosed glaucoma samples [22]. The diagnosis results are printed on the labels for each

sample (0 refers to negative glaucoma and 1 refers to suspicious glaucoma). A different approach of eye tracking is used to further label attention regions on 5824 fundus images, 2392 are glaucoma and 3432 are healthy images.

- STARE: The Structured Analysis of the Retina dataset was developed for the project called STARE and was conceived by Dr. Michael Goldbaum at the University of California. Many people have published many articles on this project [23, 24]. Images were provided by the San Diego Veterans Administration Medical. There are a total of 400 raw images in the dataset. Text files including expert commentary on the manifestations (features) visible in each image. During data collection, the experts were queried about a total of 44 potential manifestations, which were then condensed to 39 values during coding. 40 manually annotated photos of blood vessels, our results, and a demonstration. 10 photos are labeled as arteries or veins by experts 1 and 2. Work on detecting the optic nerve utilizing 80 photos of the ground truth and their findings.

Techniques for analyzing fundus images

There are two fundus images analyzing techniques manual and automated methods.

Manual analysis techniques – (i) Direct observation: Simply viewing the fundus image to identify glaucoma. (ii) Stereo photography: Involves taking two images of the same area of the retina at slightly different angles to create a three-dimensional image which can help to identify changes.

Automated analysis techniques – (i) Fundus auto fluorescence: It is a technique that uses the natural fluorescence of the retina to create images and the level of auto fluorescence shows the abnormalities in the retina, (ii) Machine learning (ML): These algorithms can be trained on large datasets to identify patterns and features associated with different diseases, and (iii) Image segmentation: This involves dividing the fundus image into different regions, such as the optic disc, macula, and blood vessels, which will help in the identification of abnormalities in specific regions of the retina.

If the database is not clearly defined, there will be a lot of problems during classification using the deep neural network concept.

Conclusions

In conclusion, this review has provided detailed information about the publicly available and commonly used fundus image datasets. This has assisted glaucoma researchers in learning about the availability and accessibility of public fundus image datasets, as well as in selecting the correct and appropriate dataset for their glaucoma-related work. These datasets have been used to develop and evaluate a range of automated glaucoma detection algorithms. However, further research is needed to address several challenges and limitations associated with fundus image datasets, such as variations in image quality, differences in image acquisition and processing techniques, and the need for larger, more diverse datasets to improve the accuracy and generalizability of glaucoma detection models. Thus, the fundus image datasets offer a promising avenue for advancing our understanding and management of glaucoma.

References

1. World Health Organization. Strengthening medical imaging. https://www.who.int/activities/strengthening-medical-imaging.
2. The University of British Colombia, Department of Ophthalmology and Visual Sciences, https://ophthalmology.med.ubc.ca/patientcare/ophthalmic-photography/color-fundus-photography.
3. National Eye Institute, https://www.nei.nih.gov/learn-about-eye-health/eye-conditions-and diseases/glaucoma/types-glaucoma.
4. Glaucoma Research Foundation. Types of Glaucoma. https://glaucoma.org/learn-about-glaucoma/types-of-glaucoma/.
5. Almazroa, A., et al. (2018). Retinal fundus images for glaucoma analysis: the RIGA dataset. *Medical Imaging 2018: Imag. Informat. Healthcare Res. Appl.*, 10579. 1–6.
6. Zhang, J., Dashtbozorg, B., Bekkers, E., Pluim, J. P., Duits, R., and ter Haar Romeny, B. M. (2016). Robust retinal vessel segmentation via locally adaptive derivative frames in orientation scores. *IEEE Trans. Med. Imag.*, 35(12), 2631–2644.
7. Sivaswamy, J., et al. (2014). Drishti-gs: Retinal image dataset for optic nerve head (onh) segmentation. *2014 IEEE 11th Inter. Symp. Biomed. Imag. (ISBI)*, 53–56.
8. Budai, A., et al. (2013). Robust vessel segmentation in fundus images. *Inter. J. Biomed. Imag.*, 2013, 1–12.
9. Bajwa, M. N., et al. (2020). G1020: A benchmark retinal fundus image dataset for computer-aided glaucoma detection. *2020 Inter. Joint Conf. Neural Netw. (IJCNN)*, 1–7.
10. Diaz-Pinto, A., Morales, S., Naranjo, V., Köhler, T., Mossi, J. M., and Navea, A. (2019). CNNs for automatic glaucoma assessment using fundus images: An extensive validation. figshare. Dataset. https://doi.org/10.6084/m9.figshare.7613135.v1.
11. Kim, U. (2018). Machine learn for glaucoma. https://doi.org/10.7910/DVN/1YRRAC, Harvard Dataverse, V1.

12. Pachade, S., et al. (2021). Retinal fundus multi-disease image dataset (RFMiD): A dataset for multi-disease detection research. *Data*, 6(2), 14.
13. Batista, F. J. F., et al. (2020). Rim-one dl: A unified retinal image database for assessing glaucoma using deep learning. *Image Anal. Stereol.*, 39(3), 161–167.
14. Holm, S., et al. (2017). DR HAGIS—A fundus image database for the automatic extraction of retinal surface vessels from diabetic patients. *J. Med. Imag.*, 4(1), 014503.
15. Tang, L., et al. (2011). Robust multiscale stereo matching from fundus images with radiometric differences. *IEEE Trans. Patt. Anal. Mac. Intell.*, 33(11), 2245–2258.
16. Niemeijer, M., et al. (2011). Automated measurement of the arteriolar-to-venular width ratio in digital color fundus photographs. *IEEE Trans. Med. Imag.*, 30(11), 1941–1950.
17. Retina Kaggle datase, https://www.kaggle.com/datasets/jr2ngb/cataractdataset.
18. Harish Kumar, J. R., Chandra Sekhar, S., Gagan, J. H., Yogish, S. K., Neetha, I. R. K., Vivekanand, U., Preeti, G., and Shilpa, P. (2022). Akshi IMAGE - A retinal fundus image database for glaucoma detection. *IEEE Dataport*. doi: https://dx.doi.org/10.21227/ktzx-a222.
19. Song, W. T., Chou, L., and Yi-Zhu, S. (2021). A statistical robust glaucoma detection framework combining retinex, CNN, and DOE using fundus images. *IEEE Acc.*, 9, 103772–103783.
20. Zhang, Z., et al. (2010). ORIGA(-light): an online retinal fundus image database for glaucoma analysis and research. *Ann. Inter. Conf. IEEE Engg. Med. Biol. Soc.*, 2010, 3065–3068.
21. Huazhu, F. (2019). Refuge: Retinal fundus glaucoma challenge. *IEEE Dataport*, 1–5.
22. Li, L., et al. (2019). Attention based glaucoma detection: a large-scale database and CNN model. *Proc. IEEE/CVF Conf. Comp. Vis. Patt. Recogn*, 1–6.
23. Hoover, V. K. and Goldbaum, M. (2000). Locating blood vessels in retinal images by piece-wise threhsold probing of a matched filter response. *IEEE Trans. Med. Imag.*, 19(3), 203–210.
24. Hoover, A. and Goldbaum, M. (2003). Locating the optic nerve in a retinal image using the fuzzy convergence of the blood vessels. *IEEE Trans. Med. Imag.*, 22(8), 951–958.

17 A survey on dynamic state estimation strategies of power system networks

Suvraujjal Dutta[1,a], *Papun Biswas*[2,b], *Sandip Chanda*[3,c] *and Abhinandan De*[4,d]

[1]Department of Electrical Engineering, Swami Vivekananda University, Kolkata, West Bengal, India

[2]Department of Electrical Engineering, Jis College of Engineering, Department of Electrical Engineering, Kalyani, West Bengal, India

[3]Department of Electrical Engineering, Ghani Khan Chowdhury Institute of Engineering and Technology, Malda, West Bengal, India

[4]Department of Electrical Engineering, IIEST, Shibpur, Howrah, West Bengal, India

Abstract

This paper discusses a few critical topics of dynamic state estimation (DSE) for control and protection applications in modern power systems like importance of robustness and security in power system DSE of a synchronous generator particularly in the face of cyber attacks. Performance comparison of different methods including extended Kalman filter (EKF), unscented Kalman filter (UKF), square-root unscented Kalman filter (SRUKF), cubature Kalman filter (CKF), and an observer-based approach for dynamic state estimation (DSE) in power systems under data integrity attacks have been presented in this paper. Both conventional and modern data-driven and probabilistic techniques have been reviewed and identified few limitations of those methods. To overcome these limitations new methodology is proposed. Hence, proposed work is comparison study on conventional method with new machine learning-based dynamic state estimation method.

Keywords: Dynamic state estimation, robust cubature Kalman filter (RCKF), extended Kalman filter (EKF), unscented Kalman filter (UKF), square-root unscented Kalman filter (SRUKF, cubature Kalman filter (CKF), Machine learning, power systems networks

I. Introduction

The advancement of dynamic state estimation (DSE) has been a gradual process over the years. The advancement of DSE is a continuous process. DSE is a valuable tool for power system operators and engineers. The benefits of using DSE include improved power system stability, improved power system control, and improved power system monitoring. The future of DSE is promising.

DSE is a technique for [15]estimating the dynamic state of a power system. Recent work has focused on developing new DSE algorithms that are more accurate, robust, and computationally efficient. Some of the most recent work[23] includes algorithms that use data from phasor measurement units (PMUs), are robust to bad data, and can be used in real time. The research on DSE is ongoing, and there are still challenges to be addressed, such as data quality, model accuracy, and computational complexity.

Despite these challenges, the recent progress that has been made in DSE is promising. As DSE technology continues to develop, it is likely to play an increasingly important role in ensuring the reliability, security, and efficiency of power systems around the world.

Here are some additional details about the recent work on DSE. One of the most promising new DSE algorithms is the [2]unscented Kalman filter (UKF). The UKF is a robust and efficient algorithm that can be used to estimate the dynamic state of a power system even in the presence of bad data[26]. Another promising new DSE algorithm is the dynamic state estimation-based fault locating (EBFL) method. The EBFL method can be used to quickly and accurately locate faults on transmission lines[49].

The research on DSE is ongoing, and there are still challenges to be addressed. However, the recent progress that has been made is promising, and it is likely that DSE will become an increasingly important tool for monitoring and controlling power systems in the future.

The paper studied by Chen et al. [1] provides a comprehensive overview of the state of the art in DSE for power systems [42]. The paper begins by motivating the need for DSE in power systems, and then provides a definition of DSE. The paper then discusses the different methodologies that have been proposed for DSE, and finally discusses

[a]Suvraujjal.phd@gmail.com, [b]head_ee.jisce@jisgroup.org, [c]sandipee1978@gmail.com, [d]abhinandan.de@gmail.com

DOI: 10.1201/9781003540199-17

the future challenges and research directions in DSE.

Dynamic state estimation (DSE) is the process of [5]estimating the state of a power system from a set of measurements. The measurements can include phasor measurements, sampled values, or other types of data. The goal of DSE is to estimate the state of the system at a specific point in time. DSE is motivated by the need to accurately estimate the state [29]of a power system in real time. The state of a power system is defined as the values of the physical variables that describe the system, such as the voltage and current at each bus, the speed of each generator, and the load on each line. Accurately [11]estimating the state of a power system is important for a number of reasons, including – control and protection. DSE can be used to improve the performance of power system control and protection schemes. For example, DSE can be used to estimate the location and severity of a fault, which can then be used to dispatch corrective actions, such as tripping breakers or redispatching generators.

System monitoring: DSE can be used to monitor the health of a power system. For example, DSE can be used to identify potential problems, such as overloaded lines or failing generators.

Operational planning: DSE can be used to support operational planning decisions, such as the scheduling of generation and transmission resources. There are a number of different methodologies that have been proposed for DSE. The most common methodology is the [45]Kalman filter. The Kalman filter is a recursive estimator that uses a set of measurements and a model of the system to estimate the state of the system. Other methodologies for DSE include the extended Kalman filter, the unscented [44]Kalman filter, and the particle filter.

The field of DSE is still evolving, and there are a number of challenges that need to be addressed in the future. Some of the key challenges include – The increasing complexity of power systems. As power systems become more complex, the task of estimating their state becomes more difficult. There is limited availability of measurements in some parts of the world. The availability of measurements is limited, which makes it difficult to accurately [12]estimate the state of the system. The presence of noise and uncertainty: The measurements[15] used in DSE are often noisy and uncertain, which can degrade the accuracy of the state estimation.

Modeling: The accuracy of DSE depends on the accuracy of the model[40] of the power system. However, power systems are complex, and it is difficult to develop accurate models.

Measurements: The accuracy of DSE also depends on the quality of the measurements. However, measurements can be corrupted by noise and other disturbances. Computational complexity – DSE can be computationally expensive, especially for large power systems.

Despite the challenges, DSE is a promising technology that has the potential to improve the performance and reliability of power systems. Future research in DSE will focus on addressing the challenges mentioned above, and on developing new DSE methodologies that are more accurate, robust, and efficient. In addition to the challenges mentioned in the paper, there are a number of other challenges that need to be addressed in the future, such as data quality. The quality of the data used for DSE is critical. However, data quality can be affected by a number of factors, such as noise, measurement errors, and cyber attacks. Security – DSE systems need to be secure from cyber attacks. Cyber attacks can be used to manipulate the measurements used for DSE, which can lead to inaccurate state estimates. Privacy – DSE systems need to protect the privacy of the data used for DSE. The data used for DSE can include sensitive information, such as the location of critical infrastructure. Despite the challenges, DSE is a promising technology that has the potential to improve the performance and reliability of power systems. Future research in DSE will focus on addressing the challenges mentioned above, and on developing new DSE methodologies that are more accurate, robust, efficient, secure, and private.

Dynamic state estimation (DSE)[21] is a technique used to estimate the state of a power system in real time. DSE takes into account the dynamic behavior of the system, which allows it to provide more accurate estimates than static state estimation (SSE) [43].

There are a number of different DSE techniques available. The most common technique is the Kalman filter. The Kalman filter[35] is a recursive estimator that uses a model of the system dynamics to update its estimates of the state. Other DSE techniques are the extended Kalman filter (EKF), the unscented Kalman filter (UKF), and the particle filter. The EKF is a more accurate version of the Kalman filter, but it is also more computationally expensive. The UKF is a less accurate version of the Kalman filter, but it is also less computationally expensive. The particle filter is a non-parametric estimator that is well-suited for systems with a large number of states.

DSE has a number of applications in power systems. It can be used to monitor the system for potential problems, such as overloads and outages.

It can also be used to control the system, such as by adjusting the output of generators. DSE is a valuable tool for power system operators. It can help them to keep the system safe and reliable. Here are some of the advantages of using DSE. DSE can provide more accurate estimates of the state of the power system than SSE.

DSE can track the dynamic behavior of the system, which allows it to provide more timely and accurate information to operators. DSE can be used to monitor the system for potential problems, such as overloads and outages. DSE can be used to control the system, such as by adjusting[42] the output of generators. Here are some of the challenges of using DSE. DSE[36] is computationally more expensive than SSE. DSE is more sensitive to noise and disturbances in the measurements. DSE requires a more accurate model of the system dynamics. Despite the challenges, DSE is a valuable tool for power system operators. It can help them to keep the system safe and reliable.

Dynamic state estimation (DSE)[41] is a technique used to estimate the state of a power system in real time. DSE takes into account the dynamic behavior of the system, which allows it to provide more accurate estimates than static state estimation (SSE). With the increasing penetration[38] of renewable energy sources (RES) in power systems, the need for DSE has become more important. RES are intermittent and unpredictable, which can cause the power system to become unstable. DSE can help to improve the stability of power systems by providing more accurate information about the system state [44].

The paper submitted by Zhou et al., (2018) states that DSE is a technique used to estimate the state of a power system in real time. DSE takes into account the dynamic behavior of the system, which allows it to provide more accurate estimates than static state estimation (SSE). With the increasing penetration of distributed energy resources (DERs) in power systems, the need for DSE has become more important. DERs [39]are intermittent and unpredictable, which can cause the power system to become unstable. DSE can help to improve the stability of power systems by providing more accurate information about the system state [45].

There are a number of different DSE techniques available. The most common technique is the Kalman filter. The Kalman filter is a recursive estimator that uses a model of the system dynamics to update its estimates of the state.

Other DSE techniques include the extended Kalman filter (EKF), [20]the unscented Kalman filter (UKF), and the particle filter. The EKF is a more accurate version of the Kalman filter, but it is also more computationally expensive. The UKF is a less accurate version of the Kalman filter, but it is also less computationally expensive. The particle filter is a non-parametric estimator that is well-suited for systems with a large number of states. DSE has a number of applications in power systems with DERs. It can be used to monitor the system for potential problems, such as overloads and outages. It can also be used to control the system, such as by adjusting the output of generators.

DSE [3]is a valuable tool for power system operators. It can help them to keep the system safe and reliable.

The paper then goes on to discuss the challenges of DSE in power systems with DERs. These challenges include the following:

The dynamic behavior of DERs is complex and difficult to model. DERs can cause disturbances in the power system that can make DSE more difficult.

The communication infrastructure for DSE in power systems with DERs is often not reliable.

The paper then discusses some of the recent advances in DSE for power systems with DERs. These advances include:

The use of machine learning techniques is to improve the accuracy of DSE. The development of DSE algorithms are specifically designed for power systems with DERs. The integration of DSE with other power system control and monitoring systems. The paper concludes by discussing the future of DSE in power systems with DERs. The authors believe that DSE will become increasingly important as the penetration of DERs in power systems continues to grow. They also believe that the challenges of DSE can be overcome through the use of advanced technologies, such as machine learning and distributed computing.

DSE is a technique used to estimate the state of a power system in real time [46]. DSE takes into account the dynamic behavior of the system, which allows it to provide more accurate estimates than static state estimation (SSE). Uncertainties in power systems can arise from a variety of sources, including measurement noise, model errors, and external disturbances. These uncertainties can degrade the accuracy of DSE. To address this issue, a number of techniques have been developed to incorporate uncertainty into DSE. One approach is to use a robust estimator. Robust estimators are designed to be less sensitive to noise and disturbances than traditional estimators. Another approach is to use a probabilistic estimator. Probabilistic estimators provide a distribution of possible states, rather than a single point estimate. This can be useful for decision-making, as it allows for the consideration of uncertainty.

The choice of uncertainty handling technique depends on the specific application. For example, robust estimators may be preferred for applications where accuracy is critical, while probabilistic estimators may be preferred for applications where flexibility is important.

In recent years, there has been growing interest in the use of DSE for power systems with uncertainties. This is [9]due to the increasing penetration of renewable energy sources (RES) and distributed energy resources (DERs) into power systems. RES and DERs are intermittent and unpredictable, which can introduce new sources of uncertainty into the power system. DSE can be used to improve the reliability and stability of power systems with uncertainties.

II. Advancement of different research on dynamics state estimation

The use of DSE[47] for control and protection applications in modern power systems, particularly in the context of increasing penetration of converter-based resources (CBRs) such as renewable energy sources and energy storage has been proposed [14]. The paper discusses the differences between DSE and observers, and presents examples and practical applications of DSE-based solutions for ensuring system stability, reliable protection, and security. Additionally, the paper discusses the use of sampled value measurements versus PMU measurements for DSE formulation (Figure 17.1).

The paper demonstrates how these challenges can be effectively addressed with DSE-enabled solutions. As a precursor to these solutions, the paper presents a reformulation of DSE that considers both synchrophasor and sampled value measurements. It also provides comprehensive comparisons of DSE and observers. DSE[27] is a technique used in power systems to accurately track the dynamics of the system and provide real-time updates on the system state. This paper focuses on the control and protection applications of DSE and presents different challenges that arise in modern power systems (Figure 17.2).

The paper[2] shows the usefulness and necessity of DSE-based solutions in ensuring system stability, reliable protection and security, and resilience by revamping control and protection methods. The example of implementation of DSE-based control architecture is likely to be a practical application of the solutions presented in the paper (Figures 17.3 and 17.4).

Without further context, it is difficult to provide a detailed explanation of the example implementation. However, it is likely that the implementation involves using DSE to monitor the power system and make real-time adjustments to ensure stability

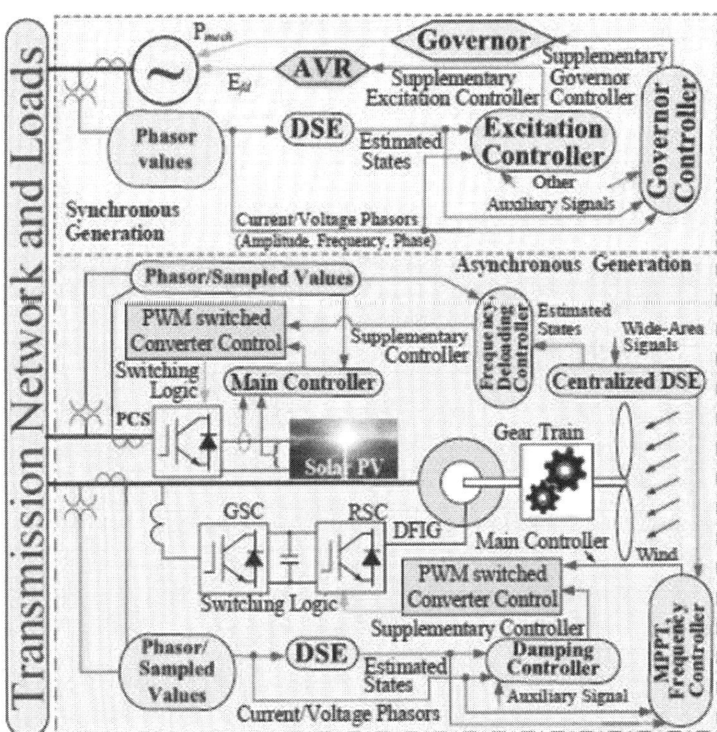

Figure 17.1 Example implementation of DSE-based control architecture

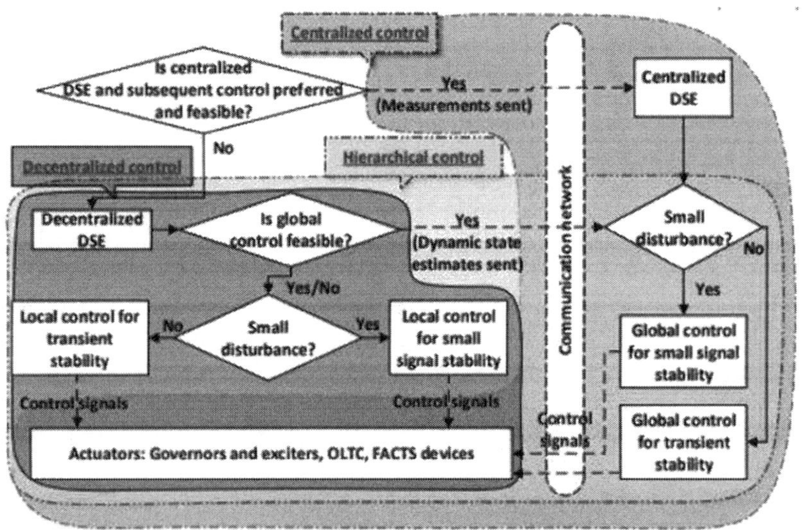

Figure 17.2 A flowchart of DSE-based control decision and design process for rotor angle stability

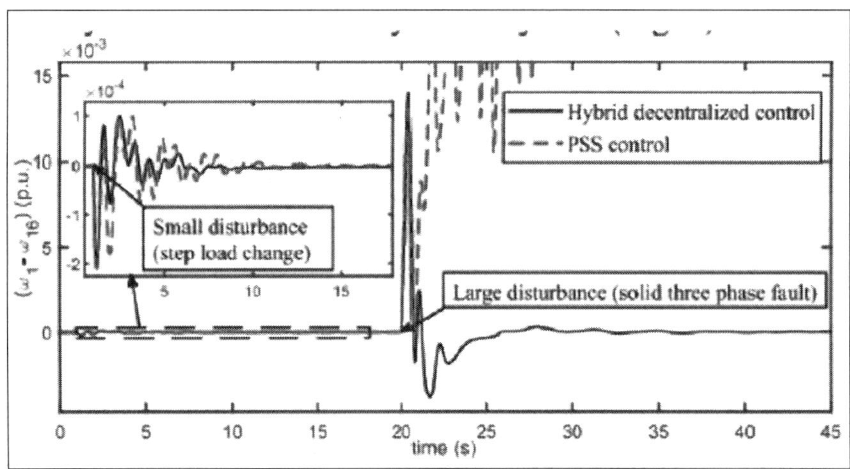

Figure 17.3 Control Performance of decentralized hybrid control

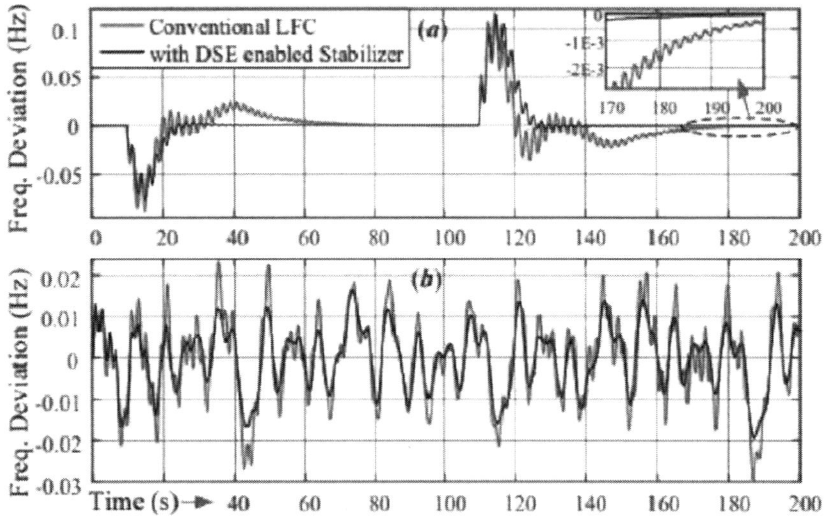

Figure 17.4 Frequency stability: DSE-stabilizer vs. conventional LFC

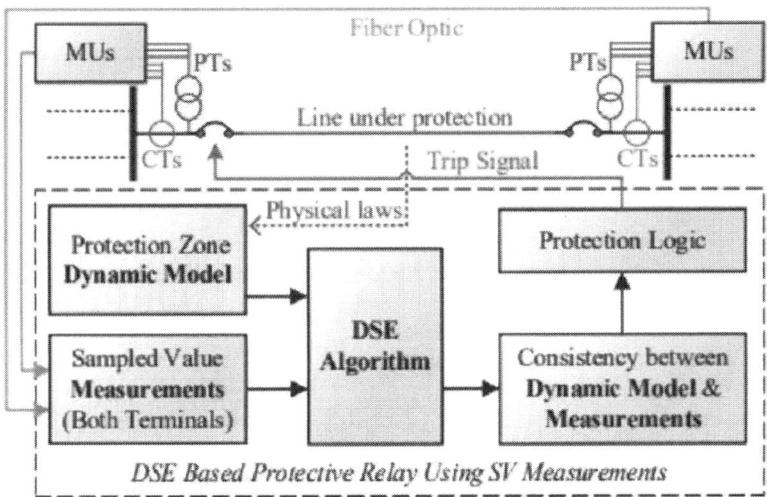

Figure 17.5 DSE-based protective relay using SV measurements

and reliability. This could involve adjusting the output of generators, switching between power sources, or taking other actions to maintain the desired system state (Figure 17.5).

Overall, the paper highlights the importance of DSE in modern power systems and provides insights into how it can be used to address control and protection challenges. The example implementation of DSE-based control architecture is likely to be a practical demonstration of the solutions presented in the paper [27] which proposes a robust observer for performing power system dynamic state estimation (DSE) of a synchronous generator. The observer is developed using the concept of L∞ stability for uncertain, nonlinear dynamic generator models.

$$\begin{cases} \dot{\delta} = \omega_0 \Delta\omega \\ \Delta\dot{\omega} = \frac{1}{T_j}\left[T_m - T_e - D\Delta\omega\right] \\ \dot{E}_q' = \frac{1}{T_{d0}'}\left[E_f - E_q' - (X_d - X_d')I_d\right] \\ \dot{E}_d' = \frac{1}{T_{q0}'}\left[-E_d' + (X_q - X_q')I_q\right] \end{cases} \tag{1}$$

$$\begin{cases} I_d = \frac{E_q' - U_t\cos(\delta - \phi)}{X_d'} \\ I_q = \frac{U_t\sin(\delta - \phi) - E_d'}{X_q'} \end{cases} \tag{2}$$

The focus of the paper [30] is to propose a robust filtering algorithm, called robust cubature Kalman filter (RCKF), for dynamic state estimation of synchronous machines in the presence of non-Gaussian measurement noise and outliers. The paper combines the Huber's M-estimation theory with the classical cubature Kalman filter (CKF) to improve the adaptability of the filter to unknown measurement noise statistics. The proposed RCKF algorithm is tested on the WSCC 3-machine 9-bus system and New England 16-machine 68-bus system, and the simulation results show that the RCKF outperforms the classical CKF in terms of robustness and adaptability. The paper also discusses the limitations and contributions of Kalman-type filtering techniques and proposes future research directions. The[5] contributions of this paper are to summarize the technical activities of the task force on power system dynamic state and parameter estimation, to propose a unified framework for DSE, to review various dynamic state and parameter estimation methods, and to identify potential applications of DSE for enhanced monitoring, protection, and control. This paper[18] also provides future research needs and directions for the power engineering community.

The paper reviews[3] the tools and methods for the implementation of DSE and discusses their pros and cons. The system observability concept is also discussed. The paper proposes a unified framework for DSE and identifies potential applications of DSE for enhanced monitoring, protection, and control.

A nonlinear dynamical system is said to be observable at a state x_0 if the nonlinear observability matrix obtained by using Lie derivatives at $x = x_0$ has a full rank. This can be explained as follows:

Consider the nonlinear system, Lie derivative of h with respect to f is defined as

$$L_f \boldsymbol{h} = \nabla \boldsymbol{h} \cdot \boldsymbol{f}. \tag{3}$$

From the definition (3)

$$L_f^0 \boldsymbol{h} = \boldsymbol{h}, \qquad L_f^k \boldsymbol{h} = \frac{\partial L_f^{k-1}\boldsymbol{h}}{\partial x} \cdot \boldsymbol{f}. \tag{4}$$

$$\Omega = \begin{bmatrix} L_f^0(\boldsymbol{h}_1) & \dots & L_f^0(\boldsymbol{h}_m) \\ L_f^1(\boldsymbol{h}_1) & \dots & L_f^0(\boldsymbol{h}_m) \\ \vdots & \dots & \vdots \\ L_f^{n-1}(\boldsymbol{h}_1) & \dots & L_f^{n-1}(\boldsymbol{h}_m) \end{bmatrix}, \quad (5)$$

$$\boldsymbol{O} = d\Omega = \begin{bmatrix} dL_f^0(\boldsymbol{h}_1) & \dots & dL_f^0(\boldsymbol{h}_m) \\ dL_f^1(\boldsymbol{h}_1) & \dots & dL_f^0(\boldsymbol{h}_m) \\ \vdots & \dots & \vdots \\ dL_f^{n-1}(\boldsymbol{h}_1) & \dots & dL_f^{n-1}(\boldsymbol{h}_m) \end{bmatrix} \quad (6)$$

The observability matrix O must have rank n in order for the system to be observable and n is the dimension of the state vector x.

One limitation of this paper is that it mainly focuses on the technical aspects of DSE and does not provide a comprehensive discussion of the economic and regulatory aspects of DSE implementation. Another limitation is that the paper does not provide a detailed comparison of the performance of different DSE methods in terms of accuracy, robustness, and computational efficiency [1]. It also discusses on various methods for dynamic state estimation in power systems, including unscented transformation, Kalman filtering, particle filtering, and neural networks. The paper also highlights the use of these methods for estimating parameters, detecting cyber attacks and unknown inputs, and the utilization of high-performance computing methods for dynamic state estimation. This paper presents: (a) alternatives that address the limitations of power system DSE routines under model uncertainty and potential cyber attacks, (b) introducing and implementing the cubature Kalman filter (CKF) and a nonlinear observer to address the challenges of effective power system DSE, (c) testing various Kalman filters and the observer on the 16-machine, 68-bus system given realistic scenarios under model uncertainty and different types of cyber attacks against synchrophasor measurements, and (d) showing that CKF and the observer are more robust to model uncertainty and cyber attacks than their counterparts, and performing a thorough qualitative comparison for Kalman filter routines and observers based on the tests.

Overall, the Lipschitz[44] condition provides a theoretical guarantee of the performance of the proposed state estimation methods under model uncertainty and cyber attacks. By satisfying this condition, the cubature Kalman filter and nonlinear observer are shown to be more robust than other Kalman filter routines and observers.

III. Conclusions

Dynamic state estimation (DSE) is a technique for estimating the dynamic state of a power system from measurements of its physical quantities. DSE is a powerful tool[16] that can be used to improve the reliability and efficiency of power systems. As power systems become more complex, the need for DSE will become even more important. The advancement of DSE is a continuous process. As power systems become more complex and interconnected, the need for accurate and reliable DSE will become even more important. The future of DSE is promising. With the development[25] of new technologies, such as machine learning, DSE will become even more accurate and reliable.

IV. References

1. Junbo, Z., Gómez-Expósito, A., Marcos, N., Mili, L., Abur, A., Terzija, V., Kamwa, I., Pal, B., Singh, A., Junjian, Q., Zhenyu, H., and Meliopoulos, A. (2019). Power system dynamic state estimation: Motivations, definitions, methodologies, and future work. *IEEE Trans. Power Sys,vol. 34, no. 4, pp. 3188-3198*
2. Binglin W., Yu, L., Dayou, L., Kang, Y., and Rui, F. (2021). Transmission line fault location in MMC-HVDC grids based on dynamic state estimation and gradient descent. *IEEE Trans. Power Delivery*,vol. 36, no. 3, pp. 1714-1725.
3. Liang, C., Peng, J., Jing, Y., Yang, L., and Yi, S. (2021). Robust Kalman filter-based dynamic state estimation of natural gas pipeline networks. *Math. Prob. Engg*,pp.1-10.
4. Zhao. (2019). Robust unscented Kalman filter for power system dynamic state estimation with unknown noise statistics. *IEEE Trans. Smart Grid*,vol. 10, no. 2, pp. 1215-1224.
5. Alhelou, H. H., Mohamad Esmail, H. G., and Nikos, H. (2020). Deterministic dynamic state estimation-based optimal LFC for interconnected power systems using unknown input observer. *IEEE Trans. Smart Grid,vol. 11, no. 2, pp. 1582-1592.*
6. Zhao, J., Netto, M., Huang, Z., Yu, S. S., Gomez-Exposito, A., Wang, S., Kamwa, I., Akhlaghi, S., Mili, L., Terzija, V., Sakis Meliopoulos, A. P., Pal, B. C., Singh, A. K., Abur, A., Bi, T., and Rouhani, A. (2021). Roles of dynamic state estimation in power system modeling, monitoring and operation. *IEEE Trans. Power Sys,vol. 36, no. 3, pp. 2462-2472.*
7. Xing, Y. and Chen, Lv. (2020). Dynamic state estimation for the advanced brake system of electric vehicles by using deep recurrent neural networks. *IEEE Trans. Indus. Elec*,vol. 67, no. 11, pp. 9536-9547.
8. Liu, H., Hu, F., Su, J., Wei, X., and Qin, R. (2020). Comparisons on Kalman-filter-based dynamic state estimation algorithms of power systems. *IEEE Acc,vol. 8, pp. 51035-51043.*
9. Kazemi, Z., Safavi, A., Naseri, F., Urbas, L., Setoodeh, P. (2020). A secure hybrid dynamic-state estimation approach for power systems under false

data injection attacks. *IEEE Trans. Indus. Informat*, vol. 16, no. 12, pp. 7275-7286.

10. Chen, Y., Yuan, Y., Lin, Y., Yang, X. (2020). Dynamic state estimation for integrated electricity-gas systems based on Kalman filter. *CSEE J. Power Energy Sys,vol. 8, no. 1, pp. 293-303.*
11. Yu, S. S., Guo, J., Chau, T. K., Fernando, T., Iu, H. H.-C., and Trinh, H. (2020). An unscented particle filtering approach to decentralized dynamic state estimation for DFIG wind turbines in multi-area power systems. *IEEE Trans. Power Sys,vol. 35, no. 4, pp. 2670-2682.*
12. Zhao, J., Zheng, Z., Wang, S., Huang, R., Bi, T., Mili, L., and Huang, Z. (2020). Correlation-aided robust decentralized dynamic state estimation of power systems with unknown control inputs. *IEEE Trans. Power Sys,vol. 35, no. 3, pp. 2443-2451.*
13. Wang, W., Tse, C., and Wang, S. (2020). Dynamic state estimation of power systems by p-norm nonlinear Kalman filter. *IEEE Trans. Circuits Sys. I Reg. Papers,vol. 67, no. 5, pp. 1715-1728.*
14. Liu, Y., Singh, A. K., Zhao, J., Sakis Meliopoulos, A. P., Pal, B. C., Ariff, M. A. M., Van Cutsem, T., Glavic, M., Huang, Z., Kamwa, I., Mili, L., Mir, A. S., Taha, A. F., Terzija, V., and Yu, S. (2021). Dynamic state estimation for power system control and protection IEEE task force on power system dynamic state and parameter estimation. *IEEE Trans. Power Sys*,vol. 36, no. 6, pp. 5909-5921.
15. Zhao, J., Netto, M., and Mili, L. (2017). A robust iterated extended Kalman filter for power system dynamic state estimation. *IEEE Trans. Power Sys,vol. 32, no. 4, pp. 3205-3216.*
16. Akhlaghi, S., Zhou, N., and Huang, Z. (2017). Adaptive adjustment of noise covariance in Kalman filter for dynamic state estimation.
17. Barnes, A. K. and Mate, A.. (2021). Dynamic state estimation for radial microgrid protection,2021 IEEE/IAS 57th Industrial and Commercial Power Systems Technical Conference (I&CPS), Las Vegas, NV, USA, 2021,pp. 1-9.
18. Vieyra, N., Maya, P., and Castro, L. M. (2020). Dynamic state estimation for microgrid structures. *Elec. Power Comp. Sys.*
19. Zhao, J. and Mili, L. (2019). A decentralized H-infinity unscented Kalman filter for dynamic state estimation against uncertainties. *IEEE Trans. Smart Grid*,vol. 10, no. 5, pp. 4870-4880.
20. Nath, S., Akingeneye, I., Wu, J., and Han, Z. (2019). Quickest detection of false data injection attacks in smart grid with dynamic models. *IEEE J. Emerg. Select. Topics Power*,vol. 10, no. 1, pp. 1292-1302.
21. Zhao, J., Mili, L., and Gómez-Expósito, A. (2019). Constrained robust unscented Kalman filter for generalized dynamic state estimation. *IEEE Trans. Power Sys,vol. 34, no. 5, pp. 3637-3646.*
22. Chen, R., Li, X., Zhong, H., and Fei, M. A novel online detection method of data injection attack against dynamic state estimation in smart grid,Volume 344,2019,Pages 73-81.
23. Li, S., Hu, Y., Zheng, L., Li, Z., Chen, X., Fernando, T., Iu, H. H.-C., Wang, Q., and Liu, X. (2019). Stochastic event-triggered cubature Kalman filter for power system dynamic state estimation. *IEEE Trans. Circuits Sys. II Exp. Briefs*,vol. 66, no. 9, pp. 1552-1556.
24. Jin, X., Yin, G., and Chen, N. (2019). Advanced estimation techniques for vehicle system dynamic state: A survey. *Sensors.*
25. Qi, J., Sun, K., Wang, J., and Liu, H. (2016).Dynamic state estimation for multi-machine power system by unscented Kalman filter with enhanced numerical stability,IEEE Transactions on Smart Grid,vol. 9, no. 2, pp. 1184-1196.
26. Nicolai, M., Lorenz-Meyer, L., Bobtsov, A. A., Ortega, R., Nikolaev, N., and Schiffer, J. (2020). PMU-based decentralized mixed algebraic and dynamic state observation in multi-machine power systems,*ArXiv* abs/2003.13996 (2020): n. pag.
27. Nugroho, S. A., Taha, A. F., and Qi, J. (2020). Robust dynamic state estimation of synchronous machines with asymptotic state estimation error performance guarantees,vol. 35, no. 3, pp. 1923-1935.
28. Zhao, J. and Mili, L. (2017). Robust power system dynamic state estimator with non-Gaussian measurement noise: Part I–Theory,IEEE Transactions on Instrumentation and Measurement,vol. 71, pp. 1-10, 2022.
29. Qi, J., Taha, A. F., and Wang, J. (2018).Comparing Kalman filters and observers for power system dynamic state estimation with model uncertainty and malicious cyber attacks,IEEE Access,vol. 6, pp. 77155-77168.
30. Li, Y., Li, J., Qi, J., and Chen, L. Robust cubature Kalman filter for dynamic state estimation of synchronous machines under unknown measurement noise statistics,IEEE Access,2019,vol. 7, pp. 29139-29148.
31. Li, Y., Li, Z., Chen, L., and Nanjing. Dynamic state estimation of generators under cyber attacks,IEEE Access,2019,vol. 7, pp. 125253-125267.
32. Durgaprasad, G. and Thakur, S. S. (1998). Robust dynamic state estimation of power systems based on M-estimation and realistic modeling of system dynamics. *IEEE Trans. Power Sys,vol. 13, no. 4, pp. 1331-1336.*
33. Sinha, A. K. and Mondal, J. (1999). Dynamic state estimator using ANN based bus load prediction. *IEEE Trans. Power Sys,vol. 14, no. 4, pp. 1219-1225.*
34. Shih, K.-R. and Huang, S.-J. (2002). Application of a robust algorithm for dynamic state estimation of a power system. *IEEE Trans. Power Sys,vol. 17, no. 1, pp. 141-147.*
35. Lin, J.-M., Huang, S.-J., and Shih, K.-R. (2003). Application of sliding surface-enhanced fuzzy control for dynamic state estimation of a power system. *IEEE Trans. Power Sys,vol. 18, no. 2, pp. 570-577.*
36. Valverde, G. and Terzija, V. (2011). Unscented Kalman filter for power system dynamic state estimation. *IET Gen. Trans. Distrib,p. 29 – 37.*

37. Leung, K. T., Whidborne, J. F., Purdy, D., and Dunoyer, A. (2011). A review of ground vehicle dynamic state estimations utilising GPS/INS. *Veh. Sys. Dyn,pp. 29–58.*
38. Qi, J., Sun, K., and Kang, W. (2014). Optimal PMU placement for power system dynamic state estimation by using empirical observabilityGramian. ARXIV-MATH.OC,IEEE Transactions on Power Systems, vol. 30, no. 4, pp. 2041-2054, July 2015.
39. Qi, J., Sun, K., Wang, J., and Liu, H. (2015). Dynamic state estimation for multi-machine power system by unscented Kalman filter with enhanced numerical stability,ARXIV-MATH.OC,IEEE Transactions on Smart Grid,vol. 9, no. 2, pp. 1184-1196.
40. Liu, Y., Sakis Meliopoulos, A. P., Tan, Z., Sun, L., and Fan, R. (2017). Dynamic state estimation-based fault locating on transmission lines. *IET Gen. Trans. Distrib,p. 4184 – 4192.*
41. Zhao, J., Gómez-Expósito, A., Netto, M., Mili, L., Abur, A., Terzija, V., Kamwa, I., Pal, B., Singh, A. K., Qi, J., Huang, Z., and Sakis Meliopoulos, A. P. (2019). Power system dynamic state estimation: Motivations, definitions, methodologies, and future work. *IEEE Trans. Power Sys,pp. 3188-3198.*
42. Kamwa, I., Mili, L., and La Scala, M. (2018). Dynamic state estimation for power systems: Motivations, definitions, methodologies, and future work. *IEEE Trans. Power Sys.* 33(5),vol. 34, no. 4, pp. 3188-3198.
43. Zhao, J., La Scala, M., and Mili, L. (2017). A review of dynamic state estimation for power systems. *IEEE Sig. Proc. Mag.*, 34(6).
44. Zhao, J., La Scala, M., and Mili, L. (2017). Dynamic state estimation for power systems with renewable energy sources. *IEEE Trans. Power Sys.*, 32(6),pp. 1-8.
45. Zhang, X., Wang, R., and Wang, X. (2018). A survey of dynamic state estimation for power systems with distributed energy resources. *IEEE Acc.*, 6,vol. 11, no. 4, pp. 1065-1074.
46. Hou, D., Sun, Y., Wang, J., Zhang, L., and Wang, S. (2022). Dynamic state estimation for power systems with uncertainties. *IEEE Trans. Power Sys.*, 3465–3476
47. Zhang, W., Feng, F., and Zhang, Y. (2021). A deep learning approach for dynamic state estimation in power systems. *IEEE Trans. Power Sys.*, 36(4), 3465–3476.
48. Zhang, Y., Li, Q., Chen, C., Zheng, X., and Tan, Y. (2022). Improved dynamic state estimation of power system using unscented Kalman filter with more accurate prediction model. *Elec. Power Sys. Res.*, 209, 112049.
49. Wang, L., Zhang, Y., Liu, Y., and Wang, X. (2021). Dynamic state estimation aided by machine learning. *IEEE Trans. Power Sys.*, 36(1), 131–142.

18 A review on load frequency control considering deregulated hybrid power system model: Recent trends and future prospects

Pralay Roy[1,a], *Pabitra Kumar Biswas*[1,b] *and Chiranjit Sain*[2,c]

[1]Department of Electrical Engineering, National Institute of Technology, Mizoram, Mizoram, India

[2]Department of Electrical Engineering, Ghani Khan, Choudhury Institute of Engineering & Technology, Malda, India

Abstract

The demand for electricity has been steadily rising over the past 10 years, along with the rapid depletion of fossil resources and the advent of electrical deregulation policies. Unpredictability, fluctuation, and shifting load requirements can all be handled by a stable power system. A reliable power system is capable of withstanding disruptions, variations, and different load requirements. Because load demand is erratic and consequently causes the tie-line power and system frequency to deviate from their nominal values, loads in power systems are especially unstable and unpredictable. This article provides an overview of various deregulated power system control methodologies/techniques that aim to alleviate various load frequency control (LFC) challenges in such systems. The discussion includes a thorough study of various control strategies based on traditional control, modern control and robust procedures. This review effort effectively addresses this issue. With the aid of the merits and shortcomings of various controllers are also examined. Researchers may be able to figure out the gap between present advancements, implementation, difficulties, and anticipated future trends in case of LFC analysis with the aid of this thorough, efficient literature evaluation.

Keywords: Interconnected power systems, load frequency control (LFC), frequency deviation ,tie line power deviation, applications of different kinds of controllers with their merits and demerits, various kind of optimization techniques for tuning the controller's parameters for getting the better dynamic responses, discussion against the application of modern energy storage arrangement in interconnected systems.

Introduction

The electrical power system refers to the tie lines that link together various control areas. The generators in the control zone also accelerate or decelerate in order to maintain the predefined values of frequency along with power angle under static as well as dynamic conditions [1, 2]. The case of changing the real power output from generators with the variations in system frequency and the interchange of tie line power within the preselected limits is known as load frequency control, or LFC [3]. Following regulation of power generation in the demand-load line to reduce the zone control error's temporal average, the connection line's power profile and frequency are stabilized by the LFC. This is accomplished by properly adjusting the controller to control the electrical power system's generation unit. Consequently, the frequency change stays within the nominal range of 0.1 Hz or less of the system frequency [4, 5].

Contributions of the work

i) This work preciously summarized the opportunities and challenges associated with LFC analysis.

ii) The benefits and drawbacks of a number of suggested controllers are evaluated in this work using a comparison table (Table 1).

iii) Researchers can bridge the gap between research, development, and implementation by using this literature review.

Using Boolean search method the databases has been collected from Web of Science and Scopus up to 2022 [6, 7]. This data clearly shows that there has been relatively less focus on the LFC of renewable sources in large multi-area power systems. Although the terms "multi-area" and "large power systems" appear in the headlines of published articles, the majority are only capable of supporting four area power systems due to complexity.

Utilization of unique controllers for energy-based LFC systems those are conventional, deregulated, and non-traditional

The activity of the controller in relation to the numerous sub-systems of any particular LFC model is what governs how the LFC scheme functions in an interconnected system. The past studies on load frequency management that have been done up to

[a]pralay07@gmail.com, [b]pabitra.eee@nitm.ac.in, [c]chiranjit@gkciet.ac.in

DOI: 10.1201/9781003540199-18

this point are listed and suggested below in this work.

Double integral and double derivative control
In addition to double derivative proportional-integral-derivative (PID) management for two area power systems, non-linearity such as governor dead band (GDB), generation rate constraint (GRC), and boiler dynamics have been discussed by Yang, et al. (2018) [8]. It is evaluated [9] in comparison to other PID controllers modified by contemporary heuristic optimization algorithms. The dynamic responses of several controllers, such as I (integral), PI (proportional-integral), PID, and PID + DD, were compared using ALO (Ant Lion Optimization) optimization in the paper [10]. The author discovered that PID + DD performs better than the other controllers in terms of settling time, peak overshoots, frequency oscillations, and tie-line powers.

Variable structure control approach
The primary objective of controller designing is to improve the dynamic response of electrical grids. In that point of view, settling times should be faster, frequency overshoot has to be reduced and the tie line power deviations becomes smaller. Thus, the control parameter optimization is done to enhance controller performances. For frequency management, the work did by Elsisi (2020) [11] suggests a novel variable structure gain scheduling (VSGS) to reduce settling time and the overshoot of frequency changes. Paper presented by Castilla (2018) [12] described about the application of variable structure control on reduced converter model for Three-Phase LCL-Filtered.

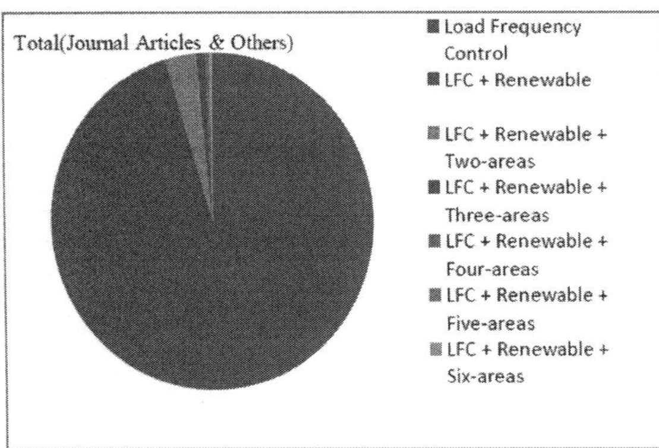

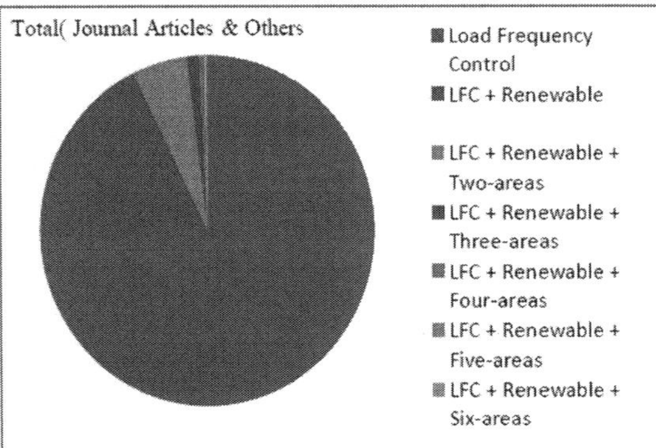

Figure 18.1 (a) Web of Science publication details of LFC in the last decade. 18.(b) Scopus publication details of LFC in the last decade

Fuzzy with proportional integral and PID control approach
For frequency regulation of a four-area power system, a unique hybrid neuro-fuzzy (HNF) is suggested by Prakash (2015) [13]. The mentioned controller is quicker than the alternative controller and can manage non-linearity. Due to the significant nonlinearity of the wind energy conversion system (WECS), a FOFPID controller is used [14] to adjust pitch angle and obtain a 2 MW output power. A fuzzy PID controller for LFC in electric power systems is proposed by Osinski et al. (2019) [15]. A hybrid fuzzy logic intelligent PID controller used for frequency regulation of an interconnected multiple-area power system has been discussed in paper presented by Haroon Gomaa et al. (2017) [16]. Due to the uncertainties in wind turbine modeling and wind speed profiles, Asgharnia et al. (2018) [17] introduced two complex controllers, fuzzy PID and fractional-order fuzzy PID, to enhance pitch control performance.

Artificial neural network (ANN) control
A work did by Yin et al. (2016) [18] uses artificial emotional reinforcement learning presents a fast and efficient adaptive load shedding solution for a disconnected power grid. In the paper submitted by Saviozzi et al., the application of covered neural structure in non-linear power systems control was explained [19]. In a two-area power system, the article by Nasiruddin et al. additionally takes into account the high-voltage direct current (HVDC) tie-line connected in parallel with the extra high-voltage direct current (EHVAC) line [20]. The controller called adaptive neuro fuzzy system (ANFIS) combines the advantages of an ANN's flexibility and fast response time with the advantages of a fuzzy controller.

Adaptive control
The real aim of the adaptive control strategy is to render the plant insensitive to changes in plant specifications by keeping its function unchanged from an operational standpoint. In order to maintain the system frequency within preselected safe limits, Marzband et al. described an efficient adaptive

control technique for figuring out the sizes of the loads and their optimal positions within the correlated power system [21]. A doubly-fed induction generator (DFIG)'s droop and inertia control loop gains can be modified adaptively using the method described by the author Verij et al. in order to solve problems and shortcomings with the interconnected power system and wind turbine modeling [22].

Model predictive control (MPC)

Forecasting future values and measuring current outputs are the basic operation of model predictive control, or MPC. Jiang et al. has shown the control rules of predictive model control are computed explicitly in order to investigate an explicit-MPC approach [23]. Zheng et al. developed a distributed MPC-based LFC technique using a discrete-time Laguerre function in order to enhance the control performances of the power system [24]. The work did by Ma et al. shows the application of the load frequency control of the multiple-area interconnected power systems in accordance with wind turbines (WTs) with the help of distributed model predictive control [25].

The researcher will probably be persuaded by this analysis to back large power networks, which will ultimately contribute to a reduction in carbon emissions. In addition, a second analytical analysis of the Web of Science was conducted to look into the interest among researchers in using metaheuristic

Table 18.1 Comparison of the benefits and drawbacks of various controllers

Sl. No.	Types of controllers	Merits	Demerits
1	P controller	The system becomes more stable by lowering the steady state error	Boost the system's maximum overshoot
2	PI controller	Significantly less steady-state error is produced	More constrained stability range
3	PD controller	System's stability is increased without the steady-state error being impacted	Amplifies noise at high frequencies
4	PID controller	Improved reactivity and stability	PID control operates differently in nonlinear systems because it is symmetric and linear
5	Classical Fractional order controller (i.e., FOI, FOPI, FOPID, etc.)	A thorough investigation has been made The design procedure is easy	Only suitable for a few operational circumstances
6	Cascaded Fractional order controller (i.e., FOI-TDN, FOPI-FOPID, PIFOD-(1 + PI), etc.)	It is simple to convert a higher order system to a lower order system Ability to endure parametric uncertainty and variation	A method using higher-order controllers Iterations require more time to finish Non-linear dynamics incompatible In order to get efficient dynamic reactions, it is essential to carefully choose the inner and outer loops
8	Fuzzy based Fractional order controller (i.e. FL-FOIDF, FOFLPID, FL-PIDNFOI, FL-FOPIFOPD, FL-PIDNFOPIDN, etc.)	Compatible with both linear and non-linear models Possess the ability to deal with both internal and external stresses Comparatively better dynamic responsiveness to classical and cascaded fractional order controllers	To choose the parameters, a high level of prior knowledge is needed 2. Compared to classical and cascaded fractional order controllers, this controller is more sophisticated 3. The tuning procedure takes longer 4. The dynamic performance depends on the membership function choice
9	Model predictive control	Dependable within the limitations of the system It is possible to use large multivariable processes	It is possible to use large multivariable processes 2. Sensitive to the forecast horizon
10	Meta heuristics control	Simple integration is possible with other techniques Suppleness of execution Problem-independent	There is no assurance of an ideal answer

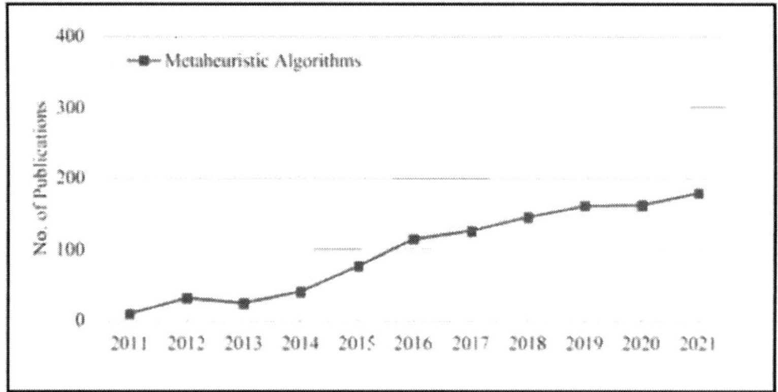

Figure 18.2 Metaheuristic algorithms have gained popularity with researchers in the past ten years for LFC investigations

algorithms and controllers for LFC investigations throughout the last ten years. The need for both controllers and metaheuristic algorithms is increasing annually, as seen in Figure 18.2 [6, 26].

Conclusions

This article examines the most recent advances in LFC methods for various conventional and renewable energy-based power systems. The application of contemporary control and other newly developed and employed controllers in the LFC field has been investigated. With the use of a comparison table, the merits and shortcomings of several proposed controllers of the pertinent study work are examined. Researchers' focus has significantly changed away from conventional PID and FOPID controllers towards hybrid and cascaded controllers in order to achieve improved stability in hybrid power systems. Similarly, it is preferred to use hybrid and modern metaheuristic optimization strategies to adjust the controller gains. Installing RESs like PV and wind has become more affordable than building a coal-fired power station in the last decade. The disadvantage of PV and wind is that they reduce the overall system inertia. To create hybrid virtual inertia, which combines the advantages of two or more energy storage systems, such as SMES, super capacitors, RFB, flywheels, aqua-electrolyzers, and fuel cells, different ESS are connected in parallel for increased stability. This paper also establishes a minimal standard for conducting great research on large-area hybrid power systems, which will help Researchers Bridge the gap between present and future LFC trends.

Declaration of competing interest

The authors declare that they have no known competing financial interests or personal relationships that could have appeared to influence the work reported in this paper.

References

1. Rajamand, S. (2021). Load frequency control and dynamic response improvement using energy storage and modeling of uncertainty in renewable distributed generators, Journals of Energy Storage,Vol. 37 (May, 2021), 102467. https://doi.org/10.1016/j.est.2021.102467.
2. Tan, K. M., Babu, T. S., Ramachandaramurthy, V. K., Kasinathan, P., Solanki, S. G., and Raveendran, S. K. (2021). Empowering smart grid: a comprehensive review of energy storage technology and application with renewable energy integration, Journals of Energy Storage,Vol. 39 (July, 2021), 102591. https://doi.org/10.1016/j.est.2021.102591.
3. Peddakapu, K., Mohamed, M. R., Sulaiman, M. H., Srinivasarao, P., Veerendra, A. S., and Leung, P. K. (2020). Performance analysis of distributed power flow controller with ultra-capacitor for regulating the frequency deviations in restructured power system.. Journals of Energy Storage,Vol. 31 (October, 2021), 101676, https://doi.org/10.1016/j.est.2020.101676.
4. Ali, H., Magdy, G., and Xu, D. (2021). A new optimal robust controller for frequency stability of interconnected hybrid microgrids considering non-inertia sources and uncertainties,International Journal of Electrical Power and Energy Systems, Vol. 128 (June, 2021), 106651 https://doi.org/10.1016/j.ijepes.2020.106651.
5. Calasan, M., Abdel Aleem, S. H. E., Bulatovi´c, M., Rube˘zi´c, V., Ali, Z. M., and Micev, M. (2021). Design of controllers for automatic frequency control of different interconnection structures composing of hybrid generator units using the chaotic optimization approach, International Journal of Electrical Power and Energy Systems, Vol. 129 (Feb., 2021), 106879 https://doi.org/ 10.1016/j.ijepes.2021.106879.
6. Clarivate, "Web of Science," 2022. https://www.scopus.com/.

7. "Scopus," 2022. https://www.scopus.com/.
8. Yang, F., He, J., Wang, J., and Wang, M. (2018). Auxiliary-function-based double integral inequality approach to stability analysis of load frequency control systems with interval time-varying delay. *IET Control Theory Appl.*, 12, 601–612. https://doi.org/10.1049/iet-cta.2017.1187.
9. Sahib, M. A. (2015). A novel optimal PID plus second order derivative controller for AVR system. *Engg. Sci. Technol. Int. J.*, 18, 194–206. https://doi.org/10.1016/ j.jestch.2014.11.006.
10. Raju, M., Saikia, L. C., and Sinha, N. (2016). Automatic generation control of a multi-area system using ant lion optimizer algorithm based PID plus second order derivative controller. *Int. J. Electr. Power Energy Syst.*, 80, 52–63. https://doi.org/ 10.1016/j.ijepes.2016.01.037.
11. Elsisi, M. (2020). New variable structure control based on different meta-heuristics algorithms for frequency regulation considering nonlinearities effects. *Int. Trans. Electr. Energ. Syst.*, Vol. 30, (July, 2020). 12428. https://doi.org/10.1002/2050-7038.12428.
12. Castilla, M., Miret, J., and De Hoz, J. (2018). Variable structure control for three-phase LCL-filtered inverters using a reduced. *IEEE Trans. Ind. Electron.*, 65, 5–15. https://doi.org/10.1109/TIE.2017.2716881.
13. Prakash, S. and Sinha, S. K. (2015). Simulation based neuro-fuzzy hybrid intelligent PI control approach in four-area load frequency control of interconnected power system. *Appl. Soft Comput. J.*, 23, 152–164. https://doi.org/10.1016/j. asoc.2014.05.020.
14. Pathak, D. and Gaur, P. (2019). A fractional order fuzzy-proportional-integral-derivative based pitch angle controller for a direct-drive wind energy system. *Comput. Electr. Engg.*, 78, 420–436. https://doi.org/10.1016/j. compeleceng.2019.07.021.
15. Osinski, C., Villar Leandro, G., and Da Costa Oliveira, G. H. (2019). Fuzzy PID controller design for LFC in electric power systems. *IEEE Lat. Am. Trans.*, 17, 147–154. https://doi.org/10.1109/TLA.2019.8826706.
16. Gomaa Haroun, A. H. and ya Li, Y. (2017). A novel optimized hybrid fuzzy logic intelligent PID controller for an interconnected multi-area power system with physical constraints and boiler dynamics. *ISA Trans.*, 71, 364–379. https://doi.org/ 10.1016/j.isatra.2017.09.003.
17. Asgharnia, R. and Shahnazi, A. J. (2018). Performance and robustness of optimal fractional fuzzy PID controllers for pitch control of a wind turbine using chaotic optimization algorithms. *ISA Trans.*, 79, 27–44. https://doi.org/10.1016/j. isatra.2018.04.016.
18. Yin, L., Yu, T., Zhou, L., Huang, L., Zhang, X., and Zheng, B. (2017). Artificial emotional reinforcement learning for automatic generation control of large-scale interconnected power grids. *IET Gener. Transm. Distrib.*, 11, 2305–2313. https://doi.org/10.1049/iet-gtd.2016.1734.
19. Saviozzi, M., Massucco, S., and Silvestro, F. (2019). Implementation of advanced functionalities for Distribution Management Systems: Load forecasting and modeling through artificial neural networks ensembles. *Electr. Power Syst. Res.*, 167, 230–239. https://doi.org/10.1016/j.epsr.2018.10.036.
20. Nasiruddin, G. S., Niazi, K. R., and Bansal, R. C. (2017). Non-linear recurrent ANN-based LFC design considering the new structures of Q matrix. *IET Gener. Transm. Distrib.*, 11, 2862–2870. https://doi.org/10.1049/iet-gtd.2017.0003.
21. Marzband, M., Moghaddam, M. M., Akorede, M. F., and Khomeyrani, G. (2016). Adaptive load shedding scheme for frequency stability enhancement in microgrids. *Electr. Power Syst. Res.*, 140, 78–86. https://doi.org/10.1016/j.epsr.2016.06.037.
22. Verij Kazemi, M., Gholamian, S. A., and Sadati, J. (2019). Adaptive frequency control with variable speed wind turbines using data-driven method. *J. Renew. Sustain. Energy*, 11, 043305. https://doi.org/10.1063/1.5078805.
23. Jiang, H., Lin, J., Song, Y., You, S., and Zong, Y. (2016). Explicit model predictive control applications in power systems: an AGC study for an isolated industrial system. *IET Gener. Transm. Distrib.*, 10, 964–971. https://doi.org/10.1049/iet-gtd.2015.0725.
24. Zheng, Y., Zhou, J., Xu, Y., Zhang, Y., and Qian, Z. (2017). A distributed model predictive control based load frequency control scheme for multi-area interconnected power system using discrete-time Laguerre functions. *ISA Trans.*, 68, 127–140. https:// doi.org/10.1016/j.isatra.2017.03.009.
25. Ma, M., Liu, X., and Zhang, C. (2017). LFC for multi-area interconnected power system concerning wind turbines based on DMPC. *IET Gener. Transm. Distrib.*, 11, 2689–2696. https://doi.org/10.1049/iet-gtd.2016.1985.
26. Ahmed Khan, H. M., Nadzirah Mansor, N., Azil Illias, H., Jamilatul Awalin, L., and Wang, L. (2023). New trends and future directions in load frequency control and flexible power system review. *Alexandria Engg. J.*, 71, 263–308. https://www.sciencedirect.com/science/article/pii/S1110016823002028.

19 Comparative performance analysis of energy management techniques for fuel cell hybrid electric vehicles

Debasis Chatterjee[1,a], *Pabitra Kumar Biswas*[1] *and Chiranjit Sain*[2]

[1]Department of Electrical and Electronics Engineering, National Institute of Technology, Mizoram, Aizawl, India

[2]Department of Electrical Engineering, Ghani Khan Choudhury Institute of Engineering & Technology, Malda, West Bengal, India

Abstract

The automotive industry is growing along with the global economy. Researchers have paid close attention to fuel cell hybrid electric vehicles (FCHEVs) because of their advantages for its maximum energy efficiency, environmental preservation, and extended driving range. Firstly, the basic structure of standard FCHEVs is presented, afterwards the latest energy management techniques for FCHEVs are discussed. Compared to the literature's currently published works, this study investigates various energy management strategies like Model Predictive Control (MPC), equivalent consumption minimization strategy (ECMS), proportional-integral (PI) control, rule-based approach and genetic algorithm-based strategy for a fuel-cell hybrid electric car are compared and overall efficiency is tabulated. Further, the performance analysis in the real-time platform reveals that the efficiency of MPC controller and fuzzy control strategy gives the highest efficiency as compared to other techniques. Additionally, this analysis illustrates the most cost-effective and fuel-efficient usage of electricity flow. Finally, the comparative performance analysis obtained through MATLAB simulation and real-time implementation claims an effective energy management strategy for FCHEV in a modern energy-efficient environment.

Keywords: Fuel cell hybrid electric vehicle (FCHEV), energy management system (EMS), equivalent consumption minimization strategy (ECMS), fuel cell power module (FCPM)

Introduction

Hybrid fuel cell vehicles commonly incorporate batteries or ultra-capacitors as additional energy sources alongside fuel cells, which serve as their primary power source. Driving circumstances for cars on the road are extremely complicated. They frequently deal with different significant variations and unexpected increases in electricity demand due to crises and changes. Effective energy management is crucial for optimizing the performance and efficiency of Fuel cell hybrid electric vehicles (FCHEVs). Various energy management techniques it has been suggested and implemented to enhance the overall performance of these vehicles. However, there is a need for a comprehensive comparative analysis to analyze the efficacy of these techniques under different driving conditions and scenarios. This paper presents a performance comparison analysis of existing energy management techniques for FCHEVs, aiming to identify the most suitable approach for maximizing energy efficiency and vehicle performance. Recently, proposals for energy management using conventional proportional integral controllers have been made. This approach is according to the use of proportional-integral (PI) controllers to regulate the primary performance variables, such as the battery level of charge or percentage of charge, super-capacitor voltage, or dc-bus voltage [1–3].

It would be wise to implement such an energy management plan to maximize fuel efficiency while ensuring that each source of energy reaches its optimum efficiency potential. In this context a multi power source FCHEV is considered and a DRL-based algorithm is proposed for effective fuel economy [2]. Also, ideally, less influence from the energy management system (EMS) is anticipated on the entire life span of the hybrid or combined power system. The sample study outline in the area of energy management methods is therefore the optimization algorithm used throughout a specific driving cycle [4-6]. Numerous fuel cells combined with various energy sources linked to energy management strategies for power systems discussed in the investigation [7]. The research on state machine control approach [8, 9] describes a simple and efficient rule-driven method for optimizing fuel efficiency.

The remaining portions of this article are organized as: Section-II represents the system description, Section-III comprising of results and discussion and Section-IV concludes the paper.

[a]chatterjeedebasis1@gmail.com

DOI: 10.1201/9781003540199-19

System description

Here we have considered one FCHEV model. Systems that use hybrid technology can be set up in series or parallel. Systems with a series configuration often serve as range extenders because the fuel cell is not electrical engine. More frequently, concurrent systems (Figure 19.1) are used. In this setup, the fuel cell is linked to the DC bus through a one-way converter, which serves to control the power flow.

The block diagram is focused on examining the connections between system components' power flows. Figure 19.1 depicts the power flow relationship and the structure of the vehicle powertrain simulation model; the solid line indicates the power flow. The structure of FCHEV is electric vehicle (EV) driven by the four-wheel motor.

The model, depicted in Figure 19.2 and implemented using MATLAB Simulink software, is an integral component of the proposed system.

A Simulink model of FCHEV is designed. It includes two sources: (i) A fuel cell of 2.4 KW, 48 V_{dc} (ii) A battery of 5.4 Ah (initial SoC 100%) (iii) DC to DC step up converter (iv) A load DC motor is used (v) Controller for fuzzy logic.

Hybrid power system design

The hybrid architecture is tailored to the energy requirements of the system.

The power module made up of fuel cell and the batteries are used in a way so that efficient energy utilization can be done. The design of hybrid power systems for comparative performance analysis of energy management techniques in FCHEVs is critical for optimizing vehicle efficiency and performance [10-12]. These systems typically integrate fuel cells, batteries, and additional energy storage devices are employed to fulfill the fluctuating power requirements of the vehicle. The design process involves selecting appropriate components, sizing

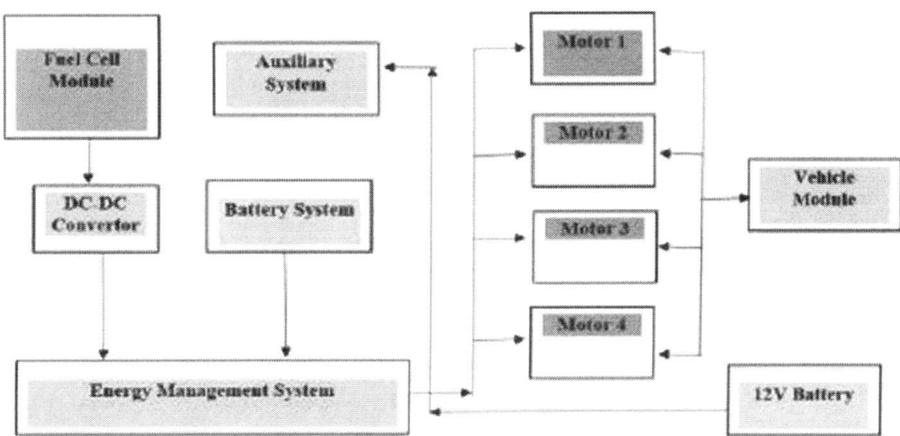

Figure 19.1 FCHEV hybrid electric vehicle block diagram

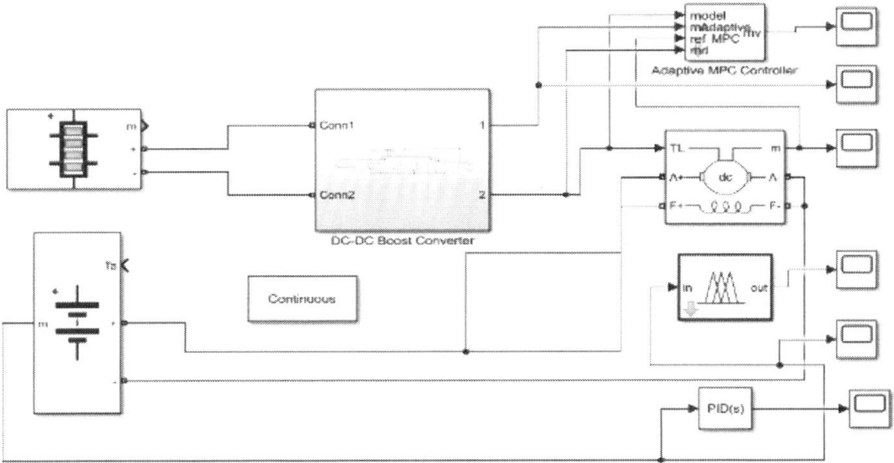

Figure 19.2 MATLAB simulation of FCHEV

them to meet the vehicle's power requirements, and developing control strategies to manage the flow of energy between the different components [13-16]. By analyzing the performance of these systems under different conditions, such as varying loads and driving patterns, researchers can evaluate the effectiveness of different energy management techniques and optimize the design for improved efficiency and performance.

The following are the primary attributes of fuel cell hybrid power system:

- **Fuel-cell power module system**
  It's a liquid-cooled, 2.4 kW, 48 Vdc FCPM. An integrated controller for interacting using the primary controller and safeguards is also included with the FCPM (H_2 low pressure, high temperature, excess current and low voltage). Under equality constraint:

$$P_{Fuel\ Cell} = P_{Load} - P_{Battery} \tag{1}$$

- $P_{Fuel\ Cell}$ is the electrical output generated by the fuel cell
  P_{Load} is the power demand by the vehicle's electrical load, and
- $P_{Battery}$ is the power delivered by the battery.
  The above Equation (1) above illustrates that the power provided by the fuel cell equals the overall power needed by the vehicle's electrical system, minus the power supplied by the battery. This equation is essential for grasping the energy equilibrium and control in FCHEVs.
  The amount power developed to finish the trip:

$$P_d = \int_0^t (P_{fc} + P_B) \tag{2}$$

where, P_{fc} = power developed by fuel cell in that duration

P_B = power developed by the battery

Equation (2) is employed in studying energy consumption and management in FCHEVs. The objective is to efficiently operate the fuel cell and battery to meet the vehicle's power demand while maximizing energy efficiency and performance.

- **The battery system**
  It is made up of series-connected 5.4 Ah, 48-volt lithium-ion battery set. Every module has an inbuilt controller for calculating SOC, monitoring cell temperature, voltage, current, and cell-to-cell balancing (also known as intro module balancing).
- **System of DC-DC converters for fuel cells**
  It comprises of five parallel-connected (40–64 V) dc–dc isolated boost converters. In each module overloading and excessive voltage protection is built into the system. For the creation of a FCEV system, this MATLAB simulation model was created. The reference document is used to help choose the parameter.

Energy management strategies

EMS plays a crucial role in the comparative performance analysis of energy management techniques in FCHEVs. These strategies determine how power is distributed among the various components of the hybrid powertrain, such as the fuel cell, battery, and electric motor, to optimize efficiency and performance. One common strategy is rule-based control, one method involves pre-established rules dictating power allocation influenced by variables like vehicle load, speed and battery state. Alternatively, MPC employs a mathematical model of the vehicle and its components to anticipate future power needs and optimize control signals accordingly. By comparing the performance of these and other energy management strategies, researchers can identify the most effective techniques for improving the efficiency and performance of FCHEVs.

The EMS is required the below mentioned necessities:

- Minimum hydrogen utilization
- Excellent overall system efficiency
- Limited range of the batteries
- Extended life span of all requirements.

The subsequent portions provide a detailed description of the EMS under discussion.

Genetic algorithm

Among the heuristic techniques is the genetic algorithm. They assist in resolving optimization issues. The research society for HEVs has proven that genetic algorithm (GA) is a legitimate technique and a useful method. For FCHEVs, a GA can optimize power distribution between the fuel cell and battery, considering factors like vehicle speed, load, and road conditions. By continuously adjusting the power allocation, the GA can enhance the overall efficiency of the FCHEV system, increasing the vehicle's range. GA has several benefits, which includes the following: (1) it does not require knowledge of the initial circumstances; (2) it targets the global optima; (3) it works with numerous design parameters; (4) it is simple to use; and (5) it does not require knowledge of the objective function.

Fuzzy logic strategy based on rules

Compared to state machine control, this method responds to changes in load faster and is more resistant to measurement inaccuracy. The fuel-cell power is determined based on load power, SOC membership operations, and a series of IF-THEN rules. The design is created using a methodology similar and employs trapezoidal membership functions. Power demand and SOC were the FLC's two inputs, while one FC power fraction was the FLC's output. A collection of rules is used to enter the user knowledge base. The FLC is given instructions on what the output should be based on a set of input conditions using a rule-base. One rule can be "if SOC is high" and "power demand is low" then "FC output is low".

Classical PI control method

This method uses a PI regulator [1] to control the battery SOC. To obtain the fuel-cell reference power, the battery power is subtracted from the load power at the output of the PI regulator. The classical PI control method is commonly used to analyze energy management in FCHEVs. This technique involves balancing power distribution between the fuel cell and battery according to the vehicle's power demand and state of charge. The PI controller computes the error between desired and actual power output, and then adjusts the control signals to minimize this error. By comparing the performance of the PI controller with other control strategies, researchers can evaluate its effectiveness in optimizing the energy management of FCHEVs. The analysis can help identify the strengths and weaknesses of the PI control method and provide insights for further improvements in energy management techniques for FCHEVs. The fuel-cell power decreases when the battery SOC is above the reference, and the battery supply all its available power. Conversely, when the SOC is below the reference, the fuel cell generates nearly the full load power. This approach is relatively easy to implement, and the PI gains are adjusted online for better responsiveness [3-4].

ECMS (equivalent consumption minimization strategy)

To implement either an adaptive or non-adaptive ECMS technique, use the method parameter. Because there is no plugin capability to recharge the battery, HEV systems are charge-sustaining, meaning the battery SOC must stay within a given range. If the change in SOC is kept to a minimum across a drive cycle, the battery serves as an energy buffer and all power originates from the fuel.

Non-adaptive: The ECMS equivalency factor used in the block is constant.

- To find the best fuel efficiency throughout a drive cycle, use this procedure.
- To maintain the ending SOC when changing the drive cycle or HEV architecture, adjust the ECMS weighting factor. The block employs a solitary constant by default.

The vehicle's longitudinal dynamics are:

$$P_d(t) = P_{load}(v(t)) + M_e \cdot v(t) \quad (3)$$

where,

M_e is the vehicle's effective mass in kilograms, P_{load} is the road load power in watts,

which accounts for the effects of gravity, rolling, and drag. The vehicle's speed, v(t), is a state variable in the Equation (3) above. This Equation (3) is used in the context of energy management in FCHEVs to calculate the total power demand considering the electrical power output of the fuel cell and the power required for auxiliary loads and losses. It helps in understanding the total energy usage and optimizing the operation of the FCHEV system

MPC (model predictive control)

MPC is a sophisticated control strategy that utilizes a mathematical model of the vehicle and its components to predict future power demands. By optimizing the control signals based on these predictions, MPC can effectively manage the power distribution between the fuel cell and the battery. This approach allows the MPC controller to adapt to changing driving conditions and optimize the overall efficiency of the FCHEV system. Through comparative analysis, researchers can evaluate the effectiveness of the MPC controller against other energy management techniques, providing valuable insights for improving the performance of FCHEVs

The MPC-EMS is developed to calculate the reference input I_{fc} to minimize fuel consumption while keeping the SoC level constant. It employs nonlinear programming to solve a short-term optimization problem at each time step k. MPC is a recognized control method with the following parameters.

Manipulated variables:

A_0: Power needed for the P_{fc} fuel cell

B_0: Demand for power to the battery P_b

C_0: Dissipated power P_w

Controlled variables:

U_d: Developed power

V_d: SOC of battery

To minimize the cost function over a particular interval is given by:

$$\int C_f = \sum_0^t P_d.t + f(A_0) \quad (4)$$

where,

C_f is cost function

P_d is taken from Equation (3) and

t is duration of running.

$f(A_0)$ is the function of cost input to the system.

To provide a more accurate interpretation or correction, we would need additional information about the context and the meaning of the symbols used in the Equation (4).

III. Results and discussion

On the 2.4-kW fuel-cell hybrid system, simulations and experimental tests were used to compare the EMS's performance. Every plan was executed into action in MATLAB/Simulink. The tests begin with identical beginning circumstances (battery SOC is equals to 70%, measured temperature of the battery unit is 30°C, fuel-cell voltage is 52 V, fuel-cell temperature is 40°C) to ensure identical conditions for comparison. Additionally, each plan was put into place in accordance with the energy management requirement. Therefore, overall effectiveness of the proposed strategy is represented in Table 19.1.

The MPC control strategy offered a little higher efficiency (88.91%) and battery stress is high compared to the other schemes (of 21.91 and 35.30). The PI method used more battery energy (SOC between 60% and 40%) and consumed less fuel (65 g of H_2). The rule-based fuzzy logic system produced, as predicted, the lowest fuel-cell stress and minimizing battery usage (SOC between 58% and 41%) leads to increased fuel consumption (63.4 g of H_2 consumed) and reduced overall efficiency (81.32%).

To verify the results, the proposed strategy is tested in a real-time energy management simulator. After the test is performed, comparative data sets are collected and tabulated in Table 19.2. The MPC control strategy offered a little higher efficiency (80.27%) and battery stress compared to the other schemes (of 21.91 and 36.7, respectively). The PI method used additional battery power (SOC between 70% and 55%) and consumed fuel (68 g of H_2). The rule-based fuzzy logic system produced, as predicted, the lowest fuel-cell stress and minimizing battery usage (SOC between 70% and 55%) results in higher fuel consumption (66 g of H_2 consumed) and lower overall efficiency (79.32%). The simulation studies' results reflect these trade-offs.

Conclusions

This paper analyses the benefits and drawbacks of various energy management systems, as well as their primary functions, then summarizes and concludes their current state. Modern energy management techniques based on intelligent vehicle connection technologies are briefly introduced here. The consumption of hydrogen, battery SOC, overall efficiency, and the amount of stress each energy source experiences are the performance comparison criteria. The latter is evaluated by use of an innovative strategy using the wavelet transform. To conclude, to optimize FCHEV performance, a multi-scheme EMS should be employed, where each scheme is selected based on specific criteria for prioritization.

Table 19.1 Overall performance of each EMS scheme in MATLAB/Simulink

EMS criteria	MPC control	ECMS	Genetic algorithm	Rule-based fuzzy logic	PI control
Battery SOC (%)	64–40	62–44	62–47	58–41	60–40
H2 consumption	67.2	72.5	88	63.4	65
Overall efficiency	88.91	80.11	73.91	81.32	78.91
Battery stress	35.30	24.26	32.77	27.22	32.03

Table 19.2 Overall performance of each EMS scheme in real-time test bench

EMS criteria	MPC control	ECMS	Genetic algorithm	Fuzzy logic-based strategy	PI control
Battery SOC (%)	70–60	70–58	70–60	70–55	70–55
H2 Consumption	58.5	63.5	74	66	68
Overall Efficiency (%)	80.27	80.20	79.12	80.01	76.23
Battery Stress	36.70	35.01	35.92	35.04	34.7

- If reducing fuel consumption is the objective, a method based on conventional MPC and fuzzy may be selected.
- The next area for research is the development of an enhanced multi-objective EMS is utilized to optimize all performance metrics.
- In addition, the stability of the MPC controller is much more acceptable than that of other controllers.

References

1. Chen, X., Li, T., Shen, J., and Hu, Z. (2017). From structures, packaging to application: A system-level review for micro direct methanol fuel cell. *Renew. Sustain. Energy Rev.*, 80, 669–678. https://doi.org/10.1016/j.rser.2017.05.272.
2. Onori, S., Serrao, L., and Rizzoni, G. (2016). Hybrid electric vehicles: Energy management strategies. Springer: Berlin/Heidelberg, Germany.
3. Pisu, P. and Rizzoni, G. (2007). A comparative study of supervisory control strategies for hybrid electric vehicles. *IEEE Trans. Control Sys. Technol.*, 15(3), 506–518.
4. Sabri, M., Danapalasingam, K., and Rahmat, M. (2016). A review on hybrid electric vehicles architecture and energy management strategies. *Renew. Sustain. Energy Rev.*, 53, 1433–1442.
5. Mukhitdinov, A. A., Ruzimov, S. K., and Eshkabilov, S. L. (2006). Optimal control strategies for CVT of the HEV during a regenerative process. *IEEE Conf. Elec. Hybrid Veh.*, 1–12.
6. Zhang, D. H., Zhou, Y., Liu, K. P., and Chen, Q. Q. (2009). A study on fuzzy control of energy management system in hybrid electric vehicle. *Power Energy Engg. Conf.* 1–4.
7. Yang, S. C., Yi, M., Xu, B., Guo, B., and Zhu, C. G. (2010). Optimization of fuzzy controller based on genetic algorithm. *Inter. Conf. Intell. Sys. Design Engg. Appl.*, 1–8.
8. Biswas and Emadi, A. (2019). Energy management systems for electrified powertrains: State-of-the-art review and future trends. *IEEE Trans. Veh. Technol.*, 68(7), 6453–6467.
9. Garcia, P., Fernandez, L. M., Garcia, C. A., and Jurado, F. (2010). Energy management system of fuel-cell-battery hybrid tramway. *IEEE Trans. Ind. Electron.*, 57(12), 4013–4023.
10. Ke, J., Xinbo, R., Mengxiong, Y., and Min, X. (2009). A hybrid fuel cell power system. *IEEE Trans. Ind. Electron.*, 56(4), 1212–1222.
11. Caux, S., Hankache, W., Fadel, M., and Hissel, D. (2010). On-line fuzzy energy management for hybrid fuel cell systems. *Int. J. Hydrogen Energy*, 35(5), 2134–2143.
12. Chun-Yan, L. and Guo-Ping, L. (2009). Optimal fuzzy power control and management of fuel cell/battery hybrid vehicles. *J. Power Sources*, 192(2), 525–533.
13. Jayakumar, A. C. and Lie, T. T. (2017). Review of prospects for adoption of fuel cell electric vehicles in New Zealand. *IET Elect. Sys. Trans.*, 7(4), 259–266.
14. Gözüküçük, M. A. and Teke, A. (2011). A comprehensive overview of hybrid electric vehicle: Powertrain configurations, powertrain control techniques and electronic control units. *Energy Convers. Manag.*, 52(2), 1305–1313.
15. Rodatz, P., Paganelli, G., Sciarretta, A., and Guzzella, L. (2005). Optimal power management of an experimental fuel cell/supercapacitor-powered hybrid vehicle. *Control Engg. Pract.*, 13(1), 41–53.
16. Veer, K., Hari, S., Bansal, O., and Singh, D. (2019). A comprehensive review on hybrid electric vehicles: Architectures and components. *J. Modern Trans.*, 27(2), 77–107.

20 Implementation of image segmentation-based stages classification technique for glaucoma infected images

Amarjit Roy[1,a], Vijaya Kumar Velpula[2,b], Ambavaram Sarihariharnadhareddy[2,c], Lakhan Dev Sharma[2,d] and Diksha Sharma[3,e]

[1]Department of Electrical Engineering, Ghani Khan Choudhury Institute of Engineering Technology, Narayanpur, Maligram, 732141, West Bengal, India

[2]School of Electronics Engineering, VIT-AP University, Amaravati, AP, 522237, India

[3]Department of Nanoscience and Technology, Central University of Jharkhand, Ranchi, India

Abstract

In the recent advancement of biomedical research, glaucoma is considered one of the important research problems with respect to its severity. The main objective of this work is a calculation of the cup-to-disk ratio by isolating the relevant parts of retinal images achieved using segmentation of optic cup and optic disk. This proposed work has utilized adaptive thresholding based segmentation, k-means clustering-based segmentation along with region based segmentation for calculation of geometrical parameters of cups and disk which is further involved to calculate cup-to-disk ratio. Once, the classification using the segmentation based multistage glaucoma detection is ready, it can be employed for higher level of research especially in the field of machine learning deep neural networks. It has been observed from performance analysis that it provides better performance than some of the state-of-art techniques.

Keywords: Cup-to-disc ratio, fundus image, different stages of glaucoma, image segmentation, deep neural network

Introduction

Glaucoma, diabetic retinopathy, and macular degeneration are considered as one of the most critical diseases related to the human eye. It is observed that the optic nerve is damaged due to Glaucoma, however, blood vessels in the retina are affected due to diabetic retinopathy to a large extent. Apart from these, in the case of macular degeneration, a thin layer of tissue is pulled away from the normal position. All these diseases may lead to permanent loss of vision if the diagnosis is not done properly and at the proper time. One of the biggest challenges in the present scenario is the rural health care system since around 70% of Indian people belong to rural places in India. The incorporation of new technologies in healthcare systems will be beneficial not only in urban areas but also in rural areas. It has been observed from a survey that awareness of diabetic retinopathy is 27% and for glaucoma is very poor around 2.3% in an urban area. This implies that awareness in rural areas will be almost at a negligible level among the people.

Cheng et al. have proposed a concept that the optic disc and optic cup segmentation by employing the classification of super pixel over glaucoma screening [1]. During optical disk segmentation, histograms and center-surround statistics are employed for the classification of super pixels as disk or non-disk. Diaz-Pinto et al. proposes deep convolutional GANs based on a new retinal image synthesizer along with a semi-supervised learning methodology for the assessment of glaucoma [2]. This system method generates synthetic images and provides automatic labeling of images. AG-CNN (attention-based CNN) proposed by Li et al. who introduces a new methodology that can avoid the drawbacks of state-of-the-art techniques [3]. Due to the high redundancy of present approaches in the classification of glaucoma, the reliability and accuracy of glaucoma detection may be reduced using this proposed work. An automatic two-stage glaucoma detection was proposed by Sreng et al. to diminish the assignment of ophthalmologists [4]. This procedure makes use of a DeepLabv3+ architecture, which segments the optic disc area before switching out the encoder module for a cluster of deep convolutional neural networks. Elangovan et al. proposes statistically segmentation-based glaucoma detection in which parameters are computed from the optic disk and cup of color fundus images [5]. This work considers an enhanced FCM algorithm (FRFCM) based on morphological reconstruction and membership filtering for optic disc and optic cup segmentation. In addition to this, we measure kurtosis, cup entropy, and rim entropy, among other statistical features. By using these,

[a]roymarjit90@gmail.com, [b]vijaya.20phd7043@vitap.ac.in, [c]sarihariharn.18bec7103@vitap.ac.in, [d]lakhan.sharma@vitap.ac.in, [e]diksha.hec@gmail.com

DOI: 10.1201/9781003540199-20

glaucoma and non-glaucoma patients can be separated. Adal et al. suggests a strategy by analyzing a routine screening program for diabetic retinopathy, which encompasses a broad spectrum of DR levels resulting from low red blood cell counts, from normal to moderate (with clinically relevant retinal lesions). [6]. Li et al. proposes a deep convolutional neural network (DCNN)-based novel algorithm in context to this research [7]. In this approach, fractional max-pooling replaces the generally used max-pooling layers for this purpose which is different from the traditional one. To generate more discriminating features, two of these DCNNs are trained and tested to get better results along with accuracy for getting a greater number of discriminative features. The suggested DR classifier categorizes the DR phases into five groups by labeling an integer ranging from 0 to 4. In their comprehensive study of diabetic retinopathy categorization. Mateen et al. suggests three main areas: the development of retinal datasets, algorithms for detecting diabetic retinopathy (DR), and metrics for evaluating performance [8]. Based on the literature proposed by Lin et al., a dataset containing 1219 fundus images (from DR patients and healthy controls) with observations of exudate lesions has been constituted in this work [9]. Besides this, left versus right eye labels, DR grade (severity scale) from three different grading protocols, the bounding box of the optic disc (OD), and fovea location-related works are also done in this methodology in their study. Qiao et al. suggest a method for analyzing fundus images for micro aneurysms [10]. The method involves convolutional neural network algorithms that incorporate deep learning as a core component that can detect and segment medical images with low latency and high performance. Apart from these, a lot of recent state-of-the-art has been surveyed in the context of image classification and machine learning [11–17].

Considering the different state-of-the-art techniques, a limited number of works based on segmentation are available in the literature on the related research topic. So, segmentation-based approaches such as adaptive thresholding segmentation, k-means clustering segmentation, and region-based segmentation have been utilized during the formulation of the cup-to-disk ratio. Based on the ratio of cup-to-disk, stages of glaucoma can be defined [18–25]. Thus, it will be very helpful from the medical aspect to determine the stage in which the patient exists. These stages of classification will be further applicable for machine or deep neural network-based research in which pre-defined classification is very helpful.

The rest of the paper is organized with datasets used in this work in Section II, segmentation-based stages classification in Section III, results and discussion in Section IV followed by conclusions in Section V.

Datasets used in this work

The following two datasets which are publicly available are used in this proposed work.

STARE: Dr. Michael Goldbaum at UC created the STARE dataset. The dataset contains 400 raw pictures (A. Hoover et al., 2000). Text files with expert commentary on each image's appearance. During data collection, specialists were asked about 44 manifestations, which were coded into 39 values. 40 hand-annotated blood vessel images, results, and demonstration. Two experts label 10 images of arteries or veins.

DRIVE: Images used in the DRIVE database were from a Dutch initiative to detect diabetic retinopathy. Four hundred people with diabetes, ranging in age from 25 to 90, made up the screening population. Of the 40 images chosen at random, seven reveal mild early diabetic retinopathy and thirty-three do not. The images were taken with a 45° field of view (FOV) using a Canon CR5 non-mydriatic 3CCD camera. The resolution of each picture was 768×584 pixels, and 8 bits were used for each color plane. There are two sets of 20 photographs each: one for training and one for testing, from a total of 40 images.

Segmentation-based stages classification

Three different types of segmentation have been considered in this proposed work such as adaptive thresholding segmentation, k-means clustering segmentation, and region-based segmentation. Using these segmentation techniques, stages of glaucoma have been defined and classified.

Segmenting an image using the adaptive threshold-based segmentation method is as simple as assigning a foreground value to each pixel whose intensity is greater than a pre-fixed threshold, and then resetting the intensities of all other pixels to their ground-level values. In adaptive threshold-based segmentation, the threshold values are varied as the operation progresses over the image, in contrast to the traditional thresholding operator.

k-means clustering algorithm based on unsupervised learning works on the images with no labeled data available in the database. Using the similarity, classes are identified to form classes or clusters within the given dataset available in the input set. Datasets with similar patterns are clustered in the same class and another dataset in the other. In this algorithm, the number of clusters is defined by k.

In the region-based algorithm, neighboring pixels are grown surrounding the similar and connected pixels to perform the segmentation. Here, the top–down approach or bottom–up approach is performed considering the seed pixel values in the region growing approach. Thus, segmentation is performed using a similar measure followed by region region-merging technique.

For the computation of CDR, the guide approach makes use of the ratio of the vertical diameter of OC and OD. Using the segmentation of OD, the minimal row coordinate is calculated as y_{min1}, and the most row coordinate is calculated as y_{max1}. The Euclidian distance among those coordinates is the vertical diameter of the OD, OD_{eu}.

$$OD_{eu} = y_{max1} - y_{min1} \quad (1)$$

Similarly, after the segmentation of OC, from OC the minimal row coordinate is calculated as y_{min2}, and most row coordinate is calculated as y_{max2}. The Euclidian distance among those coordinates is the vertical diameter of the OC, OC_{eu}.

$$OC_{eu} = y_{max2} - y_{min2} \quad (2)$$

So, the formulated CDR is as follows:

$$CDR = OC_{eu} / OD_{eu} \quad (3)$$

Based on the above calculated CDR, the classes are defined as normal, stage 1, stage 2, and stage 3 which are as follows:

- If CDR<=0.3 then it is treated as normal eye
- If CDR > 0.3 to CDR<=0.5 then it is treated as beginner level (stage 1)

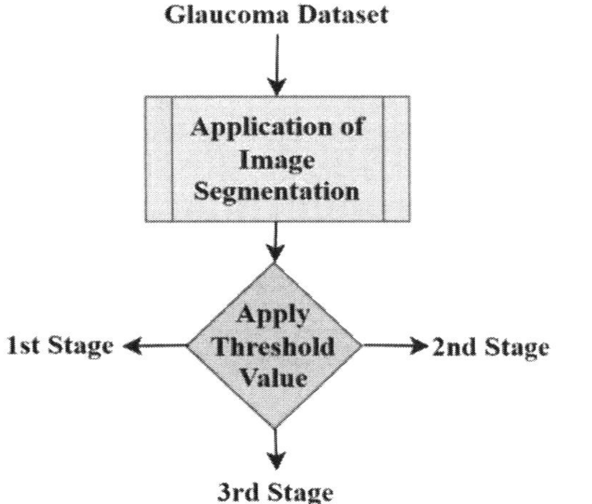

Figure 20.1 Basic flowchart of proposed technique

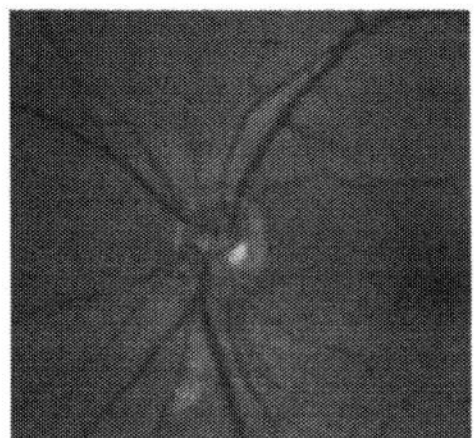

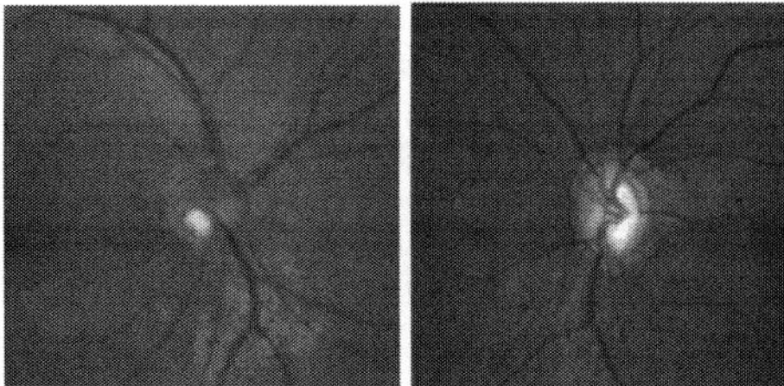

Figure 20.2 Different stages of glaucoma detection

Table 20.1 Tabular representation various stages

S. No.	Image number	CDR	Stages
1	Image 1	0.13	Normal
2	Image 3	0.14	Normal
3	Image 5	0.17	Normal
4	Image 6	0.19	Normal
5	Image 7	0.2	Normal
6	Image 9	0.10	Normal
7	Image 11	0.26	Normal
8	Image 13	0.23	Normal
9	Image 14	0.15	Normal
10	Image 19	0.13	Normal
11	Image 20	0.23	Normal
12	Image 21	0.13	Normal
13	Image 23	0.61	Stage2
14	Image 25	0.14	Normal
15	Image 27	0.27	Normal
16	Image 28	0.15	Normal
17	Image 29	0.2	Normal
18	Image 30	0.33	Stage1
19	Image 34	0.20	Normal
20	Image 39	0.31	Stage1
21	Image 43	0.22	Normal
22	Image 48	0.15	Normal
23	Image 50	0.32	Stage1
24	Image 52	0.11	Normal
25	Image 53	0.19	Normal

- If CDR > 0.5 to CDR<= 0.7 then it is treated as moderate level (stage 2)
- If CDR > 0.7 it is treated as severe level (stage 3).

The basic flowchart of the proposed technique has been presented in Figure 20.1. It is observed that starting from capturing of glaucoma dataset, the stages are classified subsequently based on the threshold values.

Results and discussions

The performance of the proposed technique has been shown in this section. The visual observation of the results has been presented in Figure 20.2. It is seen that the size of the cup has increased at the advanced stage in comparison to the early stage.

The performance of the proposed technique in terms of CDR values along with stages has been shown in Table 20.1. It has been observed that the stages have been increased with incremental values of CDR.

Conclusions

This paper has proposed stages classification of glaucoma using image segmentation. Three different techniques: adaptive thresholding segmentation, k-means clustering segmentation, and region-based segmentation are combined and applied for C/D ratio. Also, a maximum of the contemporary strategies had been examined on a constrained quantity of datasets inclusive of DRIVE and STARE. These datasets do now no longer offer pictures with many unique characteristics. Even more problematic for the segmentation process is the often poor photo resolution, which ranges from 0.4 to 0.3 megapixels. The suggested method achieved an extraction rate of 96%, a concordance rate of 86%, and an over-extraction rate of 11% compared to the disc regions determined by an ophthalmologist. In this study, we additionally offered a technique for spotting glaucoma via way of means of calculating the C/D ratio. The approach effectively recognized 71% of glaucoma instances and 83% of everyday instances. Thus, its miles cautioned that this approach is beneficial in the glaucoma evaluation of retinal fundus pictures.

References

1. Cheng, J. et al. (2013). Superpixel classification based optic disc and optic cup segmentation for glaucoma screening. *IEEE Trans. Med. Imag.*, 32(6), 1019–1032.
2. Diaz-Pinto, A. C., Naranjo, V., Morales, S., Xu, Y., and Frangi, A. F. (2019). Retinal image synthesis and semi-supervised learning for glaucoma assessment. *IEEE Trans. Med. Imag.*, 38(9), 2211–2218.
3. Li, L. et al. (2020). A large-scale database and a CNN model for attention-based glaucoma detection. *IEEE Trans. Med. Imag.*, 39(2), 413–424.
4. Sreng, S., Maneerat, N., Hamamoto, K., and Win, K. Y. (2020). Deep learning for optic disc segmentation and glaucoma diagnosis on retinal images. *Appl. Sci.*, 10(14), 4916.
5. Elangovan, P., Nath, M. K., and Mishra, M. (2020). Statistical parameters for glaucoma detection from color fundus images. *Proc. Comp. Sci.*, 171, 2675–2683.
6. Adal, K. M., van Etten, P. G., Martinez, J. P., Rouwen, K. W., Vermeer, K. A., and van Vliet, L. J. (2018). An automated system for the detection and classification of retinal changes due to red lesions in longitudinal fundus images. *IEEE Trans. Biomed. Engg.*, 65(6), 1382–1390.
7. Li, Y. H., Yeh, N. N., Chen, S. J., and Chung, Y. C. (2019). Computer-assisted diagnosis for diabetic retinopathy based on fundus images using deep convolutional neural network. *Mobile Inform. Sys.*
8. Mateen, M., Wen, J., Hassan, M., Nasrullah, N., Sun, S., and Hayat, S. (2020). Automatic detection of diabetic retinopathy: A review on datasets, methods and evaluation metrics. *IEEE Acc.*, 8, 48784–48811.
9. Lin, L., Li, M., Huang, Y., Cheng, P., Xia, H., Wang, K., Yuan, J., and Tang, X. (2020). The SUSTech-SYSU dataset for automated exudate detection and diabetic retinopathy grading. *Sci. Data*, 7(1), 1–10.
10. Qiao, L., Zhu, Y., and Zhou, H. (2020). Diabetic retinopathy detection using prognosis of microaneurysm and early diagnosis system for non-proliferative diabetic retinopathy based on deep learning algorithms. *IEEE Acc.*, 8, 104292–104302.
11. Chaitanya, M. K., Sharma, L. D., Rahul, J., Sharma, D., and Roy, A. (2013). Artificial intelligence based approach for categorization of COVID-19 ECG images in presence of other cardiovascular disorders. *Biomed. Phy. Engg. Exp.*, 9(3), 035012.
12. Chaitanya, M. K., Sharma, L. D., Roy, A., and Rahul, J. (2022). A review on artificial intelligence for electrocardiogram signal analysis. *Big Data Analyt. Artif. Intell. Healthcare Indus.*, 38–72.
13. Roy, L., Sharma, D., and Shukla, A. K. (2022). Multiclass CNN-based adaptive optimized filter for removal of impulse noise from digital images. *Visual Comp.*, 1–14.
14. Roy, A. and Laskar, R. H. (2016). Multiclass SVM based adaptive filter for removal of high-density impulse noise from color images. *Appl. Soft Comput.*, 46, 816–826.
15. Sharma, L. D. and Rahul, J. (2022). Automatic cardiac arrhythmia classification based on hybrid 1-D CNN and Bi-LSTM model. *Biocybernet. Biomed. Engg.*, 42(1), 312–324.

16. Velpula, V. K. and Sharma, L. D. (2023). Multi-stage glaucoma classification using pre-trained convolutional neural networks and voting-based classifier fusion. *Front. Physiol.*, 14, 1175881.
17. Velpula, V. K. and Sharma, L. D. (2023). Automatic glaucoma detection from fundus images using deep convolutional neural networks and exploring networks behaviour using visualization techniques. *SN Comp. Sci.*, 4(5), 487.
18. Qureshi, M., Khan, A., Sharif, M., Saba, T., and Ma, J. (2020). Detection of glaucoma based on cup-to-disc ratio using fundus images. *Inter. J. Intell. Sys. Technol. Appl.*, 19(1), 1–16.
19. Park, K., Kim, J., and Lee, J. (2020). Automatic optic nerve head localization and cup-to-disc ratio detection using state-of-the-art deep-learning architectures. *Sci. Rep.*, 10(1), 1–10.
20. Al-Bander, W. A.-N., Al-Taee, M. A., and Zheng, Y. (2017). Automated glaucoma diagnosis using deep learning approach. *2017 14th Inter. Multi-Conf. Sys. Sig. Dev. (SSD)*, 207–210.
21. Sivaswamy, J., Krishnadas, S., Joshi, G. D., Jain, M., and Tabish, A. U. S. (2014). Drishtigs: Retinal image dataset for optic nerve head (onh) segmentation. *2014 IEEE 11th Inter. Symp. Biomed. Imag. (ISBI)*, 53–56.
22. Nawaz, M., Nazir, T., Javed, A., Tariq, U., Yong, H.-S., Khan, M. A., and Cha, J. (2022). An efficient deep learning approach to automatic glaucoma detection using optic disc and optic cup localization. *Sensors*, 22(2), 434.
23. Sevastopolsky. (2017). Optic disc and cup segmentation methods for glaucoma detection with modification of u-net convolutional neural network. *Patt. Recogn. Image Anal.*, 27(3), 618–624.
24. Chinnasarn, K. (2021). Early stage glaucoma detection using adaptive geometric ellipse method. *SN Comp. Sci.*, 2(4), 1–7.
25. Panda, R., Puhan, N. B., Mandal, B., and Panda, G. (2021). Glauconet:patch-based residual deep learning network for optic disc and cup segmentation towards glaucoma assessment. *SN Comp. Sci.*, 2(2), 1–17.

21 Design and performance evaluation of cost-effective light weight dual stator five-phase PMSG for renewable energy application

Raja Ram Kumar[1,a], *Amit Mistri*[1,b], *Mrinmay Manna*[1,c], *Gopal Mondal*[1,d] *and Awani Bhushan*[2,e]

[1]Department of Electrical Engineering, Ghani Khan Choudhury Institute of Engineering & Technology, Malda, India

[2]School of Mechanical Engineering, Vellore Institute of Technology; Chennai, India

Abstract

In order to support renewable energy applications, the primary goal of this paper is to design and evaluate the performance of the dual stator five-phase light weight permanent magnet synchronous generator (DSFPLW-PMSG). The generator incorporates various components, including two sets of multi-phase windings and a dual stator, resulting in a high-power density and reliable system. Additionally, cost-effectiveness and reduced weight are achieved through the implementation of shaft and rotor ventilating ducts, which also provide ample cooling for the PMs. To design and evaluate the generator's performance, the finite element method is chosen for its superior accuracy. Magneto-static analysis enables the examination of flux lines and flux density within the generator, while transient analysis is used for assessing the no-load and load performances. Given its exceptional power density, reliability, and dynamic behavior, the proposed generator proves to be highly suitable for the aforementioned applications.

Keywords: Dual stator, FEM, multi-phase, PMSG, renewable energy

Introduction

Generating power has always been a challenging endeavor. The demand for electricity has surged due to the growing population. Primarily, electricity production relies on non-renewable energy resources like coal, crude oil, and diesel. Unfortunately, this conventional method of fossil fuel-based power generation has led to environmental pollution. Harmful gases such as sulfur-dioxide, carbon-monoxide, carbon-dioxide, sulfur-trioxide, and chlorofluorocarbons are emitted during fuel-based power generation. These gases pose significant dangers to both human health and the environment. Additionally, non-renewable energy resources are limited. Consequently, there is a concerted effort to generate electricity from renewable energy sources due to their environmentally-friendly and non-hazardous nature [1, 2]. The shift towards sustainable power generation is exemplified by renewable energy sources like wind, hydro, tidal, solar, and geothermal. Among these sources, wind energy is particularly valuable as it can be harnessed during both day and night, making it a reliable option. Furthermore, wind energy generation is more economically viable. A wind farm consists of various components, including wind turbines, towers, generators, and controlling grids [3, 4]. The generator, in particular, plays a crucial role.

There are various generator alternatives available for wind turbines, such as the permanent magnet synchronous generator (PMSG), doubly fed induction generator, and field-excited synchronous generator. Among these choices, PMSG stands out due to its exceptional advantages. These include a full speed range, brushless technology requiring minimal maintenance, and the elimination of gears [5, 6]. PMSG systems come in a variety of topologies, such as single rotor single stator, single stator dual rotor, and single rotor dual stator systems. Among them, the single rotor dual stator configuration stands out for its high-power density and suitability for radial flux generators. Furthermore, utilizing a multi-phase system enhances generator power density and reliability. Multi-phase systems, with more than three phases, provide benefits such as improved reliability and reduced per-phase current. Building upon the advantages of a dual stator structure and multi-phase system, this paper proposes a dual stator five-phased lightweight PMSG. The generator consists of various components, including dual stator, single rotor, rotor ventilation duct, shaft duct, and five-phased windings for each stator. The inclusion of multi-phase winding and dual

[a]rajaram@gkciet.ac.in, [b]amitmistrinew@gmail.com, [c]mrinmaymanna01@gmail.com, [d]gopalmondal1909@gmail.com, [e]awani.bhushan@vit.ac.in

DOI: 10.1201/9781003540199-21

stator enables the generator to achieve high power density and reliability, while the rotor ventilation duct improves cost-effectiveness and aids in cooling the permanent magnet. The design and performance evaluation of this generator are done using the finite element approach because of its high degree of precision [7, 8]. The paper is structured as follows: Section 2 offers an insight into the model structure and design specifics. Section 3 displays findings from magneto-static and transient analyses, and Section 4 concludes with final remarks.

Model overview

Figure 21.1 depicts a DSFP PMSG, consisting of an inner and outer dual stator with a single rotor. Each stator periphery is equipped with 45 slots and features two sets of five-phase 8-pole windings. The rotor, situated between the two stators, consists of two sets of 8-pole permanent magnets made of NdFeB grade material, positioned around the rotor surface. The rotor structure has been modified in order to improve the permanent magnets' thermal stability. Specifically, the rotor structure is equipped with 8 holes strategically placed to facilitate effective cooling of the generator's permanent magnets. In order to lower the generator's total weight, 4 holes are also added to the shaft. The authors introduced a DSFPLW-PMSG, illustrated in Figure 21.2. The rating and detailed specifications of the proposed generator can be found in Table 21.1 and Table 21.2, respectively.

FEM results and discussion

The design and performance evaluation of the suggested DSFPLW-PMSG involves the utilization of FEM analysis. This entails magneto-static analysis for acquiring flux line distribution and flux density information within the generator. Furthermore, transient analysis is applied to evaluate the generator's operational efficiency in scenarios of both no-load and load conditions. To achieve accurate results, the DSFPLW-PMSG model is divided into finite elements. In total, 360,529 meshes have been created for this particular PMSG.

Figure 21.3 illustrates the flux-lines in the proposed generator, where the flux lines clearly demonstrate the presence of 8 poles in the magnet. Additionally, the flux-density information in the lightweight model of the PMSG is depicted in Figure 21.4.

Figure 21.5 showcases the relationship between magnetic field-density and electrical angle (in

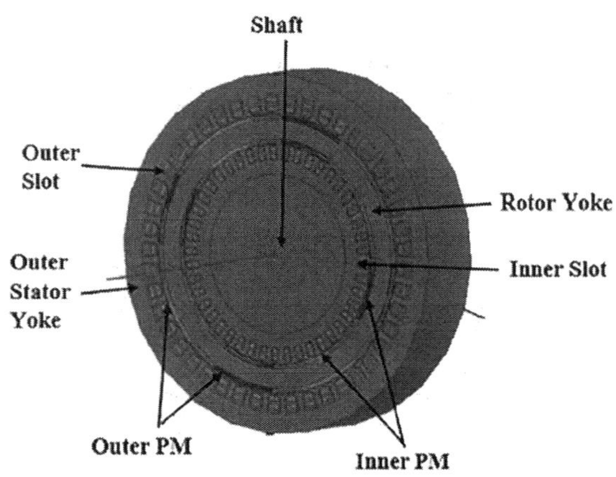

Figure 21.1 Model of DSFP PMSG

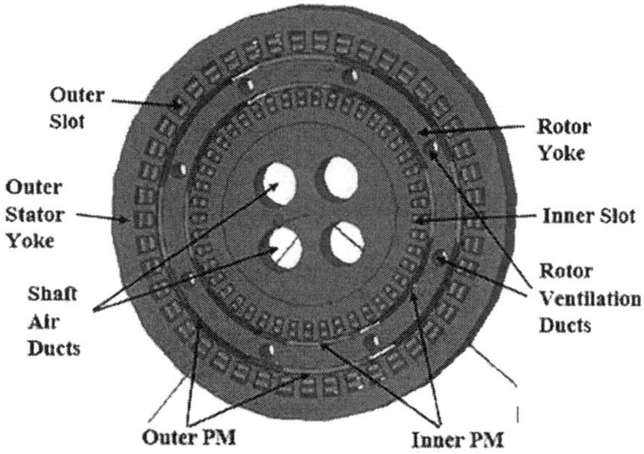

Figure 21.2 Model of DSFPLW-PMSG

Table 21.1 Rating of proposed generator

Parameters	Rating
Power (KW)	1.39
Outer-stator-voltage (Volts)	200
Inner-stator-voltage (Volts)	110
Current (Amps)	5
Phases	5
Speed (RPM)	375

Table 21.2 Dimensions of the generator

Parameters\	Value (mm)
External-radius of outer-stator	180
Internal-radius of outer-stator	140
External-radius of inner-stator	110
Thickness of the rotor-yoke	20
Thickness of the magnets	4
Air-Gap (both)	1
Shaft's radial measurement	80

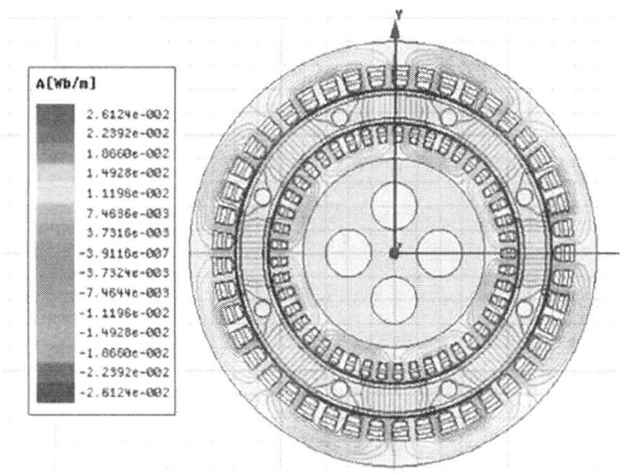

Figure 21.3 Flux line distribution in DSFPLW-PMSG

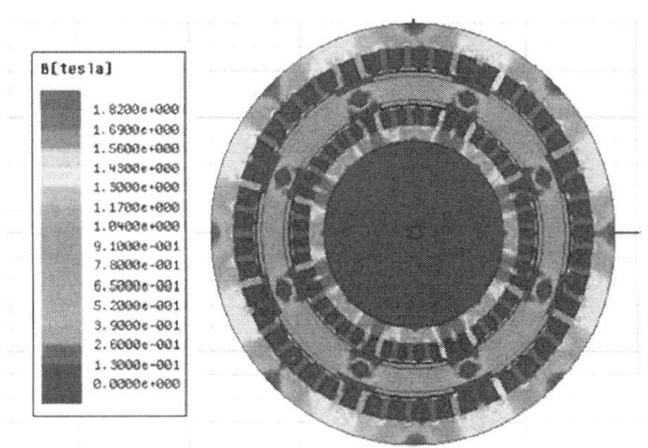

Figure 21.4 Magnetic field density distribution in DSFPLW-PMSG

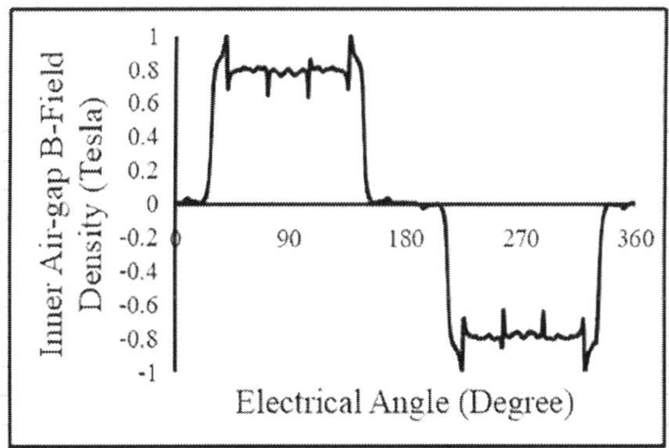

Figure 21.5 Inner B-density

degrees). Notably, the measured flux-density inside the inner air gap is 0.8 T. Furthermore, Figure 21.6 displays the outer air-gap magnetic field density as a function of the electrical angle. The calculated average air-gap flux density amounts to 0.84 T.

In Figure 21.7, there is a plot illustrating the correlation between the electromagnetic force (EMF) of the inner stator and the time period. The

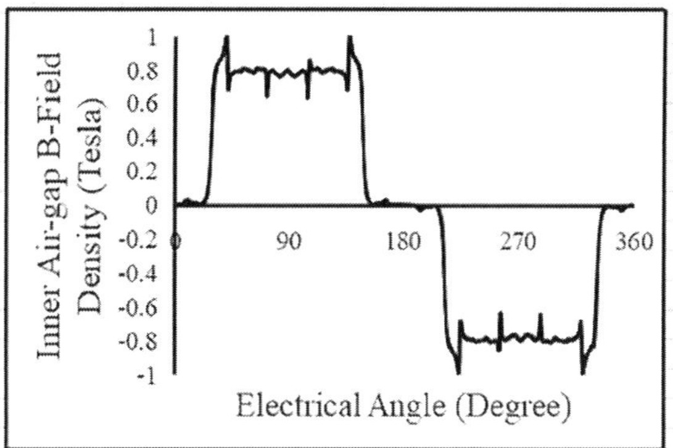

Figure 21.6 Outer B-density

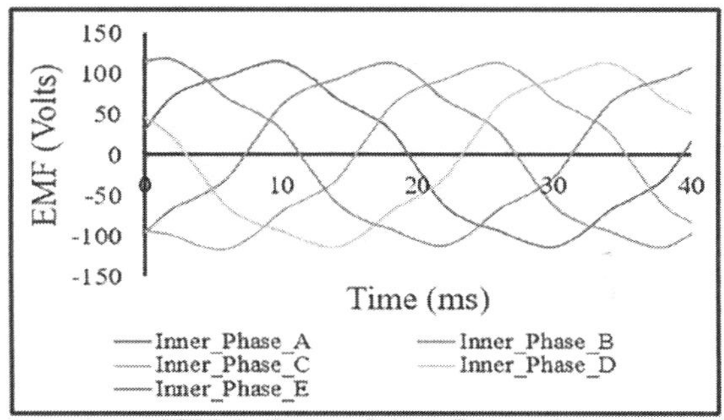

Figure 21.7 Inner stator output voltage wave form

terminal voltage of the inner stator is measured at 110 volts, with a corresponding time period of 40 ms. Consequently, the output voltage waveform has a frequency of 25 Hz. Moving on to Figure 21.8, it displays the graphical plot showcasing the outer stator electromagnetic force (EMF) as a function of the time period. Upon measurement, the voltage across the outer stator is determined to be 200 volts, and the corresponding time period is 40 ms. As a result, the frequency of the output voltage waveform is determined to be 25 Hz.

Figure 21.9 presents a plot illustrating the relationship between current and terminal voltages. The no-load terminal voltage of the inner stator is measured at 110 volts with a voltage regulation of 9.32%. The terminal voltage under no-load conditions for the outer stator is likewise established at 200 volts, demonstrating a voltage regulation of 10.86%. The examination indicates superior voltage regulation in the inner stator system compared to the outer stator system. Figure 21.10 displays the efficiency curve (% efficiency vs. load current) of the DSFPLW-PMSG, with the peak efficiency reaching 94.83% at a current of 2 amps, while the efficiency at the rated current is ascertained to be 91.79%.

A comparative study of different parameters between DSFP PMSG and DSFPLW-PMSG are

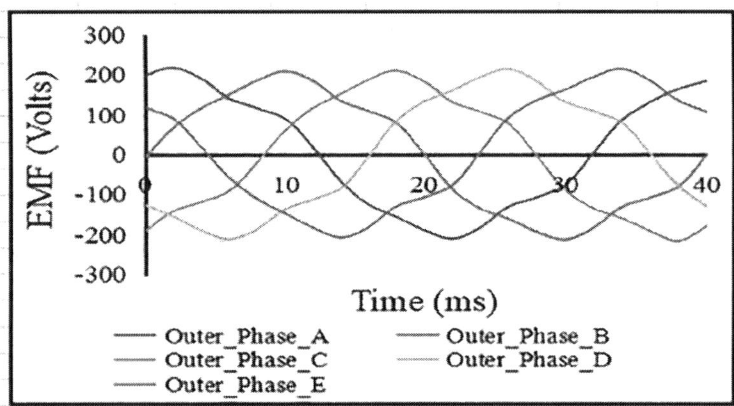

Figure 21.8 Outer stator output voltage wave form

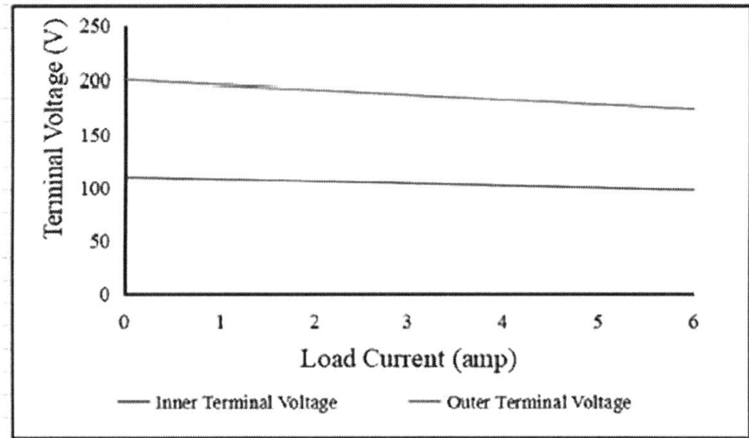

Figure 21.9 Load current vs. terminal voltage

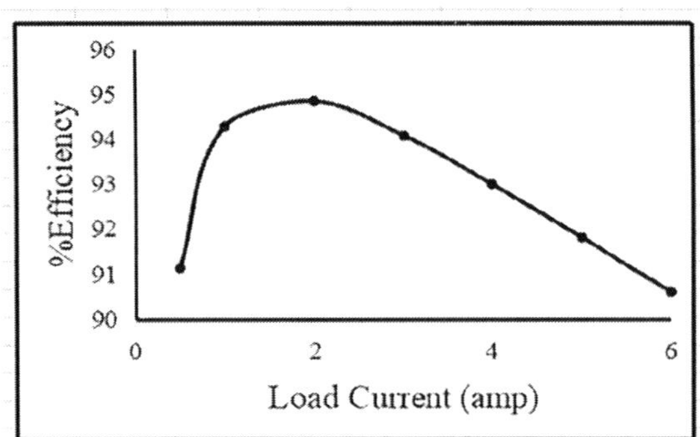

Figure 21.10 Load current vs. %efficiency

listed in Table 21.3. In comparison to DSFP PMSG the weight of proposed generator has been reduced by 3.89% whereas the power density improved by 1.42%. The proposed generators also having the 1.6785% lower value of rotor inertia (0.7604 kg.m^2) than DSFP PMSG.

Conclusions

The paper emphasizes on the design and performance analysis of a DSFPLW-PMSG, intended for renewable energy applications. FEM analysis is

Table 21.3 Comparison of DSFP PMSG and DSFPLW-PMSG

S. No.	Parameter	DSFP PMSG	DSFPLW-PMSG	% Change
1	Weight of generator	108.5834	104.3614	-3.8882
2	Power density (Kwatt/m^3)	34.408	34.894	1.41246
3	Inertia of the rotor (Kg.m^2)	0.77337	0.7604	-1.6785

opted for due to its exceptional accuracy, ensuring reliable result. The examination reveals that the outer stator tooth encounters the most elevated flux density, while staying within the bounds of magnetic material saturation. Additionally, the inner stator winding yields a no-load voltage of 110 V, whereas the outer stator winding achieves 200 V. By considering the internal parameters of the generator, the voltage regulation is found to be 9.32% for the inner stator and 10.86% for the outer stator. The proposed generator exhibits notable characteristics, including lightweight construction, high power density, and low rotor inertia, in comparison to the dual stator five-phase PMSG. The proposed generator attains a peak efficiency of 94.83%, coupled with a 91.79% efficiency at the rated current. These results lead to the conclusion that the DSFPLW-PMSG is exceptionally well-suited for renewable energy applications, particularly within the domain of wind power.

References

1. Redissi, Y., Er-rbib, H., and Bouallou, C. (2013). Storage and restoring the electricity of renewable energies by coupling with natural gas grid. *2013 Inter. Renew. Sustain. Energy Conf. (IRSEC)*, 430–435. 10.1109/IRSEC.2013.6529646.
2. Ghoneim, W. A. M., Hebala, A., and Ashour, H. A. (2018). Sensitivity analysis of parameters affecting the performance of radial flux low-speed PMSG. 2018 XIII *Inter. Conf. Elec. Mac. (ICEM)*, 968–974. 10.1109/ICELMACH.2018.8507256.
3. Hernandez, C. et al. (2022). Electromagnetic optimal design of a PMSG considering three objectives and using NSGA-III. *IEEE Trans. Magnet.*, 58(9), 1–4. 10.1109/TMAG.2022.3167306.
4. Maegaard, P. (2009). Wind energy development and application prospects of non-grid-connected wind power. *2009 World Non-Grid-Connect. Wind Power Energy Conf.*, 1–3. 10.1109/WNWEC.2009.5335847.
5. Kumar, R. R., Kumari, A., Dutta, S., and Kumar, K. (2020). Design and performance optimization of dual rotor de-coupled stator multi-phase hybrid magnetic pole permanent magnet synchronous generator for wind turbine application. *2020 IEEE Inter. Conf. Power Elec. Drives Energy Sys. (PEDES)*, 1–6. 10.1109/PEDES49360.2020.9379633.
6. Kimura, M., Kori, D., Komura, A., Mikami, H., Ide, K., Fujigaki, T., Iizuka, M., and Fukaya, M. (2012). A study of permanent magnet rotor for large scale wind turbine generator system. *2012 IEEE Inter. Conf. Elec. Mac.*, 1161–1171. 10.1109/ICElMach.2012.6350023.
7. Ahmed, M. H., Wael, A. F., and Osama, A. M. (2011). Modeling and control of direct driven PMSG for ultra large wind turbines. *2011 World Acad. Sci. Engg. Technol. Inter. J. Energy Power Engg.*, 5(11), 1269–1275. https://publications.waset.org/8556/pdf.
8. Kumar, R. R. et al. (2013). Performance analysis of dual stator six-phase embedded-pole permanent magnet synchronous motor for electric vehicle application. *2013 IET Elec. Sys. Trans.*, 13(1), 1–13. https://doi.org/10.1049/els2.12063.

22 Color image watermarking in the fusion of LWT and tensor-SVD domain in digital image watermarking framework

Saharul Alom Barlaskar[1,a], Mohiul Islam[2], Naseem Ahmad[1], Anish Monsley Kirupakaran[3], Rabul Hussain Laskar[1] and Taimoor Khan[1]

[1]Department of ECE, National Institute of Technology Silchar, Assam, India

[2]School of Electronics Engineering, Vellore Institute of Technology, Vellore, Tamil Nadu, India

[3]Department of Applied Mechanics, Indian Institute of Technology Madras, Tamilnadu, India

Abstract

In this paper, we present color image watermarking considering three RGB channels as third order tensor, upon which tensor singular value decomposition (tensor-SVD) is performed. It provides two orthogonal tensors and one diagonal tensors. As the diagonal tensor holds major energies of a color image, hence used for inserting watermark data in order to achieve adequate visual imperceptibility and robustness of watermark. Gray scale image have been used as watermark for embedding in the first diagonal tensors of SVD matrices. For embedding watermark data, 1-level lifted wavelet transformation (LWT) has been carried out in each of the separate RGB channels of the host image. LWT provide its four decomposed sub-bands, and consequently HL (horizontal) sub-band of each channels are recombine further to perform tensor-SVD operation. Grayscale watermark is transformed into principal component by SVD operation which is embedded into the first diagonal matrices. For measuring the performance, various performance matrices such as peak signal-to-noise ratio (PSNR), structural similarity index measures (SSIM), and normalized similarity ratio (NSR) have been evaluated. The experimental results reveal there is adequate imperceptibility and robustness can be achieved.

Keywords: Tensor singular value decomposition, peak signal to noise ratio, lifted wavelet transform, robustness, imperceptibility

Introduction

The tremendous growth of multi-media technology in recent decade has led to rise of copying and redistribution of digital data among users. As a result, the privacy of digital data has become a serious concerns which damages the interest of legitimate owner. Ali et al., presented image watermarking techniques to address such challenges and to protect copy right of digital data [1-3]. As the color image has drawn significant attention in real life among the users, but protecting copyright of such redistributed data is still facing challenges, which needs more research to address it. In the recent work [5-15], most of the watermarking technique has been developed in the frequency domain using discrete wavelet transform (DWT), discrete cosine transform (DCT), singular value decomposition (SVD), current transformer (CT), and discrete Fourier transform (DFT), etc., as discussed by Preda et al., compared to its counterpart such as spatial domain [4]. In frequency domain, watermark is inserted through modifications of various frequency coefficient and as a results there is sufficient flexibility of robustness possible due to its additional inherent resilient features suggested by Barlaskar et al. [10]. However in spatial domain, watermark is inserted by direct modifications of image pixels and as a result robustness is affected in diverse image processing attack.

Many of the earlier techniques were limited to a single transform domain as mentioned by Verma et al. [16], however, major limitation of such technique is the dependency of domain specific features which makes the system performance dependent on a particular domain. This drawback was addressed later by hybridizing one or two transform domain exist in many literature Singh et al. [2]. The advantage of such hybrid technique has overcome the weakness of one domain by the other and led to improve performance. The author developed a technique based on the hybrid DWT-DCT which provides adequate robustness proposed by Islam and Laskar [9] and imperceptibility. However, the technique provides poor capacity and less secure supported by Kilmer et al. [7]. Similarly, Roy et al., developed a technique based on DWT-SVD which works on grayscale image where the method provides good imperceptibility and robustness [6]. The

[a]saharul_rs@ece.nits.ac.in

DOI: 10.1201/9781003540199-22

excellent multi-resolution features of DWT and stability property of SVD utilized to perform better. However, most of the technique although perform well in the grayscale image, but the performance was not validated for color images. As a result, there is still research scope left out to test the system in the color image application.

In frequency domain, various wavelets such as DWT, lifted wavelet transformation (LWT), integer wavelet transform (IWT), quantum wavelet transform (QWT), etc., as discussed by Singh et al., used to decompose the images into multi-scale frequency band (LL, HL, LH, and HH) [2]. Modifications are done in the decomposed frequency band. George et al. [17], developed a color image watermarking technique, where blue channel was chosen for watermark insertion since blue (B) channel persists less human visual perception compared to its other two channel red (R) and green (G). The modification in blue channel may cause less visual effect. Similarly, Roy et al., presented a DWT-SVD based system for color image, where individual channels were exploited for watermark insertion [6]. Thus, mere embedding of watermark in a particular channels may look good but there is lack of robustness performance compared to grayscale framework.

Thus, in order to design a robust system there is a need of multi-dimensional transformation compared to the conventional 2-D transformation. So, it can be achieved by considering three channels together to preserve a strong correlations of three channels. Hence, the tensor-SVD is the feasible tool which can combine three channels (RGB) of the color image as the third order tensor, which can preserve the robustness features utilizing energy correlation features. In this work, we present a color image watermarking using tensor-SVD decomposition. LWT used for getting LH sub-band after 1st-level LWT transformation of each channels. Individual LH sub-bands are recombined to get tensor matrix, and which tensor-SVD is applied. Two orthogonal tensor and one diagonal tensors are so obtained. The diagonal tensor includes its first diagonal matrix followed by other two diagonal matrices. SVD is used to decompose the grayscale watermark image to get the principal components which are embedded into the first diagonal matrix of each block.

Watermarking in LWT and tensor-SVD domain

Tensor decomposition [6], is a higher dimensional extension of matrix decomposition, and usually prefers in the higher dimensional matrix analysis. The process of watermark embedding and extraction in the fusion of LWT and tensor-SVD has been

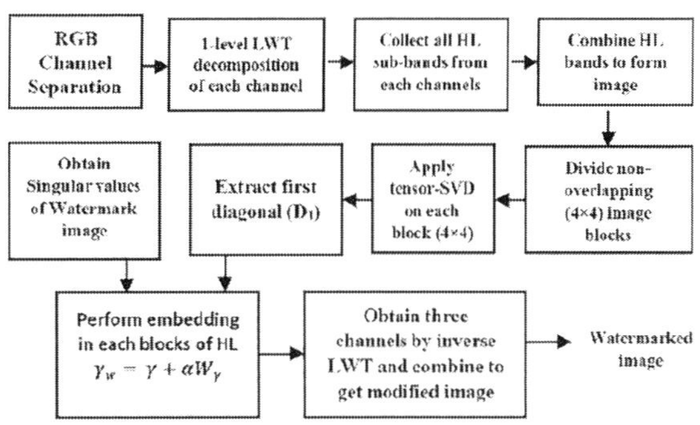

Figure 22.1 (a) Watermark embedding process

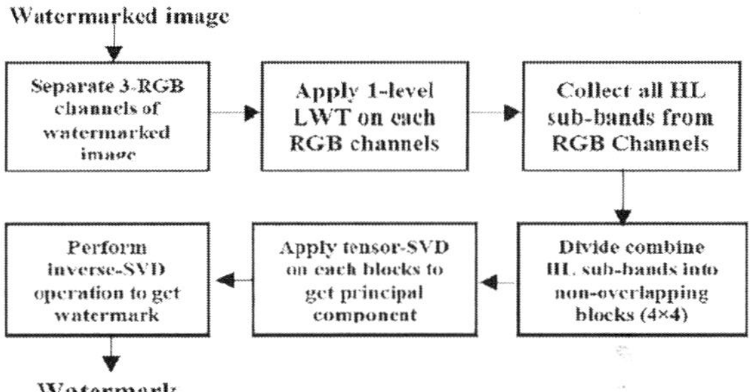

Figure 22.1 (b) Watermark extraction process

shown in Figure 22.1 (a, 22.b). The three distinct R, G, and B channels of the host image (512×512) decomposed into four multi-scale image sub-bands of size (each of 256×256). Each of HL-sub-band from three RGB channels is extracted and recombines to form HL_{RGB} (256×256×3). Further, the combine image is divided into various non-overlapping blocks (4×4×3) in order to perform tensor-SVD operation. Thus, each of the blocks provides three diagonal tensor after tensor-SVD operation. Since the energy content in the first diagonal matrix is much higher compared to other two, so the watermark is inserted in the first diagonal matrix to achieve better robustness.

Tensor-singular value decomposition

As it is well known fact that the SVD technique is a well-established mathematical tool in linear algebra that applies in any matrix. It decomposes a matrix into two orthogonal matrix and a diagonal matrix. In fact the matrix can be seen as a second order tensor. Suppose $B = R^{m \times n}$ represents an image matrix and after the SVD operation, B is decomposed to

$$B = U\sum V^T \tag{1}$$

where $U \in R^{m \times n}$ and $V \in R^{m \times n}$ are the two orthogonal matrices, $\sum \in R^{m \times n}$ represents the diagonal matrix of the SVD decomposition of the matrix B. It is to be worth mentioning that the matrix U, V holds the geometrical image properties such as horizontal and vertical details of the image whereas the matrix Σ holds brightness information of the image. For the higher order ($\rho > 2$), we can choose for tensor-SVD. Let $B \in R^{l_1 \times l_2 \times l_3}$ be the third order tensor, and B can be defied as in Equation 2

$$\mathcal{B} = \mathcal{U} \times \mathcal{S} \times \mathcal{V}^T \tag{2}$$

where $U \in R^{l_1 \times l_2 \times l_3}$ and $V \in R^{l_1 \times l_2 \times l_3}$ are the orthogonal tensor, and $S \in R^{l_1 \times l_2 \times l_3}$ diagonal tensor. The Equation (3) is called tensor-SVD and shown pictorially in Figure 22.2.

Mathematically, the tensor SVD decomposition can also be written as in Equation 3

$$\sum_{k=1}^{l_3} \mathcal{B}_{::k} = \left(\sum_{k=1}^{l_3} \mathcal{U}_{::k}\right) \times \left(\sum_{k=1}^{l_3} \mathcal{S}_{::k}\right) \times \left(\sum_{k=1}^{l_3} \mathcal{V}_{::k}\right) \tag{3}$$

where $\sum_{k=1}^{l_3} \mathcal{U}_{::k}$ and $\sum_{k=1}^{l_3} \mathcal{V}_{::k}$ are orthogonal matrix and the $\sum_{k=1}^{l_3} \mathcal{S}_{::k}$ is diagonal matrix of the tensor decomposed matrix. Since a color image composed of three channels RGB with the size $m \times n$ which can be considered as a tensor corresponds to $l_1 = m$, $l_2 = m$ and $l_3 = 3$ and $k \in \{1,2,3\}$. So after decomposition two orthogonal and one diagonal tensors are obtained in Equation 3. The diagonal tensor (S) contained main energy of the color image, moreover this diagonal ($\mathcal{S}_{::k}$) contains three separate diagonal matrices D_1, D_2, and D_3 when $k \in \{1,2,3\}$. Since three diagonals represent major energies of the color image and hence preserves strong correlations among three channels, thus suits much for embedding watermark to sustain adequate robustness. The reason behind choosing singular values of the diagonal tensor for modifications may be summarized as follows:

(i) Minor modifications in the singular values, is not reflected much in the diagonal values of D_1, D_2, and D_3 which is the stability property of SVD.

(ii) The singular values of D_1, D_2, and D_3 correspond to the brightness characteristics of the image that implies the inherent resilient properties of the SVD domain. Also the first singular values of the diagonal much higher than the corresponding values, which exhibits higher robustness property against image processing attack.

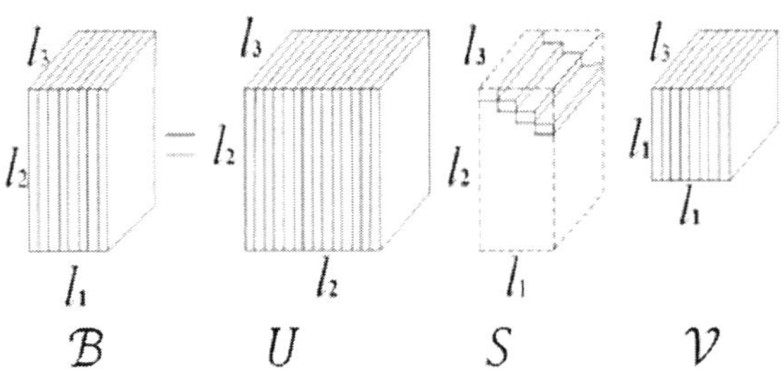

Figure 22.2 Tensor-SVD

Watermark embedding process

Let "I" be the color image of size (512×512×3) and "W" be the grayscale watermark image of size (256×256)

Step 1: Perform SVD on the watermark "W" image as

$U_w S_w V_w^T = SVD|W|$, where V_W needs to be transmitted for watermark extraction purpose. Obtain $W_W = U_W S_W$ for modifying principal component of host image.

Step 2: Separate three RGB channels of the original color host image "I". Apply 1-level LWT on each of RGB channels, and followed by extract HL_R, HL_G, HL_B sub-bands from each of RGB channels.

$$t_i^{\theta} \leftarrow LWT|I\,(:,:,i)|. \tag{4}$$

where LWT |•| is lifted wavelet transform operation on "I", $i \in \{1,2,3\}$, $i \in \{LL,LH,HL,HH\}$.

Step 3: Combine three sub-bands HL_R, HL_G, HL_B to form an image sub-bands (256×256×3). Divide into different non-overlapping blocks R^j of size (4×4×3) to perform tensor-SVD operation on each blocks using Equation 3, and extract the first diagonal matrix (D_1^j) from S^j

$$Tensor\ SVD|\mathcal{R}^j| = \mathcal{U}^j \times S^j \times (\mathcal{V}^j)^T \tag{5}$$

Step 4: Now, embed j^{th} values of the principal component of grayscale image into (D_1^j) of R^j by using Equation (6) below.

$$\gamma_w^{j,max} = \gamma^{j,max} + \alpha W_w^j \tag{6}$$

where, α is the gain factor of watermark, $\gamma_w^{j,max}$ is the modified largest singular value and so the modified diagonal tensor S_w^j is so obtained. The factor $\gamma^{j,max}$ needs to be transmitted for watermark extraction process.

Step 5: Now perform inverse tensor-SVD to get modified or watermarked block $\mathcal{R}_w^j$ using the Equation 5 shown below. The process is repeated for all the blocks to get complete modified sub-bands HL_{RGB}

$$\mathcal{R}_w^j = Tensor\ SVD^{-1}|\mathcal{U}^j \times S^j \times (\mathcal{V}^j)^T| \tag{7}$$

Step 6: Thus, all the modified sub-bands HL_{wR}, HL_{wG}, HL_{wB} are separated out and 1-level inverse LWT is carried out to get distinct modified R_w, G_w, B_w channels which are further re-combine to get watermarked image "I_W^*".

Watermark extraction process

Step 1: Initially the color watermarked image I_W^* is divided into three RGB channels such as R_W^*, G_W^* and B_W^*, followed by 1-level LWT operation is performed on each channels.

$$t_i^{\theta *} \leftarrow LWT|I_W^*(:,:,i)| \quad (8)$$

Let HL_R^*, HL_G^*, and HL_B^* are the sub-bands each of size (256x256), further combine to get HL_{RGB}^* sub-band of size (256x256x3). This combine sub-band is divided into non-overlapping blocks ($\mathcal{R}_W^{j*}$) each of size (4x4x3).

Step 2: Perform tensor-SVD on each of the block as per the Equation 6 shown below. Now, the first diagonal matrix D_1^{j*} is extracted from S^{j*} and thus the largest singular value of D_1^{j*} is $S_W^{j,*}$

$$\text{Tensor-SVD } |\mathcal{R}_W^{j*}| = \mathcal{U}^{j*} \times S^{j*} \times (\mathcal{V}^{j*})^T \quad (9)$$

Step 3: Now obtain the principal component of the grayscale watermark is as.

$$W_W^{j*} = (Y_W^{j,max} - Y^{j,max})/\alpha \quad (10)$$

Step 4: Obtain the extracted watermark by doing inverse SVD operation as follows:

$$W^* = W_W^* V_W^T \quad (11)$$

Results and discussion

In this section, experimental results are presented and discussed for evaluating the robustness and imperceptibility of watermark. For experimenta tion purpose few sample standard color image database have been used as host image as shown in Figure 22.3(a) and grayscale watermark shown in Figure 22.3(b). In order to measure imperceptibility between the original host images (I) and the watermarked image (I_W^*) peak signal to noise ratio i.e., PSNR (in dB) is calculated as Equation (12).

$$PSNR = \log_{10}\left[\frac{512\times512 \times \max\,[I(i,j)^2]}{\sum_{i=1}^{512}\sum_{j=1}^{512}[I(i,j) - I_W^*(i,j)]^2}\right] \quad (12)$$

To measure the robustness of watermark, normalized correlation coefficient (NCC) [10, 11] is computed between the original (W), and the extracted watermark image (W_W^*) using Equation (13)

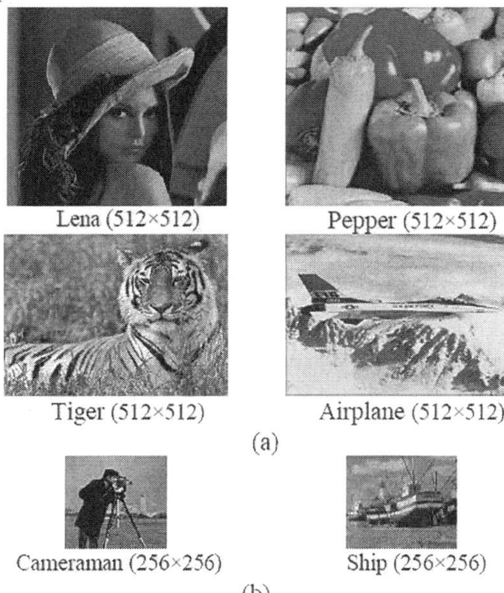

Figure 22.3 (a) Host image, (b) watermark image

$$NCC = \frac{\sum_{i=1}^{256}\sum_{1}^{256} W(i,j)\times W_W(i,j)}{\sum_{i=1}^{256}\sum_{1}^{256} W(i,j)\times W(i,j)} \quad (13)$$

And the normalized similarity ratio (NSR) [12] is computed by using Equation (14)

$$NSR = \frac{SR - \min(SR)}{1 - \min(SR)} \quad (14)$$

where NSR is depends on the similarity ratio, which given by

$$SR = \frac{S}{S + D} \quad (15)$$

where S is the total number of similar pixels and D is the total number of dissimilar image pixels in the original and extracted watermark.

Experiment 1: Robustness analysis of three diagonal matrices and imperceptibility factor

In a color image, three RGB-channel possesses strong correlations among each other because of similar texture, edge, and background information. As the first diagonal matrix (D_1) preserves most of the image energy compared to other two diagonal matrices D_2 and D_3 mentioned by Li et al., so embedding in D1 will be beneficial for maximizing the robustness [8]. In order to prove it, an experiment has been carried for different standard color images of Pepper, Lena, Tiger and airplane. Grayscale cameraman image have been embedded in the tensor diagonal matrices, and subsequent robustness values were noted down in presence of varying image processing attacks. Thus, it is evident from the Table

Table 22.1 Robustness (NCC) of watermark across three diagonal matrices under different attacks

Image	Attacks	Robustness (NCC) in diagonal		
		D_1	D_2	D_3
Pepper	Gaussian noise (0.001)	0.9675	0.9385	0.8667
	Salt & pepper (0.001)	0.9241	0.9072	0.9001
	Impulse noise	0.9551	0.9422	0.9320
	Median filter (3x3)	0.9895	0.9611	0.9408
Lena	Gaussian noise (0.001)	0.9991	0.9767	0.9799
	Salt & pepper (0.001)	0.9822	0.9776	0.9322
	Impulse noise	0.8812	0.8494	0.8355
	Median filter (3x3)	0.9643	0.9612	0.9335
Tiger	Gaussian noise (0.001)	0.8997	0.8995	0.8833
	Salt & pepper (0.001)	0.9595	0.9443	0.9376
	Impulse noise	0.9700	0.9678	0.9611
	Median filter (3x3)	0.9743	0.9676	0.9669
Airplane	Gaussian noise (0.001)	0.8895	0.8847	0.8732
	Salt & pepper (0.001)	0.9812	0.9810	0.9798
	Impulse noise	0.9192	0.9098	0.8988
	Median filter (3x3)	0.9677	0.9567	0.9510

Table 22.2 Imperceptibility measures in different standard color images in attack conditions

Image	Performance parameters		
	PSNR	SSIM	NSR
Lena	50.56	0.9998	0.9991
Pepper	49.94	0.9994	0.9995
House	49.87	0.9889	0.9843
Airplane	50.87	0.9972	0.9787
Baboon	49.98	0.9969	0.9887
Tiger	47.60	0.9787	0.9911

22.1, there is higher values of robustness possible in the first diagonal matrix embedding. Table 22.2, demonstrates the average PSNR (in dB) achieved in different standard images under attack conditions. So it can be noticed most of the cases the PSNR > 45 (dB) which is acceptable and well recognised by human visual system without visual affect as suggested by Verma et al. [16]. Other similarity measures such as SSIM and NSR were computed to judge watermarking affect and most of the cases more than 90% structurally similarity is retained.

Experiment 2: Analysis of Robustness (NCC) and Imperceptibility (PSNR) against varying gain factor (α)

In this section, a set of experiment have been carried out to understand the influence of gain factor (α) on the imperceptibility and robustness of watermark. The gain factor plays a vital role in the balancing of two conflicting parameters, which is also well proved through this sets of experimentation. A standard Lena test image has been considered to observe the variations of NCC values with varying gain factor. Few image processing attacks such as salt & pepper (0.001), Gaussian noise, compression attack (JPEG 60), impulse noise, and denoisy attacks were also imposed during watermarking process. The embedding performed with a varying gain factor values such as α=0.01, 0.02, 0.03, and 0.04. The experimental results shows higher robustness possible with higher gain factor i.e., (α=0.04) in most of the cases shown graphically in Figure 22.4. This increase of robustness is possibly due to boosting of energy content of the embedding region which means increase in resiliency of watermark against attacks. Figure 22.5 shows the variations of PSNR against the varying gain factor. Thus the trend of increase PSNR with fall in robustness is a major conflict between this two performance parameters. In the graph (shown in Figure 22.5) "pepper" image preserves highest values of PSNR~51dB. Thus, by choosing the convenient gain factor (α=0.04) an experiment was performed under different image processing attack conditions. cameraman was used as watermark. Figure 22.6

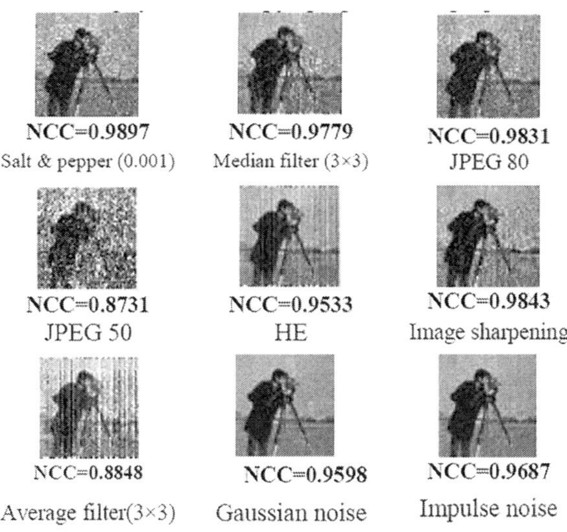

Figure 22.4 Shows different extracted watermark under different image processing attack conditions

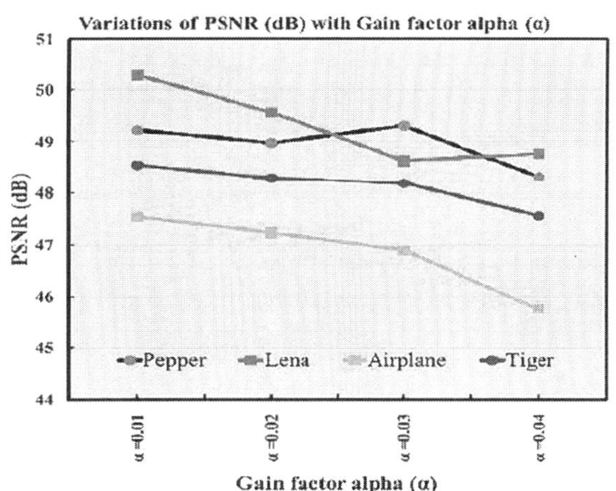

Figure 22.6 Variation of NCC values under varying gain factor (α) in image processing attack conditions

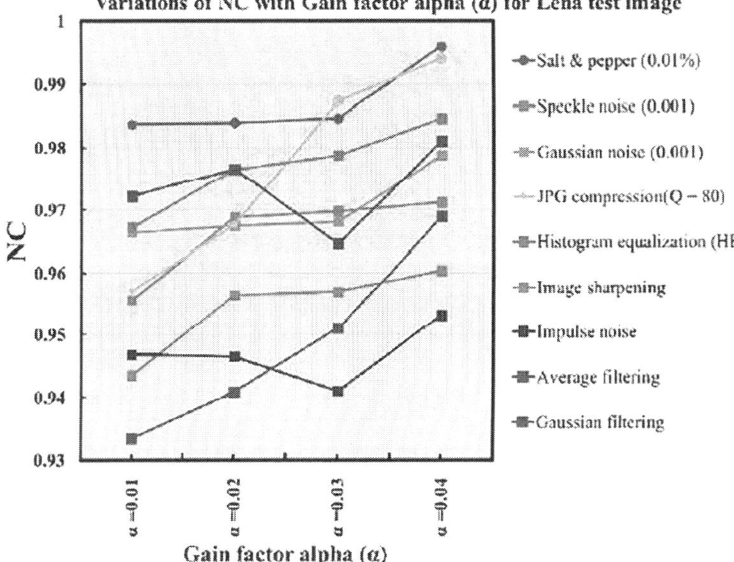

Figure 22.5 Variation of PSNR values under varying gain factor (α) under image processing attack conditions

shows the different extracted watermark versions and from the results it can be evident most of the cases, the robustness (NCC) is found to be above 90%. Thus, utilizng the first diagonal matrix of tensor based SVD decomposition, it is quite feasible to achieve a robust watermarking system for copyright protection purpose.

IV. Conclusions

A detail study of color image watermarking in the fusion of LWT and tensor-SVD have been made in this work. The results and discussion presented visualize the well proved feasibility of robustness and imperceptibility is possible to achieve under image processing attack conditions. The experiment can also be extended further utilizing other decompose image sub-bands such as LL, LH, and HH for observing robustness which have been kept at hand. The optimum balancing of NCC and PSNR can also be possible by incorporating an optimized gain factor values. This study make a clear understanding of a color image as tensor decomposition gives adequate robustness and imperceptibility compared to the conventional 2D image decomposition using SVD. Thus for further extension of work, we will be utilizing the other sub-band and other multi-scales sub-bands as well as incorporation geometric attack to understand the performance of the system.

Acknowledgement

The author would thank the members of speech and image processing laboratory, NIT Silchar for all their support.

References

1. Weng, S., Chen, Y., Ou, B., Chang, C.-C., and Zhang, C. (2019). Improved K-pass pixel value ordering based data hiding. *IEEE Acc.*, 7, 34570–34582.
2. Singh, L., Singh, A. K., and Singh, P. K. (2020). Secure data hiding techniques: A survey. *Multimedia Tools Appl.*, 79, 15901–15921.
3. Ali, S. A., Jawad, M. J., and Naser, M. A. (2017). A semi-fragile watermarking based image authentication. *J. Engg. Appl. Sci.*, 12(6), 1582–1589.
4. Preda, R. O. (2013). Semi-fragile watermarking for image authentication with sensitive tamper localization in the wavelet domain. *Measurement*, 46(1), 367–373.
5. Lin, S. D. and Chen, C.-F. (2000). A robust DCT-based watermarking for copyright protection. *IEEE Trans. Cons. Elec.*, 46(3), 415–421.
6. Roy, S. and Pal, A. K. (2019). A hybrid domain color image watermarking based on DWT–SVD. *Iran. J. Sci. Technol. Trans. Elec. Engg.*, 43, 201–217.
7. Kilmer, M. E. and Martin, C. D. (2011). Factorization strategies for third-order tensors. *Lin. Algebra Appl.*, 435(3), 641–658.
8. Li, D., Che, X., Luo, W., Hu, Y., Wang, Y., Yu, Z., and Yuan, L. (2019). Digital watermarking scheme for color remote sensing image based on quaternion wavelet transform and tensor decomposition. *Math. Methods Appl. Sci.*, 42(14), 4664–4678.
9. Islam, M. and Laskar, R. H. (2018). Robust image watermarking technique using support vector regression for blind geometric distortion correction in lifting wavelet transform and singular value decomposition domain. *J. Elec. Imag.*, 27(5), 053008–053008.
10. Barlaskar, S. A., Vir Singh, S., Anish Monsley, K., and Laskar, R. H. (2022). Genetic algorithm based optimized watermarking technique using hybrid DCNN-SVR and statistical approach for watermark extraction. *Multimedia Tools Appl.*, 81(5), 7461–7500.
11. Islam, M., Mallikharjunudu, G., Parmar, A. S., Kumar, A., and Laskar, R. H. (2017). SVM regression based robust image watermarking technique in joint DWT-DCT domain. *2017 Inter. Conf. Intell. Comput. Instrum. Control Technol. (ICICICT)*, 1426–1433.
12. Barlaskar, S. A., Kirupakaran, A. M., Ahmad, N., Yadav, K. S., Khan, T., and Laskar, R. H. (2023). Imperceptibility and robustness study in different transform domain for copyright protection in digital image watermarking in hybrid SVD-domain. *2023 3rd Inter. Conf. Artif. Intell. Signal Proc. (AISP)*, 1–5.
13. Singh, D. and Singh, S. K. (2017). DWT-SVD and DCT based robust and blind watermarking scheme for copyright protection. *Multimedia Tools Appl.*, 76(11), 13001–13024.
14. Kumar, C., Singh, A. K., and Kumar, P. (2018). A recent survey on image watermarking techniques and its application in e-governance. *Multimedia Tools Appl.*, 77, 3597–3622.
15. Gangadhar, Y., Giridhar Akula, V. S., and Chenna Reddy, P. (2016). A survey on geometric invariant watermarking techniques. *2016 IEEE Inter. Conf. Comput. Intell. Comput. Res. (ICCIC)*, 1–5.
16. Verma, V. S. and Jha, R. K. (2015). An overview of robust digital image watermarking. *IETE Tech. Rev.*, 32(6), 479–496.
17. George, J., Varma, S., and Chatterjee, M. (2014). Color image watermarking using DWT-SVD and Arnold transform. *2014 Ann. IEEE India Conf. (INDICON)*, 1–6.

23 K-means algorithm-based eye scanner for accurate prediction of diseases

Sankita Kundu[1,a], *Phularenu Das*[2,b], *Souvik Mondal*[3,c] *and Sandip Chanda*[4,d]

[1]Department of Applied Electronics & Instrumentation Engineering, RCC IIT, Kolkata, Kolkata, India

[2]Department of Electrical Engineering, NIT, Durgapur, Durgapur, India

[3]Department of Electrical Engineering, NITTTR, Kolkata, Kolkata, India

[4]Department of Electrical Engineering, GKCIET, Malda, Malda, India

Abstract

There is an interlink between various activities and risk factors of the body and the occurrence progression of numerous diseases. In particular, cotton wool spot, white ring around iris are important indications of expanded cardiovascular risks. Many infectious diseases such as HIV, bartonella enslave, leptospirosis or else diseases that causes in a bacteremia or fungal infections also may cause cotton-wool spots which can be detected from eyes. Besides this high cholesterol may cause white ring around iris and also can cause deposition of fatty, yellowish lumps around the eyes. The whitish ring around Iris also indicates that the patient may have chances of stroke. The abnormality in cup to disc ratio of eye is a sign of glaucoma which means that the nerves which are connecting eyes to brain are being damaged. Glaucoma causes high eye pressure and may cause blindness also. And yellowish eye indicates that the person may have jaundice. Thus, eye is a window to our body and various changes within eye may lead us towards a non-contact type diagnosis of various diseases. This paper presents a technique of non-contact type diagnosis of various diseases through eye scanning. The patterns like cotton wool spots, white ring around iris, abnormality in cup-to-disc ratio, glaucoma and yellowish eye have been classified. An algorithm has been developed using K-means algorithm, in this paper for clustering of patterns. The paper also reveals a possible hardware set up for the eye scanner so that maximum relevant data can be collected.

Keywords: Eye scan, cotton wool spot, whitish ring around iris, cup to disc ratio, non-contact type diagnosis, glaucoma, hypertension, clustering of patterns

Introduction

Different diseases within human body cause various abnormalities in eyes. Such as, cotton-wool-spot is the earliest sign of hypertension and diabetes. Again, white ring around iris and abnormal cup-disc ratio is respectively caused by cholesterol and glaucoma. And yellowish eye gives us sign about jaundice. So, by checking these abnormal parameters within eyes we can easily detect the presence of these diseases within human body. Ref [1] shows that the eye and heart these two organs are not linked to each other but in real they have much more similarities than someone would anticipate. Eye's vasculature and heart share many features. Various diseases like arterial hypertension, coronary artery disease, diabetes mellitus in addition to obesity all are related with structural vascular changes within the iris of eyes. Though ophthalmologists, as in ref [2] mainly focus on eye specified mechanisms and therapy of only eye diseases but if we see these eye problems in broader aspect then it may be quite revealing about our health condition. According to recent studies [3], hypertensive retinopathy is recommended as a sign for determining vascular risk and hypertension in persons. Hypertensive retinopathy anticipates the long-term danger of stroke in the people having hypertension. Retinal images were evaluated to diagnose hypertensive retinopathy and were used to be categorized. So, the occurrence of any stroke, cerebral infarction and hemorrhagic stroke were recognized by detecting hypertensive retinopathy. Regular ophthalmoscopy is suggested in the assessment of people with hypertension [4]. Hypertensive retinopathy may also be able to predict coronary heart disease (CHD). Cotton-wool spots are a symptom of placid and severe diabetic retinopathy and their arrival is also seen within the eyes of the patient with hypertension [5]. Cotton-wool spots are also the earliest sign of capillary abnormalities. Capillary abnormalities precede to develop arteriolar occlusion which also helps in the development of cotton-wool spots. Ref [6] presents a work which shows that iris of eye can provide lots of knowledge about the condition of different body organs and tissues. Through iris analysis it is possible to determine high cholesterol in the blood. Ref [7] presents

[a]sankita0499@gmail.com, [b]phularenu@gmail.com, [c]souvik1995mondal@gmail.com, [d]sandipee1978@gmail.com

DOI: 10.1201/9781003540199-23

a study on glaucoma which is a set of eye disorder that guide to progressive destruction to the optic nerve. People who are having glaucoma can lose their vision. Increment of fluid pressure inside the eyes is the most natural cause of having glaucoma. Normally, glaucoma is detected through a comprehensive examination of eyes which requires a clinical setup. As glaucoma is a progressive disease it requires a regular check-up. Few optic nerves within eyes may also resemble nerves with glaucoma, but the patients may not have any other risk components or symptoms of glaucoma. That is why the study recommends that patients should go for regular comprehensive eye examination to monitor any changes within eyes that leads to glaucoma. Ref [8] presents a work where glaucoma is detected as by the cup to disc ratio of iris. The impulses for sight are carried by optic nerves from the retina of the eyes to the brain. It is a well-known fact that detection of jaundice is possible from eye. Jaundice is not a disease itself but it is a symptom of a number of underlying circumstances that results in bile ducts, gallbladder, malfunction of liver or pancreas. In ref [9] it shows that the yellow eye or skin color during jaundice occurs due to subsequent hyperbilirubinemia, a surplus amount of bilirubin within blood. As jaundice is first noticeable within eyes it should be a clinical sign that optometrists should be on the lookout for.

All the above study depicts that by proper processing of the above-mentioned parameters, the diseases can be detected with fair accuracy. The process will certainly need data collection through simplest possible ophthalmology. The study did in ref [10] shows the usage of light has taken a major part in revealing the structural and functional details of the retina of human eyes in non-invasive nature, in past few decades. Now a-days latest scanning and imaging machineries have a huge knock on ophthalmology. Fundus photography was the first standard photographic technique introduced in the 1920's, basing on 35 mm film, and later as digital imaging. The chief interest of fundus imaging is the photography of optic nerve. And the light rays are being parallelized with the help of a focusing lens which is moveable and thus high-resolution cameras are being enabled and the retinal imaging is done. Ref [11] shows that fundus imaging is a crucial part of ophthalmic exercise with a fundus camera. But current studies tell that a reasonable quality of fundus images can be obtained using mobile phone through its in-built camera and flash light. In ref [12], it was found that there are several methods that have been used in digital image processing and Haar cascade classifier is one of them, which has been broadly applied for humans and computers interaction (HCI). The increasingly enlarging applications of machine learning (ML) in health sector is allowing us to peek at a near future where data analysis and innovation use to work hand-in-hand to aid countless patients without making them realize it. Soon, it will be quite customary to find ML-based implementations implanted with real-time patient data accessible from various healthcare systems from multiple countries which will increase the value of new ML-based therapy options as found in ref [13]. The recognition and detection of diseases and disorders which are otherwise contemplated as hard-to-diagnose is one of the main ML applications in healthcare sector. But this problem does not arise if no outputs are provided related to inputs within the model. So we can say when the label is not present we must result into unsupervised learning [14]. By monitoring their arrival minutely using high resolution camera and microscope, doctors can detect various things such as an individual's blood pressure, age, and if they smoke or not which are all major predictors of cardiovascular health. And doctors can also predict if a person is having diseases like hypertension, diabetes, cholesterol etc. or not [15].

The literature survey gives us the idea about the links within different eye parameters and health issues and also tells us about the scope of disease detection through eyes. It has been also found that the diseases can be diagnosed or can be predicted by extracting data from eye scan images using image processing and ML algorithm. This diagnosis system is cost-effective, time efficient, predicts earlier and non-contact type also so, we can say that this will not only help the patients but it will also help the doctors; as we all know early diagnosis helps to prevent the disaster and also helps to reduce the mortality rate.

Hypothesis of present work

Various deformities within eyes are earliest signs of some diseases. So, by capturing or monitoring these deformities within eyes through eye scan we can easily get to know if someone is suffering through any disease or not.

Cotton wool spots, depicted in Figure 23.1, are unnatural finding on funduscopic examination of retina of the eyes. This appears as white fluffy blotches on the retina. It occurs due to damage of nerve fibers and is an outcome of accumulations of axoplasmic substances in the nerve fiber layer. The existence of a cotton wool spot (CWS) is not contemplated as normal. The presence of a single cotton wool spot in one eye use to be the earliest ophthalmoscopic finding of diabetic or hypertensive retinopathy.

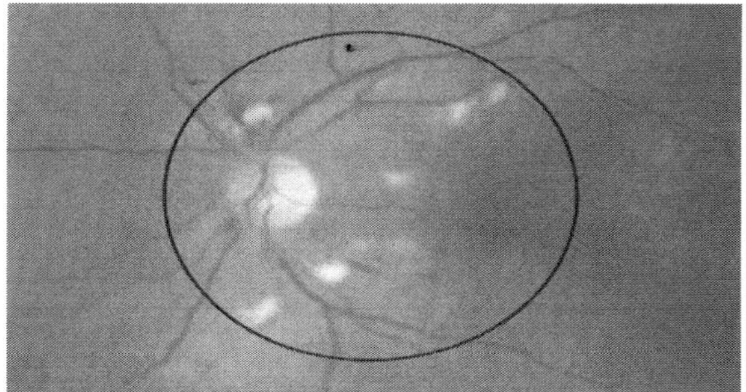

Figure 23.1 Cotton wool spot

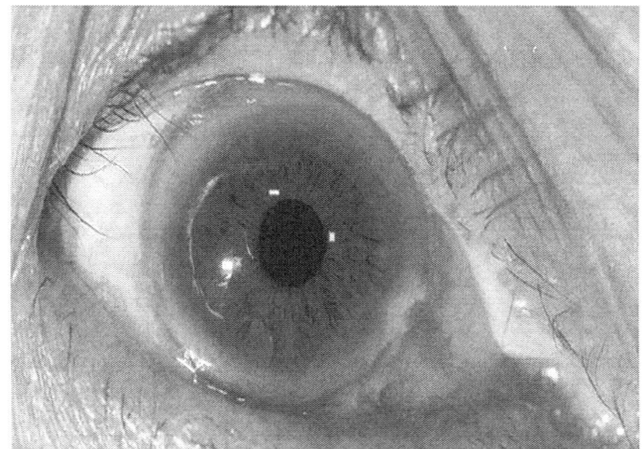

Figure 23.2 White ring around iris

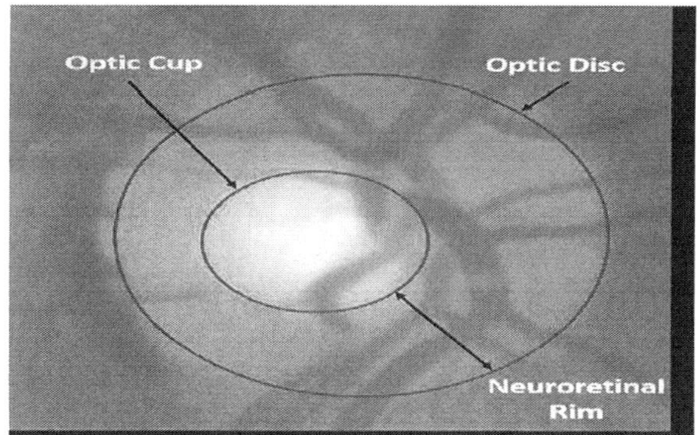

Figure 23.3 Cup and disc

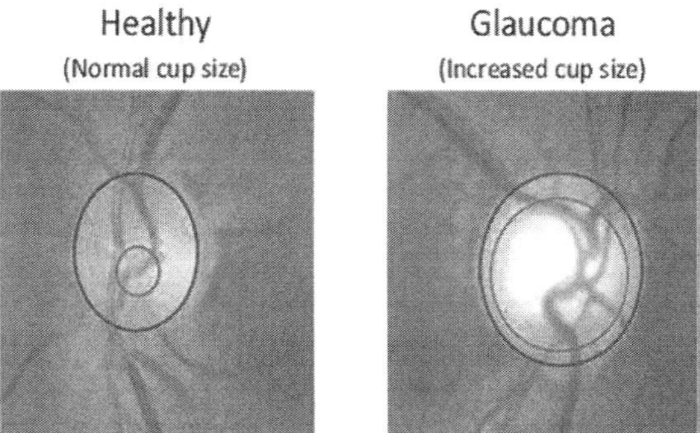

Figure 23.4 Cup and disc in healthy body vs. affected body

There is a disease of eye which is called diabetic retinopathy. This is a disease due to which increased blood sugar levels use to cause damage to blood vessels of retina. These blood vessels can inflate and also can leak otherwise they can cause stopping of blood from passing through.

White ring around iris or cloudy deposit, as depicted in Figure 23.2, on the front of the eye surrounding the pigmented iris is known as cholesterol ring. In some cases, this ring may appear with age but it is more commonly in people with high blood cholesterol levels. So, we can say the eye can give a peek at your blood cholesterol.

The cup-to-disc ratio (often notated CDR) is an estimation use to check out the progression of glaucoma. The anatomical area of "blind spot" of the eyes is optic disc. "Blind spot" is the region where the optic nerves and blood vessels enter into retina. Normally the optic disc use to be flat or it also can have a unquestionable amount of normal cupping (Figures 23.3 and 23.4).

But glaucoma, which is mostly associated with an rise in intraocular pressure, often fabricates an added pathological cupping of the optic disc. Glaucoma occurs when damage happens in the nerve which is connecting the eye with brain. This normally occurs due to high pressure within eyes and may also cause slow vision loss which may end up to blindness.

Jaundice itself is not a sickness, but it is a sign of different possible underlying diseases. Jaundice forms when there is way too much bilirubin within the body. Bilirubin is a yellowish pigment (Figure 23.5) which is formed when the dead red blood cells breakdown within the liver. The white part of the eyes can turn yellow as a result of jaundice.

Thus, we can say that different diseases use to cause different changes within eyes. And we can diagnose various diseases while scanning the eye parameters and using ML algorithms onto it.

Hardware to scan eyes

The eyes can be scanned and various changes within eyes can be determined through fundus imaging. This can be interpreted as two-dimensional photo of three-dimensional retinal tissue which one can capture by using reflected light rays. Fundus imaging or fundus photography is the standard procedure for demonstrating ocular fundus findings. The coaxial flashlight of camera of the smartphone and a hand-held power lens with high plus can generate an optical system same as indirect ophthalmoscopy which is capable of recording high resolution digital images of

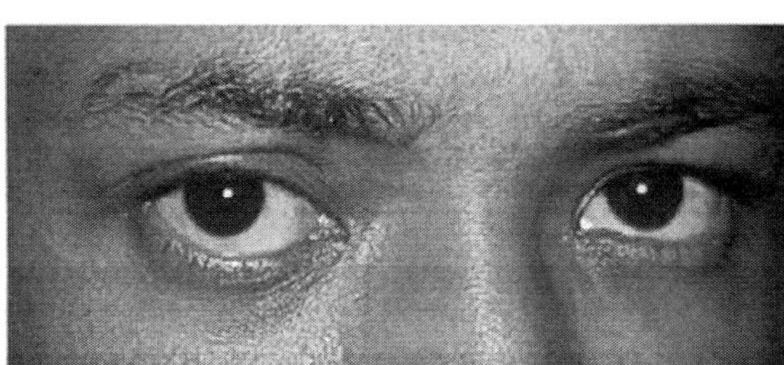

Figure 23.5 Yellowish eye while jaundice

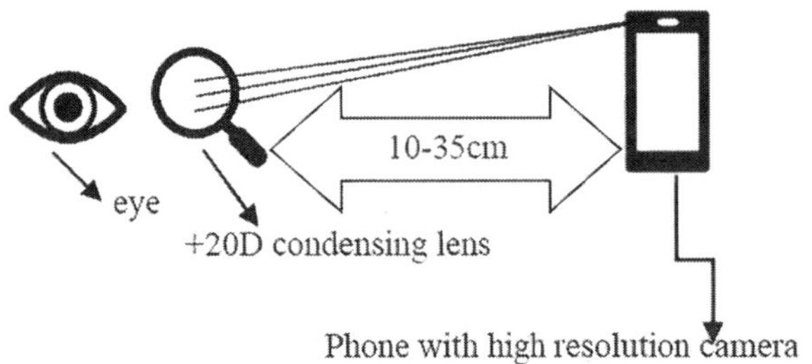

Figure 23.6 Fundus imaging hardware using smartphone's camera

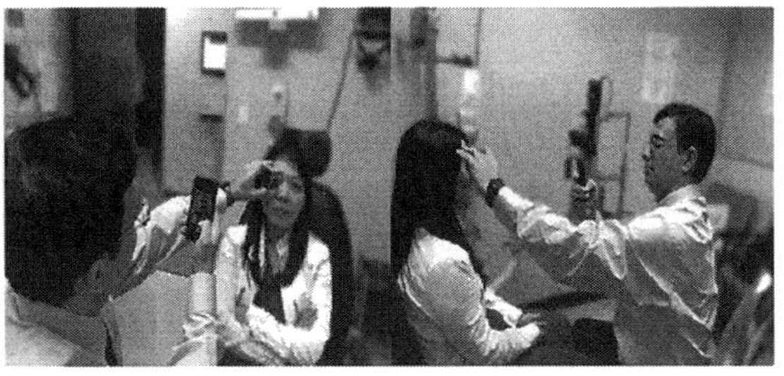

Figure 23.7 Fundus imaging (indirect ophthalmoscopy)

retina. We have followed the following steps to scan eyes of the patients (Figures 23.6 and 23.7):

A smartphone with high resolution camera and flashlight is held in one hand and has been used to take retinal images.

Using another hand, a +20D condensing lens is held in front of the patient's eye for detailed imaging.

10–35 cm distance is maintained within the smartphone and the lens.

While taking the picture using smartphone's camera, the flashlight of the phone is kept on. The lens and the phone are held steadily in required position for clear and detailed retinal imaging.

Data extraction through image processing

After scanning the patient's eye, it is needed to extract required data from it. And this data extraction can be done by digital image processing. Image processing is an algorithm where the images are given as an input and we get a particular feature or pattern as an output. Digital image processing provides a wider range to process two dimensional digital images and also provides various digital filters to sharpen the images and to clear the noise from it.

For extracting data through image processing the scanned image of patient's eye is firstly represented as a matrix and thereafter a pixel-to-pixel comparison is done with the reference image (image of healthy eye) to detect presence of any abnormality like cotton-wool spots, cholesterol ring, yellowish eye, etc., within the eyes of the patient. Through pixel-to-pixel comparison with reference image it is possible to detect difference in color intensity and this led us to an output which tells us about the presence or absence of parameters, which are the earliest signs of various diseases within the eyes of patient.

Clustering of data through ML

In this project to predict diseases from the detected abnormal parameters of eyes we mainly used overlapping clustering, as in figure strict than one clusters. Some data can also be part of no clusters and these are known as outliers.

To build a system that selects the algorithm best fitting the requirements for disease detection from eye scan some constraints need to be specified.

We used Naïve K-means algorithm for disease detection. As for working with medical data, this algorithm is the standard one for clustering. It provides the most accurate prediction through unsupervised learning. It mainly uses iterative refinement technique. This algorithm predicts by allocating data points to the closest cluster by distance, through a different distance function other than (squared) Euclidean distance which may stop or restrict this algorithm from converging. Different modifications of K-means such as spherical K-means and K-medoids is used to permit using other distance measures.

K-means algorithm mainly includes two steps:

A. Assigning step: This step includes assignment of each observational data to the cluster with nearest mean through the least squared Euclidean distance (Equation 1).

$$S_i^{(t)} = \left\{ y_P : \left\| y_P - a_x^{(t)} \right\|^2 \leq \left\| y_P - a_z^{(t)} \right\|^2 \forall z, 1 \leq z \leq m \right\} \quad (1)$$

Assuming an initial set of K-means $a_1 \cdots a_m$. In the given equation each {\display style x_{p}}y_P is assigned to exactly one {\display style

S^{(t)}}$s^{(t)}$ even if it could be allocated to two or more of them.

B. Updating step: This step includes recalculation of means (centroids) for observational data for assigning them to each cluster (Equation 2).

$$a_x^{[t+1]} = \frac{1}{s_i^t} \sum_{y_z \in s_i^t} y_z \quad (2)$$

The algorithm converges when the assigned data values change no longer.

Results

For predicting diseases from eye scan data through unsupervised learning mainly two matrixes are being formed.

These two matrixes (Tables 23.1 and 23.2) are mainly formed by the computational algorithms from the provided training dataset and by comparing these two matrices the ML algorithm will predict whether a patient has any listed diseases (diabetes, hypertension, glaucoma, cholesterol and jaundice) or probability of having this disease in near future.

As by comparing these two matrices, the algorithm will reach to the conclusion that if a person is having cotton-wool-spot then he/she might have hypertension or diabetes or both of them. And if someone is having whitish ring around iris then he/she is suffering through high cholesterol. And someone with abnormal cup-disc ratio and yellow eye is suffering through glaucoma and jaundice, respectively. This driven conclusion from training data will help the ML algorithm to predict and provide most optimum decision.

To check the accuracy and precision of the ML algorithm an N×N matrix will be formed which is known as confusion matrix. This encapsulates how efficient are the predictions of classification model is. This is a correlation matrix between the classification of label and the model. One axis of the confusion matrix is the label that is predicted by the algorithm and the other axis provides the actual label. Here, N shows the amount of classes (Table 23.3).

A true positive is a result of correctly predicting the positive class. And a true negative is a result of properly predicting the class which is negative.

Table 23.1 Assuming person vs. disease data table (matrix)

D/P	1st person	2nd person	3rd person	Person (n-1)	Person n
Diabetes	1	1	0	1	1
Hypertension	0	1	1	0	1
Glaucoma	1	1	1	0	0
Cholesterol	1	1	0	0	1
Jaundice	0	0	0	1	0

Table 23.2 Assuming parameters (abnormalities within eyes) vs. person data matrix

Parameter/person	1st person	2nd person	3rd person	Person (n-1)	Person n
Cotton wool spot	1	1	1	0	1
White ring around iris	1	1	0	0	1
Abnormal cup-to-disc ratio	1	1	1	0	0
Yellow eye	0	0	0	1	0

Table 23.3 Test case detection

Labeled/ predicted	Disease (predicted)	No-disease (predicted)
Disease (actual)	True positive	False positive
No-disease (actual)	False negative	True negative

Whereas, a false positive is a result where the model is predicting the positive class incorrectly and a true negative is a result where the model is predicting the negative class incorrectly.

Conclusions

We are developing an application-based software which can diagnose various diseases such as diabetic and hypertensive retinopathy, cholesterol, glaucoma, jaundice by determining abnormalities within eyes. This app and hardware will use the smartphone's camera to have retinal imaging and by using ML algorithms onto that image, it will provide us if the patient is having any mentioned diseases or not. This app not only leads us towards a quick diagnosis but also it leads us towards non-contact type diagnosis system. This provides an user interface (UI), that is why it is really very easy to use for laymen. This also provides an opportunity to check one's health condition from home and makes one know if he needs to go a doctor or not. This will not only help patients but also will help the doctors. As we all know early diagnosis always reduces the mortality rate.

References

1. Josef, F., Konieczka, K., Bruno, R. M., Virdis, A., Flammer, A. J., and Taddei, S. (2013). The eye and the heart. *Oxford J. Eur. Heart J.*, 34, 1270–1278. doi.org/10.1093/eurheartj/eht023.
2. Annie, S. (2015). The heart and the eye: Seeing the links. *Am. Acad. Ophthalmol,* 1–6. www.aao.org/eyenet/article/heart-eye-seeing-links.
3. Yi-Ting, O., Wong, T. Y., Klein, R., Klein, B. E. K., Mitchell, P., Sharrett, A. R., Couper, D. J., and Ikram, M. K. (2013). Hypertensive retinopathy and risk of stroke. *AHA J,* 1–6. doi.org/10.1161/HYPERTENSIONAHA.113.01414.
4. Duncan, B. B., Wong, T. Y., Tyroler, H . A., Davis, C. E., and Fuchs, F. D. (2002). Hypertensive retinopathy and incident coronary heart disease in high risk men. *Br. J. Ophthalmol,* 995–999. pubmed.ncbi.nlm.nih.gov/12185127/
5. Eva, M. K., Dollery, C. T., and Bulpitt, C. J. (1969). Cotton-wool spots in diabetic retinopathy. *Am. Diabet. Assoc. Diabet.*, 18(10), 691–704. https://doi.org/10.2337/diab.18.10.691.
6. Burak, K. G. and Kurnaz, C. (2016) . Detection of high-level cholesterol in blood with iris analysis. *20th National Biomed. Engg,* 1–7. *Meeting (BIYOMUT),* https://hdl.handle.net/20.500.12712/13569.
7. American Optometric Association (AOA). www.aoa.org/patients-and-public/eye-and-vision-problems/glossary-of-eye-and-vision-conditions/glaucoma.
8. Scott, B., Cohen, J. S., and Quigley, H. (2017). Optic nerve cupping. *Glaucoma Res. Foundation.* 1–2. www.glaucoma.org/glaucoma/optic-nerve-cupping.php.
9. Carlo, J. P. and Pizzimenti, J. J. (2017). Jaundice and the eyes. *Rev. Optom,* 1–7. www.reviewofoptometry.com/article/ro1117-jaundice-and-the-eyes.
10. Boris, I. G. (2014). Modern technologies for retinal scanning and imaging: an introduction for the biomedical engineer. *Biomed. Engg. Online,* 706–711. doi: 10.1186/1475-925X-13-52.
11. Mahesh, P. S., Mishra, D. K. C., Madhukumar, R., Ramanjulu, R., Reddy, S. Y., and Rodrigues, G. (2014). Fundus imaging with a mobile phone: A review of techniques. *Ind. J. Ophthalmol.,* 62(9), 960–962. doi: 10.4103/0301-4738.143949.
12. Fitri, U., Primaswara, R., and Sari, Y. A. (2017). Image processing for rapidly eye detection based on robust. *Inter. J. Elec. Comp. Engg. (IJECE),* 7(2), 823~830. ISSN: 2088-8708. DOI: 10.11591/ijece.v7i2.pp823-830.
13. HEALTHCARE BPO. Step by Step to K-means clustering. Data Science Blog.
14. Healthcare.ai. Google's new AI algorithm predicts heart disease by looking at your eyes.
15. The Verge by James Vincent published on Feb 19, 2018, 12:04pm EST.

24 Power generation prediction in dusty grid-tied solar panels

Debopoma Kar Ray[1,a], *Kabita Sahu*[1,b], *Tamal Roy*[1,c] *and Surajit Chattopadhyay*[2,d]

[1]Department of Electrical Engineering, MCKV Institute of Engineering, Kolkata, India

[2]Department of Electrical Engineering, Ghani Khan Choudhury Institute of Engineering & Technology, Malda, India

Abstract

In this paper, power prediction of a grid-tied PV system has been done in the presence of dirt on the surface of the panel to aid in the condition monitoring of the PV panels for both grid-tied/grid-isolated plants. This work has been advanced on a practical rooftop generating system of an institution wherein rigorous monitoring of the system data has been done for tenure of one month from the first day of cleaning to the day 21st of the cleaning and the deterioration of the system data has been reviewed from the analysis. This panel is cleaned fully at an interval of 21 days. The data obtained has been assessed using a curve fitting technique and optimized feature has been extracted. Considering the feature extraction, an algorithm has been developed for effective assessment of dirt accumulation and panel cleaning command generation in the PV arrays without the intervention of human effort. Similar work has been advanced in case studies with different connected loads with validation of the proposed algorithm.

Keywords: Algorithms, condition monitoring, curve fitting, grid-tied PV system, optimization

1. Introduction

Condition monitoring of renewable energy conversion systems is of immense concern in recent years since carbon footprint elimination is the greatest concern of modern days. In this context, current research has been seen to largely deal with predictive analysis for identifying various anomalies in the system. A detailed short circuit study on Ximeng's UHV project depicts the response of the system transient over-voltage and wind turbine power [1]. A novel algorithm has been seen with optimized doubly fed induction generator operation with laboratory prototype validation [2]. Analysis reveals the development of a protection scheme using a permanent magnet synchronous generator where over current and over voltage have been assessed within time limits with real-time data validation [3]. Analysis shows the effectiveness of power system stabilizer 4B (PSS4B) over PSS1A in damping system oscillation for a hybrid power system [4]. Normal electromagnetic relays have been seen to mal-operate during hybrid system anomalies. Hence studies depict the usage of stochastic ensemble-based classifier for effective protection of PV-wind hybrid micro-grids [5]. A comprehensive analysis of critical clearing time has been seen for a hybrid IEEE 30 bus system in the transient stability domain, wherein the system has been modeled in ETAP software with rotor angle alteration-based fault assessment with a satisfactory outcome [6]. A cost-optimized portable IoT-based control mechanism has been seen for hybrid micro-grid with multi-location authentication [7]. A low voltage ride through (LVRT) control scheme has been seen stabilizing renewable incorporated micro-grid to address various faults in both software and hardware environments [8]. A HOMER (Hybrid Optimization Model for Electric Renewables) Pro software-based eco-friendly hybrid network has been seen with 5.1 year payback period [9]. A MATLAB software-based case study has been seen for hybrid grid with the comparative study of battery-based and battery-fewer systems with satisfactory outcomes [10]. An artificial neural network-based dynamic voltage regulator has been seen for enhancing the system response for a hybrid system [11]. A grid-tied battery-storage-station-based hybrid system with MPPT technique has been used for extracting maximum power output from the system using a PID controller [12]. HOMER software-based decentralized coordinated control energy management system has been seen for hybrid systems with practical authentication using Starsim in PSCAD/EMTDC software [13]. A novel software-based hybrid DC micro-grid line-to-ground fault implementation yields satisfactory outcomes [14].

[a]debopoma86@gmail.com, [b]kabitasahu434@gmail.com, [c]tamalroy77@gmail.com, [d]pubnsc1@gmail.com

DOI: 10.1201/9781003540199-24

Series arc faults can be detected in a hybrid microgrid using Stockwell transform-based parametric analysis [15]. Unsymmetrical fault assessment has been seen in hybrid system using multi-resolution analysis of discrete wavelet transform (MRA of DWT) and Stockwell transform-based feature extraction [16]. Parallel arc faults can be detected in grid-tied/grid-isolated solar energy generation systems using Stockwell transform [17]. Assessment of nano and pico-grids has been seen with the analysis of power quality in hybrid networks [18, 19]. Robust control-oriented modeling and research have been seen for modern hybrid networks with successful outcomes [20]. None of the analyses have been observed to deal with the dirt accumulation pattern analysis in due course to automatically monitor the cleanliness of the PV panels in unmanned PV array-based grid-tied/grid-isolated power pants. Also, power prediction in dirty panels has not been seen to be assessed.

Thus, the motivation of this work is to provide predictive analysis on the quality degradation of PV panels due to the accumulation of non-uniform dust on the same. Firstly power output from the plant has been recorded on the day of panel cleaning and the average temperature of the area has been recorded. On the day of cleaning 4 consecutive data have been taken at 11 am, 1 pm, 3 pm, and 5 pm, respectively. Similar data collection has been pursued for consecutive days up to the 21st day of pane cleaning. The recorded data has been assessed using the curve fitting technique and the roots have been calculated. Thereafter the maxima, minima, and average values of the power available on each day have been found and again curve fitting technique has been used depending on which feature has been extracted for the development of an algorithm to effectively justify the degradation of panels and for effective online command generation for automatic panel cleaning of remote PV plants.

2. Data acquisition

This work has been advanced on a 5 kW grid-tied roof-top generating system of an institute during the tenure of 3rd–24th February, 2023 to during the time intervals 11 am, 1 pm, 3 pm, and 5 pm, respectively. The snapshot of the system has been given in Figure 24.1. The panel has been cleaned on 3rd February 2023 and data acquisition has been done on the following dates as given in Table 24.1 for the respective dates. The power obtained from the plant for the aforesaid days has been given in Table 24.2.

The case study aspires to explore how a curve-fitting technique is used to predict the power output of a grid-tied solar system. Curve fitting is a technique used to fit a curve or function to a set of data points. The goal is to find the best-fitting curve that accurately represents the data.

3. Curve-fitting-based data analysis

The data obtained from the field study (for D0) has been plotted in Figure 24.2. Since the plots of the other days show more or less similar responses, the data has been assessed using the curve fitting technique for decision-making, as depicted in Table 24.3 and the maxima and minima values of power obtained from the plant have been extracted from the recorded data (as provided in Table 24.4). Figure 24.2 has been assessed using the curve fitting technique in software where the nature of the response has been observed to resemble a 2nd degree polynomial as shown in Table 24.3. Similar analysis has been done for the other days (Table 24.3). In Table 24.3, the cleanest panel is a rational function whose nature retains up to 14th day of panel cleaning. On the 17th day of cleaning, the response resembles that of a cubic polynomial which again returns to rational nature on the successive days of dirt accumulation on the panel. However, the roots on each day have been seen to vary. Monitoring such responses for a long span,

Figure 24.1 System under study [Courtesy: 5kWp RTGS installed at MCKVIE, Liluah, Howrah]

Table 24.1 Data acquisition time intervals

Day	Date	Status	Time intervals
D0	3rd February, 2023	Clean panel	11 am, 1 pm, 3 pm, and 5 pm on each day
D4	7th February, 2023	4th day of cleaning	
D6	9th February, 2023	6th day of cleaning	
D7	10th February, 2023	7th day of cleaning	
D10	13th February, 2023	10th day of cleaning	
D12	15th February, 2023	12th day of cleaning	
D14	17th February, 2023	14th day of cleaning	
D17	20th February, 2023	17th day of cleaning	
D19	22nd February, 2023	19th day of cleaning	
D21	24th February, 2023	21st day of cleaning	

Table 24.2 Power obtained from the plant [Courtesy: 5kWp MCKVIE RTGS data]

Day	Time interval	Recorded data		Day	Time interval average temperature	Recorded data	
		Average temperature	Power (kW)			Average temperature	Power (kW)
D0	11 am	29	3.95	D12	11 am	29	2.69
	1 pm		2.4		1 pm		2.6
	3 pm		2.3		3 pm		1.55
	5 pm		2.29		5 pm		0.5
D4	11 am	32	3	D14	11 am	30	1.62
	1 pm		2.7		1 pm		1.6
	3 pm		2.3		3 pm		0.6
	5 pm		1.7		5 pm		0.5
D6	11 am	32	2.9	D17	11 am	32	2.2
	1 pm		2.59		1 pm		2
	3 pm		1.55		3 pm		1.3
	5 pm		0.5		5 pm		0.5
D7	11 am	31	3.25	D19	11 am	34	2.65
	1 pm		2.75		1 pm		2.4
	3 pm		2		3 pm		1
	5 pm		0.5		5 pm		0.5
D10	11 am	30	3.45	D21	11 am	34	2.19
	1 pm		3.2		1 pm		1.95
	3 pm		1.69		3 pm		0.97
	5 pm		0.5		5 pm		0.6

decision-making regarding dirt accumulation and possible cleaning of the panels is very difficult. Thus, for more specific feature extraction, the maxima, minima, and average value analysis of the data sets have been extracted as presented in Table 24.4. The data obtained has been assessed using the curve fitting technique as presented in Table 24.5 and the response has been given in Figure 24.3. From Table 24.5, it has been inferred that, the maxima value variation for the tenure of 21 days of dirt accumulation in the panel resembles an 8th degree polynomial function.

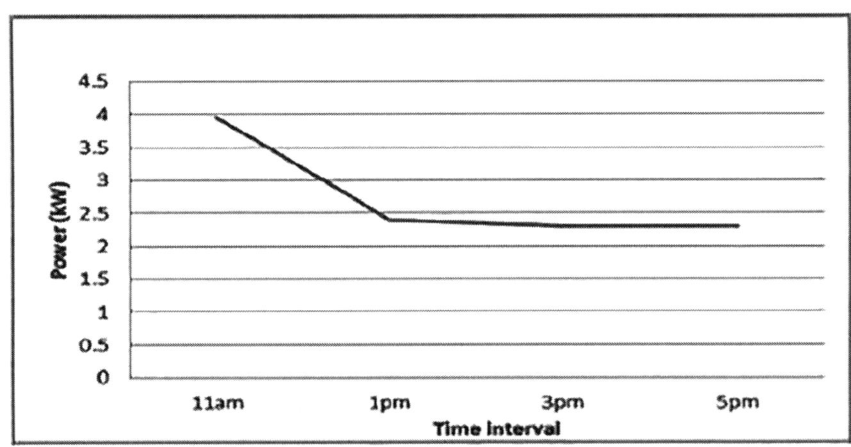

Figure 24.2 D0 power extraction for the whole day

Table 24.3 Curve fitting analysis of recorded data [Courtesy: 5kWp MCKVIE RTGS]

Day	Response analysis	Day	Response analysis	Day	Response analysis
D0	Equation: 2nd degree polynomial $f(x)=\frac{(m_1x^2+m_2x+m_3)}{(x+n_1)}$	D7	Equation: 2nd degree polynomial $f(x)=\frac{(m_1x+m_2)}{(x^2+n_1x+n_2)}$	D14	Equation: 3rd degree polynomial $f(x)=m_1x^3+m_2x^2+m_3x+m_4$
D4	Equation: 2nd degree polynomial $f(x)=\frac{(m_1x+m_2)}{(x^2+n_1x+n_2)}$	D10	Equation: 2nd degree polynomial $f(x)=\frac{(m_1x+m_2)}{(x^2+n_1x+n_2)}$	D17	Equation: 3rd degree polynomial $f(x)=m_1x^3+m_2x^2+m_3x+m_4$
D6	Equation: 2nd degree polynomial $f(x)=\frac{(m_1x+m_2)}{(x^2+n_1x+n_2)}$	D12	Equation: 2nd degree polynomial $f(x)=\frac{(m_1x+m_2)}{(x^2+n_1x+n_2)}$	D19	Equation: 2nd degree polynomial $f(x)=\frac{(m_1x+m_2)}{(x^2+n_1x+n_2)}$
Day: D21 (dirtiest day) Equation: 2nd degree polynomial $f(x)=\frac{(m_1x+m_2)}{(x^2+n_1x+n_2)}$					

Table 24.4 Maxima, minima, and average value analysis of recorded power [Courtesy: 5kWp MCKVIE RTGS]

Data	Day									
	D0	D4	D6	D7	D10	D12	D14	D17	D19	D21
Maxima (kW)	3.95	3	2.9	3.25	3.45	2.69	1.62	2.2	2.65	2.19
Minima (kW)	2.29	1.7	0.5	0.5	0.5	0.5	0.5	0.5	0.5	0.6
Average (kW)	2.74	2.4	1.89	2.13	2.21	1.83	1.08	1.5	1.64	1.43

The minima value variation resembles a Gaussian function and the average value resembles a rational function. Thus, if the responses resemble any of the above characteristic natures, cleaning of panels is suggested which can be implemented for unmanned roof-top or yard-mounted grid-tied or grid-isolated PV generating systems. Considering Table 24.4, an algorithm has been proposed in successive sections.

4. Algorithm

Depending on the feature extraction as in Table 24.5, a direction has been proposed for dirt accumulation testing and panel cleaning command for grid-tied/grid-isolated PV arrays.

Thus it can be concluded that, if the nature of the Polynomial/Gaussian/Rational functions can be monitored, adequate cleaning command generation can be given for the automatic cleaning of PV panels regularly. Also depending on the nature of the variation as in Figure 24.3 any other day power nature can be predicted in the advent of dirt accumulation.

Step 1: Record power obtained from the PV system for a chosen tenure at regular intervals.
Step 2: Compute maxima, minima, and average values on each day
Step 3: Perform curve fitting technique-based feature extraction
Step 4: Monitor the characteristic nature and extract the feature
Step 5: Assess the results and predict power generation from dusty grid-tied panels.

5. Case study

Case studies have been done on the same panel on panel in the month of March 2023 for three days keeping the panel dirty for the tenure of 21 days on the 8th, 11th, and 18th day of the cleaning of the panel and the maxima, minima and average values of the days have been recorded. The data obtained has been plotted in the same frame as the data sets in Figure 24.4.

In Figure 24.4, "D8(0)", "D11(0)" and "D18(0)" denote the 8th, 11th, and 18th day power prediction from Figure 24.3 for the previous data collection. and "D8(Case)", "D11(Case)" and "D18(Case)" denote the case studies done in the month of March 2023 on the 8th, 11th, and 18th day of the panel cleaning which has been plotted in the same frame of Figure 24.3. The results show that by this technique power extraction can be well predicted in dirty grid-tied solar panels with satisfactory outcome.

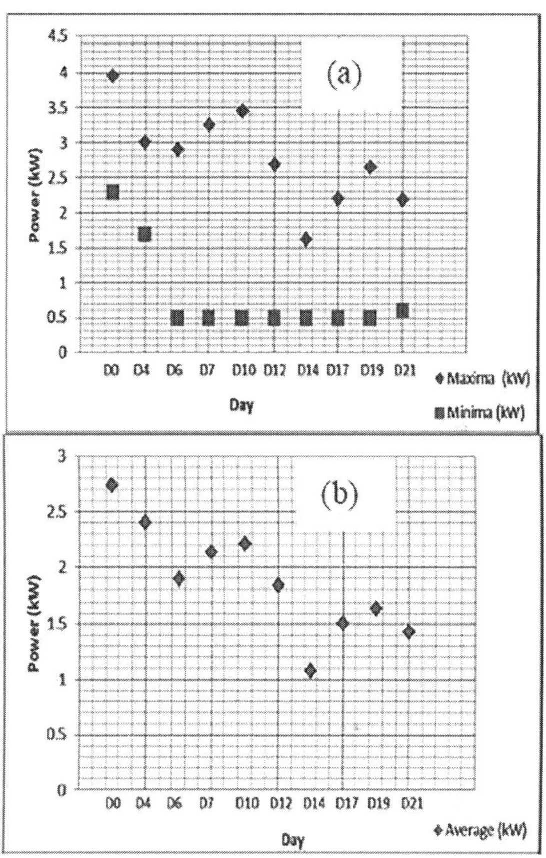

Figure 24.3 Variation in (a) maxima, minima, and (b) average values of power obtained from the grid-tied RTGS [Courtesy: 5kWp MCKVIE RTGS]

Table 24.5 Curve fitting analysis of maxima, minima, and average values [Courtesy: 5kWp MCKVIE RTGS]

Power (kW)	Response analysis
Maxima	8th degree polynomial function: $f(x) = m_1x^8 + m_2x^7 + m_3x^6 + m_4x^5 + m_5x^4 + m_6x^3 + m_7x^2 + m_8x + m_9$
Minima	Gaussian function: $f(x) = h_1e^{(-((x-p_1)/q_1)^2)} + h_2e^{(-((x-p_2)/q_2)^2)} + h_3e^{(-((x-p_3)/q_3)^2)}$
Average	Rational function: $f(x) = h_1e^{(-((x-p_1)/q_1)^2)} + h_2e^{(-((x-p_2)/q_2))} + h_3e^{(-((x-p_3)/q_3)^2)}$

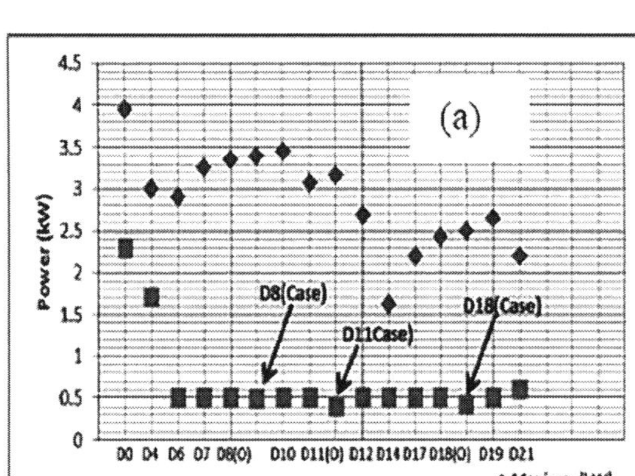

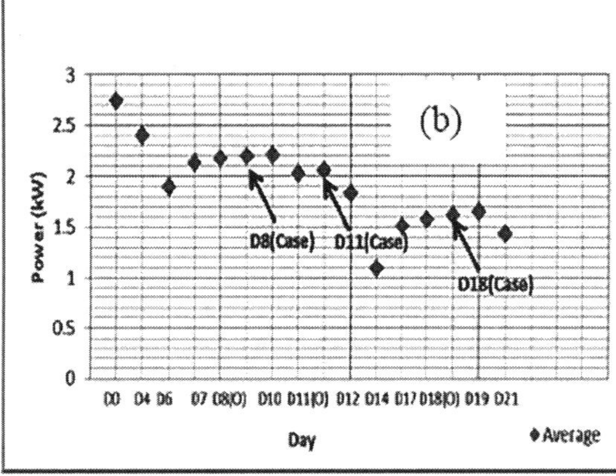

Figure 24.4 Validation of proposed technique monitoring (a) maxima, minima (b) average values of power obtained [Courtesy: 5kWp MCKVIE RTGS]

6. Conclusions

Renewable energy systems harness the power of renewable energy sources to meet the energy needs of society while reducing green-house gas emissions and mitigating climate change. Hence, the work has been advanced, wherein the concern was to provide a rule set for assisting in the automatic cleaning of PV panels in due course. For this data has been collected from a grid-tied Solar RTGS of an institute for a tenure of 21 days from the date of cleaning and the deterioration of the power obtained from the panel has been seen under the accumulation of dust on the surface of the plates. Curve fitting technique-based optimization has been pursued and depending on the nature of the curve-fitting, the best-optimized parameter has been extracted depending on which an algorithm has been introduced for proper assessment of panel degradation and the generation of cleaning command for the arrays.

Acknowledgement

The authors are grateful to MCKV Institute of Engineering, Howrah, Kolkata, India, for providing necessary facilities for analysis.

References

1. Cheng, X., Liu, H., Wu, L., Li, Y., and Sun, D. (2021). Artificial short-circuit test and result analysis of Ximeng UHV AC/DC transmission project. *IEEE 5th Conf. Energy Internet Energy Sys. Integ. (EI2)*, 2636–2642. doi: 10.1109/EI252483.2021.9713069.
2. Verma, C. A., Kumar, U., Das, B., Das, A., and Kasari, P. R. (2019). Fault tolerant control of hybrid wind-solar generation system. *IEEE Inter. Conf. Sustain. Energy Technol. Sys. (ICSETS)*, 069–074. doi: 10.1109/ICSETS.2019.8744849.
3. Farooq, U. (2019). A reliable approach to protect and control of wind solar hybrid DC microgrids. *IEEE 3rd Conf. Energy Internet Energy Sys. Integ. (EI2)*, 348–353. doi: 10.1109/EI247390.2019.9062101.
4. Li, Z. E., Tiong, T. C., and Wong, K. I. (2019). Improving transient stability of diesel-wind-solar hybrid power system by using PSS. *1st Inter. Conf. Elec. Control Instrum. Engg. (ICECIE)*, 1–6. doi: 10.1109/ICECIE47765.2019.8974702.
5. Manohar, M., Koley, E., and Ghosh, S. (2020). Stochastic weather modeling-based protection scheme for hybrid PV–wind system with immunity against solar irradiance and wind speed. *IEEE Sys. J.*, 14(3), 3430–3439. doi: 10.1109/JSYST.2020.2964990.
6. Rahman, S. R., Miah, M. A. R., Tarif, Z. N., and Jyoty, M. J. A. (2019). Determining critical clearing time in transient stability assessment for hybrid power system. *IEEE 6th Inter. Conf. Engg. Technol. Appl. Sci. (ICETAS)*, 1–5. doi: 10.1109/ICETAS48360.2019.9117282.
7. Priya, P., Narayana, P., and Reddy, P. D. K. (2021). Monitoring and controlling of IoT-based solar wind hybrid system. *3rd Inter. Conf. Adv. Comput. Comm. Control Netw. (ICAC3N)*, 569–574. doi: 10.1109/ICAC3N53548.2021.9725746.
8. He, Y., Wang, M., and Xu, Z. (2020). Coordinative low-voltage-ride-through control for the wind-photovoltaic hybrid generation system. *IEEE J. Emerg. Select. Topics Power Elec.*, 8(2), 1503–1514. doi: 10.1109/JESTPE.2019.2958213.
9. Lin, E., Phan, B. C., and Lai, Y. C. (2019). Optimal design of hybrid renewable energy system using HOMER: A case study in the Philippines. *2019 Southeast Con.*, 1–6. doi: 10.1109/SoutheastCon42311.2019.9020552.
10. Fajardo, R. R. M., Mallari, E. V., Vivo, P. J. H., Pacis, M. C., and Martinez, J. (2020). Impact study of a microgrid with battery energy storage system (BESS) and hybrid distributed energy resources using MAT-

LAB Simulink and T-test analysis. *2020 IEEE 12th Inter. Conf. Humanoid Nanotechnol. Inform. Technol. Comm. Control Environ. Manag. (HNICEM)*, 1–6. doi: 10.1109/HNICEM51456.2020.9400061.

11. Tata, H. and Veerraju, M. S. (2022). Power quality improvement of a hybrid renewable energy systems using dynamic voltage restorer with ANN controller. *2022 IEEE 2nd Inter. Conf. Sustain. Energy Future Elec. Trans. (SeFeT)*, 1–4. doi: 10.1109/SeFeT55524.2022.9908896.
12. Rakib, K., Salimullah, S. M., Hossain, M. S., Chowdhury, M. A., and Ahmed, J. S. (2020). Stability analysis of grid integrated BESS based hybrid photovoltaic (PV) and wind power generation. *2020 IEEE Region 10 Symp. (TENSYMP)*, 1717–1720. doi: 10.1109/TENSYMP50017.2020.9230650.
13. Lin, Y. and Fu, L. (2022). A study for a hybrid wind-solar-battery system for hydrogen production in an Islanded MVDC network. *IEEE Acc.*, 10, 85355–85367. doi: 10.1109/ACCESS.2022.3193683.
14. Zulu, M. and Ojo, E. (2022). Power flow and fault analysis simulation for a PV/wind hybrid DC microgrid. *2022 30th South. Afr. Universities Power Engg. Conf. (SAUPEC)*, 1–6. doi: 10.1109/SAUPEC55179.2022.9730684.
15. Kar Ray, D., Das, D., Shah, O. P., Singh, S. K., and Chattopadhyay, S. (2019). Condition monitoring of solar-wind hybrid micro-grid using stockwell transform based parametric values. *MCCS-2019*. vol. 673 pp 281–294, doi: 10.1007/978-981-15-5546-6_23.
16. Kar Ray, D. and Chattopadhyay, S. (2019). Fault analysis in solar-wind hybrid micro-grid using MRA and ST based statistical analysis. *IET Sci. Meas. Technol.*, 14(6), 639–650. doi: 10.1049/iet-smt.2019.0279.
17. Kar Ray, D., Roy, T., and Chattopadhyay, S. (2021). Adaptive time duration computation for parallel arc fault in wind-solar hybrid system. *Algorithms Intell. Sys. Book Ser. (AIS)*. pp. 145-154, doi: 10.1007/978-981-16-3368-3_15.
18. Chattopadhyay, S. (2022). Nanogrids and picogrids and their integration with electric vehicles. *IET London*. pp. 1-367, ISBN: 978-1-83953-482-9.
19. Chattopadhyay, S., Mitra, M., and Sengupta, S. (2011). Electric power quality. pp. 1-177, Springer: Netherland, ISBN: 978-94-007-0635-4.
20. Roy, T. and Barai, R. K. (2023). Robust control-oriented linear fractional transform modeling-applications for the μ-synthesis based H∞ control. Springer: Netherland, pp. 1-161, 2023. ISBN: 978-98-1197-461-8.

25 Harmonic distortion and firing angle driven optimization-based adaptive filter system topology for 12 pulse HVDC converter

Surajit Chottapadhyay[a], *Goutam Kumar Ghorai*[b] *and Dipendu Debnath*[c]

Department of Electrical Engineering, Ghani Khan Choudhury Institute of Engineering & Technology, Malda, West Bengal, India

Abstract

This paper attempts an adaptive filter system design for 12-pulse HVDC converter system. The initial triggering angle has a great influence on the output of a 12-pulse converter system. Additionally, fractional harmonics are found at the output due to voltage conversion. In this work, the influence of variation of triggering angle on DC components and on sub-harmonics has been analyzed. Then, based on their relation, optimization has been proposed for the filter network. Depending on the allowed harmonics constraints, system parameters are to be chosen for an adaptive filter unit. This method will help to optimize between triggering angle and harmonics content and is capable of suggesting the best system parameters for the filter.

Keywords: Firing angle, harmonic distortion, HVDC converter, optimization

1. Introduction

Converter system plays an important role in integrating DC lines with AC networks. Continuous operation and good performance are desired for the reliable operation of the whole AC-DC network. Lots of studies are being carried out to achieve the best performance of converter units. The performance of a converter unit depends on desired output level as well as on the quality of the output.

Input and output side impedances are important components that influence on the converter [1]. The quality of the converter depends on the harmonics present in the output of the converter. Harmonic instability [2] can occur due to the presence of harmonics at the output in large amounts. The problem can be minimized by the harmonics suppression method [3]. To improve the quality, a 12-pulse controlled rectifier can be integrated with the DC side harmonic suppression. Sometimes, a predictive current control strategy may be used. To improve the harmonics in voltage source converter and to reduce output total harmonic distortion. Harmonics mitigations are followed in converter configuration to improve the quality of the output power [4–8].

Study reveals that the output power gets distorted by low-order harmonics contents [9]. Many system models and mathematical parameters have been introduced to represent various network systems and quality issues [10–12]. Fast Fourier transform and wavelet transform-based assessment methods have been introduced to measure and analyze harmonic distortions [13–16]. A new method has been introduced for reducing harmonic content in the multilevel converter [17]. A modified space vector pulse model has been introduced for modulation used in the converter [18]. Sometimes, an auxiliary supply is used to reduce the harmonics of the main system [19]. Vector controlled model has been used for harmonics reduction and quality improvement in refs [20, 21] whereas a probabilistic model has been introduced in ref [22] for quality improvement.

Converter operation is very much dependent on the firing mechanism. The firing angle decides the performance of the converter. In this work, an attempt has been taken to observe the influence of variation of initial firing angle in a 12-pulse converter on its output and quality. The quality has been assessed in terms of low-order frequencies present in the system. After the introduction, converter modeling has been presented in Section 2. Then DC components and low-order harmonics are studied in Section 3. Section 4 describes the observation made in terms of harmonics distortions. Then, a model for designing an adaptive filter configuration has been proposed in Section 5 followed by a conclusion in Section 6.

2. 12-pulse converter modeling

DC link end terminal has been shown in Figure 25.1, it consists of a 12-pulse converter that combines two 6-pulse converters. 6-pulse configuration

[a]surajitchattopadhyay@gmail.com, [b]goutam@gkciet.ac.in, [c]dipendudebnath093@gmail.com

DOI: 10.1201/9781003540199-25

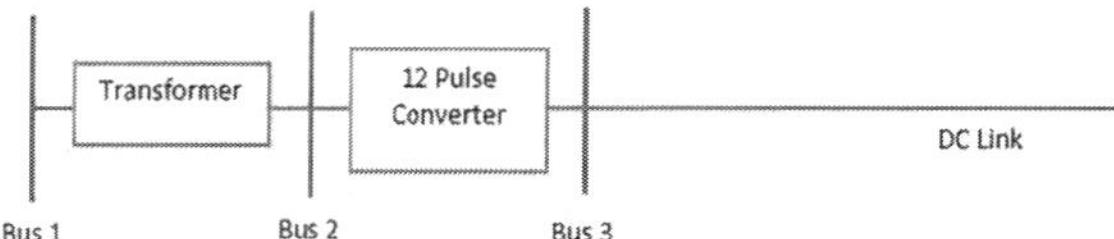

Figure 25.1 12-pulse converter-based HVDC link line

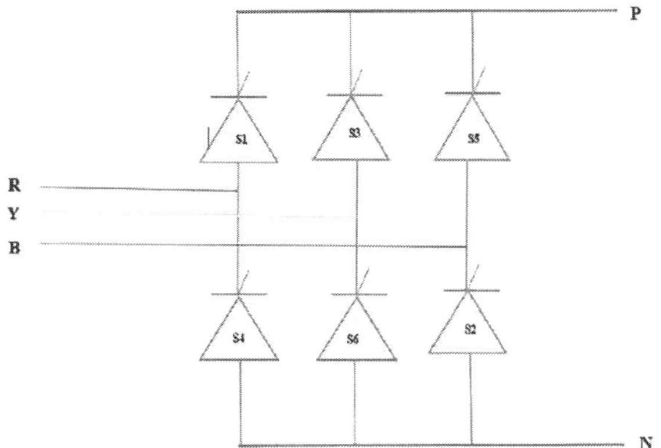

Figure 25.2 6-pulse converter

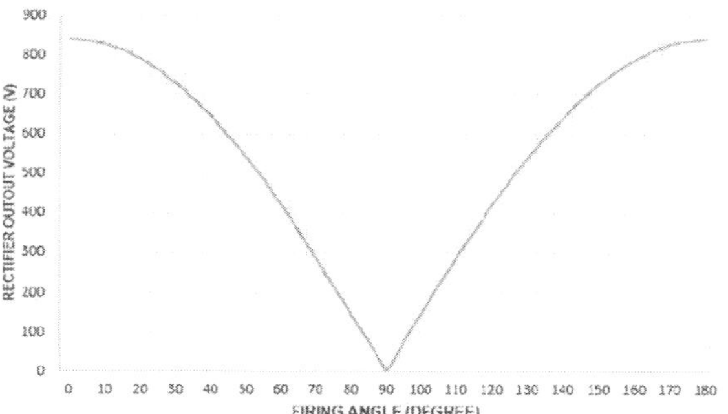

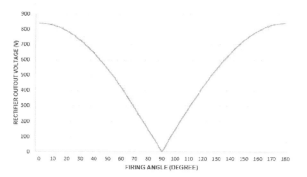

Figure 25.3 Voltage magnitude for different α

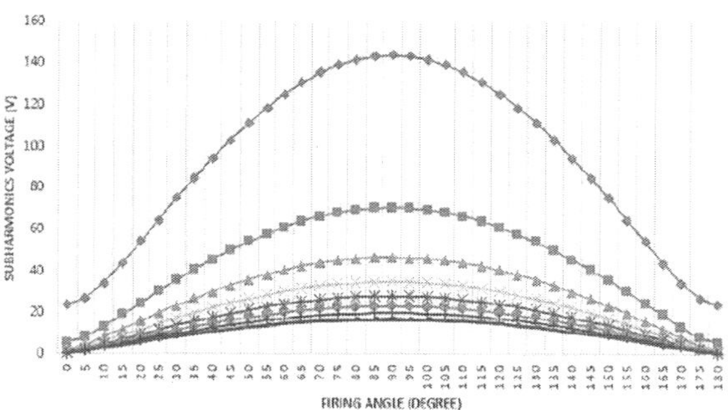

Figure 25.4 Different voltages amplitudes for different firing angle

of the converter has been shown in Figure 25.2. This converter is connected to a transformer and DC link line. Normally, in 12-pulse configuration, one 6-pulse converter is connected to the transformer in star-delta and another is connected with a 6-pulse converter connected to a delta-delta transformer. This also helps to adjust or cope with a 30° phase shift in either direction. Output depends on the triggering angle. After initial triggering, all switches follow the previous triggering after certain intervals (commonly 60°).

3. Performance study with variable initial triggering angle

Obviously, the output is not pure step and has lots of harmonics. Frequencies, available in the output waveforms, can be expressed as follows.

$h = 6n$ [Where n is an integer]

The harmonic voltages present in the DC side, also known as DC harmonics, can be written as follows:

$$V_h = V_{d0}\frac{\sqrt{2}}{h^2-1}[1 + (h^2 - 1)sin^2\alpha]^{\frac{1}{2}}$$

where, $V_{d0} = 1.35E_{LL}$

The output voltage depends on the triggering or firing angles of switches The output voltage versus firing angle has been plotted in Figure 25.3. It shows that with the increase of firing angle output decreases up to 90°, and then it increases. Then various sub-harmonics have been determined up to 50 Hz (considering it as fundamental as it is the operating frequency of the AC side of the converter unit). Sub-harmonics for different initial firing angles have been determined for 0–90° and for 90–180°. Variations of harmonic components for change of firing angle have been shown in Figure 25.4. It shows an increase of harmonic magnitude up to 90°, and then it decreases up to 180°.

4. Harmonic distortion

Total harmonic distortion (THD) refers to the ration of harmonics contribution to the fundamental. Distortion factors were determined for different firing angles considering sub-harmonics up to 50 Hz and considering 50 Hz as reference fundament components (as DC output does not have any frequency). Variation of THD with firing angle has been shown in Figure 25.5. THD increases up to 35° firing angle and then it becomes almost constant up to 90°.

5. Adaptive optimization of alpha

The results show that the output of a 12-pulse converter consists of harmonics. Two aspects have been revealed from the previous results for up to 0–90° (a) net DC output decreases with the increase of

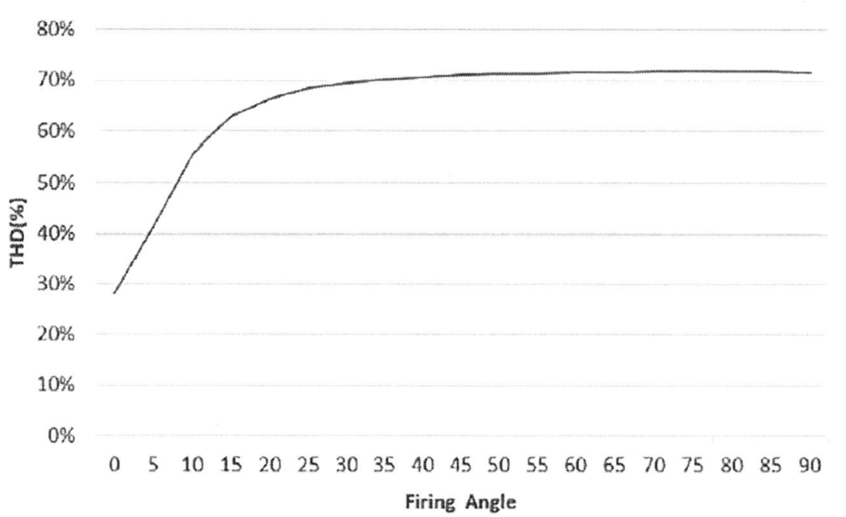

Figure 25.5 Total harmonic distortion for different α

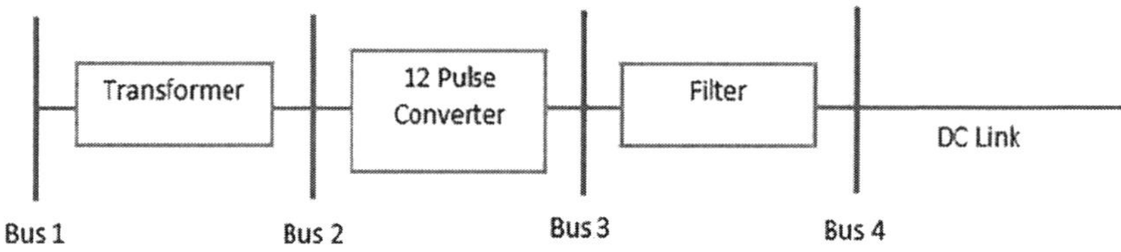

Figure 25.6 HVDC link line with DC side filter unit

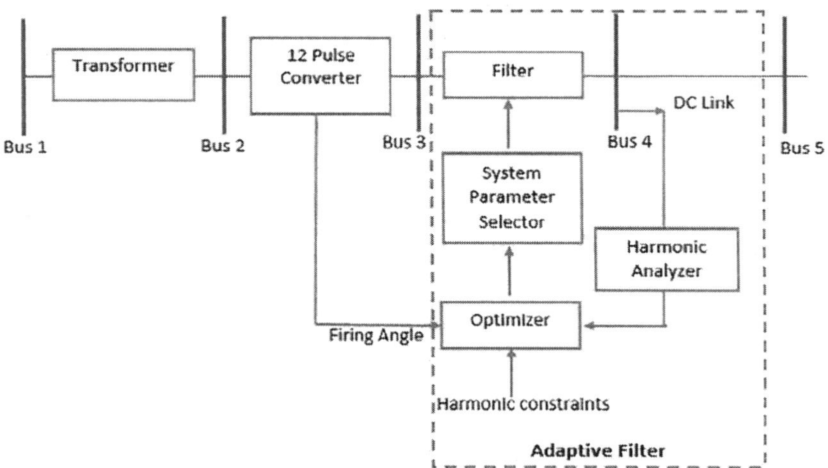

Figure 25.7 Model for adaptive filter system design for 12 pulse HVDC converter

initial firing angle and (b) THD decreases with the increase of firing angle. To minimize harmonics content, the filter is used as shown in Figure 25.6. Here, an adaptive file has been proposed as shown in Figure 25.7. It will take the firing angle and harmonics present in the output as feedback signal.

Allowable THD range will be considered as constraints. Optimization will take place prior to learnings from the relation of output and THD with firing angle and filter parameters will be adjusted. System parameters of the filter will be selected from the outcomes of optimization and will be used for filter operation. Thus, the filter configuration becomes dynamic as well as adaptive with respect to both output and harmonics content present at the output.

6. Conclusions

An adaptive filter configuration topology for 12-pulse HVDC converter system has been presented in this work. Results show that initial triggering angles influence the output of a 12-pulse converter system as well as the harmonics distortion that occurs at the output of the converter. Net DC output is seen to decrease with the increase of firing angle whereas THD de is seen to increase for the same variation of firing angle. This indicates that a fixed configuration of the filter unit cannot be the best solution. Therefore, optimization has been proposed for filter system configuration using the prior knowledge of initial firing angle and THD, and net DC components of the output.

Acknowledgement

Authors acknowledge the support received from GKCIET, Malda and thank for all their cooperation.

References

1. Riedel, P. (2005). Harmonic voltage and current transfer, and AC- and DC-side impedances of HVDC converters. *IEEE Trans. Power Delivery*, 20(3), 2095–2099. DOI: 10.1109/TPWRD.2005.848719.
2. Mao, C., He, R., and Dong, P. (2022). Research on criterion for harmonic instability of HVDC power transmission system composed of double-circuit lines bipolar 12-pulse converter considering converter transformer saturation. *2022 7th Asia Conf. Power Elec. Engg. (ACPEE)*, 601–605. DOI:10.1109/ACPEE53904.2022.9783758.
3. Xiaoqing, R., Damin, Z., Shuanglian, Y., Jianxiong, C., and Zexiang, D. (2022). Harmonic suppression strategy of 12-pulse controlled rectifier. *2022 IEEE 5th Inter. Elec. Energy Conf. (CIEEC)*, 3485–3490. DOI: 10.1109/CIEEC54735.2022.9846619.
4. Gbadega Peter, A. and Saha, A. K. (2019). The impacts of harmonics reduction on THD analysis in HVDC transmission system using three-phase multi-pulse and higher level converters. *2019 South. Afr. Universities Power Engg. Conf./Robot. Mechatron./Patt. Recogn. Assoc. South Afr. Bloemfontein, South Africa, 2019*, 444–449. DOI: 10.1109/RoboMech.2019.8704743.
5. Agarwal, R. and Singh, S. (2014). Harmonic mitigation in voltage source converters based HVDC system using 12-pulse AC-DC converters. *2014 Ann. IEEE India Conf. (INDICON)*, 1–6. DOI: 10.1109/INDICON.2014.7030590.
6. Yamashita, K. I., Morioka, T., Fukumoto, K., and Nishikata, S. (2016). A high-voltage direct current transmission system using a 12-pulse thyristor inverter without AC grids' low-order harmonic distortions. *2016 19th Inter. Conf. Elec. Mac. Sys. (ICEMS)*, 1–5.
7. Jaipradidtham, C. and Kinnares, V. (2006). Harmonic flow analysis of HVDC systems for discrete wavelet transform with advanced of the harmonic impedances and voltage source transients model of AC-DC converters. *TENCON 2006 - 2006 IEEE Region 10 Conf.*, 1–4. DOI: 10.1109/TENCON.2006.343777.
8. Nandi, S. K., Riadh, R. R., and Rahman, S. (2015). Investigation of THD on a 12-pulse HVDC transmission network and mitigation of harmonic currents using passive filters. *2nd Inter. Conf. Elec. Inform. Comm. Technol. (EICT)*, 510–515. DOI: 10.1109/EICT.2015.7392006.
9. Yamashita, K. I., Morioka, T., Fukumoto, K., and Nishikata, S. (2016). A high-voltage direct current transmission system using a 12-pulse thyristor inverter without AC grids' low-order harmonic distortions. *2016 19th Inter. Conf. Elec. Mac. Sys. (ICEMS)*, 1–5.
10. Chattopadhyay, S.. (2022). Nanogrids and picogrids and their integration with electric vehicles. *IET, London*. pp. 1-367, ISBN: 978-1-83953-482-9.
11. Chattopadhyay, S. and Das, A.. (2021). Overhead electric power lines: Theory and practice. *IET, London*. pp. 1-361, ISBN: 9781839533112.
12. Chattopadhyay, S., Mitra, S., and Sengupta, S. (2011). Electric power quality. pp. 1-177, Springer: Netherlands. ISBN: 978-94-007-0635-4.
13. Ghosh, S. S., Chattopadhyay, S., and Das, A. (2022). Fast Fourier transform and wavelet-based statistical computation during fault in snubber circuit connected with robotic brushless direct current motor. IET *Cogn. Comput. Sys.* 4(1), 31–44. https://DOI.org/10.1049/ccs2.12041.
14. Ganguly, P., Chattopadhyay, S., and Biswas, B. N. (2021). Adaptive algorithm for battery charge monitoring based on frequency domain analysis. IETE J. Res.. 69(9), 6398–6408. https://DOI.org/10.1080/03772063.2021.2000508.
15. Mukherjee, N., Chattopadhyaya, A., Chattopadhyay, S., and Sengupta, S. (2020). Discrete-wavelet-transform and stockwell-transform-based statistical parameters estimation for fault analysis in grid-connected wind power system. *IEEE Sys. J.*, 14(3), 4320–4328. 2020, DOI. 10.1109/JSYST.2020.298413224.
16. Chattopadyay, S., Chattopadhyaya, A., and Sengupta, S. (2014). Harmonic power distortion measurement in Park Plane. *Measurement*, 51, 197–205. https://DOI.org/10.1016/j.measurement.2014.02.021.
17. Kumar, A. R., Bhaskar, M. S., Subramaniam, U., Almakhles, D., Padmanaban, S., and Bo-Holm Nielsen, J. (2019). An improved harmonics mitigation scheme for a modular multilevel converter. *IEEE Acc.*, 7, 147244–147255. DOI: 10.1109/ACCESS.2019.2946617.
18. Yan, H., Yang, J., and Zeng, F. (2022). Three-phase current reconstruction for PMSM drive with modified twelve sector space vector pulse width modu-

lation. *IEEE Trans. Power Elec.*, 37(12), 15209–15220. DOI: 10.1109/TPEL.2022.3188425.
19. Fukuda, S., Ohta, M., and Iwaji, Y. (2008). An auxiliary-supply-assisted harmonic reduction scheme for 12-pulse diode rectifiers. *IEEE Trans. Power Elec.*, 23(3), 1270–1277. DOI: 10.1109/TPEL.2008.921165.
20. Singh, B., Bhuvaneswari, G., and Garg, V. (2006). Power-quality improvements in vector-controlled induction motor drive employing pulse multiplication in AC-DC converters. *IEEE Trans. Power Delivery*, 21(3), 1578–1586. DOI: 10.1109/TPWRD.2006.874660.
21. Singh, B., Bhuvaneswari, G., and Garg, V. (2006). Harmonic mitigation using 12-pulse AC-DC converter in vector-controlled induction motor drives. *IEEE Trans. Power Delivery*, 21(3), 1483–1492. DOI: 10.1109/TPWRD.2005.860265.
22. Ngandui, E. and Sicard, P. (2004). Probabilistic models of harmonic currents produced by twelve-pulse AC/DC converters. *IEEE Trans. Power Delivery*, 19(4), 1898–1906. DOI: 10.1109/TPWRD.2004.835422.

26 Design and experimental validation of a budget friendly liquefied petroleum gas detector and alarming system

Raja Ram Kumar[1,a], *Didhiti Dey*[1,b], *Sreejana Sengupta*[1,c], *Sachin Mondal*[1,d], *Pratapaditya Mondal*[1,e] *and Awani Bhushan*[2,f]

[1]Department of Electrical Engineering, Ghani Khan Choudhury Institute of Engineering & Technology, Malda, India

[2]School of Mechanical Engineering, Vellore Institute of Technology, Chennai, Tamil Nadu-600127, India

Abstract

The prime focus of this paper is to design an experimental validation of a budget friendly liquefiedpetroleum gas detector and alarming system (BFLPGDAS). The proposed system comprises several components such as MQ2 sensor, relay, buzzer, LED light and exhaust fan. In this system, MQ2 sensor senses gas leakage and relay tune on the alarming circuit as well as exhaust fan to reduce concentration level of LPG from the effected surrounding area. To check the sensitivity and the performance of BFLPGDAS, a prototype has been developed. In this connection, some performance parameters have been tested. Furthermore, a cost estimation of the proposed BFLPGDAS has been conducted, revealing its highly cost-effective and budget-friendly nature. The device's fast response time and low cost make it particularly suitable for the aforementioned application.

Keywords: Alarm system, exhaust fan, MQ2 sensor,LPG leakage, relay

Introduction

The need for energy is continuously rising as the population expands. A substantial proportion of the population in developing countries lives in rural regions.In these regions, people rely heavily on primary sources of energy such as crude oil, hard coal, natural oil, and wood to meet their energy needs. Out of all available options, wood and coal are the most readily accessible and affordable sources of energy. However, utilizing these sources as cooking fuel releases smoke, leading to health problems among rural families [1]. Also chopping wood was creating deforestation ultimately resulting global warming.Again, another primary source like coal is unavailable due to less production from mines and because of pollution, cooking fuel via coal is also not possible. Liquefied petroleum gas (LPG), being a less polluting and eco-friendly energy source, serves as a solution to this problem [2]. LPG is the mixture of ethylene, ethane, propylene, propane,normal butane, isobutene, isobutylene and butylene, Molecular formula of LPG is C3H8 or C4H10. The correct LPG flame temperature is 1967°C or 3573°F. Through various government schemes, LPG is made available to every individual in India. As of 2021, according to the Petroleum Planning & Analysis Cell, the number of LPG bottling plants had reached a cumulative count of 202. This growth aligns with the consistent increase in LPG demand within the nation. Consequently, these advancements enabled India to escalate its LPG coverage from approximately 56.2% in 2015 to an impressive 99.8% by 2021. In 2018, as per the National Statistical Office (NSO), 61% of all Indian households employed LPG for cooking purposes. Among these, 48.3% of rural households embraced LPG, whereas the prevalence was significantly greater in urban regions, reaching 86.6%. The average value of LPG consumption for India during the period 1980–2021 was 302.83 thousand barrels per day. The lowest recorded LPG consumption was 32.59 thousand barrels per day in 1980, while the highest reached 886.55 thousand barrels per day in 2021. By April 2022, the total count of active LPG domestic customers in India had surged to approximately 30.5 crores from 14.9 crore customers in 2015.

Notable companies involved in LPG distribution include Bharat Petroleum Corporation Limited, Hindustan Petroleum Corporation Limited, Indian Oil Corporation Ltd., Total Energies SE, and Reliance Petroleum Ltd. [3]. LPG derived from natural gas necessitates a gas separation facility rather than an oil refinery, resulting in cost-effective extraction. This approach stands as a more economical alternative compared to producing LPG from crude oil in refineries. The extraction of LPG from natural gas is projected to demonstrate

[a]rajaram@gkciet.ac.in, [b]didhitidey001@gmail.com, [c]sushantasengupta215@gmail.com, [d]titas8145103898@gmail.com, [e]sachinmondal646@gmail.com, [f]awani.bhushan@vit.ac.in

DOI: 10.1201/9781003540199-26

substantial growth within the Indian LPG market. But in spite of these benefits, LPG leakage can create more disaster than any other fuel. Those reasons are namely human error, knob error, cylinder burst and faulty regulator. Due to which it is very important for a handy detector fitted to cylinder to avoid the risks of ignition[4]. Keeping the above points in mind, authors have proposed a very robust, cheap LPG gas detector and alarming system. The prime focus of this paper is as follows:

- To propose a budget friendly and simple detector to reach maximum people in country.
- To test the device for different parameters to detect the range of the device.

So,a device is designed and fabricated to fulfill the above purposes. In this connection a number of tests are carried out like time response,reduction of concentration level with time and total time of operation of exhaust fan.

The structure of this paper is segmented as follows: Section 2 outlines the model description and operational principle. Section 3 provides an in-depth account of the experimental setup and outcomes. Lastly, Section 4 draws the paper to a conclusion.

2. Model description and operating principle

Figure 26.1 depicts the block diagram of the system proposed in this study. It consists of different components like MQ2 sensor, relay, buzzer, LED light and exhaust fan.Basically, MQ2 sensor is the most important component of the above system which is one of the members of MQ series sensor. This device features a MOS (metal oxide semiconductor) sensor, which incorporates a gas sensing material composed of tin dioxide semiconductor. When the tin dioxide semiconductor layer is heated to a high temperature, it attracts oxygen molecules to its surface, resulting in oxygen adsorption. In a clean air environment, oxygen molecules draw electrons from the tin dioxide's conduction band, leading to the creation of an electron depletion layer just below the surface of the tin dioxide particles. This forms a potential barrier that renders the tin dioxide film highly resistant, impeding the flow of electric current [5]. However, in the presence of reducing gases, the density of adsorbed oxygen on the surface diminishes due to reactions with the reducing gases. This reaction lowers the potential barrier, causing the release of electrons into the tin dioxide material. This electron release allows current to flow freely through the sensor, indicating the presence of these reducing gases. The operating characteristics of the MQ2 gas sensor involve a 5V DC power supply and an approximate power consumption of 800mW. Additionally, the device includes a relay component. Relays function as electronically and electromechanically operated switches. When the relay's contact is in the open state (NO), the relay remains unenergized. Here the relay is for feedback system, it conveys the message of LPG detection to the alarm system. The device also consists of buzzer. An audio signaling device, such as a buzzer, can take the form of piezoelectric, or mechanical type andelectromechanical. It is equipped with two pins: a positive pin and a negative pin. The positive terminal is indicated by the positive (+) symbol or a longer pin, and it receives power at a voltage of 6 V. On the other hand, the negative terminal is marked with the negative (-) symbol or a shorter pin, and it is linked to the ground (GND) terminal.Another component of the system is an LED (light-emitting diode), a semiconductor device that emits light when current passes through it. Within the semiconductor, electrons recombine with electron holes, generating energy that is then released as photons, producing visible light. LED in this device is used for the alerting system. The other part of the device is exhaust fan. It works by reducing the concentration of LPG gas in the leakage center.

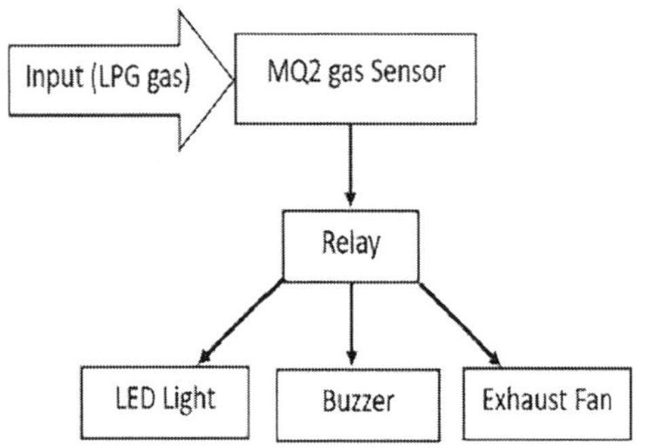

Figure 26.1 Diagram of the proposed system

Figure 26.2 shows the operating principle of the proposed device. When the LPG leakage happens, the concentration of LPG gas increases in the air. As a result, the MQ2 sensor senses the LPG gas. The sensor sends a signal to the relay. The relay processes the signal from the sensor and turn on the buzzer, LED light and exhaust fan. The buzzer and the LED light is worked for alarming purpose. The exhaust fan starts to minimize the concentration of the LPG gas.

3. Experimental setup and result

Figure 26.3 illustrates the experimental setup for proposed device. It consists of the components like MQ2 sensor, relay, buzzer, exhaust fan and LED

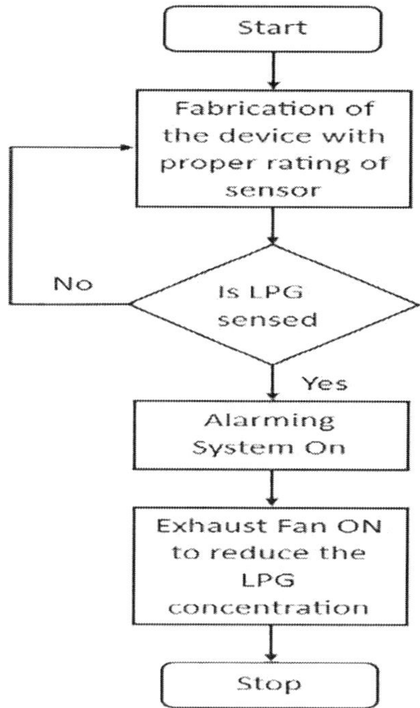

Figure 26.2 Operating principle of the proposed system

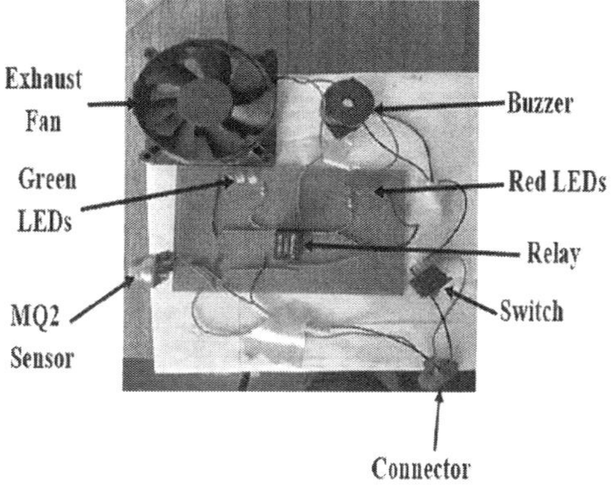

Figure 26.3 Experimental setup for LPG gas leakage detector and alarming system

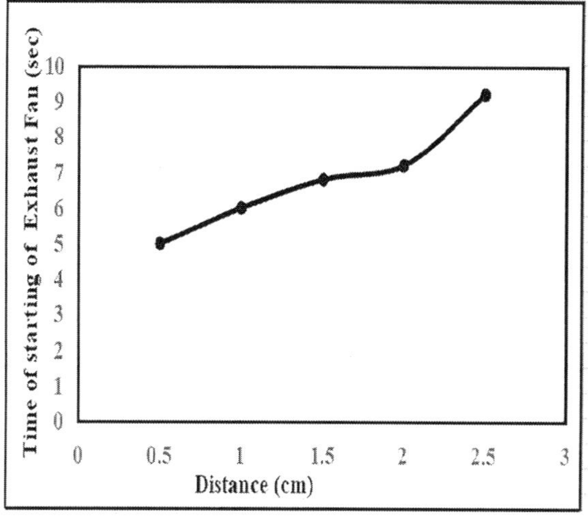

Figure 26.4 Time of stating of exhaust fan vs. distance between the sensor and LPG leakage center

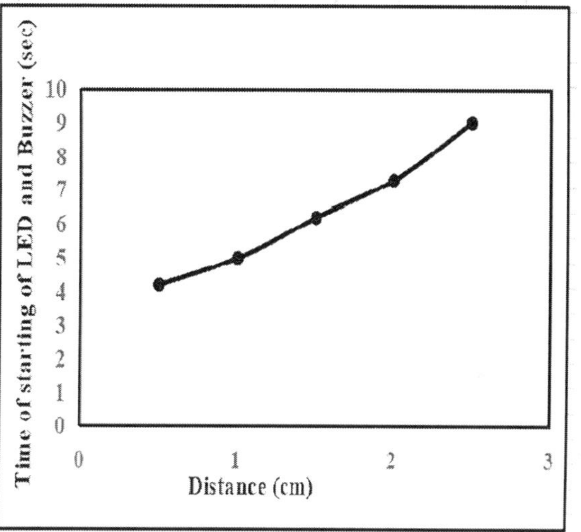

Figure 26.5 Time of starting of LED and buzzer(sec) vs.distance between the sensor and LPG leakage center

light.MQ2 sensor is used specifically for sensing the concentration of LPG gas leakage in air. The triggering of sensor sends a signal to the relay circuit. The relay processes the signal and turn on the buzzer, LED light and exhaust fan. The fan is used to minimize the gas concentration while LED light and buzzer is used for alarming process. The experiment is conducted and performances like time response of exhaust fan, reduction of concentration level with time and total time of operation of exhaust fan are carried out.

Figure 26.4 shows the time of switching on the exhaust fan vs. distance. It is found that at a distance of 0.5 cm between the sensor and LPG gas leakage, the sensor starts to work within 5 s and turns on the fan. With increasing time, the gas is diffusing in the air and after reaching a certain point, exhaust time increases very fast which leads to the conclusion that after a time the exhaust will not work because the sensor will not sense the leakage and cannot give signal to relay to turn on the exhaust fan.

Figure 26.5 exhibits the LED response time with distance. At a distance of 0.5 cm, the LED light will glow after the sensor works at 4.2 s. With increasing the distance, the time of response will gradually become slow which leads to the conclusion that after a time there will not be any response from the sensor that will not switch on the LED.

Figure 26.6 illustrates the graph showcasing the intermediate time response of the system in relation to the distance between the sensor and the LPG leakage center. When the sensor is positioned at a distance of 0.5 cm from the LPG leakage center, the exhaust fan effectively removes the gas within 40 s from a volume of space measuring 15cm × 15cm × 15cm. As the distance between the sensor and the leakage center increases, the intermediate working time also increases.

Additionally, cost estimation has been conducted for the proposed device. To accomplish this, the prices of the components were surveyed based on their market value, and the findings are presented in Table 26.1. Considering the component prices, the total estimated cost for our proposed device amounts to four hundred eighty-two rupees.

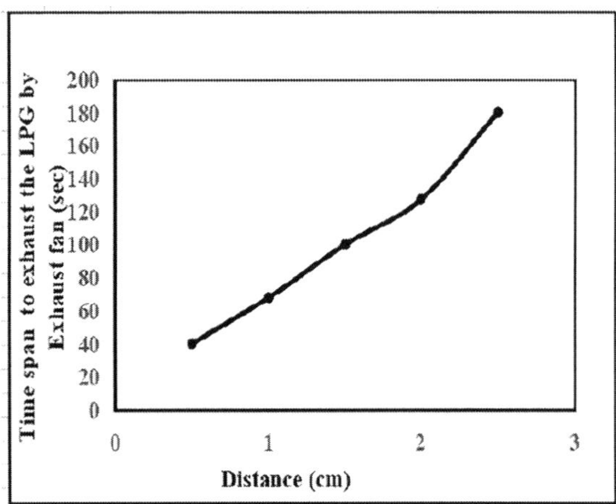

Figure 26.6 Time span to exhaust the LPG by exhaust fan vs. the distance between the sensor and LPG leakage center

4. Conclusions

The rise in livelihood threats and health issues caused by LPG leakage has sparked a growing interest in budget-friendly LPG gas detectors and alarming systems. The need for gas detectors is particularly crucial in commercial and industrial environments, as safety failures can result in tragic consequences. When it comes to gas detection, time is of the essence, and selecting the correct gas detection solution poses a significant challenge for safety professionals who cannot afford any mistakes.In this context, the authors have introduced a cost-effective LPG detector and alarming system. The device is compact, lightweight, and easily portable, allowing for effortless installation. All the necessary components for constructing the device are readily available in the market. The estimated cost of the device is only four hundred eighty-two rupees, significantly lower compared to other available devices in the market. To validate its performance and reliability, a prototype of the device has been constructed. To validate the performance and reliability of the device a prototype has been fabricated. From the test results it is concluded when MQ2 sensor of the devicesense the LPG LED and buzzer is turning on within 4.2 s and the exhaust fan is turning on within 5 s and whichis situated 0.5 cm away from leakage center.Furthermore, the device is designed to function effectively in any weather conditions, with minimal risk of damage. Due to its low cost and rapid response, the proposed device enhances safety and reduces risks.

Table 26.1 List of quantity and price of the components used in the proposed system

S. No.	Particulars	Quantity	Price (in rupees)
1	MQ2 sensor	01	150
2	Buzzer	01	50
3	Exhaust fan	01	100
4	IN4007 diode	01	3
5	6v relay	01	100
6	470-ohm resistor	02	10
7	9 Vbattery with connector	02	30
8	LED	06	30
9	Switch	01	06
10	BC557 Transistor	01	3
		Total	482

5. References

1. Rahul,R.and Bhadra, K. K. (2019). Impacts of traditional cooking fuels on the prevalence of ailments in India.*Ind. J. Human Dev.*, 13(3), 294–307.0.1177/0973703019900444.
2. Zongzheng,M.and Cheng, Y.(2011).Numerical simulation of stratified lean burn for LPG engine.*2011 Inter. Conf.Elec. Inform. Control Engg.*, 5982–5985.10.1109/ICEICE.2011.5776974.
3. Firoz,A., Saleque, K., Alam, Q., Mustary, I., Chodhury, H., and Jazar, R.(2019).Dependence on energy in south Asia and the need for a regional solution.*Energy Proc.*, 160, 26–33.10.1016/j.egypro.2019.02.114.
4. R. K. C., N. M., M. R., and H. S. (2021).LPG leakage detector with smart SMS alert using microcontroller.*6th Inter. Conf. Comm.Elec.Sys.(ICCES)*, 58–62.10.1109/ICCES51350.2021.9489037.
5. Alipour, S., Mortazavi, Y., Khodadadi, A., Medghalchi, M., and Hosseini, M. (2006).Selective sensor to LPG in presence of CO using nanogold filter, operating at low temperature, with Pt/SnO2.*Sensors*, 1089–1092. 10.1109/ICSENS.2007.355815.

27 Comfortable and safe elevator system with emergency features and smooth braking using PLC

Anuraaga Nath[a] *and Emily Datta*[b]

Department of Electrical Engineering, Heritage Institute of Technology, Kolkata, India

Abstract

Elevators provide convenience for people who need to travel multiple stories, especially when carrying large items. It also includes people with disabilities who may not be able to use stairs. Here the system is designed to provide a smooth and comfortable ride for passengers and safety in an emergency. The use of a PLC allows less cost compared to others with precise control of the elevator's movements and implementation of advanced safety features. Here we have used a ladder logic program to execute various concepts in the elevator system. It consists of different safety, and emergency measures to execute during emergency conditions like sudden power cut-off, earthquakes, fires, etc. We have shown a 3-floor elevator working system under these mentioned conditions. The system is optimized and detailed. The paper demonstrates the ladder logic program can be used to design and control an elevator system with high efficiency and reliability.

Keywords: Comfort, emergency, enhance, ladder logic programs, optimize, PLC, power cut, reliability, safety

Introduction

An elevator system consists of a lift car, a motor, a brake, a door, and a control panel. The lift car contains people or cargo which is to be uplifted by the elevator. PLC performs extraordinary work in diminishing the workload. Industrial processes have become more complex, require to be performed at more incredible speeds, and have better accuracy. An elevator system works on the essential four-quadrant operation of the electric drive. This operation introduces motors, brakes, and control panels. Various research works are being held earlier in time. We used PLC to control the whole elevator system at a very base level. This PLC can be optimized for better operation using machine learning concepts. We have discussed that part in our future work. In this paper, there is a brief explanation and definitions of different subjects that we tried to implement. Our implementation improves the cost management system as using PLC is cost-effective in both installation and maintenance. The easy maintenance process will help the system to be more usable.

An elevator system faces many challenges and limitations, such as consuming a large amount of energy and generating a high level of noise and vibration. Regular and costly maintenance is also required to ensure a proper and safe operation. Faults or failures cause the malfunction or the breakdown of the system. The system is also affected by external factors, such as the weather, traffic, or user behavior. It influences the performance of the system. We will try to explain how we can overcome these limitations.

This single elevator car system is tested and evaluated using different scenarios and performance criteria and in different braking conditions.

1. Under a sudden power cut scenario.
2. Under emergency conditions like fire, earthquakes.
3. Under overload condition.

Software implementation

To implement some modern features to the elevator system we added logic using Siemens Simatic Manager S7 Professional Software. We have introduced some innovative thoughts which we can include in a contemporary elevator system to make it more powerful and effective.

When the elevator is in a static state, the elevator will receive two kinds of signals. If the call signal from other floors is received, PLC will control the lift based on the signal response program. When the elevator is in operation if the elevator is going down, the elevator will only feedback on the received down call signal otherwise, it will only respond to the up-call signal.

The elevator system implements the four quadrant operation of the electric drive. The logic circuit works with the elevator car position detection system according to the floor. The ladder logic is made for 3 floors. As the system is generated for simulation purposes only, we have used multipoint

[a]anuraaga.nath.ee23@heritageit.edu.in, [b]emily.datta@heritageit.edu

DOI: 10.1201/9781003540199-27

interface (MPI), which is the interface of SIMATIC S7 multipoint communication. It is a network suitable for communication between a small number of sites. It is mostly used for short-distance communication between the upper computer and a small number of PLCs.

Model features

Emergency conditions

Here we are going to show the feature we have added for sudden emergency conditions like fire, earthquakes, etc. Here we have attached the lift emergency with the building emergency. So, whenever the building emergency state will be at the on state, the elevator's emergency state will also be on, and thus the logic will start running, resulting in all elevator cars reaching the ground floor (Figure 27.1) and opening elevator doors and remaining in the same condition until the emergency state gets off state (Figure 27.2). This ensures that none is trapped inside the elevator during the emergency condition and that the elevators remain operational. This also prevents any damage to the elevator due to electrical or mechanical failures during an emergency. This feature can be activated both manually and automatically by detecting building emergency systems like smoke, fire, or seismic activity in the building. In Figure 3 the ground call from the 2nd and 1st floor to the ground floor is seen to be dependent on the floor detection system.

The running condition is shown in Figure 27.3. If the elevator is on the 2nd floor, the "2nd to ground floor" (M0.4) command will run and similarly if the floor is on the 1st floor, the "1st to ground floor" (M0.3) command will run.

Sudden power cut condition

For such conditions, we have installed an auxiliary power supply just to supply the elevator to the nearest floor can be reached. The lift should not draw any more power from the auxiliary power supply. An inverter as an additional device is added to the system. The automatic auxiliary supply switching with ON logic and power-OFF logic are used. In Figure 27.4, we can see the power cut condition. The logic detected the power cut condition and enables the auxiliary power automatically (Figure 27.5). As well as when the power comes back, the auxiliary power switches off automatically as the logic condition does not fulfill.

Advantages of using an auxiliary power supply:

- Prevents users from being trapped inside the elevator in a power cut.
- Reduces the chance of damaging elevator components due to voltage fluctuations.
- Increases safety and reliability of the elevator system.

Although it can increase overall cost, complexity, and space requirement for the system it is worth it in case of the safety of the passengers.

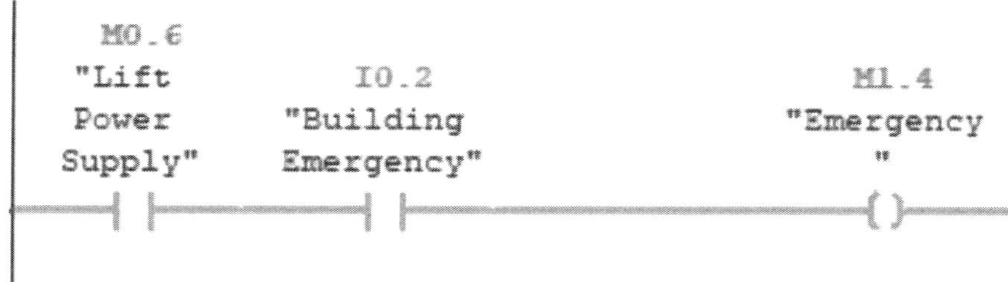

Figure 27.1 Running condition of building emergency is connected with elevator emergency sensor: Green color signifies power path (color online)

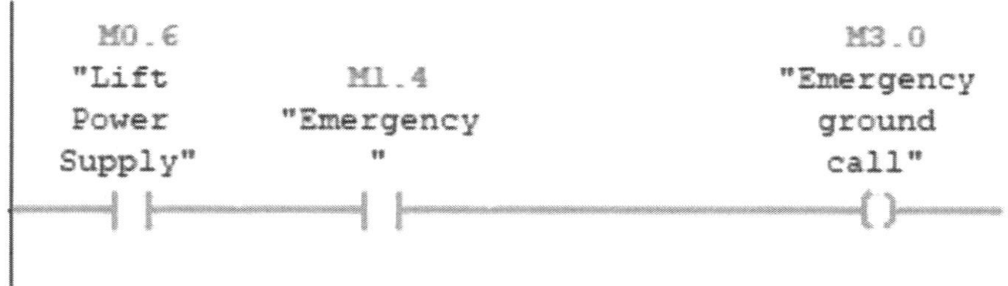

Figure 27.2 Running condition of elevator emergency system when the emergency sensor is turned on: green color signifies power path (color online)

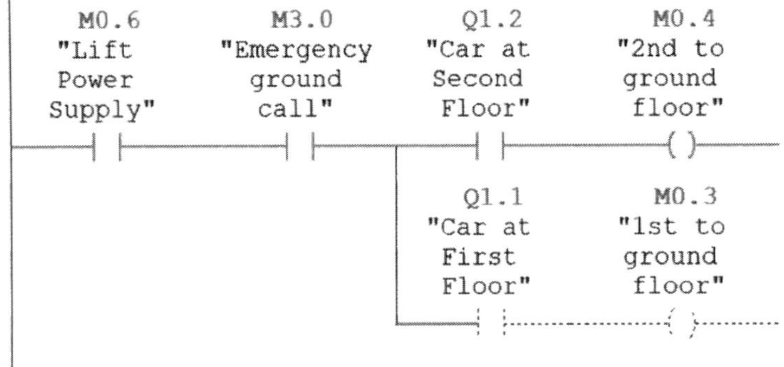

Figure 27.3 Running condition of commands with elevator emergency when car is in second floor: Green colour signifies power path (colour online)
Source: Simatic Manager S7 simulation

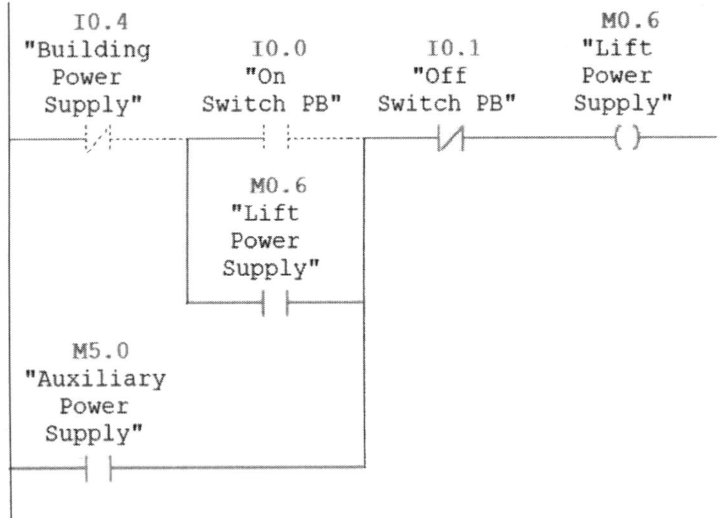

Figure 27.4 Running condition of power supply in elevator system in power cut condition, the auxiliary power supply is automatically turned on: green color signifies power path (color online)

Figure 27.5 Running condition of building power supply logic connection with the auxiliary power supply: Green color signifies power path (color online)
Source: Simatic Manager S7 simulation

Smooth braking system

The smooth braking system is introduced via a timer system. For simulation purposes, we are assuming that 5 s are needed to reach a single floor. In the case of direct calls like from ground to 2nd, timer will give signal after (5*2)-2 = 8 s. As we have implemented the simulation, we did not consider any use of safety belts as it comes with the manufacturer side as per the model and usage of the elevator car. But, for smooth braking and comfort of the passenger, we included a time-based braking system. It automatically detects the floor calls and works accordingly.

A time split is used to detect the reaching floor and engaging the braking system right away. The first 3 s is allocated for travelling from one floor to another and engaging the braking instance for last 2 s. All the cabin up commands and all cabin down commands are connected with "Cabin Up" and "Cabin Down" memory bits and those both are connected parallelly with the braking bit. This braking system command is split into 4 branches where "Empty Elevator" and "Loaded Elevator" conditions are considered.

Most importantly, the smooth braking is not only depending on PLC program but also depends on the various parts and mechanism of mechanical system and motors. We have only studied the software model of PLC which will help to achieve a smooth braking system by giving proper time of braking signal.

Conclusions

In this paper, we have presented a modern lift system using PLC which is very cost effective compared to other system and can handle emergencies such as sudden power cuts, high-intensity storms, earthquakes, or fires. We have designed and implemented an emergency power supply that can keep the lift running for a limited time until it reaches the nearest floor. We have also added an emergency ground call feature that can alert the ground staff and bring the passengers to the ground floor in case of any danger. Our system is reliable, safe, and efficient for various scenarios. Also, for the comfort of the users, we introduced a smooth braking system upon reaching the destination floor. We included a timer that detects the time to reach and initiates the braking system as required.

In future work, we are determined to improve the system using density entry, we can make assumptions about the number of people inside the elevator and the number of people waiting outside, and their corresponding input command. Density entry uses depth data, machine learning, and computer vision to anonymously count people. Moreover, we can use Zoning techniques and reinforcement learning algorithms to optimize the system more effectively.

References

1. Siemens, A. G. (2017). *Simatic* Programming with STEP 7. Germany.
2. Stephen, P. T. (2016). Programmable Logic Controller (PLC) Tutorial. Siemens Simatic, S7–1200.
3. Jürgen, K. (2011). PLC-basic course with SIMATIC S7: Vogel Business Media GmbH & Co. KG.
4. Hooper, J. (2006). *Introduction to PLCs*. 2nd ed. Carolia: Academic Press.
5. Xiaohu, H. (2020). Design and research of elevator control system based on PLC. DOI: 10.1088/1742-6596/1449/1/012101 https://iopscience.iop.org/article/10.1088/1742-6596/1449/1/012101.
6. Jadhav, A. B. and Patil, M. (2014). PLC based industrial automation system. *Inter. Conf. Recent Trends Engg. Manag. Sci.* https://www.researchgate.net/publication/311861408_PLC_Based_Industrial_Automation_System..
7. Ranganj, D. G. and Tahilramani, N. V. (2017). Automation based elevator control system. DOI: 10.1109/ICATCCT.2017.8389116. (IEEE) https://ieeexplore.ieee.org/document/8389116.
8. Singh, G., Agarwal, A., Jarial, R. K., Agarwal, V., Mondal, M. (2013). PLC controlled elevator system. DOI: 10.1109/SCES.2013.6547517 https://ieeexplore.ieee.org/document/6547517.
9. Yu, W., Xu, L., and Qiu, Q. (2011). Optimization of the elevator speed control system. *Appl. Mech. Mat.*, 101–102. https://www.scientific.net/AMM.101-102.405.
10. Chen, M., Yin, J., and Liu, F. (2018). Design of elevator control system based on PLC and frequency conversion technology. DOI: 10.1109/ICMTMA.2018.00073: IEEE. https://ieeexplore.ieee.org/document/8337384.
11. FerenčíK, N., Jadlovský, J., KopčíK, M., and Zolotová, I. (2017). PI control of laboratory model elevator via ladder logic in PLC. DOI: 10.1109/SAMI.2017.7880338. https://ieeexplore.ieee.org/abstract/document/7880338.
12. Chan, R. and Chow, K. P. (2017). Threat analysis of an elevator control system. DOI:10.1007/978-3-319-70395-4_10. https://link.springer.com/chapter/10.1007/978-3-319-70395-4_10.
13. Sale, M. D. and Prakash, V. C. (2017). Dynamic scheduling of elevators with reduced waiting time of passengers in elevator group control system: Fuzzy system approach. DOI:10.1007/978-981-10-3818-1_37.
14. Shi, D., Xu, B. 2018., Intelligent elevator control and safety monitoring system. DOI:10.1088/1757-899X/366/1/012076.

28 Deep learning-based stator winding fault detection of three phase induction motor using GoogLeNet

Riju Nandi[1,a], *Chiranjit Sain*[2], *Amarjit Roy*[2] *and Arnab Ghosh*[3]

[1]Surendra Institute of Engineering and Management, Dhukuria, Siliguri, District Darjeeling-734009, West Bengal, India

[2]Department of Electrical Engineering, GKCIET, Malda, West Bengal, India

[3]Department of Electrical Engineering, National Institute of Technology, Rourkela, India

Abstract

The detection of machine fault is a main challenge for effective machine maintenance in time. There are various methods available for identifying faults, both directly and indirectly. However, these methods can be time consuming, monotonous, and may require advanced sensors or state-of-the-art equipment. Processing of image is one of the methodologies or techniques that can lead towards the evaluation of fault directly. When used in conjunction with image processing techniques, machine learning can aid in the diagnosis of induction motor malfunctions. Lately, there has been a surge in the use of deep learning (DL) techniques for image classification. One of the most popular approaches is the application of convolutional neural networks based on DL. In present study, GoogLeNet is applied for identification of faults through various Fast Fourier Transform (FFT) plots (both healthy and faulty). A total of twenty FFT plots were acquired for three phase induction motor performance (Extended Park's Vector Approach (EPVA), a two-dimensional (2-D)-based representation) to form the input dataset for the GoogLeNet (DL) analysis. Performance of GoogLeNet was evaluated to be promising with an accuracy of 94.23%, 99.01%, 76.54% and 92.83% for faulty stator winding with no load, faulty stator winding with load, healthy stator winding with load and healthy stator winding with no load, respectively.

Keywords: GoogLeNet, artificial intelligence, deep learning, fault detection

Introduction

The arrival of this 4th industrial revolution exposed possibilities of interaction among machines with modern high-end sensors. The method has simplified analysis, real-time adaptation, and optimal decision-making. Hussain et al. (2022) has referred induction motors as the industry's workhorse which has been utilized in numerous industrial applications and household applications [1]. The induction motors are utilized in many applications i.e., pumps, exhaust fans, blowers, vacuum cleaners, and cranes. Due to severe and harsh way of operation and handling, these machines often go through different problematic conditions, which produce faults. Interaction of motors of sensors (vibration, speed and heat) and continuous monitoring might help in the detection of problem and detection of related concerns within interconnected processes and planning predictive maintenance as well. Two in induction motor faults can be faced i.e., electrical fault and mechanical fault, as reported. Hussain et al. (2022), mentioned single phasing fault and stator winding fault as two of the most significant electrical faults [1]. Stator inter-turn winding faults might cause phase winding to the ground. Single phasing has limitation during supply unbalances which may cause overheating of a motor. Faults can be avoided or repaired based on the detection timing. Artificial neural networks are mathematical algorithms which can solve problems. Layers of interconnected nodes are the building block of artificial neural networks that corresponds to the biological neural networks as seen in animal brain. There are three layers in conventional neural networks or Convolutional Neural Network (CNNs) which are: input, hidden and output layer. Since this method was technically challenging to understand before the rise of contemporary computing power—specifically, Graphics Processing Unit (GPUs) and other algorithms—in the 2010s, it has only been feasible [1]. The CNN is made up with a hidden layer while there are multiple hidden layers in deep-learning neural network. Google developed GoogLeNet which is the most popular convolutional neural network trained with millions of images, and can reportedly identify thousands of objects per day, such as computer accessories, pendrive, keyboards, pencils, other objects or even animal [2, 3]. Images with dimensions of 224 × 224 × 3 are required for its 144 layers of input. Compared to various neural networks of deep learning (viz. AlexNet).

[a]rijunandi31@gmail.com

DOI: 10.1201/9781003540199-28

GoogLeNet significantly hinders the number of variables from sixty-seventy millions to approximately four millions and is comparatively more promising while predicting. Till today, GoogLeNet is the popular most deep learning neural network that can be applied in various aspects such as diseases predictions, video processing, machine fault analysis, etc. The current study has been aimed to validate the potential of GoogLeNet on deep learning technique for the analysis of accuracy of stator winding machine fault detection which has been done in this research with a set of healthy and faulty three phase induction motors. This research will help to design methodology that can be less-costly, more accurate and less time-consuming method for machine fault estimation. Early detection of such faults may help in implementing some preventive measures so that the condition of a machine remains good.

Experimental

Designing of stator winding fault detection of three phase induction motor using GoogLeNet (deep neural network) in MATLAB application have been carried out in this research.

Material preparations for analysis with established FFT plots

A 2.5 HP, 4 pole, 400V, three phase induction motor consisting of 360 turns/phase was used to study the characteristics of stator winding faults. Variable load conditions were achieved by coupling a 2.2 kW, 230 V, 12.1 A, 1500 rpm DC shunt generator with the induction motor [4]. In these settings, a balanced supply voltage has been used for all trials:

1. Healthy condition of the motor
2. Stator winding inter turn dead short circuit fault conditions.

An inter-turn dead short circuit defect in a stator winding implies an insulation failure between particular turns of the winding. From the R-phase winding of the motor, multiple taps with particular numbers of turns were extracted and connected to a patch board in this study. Inter turn dead short circuit faults was made by shorting any two tappings of selected patch board. Relevant pictorial representation of winding is shown in Figure 28.1.

To study the characteristics of rotor faults and bearing faults, a series of experiments were performed on 1/3 HP, 380 V, three phase induction motor. Variable load conditions were achieved by coupling a 0.75 kW, 230 V, 4.6 A, 1500 rpm DC shunt generator with the induction motor. All the

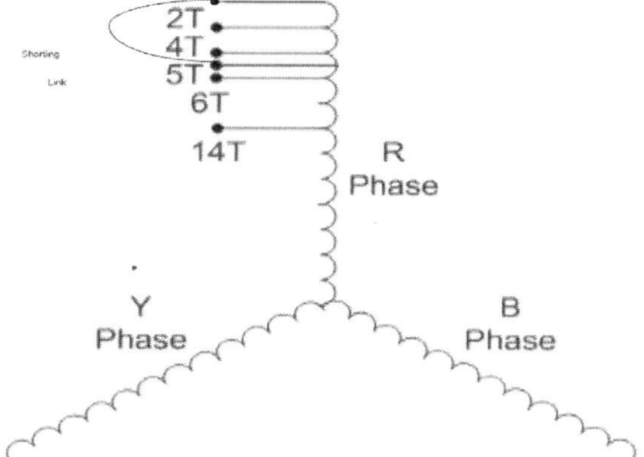

Figure 28.1 Schematic diagram of stator win ding with taps to create dead short circuit faults

experiments have been conducted at a supply voltage which was balanced. Broken rotor bar fault (to create broken rotor bar fault in induction motor, proper amount of material has been removed from the rotor bar to expose three rotor bars) and damaged bearing fault (to test characteristics of bearing fault, a motor consisting one bearing having inner race fault and one with outer race fault) were considered for this study.

To measure the currents flowing through the stator of a three phase motor under different analytical circumstances, a CRIO-9075 integrated controller and chassis system was utilized. This system included a 400 MHz PowerPC processor and a four-channel current input card. A visual examination of the three phase stator currents drawn by the motor is not sufficient to assess the information of imbalance created in the motor as a result of small stator problems. The use of Extended Park's Vector Approach (EPVA), a two-dimensional (2-D) representation was carried out for description of three phase induction motor performance [5]. After oscillating park's vector approach FFT plots were received.

GoogLeNet

Convolutional neural network or CNN makes use of an algorithm patterns to identify identical patterns in images, working similarly to the human brain. When it comes to describing models, their characteristics are often discussed during analysis of accuracy, elapsed time, loss rate, and confusion matrix. Primary feature of this model is its ability to self-train using a hierarchical representation of a dataset that contains a large number of images. The backpropagation algorithm was implemented to reset the weights with respect to input during training purposes. GoogLeNet was used in this research for identifying two classes of FFT plots, healthy and faulty stator winding current with both load and no load.

GoogLeNet increases the ability of inception block by using diverse transformation functions through splitting, followed by transforming; finally merging of big layers into smaller one. It is also referred as Inception V1 architecture [6] as it can be referred as a novel idea of inception block in deep learning which enhances the scope CNN model and capability of it. GoogLeNet also lowers the computational cost due to the tremendous reduction in the estimate because of a few quantity of neurons and the addition of a bottleneck convolution layer. GoogLeNet enhances the fully connected layers without sacrificing accuracy by expanding the size of the blocks in both width and depth. To find the mean of the channels' values, rather than using the linked layer, a global average pooling layer is employed. The GoogLeNet's property of splitting, transforming and merging enables the identification of various types of variations undetected within a single image class. Chandel et al. (2021) reported a drawback of this module i.e., customized topology that often changes according to module [6].

Deep CNN (DCNN) model framework

We have used DCNN to identify faults in stator winding of three phase induction machines in this research. DCNN detects different features of healthy and faulty FFT plots (a total of twenty plots were used) inserted as .jpg format in MATLAB programming. The comparison between healthy and faulty FFT plots denoted by deep learning models can indicate faults of stator winding. All the FFT plots have been compiled in MATLAB 2019b software which was installed in a computer (Specifications: Intel i7 processor having 8 GB RAM and windows 11 operating system with graphics card installed). The testing dataset helped to assess the models.

Deep learning model process

DCNN models rely on graphical hints; here FFT plots during training process to identify faults. The models were trained with parameters as ILR (initial learning rate), mini batch size, maxepochs in order to make an ethical comparison between the models [6]. Software- MATLAB- 2019b with a deep learning toolbox has been utilized for designing, implementation, visualization and monitoring the training process of the GoogLeNet model.

Preparation of dataset

Downscaling of FFT plots was done to suit the size specifications of GoogLeNet which was set at 224 × 224. Dataset of FFT plots stored in two different folders (healthy and faulty) were grouped into training and validation. Training was done with 70% of images and for validation 30% were used. This ratio 70:30 was similar for every FFT plots. We used an in-house MATLAB code to randomly select images for training and validation.

Model architecture

GoogLeNet was tested and evaluated for its accuracy. Each model has a unique architecture that can be determined by the amount of convolutional layers, activation functions used in each layer, and the number of hidden units in each single layer [6]. Here are the parameters that were modified: number of classes, minimum and maximum batch size, maxepochs, and frequency of validation. The images were chosen at random through programming, so each epoch had different images than the previous one. This was made possible by the continuous switching of images in these two (training and validation) stages of model formation.

Results and discussion

Using GoogLeNet, we performed a number of different FFT plots to distinguish between photos of fully functional and defective three phase induction motor stator windings (only at Minibatch size of 5 with 6 maxepochs). It can be clearly seen that the DCNN has ability to differentiate images based on classification. The deep learning models expressed varied accuracy across different FFT plots in identifying fault. Through the outcome of various accuracy rates, it can be interpreted from results that overall performance by GoogLeNet was promising. Output of FFT Plot classification using GoogLeNet has been given in Figure 28.2. The exceptional performance of GoogLeNet is due to its unique ability to apply convolutions to images across multiple stages, using filters of different sizes, and performing pooling in a single layer. Accuracy rates of selected FFT plots have been given in Table 28.1.

Hence, GoogLeNet model achieved maximum accuracy of 94.23% (Figure 28.3a) for stator winding fault with no load and maximum accuracy with stator winding fault having load was 99.01% (Figure 28.3b). While for healthy stator winding, maximum accuracy rates were 76.54% (Figure 28.3c) and 92.83% (Figure 28.3d) for "with load" and "with no load", respectively. Figure 28.3 has been prepared with the training results (of four results as discussed). After training, an increase in validation accuracy was observed while there was a decrease as depicted in the validation loss which ensured that the network was learning properly.

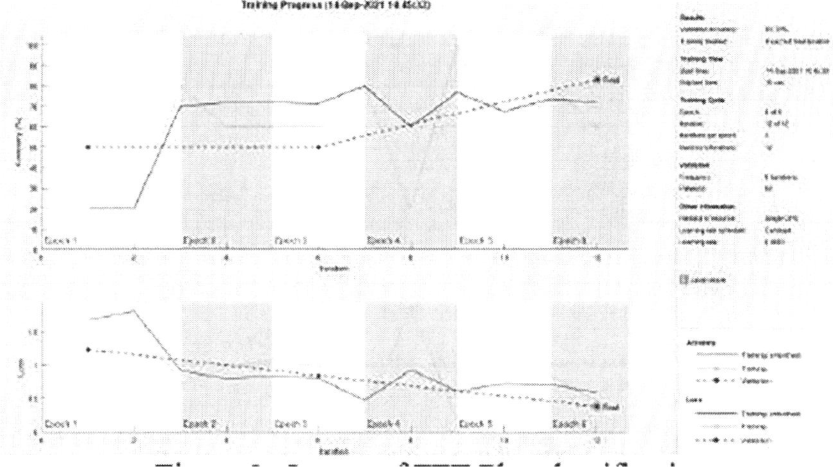

Figure 28.2 Output of FFT Plot classification

Table 28.1 Accuracy rates of selected FFT plots

FFT plots	Accuracy rate
Faulty stator winding with no load	94.23%
Faulty stator winding with load	99.01%
Healthy stator winding with no load	92.83%
Healthy stator winding with load	76.54%

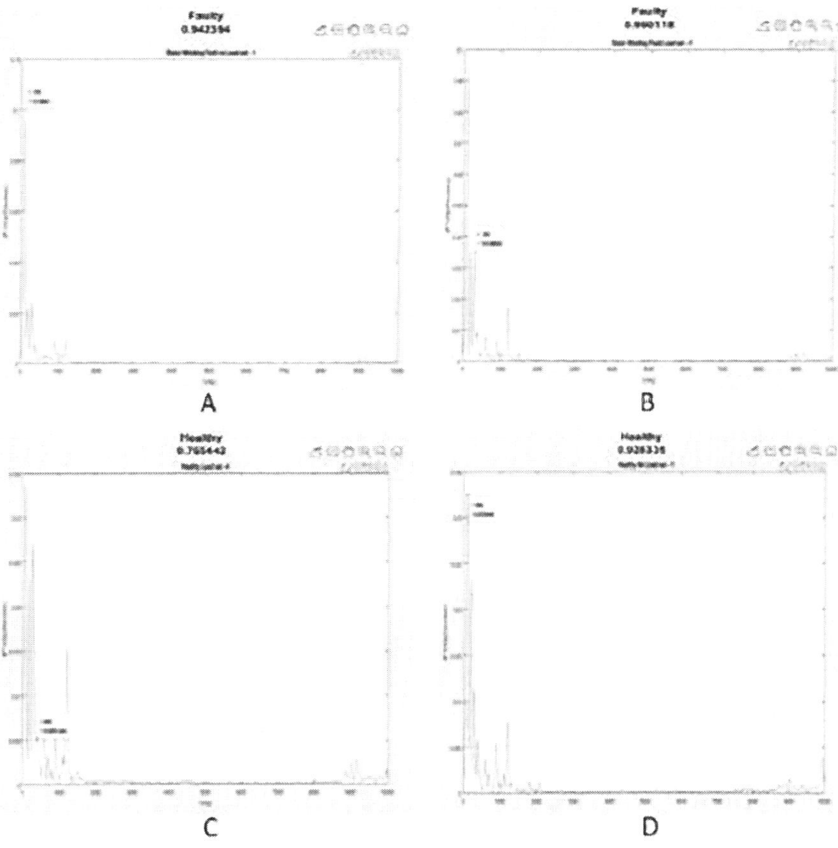

Figure 28.3 Accuracy rates of selected FFT plots

Validation accuracy rate was found to be 83.33% (Figure 28.2).

Conclusions

The aim of this research was to design stator winding fault detection of three phase induction motors using GoogLeNet (deep neural network) in MATLAB application which has been successfully carried out and outcome has been given and discussed in the results. The world is continuously upgrading itself through various technologies, where artificial intelligence (AI), more precisely–convolutional neural network (CNN) is the state-of-the-art technological study towards which the next generation is apparently tending. So, with the utilization of this technology could change the workhorse in many industries. With the help of this deep learning, faults of various induction motors can easily be detected which will be time consuming and cost effective as well.

Acknowledgement

Authors would like to thank Mr. Soumya Majumder for helping with manuscript preparation.

References

1. Hussain, M., Memon, T. D., Hussain, I., Ahmed Memon, Z., and Kumar, D. (2022). Fault detection and identification using deep learning algorithms in induction motors. CMES-Comp. Model. Engg. Sci., 133(2), 435–470.
2. Zhang, X., Pan, W., Bontozoglou, C., Chirikhina, E., Chen, D., and Xiao, P. (2020). Skin capacitive imaging analysis using deep learning GoogLeNet. *Intell. Comput. Proc. 2020 Comput. Conf.*, 2, 395–404.
3. Jahangeer, G. S. B. and Rajkumar, T. D. (2021). Early detection of breast cancer using hybrid of series network and VGG-16. *Multimedia Tools Appl.*, 80, 7853–7886.
4. Sarkar, S., Purkait, P., and Das, S. (2021). NI CompactRIO-based methodology for online detection of stator winding inter-turn insulation faults in 3-phase induction motors. *Measurement*, 182, 109682.
5. Das, S., Purkait, P., and Chakravorti, S. (2012). Separating induction motor current signature for stator winding faults from that due to supply voltage unbalances. *2012 1st Inter. Conf. Power Energy NERIST (ICPEN)*, 1–6.
6. Chandel, N. S., Chakraborty, S. K., Rajwade, Y. A., Dubey, K., Tiwari, M. K., and Jat, D. (2021). Identifying crop water stress using deep learning models. *Neural Comput. Appl.*, 33, 5353–5367.

29 Design and control of reactive power of cascaded H-bridge inverter-based STATCOM in grid-and the tied application using PV source

Somesubhra Panda[1,a], *Sanjoy Mondal*[1,b] *and Shivashis Sengupta*[2,c]

[1]Indian Institute of Technology (Indian School of Mines), Dhanbad, India

[2]National Institute of Technology, Tiruchirappalli, Tamil Nadu, India

Abstract

Distributed generation-based sustainable energy resources are the solutions to young smart grid investigator to encounter the problem of non-linear load increase in power system. Day-to-day demand for renewable energy is increasing extensively, and with the evolution of power electronics, both concepts are conjugated for PV-based grid-tied applications or islanding applications. STATCOM (Static and Synchronous Compensator) is the prime FACTS (Flexible Alternating Current Transmission System) device applicable in transmissions and distribution systems. It is used for controlling voltage and frequency as well as power flow and the transmission capacity of line. STATCOM itself is a two- or three-level VSI (Voltage Source Inverter); hence, it can generate or absorb reactive power in transmission line. STATCOM has replaced the old, aged SVC in transmission and distribution system. It has similarities with synchronous generators in terms of its utility. Apart from the conventional seven types of power quality problems, VAR control is one of the toughest tasks to meet the proper power quality during any disturbance in grid network with distributed generations. voltage source inverter is the heart of STATCOM device. GTO-based STATCOM is not cost-effective for microgrid applications. This paper has proposed new circuit-level operation of three-phase cascaded H-bridge multilevel inverter (CHB-MLI), which is incorporated inside the STATCOM. CHB-MLI is upgrading itself day-to-day under research from last 20 years. It's more popular than other multilevel inverters, as well as old-fashioned two-level inverter. Here, the sine-PWM technique is used for controlling output voltage, and two PI controllers are there as AC voltage regulator and current regulator. LCL filter is also used here. PV-based boost converter output DC. Voltage is treated here as DC input to STATCOM. The entire system modelling and simulation are performed using MATLAB.

Keywords: STATCOM, VSI, cascaded H-bridge inverter, PWM, filter, reactive power

Introduction

The present era has the offofirstm of electrical energy in way, sm.e.i.e.ansform the transformational grid grid, which is executed by extensive uses osourcesrofday-to-livesslife. As furenewable sources, provides a is gocellscellsaretinto to go in near the nearnearre and energy demand is growing rapidly throughout the wo,e have to shift our generation. From conventional resoun-conventional resource to mitiresources greenhouse gnd save our plagases from it. Towards sustaithemTo achieve aietal approach weapproach, concentrate on production of renewable energy and make it familiar in practicing engineering applications. Photovoltaic (PV) and wind energy is much more popular than tidal or geothermal resources. PV is cost-effective technology which is used for building of green energy technology. This 21st century is the era of power electronics evolution from old age thyristor to new age IGBT, SiC, GaN. FACTS device are the basic example of IGBT-based multilevel inverter with three level voltage output, these three level voltage is almost sinusoidal when it is compared with general AC. sine waveform. FACTS devices defined by IEEE "alternating current transmission systems, incorporating power electronics-based and other static controllers to enhance controllability and available power transfer capacity". General attributes of FACTS devices are—dynamic reactive power support, voltage control, control active and reactive power flow, improve power system stability. Benefits of FACTS devices are—decrease in power loss in transmission line, voltage profile increases and less voltage fluctuation, steady and transient stability limit increases, also provide power congestion management. When load changes dynamic stability is a big issue for existing power system. Dynamic issues in power system are based on transient stability, damping power swing, post-contingency voltage control, voltage stability and sub synchronous resonance (SSR). There are several FACTS devices for VAR generator and self-commutated VAR compensator. Synchronous condensers, fixed or mechanically switched capacitors,

[a]somesubhra.iitism@gmail.com, [b]sanjoymandal@iitism.ac.in, [c]senshivashissss@gmail.com

DOI: 10.1201/9781003540199-29

combined TSC and TCR falls into VAR generator category and STATCOM, SSSC, UPFC, etc., belongs to VAR compensator domain. SVC, TCSC, TCPST are variable impedance type FACTS devices they modifies the transmission or distribution line constraints, and STATCOM, SSSC, etc., are voltage source converter type FACTS devices they change equivalent impedance of the load. STATCOM is the best device for shunt compensated power network due to its better VAR compensation characteristics than other FACTS devices in conventional grid as well as micro grid [1, 2]. It was discussed by gyugyi in 1976. VAR compensation is determined in two approaches like voltage compensation and load compensation, improving power factor and active power stability voltage support can be done from perspective of load and voltage compensation can be achieved by controlling voltage sag or swell at point of common coupling (PCC) for smart grid or conventional grid tied application because presence of linear and non-linear load [3, 4]. Researchers and practicing engineer have faced problem for islanding condition of smart grid as renewable energy resource needs reactive power support [5]. A novel VAR compensation is proposed on distributed resources-based micro grid with restricted capacity of this network [6]. In wind energy-based micro grid application due to huge disturbance in non-linear load GA, ANFIS technique takes part in a pivotal role on reactive power control than conventional method [7]. There are several control strategies regarding reactive power control are conventional proportional and integral (PI) controller, different soft computing control strategies, sliding control and optimal control [8]. As photovoltaic irradiance changes frequently a battery source should couple with it to maintain the active power support throughout the day. A researcher has proposed a name to this topology as "ESTATCOM". Photovoltaic energy-based STATCOM can be treated as PV –STATCOM because in night time it gives reactive power to grid and takes active power from grid to charge the DC source, it can also be treated as partial STATCOM mode in cloudy weather, only active power can be given to grid at day time with proper solar irradiance [9]. This paper consist of six sections .section I gives the introduction to the problem findings, section II includes system description of VSI-based STATCOM, PV-based boost converter, renewable energy and LCL filter design, control technique of STACOM using PI controller has discussed in detail at section III, section IV shown simulation and modeling and simulation results in MATLAB software, section V shown simulation results, section VI includes conclusion and future works of this research.

System description

Basic idea of STATCOM

In 1970 SCR or thyristor-based reactive power compensation was introduced. Main purpose of Shunt FACTS device is to support VAR demand. Shunt compensators are placed at midpoint of transmission line because voltage sag is more at midpoint for an uncompensated transmission line. If any energy source is faulted then load demand will be increased to other sources connected in same network. That causes voltage depression, at that time shunt FACTS can supply the required capacitive VAR to the network to mitigate voltage stability limit. Here DC input of STATCOM is the output of PV-based boost converter .STATCOM I is here connected with line having 0.15 P.U tie reactance i.e. 10% to 15% of base voltage. From Equation (1) it's evident that reactive power can be controlled by means of multilevel inverter's output voltage, which can be controlled by firing pulses of IGBT switches. It has resemblance with synchronous generator tied with ac conventional grid. VDC should be 30% higher than ES, On the contrary ES is less than Et.

$$Q = \frac{\frac{(1-E_S)}{E_t}}{X} {E_t}^2 \tag{1}$$

STATCOM is fitted with PV to supply real power when irradiance is low. If ESTATCOM is less than grid voltage then VAR is essential to inject there. Hence reactive current flows from grid to inverter and inverter absorbed the VAR on the other hand inverter delivered the VAR when ESTATCOM is more than grid voltage. The power flow relation is

$$S = \frac{3E_t E_S}{X(\sin\alpha) - 3j\frac{E_S E_t}{X(\sin\alpha) - \frac{{E_S}^2}{X}}} = P - jQ \tag{2}$$

Where, S = apparent power, P = active power, Q= Re active power, ES = STATCOM's fundamental output phase voltage (r.m.s), Et = AC power grid phase voltage to neutral (r.m.s), α = Angle between ES and E (Figure 29.1).

Cascaded H-bridge multilevel inverter

Comparing with NPC inverter CHB-MLI has advantage over it in renewable energy application. Multiple H- bridges are connected in series which are supplied by low voltage DC. source and consequences in multilevel output voltages. If there are "h" no of bridges then (h+1) number of voltages at output profile. So, for 3 numbers of H- bridges then 7 levels of voltages we get in output. An example will clear a 7 level output voltages i.e., +6V, +4V,

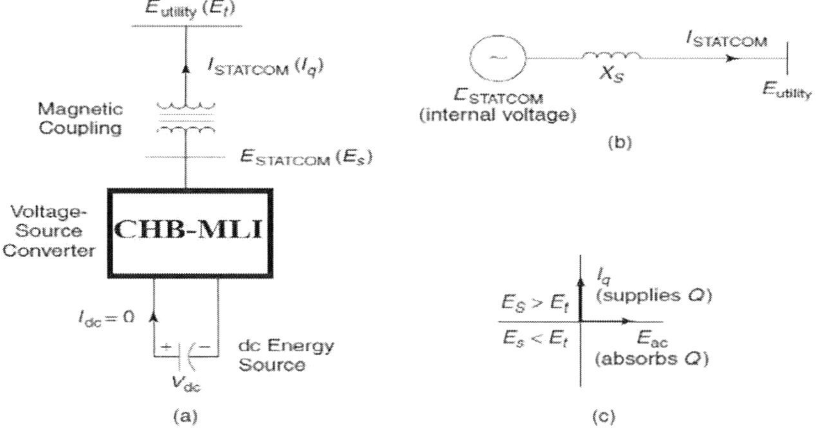

Figure 29.1 (a) Scheme of STATCOM. (b) Grid tied single line diagram of STATCOM. (c) VAR delivery and absorption

+2V, 0V; -2V, -4V,-6V. With equal and unequal dc voltage application at input CHB-MLI can be classified in two types. [15] CHB-MLI has several advantages over NPC inverter for renewable-based grid tied application. Cascaded H-bridge has separate DC source in each sub module but NPC has only one DC source, hence if any failure occurred then only that sub module is bypassed through software but NPC can't bypass that faulty section and the whole system has to be switched off [16]. If there are "N" no. of bridge per phase then maximum output levels will be equal to (2N+1). If no. of bridges is increased then output voltage level also increased but total harmonic distortion (THD) will be reduced [17]. To maintain the optimal modulation index (M.I) DC source should be selected consequently. If M.I is nearly 1 then no of levels are more high, for a three and five level CHB inverter M.I is 0.33and 0.67 consecutively. In ref [18, 19], fault tolerant and redundancy problem has discussed on PV-based CHB MLI or STATCOM (Figures 29.2 and 29.3).

Passive filter topology (LCL filter)

To connect a finite value of inductor in series or parallel with the output of multilevel VSI to create an elementary passive filter. Reactive and passive elements are used for designing passive filter topology like "L filter", "LC", "LCL filter". These reactive elements are connected through a small value resistor which increases the stability of the filter comparing with pure reactive filter. Some constraints are considerable regarding the design purpose of passive filter like that reactance of inductor be around 20% of STATCOM, ripple current should be lower than 25%, Voltage drop will be within 10% range [19, 20, 31].

Control system of STATCOM

Apart from two level inverters there are several multi-pulse and multilevel inverter applications available for STATCOM [23–27]. Comparing both sinusoidal pulse width modulation (SPWM), selective pulse width modulation (SHEPWM) technique for multilevel inverter, SPWM keep harmonics in tolerance band via IEEE standard 519 [27–29].

abc to d_q transformation

In [30, 31] "abc" stationary frame is represent, these variables a rotating two phase variable "dq" can be transformed from three phase variables "abc" stationary frame. With the space vector speed "ω" dq frame rotates.

PLL

PI controller along with Voltage control oscillator is the basic architecture of PLL. In PCC when VSD attained maximum speed then, θ = 0, V Sq = 0 therefore output of VCO.

Control of real and reactive power

Real Power (P) and reactive power (Q) output equation has shown in ref [33].

$$P = \frac{2}{3}(V_{sd}I_{dp} - V_{sq}I_{qp}) \quad (3)$$

$$Q = \frac{2}{3}V_{sq}I_{dp} - V_{sd}I_{qp} \quad (4)$$

For making V_{Sq} =0 as angle between fa and fd, this equation are remodeled as:

$$P = \frac{2}{3}V_{sd}I_{dp} \quad (5)$$

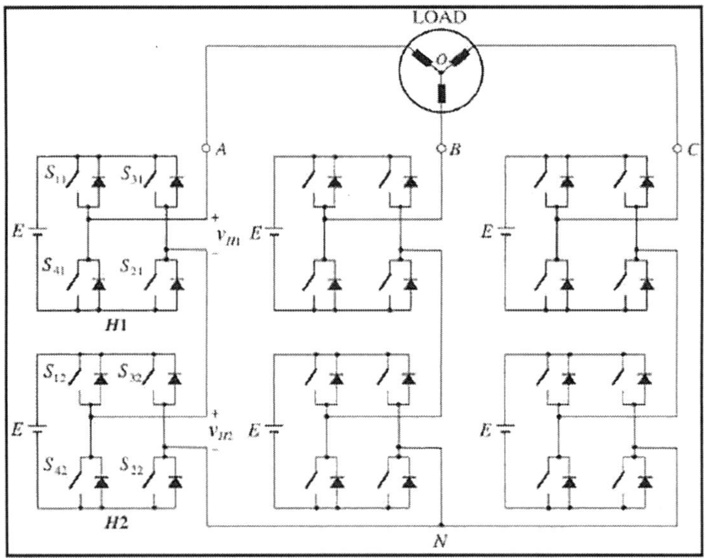

Figure 29.2 Cascaded H-bridge multilevel inverter (CHB-MLI)

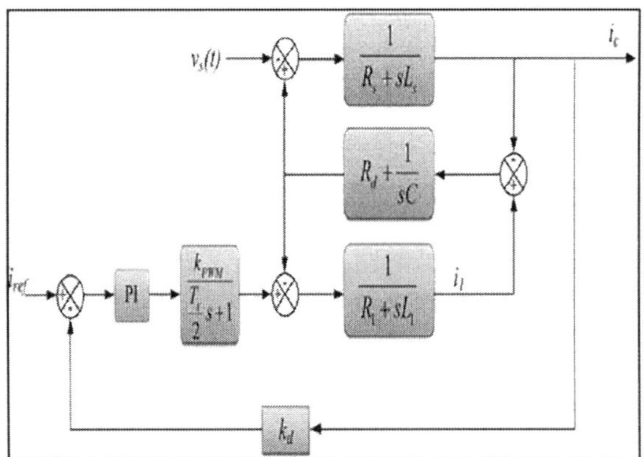

Figure 29.3 Passive filter (LCL filter) [22]

$$Q = \frac{2}{3}V_{sd}I_{dp} \tag{6}$$

Here V_{sd} is the grid voltage which does not change significantly. Therefore I_{dp} and I_{dq} can be controlled the inverter output power.

$$I_{dp_{ref}} = \frac{2}{3}V_{sd} \quad P_{ref} \tag{7}$$

$$I_{dp_{ref}} = -\frac{2}{3}V_{sd} \quad Q_{ref} \tag{8}$$

In this work STATCOM will be consist of a 3 level CHB MLI. Measurement system measures ac voltage and currents.ac voltage is measured and compared by reference voltage.ac voltage and current are in 'abc' form then they are converted into 'dq' form. Two regulator blocks are there -1.voltage regulator block (outer loop) 2. Current regulator block (inner loop). These are basically 2 PI regulators. Using Vm and Vref ac voltage regulator produce reference reactive current (Iq). (Iq ref) for current regulator. Current regulator basically compares Iqm (quadrature component of current that is measured by measurement block) and Iq ref and generates alpha (α). This alpha given to the pulse generator. Under normal condition α, this phase shift is nearly equals to zero as STATCOM is employed for reactive power control. PLL stands for Phase locked loop it synchronize the pulse within the system and produce reference angle "ωt". DC regulator control real power. Task of control system is to increase or decrease the capacitor voltage in order to generate ac voltage can be controlled or it has correct amplitude for reactive power control. Another task of control system is that to keep V0 & VPCC in phase. Main task of controller is that controller is that regulation principle. If Vm > Vref , voltage regulator ask for higher reactive current output and in order to generate more capacitive reactive power and current regulator more increase the α of inverter voltage with respect to PCC voltage. So real power flow PCC to capacitor. If Vm < Vref real power flows capacitor to PCC (Figures 29.4 and 29.5).

Simulation and discussion

Here in this work CHB–MLI-based STATCOM is implemented for reactive power and active power controlling issue in network.

There are several reasons for feasibility of VAR compensation technique

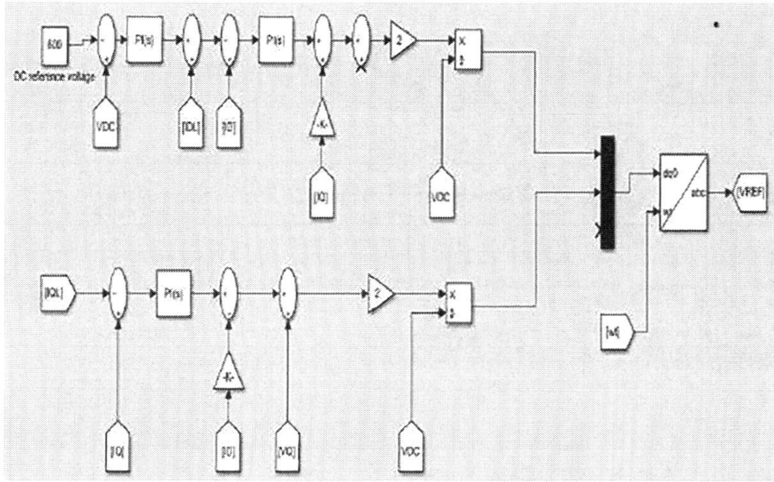

Figure 29.4 Voltage and current controller blocks

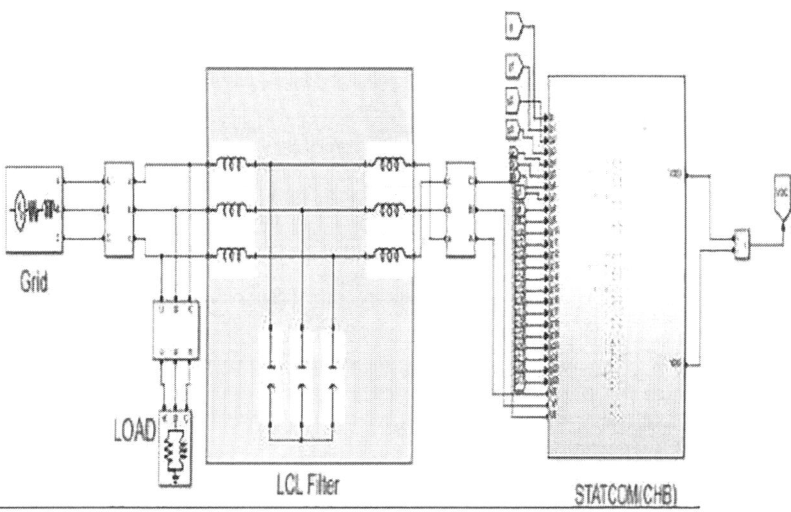

Figure 29.5 Reactive power control of CHB-MLI-based STATCOM

a) Put a stop to voltage sag and voltage swell in interconnected network
b) To increase stability of voltage
c) Better deployment of conventional or renewable resource in best possible way
d) For better voltage regulation.

The output of Dc voltage is being fed through a PV resource and a boost converter. This boost converter based on P&O MPPT output. All values are given in appendix section. Irradiance is a non-linear function, it is designed inn MATLAB by "repeating sequence interpolated" and for this work temperature is considered 25 degree as normal. After inverting the ac 5 level output using CHB MLI it comes through a passive (LCL) filter. Here we acknowledged IEEE guideline that reactive power is based on 5% of rated VoltAmp (VA) other necessary data is given in appendix. It has shown from this work active and reactive power taken from the grid, consumed by the load and supplied by this inverter. Total active power taken from the grid consumed by the load. Total reactive power supplied by inverter and consumed by the load. By changing the active power value and reactive power value of three phase load. Total active power taken from the grid consumed by this load and reactive power supplied by inverter and consumed by the load. For better performance of this PI controller it is considered that1 ms for voltage controller and 200 microseconds for current controller would be performed better. simulation outputs are given here (Figures 29.6–29.8).

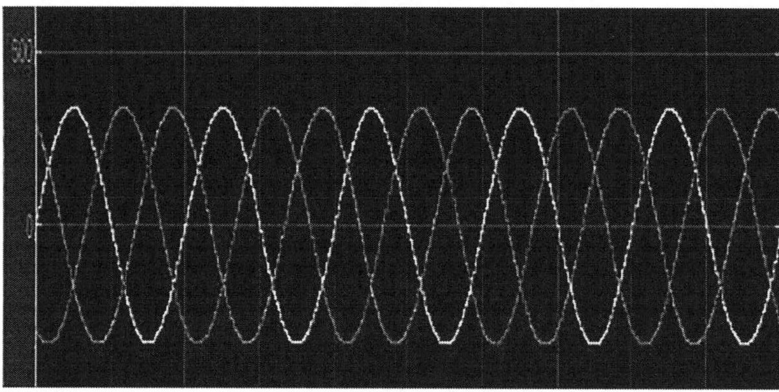

Figure 29.6 Grid voltage vs. time [3]

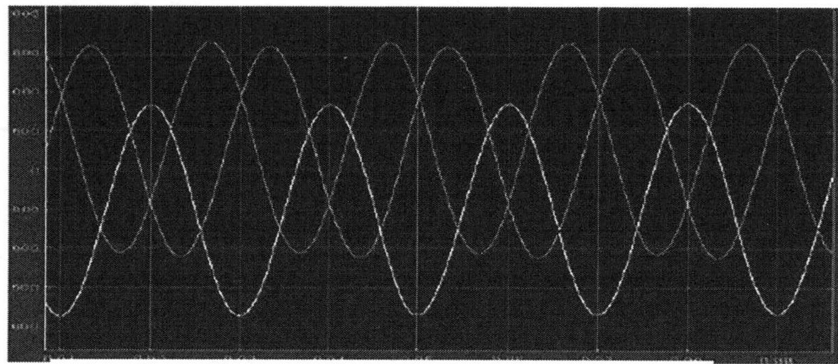

Figure 29.7 Inverter voltage vs. time

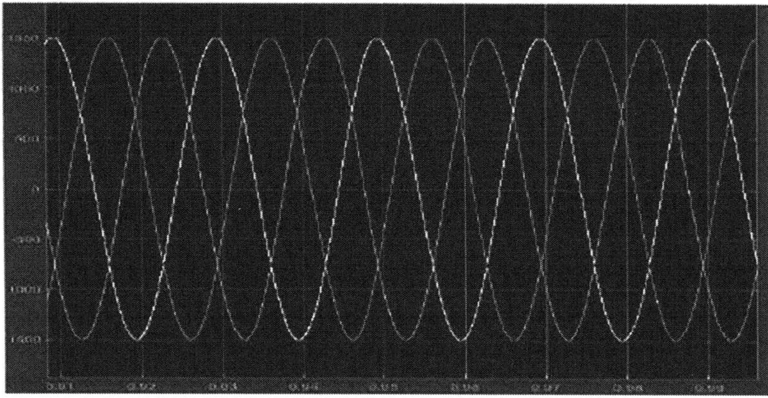

Figure 29.8 Inverter current vs. time [3]

Conclusions

It is seen that multilevel based inverter is more efficient than two level inverter. active and reactive power can be controlled by this type of arrangement. DC voltage is the primary concern for STATCOM, it should be constant .hence here Boost converter is implemented after the PV array.

References

1. Famous, O. I., Fandi, G., Švec, J., Müller, Z., and Tlusty, J. (2015) . Comparative review of reactive power compensation technologies. *16th Inter. Sci. Conf. Elec. Power Engg. (EPE)*, pp.2-7.
2. Qi, J., Zhao, W., and Bian, X. (2020). Comparative study of SVC and Statcom reactive power compensation for prosumer microgrids with Dfig-based wind farm integration. *IEEE Acc.*,vol 8, pp209878-209885.
3. Wanner, R. M. and Hausler, M. (1983). Compensation systems for industry. *Brown Boveri Rev.*, 70, 330–340.
4. Bonnard, G. (1985). The problems posed by electrical power supply to industrial installations. *Proc. IEE Part B*, 132, 335–340.

5. Neyshabouri, Y., Iman-Eini, H., Chaudhary, S. K., Teodorescu, R., and Sajadi, R. (2020). Improving the reactive current compensation capability of cascaded H bridge based STATCOM under unbalanced grid voltage. *IEEE J. Emerg. Select. Topics Power Elec.*, 8(2), pp.1466-1476.
6. Xu, Y. and Li, F. (2014). Adaptive Pi control of Statcom for voltage regulation. *IEEE Trans. Power Delivery*, 29(3),pp. 1002-1011.
7. Gayatri, M. T. L., Alivelu, M., Parimi, A. V., and Kumar, P. (2018) . A review of reactive power compensation techniques in microgrids. *Renew. Sustain. Energy Rev.*, 81 (Part 1).
8. Liu, S. D., Liu, F., Liu, B., and Kang, Y. (2008). A variable step size INC MPPT method for PV systems. *IEEE Trans. Indus. Elec.*, 55(7), 2622–2628.
9. Kumar, N. and Kumar, S. A. (2016). Reactive power control in decentralized hybrid power system with Statcom using GA, ANN and Anfis methods. *Inter. J. Elec. Power Energy Sys.*, 83, pp.175-187.
10. Foad, H. G., Abdollahahmadib, Sharaf, A. M., Pierluigisianod, J., Branislavhredzakb, and Agelidise, V. G. (2018). Review of facts technologies and applications for power quality in smart grids with renewable energy systems. *Renew. Sustain. Energy Rev.*, 82(1), pp.502-514.
11. Varma, R. and Siavashi, E. (2018). PV-STATCOM – A new smart inverter for voltage control in distribution systems. *IEEE Gen. Meet. Power Energy Soc*, pp.1681-1691.
12. Shukla, A. and Nami, A. (2015) . Multilevel converter topologies for STATCOMs, F. Shahnia, et al. (eds.). Static Compensators (STATCOMs) in Power Systems. Springer Science+Business Media Singapore.
13. Boonmee, C., Somboonkit, P., and Watjanatepin, N. (2015). Performance comparison of threelevel and multi-level for grid-connected photovoltaic systems. *Proc. IEEE Conf. Pub.*, 1–5.
14. Suresh, Y. and Panda, A. K. (2013). Investigation on hybrid cascaded multilevel inverter with reduced dc sources. *Renew. Sustain. Energy Rev.*, 26, 49–59.
15. Brückner, T. and Bernet, S. (2001). Loss balancing in three-level voltage source inverters applying active NPC switches. *Proc. IEEE Power Elec. Special Conf.*, 1135–1140.
16. Bonnard. (1985). The problems posed by electrical power supply to industrial installations. *Proc. IEE Part B*, 132, 335–340.
17. Ishida, T., Matsuse, K., Sugita, K., Huang, L., and Sasagawa, K. (2000). DC voltage control strategy for a five-level converter. *IEEE Tran Power Elec.*, 15(3), 508–515.
18. Hochgraf, E. K. and R. H. Lasseter. Chapter 1 Converter and Output Filter Topologies for STATCOMs. Statcom controls for operation with unbalanced voltages. *IEEE Trans. Power Del.*, 13(2), 538–544. http://www.energy.siemens.com/us/en/powertransmission/hvdc/hvdcplus/modularmultilevel-converter.htm.
19. Liansong, X., Fang, Z., and Minghua, Z. (2013). Study on the compound cascaded STATCOM and compensating for 3-phase unbalanced loads. *Appl. Power Elec. Conf. Expos. (APEC)*, 3209–3215.
20. Song, Q. and Liu, W. (2009) . Control of a cascade STATCOM with star configuration under unbalanced conditions. *IEEE Trans. Power Elec.*, 24(1),pp.45-58.
21. Chou, S.-F., Wang, B.-S., Chen, S.-W., Lee, C. T., Cheng, P. T., Akagi, H., and Barbosa, P. (2013). Average power balancing control of a STATCOM based on the cascaded Hbridge PWM converter with star configuration. *Proc. 2013 IEEE Energy Conv. Cong. Expos. (ECCE)*, 970–977.
22. Yazdani, A. and Irvani, R. (2013). Voltage sourced converters in power system modeling, control, and application. *IEEE Recomm. Prac. Req. Harmonic Control Elec. Power Sys.*, 1–26.
23. Guerrero, J. M., Loh, P. C., Lee, T. -L. and Chandorkar, M. (2013). Advanced control architectures for intelligent microgrids—Part II: Power quality, energy storage, and AC/DC microgrids. *IEEE Trans. Indus. Elec.*, 60(4), 1263–1270. doi: 10.1109/TIE.2012.2196889.
24. Bin, W., Lang, Y., Zargari, N., and Kouro, S. (2011) . Power conversion and control of wind energy systems. *IEEE Press*. 30-38
25. P. W. Hammond. "A new approach to enhance power quality for medium voltage ac drives," IEEETrans. Ind. Appl., vol. 33, no. 1, pp. 202–208, Jan./Feb. 1997.
26. J. S. Lai and F. Z. Peng. "Multilevel converters – A new breed of power converters," IEEE Trans.Ind. Appl., vol. 32, no. 3, pp. 509–517, 1996.
27. F. Z. Peng and J. S. Lai. "Dynamic performance and control of a static var generator using cascade multilevel inverters," IEEE Trans. Ind. Appl., vol. 33, no. 3, pp. 748–755, Mar./June 1997.
28. R. Marquardt and A. Lesnicar. "A new modular voltage source inverter topology," Proc. EPE,Sept. 2003.
29. A. Lesnicar and R. Marquardt. "An innovative modular multilevel converter topology suitable for a wide power range," Proc. IEEE Bologna Power Tech, 2003, vol. 3, pp. 23 26.
30. M. Glinka and R. Marquardt. "A new ac/ac-multilevel converter family applied to a single phase converter," Proc. IEEE-PEDS, 2003, vol. 1, pp. 16–23.
31. M. Glinka and R. Marquardt. "A new ac/ac multilevel converter family," IEEE Trans. Ind. Electron., vol. 52, no. 3, pp. 662–669, 2005.
32. H. Akagi, S. Inoue, and T. Yoshii. "Control and performance of a transformerless cascade PWM STATCOM with star configuration," IEEE Trans. Ind. Appl., vol. 43, no. 4, pp. 1041-1049, July/Aug. 2007.
33. M. Hagiwara and H. Akagi. "PWM control and experiment of modular multilevel converters," Proc. IEEE PESC, June 2008, pp. 154–161.

30 A review on mode estimation in wide area monitoring of smart grids using degraded PMU data

Subhalaxmi Satapathy[a], *Vagesh Kumar*[b] *and Shekha Rai*[c]

Department of Electrical Engineering. National Institute of Technology, Rourkela, Odisha, India

Abstract

The addition of non-conventional energy sources and the evolving landscape of the electrical industry have brought about fresh operative challenges and uncertainties within the power system. As a consequence, the power system's stability has been significantly compromised. In order to tackle these challenges, wide area monitoring systems (WAMSs) has been introduced by the power industry and to convey exact time phasor measurements it employs phasor measurement units (PMUs) in conjunction with the global positioning system (GPS). This advancement has enabled the extraction of modal information in real-time. This review paper presents a comprehensive investigation of techniques for accurate mode estimation in wide-area monitoring of smart grids using degraded PMU data. The paper discusses three efficient methods i.e., K-medoids-long short-term memory (K-medoids-LSTM), K-means-artificial neural network (K-means-ANN), and robust principal component analysis (RPCA) with probabilistic distributional clustering (PDC) algorithm, which address combines outlier detection and missing data imputation and estimation of signal parameters via rotational invariance technique (ESPRIT) is employed as a mode estimator.

Keywords: PMU, K-medoids-LSTM, K-means-ANN, RPCA, PDC, ESPRIT, modes evaluation

I. Introduction

In the recent technological era, as the demand for electricity increases, power grids experience a greater burden, resulting in the challenges like instability in the power system, recognition and observation of poorly damped low frequency oscillation. This instability can lead to issues such as blackouts, which not only impact people's daily lifestyles but also hinder a country's economic development [1]. Furthermore, as the reliance on internet-based grid monitoring grows, the risk of cyber-attacks also rises [2]. Accurately estimating low-frequency inter-area oscillations poses challenges for supervisory control and data acquisition (SCADA) systems [3]. To overcome these issues, the WAMS was introduced. Unlike SCADA system, WAMS offers improved capabilities for capturing and analyzing data from a wider geographic area, allowing for more accurate assessment and control of low-frequency oscillations [4].

Power system protection and control encounter significant challenges due to low-frequency oscillations (LFO), which can potentially result in blackouts [1]. To detect these low frequency modes, various online detection techniques have been developed, including variable projection [5], fast Fourier transform (FFT) [6], PRONY algorithm [7], sparsity [8], Kalman filter [9], ESPRIT-based algorithm [10].

The PMUs deployed on-site often experience various data quality issues that result from factors such as communication congestion, hardware failures, transmission interruptions, and other related challenges [2]. In the context of reconstructing degraded PMU data, this paper has implemented three different techniques such as K-medoids-LSTM [11], K-means-ANN [12], and RPCA [13] to address the issue of degraded PMU signals. Besides these approaches, this review paper used a two-stage technique called ESPRIT based technique [10], which focuses on mode estimation.

In the first technique, an efficient K-medoids-LSTM approach is employed. It uses K-medoids algorithm to detect and remove outliers from the signal. The LSTM model is employed to predict and impute missing values by leveraging the observed data patterns [11].

In the second technique, a K-means-ANN [12] based approach is utilized. For detecting and eliminating anomalies from the signal the K-means clustering algorithm is used. Afterwards, ANN is employed to impute missing data values.

The third technique employs a RPCA method [13]. The RPCA [14] technique is utilized to identify and remove outliers from the signal. Then, for imputation of the missing values, the PDC [15] algorithm is used.

Finally, the reconstructed signal is fed into the modified ESPRIT [10] based method for mode estimation in all three methods.

[a]522ee1007@nitrkl.ac.in, [b]222ee6309@nitrkl.ac.in, [c]rais@nitrkl.ac.in

DOI: 10.1201/9781003540199-30

All the three techniques are illustrated in Section II, afterwards the results and discussion of all the three methods are mentioned in Section III. Lastly, Section IV concludes the analysis.

II. Methodology of proposed techniques

A. K-medoids—LSTM-based algorithm

Handling of outliers by K-medoids

The steps elaborated in K-medoids clustering [11] are as follows: Initialize k clusters in the given dataset. Randomly choose k objects as medoids, assigning each object to a cluster. Compute the cost (distance) between each non-medoid entity and the medoids using the Manhattan distance formula expressed in Equation 1.

$$\text{Manhattan Distance} = \text{mod}\,(x1 - x2) + \text{mod}\,(y1 - y2) \quad (1)$$

Give each non-medoid entity to the cluster with the minimum medoid distance. Calculate the total cost, and their corresponding cluster centroids. Then allot it to d_j. Randomly select a non-medoid object (i). Temporarily swap the object (i). With its assigned medoid (j) and recalculate the total cost and give it to d_i. If $d_i<d_j$, make the swap permanent to update the set of k-Medoids. Otherwise, undo the temporary swap. Reiterate step four to step eight unit no further changes occur.

Handling of missing values by LSTM

The LSTM [19] network utilizes 3 gates (unknown gate, input gate, and output gate) to find the cell state C_t. At every stage, the input χ_t and the prior unknown state h_{t-1} are used to compute which information should be forgotten from the previous cell state i.e., C_{t-1} by the unknown gate layer. Afterwards, the unknown gate generates a digit between 0 and 1 for every component inside the cell state vector C_{t-1}, where a value of zero indicates whole disremembering and a value of one specifies retaining the information. The output of the forget gate is determined by the Equation 2.

$$f_t = \sigma\,(W_f \times [h_{t-1}, x_t] + b_f) \quad (2)$$

The input gate in the LSTM network determines which cell units must be updated. It is computed by the Equation 3.

$$i_t = \sigma\,(W_i \times [h_{t-1}, x_t] + b_i) \quad (3)$$

To establish the cell state $\bar{C}$ in an LSTM network, the computation is performed using the Equation 4.

$$\bar{C}_t = tanh\,(W_C \times [h_{t-1}, x_t] + b_C) \quad (4)$$

The cell state C_t in an LSTM network can be updated using the Equation 5.

$$C_t = f_t * h_{t-1} + (i_t * \bar{C}\,) \quad (5)$$

The output gate O_t in an LSTM network is computed using the Equation 6.

$$O_t = \sigma\,(W_O\,[h_{t-1}, x_t]) + bo \quad (6)$$

where, hidden state = C_{t-1}, input state = χ_t, sigmoid activation function = σ, output gate = O_t, next hidden state = , new cell state = C_t, W_c, W_f, W_0 represent weight matrices, while b_i, bc, and b_0 are bias vectors. The symbol * denotes element-wise multiplication.

The final step in the LSTM process is to use the updated cell state C_t and concealed state ht expressed in Equation 7 is to process the next timestep (t+1). This creates a loop where the LSTM recurrent unit takes the current cell state and hidden state, and produces the next cell state C_{t+1} and concealed state h_{t+1}. This process has to be repeated till the completion of the sequence, allowing the LSTM network to capture and process information over time.

$$h_t = O_t * tanh\,(C_t) \quad (7)$$

B. K-means—ANN-based algorithm

Handling of outliers by K-means

A popular unsupervised learning technique K-Means clustering [20] is commonly employed in data mining tasks. Its objective is to partition the data space into distinct clusters, grouping together data points that exhibit similarity while keeping dissimilar data points in separate clusters. The centroid of each cluster is used for their characterization, which represents all the data point's mean allocated to that particular cluster. The determination of the optimal number of clusters (K) is done by the elbow method by analyzing the within-cluster sum of square (W.C.S.S) distance measure (D(x,y)) which is calculated using Equation 8. This approach aids in identifying the appropriate number of clusters that effectively captures the inherent variations present within the dataset.

$$D(x,y) = \sqrt{\sum_{i=1}^{k} 1 \sum_{l=1}^{ki} (Y_{i,l} - X_i)^2} \quad (8)$$

where, suppose $Y_i = d_1, d_2, d_3, \ldots, d_n$ be the data points and initial centroids are represented as,

$X_i = x_1,\ldots, x_k$.

Handling of missing values by ANN

To handle missing values within a dataset, an advanced ANN technique is utilized. The first step is to divide the dataset into separate groups as train set and test set, with input variables identified. The train set is employed for building a model capable of predicting the missing values, while the test set is designed to assess the execution of models. By inputting the identified variables into the ANN model, prediction errors are minimized by adjusting the weights of neurons V_k expressed in Equation 9. Once the model is skilled, it is evaluated using the testing set, and the obtained outcomes are analyzed to determine the efficiency of the model in handling missing values [16].

$$V_K = \sum_{i=1}^{n} W_{Ki} Z_i + b_k \tag{9}$$

If W_{ik} is the old load, then the restructured load can be stated as shown in Equation 10.

$$\Delta_{(k)} = \mu \times x_{i(k)} \times e_k \tag{10}$$

C. Robust principal component analysis-based algorithm

The main idea behind RPCA is to decompose the dataset into two components: a low-rank component indicating the underlying structure and a sparse component indicating the anomalies or corruptions. This decomposition allows for the identification and removal of outliers while preserving the important information in the data. The RPCA algorithm iteratively estimates the low-rank and sparse components by minimizing a specific objective function. This process involves finding the optimal balance between fitting the low-rank component to the data and promoting sparsity in the sparse component [17].

Corruption model

The following equation is used to separate the clean signal and corrupted signal.

$$Z(\mathbf{t}) = \mathbf{y}(\mathbf{t}) + \mathbf{e}(\mathbf{t}) \tag{11}$$

where, z (t) signifies the measured signal, y (t) signifies the clean signal and e (t) signifies the corrupted signal.

In monitoring applications, there can be discrepancies between the measurements observed at the control center and the actual measurements due to various factors like missing data, outliers, or intentional manipulations. To account for this, a model introduces an additional term called the signal corruption, which is added to the observed measurements. The model assumes that the corruptions only impact a fraction of all signals because the monitoring devices are spread across wide geographical areas. Therefore, the corruption vector, denoted as e (t) is mostly filled with zeros, making it sparse [14].

PDC for missing values

PDC is a method which is used for missing values imputation according to their probability distributions. On the basis of possibility of their distribution appropriate to a particular cluster it uses both clustering principle and probability modeling for allocating missing values to clusters. As per the existing data, this method offers a probabilistic structure for imputation of missing values, considering the fundamental patterns and characteristics of the dataset [15].

ESPRIT-based technique for modes evaluation

The recovered data generated from all the three algorithms are used as input in the ESPRIT based method [10] to generate a robust auto-correlation matrix. This matrix is then used to estimate the model constraints accurately. The simulated real-time signal is denoted by the Equation 12.

$$S(n) = \sum_{k=1}^{m} A_{ke^{-\sigma KT_p}} \cos\left(2\pi f k n T_p + \theta_k\right) \tag{12}$$

where, amplitude (Ak), frequency (fk), phase angle (θk), and damping factor (σk). The sampling time period (Tp) is also involved in the representation of the signal.

III. Results and discussion

It is a study where different methods are applied to a besmirched test signal containing oscillations and data from multiple generators in local-area and true probing data from Western Electricity Coordinating Council (WECC) [18]. The purpose of the study is to evaluate the proposed methods (K-medoid-LSTM, K-means-ANN, and RPCA) and compare their performance. The evaluation includes a statistical analysis conducted at various levels of noise. The objective is to assess how well the proposed techniques handle missing values and outliers and determine their effectiveness in mode estimation.

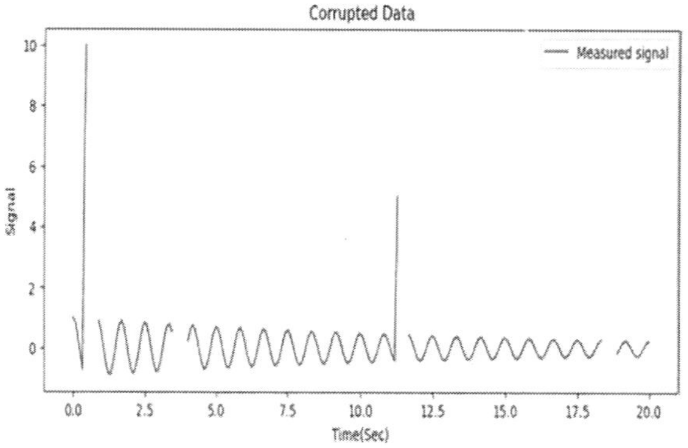

Figure 30.1 Local area oscillation mode corrupted signal obtained from simulation

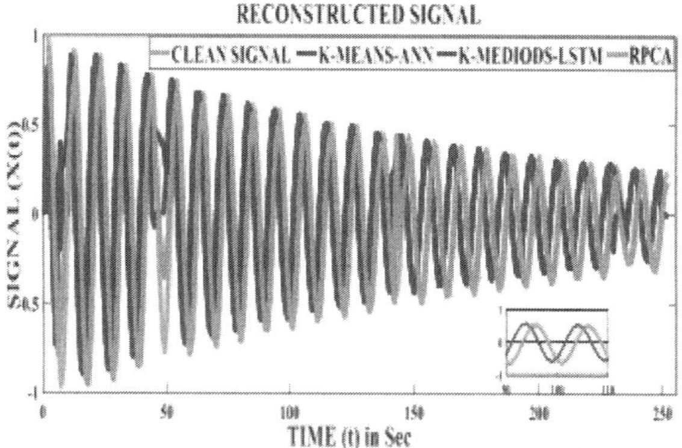

Figure 30.2 Simulated reconstructed signal at SNR 20 dB

A. Mode evaluation of local area

The respective algorithms are applied to an incomplete test signal with a frequency of 1.2 Hertz and a damping ratio of -0.07 shown in Figure 30.1. Figure 30.2 represents the signals reconstructed by all three methods. The study conducted on local-area modes, a statistical analysis is performed using 10,000 cycles of Monte Carlo simulation, and the results are presented in Table 30.1. The findings indicate that in the K-means-ANN technique the frequency range is 1.2063 Hertz as well as the damping factor is -0.0514. In the K-medoids-LSTM based technique the frequency range is 1.2068 Hertz and the damping factor is -0.0528. Whereas in the RPCA technique the frequency range is 1.1995 Hertz and damping factor is -0.0637. From the experiment, it was found that the RPCA technique offers exact estimation of the modes compared to the other two techniques, K-medoids-LSTM and K-means-ANN, at a signal to noise ratio (SNR) value of 20 dB.

Figures 30.3 and 30.4 illustrate the frequency and damping ratio plots for all 3 methods at a SNR of 20 dB. The graph clearly indicates that the RPCA technique provides highly correct damping and frequency values with reduced variation compared to the other techniques.

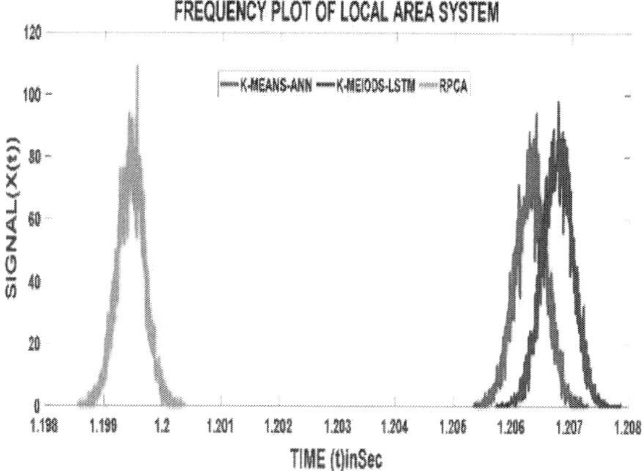

Figure 30.3 Frequency graph of local area system at SNR 20 dB representing mean and variance

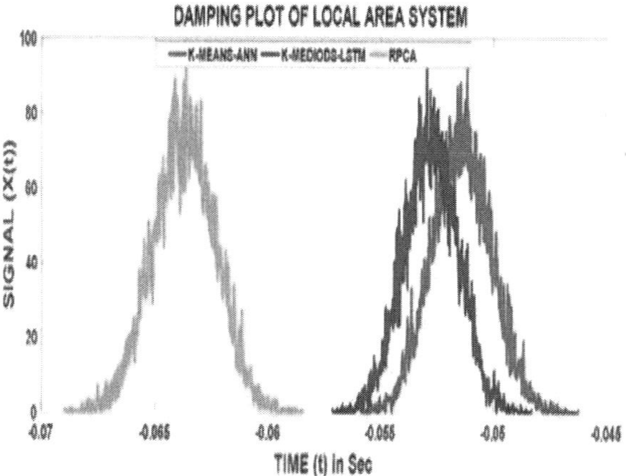

Figure 30.4 Damping graph of local area system at SNR 20 dB representing mean and variance

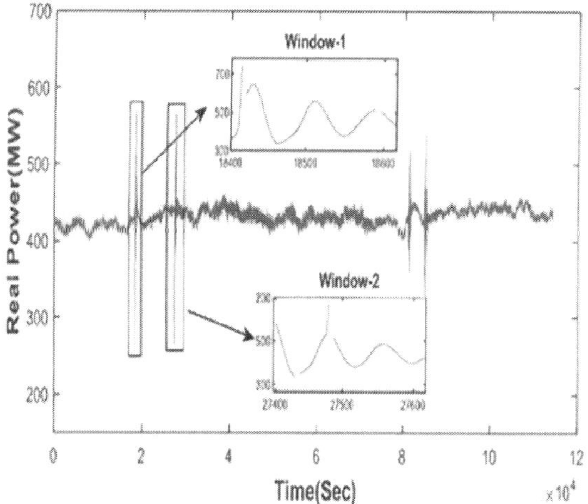

Figure 30.5 Probing data by PMU linked to WECC

Table 30.1 Predicted value of means (μ) & variance (σ) of local area mode of oscillation for test signal

Real frequency (f) = 1.2 Hertz and real damping (D)= -0.07				
		K-medoids-LSTM		
	F (Hertz)		D	
S.N.R(in dB)	μ	Σ	μ	Σ
20	1.2068	6.81×10-08	-0.0528	1.44×10-6
40	1.2068	6.86×10-10	-0.0527	1.49×10-8
50	1.2068	6.73×10-11	-0.0528	1.44×10-9
		K-means-ANN		
	F (Hertz)		D	
S.N.R (in dB)	μ	σ	μ	σ
20	1.2063	7.11×10-06	-0.0514	1.68×10-6
40	1.2063	7.01×10-10	-0.0514	1.73×10-8
50	1.2063	7.24×10-11	-0.0514	1.67×10-9
		RPCA		
	F (Hertz)		D	
S.N.R (in dB)	μ	σ	μ	σ
20	1.1995	5.50×10-08	-0.0637	1.92×10-6
40	1.1995	5.45×10-10	-0.0637	1.87×10-8
50	1.1995	5.46×10-11	-0.064	1.88×10-9

Table 30.2 Frequency and percentage of damping estimation for probing data collected from WECC

Real frequency(f)= 0.318 Hertz and real damping % = 8.3%				
WINDOW	Modes	K-meansANS-ANN	K-mediods- LSTM	RPCA
1	Frequency (HZ)	0.3340	0.3353	0.3254
	damping %	7.2999%	7.9774%	8.5438%
2	Frequency (HZ)	0.3314	0.3338	0.3142
	damping %	8.8611%	6.4013%	7.9580%

Evaluation of modes with real test signal collected from WECC

The respective techniques were verified on inclusive probing data taken from a phasor measurement unit (PMU) linked to a system of WECC retrieved on 14th Sept, 2005, for both Window 1 and Window 2 shown in Figure 30.5. The estimated mode of frequency and damping for all the algorithms at a 20 dB SNR are presented in Table 30.2.

Conclusions

This research paper discusses three novel approaches to tackle the challenges associated with degraded PMU signals. These approaches are specifically designed to address missing data imputation, outlier detection and removal. The evaluation of these techniques was conducted using real-world data (probing data of real-time) collected from the WECC, including local-area. Meant for missing data imputation, the methods employed were ANN, LSTM, and PDC algorithm. Outlier detection and removal utilized K-means, K-medoids and RPCA algorithm. Following the application of these techniques, the reconstructed signals were subjected to the ESPRIT based algorithm for estimation of frequency and damping factor of low-frequency oscillations. The evaluation considered various signal to noise ratio (SNR) values, including 20 dB, 40 dB, and 50 dB. The results revealed that among the three approaches, RPCA demonstrated high efficiency and effectiveness and this is suitable for wide area monitoring applications.

Acknowledgement

We are gratefully acknowledged the students, staff, and authority of Electrical Engineering department, NIT Rourkela for their cooperation in this research.

References

1. Lal, M. D. and Varadarajan, R. (2023). "A review of machine learning approaches in synchro- phasor technology." *IEEE Access*, vol. 11, pp. 33520-33541, 2023.
2. Adhikari, U., Morris, T., and Pan, S. (2016). WAMS cyber-physical test bed for power system, cybersecurity study, and data mining. *IEEE Trans. Smart Grid*, 8(6), 2744–2753.
3. Thomas, M. S. and McDonald, J. D. "Power system SCADA and smart grids," CRC press, 2017.
4. A. G. Phadke and T. Bi, "Phasor measurement units, WAMS, and their applications in protection and control of power systems," in Journal of Modern Power Systems and Clean Energy, vol. 6, no. 4, pp. 619-629, 2018, doi: 10.1007/s40565-018- 0423-3
5. Borden, R. and Lesieutre, B. C. (2014). Variable projection method for power system modal identification. *IEEE Trans. Power Sys.*, 29(6), 2613–2620. doi: 10.1109/TPWRS.2014.2309635.
6. Girgis, A. and Ham, F. M. (1980). A quantitative study of pitfalls in the FFT. *IEEE Trans. Aerospace Elec. Sys.*, 16(4), 434–439. doi: 10.1109/TAES.1980.308971.
7. Rai, S., Lalani, D., Nayak, S. K., Jacob, T., and Tripathy, P. (2016). Estimatn of low-frequency modes in power system using robust modified prony. *IET Gen. Trans. Distrib.*, 10(6), 1401–1409.
8. Rai, S., Tripathy, P., and Nayak, S. K. (2019). Using sparsity to estimate oscillatory mode from ambient data. *Sadhana – Acad. Proc. Engg. Sci.*, 44(4), 1–9. doi: 10.1007/s12046-019-1071-7.
9. Zhu, X., et al., "A Kalman filter based approach for markerless pose tracking and assessment" *27th International Conference on Automation and Computing (ICAC)*. pp.1-7, 2022.
10. Tripathy, P., Srivastava, S. C., and Singh, S. N. (2011). A modified TLS- ESPRIT-based method for low- frequency mode identification in power systems utilizing synchrophasor measurements. *IEEE Trans. Power Sys.*, 26(2), 719–727. doi: 10.1109/TPWRS.2010.2055901.
11. Sahoo, M., Roy, A., and Rai, S. (2023). A K-medoids-LSTM based technique for electromechanical modes identification for synchrophasor applications. *2023 Inter. Conf. Power Instrum. Control Comput. (PICC)*, 1–6.
12. Sahoo, M., Reshma, V., and Rai, S. (2023). A K-means-ANN based technique for identification of oscillatory modes in power system. 2023 IEEE 8th Inter. Conf. Converg. Technol. 1–6.
13. Chatterjee, K., Chaudhuri, N. R., and Stefopoulos, G. (2020). Signal selection for oscillation monitoring with guarantees on data recovery under corruption. *IEEE Trans. Power Sys.*, 35(6), 4723–4733.
14. Ding, Y. and Liu, J. (2017). Real-time false data injection attack detection in energy internet using online robust principal component analysis. *2017 IEEE Conf. Energy Internet Energy Sys. Integ. (EI2)*, 1–6.
15. Yuan, X., Zhao, C.-X., and Zhang, H.-F. (2012). A probability distribution-based point cloud clustering algorithm. *Inter. J. Model. Identif. Control*, 15(4), 320–330.
16. Choudhury, S. J. and Pal, N. R. (2019). Imputation of missing data with neural networks for classification. *Knowledge-Based Sys.*, 182, 104838. doi: 10.1016/j.knosys.2019.07.009.
17. Mahapatra, K. and Chaudhuri, N. R. (2018). Malicious corruption-resilient wide-area oscillation monitoring using online robust PCA. *2018 IEEE Power Energy Soc. Gen. Meet. (PESGM)*, 1–5.
18. Reportand data of WECC. Available: ftp://ftp.bpa.gov/pub/WAMS Information/.
19. Sherstinsky, A. (2020). Fundamentals of recurrent neural network (RNN) and long short-term memory (LSTM) network. *Phys. D Nonlin. Phenom.*, 404, 132306.
20. Ikotun, A. M., Ezugwu, A. E, Abualigah, L., Abuhaija, B., and Heming, J. (2022). K-means clustering algorithms: A comprehensive review, variants analysis, and advances in the era of big data. *Information Sciences*, 622, pp.178-210.

31 Instant decompression-assisted sustainable green technology for the quality improvement of rice: Machine learning and computational fluid dynamics-based process modeling and simulation

Sourav Chakraborty[1,2,a], *Sonam Kumari*[2], *Swapnil Prashant Gautam*[2] *and Manuj Kumar Hazarika*[2]

[1]Department of Food Processing Technology, Ghani Khan Choudhury Institute of Engineering and Technology, Narayanpur, Malda, West Bengal-732141, India

[2]Department of Food Engineering and Technology, Tezpur University, Assam-784028, India

Abstract

In this investigation, instant decompression-assisted sustainable green technology was developed for the quality improvement of rice. In combination with four major sections namely boiler, treatment chamber, vacuum tank fitted with vacuum pump, and instant controlled pressure release system, along with various other parts, the ICPD treatment unit was fabricated. The performance evaluation of ICPD unit based on various parameters namely pressure drop/decompression rate (PDR), operational pressure of treatment chamber (OPTC), TP-TT and TCT-TT graphs, broken rice percentage (BRP) and drying time (DT) revealed adequate operation of the system. Machine learning approach in terms of adaptive neuro-fuzzy interface system (ANFIS) was applied for the simulation of ICPD performance parameters, considering BRP and DT as input of the structure. 3-3 ANFIS structure in combination of 9 fuzzy rules and gaussmf membership function showed adequate performance for effective simulation of the unit. Further, computational fluid dynamics (CFD) was applied for the profiling of temperature modules during ICPD treatment unit process.

Keywords:ICPD, TP-TT, TCT-TT, ANFIS, CFD

Introduction

Instant decompression-assisted sustainable green technology, also known as ICPD treatment is a novel method to improve process performance and quality attributes of rice. During the parboiling process, paddy grains undergo ICPD treatment in addition to soaking and hot air drying. This form of treatment is also known as a revolutionary steaming procedure based on rapid decompression. Saturation steam is used to achieve ICPD-induced steaming in a treatment chamber. Inside the chamber, samples of paddy are exposed to saturated steam pressure for a predetermined duration. After the steam treatment, the chamber's pressure is rapidly reduced to a high vacuum state through an instantaneously controlled pressure release valve. High porosity, as well as enhanced specific surface area [5], and quick autovaporization of moisture [1–4], are the outcomes of this swelling and textural alteration. The objectives of the present study were the development of ICPD treatment unit for the quality enhancement of rice, simulation of its performance by applying machine learning approach, and profiling of temperature modules using computational fluid dynamics (CFD) modeling.

Materials and methods

Design and development of ICPD treatment unit

The ICPD treatment unit must achieve the following goals: (a) development of an introductory vacuum stage, (b) generation of high steam pressure (0.1–1.0 MPa), with pressure immediately dropping into the vacuum tank, (c) retention of vacuum stage after steaming, (d) maintaining pressure releasing rate of more than 0.25 MPa/s, and (d) releasing final pressure towards the atmosphere as the termination of the treatment.

The ICPD treatment unit was developed after careful consideration of design criteria and conceptualization. The apparatus included a boiler, a treatment chamber, a vacuum tank equipped with a vacuum pump, and an instant controlled pressure release system, as well as a number of smaller components like a moisture trapper, a pressure gauge, a temperature gauge, a control panel and an atmospheric pressure releasing valve.

[a]sourav@gkciet.ac.in

DOI: 10.1201/9781003540199-31

Performance evaluation of the ICPD treatment unit

The performance of the ICPD treatment unit was evaluated by measuring the following parameters: Pressure drop/decompression rate (PDR), operational pressure of treatment chamber (OPTC), TP-TT and TCT-TT graphs, broken rice percentage (BRP), drying time (DT).

Simulation of the performance of ICPD treatment unit

Adaptive neuro-fuzzy adaptive system (ANFIS) was used for the simulation of the performance of ICPD treatment unit. With consideration of treatment pressure (TP) and treatment time (TT) as independent and broken rice percentage (BRP) and drying time (DT) as dependent variables, ANFIS was implemented. The typical first order Takagi-Sugeno-Kang type network was used for the development of ANFIS model. ANFIS was solved by using MATLAB R2015a software. On the basis of coefficient of determination (R^2) and mean square error (MSE) values, the performance of ANFIS was evaluated. Besides, the profiling of the ICPD temperature modules was done by applying computational fluid dynamics modeling.

Results and discussions

Experimental setup of ICPD treatment unit and its operation

The ICPD treatment unit was developed with consideration of the operational parameters viz., operating pressure at treatment chamber (OPTC), and maximum pressure drop on release and pressure drop rate (PDR). The developed ICPD unit had the treatment chamber volume of 3534 cm^3 and sufficient for samples for treatment up to 1.8 kg (based on bulk density of 523 kg/m^3) The treatment unit was consisted of four major sections namely boiler, treatment chamber, vacuum tank fitted with vacuum pump, and instant controlled pressure release system, along with various other parts such as moisture trapper, pressure gauge, temperature gauge, control panel and pressure releasing valve to atmosphere, etc. The experimental setup of ICPD unit (fabricated) and its treatment chamber is shown in Figure 31.1.

Performance evaluation of the ICPD unit

The performance of the ICPD treatment unit was evaluated on the basis of various parameters namely PDR, OPTC, TP-TT and TCT-TT graphs, BRP and DT of the parboiled paddy grains. Figure 31.2

Figure 31.1 Fabricated ICPD treatment unit

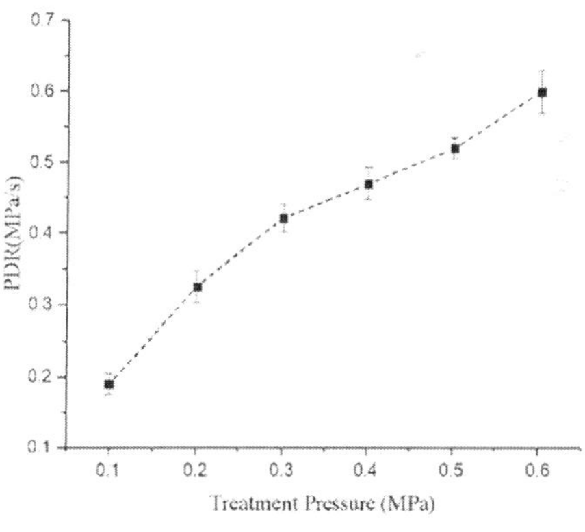

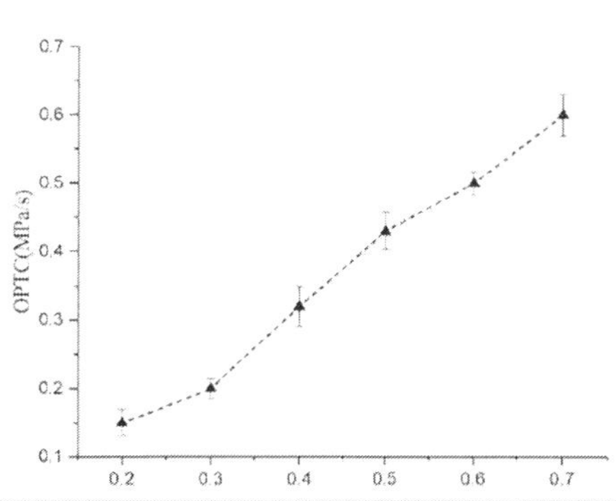

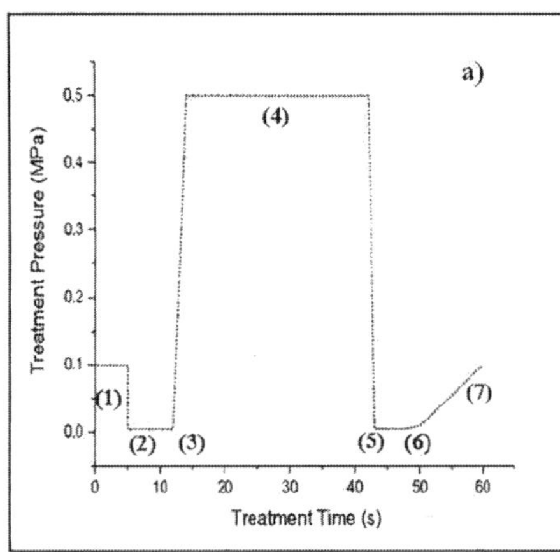

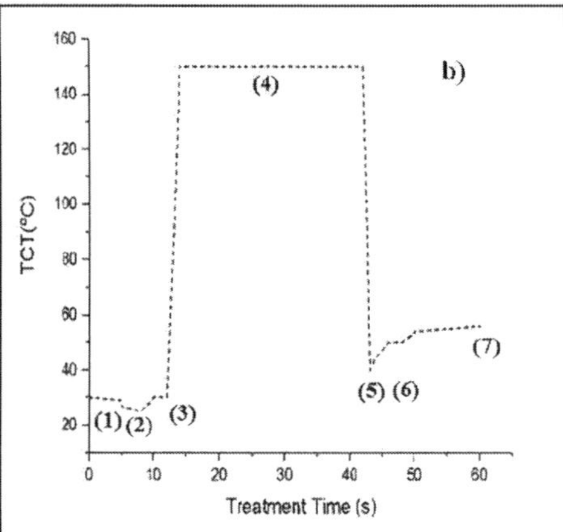

Figure 31.2 Performance evaluation of ICPD treatment

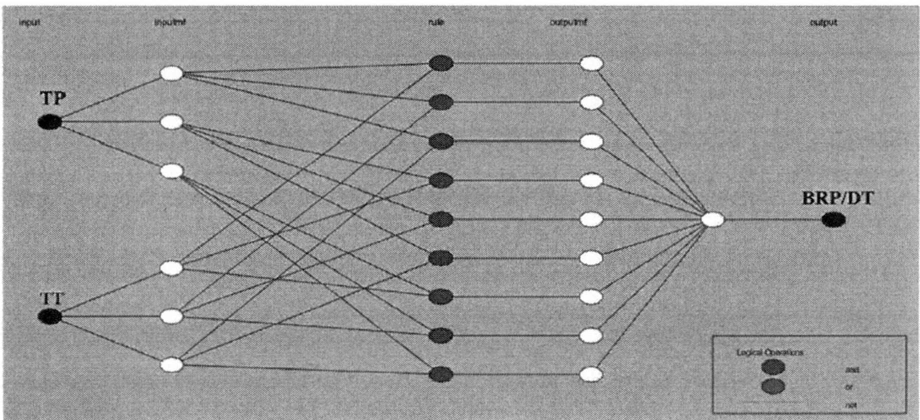

Figure 31.3 ANFIS architecture for the modeling of ICPD assisted parboiling process

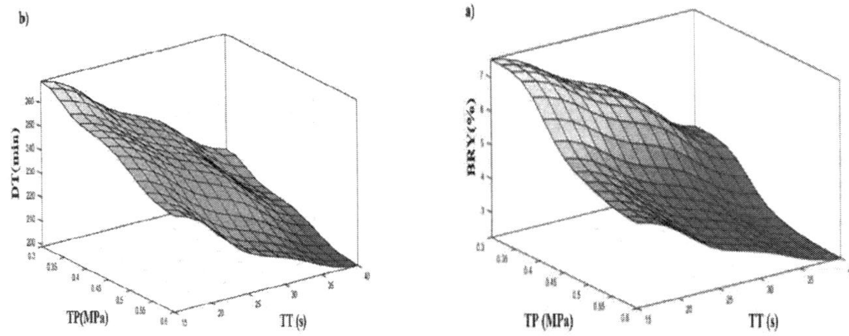

Figure 31.4 Effect of TP and TT on the broken rice

illustrates various performance parameters of the ICPD treatment unit.

Simulation of the performance of ICPD unit

The performance of the ICPD treatment was simulated by applying adaptive neuro-fuzzy interface system (ANFIS). The performance parameter, BRP and DT were simulated as a function of TP and TT. Best ANFIS architectures for BRP and DT were determined by applying Takagi-Sugeno type. After 4000 epoches, 3-3 was selected as the best models for both the ICPD treatment parameters. Total 9 fuzzy rules along with Gaussian (gaussmf) membership function were implemented for the selection of best ANFIS model. Figure 31.331.–31.5 illustrate the ANFIS based simulation of the ICPD assisted parboiling process. CFD-based temperature modules of the ICPD unit are demonstrated in Figure 31.6.

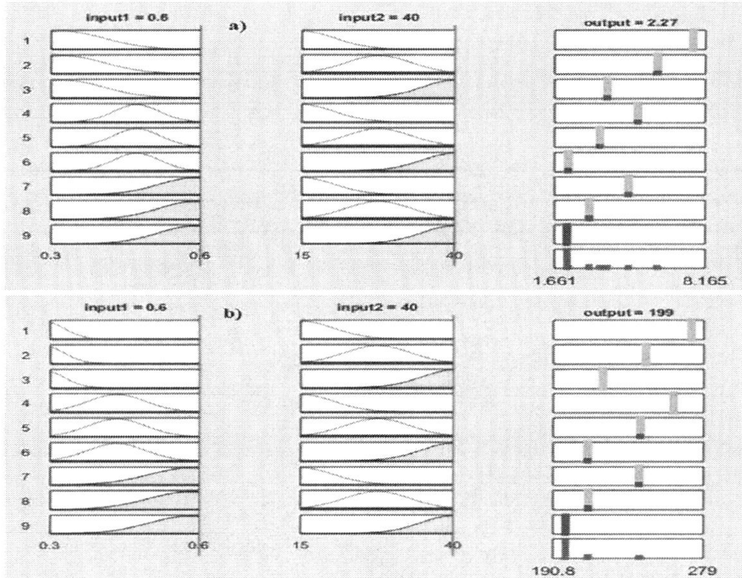

Figure 31.5 Simulation of ANFIS model for ICPD-assisted parboiling process

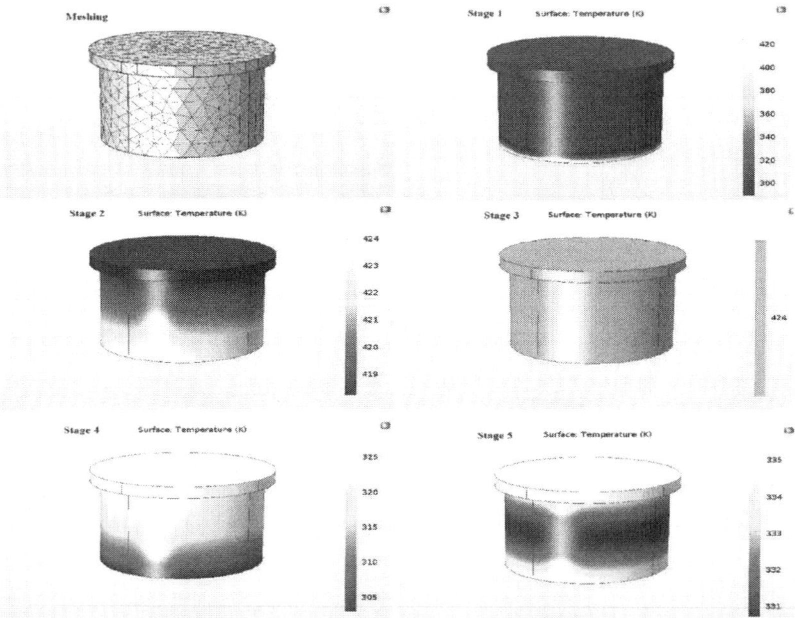

Figure 31.6 Computational fluid dynamics (CFD) modeling of ICPD treatment process

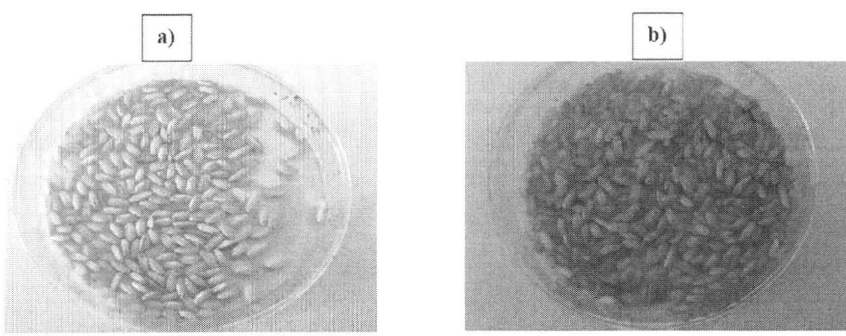

Figure 31.7 ICPD treated rice samples

Figure 31.7 represents ICPD treated samples in contrasted with control samples.

Conclusions

ICPD treatment is a sustainable green technology, which can be applied for the quality enhancement of rice and other food products during processing. In the present investigation, the developed unit showed adequate performance based on pressure drop/decompression rate (PDR), operational pressure of treatment chamber (OPTC), TP-TT and TCT-TT graphs, broken rice percentage (BRP) and post steaming drying time (DT) of the paddy grains. The results revealed that treatment pressure (TP) ranging from 0.2 to 0.6 MPa showed adequate performance based on the required criteria of ICPD treatment parameters. Machine learning approach namely 3-3 ANFIS structure in combination with 9 fuzzy rules and *gaussmf* membership function showed adequate performance for effective simulation of the unit. Further, computational fluid dynamics (CFD) modeling helped for the dynamic profiling of ICPD temperature modules.

Acknowledgement

The authors are thankful to Department of Food Engineering and Technology, Tezpur Universityfor providing all the lab facilities.

References

1. Chakraborty, S., Gautam, S. P., Das, P. P., andHazarika, M. K. (2019). Instant controlled pressure drop (DIC) treatment for improving process performance and milled rice quality.*J. Inst.Eng.(India) Ser.A*,100, 683–695.
2. Chakraborty, S., Gautam, S. P., Bordoloi, T., andHazarika, M. K. (2020). Neural network and computational fluid dynamics modeling for the gelatinization kinetics of instant controlled pressure drop treated parboiled rice.*J. Food Proc.Engg.*,43(11), e13534.
3. Chakraborty, S., Gautam, S. P., Sarma, M., andHazarika, M. K. (2021). Adaptive neuro-fuzzy interface system and neural network modeling for the drying kinetics of instant controlled pressure drop treated parboiled rice.*Food Sci.Technol. Inter.*,27(8), 746–763.
4. Pilatowski, I., Mounir, S., Haddad, J., Cong, D. T., andAllaf, K. (2010). The instant controlled pressure drop process as a new post-harvesting treatment of paddy rice: impacts on drying kinetics and end product attributes.*Food Bioproc.Technol.*,3, 901–907.
5. Thai, C. D. andAllaf, K. (2013). DIC-assisted parboiling of paddy rice. *Instant Controll. Press.Drop (DIC) Food Proc.Fundamen.Indus. Appl.*, 57–66. New York, NY: Springer New York.

32 Fusion-based noise reduction technique for CT images using block matching three-dimensional (BM3D) and total variation (TV) approach

Rita Garam[1], Marpe Sora[1], Jagdeep Rahul[2], Amarjit Roy[3,a] and Tamchi Yani[1]

[1]Department of Computer Science and Engineering, Rajiv Gandhi University, Arunachal Pradesh, India

[2]Department of Electronics and Communication Engineering, Rajiv Gandhi University, Arunachal Pradesh, India

[3]Department of Electrical Engineering, Ghani Khan Choudhury Institute of Engineering & Technology, West Bengal, India

Abstract

Noise in medical images is a prevalent issue that can significantly degrade the quality of the image and hinder accurate diagnostic interpretation. This study proposes a novel denoising filter that combines BM3D (block-matching 3D) and total variation (TV) techniques to reduce noise in lung CT images. The aim is to effectively decrease noise artifacts while preserving relevant anatomical details in the CT images. The effectiveness of the new method is evaluated by contrasting its outcomes with those attained by current cutting-edge methods. This evaluation employs well-established measurements for gauging noise, specifically the peak signal-to-noise ratio (PSNR), structural similarity index (SSIM), and mean square error (MSE). The proposed method improves the peak signal-to-noise ratio (PSNR) to 14 dB when corrupted with 10% of additive white Gaussian noise (AWGN). The experiment results suggest the proposed technique outperforms some of the traditional filters.

Keywords: Noise, CT image, BM3D, total variation denoise (TVD)

Introduction

Noise is an undesirable variation within an image that can diminish its quality. In the context of medical imaging, noise refers to artifacts that may obscure the underlying anatomical or pathological information, thereby impacting the accuracy of diagnostic interpretation [1]. Noise in medical images can significantly affect image quality, making it harder to visualize structures, detect subtle abnormalities, and accurately interpret the data. Noise reduction techniques, such as filtering algorithms, are commonly employed in medical imaging to mitigate the impact of noise and enhance image quality. These techniques aim to preserve relevant details while reducing noise artifacts effectively [2–4]. Medical imaging technologies like X-ray, computed tomography (CT), magnetic resonance imaging (MRI), and positron emission tomography (PET) are utilized for disease diagnosis. In recent times, CT scans have become increasingly popular among radiologists as the preferred method for detecting life-threatening diseases. This preference is because CT imaging allows for the clear visualization of anatomical structures in great detail due to its high spatial and temporal resolution. CT scans possess advanced capabilities that produce precise and detailed images, enabling radiologists to accurately identify and analyze abnormalities and diseases. The high spatial resolution of CT scans allows for the visualization of the smallest structures within the body, while the high temporal resolution permits the examination of dynamic processes such as blood flow and organ function [5]. These qualities make CT scans invaluable in diagnosing, monitoring, and treating various life-threatening conditions.

It is important to note that high-resolution CT scans use a high intensity of X-rays, which can raise the risk of cancer. Low-intensity CT scans are frequently used to prioritize patient health [6]. However, due to a variety of circumstances, these low-intensity CT scans are more vulnerable to noise. These CT scans employ less radiation to lower patient risk, but the signal is weaker as a result. The weaker X-ray signal exhibits the electronic noise from the CT scanner [7]. The fewer X-ray photons observed introduce statistical disturbance known as Poisson noise, which has a random distribution and can be more visible in the photographs. CT image noise makes low-contrast objects difficult to diagnose. To improve disease diagnosis, CT images

[a]amarjit@gkciet.ac.in

DOI: 10.1201/9781003540199-32

must be denoised. Medical imaging uses filtering algorithms and other noise reduction methods to reduce noise and improve image quality. These methods preserve details while reducing noise artifacts. In recent decades, researchers have devoted considerable efforts to investigating the denoising of CT images, recognizing its crucial role as a preprocessing step in image classification. Earlier studies have extensively utilized various classical filters [8], including the Gaussian filter, adaptive median filter (AMF), and adaptive Weiner filter (AWF). In ref [9], AWF was used to enhance the quality of images derived from low-resolution video frames. However, Weiner filter effectiveness relies on accurate signal-to-noise ratio (SNR) estimation, it encounters a balance between noise reduction and preserving image intricacies. In ref [10], four filter types, specifically the mean filter, median filter, adaptive median filter (AMF), and Weiner filter, were applied to mammography images, and the findings indicated that the AMF outperformed the other three filters in terms of noise reduction. The problem with the median filter is that it treats every pixel, regardless of whether it is corrupted by noise or not, leading to a blurred effect in the image. Another denoising technique [11] was introduced, based on the collaborative impact of both the Gaussian filter and the non-local means filter. In this approach, the Gaussian filter was employed to smooth the image, while the non-local means filter selected pixels similar to the current pixel for a weighted average, effectively removing noise artifacts.

Presently, the integration of wavelet-based denoising methods with other filters is extensively embraced by researchers in their applications. In method [12], a combination of wavelet and median filtering techniques was utilized for noise filtering. In a study [13], a 2D autoregressive model was proposed to analyze filtered residuals. Residuals obtained from the median filter, average filter, and Gaussian filter were considered important characteristics for the analysis. This study conducted experiments to evaluate the effectiveness of different traditional filters. The findings demonstrated that the fusion of block matching 3D (BM3D) and total variation filter (TVF) yielded the most favorable outcomes specifically for CT images. To assess the efficacy of the combined filter, we conducted a comparative analysis against established cutting-edge techniques, assessing its effectiveness using fundamental metrics like peak signal-to-noise ratio (PSNR), structural similarity index (SSIM), and mean square error (MSE). The subsequent segments of the paper are organized as follows: Section 2 outlines the proposed approach. In Section 3, the experimental outcomes are discussed and presented. Subsequent to this, Section 4 concludes the study and highlights possible avenues for further research.

Methodology

A novel fusion-based filtering technique is proposed in this work. The proposed filter was designed by combining BM3D with the TV technique. In literature, BM3D with the combination of wavelet decomposition is so far explored [14–16]. BM3D filter is very efficient in removing white additive Gaussian noise (AWGN) while TVF avoids over-smoothing of the image by preserving the sharp edges.

Hence, the fusion of BM3D and TVF yields improved performance regardless of the noise density. The fundamental schematic of the applied method is depicted in Figure 32.1.

Block matching three-dimensional (BM3D) algorithm

BM3D [15] is a popular noise-removal algorithm that has gained significant attention recently. The algorithm operates by dividing an image into non-overlapping blocks and subsequently conducting a collaborative filtering process based on the similarity of these blocks. The algorithm comprises two primary stages: block-matching and 3D filtering, respectively.

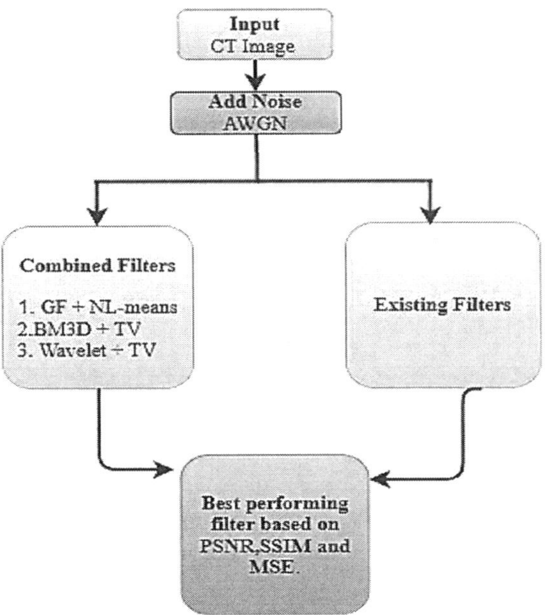

Figure 32.1 Basic block diagram of the proposed method

The noisy image is partitioned into non-overlapping blocks of dimensions B×B. Within each block, a search is performed to find similar blocks using the Euclidean distance. Let's denote R as the reference block, and S = {S1, S2, ..., Sn} as the set of similar blocks. The similarity between the reference block R and each block in the set S are measured. The best matching block from the set is then selected. Mathematically, the BM3D algorithm can be represented as:

S = {S1, S2, ..., Sn}, where Si is a similar block to R.

G = {R, Bm, S1, S2, ..., Sn}, 3D group formation

Y = T(G) Transform:

Y_thresh = T_coeff(Y) Thresholding:

G_denoised = InverseTransform(Y_thresh).

Aggregation: Merge overlapping regions from G_denoised to obtain the denoised image.

Total variation approach

TV denoising is a popular image-denoising technique that aims to preserve edges while effectively reducing noise [17]. It is based on the principle that images with smooth regions and sharp transitions between them tend to have a small total variation. Mathematically, TV denoising can be formulated as an optimization problem. Given a noisy image n, the goal is to determine an image u with reduced noise which optimizes the following expression:

$$E(u) = \|u - n\|^2 + \lambda * TV(u) \quad (1)$$

where $\|u\text{-}n\|^2$ represents the data consistency factor, quantifying the distinction between the denoised image u and the noisy counterpart n. TV(u) represents the total variation of denoised image u, and λ signifies a parameter that maintains the balance between consistency and total variation. The total variation $TV(u)$ is defined as the integration of the absolute gradient of u over the image domain:

$$TV(u) = \int\int \|\nabla u(x, y)\|\, dxdy \quad (2)$$

Here, $\nabla u\ (x, y)$ stands for the gradient of u at every pixel (x, y). TV denoising aims to determine u that minimizes the energy function by striking a balance between preserving the noisy image's fidelity and promoting the denoised image's smoothness. This is typically achieved using iterative optimization methods such as the split Bregman algorithm.

Results and discussion

To begin the experiment, an original image of a lung CT scan was converted into a grayscale image of size 277×277 and loaded into MATLAB. The loaded CT image was then intentionally corrupted with white additive Gaussian noise (AWGN) at various noise densities. The noisy image was subsequently processed using several commonly used filters, including the Gaussian filter (GF), non-local means filter (NLF), median filter (MF), total variation filter (TVF), Wiener filter (WF), and wavelet decomposition. The denoising effects of these filters were evaluated at different levels of noise densities ranging between 10% and 50%. The sample image utilized in this experiment was obtained from a publicly accessible platform, namely Kaggle (https://www.kaggle.com/datasets/mohamedhanyyy/chest-ctscan-images) (Table 32.1).

The earlier studies demonstrated that the combination of filters yielded superior results [8]. In this research, we explored several filter combinations to enhance noise removal. The noisy image was subjected to different combinations of filters, including

Table 32.1 Comparative analysis of different composite filters for AWGN reduction in CT image (277×277) based on PSNR (dB), SSIM, and MSE

Filter-combination		Noise density (%)				
		10%	20%	30%	40%	50%
Gaussian+Non Local-means	PSNR	19.2343	13.8304	10.4088	8.0261	6.3156
	SSIM	0.56432	0.50013	0.44814	0.41388	0.39594
	MSE	0.11928	6.041396	0.09017	0.15754	0.23358
Wavelet+TV	PSNR	28.8522	18.8256	13.4858	10.2589	8.0915
	SSIM	0.74136	0.63955	0.56939	0.51301	0.46185
	MSE	0.0013081	0.013105	0.044945	0.094214	0.15519
BM3D+TV	PSNR	**31.0436**	**18.9094**	**13.5260**	**10.2703**	**8.0989**
	SSIM	**0.81386**	**0.70091**	**0.62383**	**0.55629**	**0.49242**
	MSE	**0.000798**	**0.012855**	**0.044402**	**0.093967**	**0.15492**

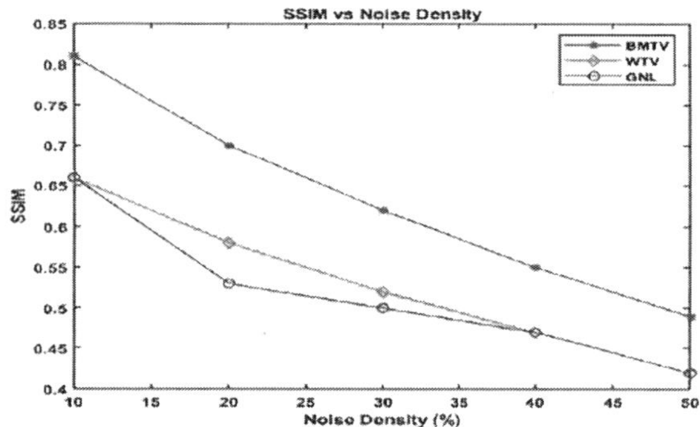

Figure 32.2 SSIM vs. noise density (%) graph of various composite filters at (10%, 20%, 30%, 40%, 50%) noise density

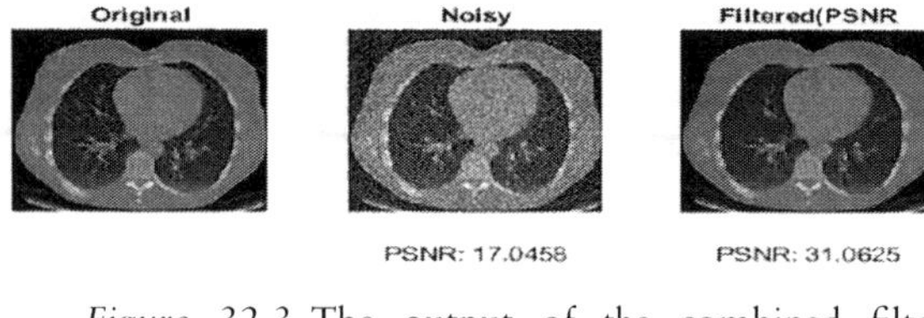

Figure 32.3 The output of the combined filter (BM3D+TV)

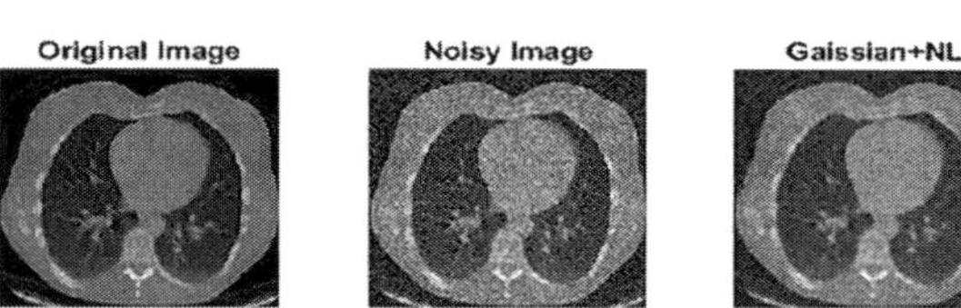

Figure 32.4 The output of the combined filter (Gaussian and NL-means)

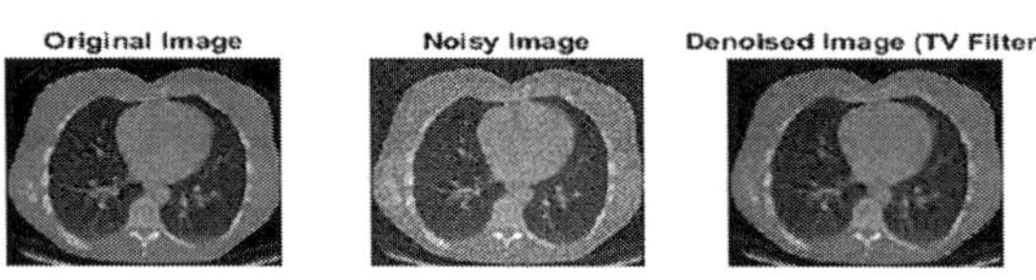

Figure 32.5 The output of the combined filter (Wavelet filter + TV)

wavelet decomposition with TV, a combination of Gaussian and non-local means, and BM3D with TV. It was observed that the fusion of BM3D and TV produced the best outcomes among the studied filter combinations. Initially, the noisy image underwent BM3D filtering, followed by the implementation of TVF on the filtered image. Figures 32.3–5 illustrates the outputs of various combined filters used in the experiment. The CT image used in the experiment is subjected to 10% of additive white Gaussian noise (AWGN). Table. 32.1 provides a comparative analysis of different combined filters in terms of performance metrics such as PSNR, SSIM, and MSE whereas Figure 32.2 represents the comparison graphs of various fusion-based filters in terms of performance metrics such as SSIM. The results of the experiments indicate that the suggested denoising approach involving fusion surpasses alternative composite filters. Notably, the combination of BM3D and TV stands out, showcasing a noteworthy enhancement of 14 dB in PSNR and thereby signifying a substantial improvement in image quality.

Conclusions

Diagnostic imaging demands high-quality, noise-free pictures. Denoising methods enhance image quality. These algorithms perform differently depending on the noise. In this study, a fusion of BM3D and TV filters is applied to enhance image quality. The assessment of both established filters and the newly proposed approach employs PSNR, SSIM, and MSE metrics. Notably, the suggested approach surpasses the amalgamated filters examined in this study, resulting in a 14dB increase in PSNR when contrasted with the initial noisy image. The potential for algorithmic enhancement in future studies dedicated to denoising remains evident.

References

1. Hendee, W. R. and Russell Ritenour, E. (2003). Medical imaging physics. John Wiley & Sons.
2. Rajni and Anutam. (2013). Image denoising techniques - An overview. *Inter. J. Comp. Appl.*, 86. 10.5120/15069-3436.
3. Diwakar, M. and Kumar, M. (2018). A review on CT image noise and its denoising. *Biomed. Sig. Proc. Control*, 42, 73–88.
4. Kaur, S., Nikita, S. J., and Singh, A. (2021). Review on medical image denoising techniques. *2021 Inter. Conf. Innov. Prac. Technol. Manag. (ICIPTM)*, 61–66. doi: 10.1109/ICIPTM52218.2021.9388367.
5. Jaffe, T. A. and Nelson, R. C. (2018). Computed Tomography. In: Reiser, M. F., Semelka, R. C., Le Bihan, D., editors. Magnetic Resonance Tomography. Berlin, Heidelberg: Springer Berlin Heidelberg. 121–162.
6. Pearce, M. S., et al. (2012). Radiation exposure from CT scans in childhood and subsequent risk of leukemia and brain tumors: A retrospective cohort study. *Lancet*, 380(9840), 499–505.
7. Smith, J. K. and Johnson, A. B. (2023). Impact of radiation dose reduction on image quality in computed tomography: A literature review. *Med. Imag. J.*, 20(2), 78–93.
8. Kavya, D., Jaswanth, K., Chethana, S., Shruti, P., and Sarada, J. (2023). Comparative analysis of

image denoising using different filters. In: Gupta, D., Khanna, A., Bhattacharyya, S., Hassanien, A. E., Anand, S., Jaiswal, A. (eds). Inter. Conf. Innov. Comput. Comm. Lec. Notes Netw. Sys., 471.

9. Hardie, R. (2007). A fast image super-resolution algorithm using an adaptive Wiener filter. *IEEE Trans. Image Proc.*, 16(12), 2953–2964. doi: 10.1109/TIP.2007.909416.
10. I. J. (2013). Image, graphics, and signal processing. 5, 47–54. Published Online April 2013 in MECS (http://www.mecs-press.org/) DOI: 10.5815/ijigsp.2013.05.06.
11. Wang, M, Zheng, S., Li, X., and Qin, X. (2014). A new image denoising method based on Gaussian filter. *2014 Inter. Conf. Inform. Sci. Elec. Elect. Engg.*, 163–167. doi: 10.1109/InfoSEEE.2014.6948089.
12. Boyat and Joshi, B. K. (2013). Image denoising using wavelet transform and median filtering. *2013 Nirma University Inter. Conf. Engg. (NUiCONE)*, 1–6. doi: 10.1109/NUiCONE.2013.6780128.
13. Yang, J., Ren, H., and Zhu, G., et al. (2018). Detecting median filtering via two-dimensional AR models of multiple filtered residuals. *Multimed. Tools Appl.*, 77, 7931–7953.
14. Hou, Y., Zhao, C., Yang, D., and Cheng, Y. (2011). Image denoising by sparse 3-D transform-domain collaborative filtering. *IEEE Trans. Image Proc.*, 20(1), 268–270. doi: 10.1109/TIP.2010.2052281.
15. Dabov, K., Foi, A., Katkovnik, V., and Egiazarian, K. (2007). Image denoising by sparse 3-D transform-domain collaborative filtering. *IEEE Trans. Image Proc.*, 16(8), 2080–2095.
16. Yahya, A. A., Tan, J., Su, B., et al. (2020). BM3D image denoising algorithm based on adaptive filtering. *Multimed. Tools Appl.*, 79, 20391–20427.
17. Chambolle, A. (2004). An algorithm for total variation minimization and applications. *J. Math. Imag. Vis.*, 20(1–2), 89–97.

33 FFT-based performance analysis and classification of E-vehicle inverters

Tapash Kr. Das[a], *Ayan Banik, Mrinmoy Nayek, Raja Kumar Das and Debdeepta Ghosh*

Department of Electrical Engineering, Ghani Khan Choudhury Institute of Engineering & Technology, Malda, West Bengal, India

Abstract

Modern life requires transportation, but the old combustion engines are quickly becoming obsolete. Petrol and diesel vehicles are very polluting and are rapidly being replaced with completely electric vehicles. Electric vehicle (EVs) emits zero carbon and is much better for the environment. An electric vehicle has substantially lower operating costs than gasoline, petrol and diesel vehicles. Electric vehicles charge their batteries with electricity rather than using fossil fuels such as gasoline or diesel. This paper presents an electric vehicle (EV) performance analysis utilizing an interior permanent magnet synchronous motor (IPMSM). The performance and analysis include the evaluation of the IPMSM torque and speed characteristics, as well as its efficiency and power factor under various operating conditions. The performance of the EV has been assessed in terms of its acceleration, maximum speed, and energy consumption. Various computer simulations have been carried out using MATLAB, and the results are compared with a conventional DC motor-based EV. It is observed that the IPMSM-based EV exhibits better torque and speed characteristics compared to the DC motor-based EV. The IPMSM's high torque density results in better acceleration and a higher maximum speed for the EV. Moreover, the IPMSM's high efficiency and power factor contribute to the reduced energy consumption of the EV. With the help of simulation results, various characteristics of IPMSM and the nature of different inverters have been demonstrated. Overall, this paper demonstrates the advantages of utilizing an IPMSM in an EV and provides a comprehensive performance analysis of the IPMSM-based EV. The results indicate that the IPMSM-based EV exhibits superior performance and efficiency compared to the DC motor-based EV, making it a promising option for future EV applications.

Keywords: Brushless direct current motor (BLDC), electric vehicle (EV), electric vehicle inverter, First Fourier Transformation (FFT), inverter, interior permanent magnet synchronous motor (IPMSM)

Introduction

Electric vehicles (EVs) have gained significant attention in recent years as a sustainable transportation solution. The performance and efficiency of EVs heavily depend on the inverter-based technologies used to convert DC power from the battery to AC power for driving the electric motor. This literature survey aims to provide an overview of the research and developments in the performance analysis and classification of e-vehicle inverter-based technologies. The performance analysis and classification of e-vehicle inverter-based technologies involve evaluating and categorizing these technologies based on factors such as power conversion efficiency, switching frequency, voltage and power rating, control strategies, cooling and thermal management, integration and packaging, and reliability and durability. By considering these aspects, researchers and manufacturers can assess the efficiency, reliability, and overall performance of electric vehicle inverters, enabling the development of more advanced and efficient systems for electric vehicles.

Literature survey

For vehicle-to-grid (V2G) applications, this research explores the decoupled dq current control method for grid-tied packed E-cell inverters. The authors offer a control technique for ensuring power quality during bidirectional power flow between EVs and the grid [1]. During grid integration and V2G activities, the decoupled dq control assures precise current tracking and improves performance. In this research, the authors of a rectifier-inverter-fed induction motor drive study the effects of dead time on the input current of the inverters, DC-link dynamics, and light-load instability [2]. They examine how standby time affects the efficiency of the motor drive system and suggest ways to fix the problem. The research shows that dead time compensation is critical to increasing system efficiency and steadiness. Future electric vehicles (EVs) face a number of obstacles and possibilities that are explored in this review article [3]. The author summarizes recent developments in the charging infrastructure, electric motor drives, power electronics,

[a]tapash@gkciet.ac.in

DOI: 10.1201/9781003540199-33

and energy storage systems for EVs. The study also discusses the difficulties of EV adoption and offers suggestions for facilitating the mainstream use of EVs [4]. This work addresses the issue of torque ripple reduction in low-switching-frequency induction motor drives supplied by three-level inverters with a neutral-point clamp. They suggest a control technique that maximizes efficiency by minimizing torque ripple by adjusting the inverter's switching angles. The results of the research show that the suggested method is useful for improving the performance and ensuring the smooth operation of low-switching-frequency motor drives [5]. This study discusses the suppression of electromagnetic interference (EMI) and the regulation of surge voltage in ecocars outfitted with PWM inverters. The authors provide a practical approach to reducing EMI noise and surge voltages at the motor's connections. The research sheds light on the factors to consider and control strategies to use in ecovehicles' design for dependable, low-noise operation. This study details the design and implementation of integrated vehicle-solar-grid systems. Solar power, electric cars, and the grid may all be integrated in this article's discussion of bidirectional power flow and energy management. Sustainable mobility and renewable energy are highlighted as the research delves into the possible advantages, constraints, and practical implementation elements of vehicle-solar-grid integration. In this study, we provide a Fourier transform-based technique for categorizing solar PV microgrids. The authors suggest utilizing the Fast Fourier Transform (FFT) to categorize the operating stages of a PV system by analyzing the frequency components of data. The research shows that the FFT-based method is useful for distinguishing between normal operation, incorrect operation, and partial shading, among other circumstances [6]. There may be uses for the suggested classification method in PV microgrid fault detection and diagnostics. Using solar and waste heat energy technologies, the authors of this research offer a novel method for cooling power transformers. The research presents a cooling method that makes use of solar energy and waste heat recovery to improve the effectiveness of cooling in power transformers. By reducing the need for traditional cooling techniques, the suggested solution helps power transformers run more efficiently and sustainably. The study shows how renewable energy sources may be used to cool transformers [7]. In this study, the authors examine solar PV arrays and how to identify string faults using sub-harmonics. They provide an approach that uses sub-harmonic analysis to detect and localize problematic PV string nodes. The research shows that the suggested method can successfully identify and diagnose string defects, including open-circuit faults, in PV arrays. The findings help make solar PV systems more durable and easier to maintain. In this study, we discuss a method for solar-powered irrigation pumping that makes use of Zeta converters and permanent magnet brushless DC (PMBLDC) motors [8]. For solar photovoltaic (PV)-based irrigation pumping systems, the authors offer an effective power conversion and control technique. The research aims to maximize the efficiency with which electricity is transferred from the PV system to the motor. The findings help pave the way for the creation of renewable energy-powered irrigation systems that are both long-lasting and efficient [9]. These provide quick glimpses into the state of the art and cutting-edge methods for harnessing the power of the sun to boost the efficiency, effectiveness, and longevity of any given system. While studying authors feel that most of the paper focus on the application of solar but very less work has been done on the type segregation or classification of solar powered e-vehicle inverter. This has significantly motivated authors to research on the above problem statement. The explicit review has not only help authors to understand the level or degree of research carried out but also motivate to explore uncertainty in this area [10].

Model layout of EV inverter

This diagram shows the project's block diagram. An IPMSM motor is included in the project's suggested model. Also included in this model are five distinct converter types: GTO, Thyristor, IGBT, MOSFET, and ideal semiconductor. The circuit needs input at first. The control circuit receives this input power and uses it to regulate the voltage and current. The IPMSM motor then transforms the control voltage and current to determine the source's velocity. A very work on electric vehicles using interior permanent magnet synchronous motors (IPMSM) has been observed from the literature survey and that has motivated the authors to do the FFT-based classification and performance analysis on the proposed model (Figure 33.1).

FFT nature of EV inverter

There we compare three types of inverters i. e., MOSFET, IGBT, Thyristor. We compare the inverter in two ways one is relative to fundamental and another is relative to DC component (Figure 33.2 and Table 33.1).

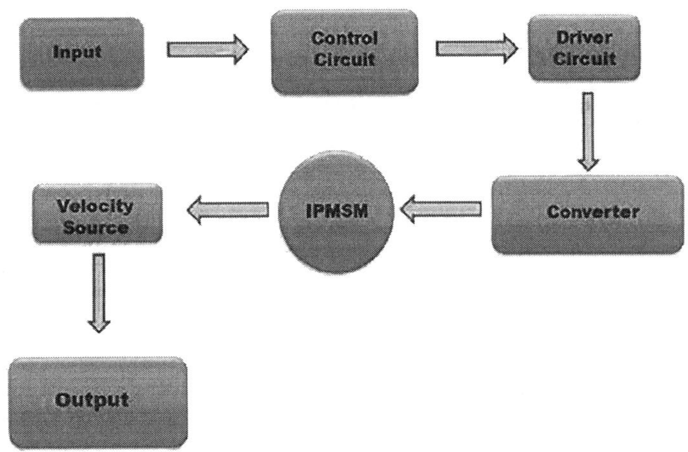

Figure 33.1 Model layout of EV inverter

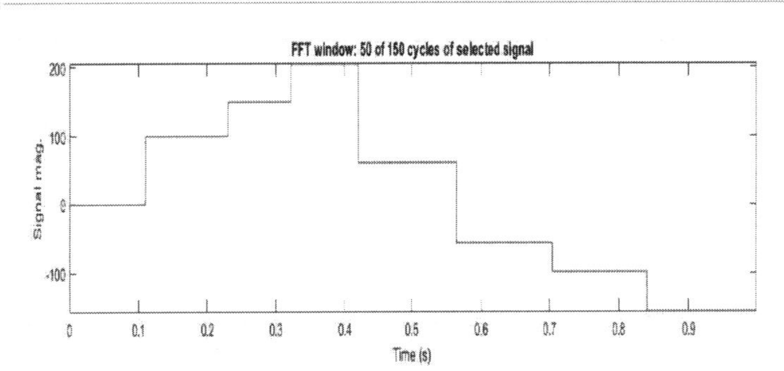

Figure 33.2 FFT nature of EV inverter

Table 33.1 FFT data driven from various EV inverters

Magnitude of fundamental				Magnitude of DC component			
Frequency	MOSFET	IGBT	Thyristor	Frequency	MOSFET	IGBT	Thyristor
0	619.62	602.56	591.22	0	100	96	98
1	11010.61	10152.69	9890.88	1	1777.02	1520.22	1369.73
2	1526.61	1492.12	1420.58	2	246.38	240.63	210.86
3	1646.98	1545.52	1502.12	3	265.81	156.02	205.45
4	1132.91	1101.98	1021.58	4	182.84	156.59	169.56
5	501.86	496.42	400.25	5	80.99	76.56	60.56
6	1299.08	1185.65	1180.26	6	209.66	185.65	162.56
7	682.39	652.26	550.36	7	110.13	85.56	96.18
8	1332.42	1305.54	1190.56	8	215.04	175.65	185.52
9	744.82	710.56	530.69	9	120.21	98.1	86.96
10	627.86	599.05	458.32	10	101.33	95.98	89.17

Magnitude of fundamental for MOSFET

Magnification (% of fundamental) vs. frequency characteristics of a MOSFET's relative fundamental are shown by the Figure 33.3. The frequency is shown by the x-axis and the magnification percentage by the y-axis. Frequency 0, the magnification percentage is 616.6, and at frequency 1 the value of Mag (% of fundamental) is 11010.61. It is a transient type of graph in which the value is constantly changing at different points.

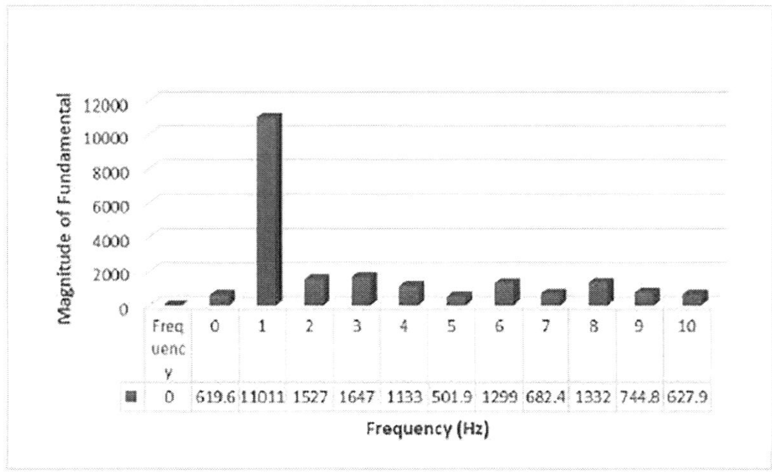

Figure 33.3 Magnitude of fundamental vs. frequency characteristic of MOSFET

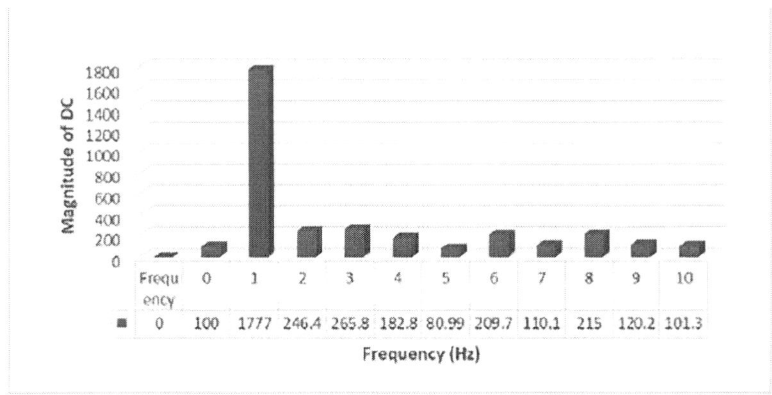

Figure 33.4 Magnitude of DC component vs. frequency characteristic of MOSFET

Magnitude of DC component for MOSFET

According to Figure 33.4, the MOSFET's relative magnification (% of DC) vs. frequency (Hz) characteristics are shown. At frequency (Hz), 0 max (% of DC) is 100; after that frequency, 1 mag (% of DC) is 1777.02. The x-axis and y-axis show frequency (Hz) and mag (% of DC), respectively. It is a transient type of graph in which the value is constantly changing at different points.

Magnitude of fundamental for IGBT

Figure 33.5 represents that the mag (% of fundamental) vs. frequency characteristics of relative of fundamental of IGBT. x-axis represent the frequency and y-axis represent Mag (% of fundamental) at frequency 0 max (% fundamental) is 602.56, after that when frequency is 1 the value of mag (% fundamental) is 10152.69. It is transient type of graph where the value always fluctuating at various point.

Magnitude of DC component for IGBT

The IGBT's relative magnification (% of DC) vs. frequency (Hz) characteristics is shown in Figure 33.6. The frequency is shown on the x-axis (Hz), and the magnification at that frequency is shown on the y-axis (% of DC). Max (% of DC) is 96 at zero frequency; after that, 1 mag (% of DC) is 1520. It is a transient type of graph in which the value is constantly changing at different points.

Magnitude of fundamental for thyristor

The IGBT's relative magnification (% of DC) vs. frequency (Hz) characteristics is shown in Figure 33.7. The frequency is shown on the x-axis (Hz), and the magnification at that frequency is shown on the y-axis (% of DC). Max (% of DC) is 96 at zero frequency; after that, 1 mag (% of DC) is 1520. It is a transient type of graph in which the value is constantly changing at different points.

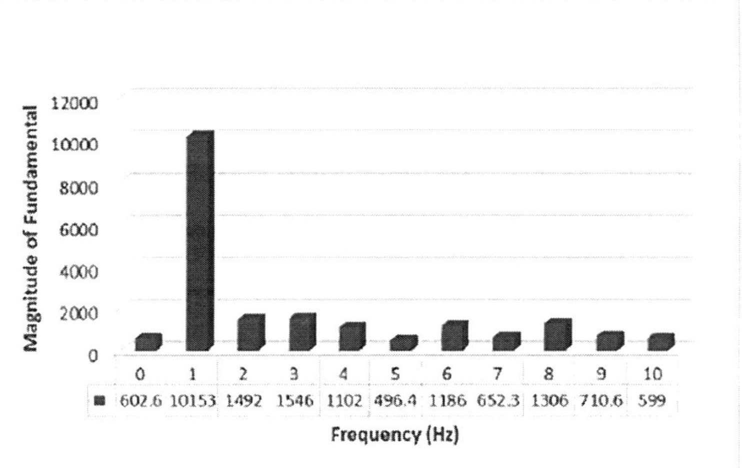

Figure 33.5 Magnitude of fundamental vs. frequency characteristic for IGBT

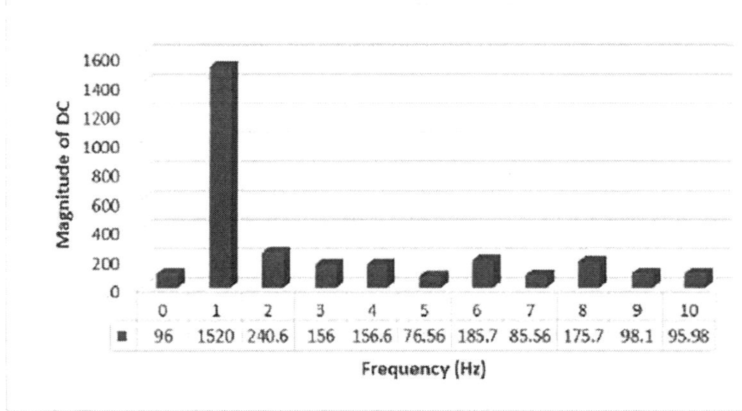

Figure 33.6 Magnitude of DC component vs. frequency nature of IGBT

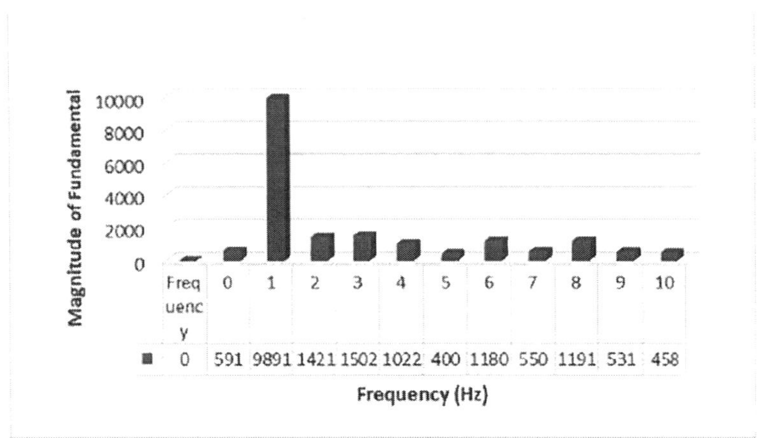

Figure 33.7 Magnitude of fundamental vs. frequency nature of thyristor

Magnitude of fundamental for MOSFET, IGBT, thyristor

The features of the relative fundamental of MOSFET, IGBT, and thyristors are shown in Figure 33.8 as a function of frequency (Hz) and magnification (% of fundamental). At frequency 0 the value of max (% of fundamental) is 619, 602, 592; thereafter, at frequency 1 mag (% of fundamental) is 11600, 10100, and 9900 on the x-axis and y-axis, respectively. It is a transient type of

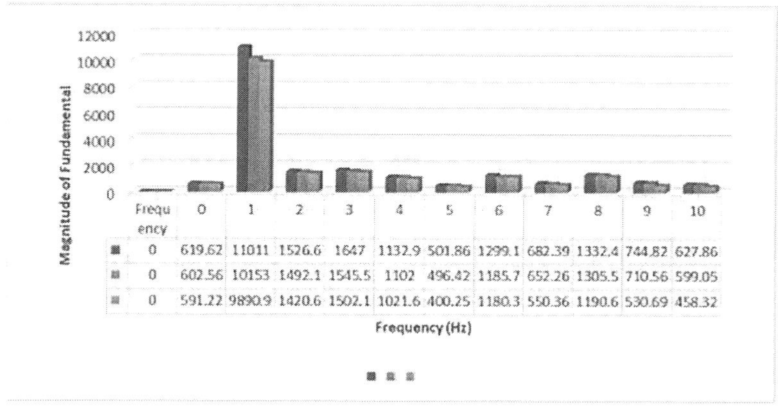

Figure 33.8 Comparison between three inverters of mag of fundamental vs. frequency char

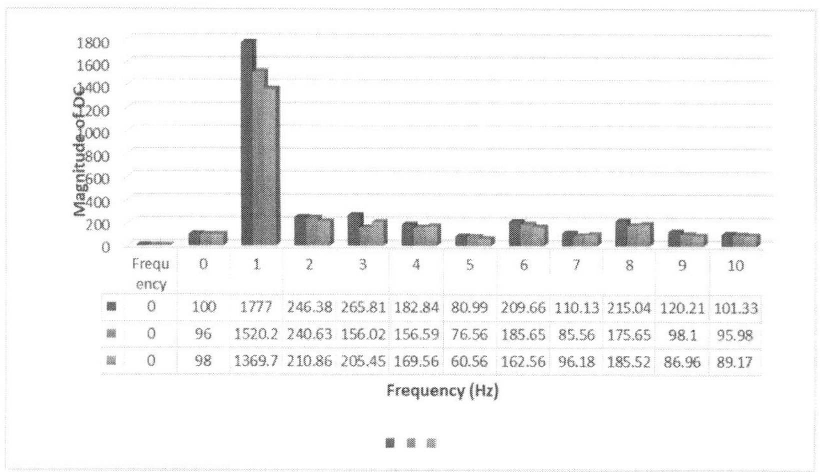

Figure 33.9 Comparison between three inverters of mag of DC component vs. frequency char

graph in which the value is constantly changing at different points.

Magnitude of DC component for MOSFET, IGBT, thyristor

In this project, inverters of the MOSFET, IGBT, and thyristors types are used, and Figure 33.9 illustrates the characteristics of the superior EV inverter. First, we compare according to fundamental criteria, and then MOSFET, IGBT, and thyristors are compared according to DC component criteria.

The blue bar in this diagram shows a MOSFET-type inverter, the orange bar an IGBT-type inverter, and the grey bar a thyristors-type inverter. This is how EV inverters are classified using FFT. The characteristics of the MOSFET, IGBT, and thyristors types of EV inverters are depicted in this diagram. This graph shows the EV inverter's THD (%) vs. frequency characteristic.

Conclusions

The study investigates the effectiveness of FFT analysis in assessing the performance of e-vehicle inverters. Researchers obtained voltage and current waveforms from various types of inverters and subjected them to FFT analysis to extract frequency-domain characteristics and identify potential performance issues. The results showed that FFT analysis is a powerful tool for evaluating inverter performance, detecting anomalies like voltage and current harmonics, switching frequency variations, and other issues related to inverter operation. The researchers developed a classification scheme based on FFT analysis results, categorizing inverters into different performance classes, allowing for quick and effective assessments of their overall health and functionality. The findings have significant implications for the design, development, and maintenance of e-vehicle inverters, aiding in early fault detection, proactive maintenance scheduling, and overall improvement in reliability and efficiency. However, FFT analysis is not a standalone solution and should be integrated with other diagnostic methods and domain knowledge to obtain a comprehensive understanding of inverter performance. Further research is needed to explore the application of FFT analysis in real-time monitoring and control systems for e-vehicle inverters. Promising approach to enhance monitoring and maintenance of critical components, identifying and addressing potential issues before they lead to significant failures, contributing to the overall reliability and performance.

References

1. Sadabadi, M. S., Sharifzadeh, M., Mehrasa, M., Karimi, H., and Al-Haddad, K. (2023). Decoupled dq current control of grid-tied packed E-cell inverters in vehicle-to-grid technologies. *IEEE Trans. Indus. Elec.*, 70(2), 1356–1366.
2. Guha and Narayanan, G. (2018). Impact of dead time on inverter input current, DC-link dynamics, and light-load instability in rectifier-inverter-fed induction motor drives. *IEEE Trans. Indus. Appl.*, 54(2), 1414–1424.
3. Husain. (2021). Electric drive technology trends, challenges, and opportunities for future electric vehicles. *IEEE*, 109(6), 1039–1059.
4. Tripathi and Narayanan, G. (2018). Torque ripple minimization in neutral-point-clamped three-level inverter fed induction motor drives operated at low-switching-frequency. *IEEE Trans. Indus. Appl.*, 54(3), 2370–2380.
5. Mutoh, N. and Kanesaki, M. (2008). A suitable method for ecovehicles to control surge voltage occurring at motor terminals connected to PWM inverters and to control induced EMI noise. *IEEE Trans. Veh. Technol.*, 57(4), 2089–2098..
6. Kydd, P. H., Anstrom, J. R., Heitmann, P. D., Komara, K. J., and M. E. (2016). Vehicle-solar-grid integration: Concept and construction. *IEEE Power Energy Technol. Sys. J.*, 3(3), 81–88.
7. Das, T. K., Banik, A., Chattopadhyay, S., and Das, A. (2019). FFT based classification of solar photo voltaic microgrid system. *2019 Second Inter. Conf. Adv. Comput. Comm. Paradigms.*
8. Das, T. K., Banik, A., Chattopadhyay, S., and Das, A. (2021). Energy-efficient cooling scheme of power transformer: An innovative approach using solar and waste heat energy technology. In: Ghosh, S. K., Ghosh, K., Das, S., Dan, P. K., and Kundu, A. (eds). Advances in Thermal Engineering, Manufacturing, and Production Management. ICTEMA 2020. *Lec Notes Mec. Engg.* Springer.
9. Das, T. K., Banik, A., Chattopadhyay, S., and Das, A. (2019). Sub-harmonics based string fault assessment in solar PV arrays. In: Chattopadhyay, S., Roy, T., Sengupta, S., and Berger-Vachon, C. (eds). Modeling and Simulation in Science, Technology and Engineering Mathematics. *Adv. Intell. Sys. Comput.*, 749.
10. Banik, A. and Sengupta, A. (2022). Irrigation pumping scheme based on solar PV and Zeta converters with PMBLDC motor. In: Kumar, S., Ramkumar, J., and Kyratsis, P. (eds). Recent Advances in Manufacturing Modeling and Optimization. *Lec. Notes Mec. Engg.* Springer.

34 A design and performance analysis of electrical parameters in electric vehicle (EV) inverter

Tapash Kr. Das[a], *Ayan Banik, Mrinmoy Nayek, Raja Kumar Das and Debdeepta Ghosh*

Department of Electrical Engineering, Ghani Khan Choudhury Institute of Engineering & Technology, Malda, West Bengal, India

Abstract

With the rapid advancement of electric vehicle (EV) technology, the design and optimization of electric powertrain components have become crucial. Among these components, the electric vehicle inverter plays a vital role in converting the DC power from the battery to AC power for driving the electric motor. The design and performance analysis of electrical parameters in the EV inverter are of paramount importance to achieve efficient power conversion, improve system reliability, and enhance overall vehicle performance. This paper aims to explore the key electrical parameters involved in the design of an electric vehicle inverter and analyze their impact on the inverter's performance. The paper begins with an introduction that highlights the importance of optimizing powertrain components in EV technology. It then provides an overview of the electric vehicle inverter, including its function, configuration, and power electronics devices used. Next, the paper delves into the various electrical parameters that play a crucial role in the design of the EV inverter. These parameters include DC link capacitance, inductor design, switching frequency, gate drive circuit, heat sink design, and control strategy. Design considerations and optimization techniques are discussed to improve efficiency, thermal management, electromagnetic interference (EMI)/electromagnetic compatibility (EMC), reliability, and fault tolerance of the inverter. The paper emphasizes performance analysis and evaluation, covering power loss analysis, efficiency analysis, thermal performance evaluation, harmonic analysis, and dynamic response analysis. Case studies and experimental results are presented to validate the design and showcase the practical implications of the discussed electrical parameters. The paper concludes by highlighting future trends and challenges in the field, including emerging technologies, grid integration, and standards and regulations.

Keywords: Inverter, parameter, electrical vehicle, harmonic assessment and sustainability

Overview

The design and performance analysis of electrical parameters in electric vehicle (EV) inverters are crucial aspects of optimizing the powertrain system. In this overview, we will delve into the importance of designing and analyzing electrical parameters in EV inverters, highlighting the key considerations and their impact on overall performance. Electric vehicle inverters serve as the bridge between the battery pack and the electric motor, converting the direct current (DC) power from the battery into alternating current (AC) power required by the motor. The efficiency, reliability, and overall performance of the inverter significantly influence the vehicle's drivability, range, and energy efficiency. To achieve optimal performance, several electrical parameters within the EV inverter must be carefully considered. These parameters include the DC link capacitance, inductor design, switching frequency, gate drive circuit, heat sink design, and control strategy. The DC link capacitance plays a crucial role in energy storage and voltage stability. Proper selection and sizing of the capacitance are essential to maintain a stable DC voltage level and minimize ripple currents. Inductor design is vital for managing the current flow and minimizing losses. The inductor's characteristics, such as inductance value and core material, influence the overall efficiency and electromagnetic interference (EMI) performance of the inverter. The switching frequency determines the speed at which the inverter operates and affects its efficiency and electromagnetic compatibility (EMC). The selection of an appropriate switching frequency is crucial to balance switching losses and harmonic content. The gate drive circuit controls the switching behavior of power devices within the inverter. Its design and implementation impact the switching speed, losses, and reliability of the inverter. Proper heat sink design is essential to manage thermal dissipation within the inverter. Efficient cooling techniques, such as heat sinks and thermal management systems, are crucial to maintain component temperatures within acceptable limits and ensure long-term reliability. The control strategy governs the overall operation and performance of

[a]tapash@gkciet.ac.in

DOI: 10.1201/9781003540199-34

the inverter. Advanced control algorithms, such as pulse width modulation (PWM), can optimize efficiency, dynamic response, and fault management in the inverter system. The design considerations and optimization techniques of these electrical parameters focus on improving efficiency, thermal management, EMI/EMC performance, reliability, and fault tolerance of the inverter. Balancing these aspects is crucial to achieving a well-designed, high-performance EV inverter. To evaluate the inverter's performance, various analysis techniques are employed. Power loss analysis helps identify areas of high losses, enabling the optimization of power device selection and driving strategies. Efficiency analysis quantifies the overall energy conversion efficiency of the inverter system. Thermal performance evaluation ensures that the inverter operates within safe temperature limits. The harmonic analysis examines the inverter's harmonic content to meet regulatory requirements and minimize grid disturbances. Dynamic response analysis assesses the inverter's ability to handle transient conditions and rapid changes in load demand.

Literature review

This paper focuses on the analysis and design of an automatic motion inverter. It discusses the development of a control strategy for an inverter used in motion control applications, including electric vehicle propulsion. The study explores the implementation of robust control algorithms to improve the inverter's performance and reliability [1]. This paper focuses on the development of digital signal processor (DSP)-based sensorless electric motor fault-diagnosis tools for electric and hybrid electric vehicle powertrain applications. It discusses the implementation of advanced algorithms to detect and diagnose faults in electric motors, providing insights into improving the reliability and performance of EV inverters. This paper presents the design and analysis of an integrated electric vehicle (EV) battery charger with retrofit capability. It explores the electrical parameters involved in the charger's design and investigates its performance, efficiency, and compatibility with different EV models [2]. The study highlights the importance of integrating the charger with the inverter system for enhanced EV charging infrastructure. This paper focuses on the analysis and design of a two-phase "on" drive operation for a permanent magnet synchronous machine (PMSM) electromechanical actuator. It discusses the electrical parameter considerations and performance analysis of the PMSM inverter system for improved efficiency and dynamic response [3]. The study presents insights into achieving optimal drive operation for electric vehicle applications. This paper presents an evaluation of a 1200-V, 800-A all-silicon carbide (SiC) dual module for electric vehicle inverter applications. It analyzes the electrical parameters of the SiC module and investigates its performance in terms of power loss, efficiency, and thermal characteristics. The study highlights the potential of SiC devices to improve the performance of EV inverters. This paper discusses the utilization of the d-q transformation technique to vary the switching frequency for interior permanent magnet synchronous motor (IPMSM) drive systems [4]. The study explores the impact of different switching frequencies on the performance and efficiency of the IPMSM drive. It demonstrates how adjusting the switching frequency through the d-q transformation can optimize the motor drive system for electric vehicle applications [5]. This paper focuses on the classification of faults in solar photovoltaic (PV) microgrid systems using the fast Fourier transform (FFT) technique. It investigates the electrical parameters and fault characteristics of PV arrays and proposes an FFT-based approach for fault diagnosis and classification. The study highlights the importance of analyzing electrical parameters to enhance the reliability and performance of solar PV microgrid systems [6]. This paper presents an innovative approach to achieve energy-efficient cooling for power transformers by utilizing solar and waste heat energy technologies. It discusses the design and analysis of a cooling scheme that optimizes the electrical parameters and cooling techniques for power transformers. The study highlights the potential of integrating renewable energy sources to improve the efficiency and sustainability of power transformer cooling systems [7]. This paper focuses on the assessment of string faults in solar PV arrays using sub-harmonic analysis. It investigates the electrical parameters and fault characteristics of PV strings and proposes a sub-harmonics-based approach for fault detection and assessment. The study emphasizes the importance of analyzing electrical parameters to enhance the reliability and performance of solar PV systems [8]. This paper presents an irrigation pumping scheme that integrates solar PV systems, Zeta converters, and permanent magnet brushless DC (PMBLDC) motors. It discusses the electrical parameter considerations and design of the system, highlighting the benefits of utilizing renewable energy sources for irrigation applications [9]. The study showcases how optimizing the electrical parameters and control strategies can enhance the efficiency and reliability of the irrigation pumping system [10].

Model architecture of inverter

The battery unit will be charged by a single-phase AC (230 V) supply to provide backup for various loads. In order to provide the necessary voltage for various circuit components, two step-down transformers with ratings of (12 V-0-12 V, 8 A) and (12 V-0-12 V, 3 A) have been utilized to step down the voltage level from 230 V/12 V. A bridge rectifier circuit that converts AC to DC for a battery charging circuit has been created using the diode 1N5408. The charge in the battery has been stored via the charging circuit. Battery-powered DC power will be supplied to the inverter circuit's input during the power outage. The inverter circuit was created using 4 no of MOSFETs (IRFP 150 N-Channel), 2 no of transistors (2N2222A NPN), 2 no resistances (470 ohm, 1 K ohm), and 2 no capacitors (1 uF 63 V, 4700 uF 25 V). These electronic parts were used in inverter circuits to operate and control various electrical parameters as well as convert DC to AC for powering the load from the secondary side. In order to provide the load in the IPMSM-based EV model, an 850 VA square wave inverter has been constructed for this suggested model. Additionally, the many attributes of the proposed inverter have been carefully observed. The layout diagram for the proposed work's inverter is shown in Figure 34.1. The classification of EV inverters is the project's main goal. As needed for the inverter, we used a variety of AC and DC components there.

A step-down transformer, a charger circuit, a battery unit, an inverter circuit, and an 850 VA step-up square wave transformer are all used in this project. For the charger and inverter circuits, various types of electronic components were used. This inverter has a 700-W maximum load capacity.

Required components of inverter

This project calls for a variety of electrical and electronic components, including a transformer, battery, resistance, capacitor, MOSFET, transistor, diode, heat sink, and bulb, among others (Table 34.1).

Single line diagram of inverter

This is the single-line diagram of our proposed model. In this project, we divided into a bus. A single-line diagram (SLD) of an inverter also sometimes called a one-line diagram is the simplest representation of an electric power system. A single line diagram typically corresponds to more than one physical conductor in a direct current system the line included the supply and return paths in a three-phases (the conductor is both supply and return due to the nature of the alternating current circuits).

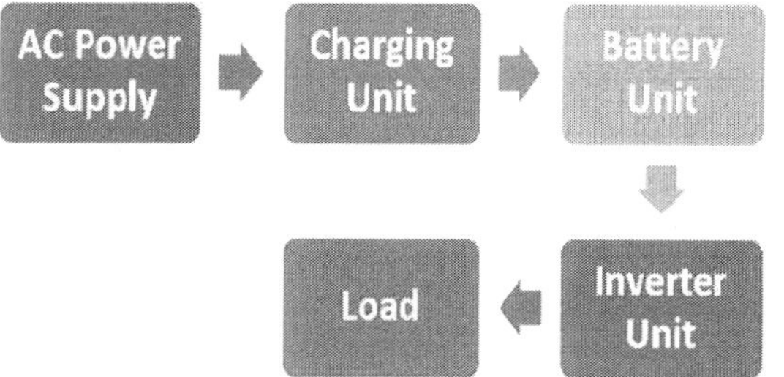

Figure 34.1 The layout diagram for the proposed work's inverter

Table 34.1 Required components name

Component name	Range	Quantity
Transformer 1	12-0-12(3 amp)	1
Transformer 2	850 VA	1
Transistor	BC 547 npn	2
Resistance 1	10 k	4
Resistance 2	470 ohm	2
Capacitor	1 uf	1
Diode 1	5408	1
Diode 2	1N4007	1
Battery	12 V	1
MOSFET	Irfp 150 N channel	4
Resistance 3	220 ohm	2

In Figure 34.2, the single-line diagram shows the left from the AC supply to the load stage. From the one-line drawing, one can easily understand that the line shows from AC supply, charger, battery unit, inverter, and load.

1. AC supply: 240 V, 50 Hz 1-phase supply.
2. Charger: Half wave rectifier circuit.
3. Battery unit: 12 V DC 7 Amp
4. Inverter: DC converter to AC
5. Load: Lamp load (100 W, 200 W, 300 W)
6. Bus: A bus is a junction where input and output current are exchanged.

Bus 1: We gave the AC supply to Bus 1 and the current passed through Bus 1 and then enter the charger.

Bus 2: Bus 1 through then the charger passes the current through B2, if the circuit is closed then the current passes to the battery unit and the battery unit is charged.

Bus 3: Again if we not close the circuit the current will not pass to the inverter, if we close the circuit the current via B3 enters the inverter.

Bus 4: Now finally inverter passes the current through B4 and then enters the load.

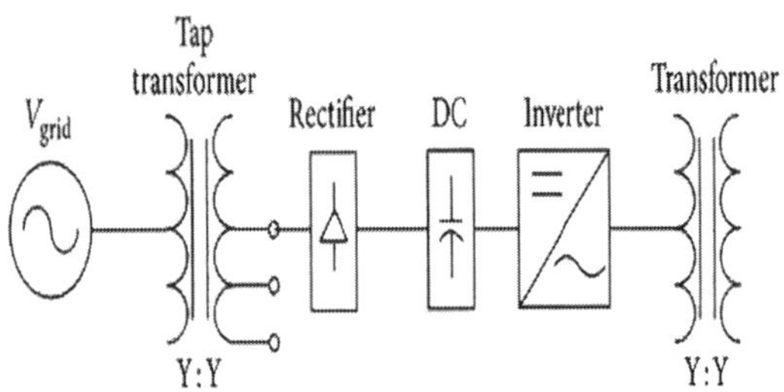

Figure 34.2 Single line diagram of inverter

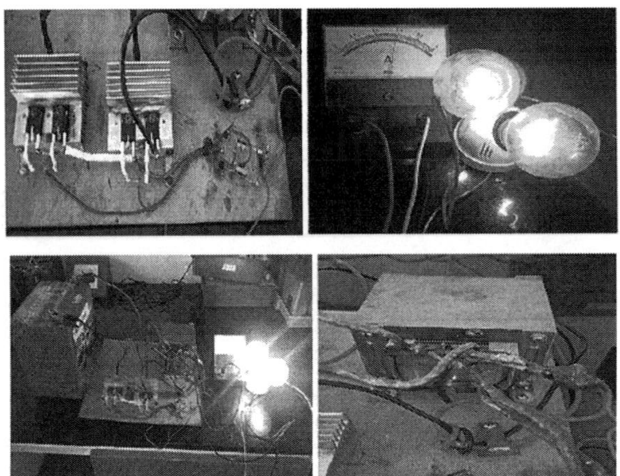

Figure 34.3 The developed prototype working model

Characteristics of all Buses (Table 34.2)

Bus 1 voltage and current at different load condition

Different types of voltage and current subject to load are displayed in Table 34.3. We use the following values for voltage and current: 9 W, 18 W, 100 W, 118 W, 218 W, 309 W, and 400 W.

Characteristic of bus 1 (B1) voltage

The voltage vs. load nature in Figure 34.4 at Bus 1, we have a supply of 220 V AC that we take into account for Bus 1, and that supply we use for domestic purposes. The voltage of the 220-V AC supply has fluctuated and is not constant. After measuring Bus 1, we obtain various voltages. We get 228 at Measure 1 and 227 at Measure 2, so we take 8 values as shown in the Figure. After the measurement, we have a max. value of 233 V and a minimum value of 225 V.

Characteristic of Bus 1 (B1) current

The current vs. load is depicted in watts in Figure 34.5. In this diagram, the current value of zero loans in watt remains zero at the initial condition, decreasing as the watt current value increases, so that B1 current at various loading conditions is inversely proportional.

Bus 2 voltage at different load condition

Different types of voltage subject to load are displayed In Table 34.4. We use the following values for voltage: 9 W, 18 W, 100 W, 118 W, 218 W, 309 W, and 400 W. This is a charging transformer's AC output voltage. At no load, the voltage is 12.7 V; at 9 W of load, it is 12.5 V; at 18 W of load, it is 12.6 V; at 100 W of load, it is 12.7 V; at 118 W of load, it is 218 W of load; it is 309 W of load; and at 400 W of load, it is 12.7 V.

Characteristic of Bus 2 (B2) voltage

The bar graph in Figure 34.6 illustrates how the voltage versus watt fluctuates between 0 and 400 W. The load is shown on the x-axis in Watts, while the voltage is shown on the y-axis. At first, when the load supply is assumed to be 12.7 V, there is no wattage. On the other hand, the load is high when the voltage supply is 12.7 V.

Bus 3 voltage at different load condition

Different types of voltage subject to load are displayed in Table 34.5. We use the following values for voltage: 9 W, 18 W, 100 W, 118 W, 218 W, 309 W, and 400 W. These charging circuit operates at a DC voltage. At no load, the voltage is 12.7 V; at 9 W of load, it is 12.5 V; at 18 W of load, it is 12.6 V; at 100 W of load, it is 12.7 V; at 118 W of load, it is 218 W; at 309 W of load, it is 12.7 V; and at 400 W of load, it is 12.7 V.

Characteristic of Bus 3 (B3) voltage

The fundamental characteristics of voltage vs. watt for load currents under various conditions are depicted in Figure 34.7. At first, when the load supply is assumed to be 12.7 V, there is no wattage. On the other hand, the load is high when the voltage supply is 12.7 V. These phenomena can be observed in a variety of switching scenarios. Furthermore, based on Figure 34.7, the proposed model may be capable of real-time remote data monitoring.

Bus 4 voltage and current at different load condition (Table 34.6)

Characteristic of Bus 4 (B4) voltage

The graph of bus 4 voltage vs. power is shown in Figure 34.8. The Bus 4 voltage will be 12.8 at start-up, and if the load continuously increases, the voltage will decrease, so the diagram showed that the Bus 4 power and voltage are inversely proportional.

Characteristic of Bus 4 (B4) current

The load vs. current graph for Bus 4 is shown in Figure 34.9. When the load is zero during the initial phase, the current will also be zero, and as the load

Table 34.2 Values of all buses

Watt	*B1*		*B2*	*B3*	*B4*		*B5*	*B6*	
	V	*I*	V_{dc}	V_{dc}	V_{dc}	*I*	*V*	*V*	*I*
0	228	0	12.7	12.7	12.8	0	12.1	235	0
9	227	0.05	12.5	12.5	12.7	1.3	10.7	231	0.11
18	225	0.06	12.6	12.6	12.2	2.2	10.6	228	0.19
100	227	0.04	12.7	12.7	12.3	7.9	9.8	192	0.5
118	230	0.05	12.6	12.6	12.2	9.6	9.6	198	0.6
218	231	0.04	12.7	12.6	11.9	14.1	8.2	180	0.9
309	233	0.06	12.5	12.7	11.8	15.1	8.1	178	1.2
400	230	0.07	12.7	12.7	11.7	15.4	8	160	1.4

Table 34.3 Bus 1 voltage and current at different load condition

Load (w)	0	9	18	100	118	218	309	400
Voltage	228	227	225	227	230	231	233	230
Current	0	0.05	0.06	0.04	0.05	0.04	0.06	0.07

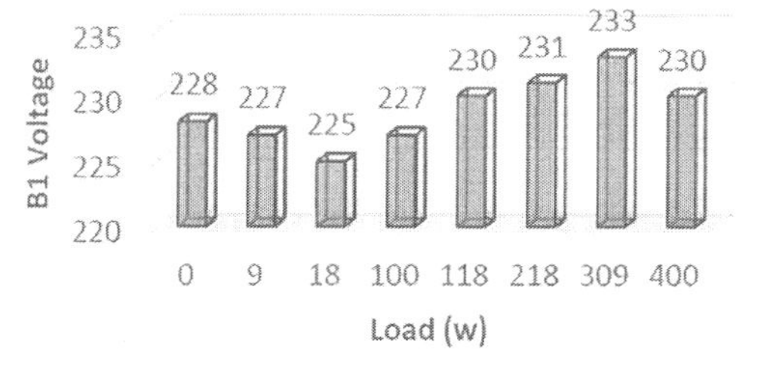

Figure 34.4 Characteristics of load vs. voltage of Bus 1

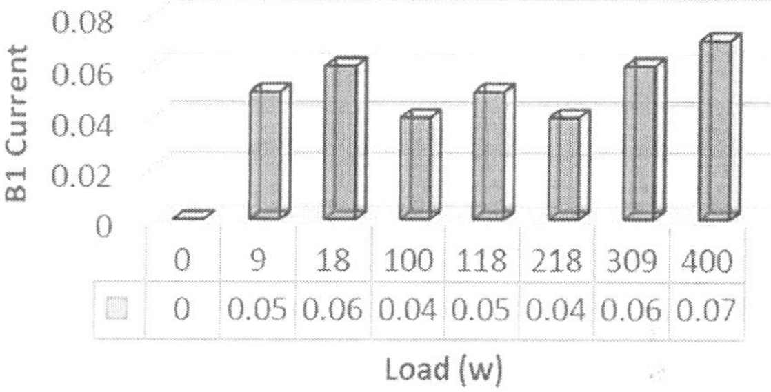

Figure 34.5 Load vs. current characteristic at Bus 1 (B1)

Table 34.4 Bus 2 voltage at different load condition

Load (w)	0	9	18	100	118	218	309	400
Voltage	12.7	12.5	12.6	12.7	12.6	12.7	12.5	12.7

increases continuously, the current value rises, making the graph's current-to-load ratio proportional.

Bus 5 voltage at different load condition

The load vs. current graph for Bus 4 is shown in Figure 34.10. When the load is zero during the initial phase, the current will also be zero, and as the load increases continuously, the current value rises, making the graph's current-to-load ratio proportional (Table 34.7).

Bus 6 voltage and current at different load condition (Table 34.8)

Characteristic of B6 voltage

In Figure 34.11 graph current vs. load, x-axis represented load and y-axis represented current. At initial condition load, zero current 235 and load increasing current value decreasing. So B6 diagram current at different loading condition load is directly proportional to the current.

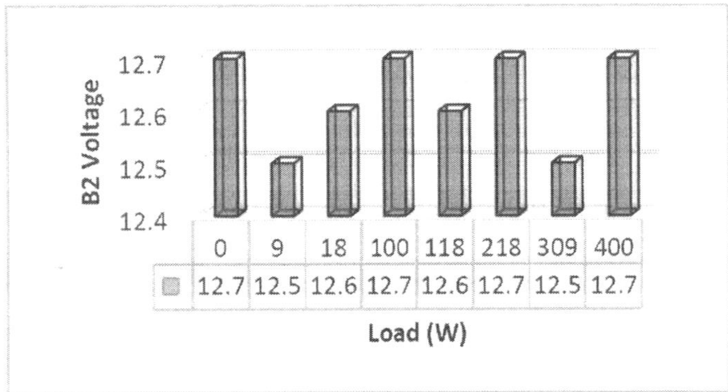

Figure 34.6 Load vs. voltage characteristics at Bus 2

Table 34.5 Bus 3 voltage at different load conditions

Load (w)	0	9	18	100	118	218	309	400
Voltage	12.7	12.5	12.6	12.7	12.6	12.6	12.7	12.7

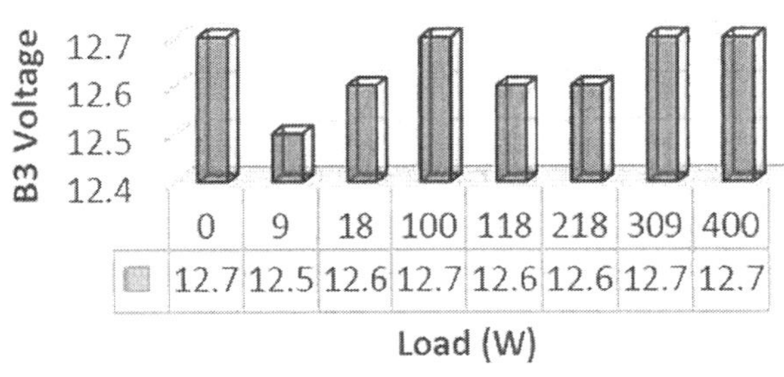

Figure 34.7 Load vs. voltage characteristic at B3

Table 34.6 Bus 4 voltage and current at different load condition

Load (w)	0	9	18	100	118	218	309	400
Voltage	12.8	12.7	12.2	12.3	12.2	12.9	128	12.7
Current	0	1.3	2.2	7.9	9.6	14.1	15.1	15.4

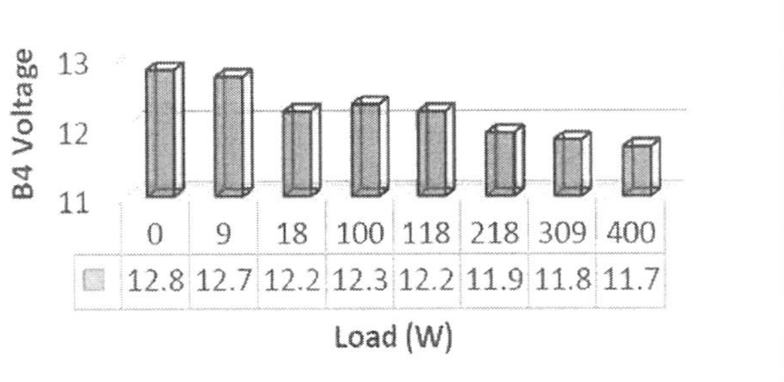

Figure 34.8 Load vs. voltage characteristic at B4

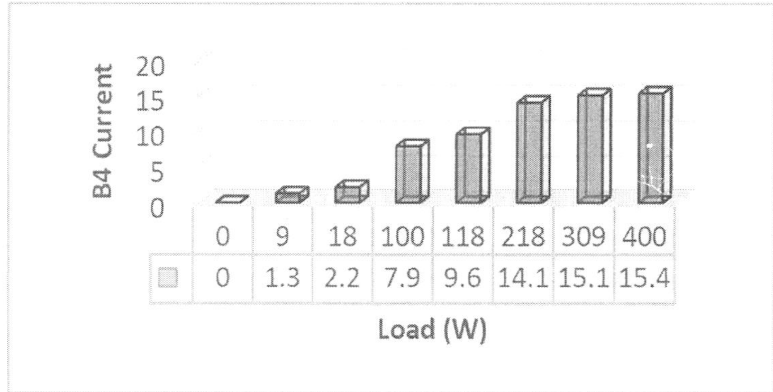

Figure 34.9 Load vs. current characteristic at B4

Table 34.7 Bus 5 voltage and current at different load condition

Load (w)	0	9	18	100	118	218	309	400
Voltage	12.1	10.7	10.6	9.8	9.6	8.2	8.1	8.0

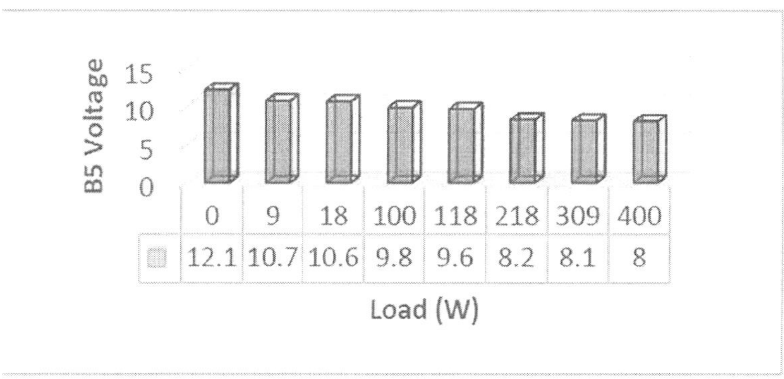

Figure 34.10 Load vs. voltage characteristic at B5

Table 34.8 Bus 6 voltage and current at different load condition

Load (w)	0	9	18	100	118	218	309	400
Voltage	235	231	228	192	189	180	178	160
Current	0	0.11	0.19	0.5	0.6	0.9	1.2	1.4

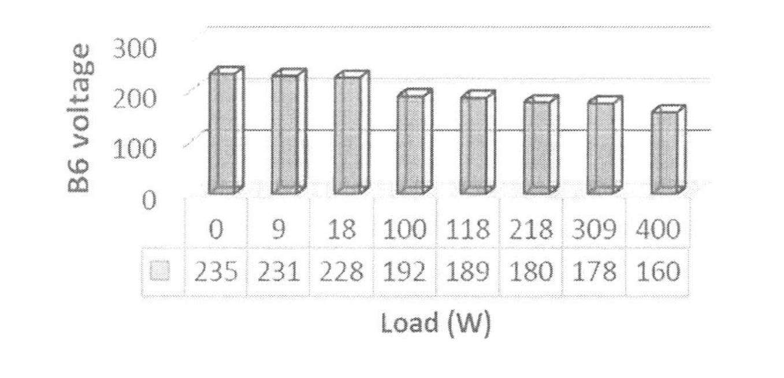

Figure 34.11 Load vs. voltage characteristic at B6

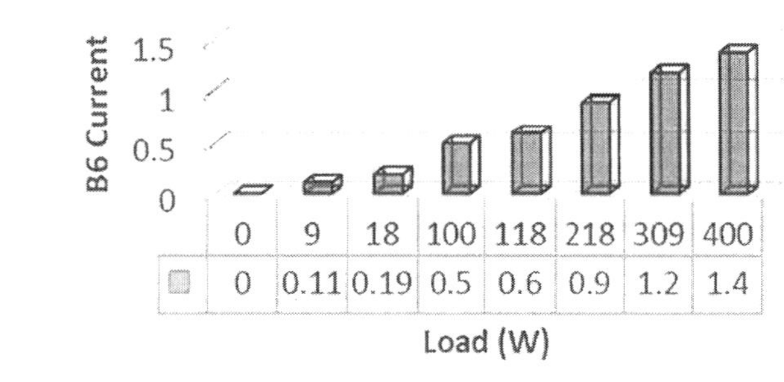

Figure 34.12 Load vs. current characteristics at B6

Characteristic of Bus 6 (B6) current

The graph in Figure 34.12 will show the voltage vs. load for Bus 5, where x-axis and y-axis stand for load and voltage, respectively. When the load is zero at start up, the voltages will be the bus voltage, and as the load increases, the load decreases, making the graph inversely proportional.

Conclusions

The paper attempts to provide a comprehensive understanding of the electrical parameters involved in the design of an electric vehicle inverter. It will discuss the design considerations and optimization techniques to enhance the inverter's performance. The performance analysis and evaluation will help in identifying areas for improvement and validate the design through case studies and experimental results. The paper has concluded with future trends and challenges, highlighting the importance of ongoing research and development in this field.

References

1. Savaresi, S. M., Tanelli, M., Taroni, F. L., Previdi, F., and Bittanti, S. (2006). Analysis and design of an automatic motion inverter. *IEEE/ASME Trans. Mechatron.*, 11(3), 346–357.
2. Akin, B., Ozturk, S. B., Toliyat, H. A., and Rayner, M. (2009). DSP-based sensorless electric motor fault-diagnosis tools for electric and hybrid electric vehicle powertrain applications. *IEEE Trans. Veh. Technol.*, 58(6), 2679–2688.
3. Ranjith, S., Vidya, V., and Kaarthik, R. S. (2020). An integrated EV battery charger with retrofit capability. *IEEE Trans. Transport. Elec.*, 6(3), 985–994.
4. Haskew, T. A., Schinstock, D. E., and Waldrep, E. M. (1999). Two-phase on drive operation in a permanent magnet synchronous machine electromechanical actuator. *IEEE Trans. Energy Conv.*, 14(2), 153–158. doi: 10.1109/60.766970.
5. Wood, R. A. and Salem, T. E. (2011). Evaluation of a 1200-V, 800-A All-SiC dual module. *IEEE Trans. Power Elec.*, 26(9), 2504–2511.
6. Yang, F., Taylor, A. R., Bai, H., Cheng, B., and Khan, A. A. (2015) Using *d–q* Transformation to vary the switching frequency for interior permanent magnet synchronous motor drive systems. *IEEE Trans. Transport. Elec.*, 1(3), 277–286.
7. Das, T. K., Banik, A., Chattopadhyay, S., and Das, A. (2019). FFT based classification of solar photo voltaic microgrid system. *2019 Second Inter. Conf. Adv. Comput. Comm. Paradigms (ICACCP)*, 1–5.
8. Das, T. K., Banik, A., Chattopadhyay S., and Das, A. (2021). Energy-efficient cooling scheme of power transformer: An innovative approach using solar and waste heat energy technology. In: Ghosh, S. K., Ghosh, K., Das, S., Dan, P. K., and Kundu, A. (eds). Advances in Thermal Engineering, Manufacturing, and Production Management. *Lec. Notes Mec. Engg.*, Springer.
9. Das, T. K., Banik, A., Chattopadhyay, S., and Das, A. (2019). Sub-harmonics based string fault assessment in solar PV arrays. In: Chattopadhyay, S., Roy, T., Sengupta, S., and Berger-Vachon, C. (eds). Modelling and Simulation in Science, Technology and Engineering Mathematics. *Adv. Intell. Sys. Comput.*, 749. Springer, Cham.
10. Banik, A. and Sengupta, A. (2022). Irrigation pumping scheme based on solar PV and Zeta converters with PMBLDC motor. In: Kumar, S., Ramkumar, J., and Kyratsis, P. (eds). Recent advances in manufacturing modelling and optimization. *Lec. Notes Mec. Engg.* Springer.

35 Analytics and computation of energy conservation through energy audit for sustainability in energy consumption development in educational building: A case study

Yogesh Prajapati[1,a], *Jayesh Pitroda*[2,b], *Jagdish Rathod*[3,c] *and Indrajit Patel*[4,d]

[1]Department of Electrical Engineering, BVM Engineering College, Vallabh Vidyanagar, India

[2]Department of Civil Engineering, BVM Engineering College, Vallabh Vidyanagar, India

[3]Department of Electronics Engineering, BVM Engineering College, Vallabh Vidyanagar, India

[4]Department of Structure Engineering, BVM Engineering College, Vallabh Vidyanagar, India

Abstract

Energy is an important factor in all aspects of a country's economy. A variety of industries, including manufacturing and agriculture, are conducting energy audits and energy conservation studies. The current work focuses on effective energy management based on an audit conducted at the BVM Engineering campus by improving performance of existing loads and estimating the amount of solar energy that can be harnessed. The energy audit consisted of two stages: preliminary and detailed. The preliminary energy audit phase completed the first 5 years of data collection and identification of energy savings potential, while the comprehensive energy audit phase completed monitoring, verification, and analysis based on the preliminary energy audit load. The detailed energy audit phase evaluated numerous operating equipment such as transformers, capacitors, lighting systems, air conditioners, fans, and pumps. Following the assessment, the inefficient load is replaced with a more efficient load, such as LED, and the fan is replaced. The energy savings are shown.

Keywords: Energy management, energy audit, utility assessment, solar rooftop

Introduction

Energy is an important factor in all aspects of a country's economy. A country's economic progress is inextricably linked to its energy consumption. Energy conservation reduces waste. Energy savings through energy efficiency and conservation reduce pollution by eliminating the fuel, mining, transportation, water, and land investments required for power plants. The energy crisis is a serious worry in today's globe, as energy use rises. To fulfill the ever-increasing necessity, the demand for energy has increased. The approach to the energy conundrum will be energy conservation and the development of energy-efficient technology [1]. Some authors have offered their views on energy conservation, various techniques, and energy audit systems based on multiple case studies. By comparing India's energy efficiency to that of other nations, certain energy-saving technologies and implementation challenges in India are studied [2]. Also, some authors have presented a case study of energy saving in a university's campus lighting [3]. The usage of electrical energy audit technologies in a college institute is also presented [4]. So, an industrial energy audit is an efficient tool for developing and implementing an extensive energy management programme. As a result, energy audits may be divided into two types: preliminary audits and detailed audits. The preliminary energy audit is a relatively quick exercise to establish energy consumption in the organization, estimate the scope for savings, identify the most likely and easiest areas for attention, identify immediate improvements/savings, set a "reference point," and identify areas for more detailed study/measurement. The detailed audit considers how each project interacts with the others, determines the amount of energy used by each key piece of equipment, and provides comprehensive cost and energy cost savings estimates. One of the crucial components of an exhaustive audit is the energy balance. Three stages make up a detailed energy audit: Phase I is the pre-audit phase, II is the audit phase and phase III is the post-audit phase.

Energy audit methodology for the Institution

The energy audit carried out in two sections: Preliminary energy audit and detailed energy audit. Primary data such as block-wise linked details, area-wise energy scopes in the connected load, diesel generator specifics, solar rooftop details, and so on

[a]yrprajapati@bvmengineering.ac.in, [b]jayesh.pitroda@bvmengineering.ac.in, [c]jmrathod@bvmengineering.ac.in, [d]inpatel@bvmengineering.ac.in

DOI: 10.1201/9781003540199-35

are gathered [5]. The detailed energy audit included the measurement, verification, and monitoring of energy-consuming equipment [6]. The computation and comparison with the rated data have been completed. Finally, energy-saving opportunities are discovered, and the implementation method is finished based on the energy-saving findings.

Preliminary energy audit

The primary aspects covered in this kind of analysis are establishing energy usage in the institute, estimating the scope for savings, and determining the most probable and easy areas for focus [7]. Determine quick improvements/savings, making a "reference point," identify areas for further investigation/measurement. The preliminary energy audit makes use of pre-existing or easily acquired data [8]. Based on the findings, data such as transformer loading condition, installed capacitor, cable size, block-wise feeder distribution, location-wise connected load details, and savings prospects are gathered. Lighting fixtures, fans, air conditioners, water pumps, computers, and other key loads are all associated with load data. Following the collection of the connected load, the monitoring, measuring, and analysis procedure was carried out as part of the thorough energy audit [9]. The block diagram of BVM Engineering College is presented in Figure 35.1. The institute's connection is subject to the HTP-1 tariff having 250 KVA contract demand. The installed transformer has a capacity of 500 KVA. BVM Engineering College has a diesel generator with a capacity of 200 KVA installed. It is operational in an emergency, such as a power outage from the grid during an examination, essential workshops/seminars, experiments, or major functions. The sizes and the connected load details are highlighted in each block as shown in Figure 35.2.

The single-line diagram in Figure 35.3 illustrates the institute campus's block-wide electrical distribution. 3/2-core aluminum cables have been installed from the transformer to the main panel is 400 sqm in size. The remaining cables, which are dispersed across the various blocks, are 120 sqm in size. The load installed in the institute in a kW and percentage is presented in Figures 35.4 and 35.5, respectively. The major load installed is the computer (PC) load which is 46% and the minimum. The following figure represents the connected load details to be considered for energy conservation.

Electricity bill analysis

BVM Engineer College is a Madya Gujarat Vij Company Limited (MGVCL). HTP-1 customer with a contract demand of 250 KVA, and 213 KVA

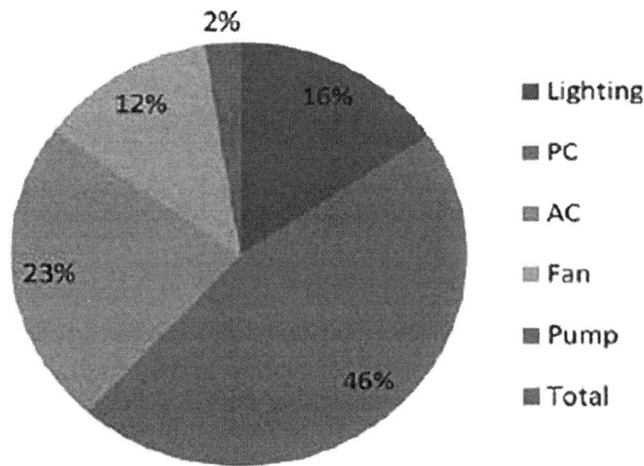

Figure 35.2 Load distribution

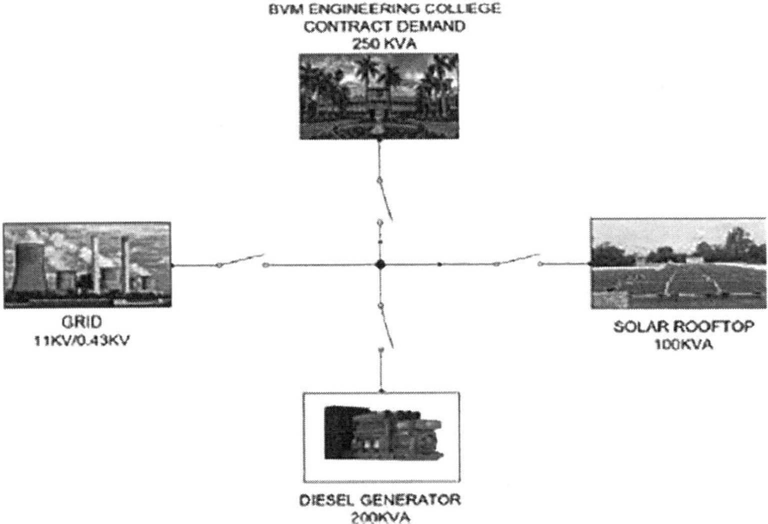

Figure 35.1 Block diagram of power supply

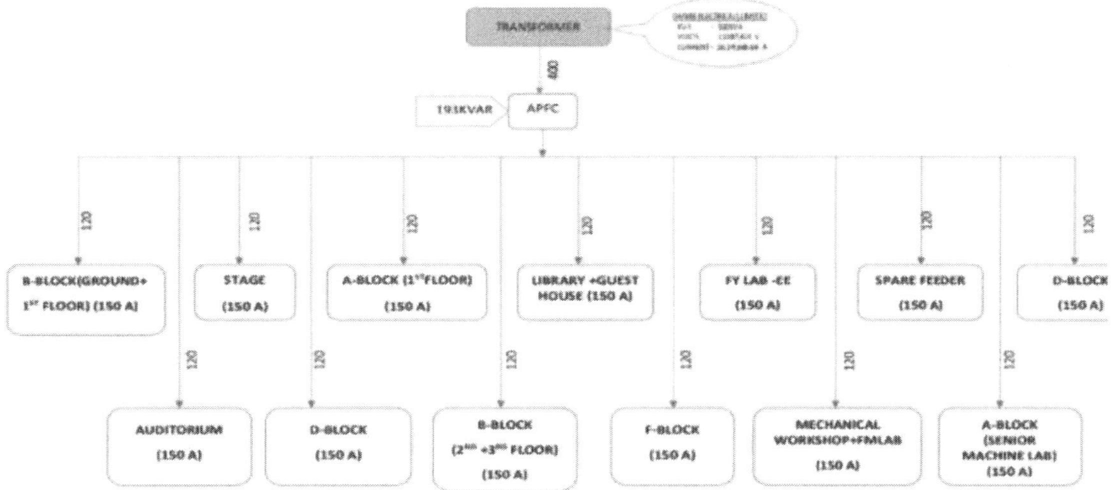

Figure 35.3 Single line diagram of feeder distribution

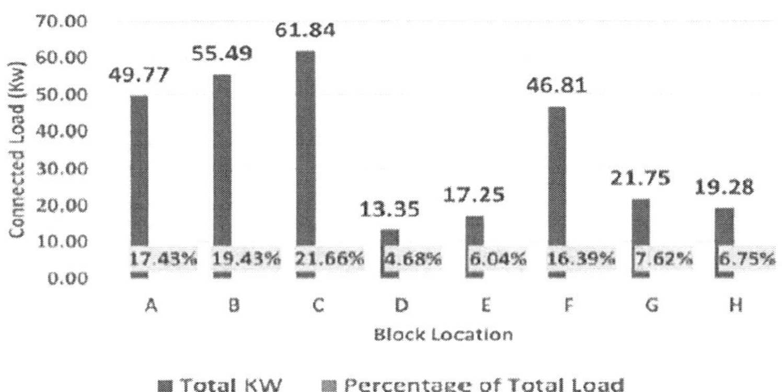

Figure 35.4 Block-wise load distribution

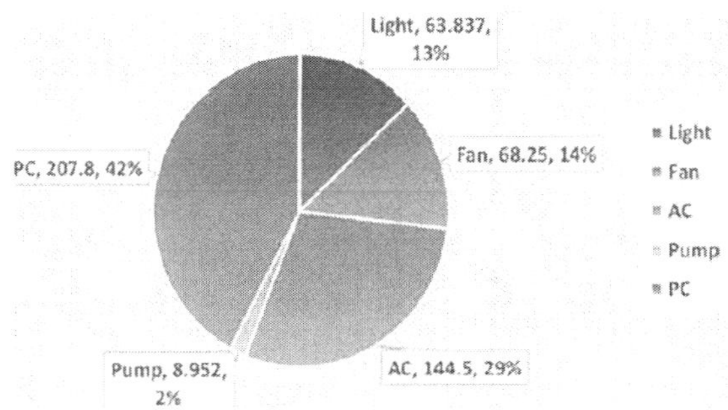

Figure 35.5 Connected load, KW

represents 85% of the contract demand. Figure 35.6 below depicts KVA usage from 2017 to 2022. The average electricity use is 131.67 KVA. The KVA in 2020 has been reduced due to the pandemic in the corona era. KVA usage began in 2021 because of online education. However, the grid-connected solar plant will be constructed in July 2021, which is one of the reasons for the drop in KVA. As shown in Figure 35.7, from the 2017 to 2022, the average unit usage is 12,276 kWh (kilo watt hours). The graph indicates that unit consumption drops dramatically in 2020 and 2021. It was due to the corona epidemic, and the students were teaching online. Similarly, Figure 35.8 depicts the unit used throughout the night and peak hours. The peak hours in Gujarat are from 6 to 10 a.m. and from 7 to 11 p.m. Peak-hour unit rates are 0.45 Rs more than off-peak rates. The night unit prices are computed at 0.43 Rs per unit. The night unit consumption is refundable [10]. Other charges such as demand charges, energy charges, and fuel prices are computed in accordance with Madya Gujarat Vij Company Limited's guidelines (MGVCL).

Transformer efficiency

BVM Engineering College has installed a 500 KVA transformer. The transformer's efficiency has been

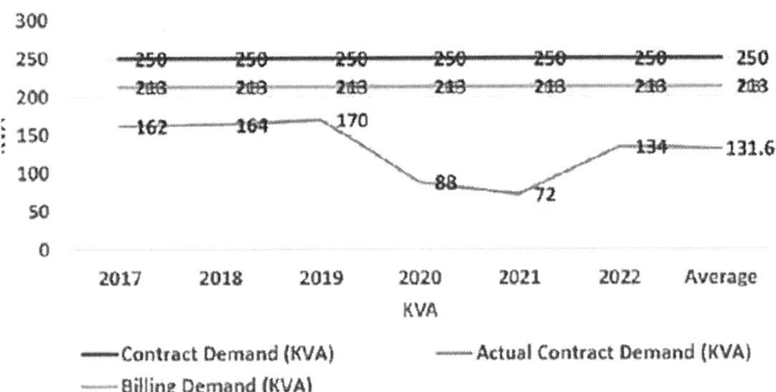

Figure 35.6 KVA demand

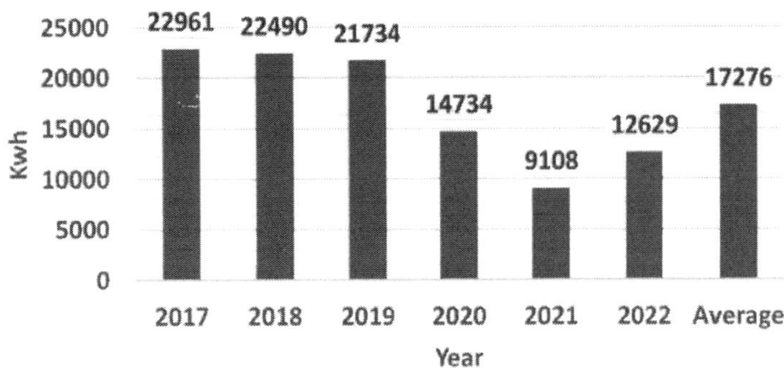

Figure 35.7 Unit consumption

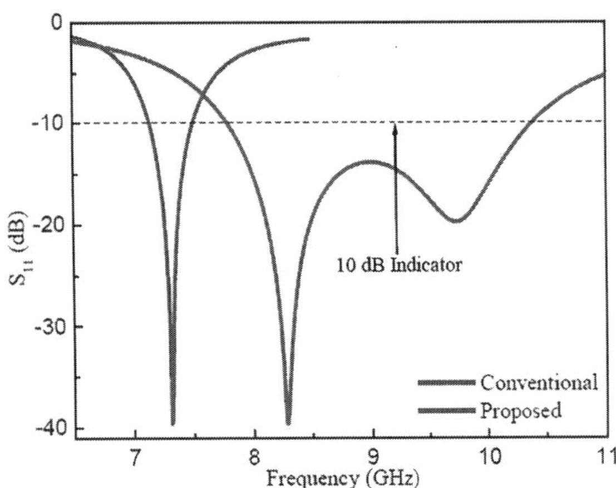

Figure 35.8 Peak hours unit and night unit consumption

verified. Figure 35.9 displays the efficiency and percentage loading from 2017 to 2022. The no load (iron) loss and load (copper) loss are computed to determine transformer efficiency [11]. A power analyzer is used to measure an average no-load power of 2.5 kW. Similarly, the total load is determined by the difference in kWh between the transformer's input and output sides. The copper load may be estimated by subtracting the iron loss from the overall loss. The transformer's efficiency is estimated from Equation (1). The output power is measured on the 440 V LT side.

$$Transformer\ efficiency = \frac{P_{out}}{P_{out} + Total\ Losses} \qquad (1)$$

where, total losses = iron loss+ copper loss.

In the automatic power factor controller (APFC) panel, a capacitor with a capacity of 193 KVAr is used. All the capacitors were found to be in good working order. The average power factor as shown in Figure 35.10 is 0.97, and all payments include refunds.

Lighting load assessment

The lighting load consumes a large amount of energy in an academic institution. The lighting source's specifications, such as wattage, fixture type, and fixture inventory, must be gathered by location or floor, and light power density (watts/m^2) must be calculated. The following criteria were used to calculate the installed load efficacy ratio (ILER) and annual energy wastage (kWh/Annum) in accordance with IS standard [12]. To measure the light intensity, the lux meter is used. The area and height are measured using the measuring tape. Figures 35.11 and 35.12 represents the existing 50

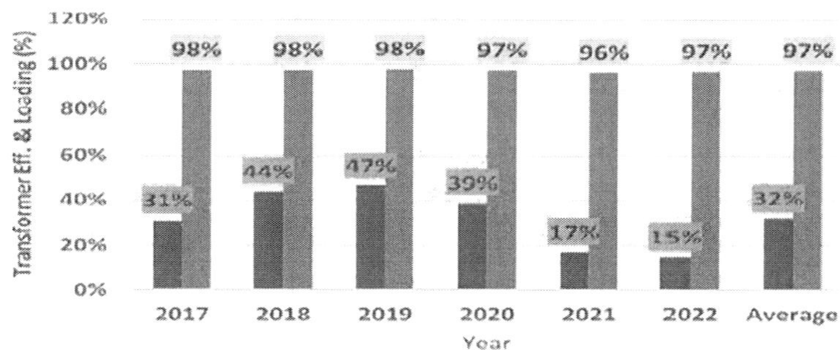

Figure 35.9 Transformer efficiency (%) and loading (%)

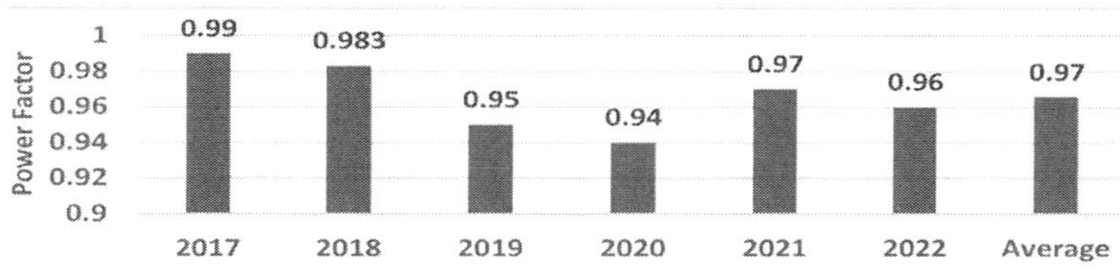

Figure 35.10 Power factor

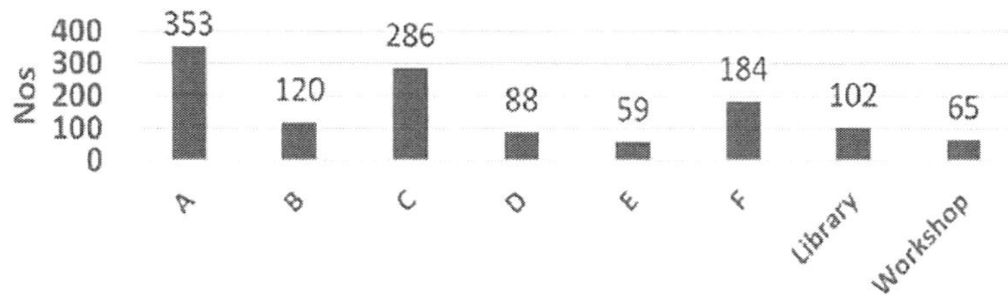

Figure 35.11 Existing old FTL details before replacement

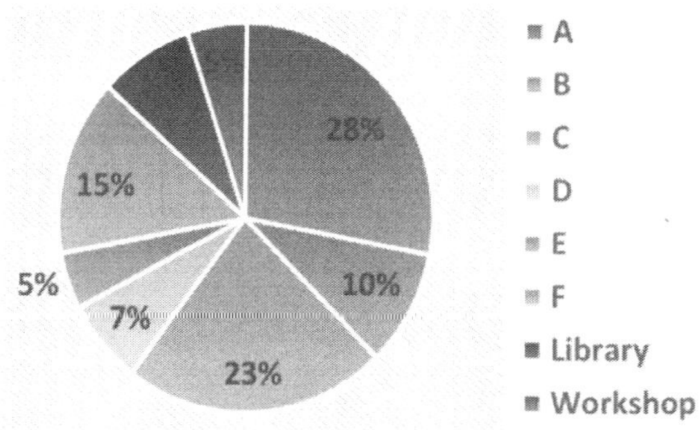

Figure 35.12 Existing old FTL in %

W florescent lighting detailed before replacement by efficient lights. The details of the replacement of the old lights by efficient lights are presented in Figure 35.13. After installing the efficient lights, the lighting load is reduced by 34% as shown in Figure 35.14. For the lighting system assessment method, the guidelines as per the Bureau of Energy Efficiency (BEE) are followed is given in ref [13]. In this the parameters such as room index (RI), Lux/Watt/m^2 and installed light efficacy ratio (ILER) are considered for the calculation of the wastage. For the ILER between 0.5 and 0.7, the lux level is satisfactory. The efficient lighting system in the rooms is assessed. Figure 35.15 represents the assessment of the lighting system for some classrooms.

Heating ventilating and air conditioning system assessment

Air conditioning and refrigeration systems require different amounts of energy depending on the load, season, operation and maintenance, ambient temperatures, and other factors. The purpose of performance evaluation is to evaluate the performance of a refrigeration system using field data and to estimate energy consumption [14]. To compute the net refrigeration capacity in tones, the following Equation (2) is applied. The velocity (m/s) is measured with a vane anemometer, while the dry and wet bulb temperatures are measured with a sling psychrometer. The enthalpy in kj/kg may be calculated using the psychometric chart.

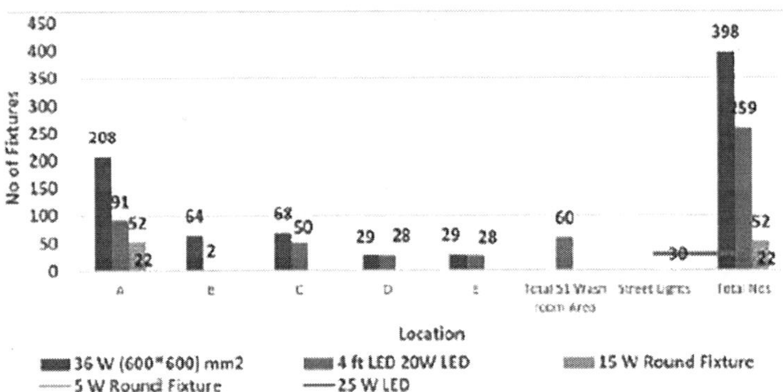

Figure 35.13 Installed New efficient lighting fixture details

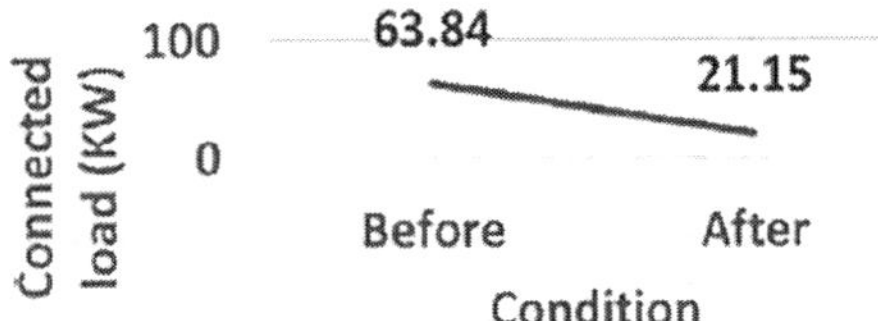

Figure 35.14 Reduction in lighting load (Kw) after replacement of by efficient lights

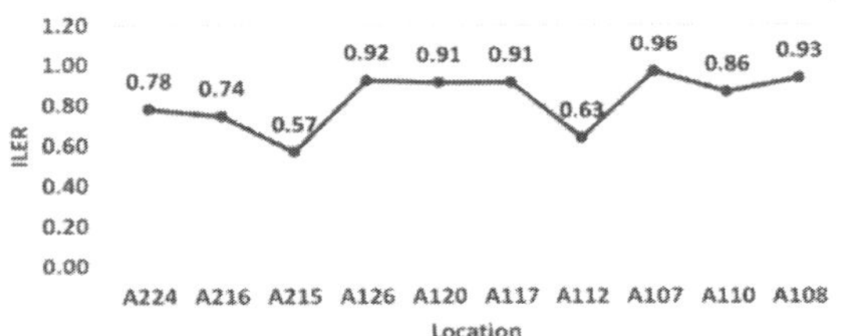

Figure 35.15 ILER assessment after installing efficient lighting fixture

$$\textbf{Heat Load}(TR) = \frac{m\times(h_{in}-h_{out})}{4.18\times3024} \tag{2}$$

where, m–mass flow rate of chilled water, kg/hr, h_{in}–enthalpy of inlet air at air conditioning unit, kJ/kg, h_{out}–enthalpy of outlet air at air conditioning unit, kJ/kg.

Figure 35.16 depicts a comparison between rated TR and actual TR for air conditioning (AC) systems installed in various locations. From May through August, the institute's air conditioners operate for 3–4 hours every day. The installed air conditioners are well-maintained and in good working order.

Fan assessment

The old fan existed in the institute with different capacities such as 120 W, 60 W, etc. After completing the data collection during the preliminary energy audit, it has been found that most of the fans are old as shown in the Table 35.1. The summary of the old fan in kW is given in Figure 35.17. These old fans are replaced by the efficient fan with capacity of 45 W. The reduction in kW is presented in Figure 35.18.

Water pump assessment

Pumping is the process of the addition of kinetic and potential energy to a liquid for the purpose of moving it from one point to another. Performance assessment of pumps would reveal the existing operating efficiencies in order to take corrective action. The purpose of the performance test of the pump is to determine the pump's efficiency [15]. The pump efficiency can be calculated based on Equation (3). The pump water flow is measured in m^3/h. using an ultrasonic meter, the head is measured in kg/cm^2 using a pressure gauge, and the power input to the motor is measured using a power analyzer. The motor's output is estimated based on its rated efficiency, which is the shaft input to the pump.

$$Pump\ Efficiency = \frac{Hydraulic\ Power, P_h}{Power\ Input\ to\ the\ Pump} \times 100 \tag{3}$$

where,

$$\text{Hydraulic Power}, P_h = Q \times (h_d - h_s) \times \rho \times g/1000$$

In this Q is volume flow rate (m^3 /s), ρ is density of the fluid (kg/m^3), g is acceleration due to gravity (m/s^2) and (hd–hs) is total head in meters. Figure 35.19 presents the pump efficiency calculated by ultrasonic flow meter method.

Solar roof top system assessment

A 100 kW grid-connected solar system with net metering is placed on the institute's roof. As shown in the Figure 35.20, approximate 47% unit's consumption of the institute is compensated by solar units. Monthly solar unit generation ranges from

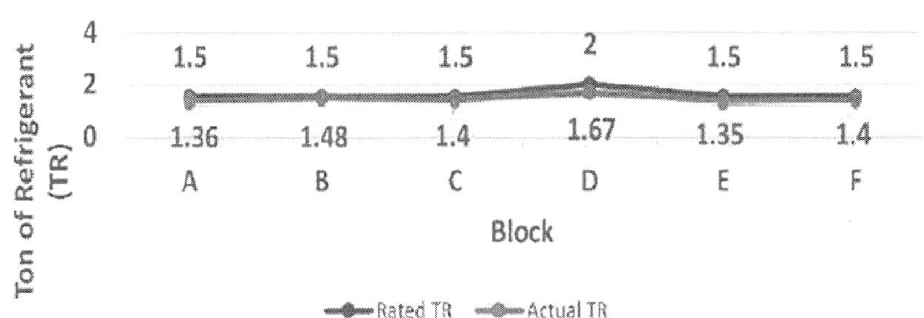

Figure 35.16 Assessment of air conditioning unit

Table 35.1 Details of the Old Fan

Sr. No	Block	Fan
1	A	All fan of100W
2	B	Ground floor - 80% fan of 100 w
3		1st floor - all fans 60 w
4		2nd floor- all fans of 60 w
5		3rd floor all fans of 60 w
6	C	100 % of the 100 w
7	D	1st & 2nd floor - 60% of 100 w
8		40% of the 60 w
9	E	80% of the 60 w
10		20% of 100w
11	F	Ground &1st floor -100% of 80 w
12		2nd & 3rd floors all fans of 60w
13	Workshop	45 Nos. of 100w & remaining 60 W.

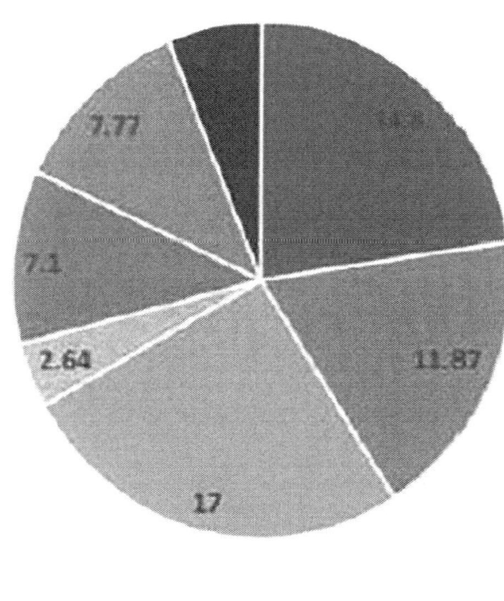

Figure 35.17 Old fan load Kw 65.2 Kw

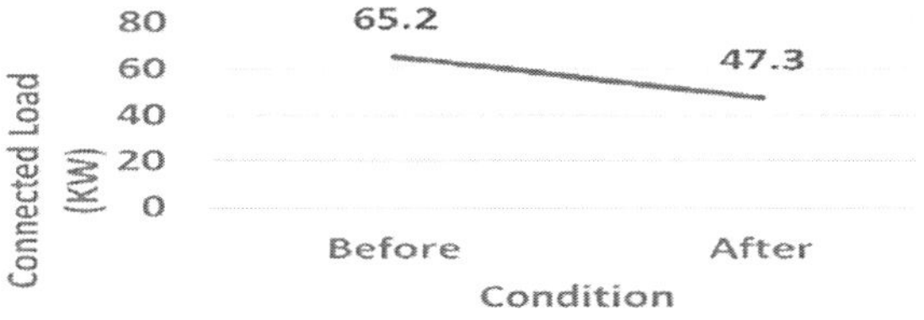

Figure 35.18 Reduction in fan load (Kw) after the replacement fan of by efficient fan (45W)

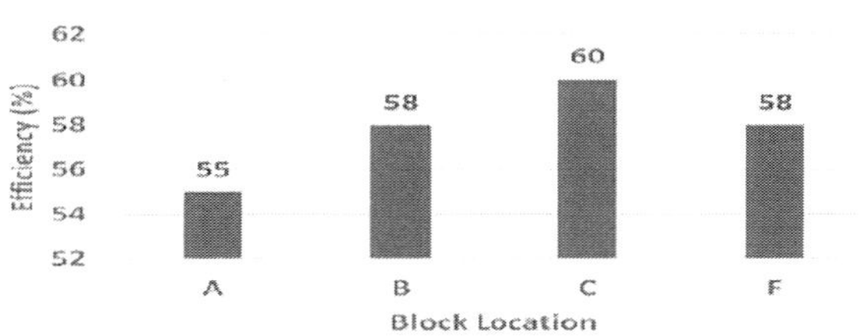

Figure 35.19 Pump efficiency

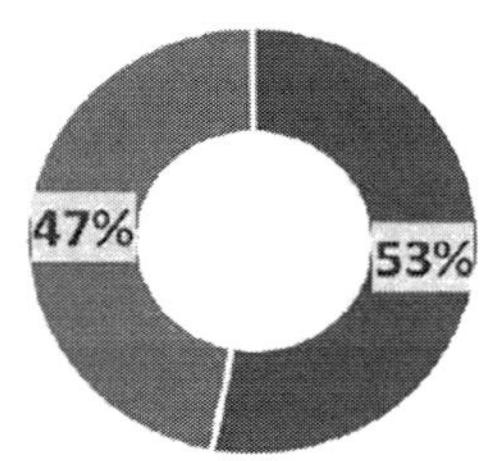

Figure 35.20 Grid unit consumption and solar unit generation

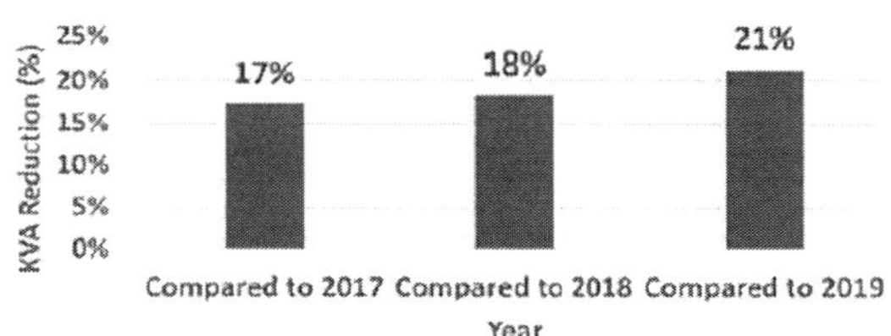

Figure 35.21 Reduction in contract demand after solar installation

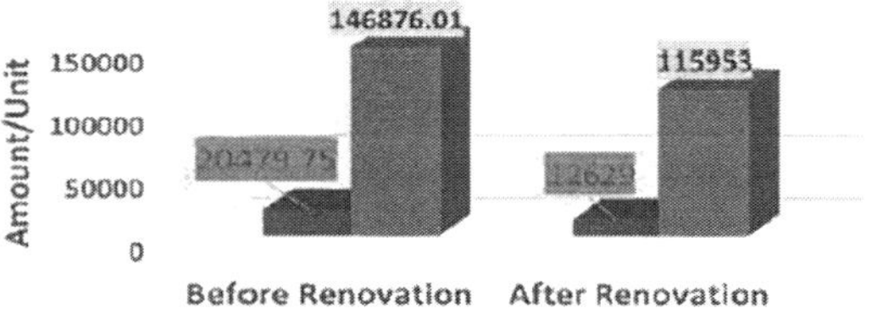

Figure 35.22 Reduction in unit consumption and amount after solar installation

450 units to 550 units in the summer, 400 units to 550 units in the winter, and 300 units to 550 units in the monsoon season. The solar panels have been cleaned twice a month on a regular basis. CO_2 emission saved is 72,999.01 kg and equivalent trees planted is 2,178.8 till December 2022.

Results and analysis

Figure 35.21 depicts a reduction in real demand (KVA) consumption of 17%, 18%, and 21% in 2022 compared to 2017, 2018, and 2019. Figure 35.22 represents the reduction in the unit consumption due to solar unit generation. The financial benefits in the unit consumption are presented in Figure 35.21 which is almost 38% in the reduction in the unit and 21% in the bill amount.

Conclusions

The energy audit has been carried out with the objective to decrease the unnecessary usage of energy as well as the energy consumption waste. During the preliminary energy audit, fundamental data such as connected load details, electricity bill, utility details such as transformer, diesel generator, and solar, scopes of energy savings in the specific area, and so on were collected. Following data collection, measurement, verification, and analysis, as well as implementation, were accomplished and provided in the article. As a result of the entire procedure, there is a 21% reduction in KVA usage. The units on the power bill are decreased by 38%, but the net bill amount is lowered by 21%. Energy and money savings are realized as a result of a 67% reduction in lighting demand, a 28% reduction in fan load, and a grid-connected solar plant via net metering.

Acknowledgement

We authors are sincerely acknowledging our parent management and our beloved chairman Er. Bhikhubhai Patel for his all kind of support and motivation throughout this solar energy concept establishment. We are thankful to BVM and other TEQIP-World Bank assisted project for establishing the lab & necessary instruments facilities for the testing. We are thankful to all who have indirectly or directly support us for this paper.

References

1. Sarangi, G. K. and Taghizadeh-hesary, F. (2020). Unleashing market-based approaches to drive energy efficiency interventions in India: An analysis of the perform, Achieve, Trade (PAT) scheme. *Asian Dev. Bank Institute*, 1177, 1–21.
2. Ullah, K. R., Thirugnanasambandam, M., Saidur, R., Rahman, K. A., and Kayser, M. R. (2021). Analysis of energy use and energy savings: A case study of a condiment industry in India. *Energies*, 14(4798), 1–24. doi: 10.3390/en14164798.
3. Kerem. (2022). Assessing the electricity energy efficiency of university campus exterior lighting system and proposing energy-saving strategies for carbon emission reduction. *Microsys. Technol.*, 28(12), 2623–2640. doi: 10.1007/s00542-022-05268-x.
4. Raj Nadimuthu, L. P., et al. (2021). Energy conservation approach for continuous power quality improvement: A case study. *IEEE Acc.*, 9, 146959–146969. doi: 10.1109/ACCESS.2021.3123153.
5. Akhtar, S. K., Suhail, M., and Jameel, M. (2021). Advanced fuzzy-based smart energy auditing scheme for smart building environment with solar

integrated systems. *IEEE Acc.*, 9, 97718–97728. doi: 10.1109/ACCESS.2021.3095413.

6. Engmann, F., Adu-Manu K. S., Abdulai, J. D., and Katsriku, F. A. (2021). Network performance metrics for energy efficient scheduling in wireless sensor networks (WSNs). *Wireless Comm. Mobile Comput.*, 2021, 1–14. doi: 10.1155/2021/9635958.
7. Patsakos, E. V. and Papakostas, G. A. (2022). A survey on deep learning for building load forecasting. *Math. Problems Engg.*, 2022, 1–25. doi: 10.1155/2022/1008491.
8. Beijing, C. (2021). Abstracts from the energy informatics. Academy Asia 2021 conference and PhD workshop. *Energy Informat.*, 4(30), 1–39. doi: 10.1186/s42162-021-00145-9.
9. Khalid, A. S. and Sarwat, A. I. (2021). Overview of technical specifications for grid-connected microgrid battery energy storage systems. *IEEE Acc.*, 9, 163554–163593. doi: 10.1109/ACCESS.2021.3132223.
10. Commission, G. E. R. (2015). Tariff for supply of electricity at low tension, high tension and extra high tension, MEGHALAYA POWER DISTRIBUTION CORPORATION LIMITED, 1-26.
11. Hashemi, M. H., Kiliç, U., and Dikmen, S. (2023). Applications of novel heuristic algorithms in design optimization of energy-efficient distribution transformer. *IEEE Acc.*, 1–14. doi: 10.1109/ACCESS.2023.3245327.
12. Hajibabaei, M., Saki, A., Golmohammadi, R., Cheshmehkhavar, M., Sarabi, M., and Isvand, M. (2014). Performance indexes assessment for lighting systems based on the normalized power density and energy losses estimation in University workrooms. *Inter. J. Occup. Hyg.*, 6(3), 131–136.
13. Thielemans, S., Di Zenobio, D., Touhafi, A., Lataire, P., and Steenhaut, K. (2017). DC grids for smart LED-based lighting: The EDISON solution. *Energies*, 10(10), 1–26. doi: 10.3390/en10101454.
14. Oladipo, S., Sun, Y., and Amole, A. (2022). Performance evaluation of the impact of clustering methods and parameters on adaptive neuro-fuzzy inference system models for electricity consumption prediction during COVID-19. *Energies*, 15(21), 1–20. doi: 10.3390/en15217863.
15. Martin-Candilejo, D. S. and Garrote, L. (2020). Pump efficiency analysis for proper energy assessment in optimization of water supply systems. *Water (Switzerland)*, 12(1), 1–18. doi: 10.3390/w12010132.

36 Implementation of symmetrical pair nonlinear defected ground structures for improvement of bandwidth, cross-polarization isolation, and stable gain in circular patch antenna

Manoj Sarkar[1], *Zonunmawii*[2], *Abhijyoti Ghosh*[2,a], *L. Lolit Kumar Singh*[2] *and Sudipta Chattopadhyay*[2]

[1]Department of Radio Physics and Electronics, Calcutta University, Kolkata, India

[2]Department of ECE, Mizoram University, Aizawl, India

Abstract

In this article, a coaxial probe feeding circular patch antenna (CPA) with a pair of nonlinear-shaped defected ground structures (DGS) has been investigated at the radiating edge on the RT-Duroid substrate (ε_r = 2.33). The suggested structure has been depicted to have a direct impact on enhancing the impedance bandwidth, polarization purity, and stable gain compared to a traditional circular patch antenna. The suggested structure offers around 31.1% impedance bandwidth, 24 dB CP-XP isolation, and steady gain over the whole operational frequency, which is quite adequate. This current structure may be excellent for use in the era of communication systems wherein high bandwidth, good polarization purity, and a steady radiation pattern abide required.

Keywords: Circular patch antenna, nonlinear shape defected ground structure, impedance bandwidth, gain, co-polarization, cross-polarization

Introduction

The patch antenna is the most useful radiator in the modern era of communication due to its simplicity of integration with active devices like (MMIC, HMMIC, MEMS), low cost, easy fabrication, and lighted. It has numerous substantial drawbacks in addition a narrow bandwidth, inadequate co-polarization gain, and weak polarization purity. Polarization purity (PP) decreases as the operating frequency of the antenna increases. [1, 2]. Conventional CPA (CCPA) emanant linearly polarized fields at the wide side direction which is the dominant TM_{11} mode. However, only a small amount of cross-polarized (XP) radiation, also known as orthogonal polarization, occurs in the first higher-order mode, TM_{21} mode [3].

Different strategies, such as defective ground structure (DGS), defected patch structure (DPS), and shorting technique, have been recommended by scientists and researchers in the area of antennas to increase the traditional CPA's input and radiation performance. CPA with circular DGS [4] and arc-shaped DGS [5] are recorded to achieve polarization purity of 5–7 dB, however, impedance bandwidth is relatively small in those studies. Several defective patch surface (DPS) models with claimed polarization purity of 20–25 dB is investigated in refs [6, 7]. With various types of rectangular slots on a circular patch surface, it has been claimed to have a bandwidth of about 22% and a gain of 3.24 dBi [8]. By completely shorting the non-radiating edges of the typical CCPA, polarization purity of roughly 25 dB has been reported in ref [9], however, the impedance bandwidth is quite small, falling between 7% and 9%. A rectangular strip with discrete shorts in another structure, [10], is said to be able to achieve 27 dB of polarization purity with only an 8% impedance bandwidth.

Different cavity-enclosed CPA types have been studied in ref [11] for bandwidth augmentation; impedance bandwidths of 7–8%, 20% bandwidth in ref [12]. Recently, radiation performance has been enhanced by shorting the wall in ref [13] and using a rectangular strip-loaded CMPA in ref [14], respectively, with no improvement in polarization purity. In the aforementioned report enhancement in either impedance bandwidth or polarization purity has been observed.

In this article, a coaxial probe feeding circular patch antenna (CPA) with a pair of nonlinear shaped defected ground structures (DGS) at the radiating edge on RT-Duroid substrate (ε_r = 2.33) with height (h) 1.58 mm has been explored to deal with the limitations of earlier studies (Figure 36.1).

[a]abhijyoti_engineer@yahoo.co.in

DOI: 10.1201/9781003540199-36

Approach towards the proposed structure

Field and cavity boundaries are disrupted when changes are made in the structure of CCPA. For the best performance of the antenna, the defect dimensions must be carefully chosen. Changes in defect size have an effect on the radiation characteristics and cavity field behavior. Larger defects result in more bandwidth, but they also tend to cause higher loss and a worse quality factor.

In the broadside direction, linearly polarized electric fields are present in the CCPA, which is a resonator. It frequently takes place henceforth in the fringing field that is radiated from the patch's radiation edge, which is the dominant mode TM_{11} [2, 15].

The radiation from the non-radiating edge of the CCPA in the first higher-order orthogonal mode (TM_{21}) is the key factor that generates high XP radiation (TM_{21}) [14]. The TM_{11} fundamental dominant mode and the TM_{21} higher-order field are orthogonal to one another.

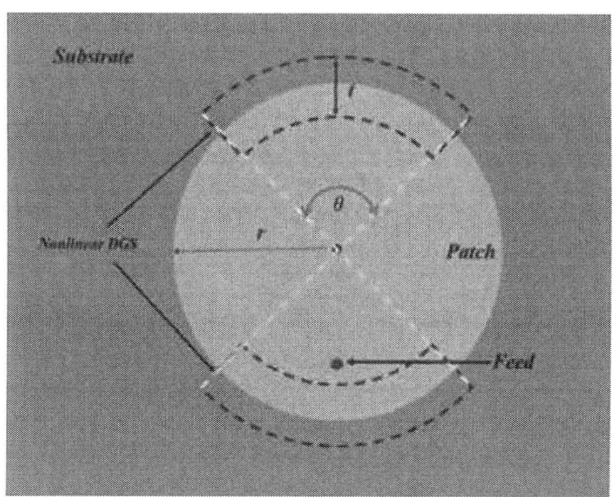

Figure 36.1 Schematic representation of suggested nonlinear shaped defective ground structure incorporated CPA

With the aid of the aforementioned concept of the structure, a symmetrical pair of nonlinear DGS profiles has been introduced in the ground plane at the radiating sides of the CCPA, which has a radius of (r) 7mm. Initially, the angle (θ) of the nonlinear DGS was kept at $\theta = 114^{\circ}$. The optimal dimension of the nonlinear DGS width (t) has been varied to get the optimum dimension.

Initially, a symmetrical pair of nonlinear DGS with width (t) 0.5 mm at $\theta = 114°$ is placed at the radiating edges of CCPA. Comparing the nonlinear DGS-integrated CPA to the CCPA, an improvement in the impedance bandwidth is observed. The CCPA (t = 0 mm) conveys a 4% impedance bandwidth, which can be boosted to 6% using a DGS structure with a width of 0.5 mm, as shown in Figures 36.2 (a, 36.b). Figure 36.2(b) also shows that the lowest order mode's resonance frequency for the proposed structure shifts to the upper side of the frequency spectrum. The encapsulation of the nonlinear shape DGS influences the substrate's effective permittivity. The DGS's width (t) is increased up to 3.5 mm. The impedance bandwidth increases with the increase of width (t) up to 2.5 mm, after which it drops for 3 mm. Figures 36.2(a) and Figure 36.2(b) have confirmed the alternation. The impedance bandwidth at t = 2.5 mm is 29.11%, which is considerably wider as compared to CCPA.

In addition to the impedance bandwidth, examined on polarization purity are also carried out. As the width (t) expansion from 0.5 mm to 3.5 mm while maintaining the nonlinear shape DGS angle (θ) at 114°. The placement of the nonlinear form

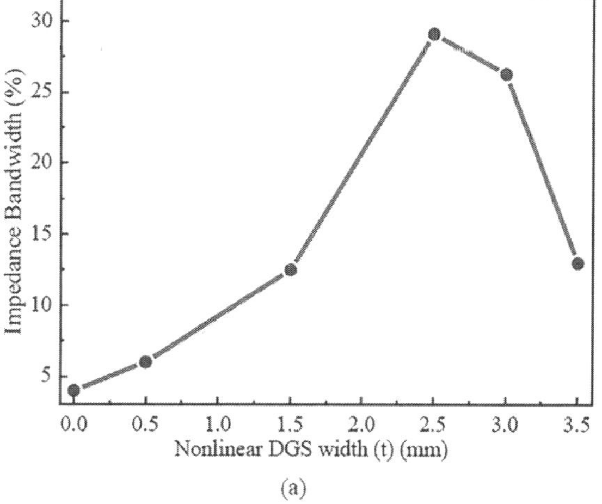

(a)

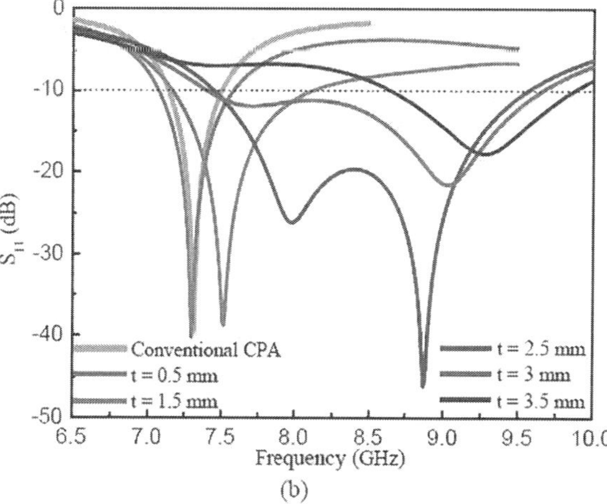

(b)

Figure 36.2 (a) Impedance bandwidth variation (%) as a function of nonlinear DGS width (t) with angle fixed at $\theta = 114^{\circ}$. (b) Reflection coefficient profile for CCPA and nonlinear DGS incorporated CPA with fixed at $\theta = 114^{\circ}$

DGS initially causes a reduction in polarization purity, but latterly it has been improved and offers the optimum polarization purity at t = 2.5 mm. In the case of a CCPA, the polarization purity is only 17 dB, but it is around 22 dB with a modified structure. The aforementioned things have been revealed in Figures 36.3 and 36.4.

Now effect of the nonlinear shape DGS angle has been varied further in order to get a good result with keeping the optimized curved shape width (t = 2.5mm) constant. The nonlinear shape DGS angle has been varied from $\theta = 80^{\circ}$ to 120° for further investigation of the impedance bandwidth, polarization purity and gain.

Therefore, nonlinear-shaped DGS has been placed at different angles in order to get optimum results. At $\theta = 98^{\circ}$ angle around 31.12% bandwidth has been achieved which can be further verified from Figure 36.5(a) whereas CCPA provides only 4% impedance bandwidth.

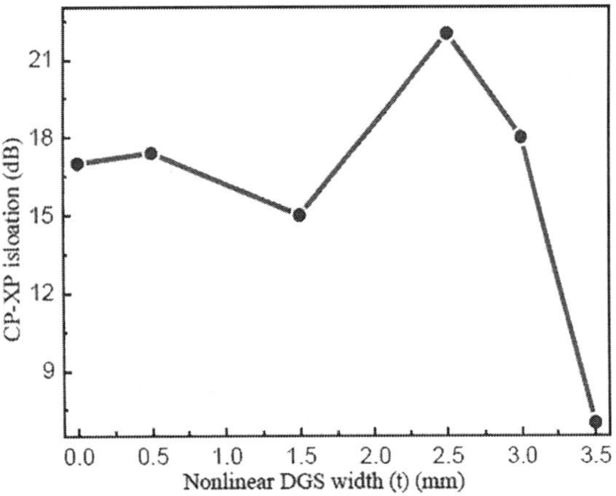

Figure 36.3 The variance of CP-XP isolation as a function of nonlinear DGS width (t)

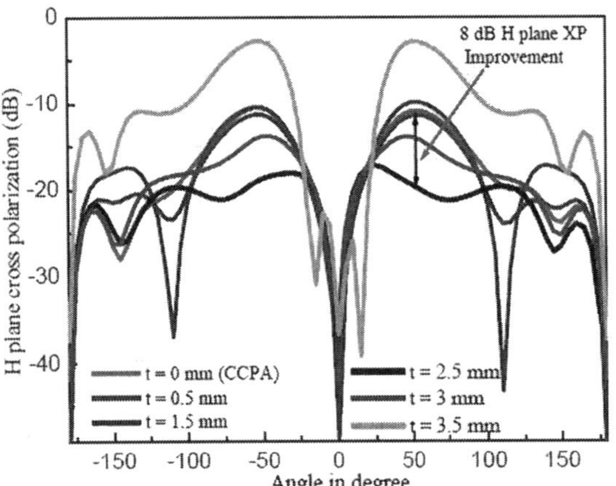

Figure 36.4 The variance of H plane cross-polarization as a function of nonlinear shape DGS width (t) at angle of $\theta = 114^{\circ}$

The S_{11} profile, which is shown in Figure 36.5(b), provides additional support for the aforementioned impedance bandwidth investigation. It is obvious that after changing the nonlinear shape of the DGS angle up to 98°, the impedance bandwidth surprisingly increased after it degrade. It exhibits quit good impedance matching.

Polarization purity is accomplished in addition to bandwidth by varying the nonlinear shape DGS structure angle (θ) up to 123°, as shown in Figure 36.6. Initially at 80° cross-polarizations has been decreased, and then it increases and attains 24 dB CP-XP isolation at 98° which is good and convenient.

The proposed nonlinear shape DGS incorporating CPA that was previously mentioned further validates the increase of polarization purity by the CP-XP isolation illustrated in Figure 36.7.

Optimum structure

Initially, a conventional circular patch antenna with a radius of 7 mm, a dielectric constant of 2.33, and a height of 1.58 mm had been designed. Then, a symmetrical pair of nonlinear shape DGS having a width (t = 2.5 mm) kept up at 98° angle is deposited in the radiating edge of CCPA. The finalized suggested structure has been designed in the way which is depicted in Figure 36.1.

Discussions and outcome

The aforementioned study explores the outcomes produced from the final suggested structure with all optimized parameters using [16]. Figure 36.8 depicts the suggested CPA and CCPA reflection coefficient (S_{11}) profiles. It has been clear evident that CCPA resonates at 7.31 GHz, however, the proposed antenna with a nonlinear shape DGS resonates at 8.29 GHz. Therefore, it is obvious that the resonance frequency has been switched into a higher frequency spectrum as a result of the incorporation of DGS structure at the radiating edge but both the models provide very fascinating impedance matching as shown in Figure 36.8. It is clearly evident that the suggested renovated structure achieved an impedance bandwidth of 31.12%, which is significantly larger than the standard CPA, whereas the CCPA has a confined impedance bandwidth of the order of 4%.

Figure 36.9(a) shows the radiation profile normalized to the E plane. The figure convincingly shows that both the traditional CPA and the suggested antenna have E plane XP values less than -40 dB. The incorporation of DGS at the radiating edges of co-polarization radiation has had minimal impact in the E plane.

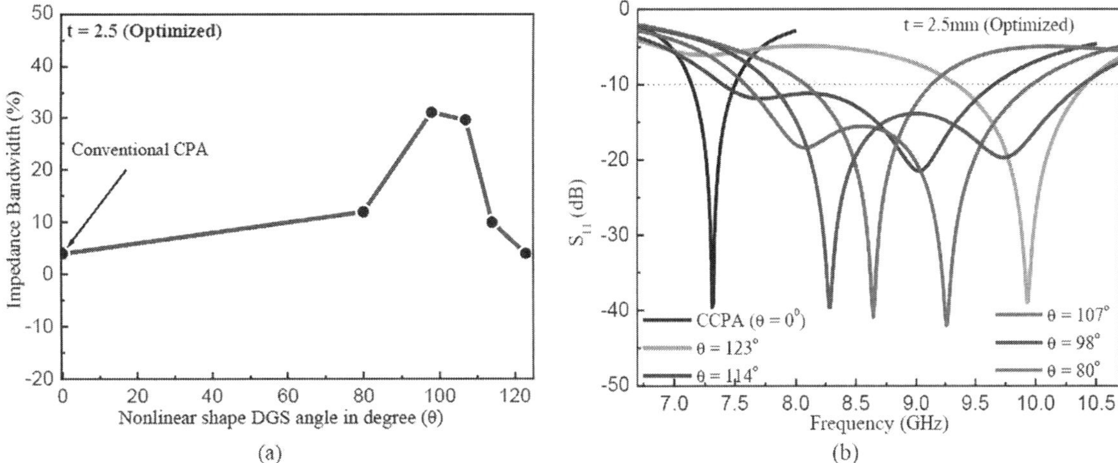

Figure 36.5 (a) Variation of impedance bandwidth (%) as a function of nonlinear DGS angle θ. (b) S_{11} profile Variation of impedance bandwidth (%) as a function angle θ (t = 2.5 mm)

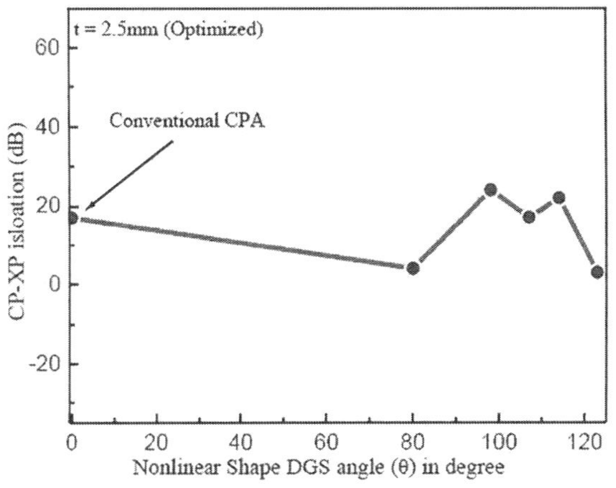

Figure 36.6 CP-XP isolation varies as a function of nonlinear DGS angle θ

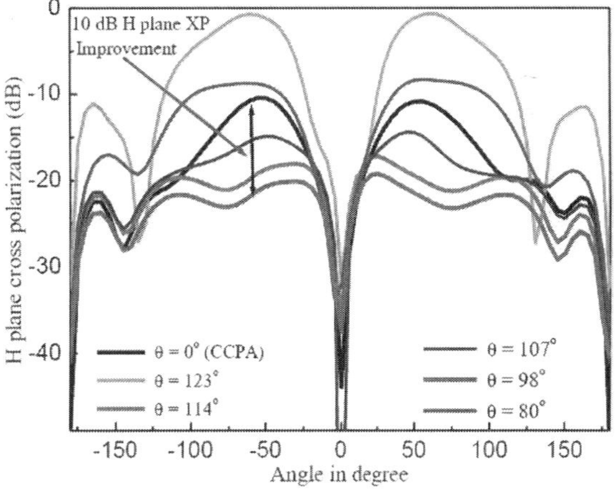

Figure 36.7 Variation of H plane cross-polarization as a function of nonlinear shape DGS angle θ

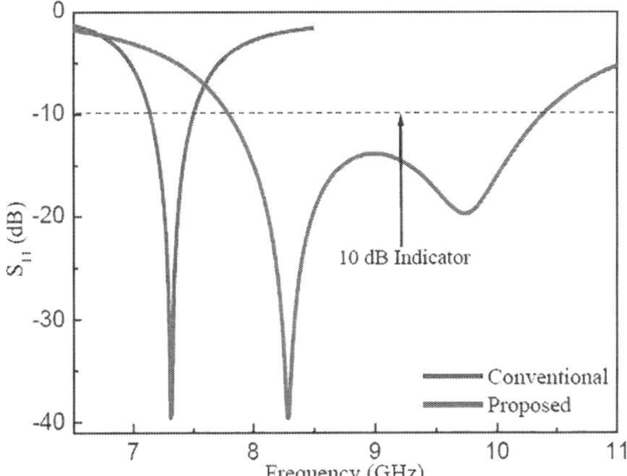

Figure 36.8 Reflection coefficient profile of standard and suggested nonlinear shape DGS incorporated CPA

The comparison between the suggested model and the standard CPA H plane radiation pattern is shown in Figure 36.9(b). In the case of the standard CPA and the suggested antenna, the maximum H plane co-polar gain is between 5 dBi and 6 dBi, respectively. Peak XP levels for the standard CPA and the proposed antenna, respectively, are -12 dB and -20 dB in the H plane.

Conclusions

A symmetrical pair of nonlinear shape defects has been implanted at the radiating edge of a conventional circular patch antenna intending to increase the impedance bandwidth, good polarization purity, and steady gain. The 31.12% operating bandwidth and almost 24 dB CP-XP isolation and 6 dB gain have been achieved from the proposed structure

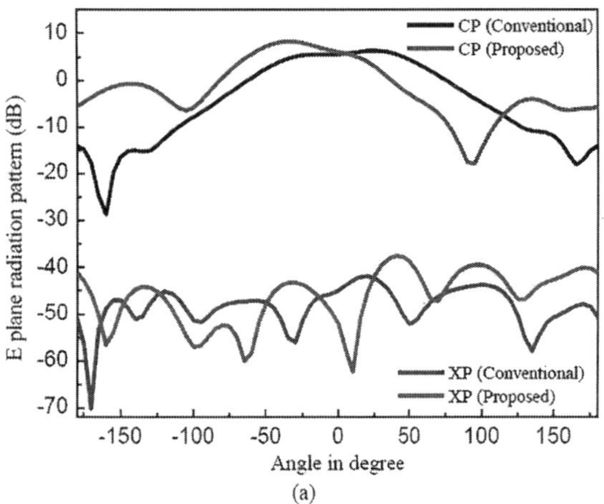

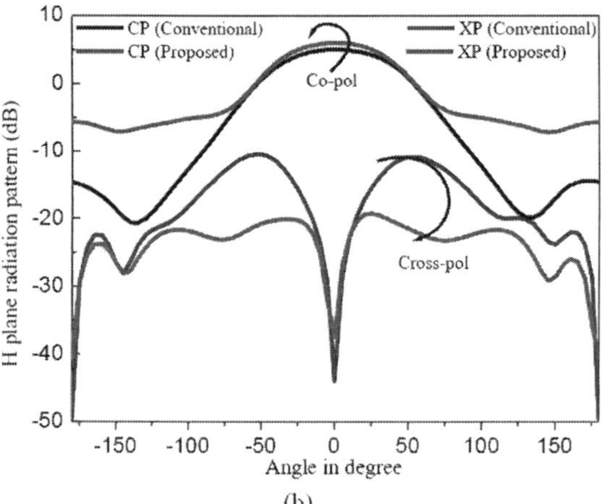

Figure 36.9 (a) The E plane radiation profile of traditional and proposed nonlinear shaped DGS integrated CPAs is normalized. (b) H plane radiation profile of traditional and proposed nonlinear shaped DGS incorporated CPAs is normalized

wherein for conventional CPA impedance bandwidth and polarization purity is only 4% & 17 dB. This research could be valuable in the present-era communication industry, wherein wide impedance bandwidth and CP-XP are required.

References

1. James, J. R. and Hall, P. S. (1992). *Handbook of microstrip antenna.* Peter Peregrinus: London.
2. Garg, R., Bhartia, P., Bahl, I., and Ittipiboon, A. (2001). *Microstrip antenna design handbook.* Artech House: Norwood, USA.
3. Lee, K. F., Luk, K. M., and Tam, P. Y. (1992). Cross polarization characteristics of circular patch antennas. *Electron Lett.*, 28(6), 587–589.
4. Guha, D., Biswas, M., and Antar, Y. M. M. (2005). Microstrip patch antenna with defected ground structure for cross polarization suppression. *IEEE Antennas Wire. Propag. Lett.*, 4, 455–458.
5. Guha, D., Kumar, C., and Pal, S., (2009). Improved cross-polarization characteristics of circular microstrip antenna employing arc-shaped defected ground structure (DGS). *IEEE Antennas Wire. Propag. Lett.*, 8, 1367–1369.
6. Ghosh, A., Ghosh, S. K., Ghosh, D., and Chattopadhyay, S. (2016). Improved polarization purity for circular microstrip antenna with defected patch surface. *Inter. J. Microw. Wire. Technol.*, 8(1), 89–94.
7. Ghosh, A., Ojha, S., Rana, Y., Anand, A., Kumar, S., Chattopadhyay, S. (2014). Improved cross polarization performance of circular microstrip antenna with arc shaped defected patch surface. *1st Inter. Sci. Technol. Cong.*, 87–92.
8. Mondal, K., Murmu, L., and Sarkar, P. P., (2017). Investigation on compactness, bandwidth and gain of circular microstrip patch antenna. *2017 Devices Integr. Circuit (DevIC)*, 742–746.
9. Zonunmawii, G., Singh, A., Chattopadhyay, L. L. K., and Sim, S. C.-Y.-D. (2021). Reduced-surface-wave-inspired circular microstrip antenna for concurrent improvement in radiation characteristics. *IEEE Antenna Wave Propag. Lett.*, 20(5), 858–862.
10. Zonunmawii, G. A., Singh, L. L. K., Chattopadhyay, S. (2022). Cavity field modulation with modulating circular patch antenna surface: A key to realize reduced horizontal radiation and omni-present improvement in radiation performance. *IEEE Acc.*, 10, 18434–18444.
11. Karmakar, N. C. (2002). Investigations into a cavity backed circular patch antenna. *IEEE Trans. Antenna Propag.*, 50, 1706–1715.
12. Singh, A. K., Gangwar, R. K., and Kanaujia, B. K. (2014). Cavity backed annular ring microstrip antenna loaded with concentric circular patch. *8th Eur. Conf. Antennas Propag. (EuCAP 2014)*, 2155–2158.
13. Zonunmawii, Singh, L. L. K., Chattopadhyay, S., and Ghosh A. (2021). Circular microstrip antenna with shorting walls for improved radiation performance. *Lecture Notes Elec. Engg.*, 937. doi: 10.1007/978-981-19-4300-3_21.
14. Zonunmawii, Singh, L. L. K., Chattopadhyay, S., and Ghosh A. (2022). Rectangular strip loaded circular patch antenna for simultaneous improvement of co polar gain and co polarization to cross polarization radiation separation. *2022 2nd Inter. Conf. Artif. Intell. Sig. Proc. (AISP)*, 1–4. https://doi: 10.1109/AISP53593.2022.9760575.
15. Hamid, S. H., Hock, G. C., Kiong, T. S., and Ferdous, N. (2019). Microstrip patch antenna design in circular topology for ultra high-frequency 900MHz radio spectrum: Size reduction technique and defected ground structure effects. *2019 IEEE Conf. Sustain. Utilization Dev. Engg. Technol. (CSUDET)*, 271–275. htpps://doi: 10.1109/CSUDET47057.2019.9214752.
16. HFSS: High frequency structure simulator, Ver. 13. Ansoft Corp., USA.

37 Low-frequency spectrum-based feature extraction for an automotive starter's winding fault diagnosis

Poulomi Ganguly[1,a], *Tirthankar Datta*[2,b] *and Surajit Chattopadhyay*[3,c]

[1]Electrical Engineering, Modern Institute of Engineering & Technology, Bandel, India

[2]Electronics and Communication Engineering, Meghnad Saha Institute of Technology, Kolkata, India

[3]Electrical Engineering, Ghani Khan Choudhury Institute of Engineering and Technology, Malda, India

Abstract

The starter motor is responsible for providing the mechanical rotating energy for automotive operation. In case of a fault on the starter motor, the vehicle will fail to operate posing problems for the driver. The starter motor and its corresponding components are prone to mechanical as well as electrical faults. In this paper, the focus has been given to the electrical fault of a starter motor and its effect on the operational characteristics of an automobile in a MATLAB Simulink environment. The response characteristics thus obtained are then converted into frequency domain through low-frequency based features analysis to obtain a continuous data set for automobile fault diagnosis. The feature analysis of the response characteristics helps in easier calculation while being effective in designing a system that can help with vehicle diagnosis and quality control.

Keywords: Armature winding fault, automotive starter, DC component, Fast Fourier transform (FFT), fault diagnosis, frequency domain analysis, MATLAB Simulink, motor diagnosis, total harmonic distortion (THD), winding fault diagnosis

1. Introduction

There are various methods and techniques for diagnosing starter motor faults in automobiles and electric vehicles. The different methods such as neuro-fuzzy, learning vector quantization (LVQ), ICA feature extraction, haar wavelet, deep learning and expert systems has used to research in order to detect and classify various fault conditions [1, 5]. These studies contribute to the development of effective diagnostic approaches for improving the performance of starter motors in vehicles. A parameter estimation technique has been suggested based on a system mathematical model for fault detection in brushless DC motor [6]. Markov model was used for the assessment of gear faults [7]. Fourier transform has become a very effective signal analysis [8]. Some more recent experiments include the development of a diagnosis scheme based on a combination of structural a priori knowledge and measured data that can be adapted for different motors like using hall effect [9], parameter of current consumption to determine starter motor efficiency [10], and other methods as researched in refs [11, 16].

Motivation

Due to a short circuit fault on the armature winding of the starter motor the current drawn increases which may not only result in starter operation failure but if allowed to flow for long the current may damage the armature winding of the starter motor [17, 18] (Table 37.1).

Contribution

In this paper, a simple automobile system has been simulated in MATLAB. The response characteristics of the system have been analyzed due to the armature winding fault in the starter motor. The data thus recorded has been then analyzed in the frequency domain through FFT analysis and the result as obtained has been demonstrated graphically.

2. System layout

Here a dynamic model of a simple automobile system has been developed in MATLAB Simulink. The system contains a battery, starter motor, alternator, gear, and starter. Figure 37.1 demonstrates the block diagram representation of the system to be simulated.

3. Vehicle response characteristics

The automotive electrical system has been modeled in MATLAB 2018 Simulink. One sample of the response characteristics and automotive start-up sequence has been shown in Figure 37.2. Maximum current variation is observed in the transient state of the current response curve. Hence the effect of decreasing starter motor armature coil inductance

[a]brinda88@gmail.com, [b]ask.tdatta@gmail.com, [c]surajitchattopadhyay@gmail.com

DOI: 10.1201/9781003540199-37

Table 37.1 Starter Motor specification

Specification Type	Component	Value/Rating
Starter Motor Electrical specifications	Armature inductance	1 mH
	Stall Torque	1 Nm
	No-Load Speed	12000 rpm
	Rated DC Supply Voltage	12 V
	Rotor inertia	0.01m2
Others	Engine Motor	Displacement - 1493 cc, Type – 4 cycles, Turbo charged
	Battery	12 V, 25 AH
	Alternator	12 V, 70 A

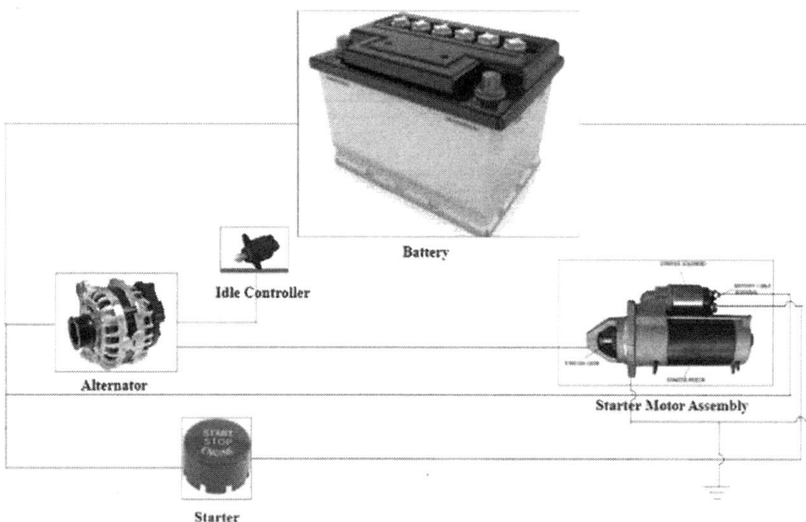

Figure 37.1 Block Diagram of the concerned system

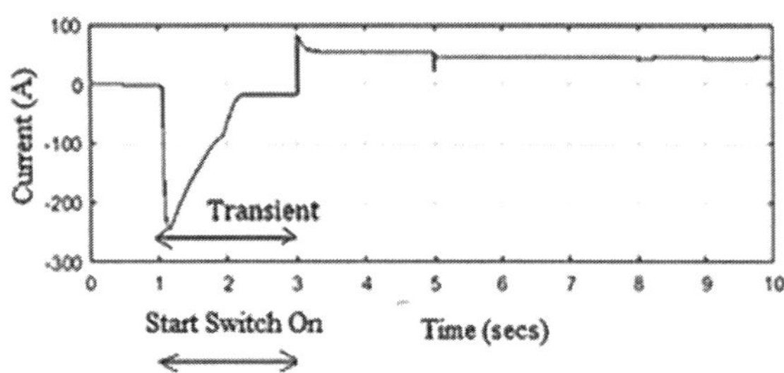

Figure 37.2 Battery current profile automotive electrical system

has been studied based on the transient part of the current response profile.

4. Low frequency-based feature extraction

Fourier analysis of a periodic function denotes the extraction of the series of sines and cosines which when combined will mimic the function. The following equation represents the decomposed form of a period function *f(t)*.

$$f(t) = a_0 + \sum_{n=0}^{\infty} a_n \cdot \cos(\frac{2\pi nt}{T}) + \sum_{n=0}^{\infty} b_n \cdot \sin(\frac{2\pi nt}{T}) \quad (1)$$

Here a_0, a_n, and b_n are Fourier coefficients and are defined as,

$$a_0 = \frac{1}{T}\int_{-T/2}^{T/2} f(t)\, dt \quad (2)$$

$$a_n = \frac{2}{T}\int_{-T/2}^{T/2} f(t) \cos\frac{2\pi nt}{T}\, dt \quad (3)$$

$$b_n = \frac{2}{T}\int_{-T/2}^{T/2} f(t) \sin\frac{2\pi n}{T}\, dt \quad (4)$$

5. Observation

Here FFT analysis of the system response has been done using powergui block in MATLAB Simulink. The parameters used to analyze the effect of starter motor armature coil degradation are %THD, DC component, and lower order harmonic frequency components.

The graphical representation of the DC component (expressed in Amps) variation with degradation of starter motor armature coil (expressed as a percentage of rated coil inductance) is shown in Figure 37.3. The overall variation in the DC component of the system response from healthy starter motor armature coil condition to short-circuited condition is about −0.2%.

The overall increase in %THD for increasing starter armature coil degradation is about 20% from healthy starter motor armature coil to total short circuit. Figure 37.4 demonstrates graphically the variation of % THD concerning starter armature coil degradation. The THD variation is considered as a percentage of the fundamental current (1 Hz in this case).

The graphical representations of harmonic frequency magnitude variation are shown in Figure 37.5. The magnitudes of the harmonic frequencies are expressed as a percentage of the fundamental current magnitude (in this case 1 Hz).

6. Algorithm for fault detection and outcome

From Figures, it is evident that the parameters considered here for monitoring undergo variation with

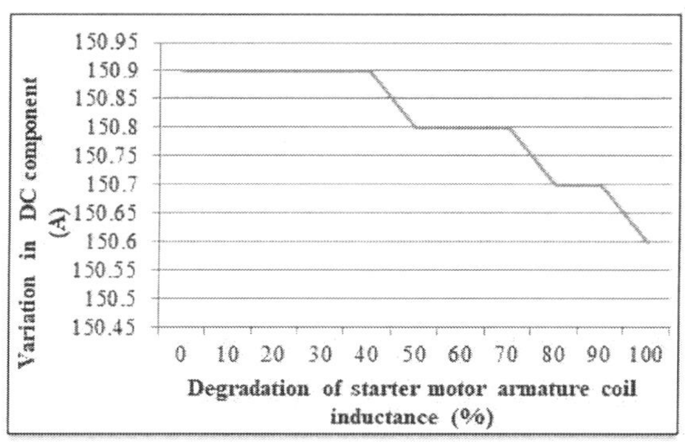

Figure 37.3 Variation in magnitude of DC Component with degradation of starter motor armature coil

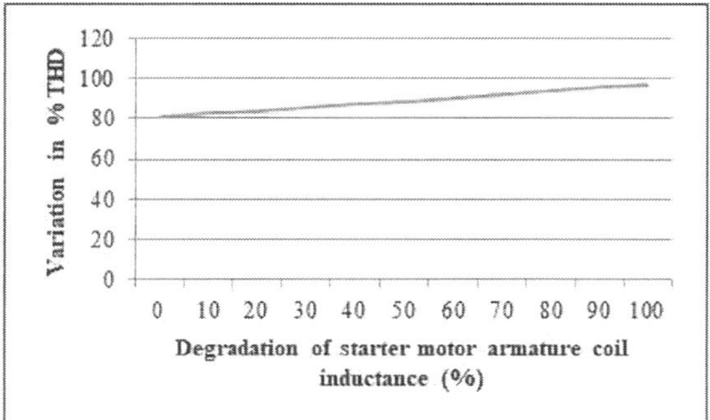

Figure 37.4 Change in THD with degradation of starter motor armature coil

degradation of starter motor coil. Comparison of the different parameter variations with degradation of starter motor armature coil is shown in Table 37.2. The magnitudes of frequencies 2 Hz and higher in Table 37.2. are expressed as a percentage of the magnitude at 1 Hz. Percentage "A" in Table 37.2. represents starter motor armature coil degradation as a percentage of rated motor inductance. The monitoring process to determine starter motor armature coil degradation can be outlined by the following algorithm shown in Figure 37.6.

7. Case Study

7.1. Automobile system

The same type of automotive system as used in simulation has been considered for two cases: (a) with healthy starter motor (b) fully armature damaged starter motor. With the turning on of the starter motor, the current is drawn from the battery to start the engine.

7.2. System characteristics for healthy starter motor

During healthy starter motor armature coil condition some amount of harmonics is present as evident from FFT analysis of the system response. The characteristics data obtained from the FFT analysis of the current response is as shown in Figure 37.7.

7.3. System characteristics for faulty starter motor

FFT analysis of the response signal during a complete short circuit of the starter motor armature coil shows an increase in the magnitude of the harmonic frequency with a reduction in the magnitude of the DC component and the signal at 1 Hz. Figure 37.8. graphically represents the FFT analysis data obtained due to a complete short circuit in the starter motor armature coil.

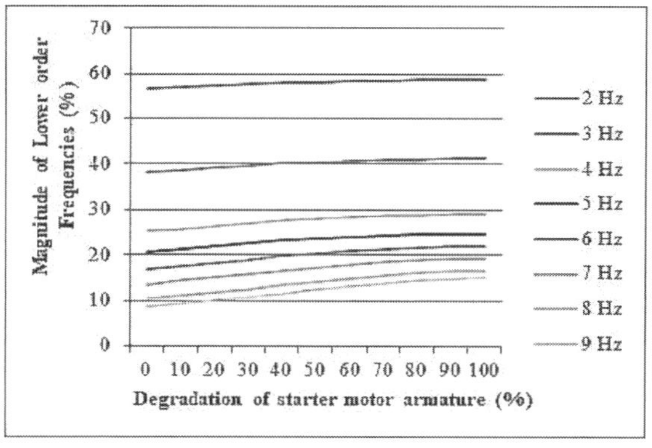

Figure 37.5 Graphical representation of variation in magnitude of harmonic frequencies with degradation of starter motor armature coil

Table 37.2 FFT analysis Parameter variation

A %	THD (%)	1 Hz (A)	2 Hz (%)	3 Hz (%)	4 Hz (%)	5 Hz (%)	6 Hz (%)	7 Hz (%)	8 Hz (%)	9 Hz (%)
0	80.82	80.79	56.46	38.2	25.17	20.68	16.84	13.58	10.38	8.75
10	82.37	80.31	56.85	38.77	25.82	21.33	17.56	14.29	11.04	9.36
20	83.95	79.81	57.21	39.3	26.45	21.96	18.3	15.04	11.75	10.03
30	85.57	79.29	57.54	39.78	27.04	22.57	19.03	15.81	12.51	10.76
40	87.22	78.77	57.82	40.2	27.58	23.13	19.74	16.58	13.3	11.53
50	88.75	78.39	58.06	40.42	27.94	23.6	20.36	17.24	14.04	12.34
60	90.47	77.84	58.27	40.72	28.34	24	20.95	17.93	14.8	13.14
70	92.24	77.27	58.44	40.95	28.66	24.31	21.42	18.52	15.48	13.88
80	94.04	76.69	58.56	41.1	28.88	24.49	21.76	18.96	16.03	14.49
90	95.89	76.1	58.64	41.18	29.01	24.54	21.93	19.23	16.39	14.9
100	97.00	75.6	58.72	41.17	29.05	24.51	21.97	19.29	16.52	15.04

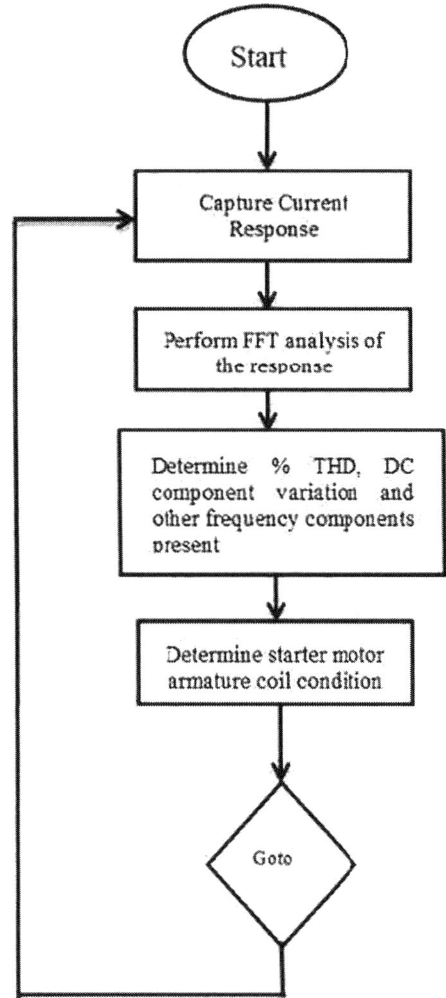

Figure 37.6 Starter motor armature coil degradation monitoring algorithm

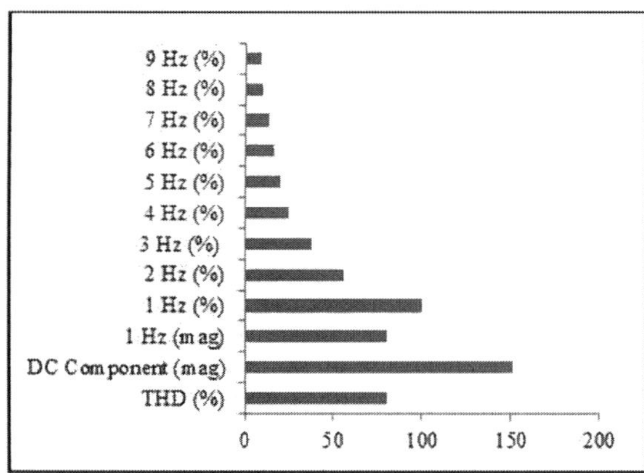

Figure 37.7 Parameters obtained during FFT analysis of the response

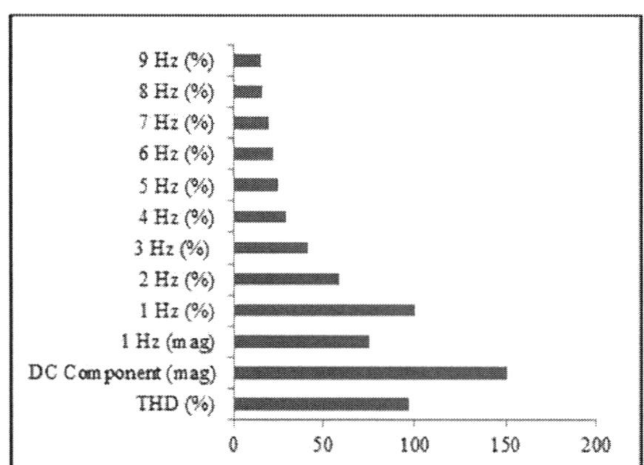

Figure 37.8 Parameters obtained during FFT analysis of response signal

7.4. FFT parameter variation and validation of proposed algorithm

While the %THD and order harmonics magnitude rise with increased degradation of starter motor armature, the value of the fundamental decreases. Figure 37.9. demonstrates graphically the variation difference of complete starter armature coil degradation with respect to normal on each of the parameters showcased in Table 37.2. Here for analysis harmonics up to 9 Hz has been considered for evaluation, as the magnitudes of higher harmonics are of less magnitude. The variation in the DC component is negligible when compared to other components obtained during FFT analysis. The variation in magnitude increases with the analysis of frequencies greater than 2 Hz. As evident from Figure 37.9, the variation in the parameters thus obtained from FFT analysis can be utilized to detect the complete short circuit of the starter motor armature coil. Maximum variation (increase) is noticed in 9 Hz frequency magnitude while %THD undergoes an increase of about 20%.

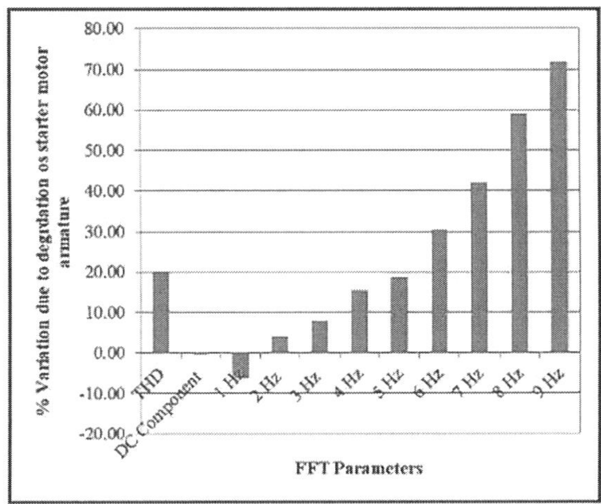

Figure 37.9 Percentage variation of parameters obtained during FFT analysis

8. Comparison with other works

The proposed method discussed in this paper has been compared with other existing methods. Table 37.3 discusses the merits and demerits of some of the methods applied so far to detect starter motor fault.

9. Conclusions

A proper monitoring system should be developed to keep in check the operational characteristics of every vehicular system on board. The paper focuses on studying the effect of the starter motor armature coil on automobile characteristic response. We have analyzed the data using the Fast Fourier Transform (FFT) method. The obtained data can be utilized as training data for developing monitoring algorithms that can accurately identify the conditions of the starter motor armature coil. By utilizing this data, it becomes possible to create an online monitoring algorithm that can assess automobile operation, particularly in dynamic conditions where operating characteristics may vary. Such an algorithm would aid in detecting any issues or anomalies related to the starter motor armature coil, allowing for timely maintenance or repairs.

Table 37.3 Comparison of different methods of starter motor fault diagnosis

Fault diagnosis method	Merit	Demerit
Learning vector quantization network [4]	1. Help detect deterioration of starter motor 2. User friendly	Has an accuracy of 91%.
Fuzzy Logic Algorithm [18]	1. User friendly 2. Can detect multiple faults	Has an accuracy of 90%.
Hall effect [9]	1. Easy diagnosis 2. Provides more information	It's more complicated construction-wise.
Current consumption [10]	1. Can successfully categorize operation of ICE with the faulty starter motor 2. Help classify different types of faults based on the current consumption of starter motor concerning mileage	Doesn't discuss the ambient temperature effect and the dependence of current consumption on it
Proposed method	1. Easy classification of faulty and normal conditions 2. It can indicate the percentage of fault 3. Accuracy of fault existence is 100%; fault measure is 97% accurate 4. A very small percentage of faults can be detected 5. Remote monitoring is possible	Other types of fault detections have not been done yet that may be carried out as future extensions

References

1. Midya, M., Ganguly, P., Datta, T., and Chattopadhyay, S. (2023). ICA-feature-extraction-based fault identification of vehicular starter motor. *IEEE Sens. Lett.*, 7(2), 1–4. 10.1109/LSENS.2023.3242814.
2. Ganguly, P., Chattopadhyay, S., and Datta, T. (2022). Haar wavelet-based statistical scanning of turn fault in vehicular starter motor. *IEEE Sen. Lett.*, 6(4), 1–4. 10.1109/LSENS.2022.3158608.
3. Xu, X., Qiao, X., Zhang, N., Feng, J., and Wang, X. (2020). Review of intelligent fault diagnosis for permanent magnet synchronous motors in electric vehicles. *Adv. Mec. Engg.*, 12(7), 1–14. 10.1177/1687814020944323.
4. Bayir, R. (2008). Condition monitoring and fault diagnosis of serial wound starter motor with learning vector quantization network. *J. Appl. Sci.*, 8(18), 3148–3156. 10.3923/jas.2008.3148.3156.
5. Dominik, F. and Rolf, I. (2000). Hierarchical motor diagnosis utilizing structural knowledge and a self-learning neuro-fuzzy scheme. *IEEE Trans. Indus. Elec.*, 47(5), 1070–1077. 10.1109/41.873215.
6. Olaf, M. and Rolf, I. (2000). Application of model-based fault detection to a brushless DC motor. *IEEE Trans. Indus. Elec.*, 47(5), 1015–1020. 10.1109/41.873209.
7. Sajjad, S., Zaidi, H., and Aviyente, S. (2011). Prognosis of gear failures in DC starter motors using hidden Markov models. *IEEE Trans. Indus. Elec.*, 58(5), 1695–1706. 10.1109/TIE.2010.2052540.
8. Cooley, J. W. and Turkey, J. W. (1965). An algorithm for the machine calculation of complex Fourier series. *Math. Comput.*, 19(90), 297–301. https://doi.org/10.2307/2003354.
9. Dziubi´nski, M., Litak, G., Drozd, A., Szydło, K., Longwic, R., and Wolszczak, P. (2018). Using the hall effect for monitoring the starter condition in motor vehicles. *Appl. Sci.*, 8(5), 1–12. 10.3390/app8050747.
10. Puzakov, A., Dryuchin, D. A., Voinash, S. A., Ariko, S. Y., Kamenchukov, A. V., and Ukraimsky, I. S. (2020). Diagnostics of automobile starters by the parameter of current consumption. J. Phy. Conf. Ser. 1679(4), 1–6. 10.1088/1742-6596/1679/4/042043.
11. Ghosh, S. S., Chattopadhyay, S., and Das, A. (2022). Snubber circuit fault detection in inverter switch of brushless DC motor. *IET Cogn. Comput. Sys.*, 14(1), 31–44. https://doi.org/10.1049/ccs2.12041.
12. Kar Ray, D., Roy, T., and Chattopadhyay, S. (2021). Switching transient-based state of Ampere-hour prediction of lithium-ion, nickel-cadmium, nickel-metal-hydride and lead acid batteries used in vehicles. *IET Nanodielectr.*, 4(3), 121–129. https://DOI.org/10.1049/nde2.12017.
13. Kar Ray, D., Roy, T., and Chattopadhyay, S. (2021). Skewness scanning for diagnosis of a small inter-turn fault in Quadcopter's motor based on motor current signature analysis. *IEEE Sens. J.*, 21(5), 6952–6961. 10.1109/JSEN.2020.3038786.
14. Kar Ray, D., Rai, A., Khetan, A. K., Mishra, A., and Chattopadhyay, S. (2020). Brush fault analysis for Indian DC traction locomotive using DWT-based multi-resolution analysis. *J. Instit. Eng. (India) Ser. B*, 101(8), 335–345. 10.1007/s40031-020-00468-3.
15. Chattopadhyay, S. (2022). Nanogrids and Picogrids and their Integration with electric vehicles. *ET, London*. pp. 1-367, 2022. ISBN: 978-1-83953-482-9.
16. Chattopadhyay, S., Mitra, M., and Sengupta, S. (2011). Electric Power Quality. pp. 1-177, Springer: Netharland. ISBN: 978-94-007-0635-4.
17. Murugesan, V. M., Chandramohan, G., Senthil Kumar, M., Rudramoorthy, R., Ashok Kumar, L., Suresh Kumar, R., Basha, D., and Vishnu Murthy, K. (2012). An overview of automobile starting system faults and fault diagnosis methods. *ARPN J. Engg. Appl. Sci.*, 7(7), 812–819.
18. Indu, K. and Aswatha Kumar, M. (2020). Electric vehicle control and driving safety systems: A review. *IETE J. Res.*, 6(6), 1–17. 10.1080/03772063.2020.1830862.

38 Using RFID technology, a cutting-edge system for traffic infringement control and punishment

Geetha Rani E.[1,a], *Bhuvaneshwari P.*[2,b], *Ch Suguna Latha*[3,c], *Jaya Lakshmi D.*[4,d], *Sethu Madhavi*[5,e], *D. Anusha*[6,f], *Dhanwand Krishna K.*[7,g] *and J. Nageswara Rao*[8,h]

[1]Assistant Professor, Department of CSE, Alliance University, Bengaluru, India

[2]Associate Professor, Department of CSE, Manipal Institute of Technology Bengaluru, Manipal Academy of Higher Education, Manipal, India

[3]Assistant Professor, Department of CSE, NRIIT, Andhra Pradesh, India

[4]Professor, Department of CSE, Saveetha school of engineering, SIMATS, Chennai, India

[5]Assistant Professor, Department of CSE, RVCE, Bengaluru, India

[6]Associate Professor, Department of CSE, SRKIT, Andhra Pradesh, India

[7]UG Student, Department of CSD, MVJCE, Bengaluru, India

[8]Associate Professor, Department of CSE, LBRCE, Andhra Pradesh, India

Abstract

A growing number of new automobiles entering the road are rising quickly, which leads to extremely congested roads. Few drivers disobey traffic laws by running red lights because they can wait so long. Due to the significant increase in traffic accidents caused by this, it is crucial to automatically identify vehicles that have violated the signal. IoT-based traffic violation detection systems can send information regarding offending automobiles to the relevant traffic law place automatically, allowing for immediate punishment of the offending vehicle. The proposed system could be implemented by utilizing an already installed vehicle ID card. If a vehicle moves forward while a red light is on, vehicle information such as the registration number, type of vehicle (two-wheeler, car, auto, truck, etc.), owner's name, vehicle color, and so on can be automatically sent to the relevant authorities via concerned mobile phones. All vehicles must be outfitted with these low-cost wireless ID cards so that vehicle data can be continuously transmitted to demonstrate the concept practically. The demonstration module includes a mock road and a miniature automatic traffic signal post. The system is set up so that if any vehicles cross zebra lines while the red signal is activated and during this time, the system will receive this information and transmit it via a Wi-Fi module. Nearby traffic police will be notified right away if anyone disregards the traffic signal. Because it regrets collecting vehicle data while there are yellow and green signals, the main processing unit must be installed close to traffic signals. One model car that will have a wireless ID card is included in the demo module.

Keywords: Vehicle ID, RFID, wireless ID, traffic laws, E-challan

Introduction

In India, the number of automobiles has expanded intensely, which has resulted in an increase in traffic accidents. The government has taken several attempts to reduce these accidents, but these initiatives have repeatedly failed, and residents' violations of traffic laws have escalated. One aspect of the cooperation is the imposition of penalties for individuals who break traffic laws, particularly signal jumps. The concept provided in this title will help to reduce road accidents and avoidance. The final accident determination is then made based on the trio main parameters for traffic accidents, which are alterations in the moving automobiles' pose, rushing, plus direction. The offender speaks with traffic police. At intersections, crosswalks, and other locations, signaling devices called traffic lights and traffic signals are used to control traffic flow. Traffic lights adhere to a universal color code with an arrangement of three light bulbs or LEDs in a standard color. Only two of the World Health Organization's (WHO) seven vehicle security recommendations are met by India in terms of vehicle safety. About 25% of complete street crashes passing involve bikes. According to crash records, about 75% of motorbike riders involved in accidents

[a]geetha.edupuganti@alliance.edu.in, [b]bhuvaneshwari.p@manipal.edu, [c]sugunachitturi@gmail.com, [d]jayalakshminandakumar2@gmail.com, [e]rallapalli.sethumadhavi@reva.edu.in, [f]d.anushasrk@gmail.com, [g]1mj21cg023@mvjce.edu, [h]nagsmit@gmail.com

DOI: 10.1201/9781003540199-38

continued to wear head protection. Authorities have a difficult time determining whether there has been a violation of traffic laws, which creates devastating conditions in which it becomes hazardous for both drivers and foot-travelers which is diverse, large, and developing quickly, governance is a challenge. India requires brand-new, cutting-edge technology to carry out massive transformation and put government plans into action. Despite being one of the world's fastest-growing developing economies, India's equitable growth is still a vital requirement. Traffic laws are frequently broken and not adhered to, which is one of the main causes of the significant number of accidents on roadways. India requires a highly regulated, impenetrable system for governance to stop these preventable accidents and control traffic. A system that forces people to drive safely and in compliance with all regulations. E-challan is that well-regulated, impenetrable system. E-challan represents an online system for e-governance that helps both traffic managers and drivers manage penalties for moving violations. E-challan offers a variety of assistance required for managing and overseeing traffic fines.

A. Motivation

The significance of the key information in this text is traffic control in urban planning and public safety, the economic benefits of improved traffic flow, the technological advancements that are transforming traffic control systems, and the integration of traffic control systems with smart city plans. Traffic control is essential for urban planning and public safety, as it reduces accidents, minimizes congestion, and improves traffic flow. Economic benefits include more efficient commuting times, reduced fuel consumption, and diminished air pollution. Technological advancements include adaptive traffic signals, smart parking systems, and systems that reduce speed and accidents. Integration with smart city plans includes sensors and cameras being used in traffic management, enforcements, and accident monitoring. The key information in this text is the role of traffic violation plus penalty systems in ensuring compliance with traffic rules and regulations, automated system solutions, public opinion and community feedback, international and national policies, and how the proposed traffic control system lines up with these policies. Traffic violation and penalty systems are used to ensure compliance with traffic rules and regulations. Automated system solutions are used to monitor and ticket vehicles automatically. Public opinion and community feedback is crucial to the success of traffic control systems. International and national policies govern traffic control and have led to improved traffic management in cities. The proposed traffic control system should align with these policies.

Literature survey

This study illustrates an automated number plate recognition (ANPR) system that uses digital images to extract number plates to identify vehicles that disobey traffic signals. Additionally, it immediately sends the actual owner of the automobile a violation SMS.[6]. The proposed system enters the vehicle's captured image. The number plate region is extracted using image segmentation techniques. The segmented image is compared to the template images using the correlation technique. [7] The resulting information can be utilized to make comparisons with the database's records [1]. Recent studies have revealed that as a result, there are 35 accidents for every 1000 vehicles in India. [8] The surveillance of traffic signal violations and their associated penalties are highlighted in this manuscript. Therefore, in this study, the benefits of the two technologies are taken advantage of while using both techniques at once to avoid the drawbacks like tampering with the license plate or hiding the Fastrack tag, etc. [10] This system employs GSM technology to notify the vehicle owner of the fine that has been imposed [2]. Recent studies have revealed that as a result, there are 35 accidents for every 1000 vehicles in India. Technology can play a significant role in overcoming these violations, which is why this work focuses on IoT and image processing-based methods. [11] The surveillance of traffic signal violations and their associated penalties are highlighted in this manuscript. [12] Therefore, in this study, the benefits of the two technologies are taken advantage of while using both techniques at once to avoid the drawbacks like tampering with the license plate or hiding the Fastrack tag, etc. This system employs GSM technology to notify the vehicle owner of the fine that has been imposed by doing this [3]. Given the significance of the outcomes achieved through deep learning concepts in AI and computer vision, the practical use of such methods for motorcycle detection and tracking is also examined. The urban motorbike dataset (UMD), which contains statistical evaluation results for various detectors, is introduced in the paper's conclusion. [13] The paper also discusses the challenges that lie ahead and offers a set of recommendations for further research in this developing field [4]. We concentrate our efforts on categorizing such violations within Indian cities as India is shifting towards automated methods to reduce violations of traffic laws and road accidents. Using computer-controlled, we describe the traffic violations in this work. We assembled a thorough dataset of more than 6 million E-challans to investigate this. In Ahmedabad, 57% of distinctive vehicles have been involved in repeat offences, according to data on how fines are paid. [14] A significant difference between the electronic challans handed down during the festivals is revealed by the temporal analysis. According to our research, severe penalties might not have a long-term effect on reducing traffic infractions [5]. Entity of new motor vehicles on the road contributes to severely congested roads and gives people license to break traffic laws. The system is additionally accurately optimized. As is already known, traffic police are the ones who photograph people who violate traffic laws, but they are unable to photograph multiple violations at once. The proposed method is faster and more effective than humans. The suggested system uses computer vision techniques to detect the most common traffic infractions in real-time and achieves good results with high accuracy [9]. One of the main issues in today's world is rush hour traffic jams. Emergency-purpose vehicles including ambulances, police cars, and fire trucks become backed up during rush hour. Because of this, human lives are lost as these emergency automobiles are unable to arrive at their destinations in a timely manner. We have created a system that allows us to give any emergency vehicle clearance by turning all red lights on its path to green, effectively giving the desired vehicle a full green wave.

Methodology

A. Problem statement

The principal objective of the system is to ensure accurate and reliable detection of vehicles that violate traffic signals. This is a critical issue that needs to be addressed promptly to prevent potential accidents, congestion, and chaos on the roads. The implementation of specialized devices and technologies for traffic violation detection becomes crucial in achieving this goal. With the system in place, the enforcement of traffic rules becomes more robust and proactive. It continually reinforces compliance with traffic regulations and actively identifies those who fail to abide by them. By promptly detecting violations, the system plays a pivotal role in maintaining order and safety on the roads. Real-time traffic violation detection is essential due to the constant vigilance of traffic authorities. The system allows for continuous monitoring of the streets, enabling traffic enforcers to stay abreast of the evolving traffic situations. This real-time capability empowers them to not only ensure the implementation of safe highways but also do so efficiently. The technology utilized in this system surpasses human observation in terms of speed and accuracy. By employing advanced algorithms and sensors, it can swiftly detect and flag instances of traffic rule violations. This capability significantly reduces the reliance on manual observation, which is prone to human errors and limitations. The system's ability to detect traffic signal breaches in real time is a game-changer. It enables authorities to promptly respond to violations, deploy necessary resources, and take appropriate actions. This proactive approach enhances overall traffic management and contributes to a safer and more orderly road environment.

B. Existing system

Recent studies have revealed that as a result, there are 35 accidents for every 1000 vehicles in India. The surveillance of traffic signal violations and their associated penalties are highlighted in this manuscript.

C. Demerits of existing system

- The existing system requires scanning the objects manually and this leads to an increase in the time consumed.
- Requires extra equipment and does not inform the user about the fines.
- Old algorithms with less image resolving capability and it require a lot of human intervention.

D. Proposed system

Using RFID to read any identified tags attached to vehicles, the initial phase is represented by data collection. Then, using an internet network communication system, some data is sent to the server. It will aid in lowering the manual labor required of the police. It will function as an automated system that maintains traffic regulations and fines offenders. The system will enforce traffic regulations, which will encourage people to drive safely. It will aid in reducing the chaos brought on by traffic law violations. Additionally, it will make pedestrians safer. The penalties will be fair and just and proportionate to the severity of the violation. Penalties will be imposed automatically via the ANPR system without human intervention.

Monitoring and enforcement: The ATCPS will be monitored and enforced using a combination of on-road patrols and automated systems. The on-road patrols will be used to monitor traffic, ensure compliance with traffic rules, and help in the event of accidents or other emergencies.

E. Advantages of proposed system

- Working of timer for the signal to turn green.
- Does not require human intervention and low-cost device and the fine details will be informed to the person via message/mail.

- Tracks the vehicles that cross the line and the data received are pretty accurate and are believable and the working model can be installed easily.

F. Vehicle detection

The methodology of a cutting-edge system for regulating and penalizing traffic infractions using radio frequency identification (RFID) technology involves a systematic approach to effectively monitor and enforce traffic regulations. This methodology integrates RFID technology into the existing traffic control infrastructure to enhance the accuracy, efficiency, and reliability of traffic violation detection and penalty management. The first step in this methodology is the installation of RFID readers at strategic locations, such as traffic signals, toll booths, and designated checkpoints. These readers are capable of wirelessly communicating with RFID tags or transponders installed in vehicles. The RFID tags can be affixed to windshields or license plates and contain unique identification information associated with each vehicle. Once the RFID readers are in place, the system continuously scans the vicinity for nearby vehicles equipped with RFID tags. As a vehicle with an RFID tag passes through a monitored area, the reader captures the tag's identification information, including the vehicle's license plate number or a unique identifier linked to the vehicle owner.

Implementation

Implementation must be clearly defined, planned carefully and systematically, otherwise it causes confusion and leads to generations of problems. Analyzing the shortcomings of current approaches and providing a solution by taking the fundamental necessities for our suggested approach into account. Considering the system's hardware specifications. We must now look at the software requirements after considering the hardware requirements. Depending on the microcontroller we choose, there are various pieces of programming, compiling, and debugging software available. Based on our requirements, we must write the source code for the proposed system, compile it, and debug it in the software. After completing all the hardware and software requirements, we must combine them to function on our system. To do this, we must write our source code onto a microcontroller. Once this is done, we must connect all the inputs as well as the output modules to the microcontroller in accordance with our needs. The implementation of the traffic violation detection system prototype focuses on effectively detecting and monitoring signal violations. The system incorporates traffic lights, which are universally recognized signaling devices that manage traffic flow at intersections and crossings, red indicating a prohibition on movement, and amber or yellow warning of an impending change to red. Despite these standardized signals, many individuals neglect to follow them due to reasons such as impatience and negligence. The working design aims to identify and seize moving violations of traffic signals to address this problem. Each vehicle is equipped with a vehicle identification card, as well as a wireless vehicle identification card reader is placed close to a traffic signal post. Overall, the implementation of this traffic violation detection system prototype automates the process of monitoring traffic and identifying violations, enabling traffic police to efficiently enforce traffic regulations. By accurately detecting signal violations and transmitting vehicle information in real-time, the system aims to improve road safety and minimize the serious repercussions of traffic rule breaches.

Results

RFID technology is used for efficient identification of violators and penalty issuance. By scanning the RFID tag of a vehicle involved in a violation, authorities can retrieve the necessary details about the owner, registration, and violation history. This simplifies the process of identifying and penalizing violators. Additionally, the RFID data, combined with violation details, can be used to automatically generate penalty notices or tickets. The penalties can then be sent to vehicle owners through email, SMS, or a dedicated mobile app, streamlining the penalty management process. By leveraging RFID technology in these ways, the system ensures accurate identification of violators and efficient issuance of penalties, contributing to effective traffic violation control. The utilization of RFID technology also enables seamless data integration and sharing (Figure 38.1). The integration of the EM-18 sensor enhances the system's capability to detect and track vehicles in real-time, ensuring effective enforcement of traffic regulations. Figure 38.1 improves the accuracy and efficiency of the violation control process by providing reliable and timely identification of vehicles involved in potential violations. In summary, the EM-18 sensor plays a pivotal role in the modern traffic violation monitoring, control, and penalty system using RFID by serving as an RFID reader module. It enables the system to detect the presence of RFID tags on vehicles, contributing to seamless vehicle identification and tracking, and enhancing the overall effectiveness of the system in enforcing traffic regulations and ensuring road safety.

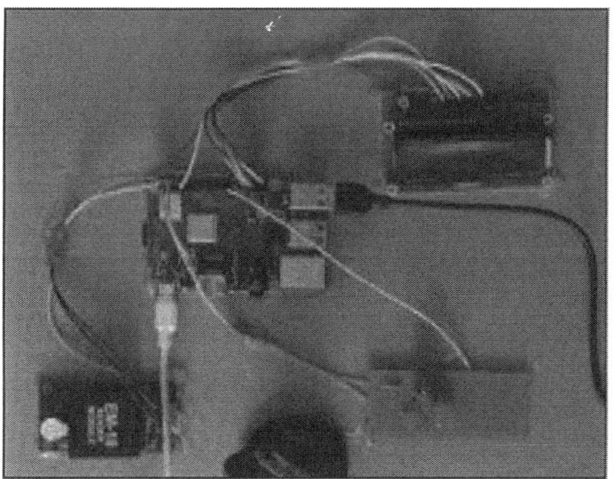

Figure 38.1 EM-18 module

Figure 38.2 Traffic signals attached with RFID scanner

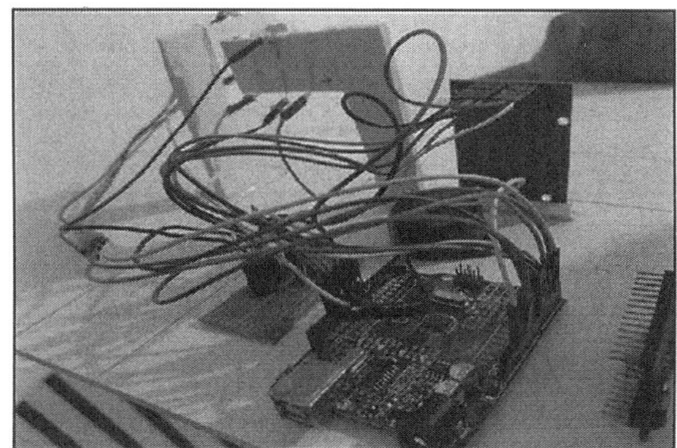

Figure 38.3 Final stage of prototyping

38.By incorporating traffic signals into the system, several key functionalities are achieved, enhancing the overall effectiveness of traffic regulation enforcement and penalty management (Figure 38.2). They regulate the flow of traffic, ensuring orderly and safe movement of vehicles at intersections and along roadways. By synchronizing the timing and sequencing of traffic signals, the system can optimize traffic flow, reduce congestion, and minimize the likelihood of violations. Additionally, traffic signals contribute to the overall efficiency of the penalty system. They provide critical information about the timing and sequencing of traffic signals during a violation, which can be correlated with violation data to determine the severity and duration of the violation. This information assists in calculating appropriate penalties and ensuring fair enforcement. Traffic signals play a crucial role in modern traffic violation monitoring, control, and penalty system using RFID. They serve as control points, capture evidence, collect data, facilitate real-time communication, contribute to penalty calculation, and provide insights for traffic management. Integrating traffic signals into the system ensures effective enforcement of traffic regulations, enhances road safety, and enables efficient penalty management.

38.The final stage of prototyping the project involves several key activities to refine and test the prototype before its implementation (Figure 38.3). During this stage, the prototype undergoes a series of refinements based on feedback received during earlier stages. The user interface, system performance, and functionalities are optimized to align with stakeholder requirements. Integration with other components of the system, such as RFID infrastructure and penalty management systems, is completed to ensure seamless communication and data exchange. Thorough testing is conducted, including performance testing to evaluate the system's responsiveness and scalability, as well as user acceptance testing to assess usability and functionality from the users' perspective. The prototype is also subjected to a security assessment to identify and address any vulnerabilities. The efforts invested in this stage ensure that the system meets the required performance standards, user expectations, and compliance with regulations, laying a solid foundation for successful implementation and operation.

Conclusions and future scope

As a result, a new system has been created to reduce traffic law breaches, which might result in more orderly traffic around the country. Drivers who break the law will now face automatic penalties, which will help to some extent minimize several problems including accidents, traffic congestion, and even pollution. The new device will only track traffic at signal poles, but it may also be used to keep an eye on one-way streets and no-entry zones.

We intend to establish order and discipline in the country's traffic system using this cutting-edge technology, making it a safer and more effective method of transportation. The research suggests a background subtraction technique for detecting moving objects (vehicles) and determining the sensitivity of the RFID tag. The calculation of the various frames is successful. When different traffic tracking movies are integrated into the application, the automobiles on both the traffic model used for simulation and the real road condition can be seen. Additionally, it's important to consider brightness variations and the intended detecting conditions when a fixed camera is keeping track of the traffic situation. Once the issue is resolved, it will be easier to accurately identify traffic violations. However, neither the development of smart cities nor the collection of data is aided by the Macau privacy law. For academic purposes, it is therefore challenging to record a traffic situation as it occurs in real time.

References

1. Srinivas Reddy, P., Nishwa, T., Shiva Kiran Reddy, R., Sadviq, Ch, and Rithvik, K. (2021). Traffic rules violation detection using machine learning techniques. *Proc. 6th Inter. Conf. Comm. Elec. Sys. (ICCES)*, ISBN: 978-0-7381-1405-7.
2. Shukla, S. (2023). Real-time monitoring and predictive analytics in healthcare: harnessing the power of data streaming. *Inter. J. Comp. Appl.*, 185, 32–37. doi: 10.5120/ijca2023922738.
3. Shukla, S. (2023). Streamlining integration testing with test containers: Addressing limitations and best practices for implementation. *Inter. J. Latest Engg. Manag. Res. (IJLEMR)*, 9, 19–26. doi: 10.56581/IJLEMR.8.3.19-26.
4. Shukla, S. (2019). Examining Cassandra constraints: Pragmatic eyes. *Inter. J. Manag. IT Engg.*, 9(3), 267–287.
5. Shukla, S. (2019). Data visualization with Python pragmatic eyes. *Inter. J. Comp. Trends Technol.*, 67(2), 12–16.
6. S. Shukla. (2019). Debugging microservices with Python. *IIOAB J.*, 10, 32–37.
7. Ruben, J F. and Mohana. (2020). Traffic signal violation detection using artificial intelligence and deep learning. *Proc. Fifth Inter. Conf. Comm. Elec. Sys. (ICCES)*, ISBN: 978-1-7281-5371-1.
8. Vyshali Rao, K. P., Srividya, R., and Dhanalakshmi, M. (2023). Reliable informational data and secured deviation notification over networks using IoT. *2023 Inter. Conf. Recent Trends Elec. Comm. (ICRTEC)*, 1–6.
9. Shukla, S. (2023). Unlocking the power of data: An introduction to data analysis in healthcare. *Inter. J. Comp. Sci. Engg.*, 11(3), 1–9.
10. Shukla, S. (2022). Developing pragmatic data pipelines using apache airflow on Google cloud platform. *Inter. J. Comp. Sci. Engg.*, 10(8), 1–8.
11. Shukla, S. (2023). Exploring the power of apache kafka: A comprehensive study of use cases suggest topics to cover. *Inter. J. Latest Engg. Manag. Res. (IJLEMR)*, 8, 71–78. doi: 10.56581/IJLEMR.8.3.71-78.
12. Shukla, S. (2023). Enhancing healthcare insights, exploring diverse use-cases with K-means clustering. *Inter. J. Manag. IT Engg.*, 13, 60–68.
13. Huang, S., Sun, D., Zhao, M., Chen, J., and Chen, R. (2022). Short-term traffic flow prediction approach incorporating vehicle functions from RFID-ELP data for Urban Road Sections. *Institut. Engg., Technol. (IET)*. doi: 10.1049/itr2.12244.
14. R. J. Franklin and Mohana, "Traffic Signal Violation Detection using Artificial Intelligence and Deep Learning," *2020 5th International Conference on Communication and Electronics Systems (ICCES)*, Coimbatore, India, 2020, pp. 839-844, doi: 0.1109/ICCES48766.2020.9137873.

39 Regression algorithm for decision trees and gradient boosting technique for eye-controlled virtual mouse

Geetha Rani E.[1,a], *Bhuvaneshwari P.*[2,b], *O. Venkata Siva*[3,c], *J. Nageswara Rao*[4,d], *Ch. Suguna Latha*[5,e], *D. Anusha*[6,f], *M. Dhana Lakshmi*[7,g] *and Dhanwand Krishna K.*[8,h]

[1]Assistant Professor, Department of CSE, Alliance University, Bengaluru, India

[2]Associate Professor, Department of CSE, Manipal Institute of Technology Bengaluru, Manipal Academy of Higher Education, Manipal, India

[3]Assistant Professor, Department of CSE, LBRCE, AP, India

[4]Associate Professor, Department of CSE, LBRCE, AP, India

[5]Assistant Professor, Department of CSE, NRIIT, AP, India

[6]Associate Professor, Department of CSE, SRKIT, AP, India

[7]Professor, Department of CSE, NHCE, Bengaluru, India

[8]UG Student, Department of CSD, MVJCE, Bengaluru, India

Abstract

This paper will go over how we can manipulate with our eye movements the mouse pointer. It uses a webcam to identify faces, facilitating easier human-computer interaction for individuals with physical disabilities. Then, we use the decision tree algorithm to track the person's eye movement. By blinking their eyes, the apps can be opened and closed by the user and carry out mouse functions without a real mouse. For most people with physical disabilities, human-computer interaction (HCI) requires the use of eye movements and hand gestures as a dependable way of input. To be able to increase the effectiveness then use of using eye tracking in human-computer interaction, a new eye control technique in this system is suggested that uses a webcam lacking somewhat additional computer hardware. The suggested system's primary objective is to offer a straightforward but practical interactive method that only uses the handler's eyes. The wished-for makes usage of iris also moves the pointer according to where the iris is, using Python and a webcam to manipulate the screen's cursor.

Keywords: Decision tree algorithm, mouse operations, mouse cursor, human-computer interaction

Introduction

The design of computers makes them easy for the average person to use. But utilizing a data center can be especially hard for those who have bodily impairments like cerebral palsy. To enhance how users interact with computers, extensive research has been studied on human-computer interaction (HCI). Nearly all of these, sadly, apply just to common people. [1-3] This covers technologies like touch-sensitive screens and voice recognition. There are numerous others. These methods were efficient, but they weren't suitable for people with physical disabilities. We employ a straightforward mechanism. Additionally, the operator is not required to physically engage with the framework. [4] For those with physical limitations, this eye-tracking workaround is straightforward. All that is required is a computer or laptop that has an integrated webcam. The following output frame shows the execution: To enable physically disabled people to use computers independently, an eye-controlled virtual mouse cursor has been developed. allowing them access to and use of contemporary computers. [5-8] To move from making observations about an item to drawing conclusions about the item's target value, the decision tree technique uses a decision tree as a predictive model. With observations acting as branches and values as leaves, this method is applied in machine learning, data mining, and statistics. In this kind of tree structure, the branches stand in for the combinations of traits that give rise to class labels; the leaves, however, stand for category names. [9] Models of trees called classification trees have as their target variable a limited range value set. Using a decision

[a]geetha.edupuganti@alliance.edu.in, [b]bhuvaneshwari.p@manipal.edu, [c]orsuvenkatasiva@gmail.com, [d]nagsmit@gmail.com, [e]sugunachitturi@gmail.com, [f]d.anushasrk@gmail.com, [g]mdhanalak72@gmail.com, [h]1mj21cg023@mvjce.edu.in

DOI: 10.1201/9781003540199-39

tree is also called a regression tree when the values of the target variable are continuous. A decision tree in data mining is an effective visual representation of decision-making and the procedures involved in making decisions, the data is represented by a decision tree, nonetheless, the classification tree generated can serve as decision-making input. The sole objective of this research is to create a structure that can classify in well-lit surroundings, motions automatically. [10] We are creating a synchronous gesture in a real time recognition system to achieve this objective. The objective of this project is to develop a comprehensive system that uses computer vision to recognize, pick up on, and describe eye motion. This structure will aid in the visualization of user interaction, artificial intelligence, and computer vision. [11-14] It creates an eye motion recognition function based on several arguments. The main goal of the structure is to simplify it such that operators are accessible, and simple to control. The objective of this research is to create a thorough system that uses computer vision to recognize, detect, and explain eye motion. This framework will aid in the visualization of user interaction, artificial intelligence, and computer vision. It creates an eye motion recognition function based on several arguments. The main goal of the structure is to make it straightforward, user-friendly, and manageable.

Literature survey

Initially, personal computers were used for text processing and mathematical problem-solving. However, in recent years, computers have become indispensable for all our daily tasks. Shopping, socializing, and internet entertainment are examples of work-related and personal activities. [15] Computers are built to be user-friendly for the public. individuals with significant physical disabilities, such as amyotrophic lateral sclerosis or cerebral palsy, on the other hand, using a computer can be extremely difficult. Human-computer interface (HCI) research is being conducted. [16] A screen that responds to touch and using speech recognition software is among the interfaces available, and a variety of other approaches. [17-19] Despite the success of people with physical disabilities were unable to use these approaches. Computers are built to be user-friendly for the public. Individuals who have serious physical disabilities, such as amyotrophic lateral sclerosis or cerebral palsy, on the other hand, utilizing a computer can be very challenging. [20] Human-computer interface (HCI) research is being conducted to increase the person's interaction with the computer system. Furthermore, the user is not required to physically engage the system. As a result, eye tracking is used to provide a straightforward fix for people who are bodily disabled. We only require a computer or laptop with a built-in webcam or a pre-installed webcam. The eye-controlled virtual mouse cursor is primarily intended to assist physically disabled people in using computers on their own and to provide them with the opportunity to use and work with modern computers. [21-23] Many methods have previously been developed to assist the disabled in working signals from the brain and facial muscles, such as electrooculogram (EOG) and electroencephalography (EEG), which are used on computers. Using image processing techniques, For the backdrop, we convert the image to grayscale. removal and to detect the presence of the frame's eye edges and counter. Hand motions are an automatic and natural way of communicating dependable mode of communication for HCI. Computer input devices like examples include the touch screen, joystick, keyboard, and mouse, but they do not offer a suitable interface. The hand gestures used in the current system, however, are typically produced with a webcam's assistance that captures images of the hands, while wearing data gloves, or both. It will be interfaced by a desktop or laptop. The initial stage of gesture recognition procedure is the capture and evaluation of the hand. using data-glove methods to initialize programmable hand gestures and finger movement using sensors. The technique based on vision, in contrast, does not call for any additional hardware to detect human-computer interaction; all that is needed is a camera. The method's drawbacks include the consistent background, the occasional presence of people, and the lighting. The numerous procedures and algorithms used in this methodology are described below along with the popularity strategies. Segmentation is the search for a connecting region within an image that meets certain criteria, such as color or concentration, and where an algorithm is present modifiable. It is difficult to develop recognition systems that work well in several contexts; nonetheless, it is further doable because these difficulties really exist. There are various compound backgrounds and lighting effects in addition to constraining several scaling effects in terms of translation, rotation, and angle. Another is the price calculation. element It needs to be accounted for. A few feature abstraction techniques, like the combination of Gabor filters which restrict using them in real-world scenarios. The use of hand gestures should, however, thinking about the cost-accuracy trade-off in 2003, Chen, Fu, and Huang. The majority of hand gesture recognition, though.

Methodology

The concept behind this paper is that there is currently no classic for hand movements to switch the cursor. The plan is to demonstrate how an incapacitated person can use a mouse by using only their eyes. We operate with a mechanical mouse in the current system. People must wear heavy equipment and participate in activities to activate the system. Webcam gathers data from the eye. After receiving these camera inputs, it will convert them into frames. It will look out for adequate because webcams and illumination after converting them to frames require adequate lighting after external bases. The focused eye images from the input source are analyzed for iris detection. The next step is to pinpoint the precise address of the iris within the eye frame. and the iris position is transferred to a screen position. The center of both the right and left eyes is then averaged to determine a mid-point. Finally, the rodent shall perform the essential screen actions based on arranged tracked movements. To begin, the video is captured using a webcam. The recorded video is used to create frames. Finding the image's edges and counter requires a good face image with adequate lighting. which locate the mouth and eye in the frame, and use using them as a guide, after locating the benchmarks, eye blinks and head movement is picked up. Here the use of both eyes allows for quicker and more precise processing time. The decision tree algorithm that we employed combined with gradient boosting. A quick algorithm that resembles techniques that take tree structure's form is the decision tree regression. It involves breaking up a large data set into more manageable portions. This division results in a tree with leaf nodes and decision nodes. Because they determine the critical positions of the face directly from pixel intensity, regression trees are used to produce images of high quality. Gradient boosting is also used to minimize mistakes made when recording the film and making the point predictions. For learning all regression trees, the gradient boosting technique creates a prediction model.

Implementation

A. Expression detection

Get a true perspective of the subject's face, scan the users face and take entry. The space ought to be well-lit, so that obtaining input is simple. We must find a face. Before beginning image handler use the frontal face detector to find faces and locate the eyes including with DLIB.

B. Ear

The more fundamental than the alluring representative because it effectively utilizes using facial region. In the ear, it was helpful aimed at spotting movements like intermittent and sparkling. We can see when the eye is closed. See a drop in the ear rate. It is possible to train a straightforward classifier to recognize the eye-closing drop.

C. Mouth aspect ratio

To create a metric capable of distinguishing between a closed and open mouth, we created a formula that has a small amount of effect from the prior ear representative. Despite being unoriginal, it works. The MAR value. Like the ear, it rises when the mouth opens. The same fundamental rules when the mouth opens. The same fundamental rules apply to this MAR measure.

D. Click activity

The mouse operations are initiated, and gestures are used to move the cursor following the location of the iris and eye, accessing the reference points. We employ "Pyautogui" as a mouse action implementation library. Simply opening our mouth for a short period of time will activate the scroll mode and closing it will do the opposite. Our heads are used for scrolling, and our eyes are used to open and close applications or folders. Clicking utilizing the mouse's left mouse button, we wink our left eye, and to click utilizing the mouse's right mouse button, we wink our correct eye. The operations' actions are briefly described.

Results

We successfully created and tested a tree-based decision method for cursor control, extraction of the eye region, iris tracking, and face recognition. With the aid of our technology, users can virtually eliminate the need for traditional input devices when entering data into laptops. Our experiments demonstrate the ease of learning our eye movement system. We tested the system for the anticipated results as part of the project's development and testing, and the results are displayed below. Figure 39.1 shows, our paper has been trained to locate the user's eyes even when they are wearing eyeglasses. Anyone who doesn't wear glasses can easily implement our paper. As shown above, our project functions flawlessly with or without glasses, so it can be used in either situation. The recorded frames are shown below for clarity, and the cursor moves in response to head movements.

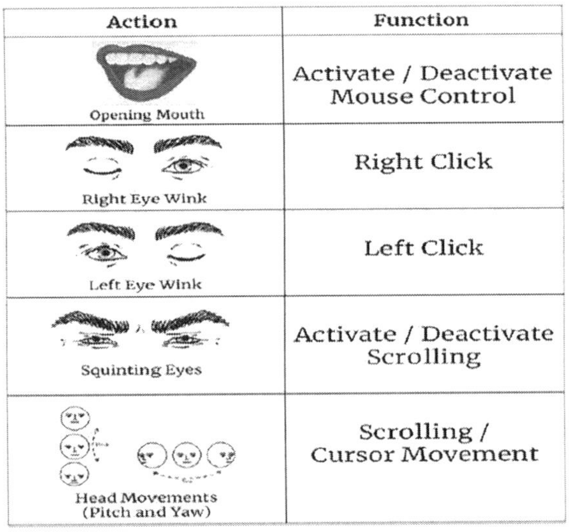

Action	Function
Opening Mouth	Activate / Deactivate Mouse Control
Right Eye Wink	Right Click
Left Eye Wink	Left Click
Squinting Eyes	Activate / Deactivate Scrolling
Head Movements (Pitch and Yaw)	Scrolling / Cursor Movement

Figure 39.1 Actions

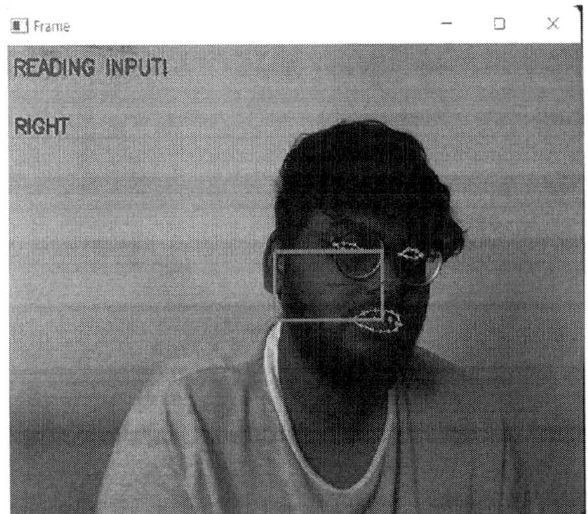

Figure 39.3 Cursor movement to the right

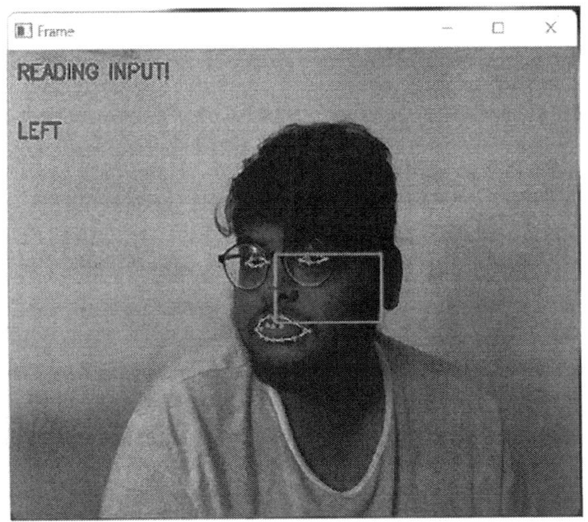

Figure 39.2 Left movement of the cursor

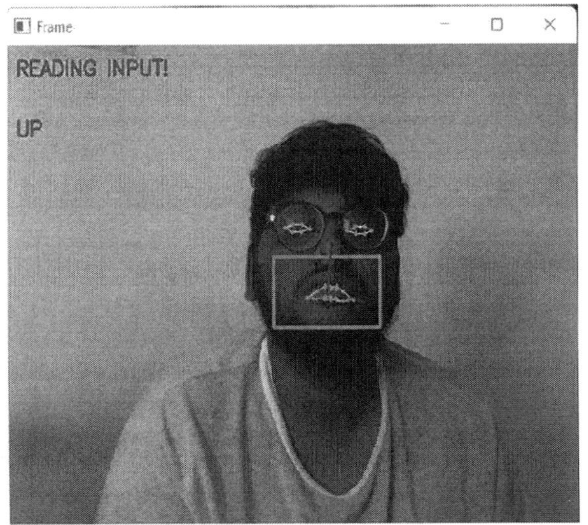

Figure 39.4 Cursor moving upwards

Figure 39.2 discusses that the frame can be seen in the image above, where the left-moving cursor is visible as shown in the image; the movements are made with our heads. The frame in the next image, which is displayed below, shows the cursor moving right as shown; we are using our heads to make the movements.

Figures 39.3 and 39.4 illustrates depicts the cursor descending as shown in the image; we can move our heads or make movements. The frame in the next image, which is displayed below, shows how the cursor is moving up as shown; we are using our heads to make the movements.

Conclusions

An interaction system for computers' design and testing for disabled individuals was successful. The method can be strengthened even more and used for different purposes. The system can be altered to help people with disabilities cutting-edge using residential home electronics including, (but not limited to) doors, lights, and televisions. Technology can also be altered to allow use by those who are totally paralyzed. Eye monitoring can also be used to identify drowsy driving, which help to avoid car accidents. They are used for tracking and detecting eye movements in virtual reality and gaming. The mouse is now simulated by this system for everyday computer communication. We intend to test new functionality for more situations where users can interact with the media in the future, as well as modify our system for new stands like phones and tablets. Numerous operation modules will also be developed to give users consistent operating experience from computer boot-up to shut down.

References

1. Sun, L., Liu, Z., and Sun, M.-T. (2015). Real time gaze estimation with a consumer depth camera. *Inform. Sci.*, 320, 346–360.
2. Cheung, Y.-M. and Peng, Q. (2015). Eye gaze tracking with a web camera in a desktop environment. *IEEE Trans. Human Mac. Sys.*, 45(4), 419–430.
3. Holzman, P. S., Proctor, L. R., and Hughes, D. W. (1973). Eye-tracking patterns in schizophrenia. *Science*, 181(4095), 179–181.
4. Donaghy, M. J. T., Pioro, E. P., Gibson, J. M., and Leigh, R. J. (2011). Eye movements in amyotrophic lateral sclerosis and its mimics: A review with illustrative cases. *J. Neurol. Neurosurg. Psychiat.*, 82(1), 110–116.
5. Shukla, S. (2023). Real-time monitoring and predictive analytics in healthcare: Harnessing the power of data streaming. *Inter. J. Comp. Appl.*, 185, 32–37. doi: 10.5120/ijca2023922738.
6. Shukla, S. (2023). Streamlining integration testing with test containers: Addressing limitations and best practices for implementation. *Inter. J. Latest Engg. Manag. Res. (IJLEMR)*, 9, 19–26. doi: 10.56581/IJLEMR.8.3.19-26.
7. Shukla, S. (2019). Examining Cassandra constraints: Pragmatic eyes. *Inter. J. Manag. IT Engg.*, 9(3), 267–287.
8. Shukla, S. (2019). Data Visualization with Python pragmatic eyes. *Inter. J. Comp. Trends Technol.*, 67(2), 12–16.
9. Shukla, S. (2019). Debugging microservices with Python. *IIOAB J.*, 10, 32–37.
10. Shukla, S. (2023). Unlocking the power of data: An introduction to data analysis in healthcare. *Inter. J. Comp. Sci. Engg.*, 11(3), 1–9.
11. Shukla, S. (2022). Developing pragmatic data pipelines using apache airflow on Google cloud platform. *Inter. J. Comp. Sci. Engg.*, 10(8), 1–8.
12. Shukla, S. (2023). Exploring the power of apache kafka: A comprehensive study of use cases suggest topics to cover. *Inter. J. Latest Engg. Manag. Res. (IJLEMR)*, 8, 71–78. doi: 10.56581/IJLEMR.8.3.71-78.
13. Shukla, S. (2023). Enhancing healthcare insights, exploring diverse use-cases with K-means clustering. *Inter. J. Manag. IT Engg.*, 13, 60–68.
14. Kumar, K. and Dwivedi, A. (2017). Big data issues and challenges in 21st century. *Inter. J. Emerg. Tech. (Special Issue NCETST-2017)*, 8(1), 72–77.
15. Dwivedi, R. P. P., Pandey, S., and Kumar, K. (2018). Internet of Things' (IoT's) impact on decision oriented applications of big data sentiment analysis. *2018 3rd Inter. Conf. Internet of Things Smart Innov. Usages (IoT-SIU)*, 1–10. doi: 10.1109/IoT-SIU.2018.8519922.
16. Bhondge, S. K., Bhoyar, D. B., and Mohod, S. (2016). Strategy for power consumption management at base transceiver station, *2016 World Conf. Futur. Trends Res. Innov. Soc. Welfare (Startup Conclave)*, 1–4. doi: 10.1109/STARTUP.2016.7583988.
17. Raut, P. W. and Bhoyar, D. B. (2009). Design FPGA based implementation of MIMO decoding in a 3G/4G wireless receiver. *2009 Second Inter. Conf. Emerg. Trends Engg. Technol.*, 1167–1172. doi: 10.1109/ICETET.2009.7.
18. Nanare, K. H. S., Bhoyar, D. B., and Balam War, S. V. (2021). Remote sensing satellite image analysis for deforestation in Yavatmal district, Maharashtra, India. *2021 3rd Inter. Conf. Sig. Proc. Comm. (ICPSC)*, 684–688. doi: 10.1109/ICSPC51351.2021.9451744.
19. Raut, P. W., Bhoyar, D., Kulat, K. D., Deshmukh, R. (2020). Design and implementation of MIMO-OFDM receiver section for wireless communication. *Inter. J. Adv. Res. Engg. Technol. (IJARET)*, 11(12), 44–54.
20. Yadav, A., Kakde, S., Khobragade, A., Bhoyar, D., and Kamble, S. (2018). LDPC decoder's error performance over AWGN channel using min-sum algorithm. *Inter. J. Pure Appl. Math.*, 118(20), 3875–3879. ISSN: 1314-3395.
21. Bondare, R., Bhoyar, D. B., Dethe, C. G., and Mushrif, M. M. (2010). Design of high frequency phase locked loop. *2010 Inter. Conf. Comm. Control Comput. Technol.*, 586–591. doi: 10.1109/ICCCCT.2010.5670772.
22. Pande, N. A. and Bhoyar, D. (2022). A comprehensive review of lung nodule identification using an effective computer-aided diagnosis (CAD) system. *2022 4th Inter. Conf. Smart Sys. Inven. Technol. (ICSSIT)*, 1254–1257. doi: 10.1109/ICSSIT53264.2022.9716327.
23. Bhoyar, D. B., Dethe, C. G., and Mushrif, M. M. (2014). Modified step size LLMS channel estimation method for MIMO-OFDM system. *Rec. Adv. Comm. Netw. Technol.*, 3(2), 98–105.

40 A study on tuning criterion, design parameters and performance of a nonlinear energy harvesting dynamic vibration absorber attached to SDOF system

Soumi Bhattacharyya[1,a] *and Shaikh Faruque Ali*[2,b]

[1]Department of Civil Engineering, Ghani Khan Choudhury Institute of Engineering and Technology, Malda, India

[2]Applied Mechanics Department, Indian Institute of Technology Madras, Chennai, India

Abstract

This work studies an EHDVA, i.e., nonlinear energy harvesting dynamic vibration absorber, for possible energy generation and simultaneous vibration attenuation of a single degree of freedom (SDOF) system subjected to base excitation. A nonlinear EHDVA, duffing-type nonlinear DVA with a piezoelectric patch, is proposed to harvest electricity while mitigating the vibration. The electromechanical interactions between the harvester-DVA and structural system are considered in the total system's modeling and time domain formulation. A non-dimensional simulation study is carried out for harmonic base excitation to understand the effect of nonlinearity. A comparative study between nonlinear and linear EHDVAs and DVAs is reported. The tuning criterion of nonlinear EHDVA for optimum vibration mitigation is identified. Further, a parametric study is carried out to obtain the optimal values of design parameters for EHDVA. The design of an EHDVA is also presented for optimal performance in energy generation and vibration control with an example SDOF system.

Keywords: Design parameters, nonlinear dynamic vibration absorber (nonlinear DVA), nonlinear energy harvesting dynamic vibration absorber (nonlinear EHDVA), nonlinear tuned mass damper (nonlinear TMD), piezoelectric transduction, tuning criterion

Introduction

Over the past decades, different vibration control techniques such as base isolation, passive energy dissipation devices, and active control systems have been investigated to safeguard structures [11]. Various passive, active, semi-active, and hybrid systems are used for structural control. A dynamic vibration absorber (DVA), popular as tuned mass damper (TMD), is a passive control device that reduces vibration at a particular design frequency of a vibrating host or the primary structure. In a typical application, a DVA shifts energy from a host vibration to itself thereby reducing the vibration of the primary system. This manuscript focuses on generating electrical energy from the vibrating DVA. This technique of tapping surrounding energy to energize the sensors is known as energy harvesting.

An efficient structural health monitoring system ensures proper application of the control techniques. In an efficient monitoring system, sensors should be placed all throughout the structure including the remote, inaccessible locations. For such applications remote and wireless sensing is desirable. Presently, wireless sensors work on battery power. However, batteries are not environment friendly and their replacement at remote locations is uneconomical and very risky sometimes. A way out of this problem is to use the energy harvested from the host vibrations to power up wireless sensors.

There are different types of energy harvesting devices or energy harvesters that can scavenge energy from several sources namely temperature differences, light radiation, electromagnetic fields, kinetic energy, etc. [6, 13, 14, 18]. For vibration control and health monitoring, vibration-based energy harvesters can be used which convert unwanted and available vibration energy to electrical energy. Since the last decade, harvesting energy from surrounding vibration has gathered popularity owing to its advantages in energizing wireless sensors [4, 20]. Three basic vibration energy harvesting mechanisms are, electromagnetic [12, 20], electrostatic [17] and piezoelectric transductions [2, 8]. In piezoelectric harvesters (PEHs), the mechanical strain, generated by surrounding vibration, is transformed into electrical charge by the piezoelectric material. This method is simple to design and able to have higher power densities [2, 8]. Review articles on piezoelectric transduction are reported in refs [4, 19].

It is noted that linear energy harvesting devices have been investigated widely and the

[a]soumibhttchr86@gmail.com, soumi@gkciet.ac.in, [b]sk.faruque.ali@gmail.com

DOI: 10.1201/9781003540199-40

optimal parameters are extensively defined [1, 2, 8]. However, the efficiency of linear energy harvesters is observed only around resonance [7], and while designing mostly harmonic excitations are considered. However, in case of unknown or random excitations harvesters with a broadband or adaptive response are likely to be beneficial [1, 10]. In one approach, multiple harvesters tuned to different frequencies can also be used for broadband energy harvesting [9, 16]. Alternatively, to maximize the harvested energy over a wide range of excitation frequencies, nonlinear structural systems have been used [1–3, 10], typically known as broadband energy harvesters. This paper contains a study on these nonlinear energy harvesters.

Existing literature has detailed investigations on structural control and energy harvesting separately. However, combined nonlinear controller-harvesting systems are less explored. This paper investigates a nonlinear device that controls the vibration of a primary SDOF system while generating energy from surrounding vibration. In order to develop the device, a conventional nonlinear dynamic vibration absorber (DVA) and the mechanism of piezoelectric transduction are combined. DVA, technically known as a passive vibration control device, dissipates energy from the primary structure and undergoes large vibrations which are utilized to produce energy by using piezoelectric material. It is to be noted that any real structures can be modeled as an SDOF system and the EHDVA device under study can be designed for that structure. Specifically, a nonlinear energy harvesting dynamic vibration absorber (EHDVA) [5] is studied through a detailed numerical study with non-dimensional equations of motion. The model of the SDOF-EHDVA system used in this study is explained. Then the characteristics of the nonlinear EHDVA are compared with those of linear EHDVA and DVA. Further, the optimal values of critical design parameters are evaluated for controlling the vibration of a base excited single degree of freedom (SDOF) primary structure. Then the design procedure of the EHDVA system attached to an example SDOF primary system is developed and a parametric study is carried out. Also, the performance of the designed EHDVA system is analyzed numerically.

Linear and nonlinear EHDVA-SDOF system and comparative study

The linear and nonlinear EHDVA model attached to an SDOF primary system, described in ref [5], is used for the present study.

In order to compare the effect of nonlinearity, the non-dimensional displacement responses of the linear DVA and EHDVA over a range of non-dimensional frequency is plotted and compared with the responses of nonlinear DVA and EHDVA with reverse sine sweep and shown in Figure 40.1a (also see Figure 40.1b, enlarged from Figure 40.1a). Figure 40.1a shows that the response amplitude is much higher in the case of linear DVAs when it is tuned i.e. $\Omega_h \cong 1$ [5] though for the nonlinear DVAs the response is higher over a wider frequency range than that of the linear DVAs. For nonlinear DVAs the responses show the hardening effect of the Duffing oscillator. Moreover, from Figure 40.1 it is also observed that the displacement amplitudes of EHDVAs are less than that of DVAs for both linear and nonlinear cases. The presence of the energy harvester induces this additional damping to the system.

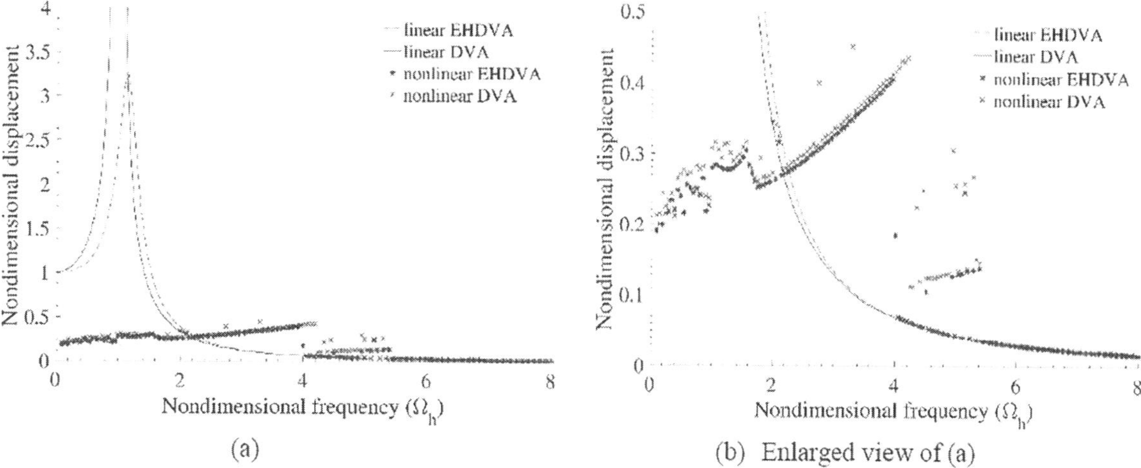

Figure 40.1 Comparison between non-dimensional displacement amplitudes of linear and nonlinear DVAs and EHDVAs over a range of non-dimensional excitation frequency ()

SDOF-EHDVA system: tuning, parametric study and design

The tuning criterion of nonlinear EHDVA, evaluated by Bhattacharyya and Ali [5], is adopted in the present work. The value of the optimum tuning ratio $\nu = 0.7$, where the response reduction has the maximum value [5], is used to design the EHDVA. In order to design an efficient EHDVA system, that can control the vibration of an SDOF system and also generate power, it is important to obtain the optimal values of design parameters.

The non-dimensional voltage generated with the variation in ▯ value is presented in Figure 40.2. Figure 40.2 shows the presence of a different optimum tuning ratio in terms of maximum power generation at $\nu = 0.5$. Now, the designer has to choose between the two optima ▯ values as per the requirement. For the present study, two design objectives are there - control of the primary SDOF structure and generation of power. $\nu = 0.7$ can be provided if control is the primary requirement (displacement reduction of 87.95% and a velocity reduction of 88.00%) but the generated power is much less (0.008 W) at this tuning ratio value. However, at $\nu = 0.5$ a maximum power can be obtained (0.0278 W) without much compromise in control performance (displacement reduction of 75.05% and a velocity reduction of 74.21%). Hence, it is better to provide optimal parametric values from the consideration of maximum harvested power.

Further, it is relevant to investigate the influence of nonlinear coefficient ▯ on the performance of the EHDVA system. The generated power is plotted over a range of ▯ and ▯ values in Figure 40.3. Figure 40.3 shows the presence of an optimum tuning ratio for which power generation is maximum. With the increase in the imposed nonlinearity in the system, the optimum ▯ shifts towards a smaller value. The inclusion of the multiple peaks is also observed in the widened power generation curve for higher values of ▯. Observing Figure 40.3 it may also be said that the maximum value of ▯ should be provided as 0.8, beyond which the fluctuation in the peak value of generated power increases. It is to be noted that, though the power generation curve is widened over the tuning ratio range for higher ▯ values, the maximum power is generated at $\alpha = 0.38$ ($\nu = 0.55$) and not at $\alpha = 1$. Hence, $\alpha = 0.38$ and $\nu = 0.55$ are considered as the optimal values of design parameters.

Performance of designed SDOF-EHDVA system

For the optimum values of the design parameters, i.e. $\nu = 0.55$ and $\alpha = 0.38$, the time histories are presented for the responses of the SDOF-EHDVA system. Figures 40.4 (a and b) shows the uncontrolled and EHDVA-controlled displacement and velocity of the SDOF primary system, respectively. From Figure 40.5 the occurrence of beat phenomenon is observed for EHDVA controlled responses. This indicates a near resonance solution for the SDOF- EHDVA system. A displacement and velocity reduction of 72.18% and 71.48% are achieved, respectively which is a satisfactory controlling performance obtained by the designed EHDVA.

The time histories for non-dimensional voltage and actual power generated by the EHDVA are presented in Figures 40.5 (a and b). From Figure 40.5a it is observed that a peak non-dimensional voltage of 0.548 is achieved by the EHDVA system. Also, 38.2 mW peak power amplitude is generated (see Figure 40.5b).

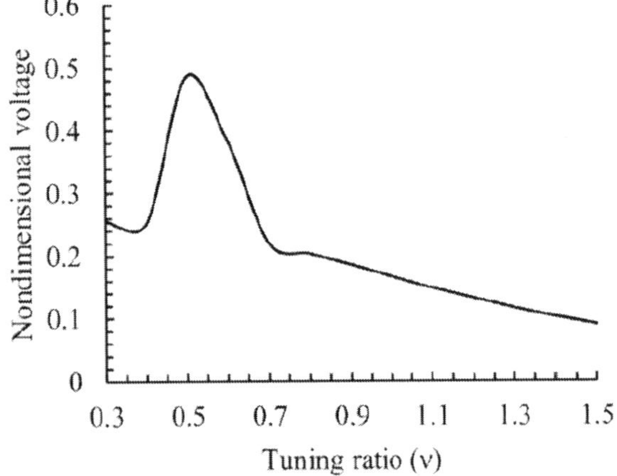

Figure 40.2 Variation in non-dimensional voltage generated by EHDVA attached to SDOF system over a range of tuning ratios ()

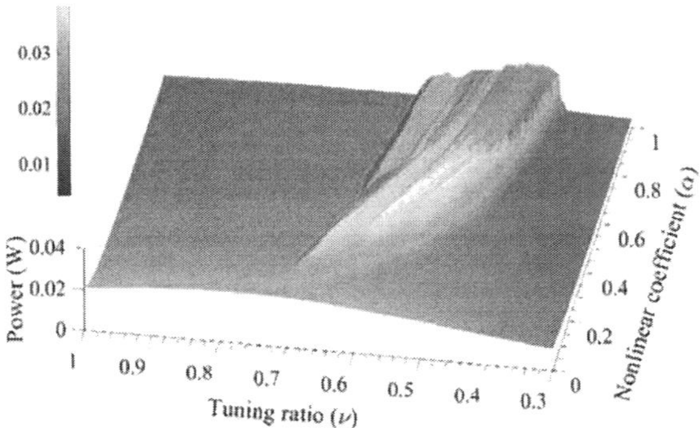

Figure 40.3 Values of power generated by EHDVA over a range of tuning ratios () and nonlinear coefficients ()

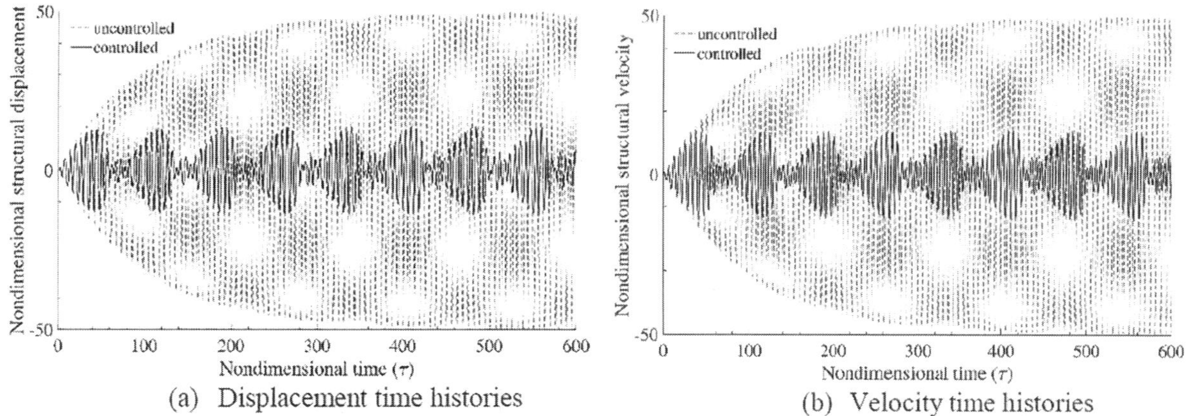

Figure 40.4 Time histories - uncontrolled and controlled non-dimensional displacement and velocity of primary SDOF system for 0.55 and 0.38

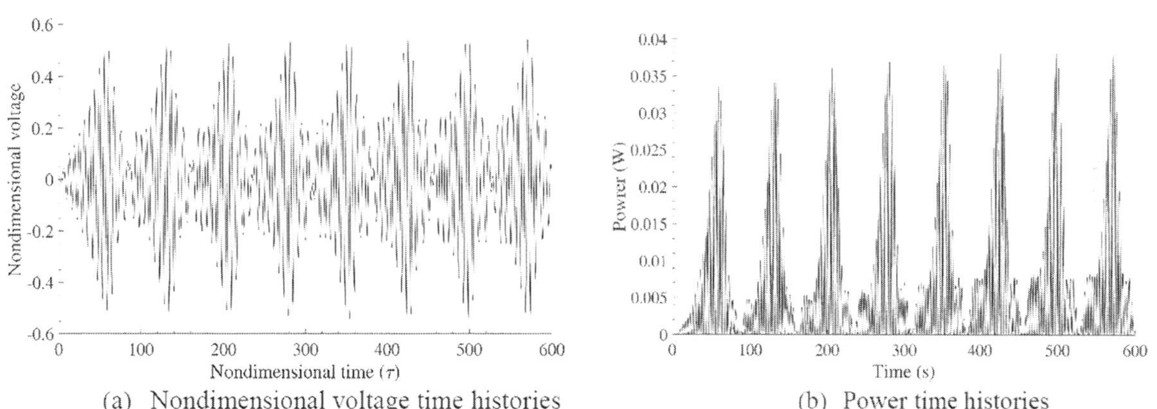

Figure 40.5 Time histories of non-dimensional voltage and power generated by EHDVA attached to primary SDOF system for 0.55 and 0.38

Conclusions

Conventional DVAs mitigate structural vibration by dissipating energy from primary structure through large displacements. In the case of an energy harvester, the higher the vibration higher the energy generated. Therefore, in this study an energy harvester is attached to the DVA, and the primary structure–EHDVA interaction is utilized for both the purposes of vibration control and energy harvesting. A sine sweep study reveals the characteristics of EHDVA modeled with hardening type Duffing nonlinearity and locates the frequency band with consistently higher system responses. The advantages of nonlinear EHDVA over the linear ones are reported. It is observed that the nonlinear harvester provides power over a broad range of frequencies and it is able to dissipate energy from the primary structure subjected to wideband excitations. Results reveal that the tuning ratio is the most important design parameter for EHDVA. The value of the tuning ratio is selected on the basis of the broadband response of the EHDVA system subjected to non-dimensional sine sweep frequency excitation. However, the sensitivity study on tuning ratio and nonlinearity coefficient reveals that the generated power is an important performance criterion to obtain the designed optimal tuning ratio. The performance analysis of the designed system in the time domain shows the efficiency of the same by attaining of 72.18% displacement reduction of the primary structure while generating 38.2 mW peak power amplitude. This shows the competency of the designed EHDVA from both control and harvesting aspects. This study also provides the basis for further analytical and experimental study on the performance of nonlinear EHDVA systems.

Acknowledgement

The authors gratefully acknowledge the financial and technical support from the Indian Institute of Technology Madras, Chennai - 600036, India through the Institute Post Doctoral Fellowship scheme.

References

1. Adhikari, S., Friswell, M., and Inman, D. (2009). Piezoelectric energy harvesting from broadband random vibrations. *Smart Mat. Struct.*, 18(11), 1–7.
2. Ali, S., Adhikari, S., Friswell, M., and Narayanan, S. (2011). The analysis of piezomagnetoelastic energy harvesters under broadband random excitations. *J. Appl. Phy.*, 109(7), 074904.
3. Ali, S., Friswell, M., and Adhikari, S. (2010). Piezoelectric energy harvesting with parametric uncertainty. *Smart Mat. Struct.*, 19(10), 105010.
4. Anton, S. and Sodano, H. (2007). A review of power harvesting using piezoelectric materials (2003–2006). *Smart Mat. Struct.*, 16, R1–R21.
5. Bhattacharyya, S. and Ali, S. (2020). Design of a nonlinear energy harvesting dynamic vibration absorber. *Recent Adv. Comput. Mec. Simulat.*, II, 48.
6. Chongfeng, W. and Xingjian, J. (2017). A comprehensive review on vibration energy harvesting: Modelling and realization. *Renew. Sustain. Energy Rev.*, 74, 1–18.
7. Daqaq, M. (2010). Response of a uni-modal duffing type harvesters to random force excitations. *J. Sound Vibrat.*, 329, 3621–3631.
8. Erturk, A. and Inman, D. (2011). Piezoelectric energy harvesting. UK: John Wiley and Sons Ltd., 1-48.
9. Ferrari, M., Ferrari, V., Guizzetti, M., Marioli, D., and Taroni, A. (2008). Piezoelectric multi-frequency energy converter for power harvesting in autonomous micro-systems. *Sens. Actuat. A Phy.*, 142, 329–335.
10. Friswell, M., Ali, S., Adhikari, S., Lees, A., Bilgen, O., and Litak, G. (2012). Nonlinear piezoelectric vibration energy harvesting from an inverted cantilever beam with tip mass. *J. Intell. Mat. Sys. Struct.*, 23(3), 1505–1521.
11. Housner, G., Bergman, L., Caughey, T., Chassiakos, A., Claus, R., Masri, S., Skelton, R., Soong, T., Spencer, B., and Yao, J. (1997). Structural control: Past, present, and future. *J. Engg. Mec.*, 123, 897–971.
12. Jones, P., Tudor, M., Beeby, S., and White, N. (2004). An electromagnetic, vibration powered generator for intelligent sensor system. *Sens. Actuat. A Phy.*, 110, 244–349.
13. Kumar, A., Ali, S. F., and Arockiarajan, A. (2019a). Influence of piezoelectric energy transfer on the inter-well oscillations of multi-stable vibration energy harvesters. *ASME J. Comput. Nonlin. Dynam.*,14(3), 031001.
14. Kumar, R., Gupta, S., and Ali, S.F. (2019b). Energy harvesting from chaos in base excited double pendulum. *Mec. Sys. Sig. Proc.*, 124, 49–64.
15. Li, L. and Cui, P. (2017). Novel design approach of a nonlinear tuned mass damper with duffing stiffness. *J. Engg. Mec.*, 143(4), 1-8.
16. Malaji, P. and Ali, S. (2017). Magneto-mechanically coupled electromagnetic harvesters for broadband energy harvesting. *Appl. Phy. Lett.*, 111, 083901.
17. Mitcheson, P., Milao, P., Start, B., Yeatman, E., Holmes, A., and Green, T. (2004). Mems electrostatic micro power generator for low frequency operation. *Sens. Actuat. A Phy.*, 115, 523–529.
18. Rajarathinam, M. and Ali, S. F. (2019). Investigation of a hybrid piezo-electromagnetic energy harvester. *Technisches Messen*, 85(9), 541–552.
19. Sodano, H., Park, G., and Inman, D. (2004). A review of power harvesting from vibration using piezoelectric materials. *Shock Vibrat. Digest*, 36, 197–205.
20. Williams, C. B. and Yates, R. B. (1996). Analysis of a micro- electric generator for microsystems. *Sens. Actuat. A Phy.*, 52, 8–11.

41 Low cost IoT-based monitoring and control system for aeroponic vertical farming

Dinesh B. Bhoyar[1,a], *Swati K. Mohod*[2,b], *D. G. Bhalke*[3,c], *J. W. Chawhan*[4,d] *and Rajesh H. Khobragade*[5,e]

[1]Department of ETC, Yeshwantrao Chavan College of Engineering, Nagpur, Maharashtra, India

[2]Department of EE, Yeshwantrao Chavan College of Engineering, Nagpur, Maharashtra, India

[3]Department of ETC, D.Y. Patil College of Engineering. Pune, Maharashtra, India

[4]Department of ETC, K.D.K. College of Engineering, Nagpur, Maharashtra, India

[5]Department of E&CS, Shri Ramdeobaba College of Engineering, Nagpur, Maharashtra, India

Abstract

This paper introduces a low-cost monitoring and control system designed for aeroponic vertical farming, focusing on the growth of tomato plants. The system tracks temperature and humidity in a closed chamber and regulates the supply of water and nutrients to optimize plant growth. Experiments were conducted to evaluate the system's performance, and the results showed its effectiveness in maintaining ideal growing conditions and enhancing tomato plant health and productivity. Implementing this affordable system in aeroponic vertical farming can potentially increase tomato yields and promote sustainable agricultural practices.

Keywords: Aeroponic, vertical farming, monitoring system, control system

Introduction

Aeroponic vertical farming has emerged as a sustainable and efficient method of crop production, offering advantages such as optimized space utilization, reduced water consumption, and controlled environmental conditions. In aeroponics, plants are grown without soil, with their roots suspended in the air and nourished by a nutrient-rich mist. To maximize the potential of aeroponic systems, it is crucial to monitor and control the growth conditions to ensure the plants receive the necessary water and nutrients. This study presents a low-cost monitoring and control system designed specifically for aeroponic vertical farming, focusing on the cultivation of tomato plants. The system combines real-time monitoring of temperature and humidity within the growing chamber with precise control over water and nutrient distribution. By integrating these components, our proposed system aims to optimize the growth conditions for tomato plants and enhance their productivity

Monitoring the environmental conditions is essential for maintaining optimal plant growth and health. Temperature and humidity play critical roles in plant development, nutrient uptake, and photosynthesis. Deviations from the desired ranges can negatively impact plant growth and lead to stress-related issues. Therefore, continuous monitoring allows for prompt adjustments to create an ideal microclimate for tomato plants. It ensures that the plants thrive in an environment that promotes their growth and minimizes the risk of physiological disorders or disease outbreaks. The control system in our setup enables precise regulation of water and nutrient supply to the tomato plants. By automating this process, human intervention is minimized and consistent and accurate delivery of essential resources is ensured. The control system adjusts the nutrient concentration and timing of nutrient delivery, tailored to the specific requirements of tomato plants at different growth stages. This optimization helps prevent nutrient deficiencies or excesses, ultimately enhancing plant health, yield, and overall productivity. To validate the effectiveness of our low-cost monitoring and control system, we conducted a case study focusing on tomato plants. The performance of the system was evaluated by monitoring the growth, yield, and overall health of the tomato plants throughout their growth cycle. Several prior studies have highlighted the benefits of monitoring and control systems in aeroponic farming. Environmental monitoring and automated nutrient supply in aeroponic systems have a positive impact on lettuce growth and quality. By monitoring conditions and providing

[a]dinesh.bhoyar23@gmail.com, [b]swatimohod6882@gmail.com, [c]dg.bhalke@dypvp.edu.in, [d]jyotsna.chauhan@kdkce.edu.in, [e]khobragaderh358@gmail.com

DOI: 10.1201/9781003540199-41

nutrients automatically, growers can create an optimal environment for lettuce, leading to improved growth and higher quality harvests. The research underscored the importance of maintaining optimal environmental conditions, such as temperature and humidity, for optimal plant performance in aeroponics. Additionally, precise control over nutrient delivery was found to be significant in maximizing crop productivity and enhancing nutrient utilization efficiency. By integrating our low-cost monitoring and control system into aeroponic vertical farming, we aim to provide small-scale farmers with an affordable solution to optimize tomato plant growth and yield. The findings of this study have the potential to contribute to the advancement of sustainable agricultural practices, promoting resource-efficient cultivation and enhancing food production in controlled environments.

Literature review

Aeroponic vertical farming has gained significant attention as a sustainable and efficient method of crop production. Smith et al. conducted a study on lettuce cultivation in an aeroponic system with an integrated monitoring and control system. They found that maintaining specific temperature and humidity ranges significantly improved lettuce growth and quality. The study highlighted the importance of real-time environmental monitoring in creating an ideal growth environment [1].

Nutrient management is another crucial aspect of aeroponic vertical farming. Precise control over nutrient delivery is necessary to maximize plant growth and yield while minimizing resource waste. Johnson and Johnson (2020) emphasized the significance of nutrient delivery control in aeroponic cultivation [2]. They demonstrated that adjusting nutrient concentration and timing based on plant growth stage and demand improved crop productivity and nutrient utilization efficiency. Implementing a low-cost monitoring and control system for nutrient management can provide small-scale farmers with a cost-effective solution for optimizing plant nutrition.

The integration of low-cost monitoring and control systems in aeroponic vertical farming has been the subject of several studies. Li et al. (2018) proposed an automated monitoring and control system for an aeroponic strawberry production system. Their low-cost system enabled real-time monitoring of environmental conditions, such as temperature, humidity, and CO_2 levels, and automated control of nutrient supply. The study reported improved strawberry growth and yield, demonstrating the feasibility and benefits of implementing cost-effective monitoring and control systems in aeroponics.

Tomato plants are one of the most commonly grown crops in aeroponic systems due to their economic significance and market demand. Chone et al. (2019) conducted a study on tomato cultivation in a controlled aeroponic environment. They found that precise control over nutrient delivery, including adjusting nutrient concentration and timing, resulted in higher tomato yields compared to traditional cultivation methods. The study highlighted the potential of monitoring and control systems in optimizing tomato production in aeroponic vertical farming.

In 2019 author developed smart monitoring and control system for aeroponic farming of tomato plants, including the use of sensors for monitoring environmental parameters, such as temperature, humidity, and pH, as well as automated control of nutrient solution and lighting [1].

Johnson and Johnson, summarizes the various control and monitoring systems that have been developed for vertical farming, including aeroponic systems [2]. The paper covers topics such as environmental monitoring, nutrient control, and automation in 2016. The paper studied by El-Sayed et al. describes the development of an automated control system for hydroponic tomato production, which includes monitoring of environmental parameters, such as temperature and humidity, as well as automated control of nutrient solution and water flow [3].

In 2020, Lim discusses the use of IoT technology for monitoring and control of vertical farming systems, including aeroponic systems. The paper covers topics such as sensor selection, data transmission, and control algorithms [4].

In 2019, Kumar et al., describes the design of a control system for aeroponic crop production, including the use of sensors for monitoring environmental parameters and automated control of nutrient solution and lighting [5].

Aeroponic vertical farming has shown promising results in terms of crop production, resource efficiency, and overall plant health. Here are some sample results from previous studies and real-world applications of aeroponic vertical farming:

Increased crop yield: Studies have reported significant increases in crop yields when using aeroponic vertical farming compared to traditional cultivation methods. For example, a study by M. S. Raju [6] (2018) demonstrated a 30% increase in strawberry yield compared to soil-based cultivation. Similarly, tomato plants grown in aeroponic systems have shown higher yields compared to conventional methods [7].

Improved resource efficiency: Aeroponic vertical farming is known for its efficient use of resources, particularly water. This cultivation method reduces water consumption by up to 95% [13] compared to traditional soil-based farming [8]. The precise control over nutrient delivery in aeroponic systems also allows for optimized resource utilization, minimizing waste and maximizing plant nutrient uptake.

Enhanced plant health: The controlled environment in aeroponic vertical farming helps create optimal conditions for plant growth, resulting in healthier plants. Studies have shown that plants grown in aeroponic systems exhibit fewer instances of pests, diseases, and nutrient deficiencies [14] compared to traditional farming methods [9]. The absence of soil-borne pathogens in aeroponics also contributes to improved plant health.

Faster growth and shorter crop cycles: Aeroponic vertical farming has been found to promote accelerated plant growth and shorter crop cycles. With precise control over environmental factors and nutrient supply, plants in aeroponic systems experience fewer growth limitations and can reach maturity faster [10]. This allows for more frequent harvests and higher overall crop production.

Year-round cultivation: The controlled environment in aeroponic vertical farming enables year-round cultivation, independent of seasonal limitations. By providing optimal temperature, humidity, and light conditions, growers can produce crops consistently throughout the year, ensuring a steady supply of fresh produce [11].

It is noted that the specific results of aeroponic vertical farming can vary depending on factors such as crop type, system design, environmental conditions, and management practices. However, the overall consensus from various studies and practical applications is that aeroponic vertical farming offers numerous benefits in terms of increased crop yields, resource efficiency, plant health, faster growth, and year-round cultivation. These results highlight the potential of aeroponic vertical farming as a sustainable and productive agricultural technique [12].

Problem statement

The most existing plant health monitoring systems are expensive and require specialized equipment and expertise, making them inaccessible to small and medium-sized vertical farms. This creates a significant challenge for farmers in maintaining optimal growing conditions and preventing plant stress, which can affect crop growth and yield.

As a result, there is a need for a low-cost plant health monitoring system that is accessible to small and medium-sized vertical farms, which can help farmers identify and address plant stress early, before it leads to significant yield losses. The system should incorporate various sensors, such as temperature, humidity, and light sensors, as well as machine learning and image processing algorithms for real-time monitoring and analysis of plant health parameters. It should also enable remote access to the data through wireless communication and cloud computing, making it easier for farmers to make informed decisions about crop management. The design and development of such a system can contribute to the advancement of sustainable and efficient agriculture practices [15] in urban areas, with significant environmental, social, and economic benefits.

Aeroponic system

Aeroponics is a type of hydroponic farming that involves growing plants in a mist or air environment instead of soil or water. Here are some of the requirements for aeroponics farming:

- Growing chamber: A growing chamber is required to hold the plants and the equipment needed for aeroponics farming. The chamber can be made of plastic, metal, or other materials.
- Mist generator: A mist generator is needed to produce the fine mist of nutrient solution that is sprayed onto the roots of the plants. The mist generator can be a high-pressure pump, ultrasonic nebulizer, or other device.
- Nutrient solution: A nutrient solution is required to provide the plants with the necessary nutrients. The solution can be mixed with water and then sprayed onto the plant roots.
- Lighting: Lighting is required to provide the plants with the energy they need to grow. LED lights are often used in aeroponics farming, as they are energy-efficient and can provide the right spectrum of light for plant growth.
- Air circulation: Air circulation is important to keep the plants healthy and prevent the growth of mold or bacteria. Fans can be used to circulate the air within the growing chamber.
- Temperature and humidity control: Temperature and humidity control is important to ensure that the plants grow in the right conditions. The growing chamber may need to be equipped with a heating or cooling system, as well as a humidifier or dehumidifier.

Design and implementation

This system's primary responsibility is to regulate the growing chamber box's temperature and air humidity within the ranges necessary for the tomato plants to thrive. These parameters will be measured by an electronic device utilizing sensors, and the results will be processed and compared to the required parameter values (set point). If there is a discrepancy, a matching action will be taken to change the surrounding circumstances such that the desired state is achieved (Figure 41.1).

There are two primary divisions of the system:

1. Monitoring system, this measures the humidity and temperature inside the growing chamber and displays the results on an LCD and a webpage.
2. A timer-based control system for the water and nutrient actuators inside the growing chamber.

A nozzle and pump system will be used to deliver water and nutrients throughout the plant development chamber. In this instance, water is supplied through pump nozzle systems, wetting the plant roots to make it simpler for the nutrient fog to adhere to the roots. Each plant has a unique area for its roots and stems to flourish. With respect to the effectiveness of nutrient uptake, the root zone should have a humidity level of between 80% and 90% RH. The air humidity should be kept between 60% and 80% RH, and the temperature should be between 20°C and 30°C, to avoid the spread of pests and diseases. When constructing systems for monitoring and control, this information serves as a guide. The system will automatically operate in accordance with the time. A pump nozzle system will be used to provide water to the plant at first. Following that, the plants will have thirty seconds to soak up the water. For the subsequent 30 s, some of the water will be absorbed by the roots and some will remain on the roots, making the roots of the plant slightly damp. Continuously, this cycle is repeated.

A. Monitoring system

The sensor system will be in charge of determining the environmental parameter values before sending data to the control system.

A sensor system comprises of two types of sensors: temperature and humidity

For the humidity and temperature measurement we use The DHT11 temperature and humidity sensor. It is a low-cost and easy-to-use device for measuring ambient temperature and humidity levels. It uses a capacitive humidity sensing element and a thermistor to measure the relative humidity and temperature, respectively. The sensor outputs a 16-bit digital signal over a single-wire interface, making it simple to use with microcontrollers such as an Adriano.

B. Control system

The results of each sensor's measurements are sent to the control system, where they are compared to the set point, or the value of the desired environmental parameters. If there is a discrepancy, the value will be converted into control signals. The microcontroller will then send a signal to the driver to activate the relay, which will turn on the water pump and supply the plants with water and nutrients. The sensor and signal conditioning circuit will read parameters from the growing chamber.

The LCD and a webpage powered by NodeMCU will both display temperature and humidity data. We use Webpage for user setting by these we use our system remotely, as shown in Figure 41.2, system will receive the temperature and humidity parameter and compare with set parameters according to that fans, and light works. If temperature is above the set range fans are get turned on, and turn off

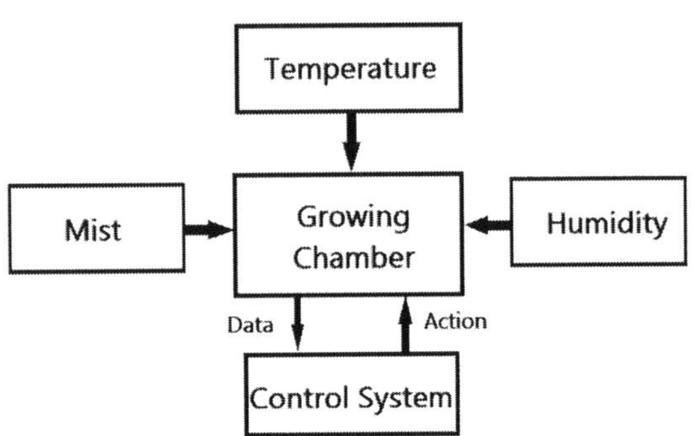

Figure 41.1 Block diagram of monitoring and control system for potato growing chamber

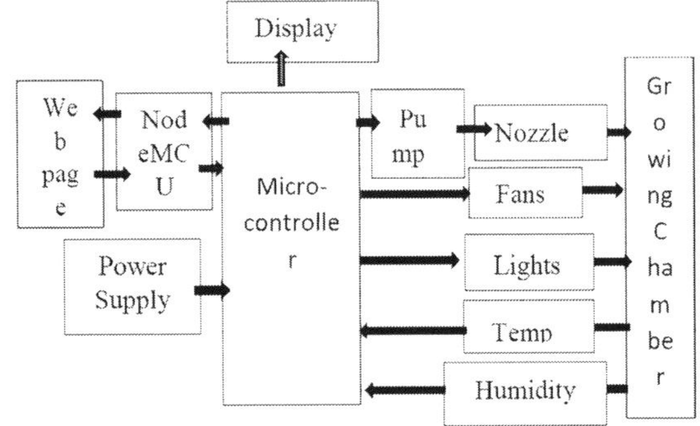

Figure 41.2 Block diagram of monitoring and control system for tomato growing chamber

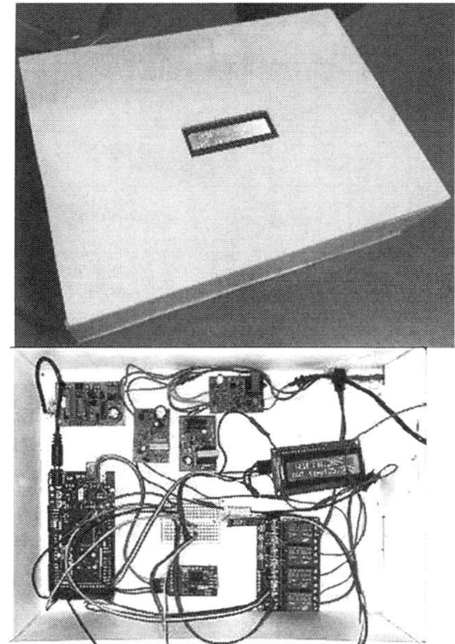

Figure 41.3 Circuit of the proposed system

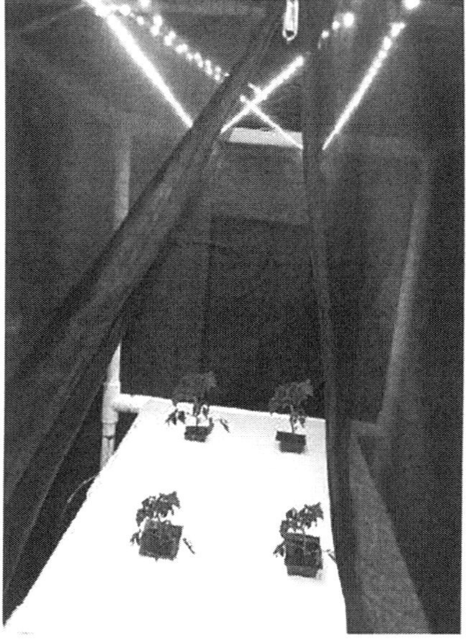

Figure 41.4 Aeroponics growing chamber

when temperature is in set range. Lights are used for photosynthesis process.

- width and length of the growth chamber = 0.7 m × 0.7 m
- Growth chamber area = Length × width × height = 0.7 × 0.7 × 1 m = 0.79 sq. m.
- Total capacity of growth chamber is 4 plants.

As shown in Figure 41.3, the control system is integrated into a single unit and packed and the structural design of PVC growth chamber as shown in Figure 41.4.

Table 41.1 Sensor testing results

Condition	Thermometer (°C)	Sensor (°C)
Chamber temp.	26	26
Fans turned off	30	30
Fans turned on	28	28

Height of the growth chamber = 1 m,

Testing of temperature sensors

As shown in the Table 41.1, the output data from temperature sensors and thermometer readings are largely identical. This demonstrates that the temperature sensor has been operating as intended and has provided accurate temperature measurements.

Result and discussion

It has been successfully constructed, put into practice, and field tested to distribute water and nutrients in an aeroponic growing chamber for the development of tomatoes. It has the ability to measure and keep track of the temperature and humidity of an in-situ chamber. The actuator (nozzle pump, mist maker, fan, and reservoir pump) that distributes water and nutrients on an aeroponic growth system for producing potatoes can be controlled by it. Customized user settings allow the user to manage system operation. The created system was able to transfer temperature and humidity data from the sensor to the display in real time, providing a growing chamber monitoring parameter.

Conclusions and future scope

Aeroponic vertical farming holds significant promise in mitigating the challenges associated with traditional agriculture. By utilizing vertical space and advanced cultivation techniques, it offers a solution for increasing food production in the face of population growth, urbanization, and climate change. However, it is crucial to thoroughly evaluate the costs and benefits of this technology, including energy consumption, water usage, and economic feasibility. Continued research and development are essential to optimize aeroponic vertical farming's efficiency, sustainability, and scalability, thereby maximizing its potential as a valuable tool in future food production systems.

Thus, the future of monitoring and control systems for aeroponics farming is bright. As the technology advances, these systems will become more sophisticated and more widely used, leading to increased efficiency, higher yields, and lower costs.

The present scenario has shown the increased threats represented by respiratory illness like chronic obstructive pulmonary disease (COPD), asthma, etc. That risk has increased due to increase in air pollutants like PM2.5, PM10, etc. a respirator can be utilized as an immediate countermeasure on an individual level safety measure as bringing down pollution levels require much longer time than the severity of the problem is allowing.

References

1. Smith, A., Johnson, B., and Thompson, B. (2019). Implementing control systems for nutrient delivery in aeroponics. *Hort. Sci.*, 54(9), S321.
2. Johnson, B. and Johnson, C. (2020). Aeroponic cultivation: Maximizing productivity through control systems. *Proc. Inter. Symp. Grow. Media Soil. Cult. Compost Utiliz. Horticul.* 55–61.
3. Riswandi, M. Niswar, Z. Tahir, Zainal and C. Y. Wey, (2022).Design and Implementation of IoT-Based Aeroponic Farming System *IEEE International Conference on Cybernetics and Computational Intelligence (CyberneticsCom)*, Malang, Indonesia, 2022, pp. 308-311, doi: 10.1109/CyberneticsCom55287.2022.9865284.
4. Benedetta Fasciolo, Ahmed Awouda, Giulia Bruno, Franco Lombardi (2023). A smart aeroponic system for sustainable indoor farming,Procedia CIRP, Volume 116, Pages 636-641, ISSN 2212-8271, https://doi.org/10.1016/j.procir.2023.02.17.
5. Lloret, J.; Sendra, S.; Garcia, L.; Jimenez, J.M (2021). A Wireless Sensor Network Deployment for Soil Moisture Monitoring in Precision Agriculture. *Sensors* **2021**, *21*, 7243. https://doi.org/10.3390/s21217243
6. Méndez-Guzmán, H.A.; Padilla-Medina, J.A.; Martínez-Nolasco, C.; Martinez-Nolasco, J.J.; Barranco-Gutiérrez, A.I.; Contreras-Medina, L.M.; Leon-Rodriguez, M. (2022) IoT-Based Monitoring System Applied to Aeroponics Greenhouse. *Sensors* **2022**, *22*, 5646. https://doi.org/10.3390/s22155646
7. Hati, Anirban Jyoti, and Rajiv Ranjan Singh. (2021). "Smart Indoor Farms: Leveraging Technological Advancements to Power a Sustainable Agricultural Revolution" *AgriEngineering* 3, no. 4: 728-767. https://doi.org/10.3390/agriengineering3040047.
8. Méndez-Guzmán HA, Padilla-Medina JA, Martínez-Nolasco C, Martinez-Nolasco JJ, Barranco-Gutiérrez AI, Contreras-Medina LM, Leon-Rodriguez M.(2022) IoT-Based Monitoring System Applied to Aeroponics Greenhouse. Sensors (Basel). 2022 Jul 28;22(15):5646. doi: 10.3390/s22155646. PMID: 35957199; PMCID: PMC9371135.
9. Friha, M. A. Ferrag, L. Shu, L. Maglaras and X. Wang, (2021)"Internet of Things for the Future of Smart Agriculture: A Comprehensive Survey of Emerging Technologies," in *IEEE/CAA Journal of Automatica Sinica*, vol. 8, no. 4, pp. 718-752, April 2021, doi: 10.1109/JAS.2021.1003925.
10. Yaw Long Chua1 , Shaik Mohamed Faizal Mohamed Bharuddin1 , Yit Yan Koh (2023)Design and Development of Automated Indoor Farming with Alert System.Journal of Advanced Research in Applied Sciences and Engineering Technology 31, Issue 2 (2023) 210-219
11. Chowdhury, Muhammad E. H., Amith Khandakar, Saba Ahmed, Fatima Al-Khuzaei, Jalaa Hamdalla, Fahmida Haque, Mamun Bin Ibne Reaz, Ahmed Al Shafei, and Nasser Al-Emadi. (2020). "Design, Construction and Testing of IoT Based Automated Indoor Vertical Hydroponics Farming Test-Bed in Qatar" *Sensors* 20, no. 19: 5637. https://doi.org/10.3390/s20195637.
12. Rashmi , Dr. Shilpa Shrigiri S (2023) Smart Cultivation: An Arduino-based IoT Aeroponics System for Indoor Farming, International Research Journal of Engineering and Technology (IRJET)Volume: 10 Issue: 08|Aug-2023 PP-271-275
13. Raut, P. W., Bhoyar, D., Kulat, K. D., and Deshmukh, R. (2020). Design and implementation of MIMO-OFDM receiver section for wireless communication. *Inter. J. Adv. Res. Engg. Technol. (IJARET)*, 11(12), 44–54.
14. Bhondge, S. K., Bhoyar, D. B., and Mohad, S. (2016). Strategy for power consumption management at base transceiver station. *2016 World Conf. Futur. Trends Res. Innov. Soc. Welfare (Startup Conclave)*, 1–4. doi: 10.1109/STARTUP.2016.7583988.
15. G. R. E., Bellam, T., M. E., B. P., G. K. C., and A. D. (2022). A practical approach of recognizing and detecting traffic signs using deep neural network model. *2022 Fourth Inter. Conf. Emerg. Res. Elec. Comp. Sci. Technol. (ICERECT)*, 1–5. doi: 10.1109/ICERECT56837.2022.10060522.

42 The identification and classification of images for the predictive diagnosis of brain diseases: A review

Swati K. Mohod[a] *and R. D. Thakare*[b]

Department of Electronics Engineering, Yeshwantrao Chavan College of Engineering, Nagpur, India

Abstract

The identification and classification of images for the predictive diagnosis of brain diseases has been an important research area in recent years. This review paper provides an overview of the current research on imaging techniques, such as X-ray, MRI, CT, PET, and SPECT, for diagnosing brain diseases. It discusses the principles, applications, strengths, limitations, recent advancements, and future directions in the field. We also examine the use of machine learning algorithms in the identification and classification of brain disease images, and discuss the challenges and opportunities in this area.

Keywords: Brain, X- ray, MRI, CT, PET, SPECT, machine learning

Introduction

Brain diseases can have a significant impact on a person's quality of life, and early detection and treatment are crucial to improving outcomes. Imaging techniques such as X-rays, MRI, CT, PET, and SPECT have been used for many years to aid in the diagnosis of brain diseases. However, the accurate interpretation of these images can be challenging, and radiologists may miss subtle changes that indicate the presence of a disease. Machine learning algorithms hold promise for brain disease image identification and classification, but challenges remain in areas such as dataset diversity, feature extraction, interpretability, and generalizability. Further research is needed to overcome these obstacles.

Classification of brain tumors

Brain tumors are classified into two types which are as follows:

Malignant tumor

Malignant tumors can be challenging to detect and diagnose accurately using traditional methods. However, digital image processing and medical image segmentation techniques have greatly improved the efficiency and reliability of identifying these tumors using MRI and CT scans. Radiologists can analyze the size, position, and characteristics of the tumor based on the images, which can help in determining the best course of treatment, including surgery.

Benign tumor

It is important to note that although benign tumors are not cancerous and do not have the ability to invade neighboring tissue or spread to other parts of the body, they can still cause health problems if they grow large enough or if they are in a critical location. Treatment for benign tumors may involve surgical removal, especially if they are causing symptoms or affecting the function of nearby organs. Additionally, regular monitoring may be necessary to ensure that the tumor does not grow or become cancerous over time.

Non-cancerous brain tumors can present a wide range of symptoms depending on their size and location. Common symptoms include headaches, seizures, vision or hearing issues, coordination difficulties, and changes in mood or personality. However, many non-cancerous brain tumors are slow-growing and may not cause any symptoms at all, or may only cause mild symptoms that are easily overlooked. It is vital to consult a doctor if you have persistent or concerning symptoms, as early detection and treatment can greatly impact the outcomes for non-cancerous brain tumors. Other common symptoms of brain tumors may include difficulty with balance, coordination, or walking, numbness or tingling in the arms or legs, problems with memory, concentration, or thinking, changes in hearing, difficulty swallowing, and changes in mood or emotions.

Imaging techniques

MRI is a commonly employed imaging technique for diagnosing brain diseases due to its superior soft tissue visualization and multi planar imaging capabilities. It provides high- resolution images of the brain and can detect changes in brain tissue that

[a]swatimohod6882@gmail.com, [b]rdt2909@gmail.com

DOI: 10.1201/9781003540199-42

may indicate the presence of a disease. CT scans are also commonly used in the diagnosis of brain diseases, as they can quickly produce detailed images of the brain. PET and SPECT scans use radiotracers to identify changes in metabolic activity or blood flow in the brain, which can be indicative of a disease.

Challenges and opportunities

Despite the promise of machine learning algorithms in the diagnosis of brain diseases one of the biggest challenges is the lack of large, annotated datasets of brain disease images. Another challenge is the need for robust algorithms that can generalize to new cases and imaging modalities. There are also ethical considerations surrounding the use of machine learning algorithms in medical diagnosis, particularly with regard to bias and interpretability.

Problem formulation

Computer-aided detection (CAD) systems can enhance efficiency and accuracy in brain tumor detection by assisting radiologists in interpreting medical images, reducing the risk of missed or misdiagnosed tumors. This study explores the feasibility and effectiveness of CAD systems using image processing and machine learning techniques to improve radiologists' diagnostic abilities in brain tumor detection and quantification.

Literature review

In-depth analysis of the mentioned studies reveals various insights into the application of machine learning and deep learning techniques in processing MRI scans for medical diagnosis and analysis. Let's delve into some key aspects of these studies:

In the study [1] focused on high-grade glioma (HGG), the authors highlight the importance of comprehensive genomic profiling to identify molecular markers, such as mutations in IDH, 1p/19q co-deletion, and changes in MGMT, for making early clinical decisions and improving survival rates. They propose using multi-modal MRI-based phenotypic tools, such as CNNs and radiomics-based multi-variate models, to detect these genotypes. To address the drawbacks of current methods and the challenge of using unlabeled data, the authors suggest a semi-supervised hierarchical multi-task model that can incorporate unlabeled glioma data and predict various molecular markers simultaneously. The proposed methodology achieves an average test accuracy of 82.35% using a dataset of labeled, partially labeled, and unlabeled data. They validate the results using task- and modality-wise ablation analysis and class activation maps, which provide clinical interpretability. Ahmad and Choudhury emphasize the importance of early brain tumor detection to improve prognosis and recovery chances [2]. They explore deep learning-based classification techniques using 2D MR images for brain tumor identification. By combining traditional classifiers and transfer learning-based deep learning techniques, they achieve high accuracy, precision, recall, F1-score, Cohen's kappa, AUC, Jaccard, and specificity. The top-performing model achieves an impressive 10-fold cross-validation-achieved accuracy of 99.39%, providing valuable guidance for selecting deep transfer learning-based approaches for brain tumor identification. In their research, Zhu et al., investigate the dynamic nature of functional connectivity (FC) in the brain network and its relation to self-construal, a cultural marker [3]. They propose using machine learning-based models to explore the dynamic connection patterns of independence and dependency in the brain. The authors employ XGBoost regression and demonstrate that the efficiency-based dynamic FC successfully distinguishes between these two orientations. They find increased prediction accuracy compared to traditional static FC methods, revealing that self-construals are linked to disperse neural networks throughout the brain.

The authors focus on Alzheimer's disease and advocate for the integration of MRI and PET [4] to capture valuable information. They propose a novel machine learning architecture, SSMI, which incorporates same-subject-modalities-interactions to extract supplementary data and insights from MRI and PET scans. The ridge-classifier model applied to an ADNI dataset shows high accuracy, precision, specificity, recall, F1-score, and AUC, with regions chosen by ridge classifier demonstrating relevance to Alzheimer's disease biomarkers. The study concentrates on brain lesion classification using machine learning and deep learning approaches. The authors suggest feature extraction using the 2D DOST [5] and dimension reduction with (2D) 2PCA, followed by training convolutional neural networks (CNNs) for MRI classification. Their proposed CAD tool outperforms modern approaches and correctly diagnoses MRI scans. Kuo et al., propose using transfer-learning machine learning models to predict dementia from MRI data [6]. Their model achieves an accuracy of 90.7% on a dataset using k-fold cross-validation and synthetic minority oversampling. This paradigm offers potential for diagnosing dementia in underserved areas and facilitating timely treatment and rehabilitation planning. Nath et al., investigate active learning for medical imaging data segmentation and present a query-by-committee strategy using Stein variation gradient descent [7]. They achieve enhanced data reduction and complete accuracy on two medical imaging datasets. The study introduces DeepCurvMRI, a curvelet transform based-CNN model [8], for early-stage AD diagnosis using MRI images. The model performs better than comparable techniques and accurately differentiates distinct stages of Alzheimer's disease. Majib et al., discuss early detection of brain tumors and suggest transfer learning approaches based on neural networks for brain tumor classification [9]. They propose a stacked classifier with excellent precision, recall, and F1 scores. The study focuses on grading Autism Spectrum Disorder (ASD) using a computer-assisted system [10] based on brain activity in response to a speech experiment. They create a two-stage system with various classifiers using region of interest (ROI) zones for feature extraction, providing encouraging results for early computer-aided grading of ASD in newborns and toddlers.

Each of these studies contributes to the advancement of medical imaging and its application in diagnosing and understanding various neurological conditions, including glioma, brain tumors, Alzheimer's disease, brain lesions, dementia, and ASD. The use of machine learning and deep learning approaches in these studies demonstrates their potential in assisting clinicians with accurate and early diagnosis, leading to improved patient outcomes.

Comparative analyses

The in-depth review of existing methods used for processing MRI scans reveals a wide array of models proposed for various applications. The extensive empirical survey presented in Table 42.1 shows the diverse range of cases where these models find application, making it easier for readers to identify the most suitable model for their specific scenarios. Table 42.1 provides a comprehensive overview of the applications of machine learning and deep learning models [11] in processing MRI scans. Each study highlights the specific use case it targets and the key techniques or models employed. This survey empowers researchers and medical professionals to efficiently choose the appropriate model that best aligns with their requirements and medical imaging tasks. Contrasting with the comparison mentioned earlier, Table 42.2 aims to help readers identify performance-specific models suitable for their specific contextual use cases. In this evaluation, we directly inferred the values of Precision (P), Accuracy (A), and Recall (R). However, for Complexity (C), Delay (D), and Scalability (S), we transformed the values into empirical fuzzy ranges, categorized as low (L=1), medium (M=2), high (H=3), and very high (VH=4) based on their operating characteristics. This approach allows for a more nuanced and comprehensive understanding of the models' capabilities and limitations, enabling readers to make informed decisions when selecting the most appropriate model for their needs.

For accuracy, the following models have demonstrated better performance: Prostate cancer MRI segmentation [1], brain tumor segmentation [3], These models can be utilized in scenarios where high accuracy is crucial. Prostate cancer MRI segmentation [1]. These models are well-suited for scenarios that demand high recall, ensuring comprehensive identification of relevant features.

Moving on to computational complexity, the following models outperform others: Prostate cancer MRI segmentation [1], MRI brain tumor segmentation [2], brain tumor segmentation [3], MRI brain

Table 42.1 Analysis of applications for different models

Method used for MRI analysis	Application area
[1] Prostate cancer MRI seg	Medical imaging – Prostate cancer
[2] Cerebral ischemic stroke seg	Medical imaging – Stroke
[3] Brain tumor and stroke classification	Medical imaging – Brain abnormalities
[4] MRI brain tumors segmentation	Medical imaging – Brain tumor
[5] Radiogenomic approaches	Medical imaging – Glioma
[6] Alzheimer's disease classification	Medical imaging – Alzheimer's disease
[7] Predicting Alzheimer's disease stage	Medical imaging – Alzheimer's disease
[8] Deuterium MR spectroscopic imaging	Medical imaging – Brain tumor
[9] Uncertainty-aware brain age prediction	Medical imaging – Brain age prediction
[10] Multitask ensemble learning	Medical imaging – Alzheimer's disease

Table 42.2 Empirical comparison of different models and processes [1–10]

Work	P	A	R	C	D	S
[1] Prostate cancer MRI seg	92.43	95.85	90.02	3	3	3
[2] Cerebral ischemic stroke seg	91.28	86.98	86.57	3	3	3
[3] Brain tumor and stroke classification	94.28	94.42	87.02	3	3	3
[4] MRI brain tumors segmentation	86.47	86.84	86.11	3	3	3
[5] Radiogenomic approaches	86.25	87.85	87.07	3	3	3
[6] Alzheimer's disease classification	86.74	86.41	87.11	3	3	3
[7] Predicting Alzheimer's disease stage	86.31	87.98	87.50	3	3	3
[8] Deuterium MR spectroscopic imaging	86.07	87.95	87.96	4	4	2
[9] Uncertainty-aware brain age prediction	86.52	87.81	87.05	4	4	2
[10] Multitask ensemble learning	87.53	86.05	86.30	3	3	3

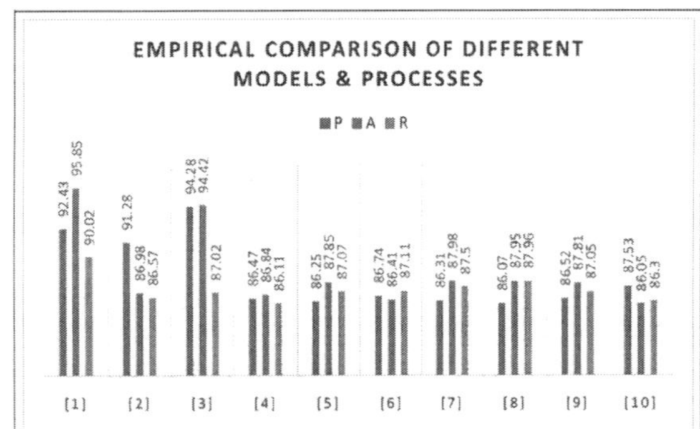

Figure 42.1 Precision, accuracy and recall of different models and processes

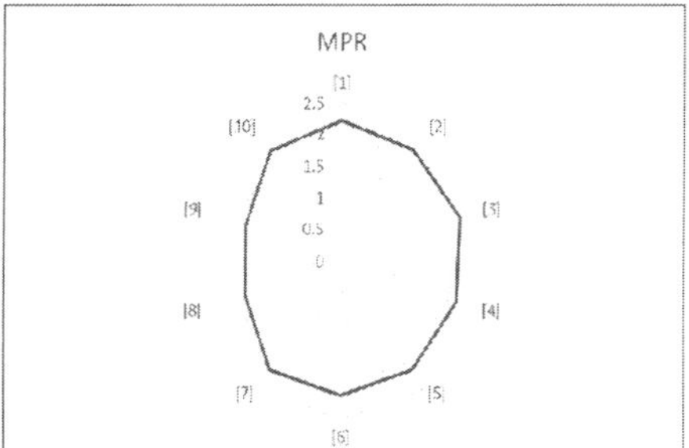

Figure 42.2 MRI processing rank (MPR) of different models and processes

tumor segmentation [4], brain tumor segmentation [5], MRI brain tumor

segmentation [6], MRI brain tumor segmentation [7], brain tumor segmentation [10]. These models are recommended for deployment in scenarios where low computational complexity is essential.

Lastly, for computational delay and scalability, the following models exhibit superior performance: Prostate cancer MRI segmentation [1], MRI brain tumor segmentation [2], brain tumor segmentation [3], MRI brain tumor segmentation [4], and brain tumor segmentation [5]. These models are well-suited for high-speed processing scenarios (Figures 42.1 and 42.2).These metrics were combined via Equation 1 to evaluate an augmented MRI processing rank (MPR) as follows:

$$MPR = \frac{P+A+R}{300} + \frac{1}{D} + \frac{1}{C} + \frac{S}{5} \quad (1)$$

As per this evaluation, it can be observed that prostate cancer MRI seg [1], brain tumor seg [3], MRI brain tumor seg [2], deep transfer ensemble (DTE) [7] outperforms other methods [12], thus can be used for high precision, high accuracy, high recall, low delay, low complexity, and high scalability scenarios. Researchers can extend these models to develop high efficiency methods for their context-specific application scenarios.

Conclusions

The identification and classification of images for the predictive diagnosis of brain diseases is an important research area with many challenges and opportunities. Imaging techniques such as MRI, CT, PET, and SPECT have been used for many years to aid in the diagnosis of brain diseases. Machine learning algorithms have shown great promise in improving the accuracy and efficiency of this process. Addressing the challenges in this area will require collaboration between clinicians, researchers, and machine learning experts, but the potential benefits are substantial.

References

1. Tupe-Waghmare, P., et al. (2021). Comprehensive genomic subtyping of glioma using semi-supervised multi-task deep learning on multimodal MRI. *IEEE Acc.*, 9, 167900–167910. doi: 10.1109/ACCESS.2021.3136293.
2. Ahmad, S. and Choudhury, P. K. (2022). On the performance of deep transfer learning networks for brain tumor detection using MR images. *IEEE Acc.*, 10, 59099–59114. doi: 10.1109/ACCESS.2022.3179376.
3. Zhu, Y., Li, X., Sun, Y., Wang, H., Guo, H., and Sui, J. (2022). Investigating neural substrates of individual independence and interdependence orientations via efficiency-based dynamic functional connectivity: A machine learning approach. *IEEE Trans. Cogn. Dev. Sys.*, 14(2), 761–771. doi: 10.1109/TCDS.2021.3101643.
4. Guelib, B., Zarour, K., Hermessi, H., Rayene, B., and Nawres, K. (2023). Same-subject-modalities-interactions: A novel framework for MRI and PET multi-modality fusion for Alzheimer's disease classification. *IEEE Acc.*, 11, 48715–48738. doi: 10.1109/ACCESS.2023.3276722.
5. Soleimani, M., Vahidi, A., and Vaseghi, B. (2020). Two-dimensional stockwell transform and deep convolutional neural network for multi-class diagnosis of pathological brain. *IEEE Trans. Neural Sys. Rehab. Engg.*, 29, 163–172. doi: 10.1109/TNSRE.2020.3040627.
6. Kuo, P.-H., Huang, C.-T., and Yao, T.-C. (2023). Optimized transfer learning based dementia prediction system for rehabilitation therapy planning. *IEEE Trans. Neural Sys. Rehab. Engg.*, 31, 2047–2059. doi: 10.1109/TNSRE.2023.3267811.
7. Nath, V., Yang, D., Landman, B. A., Xu, D., and Roth, H. R. (2021). Diminishing uncertainty within the training pool: Active learning for medical image segmentation. *IEEE Trans. Med. Imag.*, 40(10), 2534–2547. doi: 10.1109/TMI.2020.3048055.
8. Chabib, C. M., Hadjileontiadis, L. J., and Shehhi, A. A. (2023). DeepCurvMRI: Deep convolutional curvelet transform-based MRI approach for early detection of Alzheimer's disease. *IEEE Acc.*, 11, 44650–44659. doi: 10.1109/ACCESS.2023.3272482.
9. Majib, M. S., Rahman, M. M., Sazzad, T. M. S., Khan, N. I., and Dey, S. K. (2021). VGG-SCNet: A VGG net-based deep learning framework for brain tumor detection on MRI images. *IEEE Acc.*, 9, 116942–116952. doi: 10.1109/ACCESS.2021.3105874.
10. Bhondge, S. K., Bhoyar, D. B., and Mohad, S. (2016). Strategy for power consumption management at base transceiver station. *2016 World Conf. Futur. Trends Res. Innov. Soc. Welfare (Startup Conclave)*, 1–4. doi: 10.1109/STARTUP.2016.7583988.
11. Pande, N. A. and Bhoyar, D. (2022). A comprehensive review of lung nodule identification using an effective computer-aided diagnosis (CAD) system. *2022 4th Inter. Conf. Smart Sys. Inven. Technol. (ICSSIT)*, 1254–1257. doi: 10.1109/ICSSIT53264.2022.9716327.
12. Geetha Rani, E., Afeef Hussain, M., Azeezulla, M., Shandilya, M., and Susan Varughese, P. (2023). Skin disease diagnosis using VGG19 algorithm and treatment recommendation system. *2023 IEEE 8th Inter. Conf. Converg. Technol.*, 1–8. doi: 10.1109/I2CT57861.2023.10126212.

43 Various linear regression models are subjected to a comparison analysis of parameter estimation techniques

Geetha Rani E.[1,a], *Bhuvaneshwari P.*[2,b], *D. Anusha*[3,c], *M. Dhana Lakshmi*[4,d], *Jayalakshmi D.*[5,e], *Sethu Madhavi*[6,f], *Ch Suguna Latha*[7,g] *and Kshama A. S.*[7,h]

[1] Assitant Professor, Department of CSE, Alliance University Bengaluru, India

[2]Associate Professor, Department of CSE, Manipal Institute of Technology Bengaluru, Manipal Academy of Higher Education, Manipal, India

[3]Associate Professor, Department of CSE, SRKIT, Bengaluru, India

[4]Professor, Department of CSE, NHCE, Bengaluru, India

[5]Professor, Department of CSE, Saveetha school of engineering, SIMATS, Chennai, India

[6]Assistant Professor, Department of CSE, REVA University, Bengaluru, India

[7]Department of CSD, MVJCE (VT of Affiliation), Bengaluru, India

Abstract

Regression analysis is a popular analytical technique used to find and measure correlations among a few independent and dependent factors. Models of regression have evolved into many different subtypes over time, each with unique strengths and weaknesses. The linear regressions, least absolute operator for selection and shrinkage (LASSO), quantile, net elasticity, and Poisson are among the commonly used regression models that are evaluated and compared in this comparative study. Simple and often used, the quantile method of regression can be utilized for calculating the variable's conditional quantiles that are a dependent association among the if the factor isn't standard. When working with highly correlated predictors, the elastic network regression technique involves the outcome of regression using LASSO and ridges—which is helpful. The frequency of an event is a dependent variable when counting data, and Poisson regression is used to model this frequency. Based on the precision, robustness, and generalizability of these regression models, we assess and compare their performance in this comparative study. In this study, we examine the effectiveness of different models in forecasting prices based on available independent factors using a dataset of housing prices. This study identifies the most effective regression models for forecasting home prices and discusses the advantages and disadvantages of each model. This study aims to assist both academics and professionals in selecting the best depending on the parameters of the data collection and the intended research issue, regression model for a specific application. Numerous sectors, encompassing engineering, social sciences, economics, and finance, may be affected by the findings.

Keywords: Poisson regression, net elasticity regression, LASSO regression, quantile regression

Introduction

A popular analytical method for determining and quantifying correlations between numerous independent and dependent variables is regression analysis. In many disciplines, including engineering, the social sciences, finance, and economics, regression models are frequently used to forecast. Based on the values of the independent variables, the dependent variable's future values will be. [2-5] Questioning regression models seek to identify the core functional model that captures how variables relate to one another and allows for prediction and inference. Regression model selection for a given dataset is influenced by several variables, such as the kind and distribution of data, the nature of the correlations between variables, and the specific objectives of the research. Regression models have taken many different shapes throughout the years, each with their benefits and drawbacks. Along with linear and logistic regression, frequent models of random forests, polynomials, ridges, and logistics are some commonly used regression models and a dependent variable is said to be linearly related in a linear regression model. It is still a commonly utilized, straightforward strategy that is simple to understand and apply. [6-9] It can be applied to dependent variables that are categorical or continuous. Complex

[a]geetha.edupuganti@alliance.edu.in, [b]bhuvaneshwari.p@manipal.edu, [c]d.anushasrk@gmail.com, [d]mdhanalak72@gmail.com, [e]jayalakshminandakumar2@gmail.com, [f]sethumadhavi@reva.edu.in, [g]sugunachitturi@gmail.com, [h]1mj21cg033@mvjce.edu.in

DOI: 10.1201/9781003540199-43

relationships between variables might not be captured, and they might not function properly when presumptions are broken. Least absolute operator for selection and shrinkage (LASSO) regression: To choose crucial characteristics and lower the model's predictor count, a LASSO regression model is utilized. [10-12] By essentially reducing the magnitude of a few variables' coefficients to 0 and removing those var from the model, it decreases the regression coefficients towards zero. When there are a lot of predictors, it can be computationally expensive. Compared to other regression models, it is less susceptible to outliers. The elastic net regression comes next. This regression is produced by combining the ridge and LASSO regressions. Compared to LASSO regression, it is better able to handle strongly correlated predictors. When analyzing extremely big datasets with far more predictors than observations, it is useful. Poisson regression: To mimic count data, a Poisson regression model is used, with the variable that is dependent being the number of times a situation has occurred. It is predicated on the notion that the distribution of the dependent variable is Poisson. [13-15] Once interacting with dependent variable values that are non-negative integers, it is helpful. employing a model of negative binomial regression, it may be expanded to handle over-distributed count data. Based on the house price information, this comparison study assesses how well these three regression models performed and identifies the advantages & disadvantages of each model. This study aims to assist practitioners and academics in choosing an appropriate regression type for a specific application of need.

Literature survey

Refenes et al., compare neural network stock performance modeling versus regression models. We investigate applying neural networks in place of traditional statistical forecasting techniques when implemented to have an order of magnitude higher in-sample model fitness and forecasting precision than other statistical methods. We describe the parameters of the network whose values these performance indicators statistically stabilize. Neural networks have come under scrutiny for their inability to do sensitivity analysis and more convincingly simulate their environment compared to regression models and may offer a suitable justification for their predictive behavior. Support vector and linear regression are contrasted [1]. Using current or projected data, one of the business insights needed to forecast the customer interest, customer behavior, and product returns are examples to aid in forecasting. The mathematical techniques for the analytical model may be evaluated depending on commonly made use of time series data. Most apps included predict the weather, the economy, and the financial markets. In this work, two independent approaches to modeling freight generation—cross-classification and ordinary least squares (OLS).[16-18] Cross-classification tables were made to do this by finding groups of independent variables that describe freight generation using multiple classification analysis. Both times, corporate characteristics are mentioned as the reason for the production of goods. [19-21] Data gathered from companies in to estimate the model, New York's Brooklyn and Manhattan were used. As possible indicators of the daily deliveries received or made, more than 190 different variables were examined. We cover a total of six conceptually good and statistically significant linear regression models. James et al., compares different techniques. We explore the relationship between the density and shape of outliers, the magnitude of the regressor parameter, and the distance between outliers to determine the likelihood of swamping, or false alarms, and our ability to identify them. The outlier analysis configurations that, when using a planned experiment, are more likely to be seen in practice techniques are the subject of the simulation scenarios. The results for each scenario highlight the performance restrictions for each strategy and provide knowledge. To forecast the parameters of several This study examined least squares, robust, non-parametric, and regression models. techniques. These techniques were evaluated using fish collected from the Serranus cabrilla and measured the length overall, fork, and body. In these regression models, fork length served as the dependent variable, while body weight and overall distance served as liberated variables. acronyms and abbreviations and acronyms should be defined even if they had already been defined in the abstract, they are only employed for the first time in the text.

Methodology

1. Variations in regression models

1. Linear regression

Using the linear model, association, the modeling method most frequently employed is linear regression as shown in Figure 43.1. Regression lines, often known as the most desirable line, are employed. The linear connection's equation is $Y = c + (m * X) + e$, where "c" stands for the intercept, "m" for the line's range, and "e" for the error factor.

2. LASSO regression

The regression coefficient's absolute magnitude is ascertained using the LASSO method which is shown in Figure 43.2. Another method in LASSO

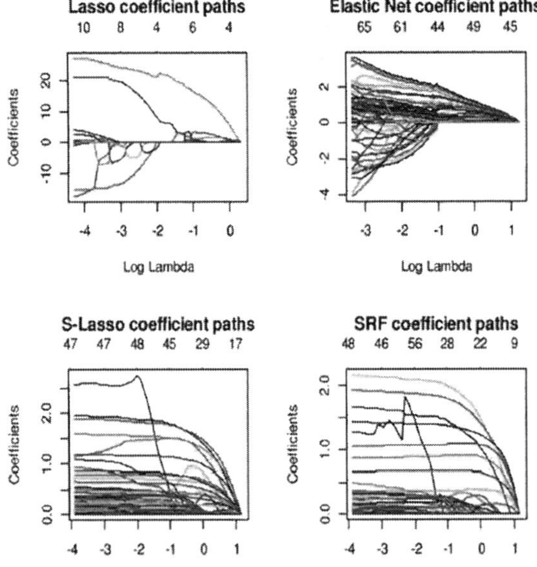

Figure 43.1 Linear regression

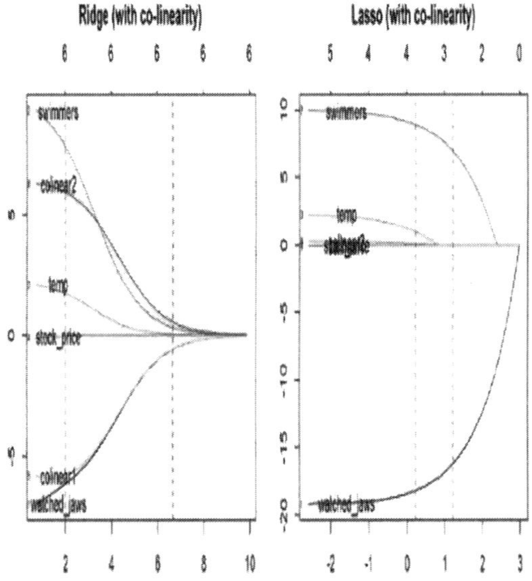

Figure 43.3 Quantile regression

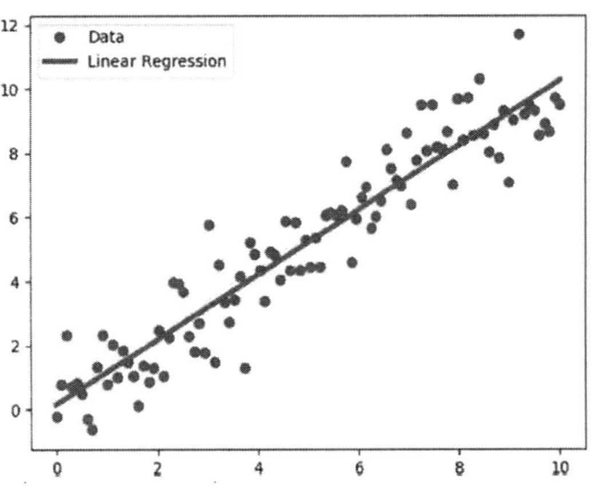

Figure 43.2 Lasso regression

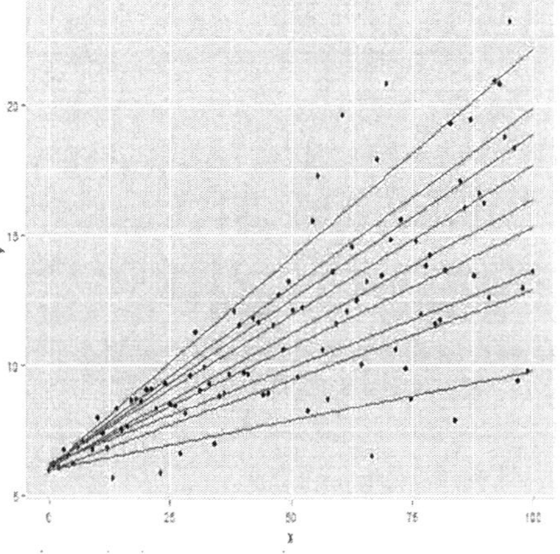

Figure 43.4 Elastic net regression

regression that produces completely zero coefficient values is variable selection.

3. *Quantile regression*

Quantile regression is a technique used as part of the linear regression approach. As the requirements for linear regression are not fulfilled or when there are outliers in the figures, it is utilized. It is employed in values for econometrics (Figure 43.3).

4. *Elastic net regression*

When working with highly correlated data, elastic net regression combines the LASSO and ridge regression techniques (Figure 43.4). Regression models are regularized by using the drawbacks associated with the LASSO and ridge regression routines. Change the number of columns: After snapping the columns button on the MS Word Standard toolbar, select the necessary number of columns from the selection palette.

5. *Poisson regression*

In arithmetical analysis, tables of incidents and data amounts are modeled using. The underlying premise of Poisson regression is that a Poisson distribution is produced by the outcome variable Y, and that a linear organization of unknown factors may accurately replicate the expected value's logarithm (Figure 43.5).

2. *Mathematical representation*

$$Y=a + bX \tag{1}$$

$$a= \frac{(\sum_Y)(\sum_{X^2}) - (\sum_X)(\sum_{XY})}{n(\sum_{x^2}) - (\sum_x)^2}$$

$$b= \frac{n(\sum_{XY}) - (\sum_X)(\sum_Y)}{n(\sum_{x^2}) - (\sum_x)^2}$$

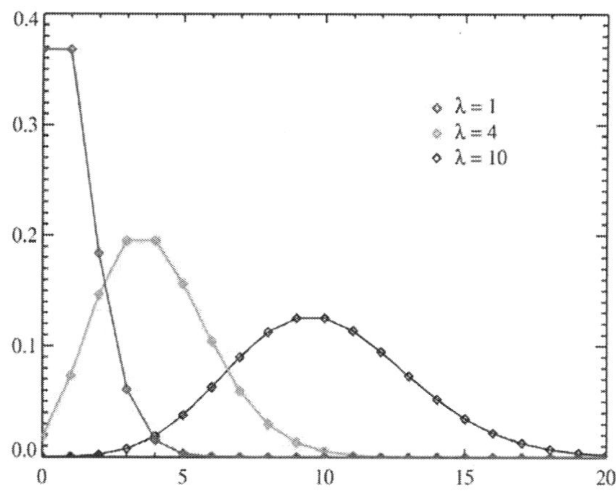

Figure 43.5 Poission regression

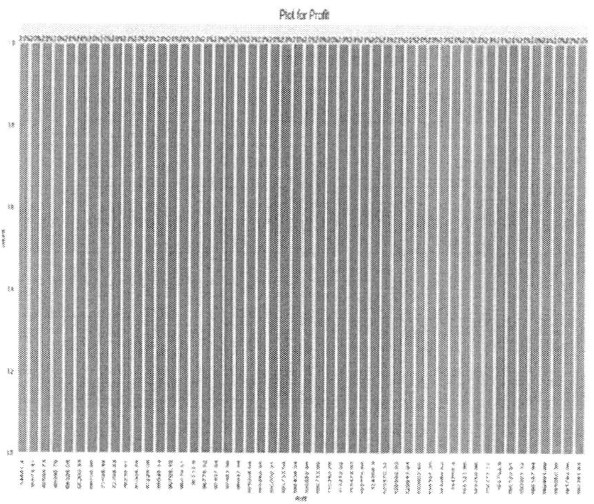

Figure 43.6 Comparing the efficiencies of regression models

3. Regression steps

A simple (just one variable each that is dependent and independent) linear regression ideal can use a single needy variable together with several independent variables.

Data collection

Gather information about a relevant subject of your choice. Make sure the data contains one or more independent variables as well as a single dependent variable or several. Additionally, it is crucial to make sure the data is adequate in terms of observations and is representative.

4. Data pre-processing

To guarantee that the regression models are precise, remove any missing values, outliers, and other mistakes from the data by pre-processing and cleaning. Data transformations like scaling, centering, or normalization may be a part of this.

5. Model fitting

Using a suitable algorithm, fit each of the chosen regression models to the preliminary data. The linear regression model is fitted using the ordinary least squares (OLS) method. The L1 regularization approach is used to fit the model for LASSO regression. The implementation of the typical is done using the quantile regression procedure in quantile regression.

Model evaluation

Figure 43.6 shows that using the chosen evaluation metric or metric(s), evaluate the performance of each model. Common metrics include R squared, root mean squared, mean squared, and mean absolute errors. Based on your study question, you might additionally take into account other metrics.

COMPARING THE MODELS:

Model Name	MSE	RMSE	R2
Linear Regression	7.79603e+07	8829.51	0.954676
Lasso Regression	7.79603e+07	8829.51	0.954676
Quantile Reression	7.79603e+07	8829.51	0.954676
Elastic Net Regression	7.79603e+07	8829.51	0.954676
Poisson Regression	1.72008e+09	41473.8	2.22045e-16

Figure 43.7 Regression model evaluation and efficiencies

6. Model comparison

Utilize statistical tests like the F-test or cross-validation to compare the performance of each model. This will enable you to decide which model predicts the dependent variable the most accurately.

Results and interpretation

Figure 43.7 describe the analyses' findings and explain how they relate to your research topic. Highlight each model's advantages and disadvantages while making suggestions for further study. Making sure the technique is transparent and the study can be replicated is essential overall. This can be accomplished by offering thorough documentation and granting public access to the code and data.

Implementation

5. Data description

R&D, marketing, and administrative finances for 50 start-ups are included in the datasheet used for the study. The observations in this compilation, which is most likely a panel dataset, each represent a corporation through time. The amount that each company funds in research and development activities is tracked by the variable R&D spending. Costs associated with the development of the product, innovation, and research may be included. The volume that each business spends on overhead expenditures like rent, payroll, and utilities is measured by control variables. The marketing spending variable tracks how much money each business spends on advertising, sponsorships, and media placements, among other marketing and promotional initiatives. We may investigate the connection between the dependent and independent variables in these three using this dataset. what way your business may save costs while achieving better financial outcomes. Also, the dataset might include specific data such as firm name, location, industry, and surveillance year in addition to the primary variables. This data can be utilized to detect patterns and trends unique to a certain group of organizations and conduct further research into the variables that affect company performance. The data must be cleaned and error-free before being used with this dataset for the regression model to be as accurate as possible. Removal of missing data, outliers, and other abnormalities may be part of this process. In addition to applying data transformations like normalization and standardization, make sure your data is appropriately sized for regression analysis. In general, this dataset provides a plethora of knowledge. for comparing regression techniques, and it may be used to learn more about the elements that contribute to a start-up's success.

Conclusions and future scope

In conclusion, our study attempted to compare and contrast the performance of five distinct regression algorithms: LASSO regression, Poisson regression, elastic net regression, and quantile regression. We investigated the association between these characteristics and firm success utilizing a dataset that includes information on the marketing, administrative, and R&D costs for 50 startups. Our investigation revealed that each regression model had distinct advantages and disadvantages. While LASSO and quantile regression showed moderate performance, linear and elastic net regression showed strong performance in predicting the dependent variable from the independent factors. Poisson regression, which was created specifically for count data, performed poorly when used to forecast the dataset's dependent variable. Additionally, we discovered that management only had a minor impact on marketing and R&D investment, which were the greatest significant independent elements in forecasting firm accomplishment. These findings help academics and experts make well-up-to-date choices regarding the regression models they use for research and application by shedding light on how businesses might optimize spending to better financial performance. aids in your descent. In conclusion, by contrasting the effectiveness of several regression models on actual datasets, our study has advanced the field of regression analysis. To assist practitioners in making more accurate job predictions, we hope that our study will stimulate further research in the field.

References

1. Kavitha, S., Varuna, S., and Ramya, R. (2016). A comparative analysis on linear regression and support vector regression. *2016 Online Inter. Conf. Green Engg. Technol. (IC-GET)*, 1–5. doi: 10.1109/GET.2016.7916627.
2. Shukla, S. (2023). Real-time monitoring and predictive analytics in healthcare: Harnessing the power of data streaming. *Inter. J. Comp. Appl.*, 185, 32–37. doi: 10.5120/ijca2023922738.
3. Shukla, S. (2023). Streamlining integration testing with test containers: Addressing limitations and best practices for implementation. *Inter. J. Latest Engg. Manag. Res. (IJLEMR)*, 9, 19–26. doi: 10.56581/IJLEMR.8.3.19-26.
4. Shukla, S. (2019). Examining Cassandra constraints: Pragmatic eyes. *Inter. J. Manag. IT Engg.*, 9(3), 267–287.
5. Shukla, S. (2019). Data visualization with Python pragmatic eyes. *Inter. J. Comp. Trends Technol.*, 67(2), 12–16.
6. Shukla, S. (2019). Debugging microservices with Python. *IIOAB J.*, 10, 32–37.
7. Shukla, S. (2023). Unlocking the power of data: An introduction to data analysis in healthcare. *Inter. J. Comp. Sci. Engg.*, 11(3), 1–9.
8. Shukla, S. (2022). Developing pragmatic data pipelines using Apache airflow on Google cloud platform. *Inter. J. Comp. Sci. Engg.*, 10(8), 1–8.
9. Shukla, S. (2023). Exploring the power of Apache Kafka: A comprehensive study of use cases suggest topics to cover. *Inter. J. Latest Engg. Manag. Res. (IJLEMR)*, 8, 71–78 doi: 10.56581/IJLEMR.8.3.71-78.
10. Shukla, S. (2023). Enhancing healthcare insights, exploring diverse use-cases with K-means clustering. *Inter. J. Manag. IT Engg.*, 13, 60–68.
11. Vyshali Rao, K. P., Ramisetty, S., and Dhanalakshmi, M. (2023). Reliable informational data and

secured deviation notification over networks using IoT. *2023 Inter. Conf. Recent Trends Elec. Comm. (ICRTEC)*, 1–6.

12. Bhondge, S. K., Bhoyar, D. B., and Mohod, S. (2016). Strategy for power consumption management at base transceiver station. *2016 World Conf. Futur. Trends Res. Innov. Soc. Welfare (Startup Conclave)*, 1–4. doi: 10.1109/STARTUP.2016.7583988.
13. Raut, P. W. and Bhoyar, D. B. (2009). Design FPGA based implementation of MIMO decoding in a 3G / 4G wireless receiver. *2009 Second Inter. Conf. Emerg. Trends Engg. Technol.*, 1167–1172. doi: 10.1109/ICETET.2009.7.
14. Nanare, K. H. S., Bhoyar, D. B., and Balam War, S. V. (2021). Remote sensing satellite image analysis for deforestation in Yavatmal district. *2021 3rd Inter. Conf. Sig. Proc. Comm. (ICPSC)*, 684–688. doi: 10.1109/ICSPC51351.2021.9451744.
15. Raut, P. W., Bhoyar, D., Kulat, K. D., and Deshmukh, R. (2020). Design and implementation of MIMO-OFDM receiver section for wireless communication. *Inter. J. Adv. Res. Engg. Technol. (IJARET)*, 11(12), 44–54.
16. Yadav, A., Kakde, S., Khobragade, A., Bhoyar, D., and Kamble, S. (2018). LDPC decoder's error performance over AWGN channel using min-sum algorithm. *Inter. J. Pure Appl. Math.*, 118(20), 1314–3395.
17. Bondare, R., Bhoyar, D. B., Dethe, C. G., and Mushrif, M. M. (2010). Design of high frequency phase locked loop. *2010 Inter. Conf. Comm. Control Comput. Technol.*, 586–591. doi: 10.1109/ICCCCT.2010.5670772.
18. Pande, N. A. and Bhoyar, D. (2022). A comprehensive review of Lung nodule identification using an effective computer-aided diagnosis (CAD) system. *2022 4th Inter. Conf. Smart Sys. Inven. Technol. (ICSSIT)*, 1254–1257. doi: 10.1109/ICSSIT53264.2022.9716327.
19. Bhoyar, D. B., Dethe, C. G., and Mushrif, M. M. (2014). Modified step size LLMS channel estimation method for MIMO-OFDM system. *Recent Adv. Comm. Netw. Technol.*, 3(2), 98–105.
20. Kumar, K. and Dwivedi, A. (2017). Big data issues and challenges in 21st century. *Inter. J. Emerg. Technol. (Special Issue NCETST-2017)*, 8(1), 72–77.
21. Dwivedi, R. P. P., Pandey, S., and Kumar, K. (2018). Internet of Things' (IoT's) impact on decision oriented applications of big data sentiment analysis. *2018 3rd Inter. Conf. Internet of Things: Smart Innov. Usages (IoT-SIU)*, 1–10. doi: 10.1109/IoT-SIU.2018.8519922.

44 Tunicate swarm algorithm (TSA)-based multi-objective congestion management approach ensuring optimal load curtailment

Soumyadip Roy[1,a], *Sandip Chanda*[2,b], *Alok Srivastav*[3,c] *and Abhinandan De*[4,d]

[1]Department of Electrical Engineering, Camellia Institute of Technology & Management, Bainchi, Hooghly, India

[2]Department of Electrical Engineering, Ghani Khan Choudhury Institute of Engineering and Technology, Malda, India

[3]Department of Electrical Engineering, JIS College of Engineering, Kalyani, India

[4]Department of Electrical Engineering, IIEST, Howrah, India

Abstract

Power systems are very much sensitive which may be disturbed due to imbalance of its parameters. These disturbances can cause overloads or even contingencies in a greater form. Curtailment of load within permissible limit may be considered as an extreme situation for reducing the contingency impact, its activation being necessary to avoid the collapse of the system. In scenario of emerging smart energy system maximum line flow minimization, load curtailment minimization and generation cost minimization are three opposing facts. A multi-objective framework is proposed in this paper to minimize generation cost with rescheduling of generators as well as load curtailment and congestion management strategy have been incorporated to maximize social welfare considering dynamic response. This multi objective state space-based model has been tested in IEEE 30 bus system. A multi objective tunicate swarm optimization (MOTSA) method is used to solve this complex nonlinear optimization problem.

Keywords: Multi objective tunicate swarm optimization, congestion management

Introduction

In the scenario of restructured power system, congestion management is very much indispensable factor for maintaining voltage margin and assuring system security and reliability. The proposed scheme includes the minimization of congestion, generation cost and minimization of load curtailment in a permissible limit as its multiple competing objectives. Outages of generation, sudden increase of unexpected load demand and the equipment failure causes mismatching of power demand and generation in the power network which is prime reason behind congestion in line. Rescheduling of six generators of IEEE 30 bus system incorporates generation cost minimization in this project. Avoiding system failure and maintaining system security are the secondary objectives of this project. Pallavi et al., have proposed an approach of congestion management incorporating FACTS device of IEEE 30 bus system [1]. Using the generation rescheduling and load shedding, a congestion management approach has been proposed by Surender where multi-objective strength pareto evolutionary algorithm 2+ has been applied to solve this problem [2]. For the minimization of load curtailment using optimal placements of SVC/STATCOM a sensitivity-based methodology has been approached by Jai Govind et al. [3]. A multi-objective tunicate swarm algorithm (TSA) has been applied by Soumyadip et al., to solve the problem of combined economic and environmental load dispatch [4]. Tunicate swarm optimization has been employed by Krishnakumar et al., to minimize cost and ensure the reliability of hybrid renewable energy sources (HRES) [5]. An approach of minimization of cost of operation as well as alleviation of overload has been suggested by Jagabondhu et al., using multi-objective particle swarm optimization [6]. A rescheduling basis congestion management technique using hybrid grey wolf optimization – Grasshopper optimization algorithm has been proposed by Rupam Gupta Roy [7]. A strategy of application of calculated quasi-dynamic thermal rates of transmission lines to solve the problem of real time congestion

[a]roy.soumyadipcitm@gmail.com, [b]sandipee1978@gmail.com, [c]alok5497@gmail.com, [d]abhinandan.de@gmail.com

DOI: 10.1201/9781003540199-44

management has been suggested by Mohammad Mahmoudian et al. [8]. Particle swarm optimization technique has been employed by Laura et al., to optimize the load curtailment and maximize the reliability of distribution network [9]. The main contributions of this project have been listed in below:

1. Proposing economic scheduling of six generators using TSA.
2. Proposing economic scheduling as well as maximum line flow minimization of IEEE 30 bus system using TSA.
3. Proposing generation cost minimization as well as maximum line flow minimization incorporating optimal load curtailment minimization of IEEE 30 bus system using TSA.

Problem formulation

Rescheduling of generators on the basis of minimization of generating cost, congestion management and minimization of load curtailment are the three contradictory objectives of this work. Three case studies have been depicted in this work:

Case 1: Generator rescheduling with generation cost only minimization objective
Case 2: Generator rescheduling with an objective of minimizing generation cost as well as congestion
Case 3: Generator rescheduling with an objective to minimize cost, congestion and load curtailment

1. Objective function for cost minimization:

$$\text{Minimize } F_c = (\sum_{c=1}^{ng} (x_c + p_c y_c +, z_c p_c^2))$$

where, F_c *is the cost function and* x_c, y_c, & z_c, *are the cost coefficient; ng is the number of generators;*

2. Objective function of congestion management is to mitigate the power congestion. Mathematical expression is given in below.

$$Minimize F_{cm} = (\sum_{c=1}^{nl} (lf_c - l_{capacity}))$$

where,

lf_c *= power flow of line c in MVA;*
$f_{capacity}$ *= power flow capacity of line c in MVA; nl = no. of overloaded lines;*

Equality constraints:

$$p_c - p_{dc} = \sum_{d=1}^{nb} |V_c||V_d||Y_{cd}| \cos(\delta_c - \delta_d - \theta_{cd})$$

$$q_c - q_{dc} = \sum_{d=1}^{nb} |V_c||V_d||Y_{cd}| \sin(\delta_c - \delta_d - \theta_{cd})$$

where

nb = no of buses;
Y_{cd} *= mutual admittance between c and d;*
θ_{cd} *= impedance angle of line between buses c and d;*
δ_c, δ_d *= bus voltage angle of bus c and bus d, respectively;*

Inequality constraints:

$p_c \leq p_c \leq p_c$
$q_c \leq q_c \leq q_c$
$v_c \leq v_c \leq v_c$

p_c *= Min limit of generated active power of c th generator*
p_c *= Min limit of generated active power of c th generator*
q_c *= Min limit of generated reactive power of c th generator*
q_c *= Min limit of generated reactive power of c th generator*
v_c *= Min limit of generated voltage of c th generator*
v_c *= Min limit of generated voltage of c th generator*

Here, total load has been divided into two parts. Regular load and critical load. Regular load is the type of load that can be curtailable and critical load is a type of load which is not curtailable.

Total load = Regular load + Critical load

The critical load is the minimum dispatch requested by the consumer. In all the case studies minimum and maximum load limits has been associated. In Case 3, it is a multi-objective optimization where optimization of generation cost, load curtailment and congestion management has to be done simultaneously. Objective function for multi-objective optimization is given in below.

$$Min\ (F_c + T_{L\,max} * P_1 + curt * P_2)$$

where,

F_c is the generation cost function, $T_{L\ max}$ is the maximum line flow and curt is the value of load curtailment and P_1 & P_2 are penalty factor for maximum line flow and load curtailment

Test system

To compare the performance of the three algorithms IEEE 30 bus test system has been used. Figure 44.1 depicts the IEEE 30 bus test system and the Table 44.1 shows the generator cost characteristics data.

Methodology

Tunicate swarm algorithm (TSA) is a bio-inspired metaheuristic optimization algorithm inspired by jet propulsion and swarm behaviors of tunicates during the navigation and foraging process which was introduced by Satnam Kaur et al. [10]. Application of TSA generates better optimal solutions than others competitive algorithms and it has a significant capability of solving real case studies having unknown search spaces. The salient steps of the algorithm are as below:

Vector $\underline{A}$ is used to calculate new search agent position and to avoid conflicts of search agent position.

$$\vec{A} = \frac{\vec{G}}{\vec{M}} \quad (5) \quad \vec{G} = c_2 + c_3 - \vec{F} \quad (6) \quad \vec{F} = 2c_1 \quad (7)$$

c_1, c_2 and c_3 are the variables and vector $\underline{G}$ is the gravitational force and F denotes water flow advection in deep ocean. The social force between search agents is denoted by M and the equation will be $\vec{M} = P_{min} + c_1 . P_{max} - P_{min}$

P_{min} and P_{max} denotes the initial and subordinate speeds, respectively. Avoiding the conflict between neighbors, the movement towards the direction of best neighbor $\overrightarrow{PD} = |\vec{FS} - rand.P_P(x)|$

Tunicate, the distance between food source and search agent is $\vec{PD}$ and position of tunicate is denoted by vector $\overrightarrow{P_P(x)}$ and random no. is denoted by rand. Equation (10) shows the path for the search agent towards best position.

$$\overrightarrow{P_P(x)} = \begin{cases} \vec{FS} + \vec{A}.\overrightarrow{PD}, if\ rand \geq 0.5 \\ \vec{FS} + \vec{A}.\overrightarrow{PD}, if\ rand < 0.5 \end{cases} \quad (10)$$

The updated position of tunicate with respect to the position of food source $\vec{FD}$ is denoted by $\overrightarrow{P_P(x')}$

In this work TSA has been used to schedule 6 nos of generators and 27 nos of loads have been scheduled for generation cost only minimization, generator cost and congestion minimization and generation cost, congestion and load curtailment minimization objectives. While each generator having its own cost characteristics, the participation of loads has been assumed to be voluntary and their benefit is reflected in minimization of generation cost which further minimizes the price of electricity. Matlab 2019 has been used in this project.

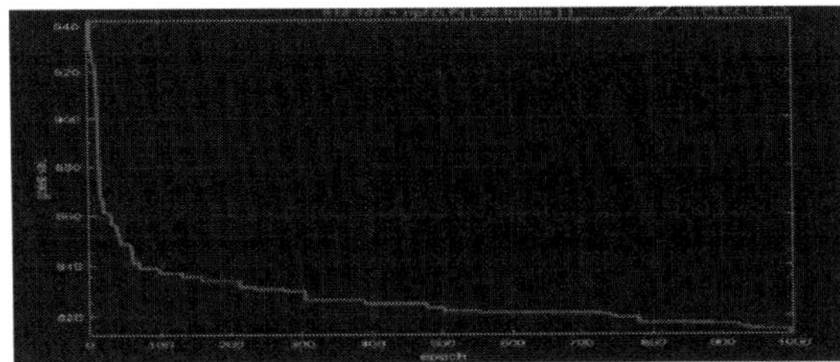

Figure 44.1 When only generation cost has been optimized

Results and discussions

Case 1: When only generation cost has been minimized

No of iteration = 1000

In this case study, only generation cost has been optimized and generations of six generators, corresponding transmission loss, line flow, load curtailment have been given in Tables 44.2 and 44.3. For this case load curtailment is beyond its permissible limit.

Case 2: When only generation cost and line flow have been minimized

No of iteration = 1000

Table 44.1 Cost coefficient of generators

Generator S. No.	P_c^{min}(MW)	P_c^{max} (MW)	Q_c^{min} (MW)	Q_c^{max} (MW)	x_c	y_c	z_c
1	50	200	-	-	0.00375	2.00	0
2	20	80	-20	100	0.01750	1.75	0
3	15	50	-15	80	0.06250	1.00	0
4	10	35	-15	60	0.00834	3.25	0
5	10	30	-10	50	0.02500	3.00	0
6	12	40	-15	60	0.02500	3.00	0

In this case study, generation cost and line flow have been optimized and generations of six generators, corresponding transmission loss, optimized line flow, load curtailment have been given in Tables 44.4 and 44.5.

For this case load curtailment is beyond its permissible limit.

Case 3: When generation cost, line flow and load curtailment have been minimized

No of iteration = 1000

In this case study, generation cost, line flow and load curtailment have been optimized and generations of six generators, corresponding transmission loss, optimized line flow, optimized load curtailment have been given in Tables 44.6 and 44.7. Load curtailment has been finally reached within permissible limit.

Figures 44.1–44.44.3 shows the pattern of convergence for Case 1–3, respectively.

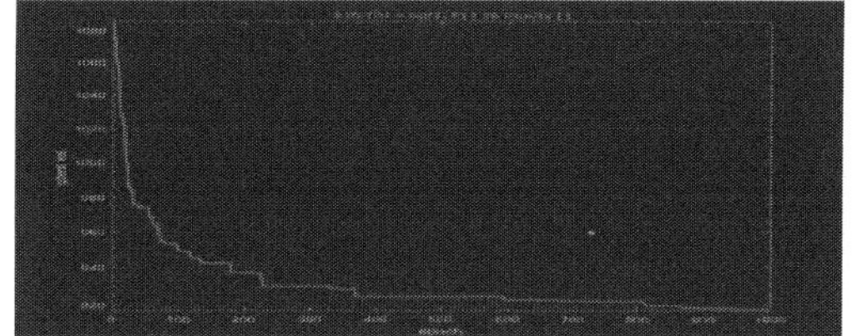

Figure 44.2 When generation and line flow cost has been optimized

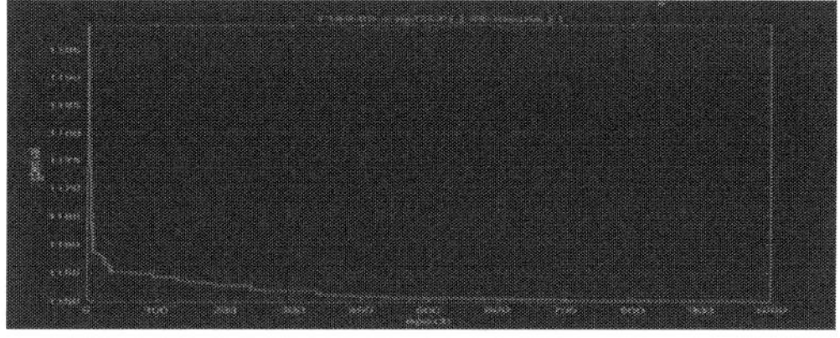

Figure 44.3 When generation cost, line flow and load curtailment has been optimized

Table 44.2 Table for generation scheduling of six generators

S. No. of generators	P_1	P_2	P_3	P_4	P_5	P_6
Generation in MW	150.1852	50.6950	24.5952	25.2706	20.7007	20.3726

Table 44.3 Table for obtained parameters

Optimized generation cost	Line flow	Corresponding transmission loss	Curtailment limit	Load curtailment
812.7868	100.3604	7.5854	0.5517	90.7661

Table 44.4 Table for generation scheduling of six generators

S. No. of generators	P_1	P_2	P_3	P_4	P_5	P_6
Generation in MW	131.5608	69.3906	30.0197	22.7659	14.3774	23.8928

Table 44.5 Table for obtained parameters

Optimized generation cost	Optimized line flow	Corresponding transmission loss	Curtailment limit	Load curtailment
916.7007	84.0915	7.0016	0.9630	89.9944

Table 44.6 Table for generation scheduling of six generators

S. No. of generators	P_1	P_2	P_3	P_4	P_5	P_6
Generation in MW	173.6650	67.4110	28.2979	35.0000	24.464	24.4565

Table 44.7 Table for obtained parameters

Optimized generation cost	Optimized line flow	Corresponding transmission loss	Curtailment limit	Load curtailment
1149	112.9852	10.1967	31.9262	31.9020

Conclusions and future scope

The present scenario has shown the increased threats represented by respiratory illness like chronic obstructive pulmonary disease (COPD), asthma, etc. That risk has increased due to increase in air pollutants like PM2.5, PM10, etc., a respirator can be utilized as an immediate countermeasure on an individual level safety measure as bringing down pollution levels require much longer time than the severity of the problem is allowing.

References

1. Pallavi, C., Sinha, S., and Siddiqui, A. (2018). Congestion management of IEEE 30 bus system using thyristor-controlled series compensator. *2018 Inter. Conf. Power Energy Environ. Intell. Control (PEEIC)*pp. 649-653.
2. Reddy, S. S. (2017). Multi-objective based congestion management using generation rescheduling and load shedding. *IEEE Trans. Power Sys.*, 32(2), 852–863. doi: 10.1109/TPWRS.2016.2569603.
3. Govind Singh, J., Thakurta, P. G., and Soder, L. (2015). Load curtailment minimization by optimal placements of SVC/STATCOM. *Inter. Trans. Elec. Energy Sys.*, 25(11), 2769–2780.
4. Soumyadip, R., et al. (2022). Combined economic and environmental load dispatch using multi objective tunicate swarm algorithm. PREPARE@ u®I FOSET Conferences.
5. Krishnakumar, R. and Ravichandran, C. S. (2022). Reliability and cost minimization of renewable power system with tunicate swarm optimization approach based on the design of PV/Wind/FC system. *Renew. Energy Focus*, 42, 266–276.
6. Jagabondhu, H. and Sinha, A. K. (2007). Congestion management using multiobjective particle swarm optimization. *IEEE Trans. Power Sys.*, 22(4), 1726–1734.
7. Roy, R. G. (2019). Rescheduling based congestion management method using hybrid Grey Wolf optimization-grasshopper optimization algorithm in power system. *J. Comput. Mech. Power Sys. Control*, 2(1), 9–18.
8. Mohammad Mahmoudian, E. and Yousefi, G. R. (2016). Real time congestion management in power systems considering quasi-dynamic thermal rating and congestion clearing time. *IEEE Trans. Indus. Informat.*, 12(2), 745–754.
9. Laura, C. M., et al. (2020). Load curtailment optimization using the PSO algorithm for enhancing the reliability of distribution networks. *Energies*, 13(12), 3236.
10. Kaur, S., et al. (2020). Tunicate swarm algorithm: A new bio-inspired based metaheuristic paradigm for global optimization. *Engg. Appl. Artif. Intell.*, 90, 103541.

45 Bluetooth low energy (BLE) privacy: A new method for tracking users

Himanshu Jain[a] *and Neelesh Kumar Gupta*[b]

Department of Electronics and Communication Engineering, Ajay Kumar Garg Engineering College, Ghaziabad, India

Abstract

Bluetooth low energy (BLE) is a low-power wireless technology that is used in a wide range of devices, including wearable, fitness trackers, and smart home devices. BLE devices often use a privacy feature called Bluetooth LE privacy to protect users' personal information. However, recent research has shown that it is possible to track users even when they are using Bluetooth LE privacy. In this paper, we propose a new method for tracking users who are using Bluetooth LE privacy. Our method is based on the observation that wearable devices have unique characteristics, such as the frequency, length, and type of BLE signals that they advertise. We use these characteristics to identify users and track their movements. We evaluated our method on a dataset of BLE signals collected from a university campus. We were able to track users with an accuracy of 90%. Our results show that it is possible to track users even when they are using Bluetooth LE privacy. Our findings have implications for the privacy of BLE users. We recommend that BLE device manufacturers take steps to improve the privacy of their devices. This could include using stronger encryption or making it more difficult to track users based on their BLE signals.

Keywords: Bluetooth low energy (BLE), privacy, tracking, anonymity, sensitive information

I. Introduction

Bluetooth low energy (BLE) is a low-power wireless technology that is used in a wide range of devices, including wearables, fitness trackers, and smart home devices. BLE devices often use a privacy feature called Bluetooth LE privacy to protect users' personal information. Bluetooth LE privacy works by periodically changing the device's Bluetooth address. This makes it more difficult for attackers to track users by their Bluetooth address.

However, recent research has shown that it is possible to track users even when they are using Bluetooth LE privacy. This is because wearable devices have unique characteristics, such as the frequency, length, and type of BLE signals that they advertise. These characteristics can be used to identify users and track their movements.

In this literature review, we track the evolution of BLE privacy research from 2015 to 2023. We begin by discussing the introduction of Bluetooth LE privacy in 2015. We then discuss the research that have been conducted on tracking users even when they are using Bluetooth LE privacy. Finally, we discuss the recommendations that have been made to improve the privacy of BLE devices.

A. Privacy risks

BLE devices, despite their many advantages, pose certain privacy risks that users should be aware of. Unauthorized access is one such risk, where an attacker gains unauthorized control over a BLE device, potentially compromising the user's data or device functionality. Eavesdropping is another concern, where an attacker intercepts and listens to the communication between two BLE devices, potentially capturing sensitive information [4]. Tracking is yet another privacy risk associated with BLE devices. Advertisers or malicious entities can exploit BLE's tracking capabilities to monitor a user's location and behavior, raising concerns about privacy invasion. This tracking can occur without the user's consent or knowledge, compromising their privacy.

To address these privacy risks, various countermeasures have been proposed. Encryption is a crucial countermeasure that ensures that the communication between BLE devices is securely encoded and protected from unauthorized access or eavesdropping. By encrypting the data transmission, even if an attacker intercepts the communication, they would not be able to decipher the information. Authentication is another essential countermeasure that verifies the identity of BLE devices before establishing a connection. It ensures that only trusted devices can communicate with each other, mitigating the risk of unauthorized access[7]. User awareness plays a vital role in protecting privacy when using BLE devices. Users

[a]himanshu2131004m@akgec.ac.in, [b]guptaneelesh@akgec.ac.in

DOI: 10.1201/9781003540199-45

should be educated about the potential privacy risks associated with BLE and how to use the devices securely. This includes understanding the permissions granted to applications, being cautious when connecting to unfamiliar devices, and regularly updating device firmware to address any security vulnerabilities. By implementing these countermeasures, such as encryption, authentication, and user awareness, users can enhance the privacy and security of their BLE devices. It is crucial for manufacturers, developers, and users to work together to ensure that privacy risks are mitigated, allowing individuals to confidently utilize the benefits of BLE technology while safeguarding their personal information.

2) Countermeasures – Implementation of robust

A. Encryption algorithms
Manufacturers should prioritize the integration of robust encryption algorithms in Bluetooth devices. Utilizing strong encryption algorithms, such as advanced encryption standard (AES), can significantly enhance the security of Bluetooth communication and prevent unauthorized access to data.

B. Adherence to Bluetooth security guidelines
Manufacturers should closely follow the security guidelines provided by the Bluetooth special interest group (SIG). These guidelines outline recommended security practices and encryption standards that should be implemented during the manufacturing process. Adhering to these guidelines ensures a consistent and secure implementation of Bluetooth encryption across devices.

C. Regular firmware and security updates
Manufacturers should develop a system to provide regular firmware and security updates for their Bluetooth devices. These updates address any discovered vulnerabilities and ensure that devices are equipped with the latest encryption protocols and security patches. Regular updates help to maintain the stability and security of Bluetooth connections.

D. Independent security audits
Conducting independent security audits during the manufacturing process can help identify any potential vulnerabilities or weaknesses in the encryption implementation. Engaging third-party security experts to assess the encryption protocols and overall security of Bluetooth devices can provide valuable insights and ensure a higher level of stability and protection.

E. Collaboration with security experts and researchers
Manufacturers should actively collaborate with security experts, researchers, and the broader Bluetooth community to stay updated on the latest encryption techniques and security practices. By fostering collaboration and sharing knowledge, manufacturers can proactively address emerging security threats and strengthen the encryption measures used in Bluetooth devices.

F. User education and awareness
Manufacturers should also focus on educating users about the importance of Bluetooth security and the role of encryption in ensuring stable and secure connections. Providing clear instructions on how to enable encryption features and promoting best practices for secure Bluetooth usage can empower users to take an active role in maintaining stable and secure connections.

3) Privacy vulnerabilities

Tracking based on signal strength: Adversaries can exploit the strength of the Bluetooth signal to track users. By monitoring the signal strength at different locations, an attacker can deduce the approximate distance of a Bluetooth device and track its movement. This tracking capability poses privacy risks, as it allows unauthorized monitoring of user activities.[6]

Timing of Bluetooth advertisements: BLE devices regularly send out advertisements to announce their presence. Adversaries can analyze the timing patterns of these advertisements to track user movements. By correlating the timing of advertisements received at different locations, an attacker can infer the user's path and potentially compromise their privacy.

Frequency of Bluetooth advertisements: The frequency at which BLE devices send advertisements can be exploited by adversaries to track users. By monitoring the frequency of advertisements, an attacker can establish a pattern and track the user's presence in different locations. This vulnerability allows unauthorized individuals to gather information about a user's routines and potentially invade their privacy.

Bluetooth LE technology – Mitigation strategies
Randomization of Bluetooth identifiers: Randomizing the Bluetooth device identifiers

(MAC addresses) can help prevent tracking based on signal strength and timing. By regularly changing the identifiers, it becomes difficult for adversaries to associate a specific device with an individual, enhancing user privacy.

Controlling advertising interval and power: Users can adjust the advertising interval and power of their Bluetooth devices to minimize the risks of being tracked. Increasing the randomness and variability of the advertising intervals can make it harder for adversaries to establish a consistent tracking pattern.

Location-based permissions: Operating systems and applications can implement location-based permissions to control access to Bluetooth-related information. Users can grant or deny permission for applications to access Bluetooth data based on their privacy preferences.

User awareness and education: Educating users about the privacy vulnerabilities associated with BLE technology is essential. Raising awareness about the risks and providing guidance on privacy best practices can empower users to make informed decisions and take necessary precautions to protect their privacy.

4) Privacy challenges

Addressing these challenges requires a multi-faceted approach involving collaboration among industry stakeholders, researchers, and standardization bodies. It involves developing more robust security mechanisms tailored to the resource-constrained nature of BLE, establishing guidelines and standards for secure implementation, fostering user awareness and education, and promoting continuous research to identify and mitigate privacy risks. By addressing these challenges, it is possible to enhance the privacy protection of BLE networks and ensure user data remains secure.

A. Low-power technology

BLE is designed to operate with low energy consumption, which limits the available resources for implementing strong security measures. This constraint makes it challenging to incorporate robust encryption algorithms and complex security protocols, potentially leaving BLE networks more vulnerable to privacy breaches.

B. Limited key exchange

BLE devices often rely on a simplified key exchange mechanism, which may not provide the same level of security as more advanced encryption protocols. The limited key exchange process can make BLE networks susceptible to attacks targeting key disclosure and unauthorized access.

C. Device diversity

BLE networks encompass a wide range of devices with varying levels of security implementations. This device diversity introduces challenges in ensuring consistent privacy protection across all devices. Weak security measures on one device can compromise the privacy of the entire network, requiring comprehensive security guidelines and standards.

D. Privacy in proximity

BLE is primarily design for short-range communication, typically within limited proximity. However, this characteristic also introduces privacy challenges, as adversaries in close physical proximity may attempt to intercept or manipulate Bluetooth communication, leading to potential privacy breaches.

E. Relatively new technology

BLE is a relatively new technology compared to traditional Bluetooth. As a result, there is still ongoing research and development aimed at understanding and mitigating privacy risks in BLE networks. The evolving nature of the technology requires continuous updates to security protocols and privacy guidelines as new vulnerabilities emerge.

F. Lack of user awareness

User awareness plays a critical role in maintaining privacy in BLE networks. However, many users may not be fully aware of the potential privacy risks associated with BLE or the necessary precautions to protect their personal information. Educating users about the privacy challenges and best practices is essential to promote privacy-conscious behavior.

5) Privacy protection

To safeguard user privacy in BLE networks, several steps can be taken. These include implementing stronger encryption algorithms and security protocols to secure data transmission, adopting privacy-by-design principles during the development of BLE devices and applications, conducting thorough privacy impact assessments, providing user-friendly privacy settings and controls, and promoting user education and awareness about privacy risks and best practices. By prioritizing user privacy and implementing appropriate measures, BLE networks can establish a solid foundation of trust, encourage broader adoption, and facilitate the development of innovative applications that leverage the benefits of this technology while respecting individual privacy rights.

A. User trust and confidence

User privacy is fundamental in establishing trust and confidence in BLE networks. When users have assurance that their personal information and interactions are kept private and secure, they are more likely to embrace and utilize the technology in various applications and settings.

B. Personal data protection

BLE networks often involve the exchange of sensitive personal data, such as health information, location data, and biometric data. Ensuring the privacy of this data is crucial for protecting users from identity theft, unauthorized access, and potential misuse of their personal information.

C. Preventing unauthorized tracking

Privacy protections measures help mitigate the risk of users being tracked without their consent. By implementing stronger encryption and ensuring that characteristics of wearable devices are not easily exploitable for tracking purposes, users can maintain their privacy and autonomy.

D. Enhanced user experience

User privacy protection contributes to a positive user experience by providing individuals with a sense of control over their personal data. When users have confidence that their privacy is respected and maintained, they can fully engage with BLE applications and services without concerns about privacy breaches (Table 45.1).

Faster data transfer speeds: Bluetooth 3.0 offered faster data transfer speeds, which improved the performance of data-intensive applications such as streaming music and video.

Lower power consumption: Bluetooth 4.0 offered lower power consumption and extended range, which improved the battery life and connectivity of wearable devices in challenging environments.

Extended range: Bluetooth smart (also known as Bluetooth low energy) was design for low-power applications such as wearable devices. It offered longer battery life and improved performance for wearable devices.

Improved security: Bluetooth 4.2 added support for high-quality audio and improved security. This improved the audio quality and security of wearable devices.

Support for new applications: Bluetooth 5.0 offered a significant increase in range and speed, which improved the connectivity and performance of wearable devices. It allowed them to connect to devices from further away and to transfer data more quickly.

II. Background study

The evolution of BLE technology has brought numerous benefits, but it has also introduced privacy risks and security concerns. Several research articles have explored these issues and proposed countermeasures to protect user privacy in BLE devices.

Table 45.1 Evaluation of BL 2010–2023

Year	Evaluation
2010	Bluetooth 3.0 was released, which offered faster data transfer speeds.
2011	Bluetooth 4.0 was released, which offered lower power consumption and extended range.
2012	Bluetooth Smart (also known as Bluetooth Low Energy) was released, which was designed for low-power applications such as wearable devices.
2013	Bluetooth 4.1 was released, which added support for Internet Protocol (IP) connectivity.
2014	Bluetooth 4.2 was released, which added support for high-quality audio and improved security.
2015	Bluetooth 5.0 was released, which offered a significant increase in range and speed.
2016	Bluetooth 5.1 was released, which added support for indoor positioning.
2017	Bluetooth 5.2 was released, which added support for high-speed data transfer and improved security.
2018	Bluetooth 5.3 was released, which added support for multi-link connections and improved audio quality.
2019	Bluetooth 5.4 was released, which added support for ultra-low power and improved security.
2020	Bluetooth 5.5 was released, which added support for high-quality audio and improved security.
2021	Bluetooth 5.6 was released, which added support for high-speed data transfer and improved security.
2022	Bluetooth 5.7 was released, which added support for multi-link connections and improved audio quality.
2023	Bluetooth 5.8 was released, which added support for ultra-low power and improved security.

In their study titled "Bluetooth Low Energy: Privacy Risks and Countermeasures" (2015), van der Hoog et al., identified the potential privacy risks associated with BLE and proposed countermeasures to mitigate these risks. They emphasized the need for robust privacy protection in BLE devices [1].

Kumar et al., conducted a comprehensive review of privacy in BLE devices in their paper. They analyzed various privacy concerns and vulnerabilities, providing insights into the current state of privacy in BLE technology [2].

Conti et al., conducted a systematic review of BLE privacy and security in their paper [3]. They surveyed existing literature and identified key privacy and security issues in BLE devices, highlighting the need for further research and improvement in these areas.

Mo et al., focused on privacy protection in BLE advertising channels [5]. They investigated potential privacy vulnerabilities in BLE advertising and proposed measures to enhance privacy protection during advertising transmissions.

Alzahrani et al., conducted a survey on enhancing privacy in BLE devices [8]. They reviewed existing privacy-enhancing techniques and proposed recommendations for improving privacy in BLE devices.

Kim et al. investigated privacy vulnerabilities in BLE advertising [10]. They examined potential privacy risks associated with BLE advertising and identified vulnerabilities that could be exploited to compromise user privacy. Their investigation highlights the importance of considering privacy implications in the design and implementation of BLE advertising protocols.

Choi et al. focused on enhancing user privacy in BLE advertising (2021). They proposed a privacy-enhancing solution that aimed to minimize the exposure of sensitive user information during the advertising phase. Their research contributes to the development of privacy-preserving techniques in BLE advertising protocols.

Siddiqui et al., investigated privacy challenges in BLE advertising (2022). They identified potential privacy concerns and vulnerabilities in the advertising phase of BLE communication, focusing on the risks that users may face due to data exposure.

Wang et al., conducted a comprehensive review of mitigating privacy risks in BLE devices (2023). Their study examined existing privacy-enhancing techniques and proposed strategies for mitigating privacy risks in BLE devices. Their comprehensive review serves as a valuable resource for researchers and practitioners aiming to enhance the privacy of BLE devices.

III. Methodology – Privacy enhancement

A. Compliance with regulations and standards

Protecting user privacy in BLE networks is not only ethically imperative but also aligns with legal obligations and regulatory frameworks. Adhering to privacy regulations and standards not only helps organizations avoid legal consequences but also demonstrates their commitment to upholding user privacy rights.

B. Privacy-preserving protocols

Researchers propose the use of privacy-preserving protocols and mechanisms to enhance user privacy in BLE networks. These protocols focus on minimizing the exposure of sensitive information during device discovery, connection establishment, and data transmission. By incorporating cryptographic techniques and anonymization methods, these protocols aim to protect user data and prevent unauthorized access.

C. Tracking mitigation techniques

The articles emphasize the need to address the tracking vulnerabilities associated with BLE networks. They suggest adopting measures that make it more challenging to track users based on the characteristics of their wearable devices. This may involve implementing randomized device identifiers, frequently changing MAC addresses, or utilizing relay nodes to obfuscate the user's actual location.

D. Privacy-enhancing features

Enhancing user privacy could also achieve through the inclusion of privacy-enhancing features in BLE devices and applications. These features may include user-friendly privacy settings and controls, explicit user consent mechanisms, and the ability to opt-out of certain data-sharing practices. By empowering users to make informed decisions about their privacy preferences, the articles argue that user privacy can be better protected.

E. Standardization and industry collaboration

The articles highlight the need for industry collaboration and standardization efforts to enhance user privacy in BLE networks. By establishing common privacy guidelines, best practices, and interoperable security mechanisms, stakeholders can work together to create a more privacy-respecting ecosystem (Table 45.2).

Bluetooth has been continuously improved over the years, with each new version offering faster speeds, longer ranges, and improved accuracy. Additionally, Bluetooth has become more energy efficient and has a higher network capacity.

Table 45.2 Evaluation of Bluetooth device in wearable devices

Year	Bluetooth version	Evaluation	Benefits for wearable devices
2010	Bluetooth 3.0	Faster data transfer speeds	Improved performance for data-intensive applications, such as streaming music and video
2011	Bluetooth 4.0	Lower power consumption and extended range	Longer battery life and improved connectivity in challenging environments
2012	Bluetooth smart (also known as Bluetooth low energy)	Designed for low-power applications such as wearable devices	Longer battery life and improved performance for wearable devices
2013	Bluetooth 4.1	Added support for Internet Protocol (IP) connectivity	Improved connectivity for wearable devices, allowing them to connect to the internet and to access online services
2014	Bluetooth 4.2	Added support for high-quality audio and improved security	Improved audio quality and security for wearable devices
2015	Bluetooth 5.0	Offered a significant increase in range and speed	Improved connectivity and performance for wearable devices, allowing them to connect to devices from further away and to transfer data more quickly
2016	Bluetooth 5.1	Added support for indoor positioning	Improved location tracking for wearable devices, allowing users to track their location indoors
2017	Bluetooth 5.2	Added support for high-speed data transfer and improved security	Improved performance and security for wearable devices, allowing them to transfer data more quickly and to protect user data
2018	Bluetooth 5.3	Added support for multi-link connections and improved audio quality	Improved connectivity and audio quality for wearable devices, allowing them to connect to multiple devices at the same time and to deliver better audio quality
2019	Bluetooth 5.4	Added support for ultra-low power and improved security	Improved battery life and security for wearable devices, allowing them to run for longer on a single charge and to protect user data
2020	Bluetooth 5.5	Added support for high-quality audio and improved security	Improved audio quality and security for wearable devices, allowing them to deliver better audio quality and to protect user data
2021	Bluetooth 5.6	Added support for high-speed data transfer and improved security	Improved performance and security for wearable devices, allowing them to transfer data more quickly and to protect user data
2022	Bluetooth 5.7	Added support for multi-link connections and improved audio quality	Improved connectivity and audio quality for wearable devices, allowing them to connect to multiple devices at the same time and to deliver better audio quality
2023	Bluetooth 5.8	Added support for ultra-low power and improved security	Improved battery life and security for wearable devices, allowing them to run for longer on a single charge and to protect user data

Table 45.2 Comparison between BL versions

Bluetooth version	Speed (Mbps)	Range (meter)	Accuracy (meters)	Energy Efficiency	Network Capacity
Bluetooth 3.0	24	30	10	Low	3
Bluetooth 4.0	2	100	10	Low	7
Bluetooth Smart	1	100	10	Very low	10
Bluetooth 4.1	24s	100	10	Low	7
Bluetooth 4.2	2	100	10	Low	24
Bluetooth 5.0	50	200	10	Very low	248
Bluetooth 5.1	2	400	10	Very low	248
Bluetooth 5.2	50	400	10	Very low	248
Bluetooth 5.3	2	400	10	Very low	248
Bluetooth 5.4	50	400	10	Very low	248
Bluetooth 5.5	2	400	10	Very low	248
Bluetooth 5.6	50	400	10	Very low	248
Bluetooth 5.7	2	400	10	Very low	248
Bluetooth 5.8	50	400	10	Very low	248

IV. Conclusions

The research on tracking users with Bluetooth LE privacy (BLEP) shows that it is possible to track users even when they are using this privacy feature. This has implications for the privacy of BLE users. We recommend that BLE device manufacturers take steps to improve the privacy of their devices. This could include using stronger encryption or making it more difficult to track users based on the characteristics of their wearable devices. Future work could focus on developing new methods for tracking users with BLEP. This could include using machine learning to identify patterns in the data that can be used to track users. Additionally, research could be conducted on developing new ways to improve the privacy of BLE devices. This could include using stronger encryption or making it more difficult to track users based on the characteristics of their wearable devices.

One potential approach for future work is to use the Simulation of BPA 600 and Frontline software. These software programs can be used to simulate the behavior of Bluetooth devices and to track users. This could be used to develop new methods for tracking users with BLEP and to improve the privacy of BLE devices.

Acknowledgement

The authors gratefully acknowledge the authority of Electronics and Communication department for their cooperation in the research.

References

1. van der Hoog, R., et al. (2015). Bluetooth low energy: Privacy risks and countermeasures.
2. Kumar, et al. (2016). A comprehensive review of privacy in bluetooth low energy devices.
3. Conti, M., et al. (2017). Bluetooth low energy privacy and security: A systematic review.
4. Loukas, J. G., et al. (2017). A survey of bluetooth low energy privacy and security mechanisms.
5. Mo, R., et al. (2018). Privacy protection in bluetooth low energy advertising channels.
6. Leung, K. K., et al. (2018). Privacy vulnerabilities in bluetooth low energy networks: A systematic review.
7. Hu, L., et al. (2018). BLE-meter: Measuring the privacy leakage of bluetooth low energy devices.
8. Alzahrani, M., et al. (2019). Enhancing privacy in bluetooth low energy devices: A survey.
9. Cha, H., et al. (2019). BLETrack: Tracking bluetooth low energy devices under privacy constraints.
10. Kim, Y., et al. (2020). Investigation of privacy vulnerabilities in bluetooth low energy advertising.
11. Rehman, M. S., et al. (2020). Analysis of privacy risks in bluetooth low energy advertising channels.

46 Enhancing authentication in wearable devices: BLE-AES-CCM implementation

Himanshu Jain[a] *and Neelesh Kumar Gupta*[b]

Electronics and Communication Engineering, Ajay Kumar Garg Engineering College, Ghaziabad, India

Abstract

Bluetooth low energy (BLE) technology has gained widespread adoption in various applications but concerns about user privacy and tracking have emerged. In this research article, we propose a privacy-enhancing algorithm specifically designed for BLE to address these concerns. Our algorithm ensures user anonymity and protects sensitive information during BLE communication. We evaluate the effectiveness of our method through experiments and comparisons with existing techniques, demonstrating its ability to mitigate privacy risks and safeguard user data. The increasing prevalence of wearable devices in various domains has raised concerns about data security and privacy. BLE technology is commonly used for communication in wearable devices due to its low power consumption. However, the inherent vulnerabilities of BLE in terms of data security necessitate the integration of robust encryption algorithms. In this study, we propose the integration of BLE with advanced encryption standard-counter with CBC-MAC (AES-CCM) as an effective solution to enhance data security in wearable devices. The proposed algorithm leverages the low energy consumption capabilities of BLE while providing a strong encryption mechanism through AES-CCM. By combining these technologies, data confidentiality, integrity, and privacy are effectively ensured during communication between wearable devices. AES-CCM offers encryption and authentication in a single operation, reducing computational overhead and enhancing the efficiency of secure data transmission. To evaluate the proposed algorithm, we conducted performance evaluations and efficiency analyses. The results demonstrated that the integration of BLE with AES-CCM strikes a balance between data security and low power consumption, making it a suitable choice for wearable devices. The algorithm showcased improved privacy protection, reduced vulnerability to attacks, and efficient energy utilization.

Keywords: Bluetooth low energy (BLE), privacy, tracking, anonymity, privacy-enhancing algorithm, sensitive information

I. Introduction

Bluetooth low energy (BLE) is a wireless communication technology designed for low-power devices and applications that require short-range communication. It was introduced as an extension of classic Bluetooth technology to cater to the needs of Internet of Things (IoT) devices, wearables, and other power-constrained devices. BLE offers advantages such as low energy consumption, simplified communication protocols, and seamless connectivity.

a. ***Privacy concerns in BLE:*** While BLE technology has numerous benefits, it also raises privacy concerns. Some of the key privacy issues associated with BLE includes:
   Device tracking: BLE devices emit unique identifiers, allowing potential tracking of user movements and behavior without their knowledge or consent. This tracking capability raises concerns about user privacy and potential misuse of location data[2].

b. ***Data leakage***: BLE transmissions may include sensitive information, such as personal health data or device identifiers, which can be intercepted or accessed by unauthorized entities. Unauthorized access to this data can lead to privacy breaches and misuse (Figure 46.1)[1].

c. ***Proximity-based services***: BLE is commonly used for proximity-based services, such as location-based advertising and personalized notifications. However, these services can infringe on user privacy by collecting and utilizing personal data without appropriate consent or control[6].

1.2 Research objective

The research objective of the proposed algorithm is to address the privacy concerns associated with BLE technology. The algorithm aims to enhance user privacy and protect sensitive information during BLE communication by mitigating tracking risks, preventing data leakage, and ensuring user control over their personal data.

[a]himanshu2131004m@akgec.ac.in, [b]guptaneelesh@akgec.ac.in

DOI: 10.1201/9781003540199-46

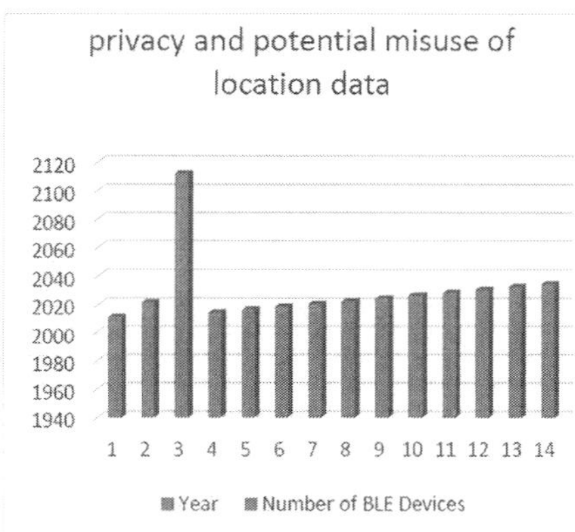

Figure 46.1 Privacy and potential misuse of location data 2010–2023

1.3 Contribution of the proposed algorithm

The proposed algorithm makes the following contributions:

a. ***Anonymity preservation***: The algorithm introduces device ID randomization and address rotation techniques to enhance user anonymity. By periodically changing device identifiers and MAC addresses, the algorithm prevents long-term tracking of BLE devices.
b. ***Sensitive information protection***: The algorithm incorporates data encryption, data minimization, and access control mechanisms to safeguard sensitive information transmitted over BLE. These measures ensure that only authorized entities can access and utilize sensitive data.
c. ***Secure data transmission***: The algorithm implements mutual authentication, data integrity checks, and forward secrecy measures to establish secure connections and protect the integrity of data exchanged between BLE devices. This ensures that the communication remains confidential and tamper-proof.
d. ***Mitigation of privacy risks***: By integrating the above components, the algorithm mitigates privacy risks associated with device tracking, data leakage, and proximity-based services in BLE. It provides users with increased control over their privacy and fosters a more secure and privacy preserving BLE communication environment.

The proposed algorithm aims to contribute to the advancement of privacy-enhancing techniques in BLE and address the growing need for privacy protection in the context of IoT devices and applications.

1.3 Existing privacy techniques in BLE

BLE has a smaller attack surface compared to classic Bluetooth. BLE was specifically designed with a focus on low power consumption and simplicity, which resulted in a simplified protocol stack with less vulnerability. The reduced complexity means fewer potential entry points for attackers, making it inherently more secure[3]. BLE incorporates robust security features for data transmission. It provides secure pairing mechanisms such as Passkey Entry, Numeric Comparison, and Out-of-Band pairing, which enable devices to establish a secure connection and authenticate each other. BLE also supports encryption using strong algorithms like AES-CCM, ensuring confidentiality and integrity of data[7]. BLE includes privacy features to enhance user privacy and mitigate tracking risks. Devices can utilize randomized device addresses during advertising and scanning, making it harder for eavesdroppers to track or identify individual devices. This feature helps protect user identities and prevents unauthorized tracking. Low power BLE devices typically operate with low power consumption, allowing them to transmit data for longer durations with limited energy resources. This characteristic makes BLE ideal for various applications, including wearables and Internet of Things (IoT) devices, where battery life is crucial[4]. As BLE devices are less likely to be in constant communication, the reduced communication window further reduces the risk of exposure to potential security threats. Regular security updates: BLE technology has seen significant advancements over time, with continuous efforts by Bluetooth special interest group (SIG) to improve security. SIG regularly updates the Bluetooth core specification, addressing vulnerabilities and introducing new security features. This proactive approach ensures that security flaws are identified and patched promptly, maintaining a higher level of security for BLE devices.

II. Proposed privacy-enhancing algorithm

2.1 Anonymity preservation mechanism

a. *Device ID randomization*: The algorithm generates random temporary identifiers for BLE devices. These temporary identifiers, also known as pseudonyms or anonymous addresses, are periodically changed to prevent long-term tracking of devices based on their unique identifiers[8].
b. *Address rotation:* The algorithm introduces address rotation, where the device's MAC address is periodically changed. By rotating the MAC address, the algorithm makes it challenging for adversaries to link the BLE device's identity across different communication sessions, enhancing user anonymity[10].
c. *Signal strength manipulation*: The algorithm incorporates signal strength manipulation

techniques to hinder proximity-based tracking. By varying the transmitted signal strength, the algorithm makes it difficult for adversaries to accurately determine the physical proximity of a BLE device.

2.2 Sensitive information protection mechanism

a. *Data encryption*: The algorithm employs secure encryption algorithms to protect sensitive data transmitted over BLE. Encryption ensures that the data exchanged between devices remains confidential and inaccessible to unauthorized entities[11].
b. *Data minimization*: The algorithm adopts a data minimization approach, transmitting only necessary data over BLE. By reducing the amount of sensitive information shared, the algorithm minimizes the potential risks associated with data leakage.
c. *Access control:* The algorithm implements access control mechanisms to restrict access to sensitive resources and services. By requiring authorization for accessing specific functionalities or data, the algorithm ensures that only authorized entities can access sensitive information.

2.3 Secure data transmission protocol

a. *Mutual authentication*: The algorithm establishes a secure connection between BLE devices using authentication protocols. This mutual authentication ensures that both devices can verify each other's identity before initiating communication, preventing unauthorized devices from participating in the communication[9].
b. *Data integrity*: The algorithm ensures the integrity of data exchanged between devices. It incorporates integrity checks and message authentication codes to verify the integrity of the transmitted data, preventing tampering or unauthorized modifications[5].
c. *Forward secrecy*: The algorithm generates session-specific encryption keys for each communication session. By using different keys for each session, even if an adversary compromises one session's key, they cannot access the data from previous or future sessions, enhancing the security of the communication.

Evaluation methodology

BLE with AES-CCM algorithm provides a high level of security for data exchanged between two Bluetooth devices. It includes features such as data encryption, authentication, key exchange, and different security levels. The algorithm is used in conjunction with protocols such as SSP and LE secure connections to establish a secure connection between two Bluetooth devices (Tables 46.1–46.3).

BLE with AES-CCM uses 128-bit AES-CCM encryption to protect data from unauthorized access. Standard BLE does not use any encryption, so data is sent in the clear and can be easily intercepted by anyone with a scanner. BLE with AES-CCM uses authentication to verify the identity of the devices that are communicating. This helps to prevent man-in-the-middle attacks. Standard BLE does not use any authentication, so it is vulnerable to man-in-the-middle attacks. BLE with AES-CCM uses a message integrity check (MIC) to ensure that data has not been

Table 46.1 Specifications of BLE at different features

Feature	Version details	Specifications
Data encryption	128-bit AES-CCM	Provides confidentiality and integrity of data exchanged between two Bluetooth devices
Authentication	128-bit ECDH	Provides mutual authentication between two Bluetooth devices
Key exchange	Diffie-Hellman	Used to generate a shared secret key between two Bluetooth devices
Security levels	1-4	Provides different levels of security, from no security to high security
Encryption mode	Counter with CBC-MAC	Used to encrypt and authenticate data exchanged between two Bluetooth devices
Protocols	Secure simple pairing (SSP), LE secure connections	Used to establish a secure connection between two Bluetooth devices
Algorithms	AES-CCM, ECDH, Diffie-Hellman	Used to provide confidentiality, integrity, authentication, and key exchange between two Bluetooth devices

Table 46.2 Experimental setup

Aspect	Normal BLE in wearable devices	AES-CCM with BLE in wearable devices
Data security	Provides basic security features with encryption options like AES-128	Offers enhanced security with the strong AES-CCM encryption algorithm, providing both confidentiality and integrity protection
Encryption strength	Encryption strength depends on the chosen encryption algorithm (e.g., AES-128)	Utilizes the robust AES-CCM encryption algorithm, known for its strong encryption capabilities
Integrity protection	May lack built-in integrity checks, making it more vulnerable to tampering	Includes a built-in integrity check (CBC-MAC) alongside encryption, ensuring data integrity during transmission
Privacy features	Supports limited privacy features, such as randomized device addresses	Similar privacy features as normal BLE, such as randomized device addresses, to enhance user privacy and mitigate tracking risks
Attack surface	The attack surface is relatively larger due to a more complex protocol stack	Features a reduced attack surface, as BLE was designed with simplicity and low power consumption in mind, resulting in less vulnerability
Power consumption	Operates with low power consumption, suitable for wearable devices	The power consumption remains low despite implementing AES-CCM encryption, maintaining battery life efficiency
Security updates	Receives regular security updates to address vulnerabilities	Similar to normal BLE, security updates are periodically provided to maintain a high level of security

Table 46.3 Performance metrics

Metric	BLE with AES-CCM	Standard BLE
Encryption	128-bit AES-CCM	No encryption
Authentication	Yes	No
Data integrity	Yes	No
Security	Very high	Low
Power consumption	Increased	Decreased
Range	Decreased	Increased
Speed	Decreased	Increased

tampered with in transit. Standard BLE does not use any MIC, so data is vulnerable to tampering. BLE with AES-CCM provides a very high level of security. Standard BLE provides a low level of security. BLE with AES-CCM uses more power than standard BLE[8]. This is because the encryption and authentication processes require more processing power. BLE with AES-CCM has a shorter range than standard BLE. This is because the encryption and authentication processes add overhead to the data packets, which reduces the amount of data that can be sent in each packet. BLE with AES-CCM is slower than standard BLE. This is because the encryption and authentication processes add overhead to the data packets, which increases the time it takes to send and receive data.

IV. Experimental results and analysis

V. Discussion

Strengths of the proposed algorithm

a. *Robust encryption*: AES-CCM is a strong encryption algorithm widely recognized for its security and reliability. By integrating AES-CCM with BLE, the algorithm provides robust encryption capabilities, ensuring the confidentiality of data transmitted between wearable devices. It protects sensitive information from unauthorized access, maintaining data privacy (Tables 46.4–46.6).

b. *Data integrity protection*: The integration of AES-CCM with BLE includes built-in integrity checks, such as CBC-MAC (Cipher Block

Table 46.4 Comparison with existing techniques

Aspect	Data security protocols in wearable devices	BLE integrated with AES-CCM encryption
Energy consumption	May not be optimized for low energy consumption, potentially draining battery life	Specifically designed for low energy consumption, prolonging battery life in wearable devices
Encryption strength	Encryption algorithms vary (e.g., AES, RSA), providing different levels of security	Utilizes the strong AES-CCM encryption algorithm, known for its robust data protection capabilities
Simplified protocol stack	Not necessarily streamlined for resource-constrained wearable devices	Offers a simplified protocol stack, reducing complexity and enabling efficient operation in resource-constrained wearable devices
Privacy features	Limited privacy features, depending on the specific protocols implemented	Includes privacy features like randomized device addresses to enhance user privacy and mitigate tracking risks
Data integrity	Security protocols may or may not provide built-in integrity checks	AES-CCM encryption includes built-in integrity checks (CBC-MAC), ensuring data integrity during transmission
Processing power and efficiency	Processing power requirements may vary, potentially impacting device performance	Optimized for low processing power and memory consumption, maintaining device efficiency while providing strong data security
End-to-end security	Depends on the specific protocols implemented; may lack end-to-end security	Provides end-to-end security through encryption and integrity checks, ensuring data confidentiality and integrity during transmission

Table 46.5 Privacy improvement achieved by the algorithm

Privacy improvement	Description
Randomized device addresses	BLE with AES-CCM integration allows the use of randomized device addresses during communication, making it difficult to track and identify individual devices. This helps enhance user privacy and mitigates tracking risks
Confidentiality of data	AES-CCM encryption ensures that data transmitted between wearable devices remains confidential. The encryption algorithm protects sensitive information from unauthorized access and helps maintain data privacy
Integrity protection	BLE integrated with AES-CCM includes built-in integrity checks, such as CBC-MAC, ensuring data integrity during transmission. This feature helps detect and prevent unauthorized modifications to the data, maintaining privacy and trustworthiness

Chaining Message Authentication Code). These checks ensure data integrity during transmission, preventing unauthorized modifications or tampering of the data. This feature enhances trustworthiness and ensures the received data remains intact.

c. *Low energy consumption*: BLE is specifically designed for low energy consumption, making it ideal for battery-powered wearable devices. The integration of AES-CCM with BLE maintains the low power consumption characteristics while providing strong data security. This combination allows wearable devices to operate efficiently without significantly draining the battery.

d. *Privacy enhancements*: BLE integrated with AES-CCM offers privacy enhancements such as randomized device addresses. By using randomized addresses during communication, it becomes more challenging to track and identify individual devices. This feature helps enhance user privacy and mitigates the risks associated with device tracking and identification.

Table 46.6 BLE data leakage prevention in wearable devices from 2010 to 2023

Year	Method	Prevention
2010	Data encryption	Data is encrypted before being transmitted over BLE, making it more difficult to intercept and read
2011	Device authentication	Devices must be authenticated before they can connect to each other, making it more difficult for unauthorized devices to access data
2012	Data access control	Users can control who has access to their data, making it more difficult for unauthorized users to access it
2013	Data minimization	Only the necessary data is collected and transmitted, making it less likely that sensitive data will be leaked
2014	Data pseudonymization	Data is made anonymous by replacing identifying information with pseudonyms, making it more difficult to link data to specific individuals
2015	Data obfuscation	Data is made difficult to understand by scrambling it or adding noise, making it more difficult to extract useful information from it
2016	Data destruction	Data is destroyed after it is no longer needed, making it less likely that it will be leaked
2017	Data leakage prevention (DLP) software	DLP software can be used to monitor and control data traffic, identify and block unauthorized access to data, and prevent data from being leaked
2018	Security best practices	Security best practices such as strong passwords, regular security updates, and awareness of phishing attacks can help to prevent data leakage
2019	Hardware-based security	Hardware-based security features such as secure elements and trusted execution environments can help to protect data from unauthorized access
2020	Cloud-based security	Cloud-based security solutions can help to protect data from unauthorized access and malicious attacks
2021	Artificial intelligence (AI)	AI can be used to detect and prevent data leakage by identifying patterns of suspicious activity
2022	Blockchain	Blockchain can be used to create a secure and tamper-proof record of data transactions, making it more difficult for data to be leaked
2023	Quantum-resistant cryptography	Quantum-resistant cryptography can be used to protect data from being accessed by quantum computers, which could pose a threat to current encryption methods

e. *Compliance with security standards*: The proposed algorithm aligns with industry standards and best practices for data security. It ensures that the implementation of BLE integrated with AES-CCM follows recognized security protocols and recommendations. This compliance enhances compatibility and interoperability with other Bluetooth devices while maintaining a high level of security.

f. *Simplified protocol stack*: BLE with AES-CCM integration provides a simplified protocol stack, reducing complexity and minimizing the attack surface. The simplified design improves efficiency, lowers the risk of vulnerabilities, and optimizes performance in resource-constrained wearable devices.

Conclusions

a. ***Summary of the research findings:*** The research findings indicate that the integration of BLE with AES-CCM encryption offers significant improvements in data security and privacy for wearable devices. By leveraging the low energy consumption capabilities of BLE and the robust encryption provided by AES-CCM, the proposed algorithm addresses the challenges of securing data transmission while maintaining efficient power utilization. The research confirms that the algorithm effectively ensures data confidentiality, integrity, and privacy in wearable device applications.

b. ***Contributions and significance of the algorithm***: The algorithm of BLE integrated with AES-CCM makes several notable contributions. Firstly, it provides a strong encryption solution tailored for low-power wearable devices, maintaining a balance between security and energy efficiency. Secondly, the integration offers privacy enhancements through randomized device addresses, mitigating tracking risks and protecting user privacy. Thirdly, the algorithm aligns with industry standards and best practices, ensuring compatibility and interoperability with other Bluetooth devices. These contributions are significant as they address the specific security and privacy requirements of wearable devices, enhancing user trust and promoting the widespread adoption of wearable technology.

c. ***Future directions and enhancements:*** While the proposed algorithm of BLE integrated with AES-CCM shows promising results, there are potential future directions and enhancements to consider. Some possible areas for further research and improvement includes: Performance optimization – Further optimizations can be explored to reduce the processing overhead and latency introduced by encryption and decryption operations. This would help maximize the efficiency of the algorithm while minimizing any impact on data transfer rates and response times. Key management – Investigating efficient key management techniques for wearable devices, such as dynamic key exchange or key derivation protocols, can enhance the overall security and simplify the management of encryption keys.

d. ***Authentication mechanisms:*** Integrating stronger authentication mechanisms, such as mutual authentication or multi-factor authentication, can enhance the overall security posture of the algorithm and provide additional layers of protection against unauthorized access.

e. ***Post-quantum cryptography:*** As quantum computing advances, exploring the integration of post-quantum cryptographic algorithms within the BLE and AES-CCM framework can ensure long-term security against quantum threats.

By focusing on these future directions and enhancements, the algorithm of BLE integrated with AES-CCM can continue to evolve and adapt to emerging security challenges, further strengthening the data security and privacy in wearable devices.

Acknowledgement

The authors gratefully acknowledge the authority of Electronics and Communication department for their cooperation in the research.

References

1. van der Hoog, R., et al. (2015). Bluetooth low energy: Privacy risks and countermeasures.
2. Kumar, et al. (2016). A comprehensive review of privacy in bluetooth low energy devices.
3. Conti, M., et al. (2017). Bluetooth low energy privacy and security: A systematic review.
4. Loukas, J. G., et al. (2017). A survey of bluetooth low energy privacy and security mechanisms.
5. Mo, R., et al. (2018). Privacy protection in bluetooth low energy advertising channels.
6. Leung, K. K., et al. (2018). Privacy vulnerabilities in bluetooth low energy networks: A systematic review.
7. Hu, L., et al. (2018). BLE-meter: Measuring the privacy leakage of bluetooth low energy devices.
8. Alzahrani, M., et al. (2019). Enhancing privacy in bluetooth low energy devices: A survey.
9. Cha, H., et al. (2019). BLETrack: Tracking bluetooth low energy devices under privacy constraints.
10. Kim, Y., et al. (2020). Investigation of privacy vulnerabilities in bluetooth low energy advertising.
11. Rehman, M. S., et al. (2020). Analysis of privacy risks in bluetooth low energy advertising channels.

47 Implementation of secure and efficient file exchange platform using blockchain technology and IPFS

Shivam Sharma[1,a], *Binduswetha Pasuluri*[2,b], *Mayuri Kundu*[1,c], *Argha Sarkar*[1,d] *and Ashwinkumar U. M.*[1,e]

[1]School of Computer Science and Engineering, REVA University, Bengaluru, India

[2]Electronics and Communication Engineering, Vardhaman College of Engineering, Hyderabad, India

Abstract

In this era of data, it is essential to develop a mechanism to facilitate people to exchange data with each other. However, sharing data among different party could become an impossible task due to the risk of security concerns. Currently data sharing occurs with the help of a trusted third party which acts as a mediator between the two parties but having a third party risks the security and transparency between the two parties. So, in this paper we try to propose a method which is de-centralized, secure and fast which can help us to deal with all the limitations we currently face in data sharing technology. To establish a decentralized connection between the two parties we make use of blockchain technology and add several encryptions to the data to make it nearly impossible for the hackers to decode and tamper with the data. Hence in this paper we will be trying to highlight the importance of blockchain in the field of data sharing thus allowing us to build a safe pathway for exchanging data among peers.

Keywords: File exchange, blockchain, advanced encryption standard, interplanetary file system

Introduction

The amount of data in this world has been increasing exponentially by the. years. According to a study the amount of data in the year 2020 was 59 zettabytes the which is expected to grow up to 175 zettabytes by the year 2075. Working with such a large amount data has always been a tedious task specially transferring data from to another. Data sharing is an important aspect of data handling. Every day, millions of bytes of data are shared. Data sharing although it might seem as a simple task but is a very sensitive task.

Currently data sharing among peers is possible only with the help of trusted third-party applications. This method of using third party applications involves several disadvantages such as the lack of decentralized network, proper encryptions, and lack of trustworthiness. Third party applications also lack transparency which gives those apps the monopoly to access the data and even tamper with it. All of these issues scare the consumer and does not give them confidence to share their confidential data through these apps.

In this paper we try to find a solution for the data sharing problem that can help us avoid all the limitations of these previous mechanisms which hinders the secure communication. In our project we make use of blockchain technology. Blockchain is a collection of nodes which collectively forms a decentralized network which connects different parties without need of a third-party application. Another advantage of using blockchain is that it is a secure network which makes it nearly impossible to tamper with [1-2].

The next most essential step in data sharing is encryption of the data. Having a secure connection is not just enough so we need to provide a second layer of protection. There are various encryption techniques, but we tend to use advanced encryption standard (AES) because it is the most secure version and updated form of data encryption standard (DES) & 3DES encryption algorithm. Along with AES we also make use of secure hash algorithm (SHA) which is an inbuilt feature of interplanetary file system (IPFS) [3-5].

All together combination of all these mechanisms helps us to achieve our aim of creating a platform which can facilitate users to securely share their confidential data with their peers without having to worry about issues of data leaking and tampering.

Methodology

The blockchain can be regarded as a record where all the transactions between the parties is stored. It is a decentralized system where the records are kept in blocks which are linked together and is secured

[a]shivmshrm2022@gmail.com, [b]bswetha.pasuluri@gmail.com, [c]kundu.mayuri@gmail.com, [d]argha15@gmail.com, [e]ashwinkumar.um@reva.edu.in

DOI: 10.1201/9781003540199-47

using cryptography. These blocks use the hash value to identify the data in the blocks This hash is used to link the previous blocks which is very difficult to temper. Blockchains are generally used to record transactions for cryptocurrency or storing votes to prevent it from being tempered.

IPFS is a network for storing sharing and distributing files. It as a decentralized system which is more efficient and resilient then the HTTP-based web protocols. Nodes are used in IPFS as the files are split into small pieces and distributed across the nodes. A unique cryptographic hash for each file is required for the verification and retrieval of the data in the file. This is possible because for any two different contents of same data has the same hash.

Advanced encryption standard (AES) is a widely used symmetric key encryption algorithm, which means that the same key is used to encrypt and decrypt data. AES works by dividing plain text into 128-bit blocks, then applying a series of mathematical operations to each block using a 128, 192, or 256-bit key. The length of the key determines the number of encryption cycles performed. For example, AES performs 10 rounds of encryption for a 128-bit key and 14 rounds for a 256-bit key. Decryption is just the reverse process, and each encryption cycle takes place in reverse order. AES is considered very secure and there are no known practical attacks that can break the encryption when strong keys are used [6-10].

Secure hash algorithm 256 (SHA-256) is a widely used cryptographic hash function that generates a fixed-size 256-bit (32-byte) message digest or hash value from an input message of arbitrary length. The SHA-256 algorithm operates on a 512-bit message block and uses a series of logical and arithmetic operations to produce a hash value. Input messages are first padded into multiples of 512 bits, then divided into 512-bit blocks. The SHA-256 algorithm has several properties that make it useful in cryptographic applications. First, it is a one-way function, which means that it is impossible to reverse the hash and determine the original input message. Second, even small changes in the input message can result in completely different hashes, which is useful for tracking data changes. Finally, the algorithm produces a fixed-length hash value, which is useful for verifying data integrity. SHA-256 has a wide range of applications, including digital signatures, message authentication codes, and password safekeeping [11-12].

Architectural diagram

The first step in data encryption is to upload the required files to the IPFS platform. Once a file is uploaded to IPFS, IPFS then generates a hash of the data and returns it to the owner. Herein, the system nodes are created, and public-private key pairs are generated and stored in the smart contract. When the owner receives the hash function of the original data from the IPFS, it starts searching the smart contract for the authenticated nodes which have the responsibility for delivering the decryption facilities to clients. Only the data sharing platform is capable of providing the requested services to which are assigned by the owner. The moment owner starts receiving the hash function, it then starts dividing it into "k" shares. After creating these shares, the owner then creates "n" numbers of random keys to use for encryption. On encrypting all these shares with their respective keys, they are then stored on the blockchain and appended with the authorized file recipients and other significant information. Data security is based on cryptographic hashing. Hashing is a unique fingerprint which characterizes a set of data. If an unsanctioned client without submitting digital content may have access to the hash. So, the chances remains that the whole data can be recovered from IPFS. This way, the owner will lose the business entirely. In the proposed scenario, only clients who have deposited funds and are authorized by worker nodes can decrypt the hash. The overall owner process that uploads files to IPFS is shown in Figure 47.1. This whole process marks the completion of complete architecture of the data sharing process using our platform.

Sequence diagram

Figure 47.2 depicts the sequence diagram. Herein, the process starts when the client wants to access the data, so it directs a request to the system. The request made by the user is combined with the customer's digital signature. The new applicants are first verified by the system using an Rivest-Shamir-Adleman (RSA) algorithm. RSA is an asymmetric algorithm which creates a public key and a private key pair where the private key is kept secret and confidential whereas on the other hand public key is shared to all other authorized users. If the recipient is not an authorized user or recognized as a non-valid requester, then the smart contract is terminated automatically, the system ignores the received request. While in the other case when the requestor is a valid user, the system starts retrieving the encrypted data from the blockchain and decrypts utilizing the private keys. The most essential point to note here is that if the owner chooses their public key when sharing the data over IPFS, the system can only decrypt on behalf of the recipient. After decryption, the recipient can obtain the hash value of the original IPFS file by reconstruction. This marks the completion of the whole process with user obtaining the desired data they requested for to the system.

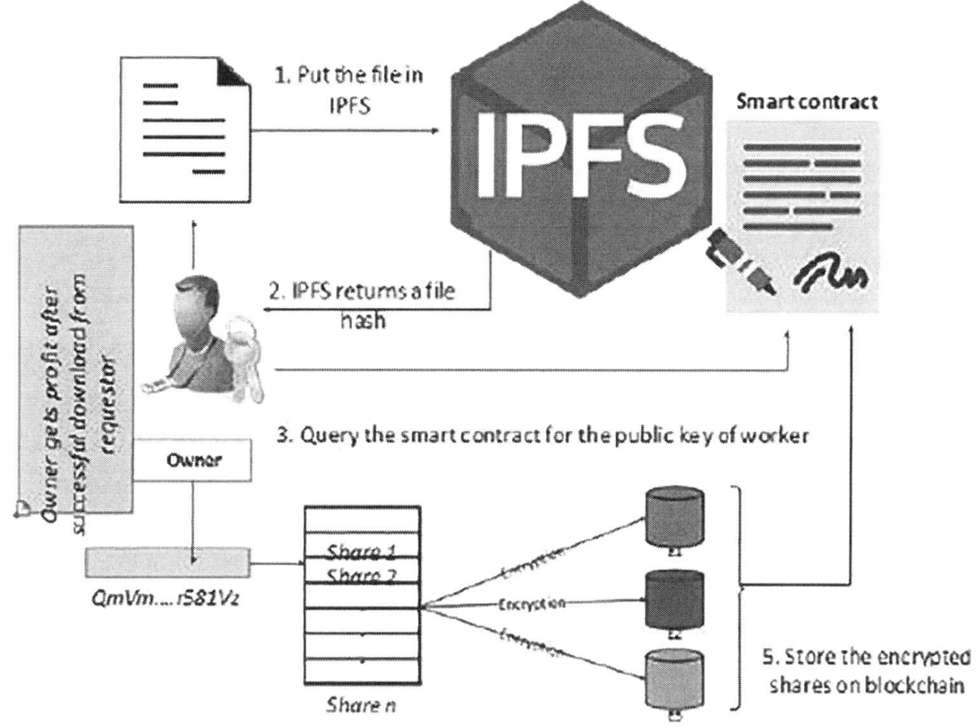

Figure 47.1 Architecture to encrypt data

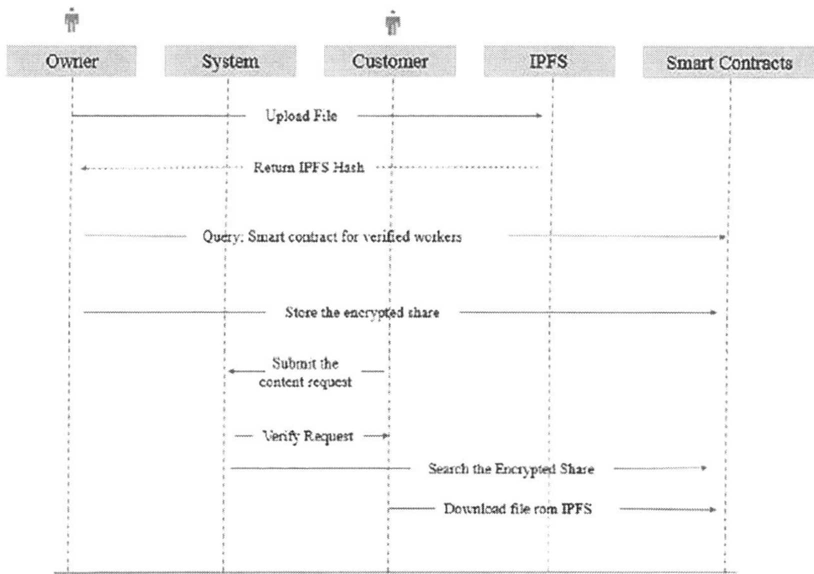

Figure 47.2 Sequence diagram

Result and analysis

The result of sharing data using blockchain and IPFS is a more secure, transparent, and efficient method of exchanging data. Blockchain technology ensures data integrity and authenticity by delivering a decentralized and immutable ledger, while IPFS enables fast and reliable access to shared data through a distributed file system. The technology offers several advantages over traditional methods of data sharing, including reduced cost and complexity, improved data privacy and control, and the elimination of middlemen and central authorities. Therefore, blockchain IPFS technology can be applied to various fields such as finance, healthcare, education and supply chain management.

Additionally, sharing data using blockchain IPFS has the potential to reduce the risk of data breaches and cyberattacks, which are increasingly common in today's digital age. today. With more secure and transparent data sharing methods, organizations and individuals can have more confidence in their data sharing practices, which ultimately leads to better decision-making and more innovation.
Overall, the result of using blockchain IPFS for data sharing is a more efficient and secure way of exchanging data that has the potential to transform the way organizations and individuals interact with each other. At the end of this project, we are able to develop a platform called SAEFE (Secure and Efficient File Exchange) Platform which facilitates peer to peer transfer of data.

Conclusions

Data sharing using blockchain and IPFS has become a safe and efficient way to share data over networks. Blockchain technology offers immutable and decentralized ledger that guarantees data integrity and authenticity, while IPFS provides a distributed file system that enables fast and reliable access to shared data. By combining these technologies, data sharing becomes more secure, transparent, and efficient, allowing organizations and individuals to exchange data without the need for intermediaries or central authorities. This not only reduces the cost and complexity associated with traditional methods of data sharing, but also improves data privacy and control. Adding all these technologies in the field of data sharing can aid the users but it also brings a few limitations with it. Currently our project cannot support the transfer of video files and can transfer only text files and images. Another limitation is that transferring huge file sizes can lead to slowing the blockchain as cryptographic encryptions require huge computational powers.

Moreover, blockchain and IPFS technology can be applied in various fields, such as finance, healthcare, supply chain management, and more, where data security and privacy are crucial. As the demand for secure and reliable data sharing continues to grow, the use of blockchain IPFS technology is expected to increase, providing more innovative solutions to the data sharing challenges of today and tomorrow.

Acknowledgement

The authors gratefully acknowledge REVA University for infrastructural support to carry on the work.

References

1. Peck, M.. (2018). Understanding blockchain technology: Abstracting the blockchain. *IEEE*.
2. Peck, M. (2018). Understanding blockchain technology: The costs and benefits of decentralization. *IEEE*.
3. Zheng, Z., et al. (2017). An overview of blockchain technology: Architecture, consensus, and future trends. *2017 IEEE Inter. Cong. Big Data (BigData congress)*.
4. Liang, X., et al. (2017). Integrating blockchain for data sharing and collaboration in mobile healthcare applications. *2017 IEEE 28th Ann. Inter. Symp. Pers. Indoor Mobile Radio Comm. (PIMRC)*.
5. Chen, Y., et al. (2017). An improved P2P file system scheme based on IPFS and Blockchain. *2017 IEEE Inter. Conf. Big Data (Big Data)*.
6. Kundu, M., Binduswetha, P., and Argha, S. (2023). Vehicle with learning capabilities: A study on advancement in urban intelligent transport systems. 2023 Third Inter. Conf. Adv. Elec. Comput. Comm. Sustain. Technol. (ICAECT).
7. Dai, M., et al. (2018). A low storage room requirement framework for distributed ledger in blockchain. *IEEE Acc.*, 6, 22970–22975.
8. Li, Y., et al. (2017). Big data model of security sharing based on blockchain. *2017 3rd Inter. Conf. Big Data Comput. Comm. (BIGCOM)*.
9. Kundu, M. and Nagendra Kumar, D. J. (2021). Route optimization of unmanned aerial vehicle by using reinforcement learning. *J. Phy. Conf. Ser.*, 1921(1). IOP Publishing, 2021.
10. Zhetao, L., et al. (2017). Consortium blockchain for secure energy trading in industrial internet of things. *IEEE Trans. Indus. Informat.*, 14(8), 3690–3700.
11. Yong, Y. and Fei-Yue, W. (2016). Towards blockchain-based intelligent transportation systems. *2016 IEEE 19th Inter. Conf. Intell. Trans. Sys. (ITSC)*.
12. Yang, Z., et al. (2018). Blockchain-based decentralized trust management in vehicular networks. *IEEE Internet Things J.*, 6(2), 1495–1505.

48 MEMS-based gesture controlled robot

Dhiraj Thote[a], *Nandini Farkade*[b], *Vedant Khorgade*[c], *Mehul Mahakalkar*[d] *and Shubhangi Tayade*[e]

Electronics and Telecommunication Engineering, Yeshwantrao Chavan College of Engineering, Nagpur, India

Abstract

In addition to presenting a concept for a gesture controlled user interface (GCUI), this article also analyses usability, technological, and application trends. We provide an integrated method for real-time tracking, and gesture-based data gauntlet technology that uses hand movements to control the wheelchair. In the paper, a microcontroller-based 3-axis wireless accelerometer system for low-cost and tiny wheelchair control was developed. A gesture identification module with micro-electromechanical systems (MEMS) sensor and wheelchair control makes up the system's two primary parts. The system's brain is microcontroller, found in the gesture recognition module. The hand-attached MEMS sensor is a 3-axis accelerometer with digital output (I2C) that detects the angle of the hand; in other words, it provides voltages to the microcontroller in accordance with the tilt of the hand. "PIC18F25K22" controller is used to operate the wheelchair control unit. Stop, backward, forward, left, and right are the four suggested motions that are successful. The outcomes of a few tests conducted using the controlled system are then presented and discussed. The system can acknowledge the input motions fast and with a valid recognition rate, according to experimental data.

Keywords: MEMS, brushless DC motor, RF transceiver HC-05, motor driver IC LM2938, PIC 18F25K22 microcontroller

I. Introduction

Today, 1% of the world's population is thought to use a wheelchair a rising number of sensors and persons with disabilities who seek to improve their own mobility, The wheelchair is the best aid available to them [3, 8]. A person who is disabled or infirm (often having a lower-body disability) may discover how easy it is to move and maneuver with a wheeled chair that can be pushed by another person or by physical power or electrically [1]. Our method enables users to coordinate the wheelchair's movement with their own movements, such as hand gestures, so they may maneuver it on a variety of surfaces without difficulty, cardiovascular issues, or tiredness. Some wheelchairs now in use have computers installed for gesture detection [2]. Utilizing the MEMS accelerometer, whose size is quite small and which the patients can place on the tip of their finger, reduces this complication[6]. The intricacy of the system is further increased by the fact that comparable sensors are used in other current systems that are also wired [7].

Additionally, they restrict long-distance communication. Through the use of RF transmission, this complexity is eliminated. RF allows signals to go farther. When the impediment is present between the transmitter and acceptor, RF signals still travel regardless of the line of sight [5]. The system can simultaneously detect obstacles in front of the wheelchair through (ultrasonic sensor) and make ease in breaking function of the wire, also the system is enabled for bearing outdoor conditions, so it has (proximity sensor) which could detect potholes and have precaution for already vulnerable being, and dark light detection provision is also made in the wheelchair where if the wheelchair is in low light or dark area it will automatically switch on the lights through (LDR sensor) [10, 11]. Due to these increased functionalities such as obstacle detection, pothole detection, and automatic switching of lights, this system becomes extremely efficient and user friendly [4] (Figures 48.1 and 48.2).

II. Proposed system

The setup comprises two main components: the transmitter and the receiver. The transmitter consists of the PIC 18F25K22 microcontroller, the MEMS accelerometer IC ADXL 335, the RF transceiver HC-05, and the voltage divider IC 7805. The system is powered by a battery providing 9 V supply, which is then reduced to 5 V using the voltage divider to meet the other components' 9 V standard supply requirement. Upon powering on, the transceiver modules automatically pair up, enabling communication between the transmitter and receiver. The 3-axis accelerometer in the transmitter detects

[a]dhiraj.thote@gmail.com, [b]farkadenandini08@gmail.com, [c]vedantkhorgade2@gmail.com, [d]mahakalkarmehul285@gmail.com, [e]tayadeshubhangi23@gmail.com

DOI: 10.1201/9781003540199-48

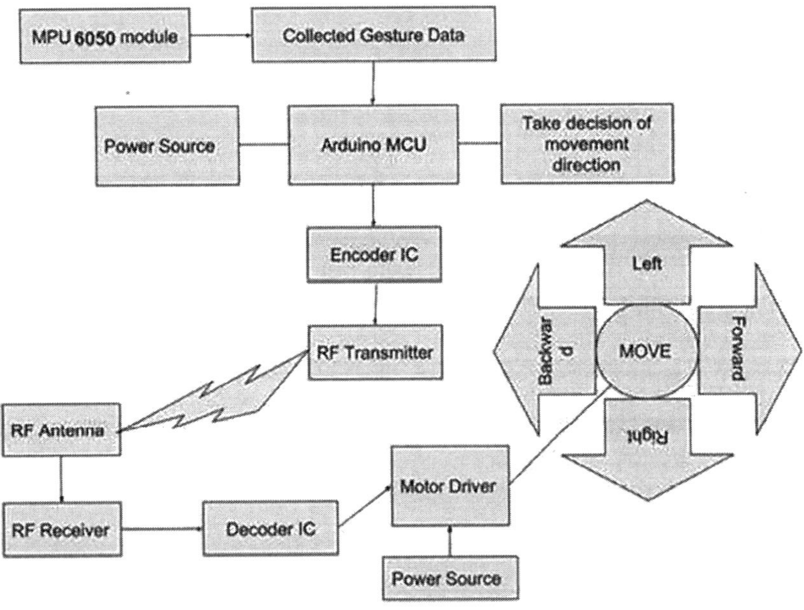

Figure 48.1 Block diagram of the proposed system

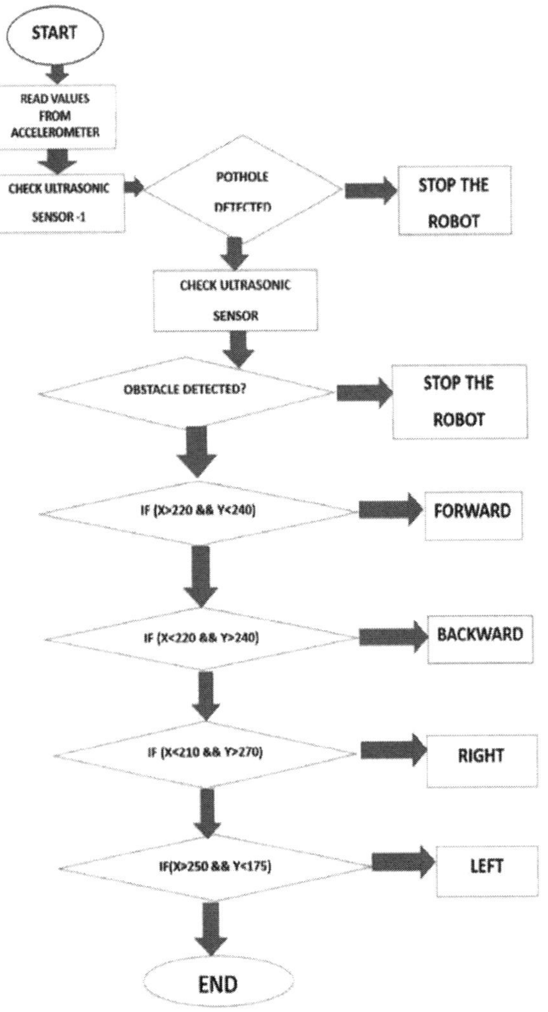

Figure 48.2 Flow diagram

the hand's tilt relative to the ground. Based on the tilt, electrical signals are sent to the microcontroller, which converts them using an internal analog-to-digital converter and transmits the digital signal to the transceiver module, acting as a master. To prevent overheating and regulate current, filter capacitors are connected in this section.

The receiver component includes the PIC 18F25K22 microcontroller, RF transceiver HC-05, motor driver IC LM2938, and two ultrasonic sensors HC-SR04 to detect obstacles and potholes. The transceiver (slave) transmits input to the microcontroller upon receiving signals from the transceiver (master). When obstacles or potholes are detected within the specified range by the ultrasonic sensors, the wheelchair halts. Otherwise, normal operation continues. The microcontroller sends digital input signals to the motor driver and gear motors if an obstacle is not recognized. Additionally, an LDR sensor is used to control the lights based on ambient light intensity. A transistor acts as a driver, converting the 5 V supply to 12 V, functioning as a switch for the LDR. When the LDR resistance drops, the LED is switched on. Two capacitors are also used to control excess supply current. Overall, the system uses gesture control and ultrasonic sensing to operate a wheelchair safely, avoiding obstacles and potholes while also providing a light control feature based on ambient light conditions.

Result and discussion

In this section, Tables 48.1–48.3 are discussed.

Table 48.1 MEMS accelerometer observation

Sr. No.	Direction motion	x-axes [tilting angle]	y-axes [tilting angle]
1	Forward	279	310
2	Reverse	359	325
3	Left	324	270
4	Right	331	369

Table 48.2 LCD display and description

LCD display of transmitter and receiver	Description of figures
	The x- and y-axis values on the transmitter side are used to control the forward movement of the wheelchair. When the x and y values from the accelerometer on the transmitter side fall within the specified threshold range, the microcontroller interprets it as a command for forward movement
	The x- and y-axis values on the transmitter side for reverse movement are used to manage the backward movement of the wheelchair. By setting appropriate threshold values for the x- and y-axes, the system ensures that only specific hand orientations or gestures trigger the reverse movement of the wheelchair
	By setting appropriate threshold values for the x- and y- axes, the system ensures that only specific hand orientations or gestures trigger the right movement of the wheelchair. This precision prevents accidental or unintended right movements, allowing the user to control the wheelchair with greater accuracy and confidence
	The x- and y-axis values on the transmitter side for left movement are used to control the wheelchair's movement in the left direction. The gesture-based control system allows for intuitive operation of the wheelchair in all directions, providing enhanced mobility and ease of use for the user
	On the receiver side, the system displays two key values: forward and depth values. These values are crucial for the system to detect potholes and obstacles effectively
	On the receiver side, when the system starts, the message "Gesture Controlled Wheelchair" is displayed. This message serves as an indication that the system has successfully initialized and is ready to be controlled through gestures. It acts as a visual confirmation for the user, letting them know that the gesture-based control functionality is activated and operational

Table 48.3 Sensors observation chart

Sr. No.	Sensor	Function	Values
1	HC - SR04 (i)	Forward direction obstacle detection	16.88
2	HC – SR04 (ii)	Downward direction pothole detection	5.58

Future scope

1) By utilizing hand movement as a form of acceleration, we can control the wheelchair's various directions using an optical sensor that tracks eye retina movement. This innovation allows us to operate the wheelchair through retinal motions effectively.
2) Researchers are actively working on creating a specialized wheelchair that interfaces with the human nervous system. This cutting-edge technology aims to enhance mobility for individuals with disabilities by leveraging nervous system signals to control the wheelchair's movements.
3) The monitoring system employs voice signals to aid differently-abled individuals in detecting obstructions. The system triggers an alarm signal with slight power section adjustments, monitoring battery voltage levels to gauge when action should be taken. To improve speed control, DC motors can be replaced with servomotors.
4) Incorporating a voice command IC allows direct connection between vocal signals and the microcontroller. Eliminating the need for computer connectivity, the system can analyze voice commands stored in the IC to interpret the user's instructions effectively.
5) To enhance user safety and security, GPS and GSM modules can be installed, enabling live location tracking of the wheelchair user. Additionally, the system can send alerts to caregivers or attendants to ensure their prompt attention and support in case of emergencies or unexpected situations

Conclusions

The main motive goal of this research is to create an electric wheelchair equipped with advanced technologies, including acceleration sensors, obstacle sensors, and computer systems, to enable persons with disabilities to move independently, especially those who are unable to move on their own. The wheelchair is designed to assist less able drivers in achieving a certain level of independent mobility. The acceleration sensor allows the user to manage the wheelchair's movement in four different directions by tilting the sensor. Additionally, to aid users who may have difficulty steering or avoiding obstacles, the obstacle sensor helps by partially handling these tasks.

One of the key advantages of this research is its low installation costs, making it accessible to a wider range of people. This increased accessibility empowers physically challenged individuals to be more independent. The wireless system not only enhances their self-assurance but also boosts their confidence in using the wheelchair effectively. The research has demonstrated competitive performance in terms of computing capabilities and gesture recognition accuracy, which ensures smooth and efficient operation of the wheelchair. With the successful implementation and widespread adoption of this technology, it has the potential to positively impact the lives of many individuals with disabilities, improving their overall quality of life and independence.

Acknowledgement

The authors gratefully acknowledge the staff and authority of Electronics and Telecommunication Engineering department for their cooperation in the research.

References

1. Vishal, P. V., Nikita, U. S., Darshana, M. P., Nikita, I. R., and Pragati, M. P. (2014). Hand gesture based wheelchair movement control for disabled person using MEMS, *Int. J. Engg. Res. Appl.*, 4, 152–158.
2. Shreedeep, G., Somsubhra, M., and Soumya, C. (2013). Intelligent gesture controlled wireless wheelchair for the physically handicapped. *Proc. Fifth IRAJ Inter. Conf. ISBN: 978-93-82702-29-0, page 40-45*
3. Diksha, G. and Saini, S. P. S. (2013). Accelerometer based hand gesture controlled wheelchair. *Inter. J. Emerg. Technol.*, 4(2), 15–20. ISSN No. (Online) : 2249–3255.
4. Rakhi, K. A. and Chitre, D. K. (2013). Automatic wheelchair using gesture recognition. pp 146-150, *Inter. J. Engg. Innov. Technol. (IJEIT)*, 2.

5. Manisha, D. and Kumar, B. A. (2014). Accelerometer based direction controlled wheelchair using gesture technology. *Inter. J. Sci. Engg. Technol.*, 3, 1065–1070.
6. Ameet, Z., Gurpreet, K., and Mashhuda, S. (2018). Use of ultrasonic sensor for detecting potholes - An IoT based system. *J. Emerg. Technol. Innov. Res.*, *5.*, *P312-314*
7. Vaidya, O. J., Ingale, K. L., and Chaudhari, R. (2019). Hand gesture based music player control in vehicle. *IEEE 5th Inter. Conf. Converg. Technol. (I2CT)*, page 1–5. doi: 10.1109/I2CT45611.2019.9033708.
8. Iyer, P., Tarekar, S., and Dixit, S. (2019). Hand gesture controlled robot. *2019 9th Inter. Conf. Emerg. Trends Engg. Technol. Sig. Inform. Proc. (ICETET-SIP-19), page* 1–5. doi: 10.1109/ICETET-SIP-1946815.2019.9092032.
9. Taniguchi, Y., Nishii, K., and Hisamatsu, H. (2015). Evaluation of a bicycle-mounted ultrasonic distance sensor for monitoring road surface condition. *2015 7th Inter. Conf. Comput. Intell. Comm. Sys. Netw.*, 31–34. doi: 10.1109/CICSyN.2015.16.
10. Baiju, P. V., Varghese K., Alapatt, J. M, Joju, S. J., and Sagayam, K. M. (2020). Smart wheelchair for physically challenged people. *2020 6th Inter. Conf. Adv. Comput. Comm. Sys. (ICACCS)*, 828–831. doi: 10.1109/ICACCS48705.2020.9074188.
11. Dhiraj, T., Harsha, T., Nitin, C., Manthan, N., Pratyush, P., and Punit, F. (2022). EV vehicle mileage booster using solar technology. *2022 pp 41-46 IEEE Inter. Women Engg. Conf. Elec. Comp. Engg. (WIECON-ECE).*

49 Empirical assessment of security models for cloud attack detection and mitigation

Sachin A. Kawalkar[a] *and Dinesh B. Bhoyar*[b]

Yeshwantrao Chavan College of Engineering, Nagpur, India

Abstract

Cloud security is one of the major concerns for scalable network deployments. With the increasing use of cloud computing, the need for effective security measures to protect against cyber-attacks is more important than it was in previous decades. Various models have been proposed for cloud attack detection and mitigation, including deep learning, machine learning, bio inspired, and prediction models. However, there is a lack of empirical analysis that assesses the effectiveness of these models in real-world scenarios. This review paper aims to address this gap by providing an empirical perspective on the statistical analysis of security models for cloud attack detection and mitigations. The paper surveys the state-of-the-art in cloud security models and critically evaluates their performance based on empirical dataset samples. The paper discusses the advantages and disadvantages of each model and highlights their suitability for different types of attacks. It also explores the challenges involved in developing accurate and robust security models, such as the need for large and diverse training datasets, dealing with imbalanced data, and the trade-off between accuracy and speed levels.

Keywords: Cloud, attack, scenarios, bio inspired, deep learning, machine learning, mitigation, delay, accuracy, precision, recall, scalability, metrics

I. Introduction

Cloud computing has become a ubiquitous part of modern information technology infrastructure due to its many advantages, including scalability, flexibility, and cost savings. However, as more and more organizations move their data and applications to the cloud, the risk of cyber-attacks increases. These attacks can result in data breaches, loss of intellectual property, and service disruptions, among other negative consequences. As a result, security has become a critical concern in cloud computing, and there is a growing need for effective security measures to protect against such attacks.

Various models have been proposed to detect and mitigate cloud attacks, including deep learning, machine learning, bio inspired, and prediction models. These models use statistical analysis and machine learning algorithms to identify patterns and anomalies in cloud data, thereby enabling the detection and prevention of attacks in real-time via the following operations,

Deep learning models: These models use neural networks to detect and mitigate cloud attacks. They can identify patterns and anomalies in cloud data and learn from large and diverse training datasets. However, they can be computationally expensive and require large amounts of data to achieve high accuracy.

Machine learning models: These models use algorithms to analyse cloud data and detect attacks. They can learn from past attacks and adapt to new threats. However, they can be limited by the quality and quantity of the training data and may not perform well on new and unknown types of attacks.

Bio-inspired models: These models are based on natural systems, such as the immune system, and use algorithms to detect and respond to attacks. They can adapt to changing threats and are less dependent on large training datasets. However, they can be complex and difficult to implement.

Prediction models: These models use statistical analysis to predict future attacks based on historical data. They can identify trends and patterns in cloud data and provide early warning of potential attacks. However, they may not be effective against new and unknown types of attacks.

However, despite the promising results of these models, there is a lack of empirical analysis that evaluates their performance in real-world scenarios.

This review paper aims to fill this gap by providing an empirical perspective on the statistical analysis of security models for cloud attack detection and mitigation. The paper surveys the state-of-the-art in cloud security models and critically evaluates their performance based on empirical data. By providing insights into the strengths and limitations of different models, this review paper can help inform the

[a]sachin.kawalkar1011@gmail.com, [b]dinesh.bhoyar23@gmail.com

DOI: 10.1201/9781003540199-49

development of more effective cloud security solutions to mitigate the growing threat of cyber-attacks.

In the following sections, we will provide an in-depth analysis of various models used for cloud attack detection and mitigation, including deep learning, machine learning, bio inspired, and prediction models. We will discuss the advantages and disadvantages of each model and their suitability for different types of attacks. We will also explore the challenges involved in developing accurate and robust security models, such as the need for large and diverse training datasets, dealing with imbalanced data, and the trade-off between accuracy and speed. Finally, we will discuss the future direction of research in this field and highlight the need for continued efforts to improve cloud security models.

Detailed literature review

A wide variety of models are proposed by researchers for improving security of cloud deployments. Each of these models vary in terms of their attack detection and mitigation performance under real-time scenarios. In this section, a detailed survey of these models is discussed, which will assist readers to identify optimal models for their application-specific use cases. According to one study [1], the flexible, on-demand, and granularly managed access to computing resources and services is a significant selling point. Due to its decentralized and ever-changing nature, as well as virtualization implementation faults, the cloud environment is vulnerable to a wide variety of cyber-attacks and security concerns that are unique to the cloud model.

As described in ref [2], the expansion of cloud computing has led to a rise in intrusions. The distributed denial of service (DDoS) attack aims to exhaust system resources to the point where authorized users are unable to access the system. DDoS attacks and defenses are ultimately motivated by resource competition. Numerous efforts have been devoted to investment and resource management. However, the cost of these defensive measures is not taken into account because it is assumed that an infinite quantity of resources will be available to repel the attacks. Due to the coarse granularity of these defensive strategies, the defender may waste resources. Researchers conduct extensive research on the topic and propose a horizontally and vertically scalable strategy for fine-grained resource management based on the concept of "birth and death" The proposed method adaptively selects the optimal resource leasing (ORL) option to enable cloud service consumers to defend themselves against DDoS attacks at the lowest feasible cost.

According to research in ref [3], which examined data privacy in cloud storage, sensitive data is encrypted before being transmitted to a cloud server. The secure and trustworthy extraction of cipher text has become more difficult. Public key encryption with keyword search (PEKS) permits the secure retrieval of cloud-based cipher texts. Unfortunately, the preponderance of PEKS protocols are vulnerable to keyword guessing attacks (KGAs) carried out by dishonest cloud storage providers.

In their research, the authors of ref [4] analyses a networked multi-agent system (NMAS) employing quantized signals for use in a predictive cloud management strategy against DoS attacks. Control signals are quantized using an arbitrary area quantizer to ascertain their useful values. To simulate denial-of-service assaults, attack-induced packet losses are modeled using a Stackelberg game. Predictive cloud control is a technology that combines cloud computing with predictive control in order to actively compensate for network or attack-caused delays and packet failures.

According to research published in ref [5], cloud service providers strive to maintain co-located virtual machines (VMs) and user actions as distinct as feasible. By constructing a distinct internal virtual network, physical network-sharing VMs can be logically isolated from one another. Coexisting VMs are susceptible to assaults that target other VMs because they share the same virtual machine monitor (VMM), virtual network, and hardware. A malicious VM may gain access to or control over other VMs via the network, shared memory, or other shared resources, or by acquiring the privilege level of its non-root host system. This investigation exposes two novel zero- day attacks on network channels between virtual machines. A hostile virtual machine (VM) could redirect network traffic intended for other VMs by masquerading as their virtual network interface controllers (VNIC) in the first type of attack. Using open-source decryption tools such as air cracks, the malicious VM may obtain encrypted data from the target VMs.

According to ref [6], distributed denial of service (DDoS) attacks are prevalent in the modern cloud computing environment. Legitimate consumers are unable to access the services due to the excessive volume of traffic, resulting in revenue losses. Despite the fact that a number of researchers have developed a variety of mitigation strategies, complications may still arise. Software-defined networking was initially deployed to defend businesses against DDoS attacks. Direct consequences of distributed denial of service attacks are revenue losses and server downtime. Still, it is difficult to satisfy service level agreements with clients.

According to research published in ref [7], heterogeneous multi-agent systems are susceptible to denial-of-service (DoS) assaults and transmission delays; this study examines a novel compensatory control technique to address these issues. This control system integrates the most beneficial aspects of cloud computing, adaptive event-triggered strategies, and predictive control schemes. The adaptive event-triggering mechanism can modify the event count adaptively, and the cloud-computing technique can eradicate the negative effects as if DoS attacks and transmission delays had never occurred. Predictive control can be utilized to actively mitigate or eliminate DDoS attacks and transmission delays, respectively.

According to research published in ref [8], cryptographic libraries such as OpenSSL and Libgcrypt are required for cloud service security. These libraries may be vulnerable to cache side-channel attacks, which are particularly prevalent and dangerous in cloud environments due to the inevitable cache competition between tenants. Due to deployment and security issues, earlier techniques for mitigating cache side-channel attacks were ineffective in cloud applications.

The research presented in ref [9] dealt with this subject. Users with limited resources frequently turn to the cloud due to the prohibitive cost and time required to train sophisticated deep neural networks. If the user suspects that an unreliable cloud service provider implanted backdoors into the returned model, they may investigate it using cutting-edge defensive techniques. From the perspective of nefarious cloud service providers, the goal of this research is to develop robust covert attacks (called RobNet) that can circumvent existing security measures.

In ref [10], cloud computing was identified as a crucial technology for the future of the IT industry. The level of cloud security is viewed as the greatest challenge for cloud service providers and a major concern for cloud consumers. EDoS attacks pose a substantial threat to cloud infrastructure. In an EDoS attack, the auto scaling and elasticity features of the cloud provider are exploited to increase the cost of using the service unjustly, resulting in the closure or insolvency of the targeted business. The EDoS attack defense shell (EDoS-ADS) is an innovative reactive method designed to defend against EDoS attacks while compensating for the shortcomings of existing countermeasures

Comparative statistical analysis

Below is the statically analysis of attack simulation and time take by existing encryption algorithms, security tools deployed on cloud network to detect and prevent the attached 256 encryption algorithm. Figure 49.1 shows the AES 256 encryption algorithm to generate the token and time taken to pass the network. Total time required to is 198.73 ms. From the request send. This is at code-level.

Figure 49.2 shows the time taken by algorithm deployed on security tool at perimeter level. Above snapshot is of VM machine from the time the query is fired and 168 ms. This time is the length of time from external network till arrival on internal network / device level.

In Figure 49.3 time taken by algorithms when the traffic is flowing from external network to internal network.

Figure 49.3 showcase various time taken by algorithms when the traffic is flowing from external network to internal network. We are comparing the various time frames across various layers and then final analysis and design of algorithm can be designed and deployed. Aim of analysis is to reduce the time taken from external traffic coming to internal network at application and database level and then if any attack surface, to be detected and prevented. Based on the detailed review of existing models, it can be observed that deep learning, bio inspired, and incremental learning models outperform linear classification models. To further identify optimal deep learning models for identification & mitigation of cloud attacks, this section evaluates the discussed models in terms of different evaluation metrics. These include, accuracy (A) of attack detection, precision (P) with which attacks are detected. Delay needed to identify attacks (D), cost of deployment (CD), and scalability (S) levels. While, accuracy and precision are directly available in the reviewed texts, estimation of delay, deployment cost and scalability was done empirically, and these values were converted into fuzzy ranges of low (L=1), medium (M=2), high (H=3), and very high (VH=4), which will assist readers to identify optimal models for

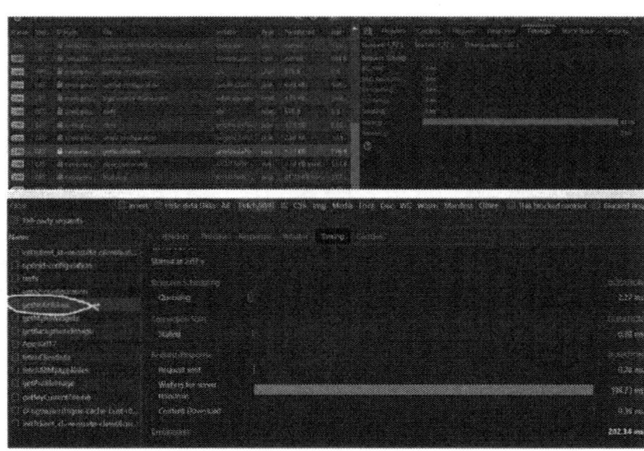

Figure 49.1 AES 256 encryption algorithm

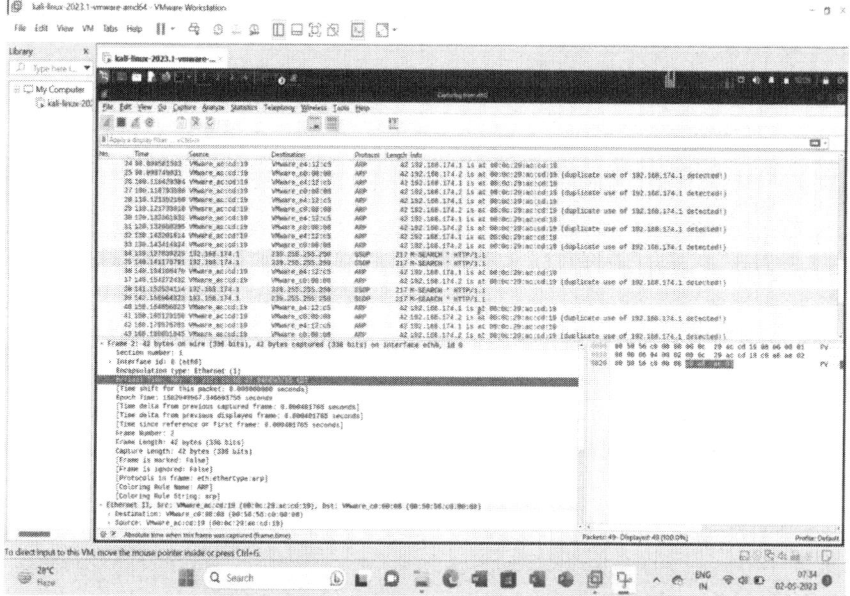

Figure 49.2 The time taken by algorithm deployed on security tool

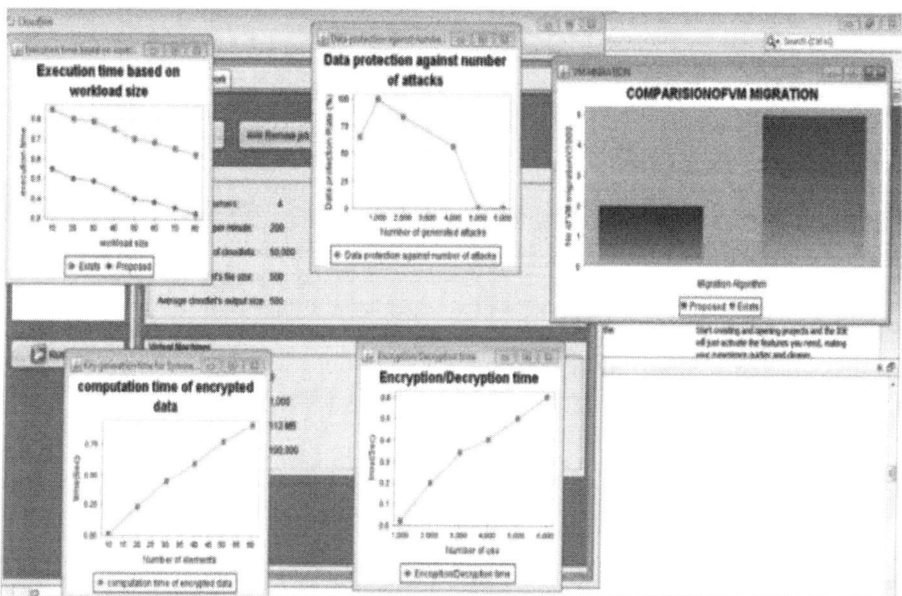

Figure 49.3 Time taken by algorithms when the traffic is flowing from external network to internal network

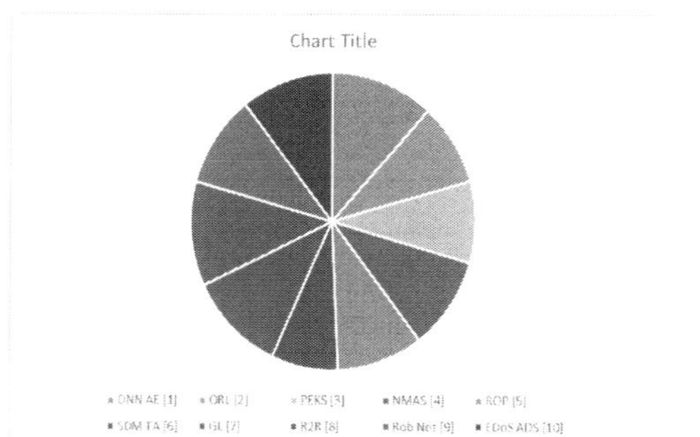

Figure 49.4 CARM of different models for identification of cloud attacks

their performance specific use cases. As per this strategy, these estimations can be observed from Table 49.1 as follows:

$$CARM = (A+P)/200 + 1/D + 1/DC + S/4 \quad (1)$$

Based on this evaluation for all models, and Figure 49.1, it can be observed that SVM, DAD, Point Net++ , TL , DNN AE [1], R2R [8] outperform other models in terms of accuracy, precision, delay, deployment cost, and security levels. Based

Table 49.1 Parametric analysis of the reviewed models under different attack scenarios

Model	A	P	CD	D	S
DNN AE [1]	97.5	78.10	VH	M	VH
ORL [2]	51.5	73.53	H	H	H
PEKS [3]	85.3	84.20	H	H	M
NMAS [4]	83.8	89.00	H	H	H
ROP [5]	83.5	91.20	H	VH	H
SDM TA [6]	99.7	98.30	VH	VH	L
GL [7]	90.4	91.93	H	M	H
R2R [8]	89.5	92.63	H	H	VH
Rob Net [9]	95.9	95.63	VH	H	H
EDoS ADS [10]	92.5	88.50	VH	M	H

on these estimations, a novel cloud attack rank metric (CARM) was estimated via Equation 1.

Thus, these models must be used for real-time cloud deployments, and can be further extended via use of incremental learning and fusion techniques.

Conclusions

The paper evaluates various statistical models for cloud attack detection and mitigation from an empirical perspective. The study focuses on accuracy, precision, deployment cost, speed, scalability, and security levels of different models. The results of the evaluation suggest that various models showcase higher accuracy, precision, and speed, while others are more cost-effective and scalable for different use cases.

It can be concluded that SDM TA, GAN, DAD, ResNet 18, BAC APP PA, TL, and Bi TCCS showcase higher accuracy levels and can be used for a wide variety of high-performance cloud deployments. Similarly, ResNet 18, PSB PN, SDM TA, DG CNN, Point Net++, and BAC APP PA are capable of showcasing higher precision levels and can be used for consistency-aware cloud deployments and scenarios.

References

1. Bhardwaj, V. M. and Vig, R. (2020). Hyperband tuned deep neural network with well posed stacked sparse autoEncoder for detection of DDoS attacks in cloud. *IEEE Acc.*, 8, 181916–181929. doi: 10.1109/ACCESS.2020.3028690.
2. Yuan, B., et al. (2020). Minimizing financial cost of DDoS attack defense in clouds with fine-grained resource management. *IEEE Trans. Netw. Sci. Engg.*, 7(4), 2541–2554. doi: 10.1109/TNSE.2020.2981449.
3. Zhou, Y., Hu, Z., and Li, F. (2023). Searchable public-key encryption with cryptographic reverse firewalls for cloud storage. *IEEE Trans. Cloud Comput.*, 11(1), 383–396. doi: 10.1109/TCC.2021.3095498.
4. Yang, H., Ju, S., Xia, Y., and Zhang, J. (2019). Predictive cloud control for networked multiagent systems with quantized signals under DoS attacks. *IEEE Trans. Sys. Man Cybernet. Sys.*, 51(2), 1345–1353. doi: 10.1109/TSMC.2019.2896087.
5. Saeed, P. G. and Hussain, S. A. (2020). Cross-VM network channel attacks and countermeasures within cloud computing environments. *IEEE Trans. Dependable Secure Comput.*, 19(3), 1783–1794. doi: 10.1109/TDSC.2020.3037022.
6. Kautish, S., R. A., and Vidyarthi, A. (2022). SDM-TA: Attack detection and mitigation mechanism for DDoS vulnerabilities in hybrid cloud environment. *IEEE Trans. Indus. Inform.*, 18(9), 6455–6463. doi: 10.1109/TII.2022.3146290.
7. Yin, X., Gao, Z., Yue, D., and Hu, S. (2022). Cloud-based event-triggered predictive control for heterogeneous NMASs under both DoS attacks and transmission delays. *IEEE Trans. Sys. Man Cybernet. Sys.*, 52(12), 7482–7493. doi: 10.1109/TSMC.2022.3160510.
8. Bhondge, S. K., Bhoyar, D. B., and Mohad, S. (2016). Strategy for power consumption management at base transceiver station. *2016 World Conf. Futur. Trends Res. Innov. Soc. Welfare (Startup Conclave)*, 1–4. doi: 10.1109/STARTUP.2016.7583988.

9. Gong, X., et al. (2021). Defense-resistant backdoor attacks against deep neural networks in outsourced cloud environment. *IEEE J. Select. Areas Comm.*, 39(8), 2617–2631. doi: 10.1109/JSAC.2021.3087237.
10. Shawahna, M. A.-A., Mahmoud, A. S. H., and Osais, Y. (2018). EDoS-ADS: An enhanced mitigation technique against economic denial of sustainability (EDoS) attacks. *IEEE Trans. Cloud Comput.*, 8(3), 790–804. doi: 10.1109/TCC.2018.
11. Bhoyar, D., Katey, B., and Ingle, M. (2018). LoRa technology based low cost water meter reading system. *Proc. 3rd Inter. Conf. Internet Things Connect. Technol. (ICIoTCT)*, 26–27.

50 Lifting wavelet domain based watermarking technique using SVM

Mohiul Islam[1,a], *Saharul Alom Barlaskar*[2], *Sushanta Debnath*[3], *Amarjit Roy*[4] *and Rabul Hussain Laskar*[2]

[1]School of Electronics Engineering, Vellore Institute of Technology, Vellore, Tamil Nadu, India

[2]Department of ECE, National Institute of Technology Silchar, Assam, India

[3]Department of Electrical Engineering, Gomati District Polytechnic, Udaipur, Tripura, India

[4]Department of Electrical Engineering, Ghani Khan Choudhury Institute of Engineering and Technology, India

Abstract

Digital watermarking can be used as a means of safeguarding digital images from unauthorized usage and distribution. It introduces a robust image watermarking technique in the lifting wavelet domain. To enhance its resilience against a various potential attacks, the method integrates support vector machine (SVM) technology during watermark extraction. The strategy involves modifying 3-level LWT coefficients to seamlessly embed specific binary watermark bits within the host image. Across a dataset of 300 images, the watermarking system demonstrates an average imperceptibility level of approximately 43.59 dB. Impressively, the algorithm exhibits heightened robustness against a variety of attack types, spanning geometric and non-geometric variations. Imperceptibility is quantified using peak signal-to-noise ratio (PSNR), while robustness is assessed through normalized cross-correlation (NC). Empirical evidence strongly suggests the potential suitability of this developed technique for fortifying copyright protection in the realm of digital imagery.

Keywords: Discrete wavelet transform (DWT), lifting wavelet transform (DCT), image watermarking, support vector machine (SVM)

1. Introduction

Watermarking in the wavelet domain has gained significant traction for its multi-resolution capabilities. Digital image watermarking proves effective in authenticating and protecting image copyrights, especially in vulnerable communication environments like the Internet. This technique involves concealing ownership data within host images, either in spatial or transform domains. Despite their simplicity, spatial domain techniques are found to offer lower robustness compared to transform domain methods, as suggested by Wang et al. [1].

The discrete wavelet transform (DWT) is renowned in image processing due to its time and frequency multi-resolution properties. The discrete cosine transform (DCT) achieves remarkable energy compaction for correlated image data. The combined utilization of DWT and DCT in watermarking typically demonstrates strong performance in terms of both invisibility and robustness. Similarly, the lifting wavelet transform (LWT) possesses energy compaction and multi-resolution features. Beyond these transformations, the singular value decomposition (SVD) stands as a potential numerical tool with applications in data hiding and image compression. Recent years have witnessed the emergence of various transform domain techniques developed by Lai et al encompassing DWT, LWT, DCT, DFT, etc. [3]. These techniques leverage signal characteristics and human perception properties to outperform spatial domain methods. Integrating different decomposition approaches like SVD and QR decomposition further enhances performance achieved by Singh and Singh [13].

This paper introduces a watermarking technique within the lifting domain. While prior methods used statistical algorithms for watermark extraction, this approach based on coefficient correlation showed limited robustness against various attack conditions, including common image processing operations like JPEG, noise, and geometric alterations. To address this, a machine learning-based approach using support vector machines (SVM) is proposed to enhance resilience against both geometric and non-geometric attacks. The SVM is seamlessly integrated into the wavelet domain algorithm, striking a balance between imperceptibility and robustness. The paper is structured as follows: Section II outlines watermark embedding and extraction processes, Section III presents experimental results and analysis, and Section IV concludes the discussion on the watermarking technique's efficacy.

[a]mohiul.islam@vit.ac.in

DOI: 10.1201/9781003540199-50

2. LWT-SVD-based watermarking scheme

This study employs an LWT sub-band to embed watermarks, with the integration of SVM during watermark extraction for bit classification. SVM is harnessed as a watermark detector due to its strong generalization capability, enhancing the overall robustness of the watermarking system. The subsequent section outlines the embedding and extraction procedure of the LWT-SVM watermarking technique.

A. Watermark embedding

The embedding process of this algorithm is similar to embedding of as performed in [JEI paper]. In both the algorithm, same embedding domain (LWT) has been utilized. However, the sub-band utilized in this method is different from the algorithm utilized in ref [15]. In this algorithm, a 3-level sub-band present in LH has been used. While in this algorithm, a sub-band present in HL as mentioned by Islam et al, which has been utilized [15].

The watermark W has been generated using RW and SW similarly as discussed by Islam et al [16]. The host image is decomposed three times using LWT to obtain a 3-level LWT sub-band. The host image undergoes 3-level LWT decomposition, shuffling coefficients with key 1 and randomizing 2×2 blocks with key 2. Binary watermark bits are embedded using non-overlapping 2×2 blocks, modifying the largest coefficient based on the watermark bit. Refer to Figure 50.1 for the flowchart of the embedding algorithm.

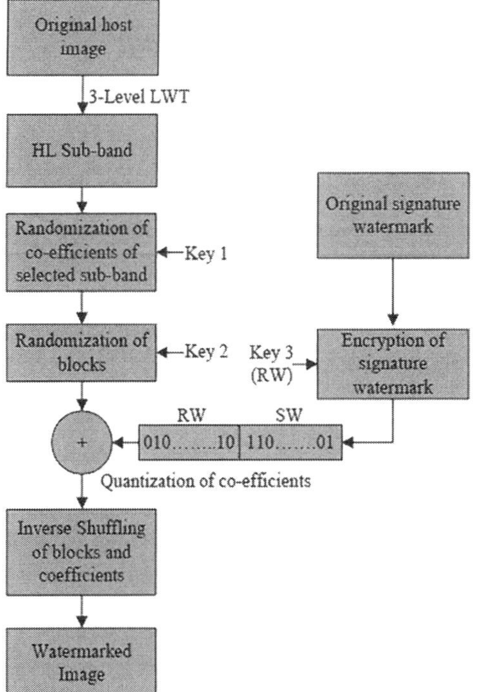

Figure 50.1 Watermark embedding procedure

B. Watermark extraction

This algorithm incorporates SVM during extraction to enhance robustness against various attack scenarios. SVM functions as a binary classifier in this method to detect watermark bits for specific reasons.

i During embedding, the watermark bit (0 or 1) corresponds to a distinct pattern of LWT coefficients, showcasing a nonlinear relationship with the two largest coefficients in the ith block, i.e., $W_i = f(c_i(n), c_i(n\text{-}1))$. SVM's potent nonlinear mapping capability makes it apt for learning this relationship.

ii Signal processing attacks on watermarked images, like noise addition and compression, alter LWT coefficients, akin to introducing diverse noises. The watermark detector needs noise resistance, resembling a generalization skill in machine learning terms. In this scheme, SVM serves as a watermark detector due to its robust generalization ability, enhancing overall robustness.

To detect the embedded watermark bits, the process begins with conducting a 3-level LWT on the compromised image, focusing on the HL3 sub-band. The coefficient blocks from the HL3 sub-band, where the watermark was embedded, are then isolated. Subsequently, a range of statistical parameters for these specific blocks are computed, following the methodology outlined by Islam et al [17]. The schematic representation of the proposed extraction technique can be seen in Figure 50.2. The detailed process of watermark extraction in step by step may be understood from the research work contributed by Islam et al. [17]. The flow chart of the proposed extraction technique is shown in Figure 50.2.

3. Results and discussion

The performance of the proposed LWT-SVM technique was assessed using a dataset of 300 images, encompassing various categories such as standard, texture, medical, and satellite images. Evaluation criteria included imperceptibility (measured via PSNR) and robustness (quantified using NC). Further elaboration on the experimental analysis is provided in the subsequent sections.

A. Imperceptibility analysis

Imperceptibility was assessed across the entire image database, revealing a trade-off between embedding threshold ('T'), imperceptibility, and robustness. Optimal balance occurred at "T=35," yielding

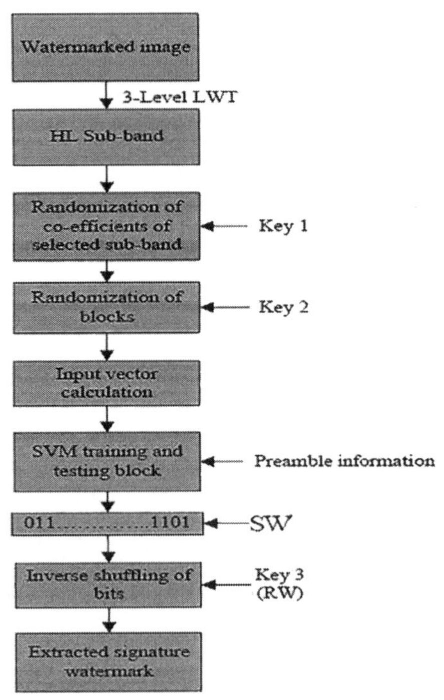

Figure 50.2 Watermark extraction procedure

satisfactory imperceptibility and robustness. PSNR results for 15 standard images, depicted in Figure 50.3, affirmed consistent imperceptibility across diverse image types. The algorithm maintained an average imperceptibility of 43.59 dB across the entire image dataset.

B. Robustness analysis

Robustness analysis encompassed diverse attacks on various image types, including JPEG compression at different quality factors, histogram equalization (HE), salt and pepper noise (SPN), speckle noise (SN), image sharpening (IS), average filtering (AF), Gaussian filtering (GF), and cropping (CR). While the comprehensive analysis spanned the entire image database, illustrative outcomes are exhibited for 10 images in Table 50.1. The algorithm's average robustness across 300 images against diverse attacks is outlined in Table 50.2, which also showcases the extracted watermark from the Lena watermarked image under different attack conditions. Experimental findings affirm the proposed algorithm's notable robustness against a range of attacks.

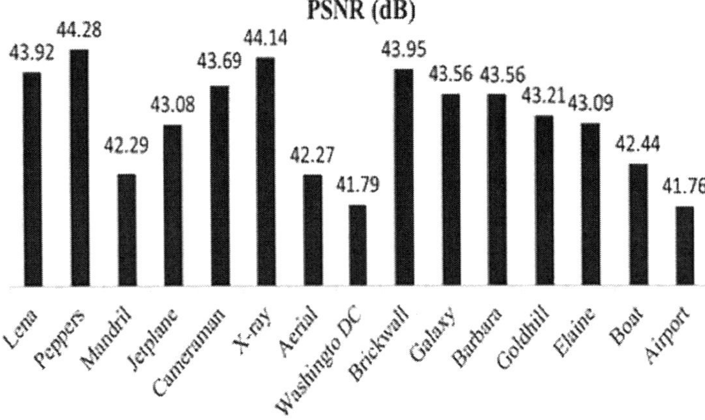

Figure 50.3 The imperceptibility for different images

Table 50.1 Average robustness in terms of NC for different attacks over 10 image

Images	JPEG 30	JPEG 50	SPN (0.01)	SN (0.01)	AF (3×3)	GF (3×3)	HE	IS	CR (15%)
Lena	0.9921	0.9921	0.8447	0.9019	0.7836	0.9961	0.9765	1.0000	0.9066
Peppers	0.9803	0.9882	0.8541	0.8856	0.8937	0.9842	0.9611	1.0000	0.8651
Mandril	0.9411	0.9723	0.8467	0.8818	0.7203	0.9686	0.9451	1.0000	0.8638
Jetplane	0.9842	0.9961	0.8429	0.6748	0.8988	0.9686	0.8479	0.9961	0.8651
Cameraman	0.9724	0.9763	0.8420	0.8831	0.8149	0.9802	0.9187	0.9921	0.8901
X-ray	0.8888	0.9570	0.8253	0.6287	0.8383	0.9802	0.8045	0.9921	0.9124
Aerial	0.8739	0.9449	0.8032	0.7162	0.6821	0.9802	0.8473	0.9921	0.9025
Washington DC	0.7953	0.8933	0.8339	0.8044	0.4051	0.9408	0.8899	0.9645	0.8804
Brickwall	0.9369	0.9803	0.8776	0.7904	0.7661	0.9921	0.9566	0.9843	0.9031
Galaxy	0.9526	0.9764	0.8604	0.9609	0.8190	0.9961	0.7782	1.0000	0.9239

Table 50.2 Average robustness (NC) for different attacks and extracted watermark on Lena image

Attacks	JPEG 30	JPEG 50	SPN (0.01)	SN (0.01)	AF (3×3)	GF (3×3)	HE
NC (average)	0.9429	0.9675	0.8477	0.8846	0.6842	0.9680	0.8854
Extracted watermark							

IV. Conclusions

This paper introduces a watermarking scheme based on LWT-SVM. The algorithm demonstrates commendable robustness against prevalent signal processing attacks while maintaining an acceptable level of imperceptibility. Yet, there's a need to bolster robustness against specific attacks like noise and denoising. Furthermore, the algorithm's vulnerability to de-synchronization attacks such as rotation, translation, and scaling is acknowledged. Future endeavors could involve integrating a geometric distortion correction algorithm into the LWT-SVM approach to bolster robustness against diverse geometric attacks.

Acknowledgement

The author would thank the School of Electronics Engineering (SENSE), VIT Vellore for all the necessary support.

References

1. Wang, R. Z., Lin, C. F., and Lin, J. C. (2001). Image hiding by optimal LSB substitution and genetic algorithm. *Patt. Recogn.*, 34(3), 671–683.
2. Mukherjee, D. P., Maitra, S., and Acton, S. T. (2004). Spatial domain digital watermarking of multimedia objects for buyer authentication. *IEEE Trans. Multimed.*, 6(1), 1–15.
3. Lai, C. C. and Tsai, C. C. (2010). Digital image watermarking using discrete wavelet transform and singular value decomposition. *IEEE Trans. Instrum. Meas.*, 59(11), 3060–3063.
4. Mehta, R., Rajpal, N., and Vishwakarma, V. P. (2017). A robust and efficient image watermarking scheme based on Lagrangian SVR and lifting wavelet transform. *Inter. J. Mac. Learn. Cybernet.*, 8(2), 379–395.
5. Verma, V. S., Jha, R. K., and Ojha, A. (2015). Digital watermark extraction using support vector machine with principal component analysis based feature reduction. *J. Visual Comm. Image Represen.*, 31, 75–85.
6. Verma, V. S., Jha, R. K., and Ojha, A. (2015). Significant region based robust watermarking scheme in lifting wavelet transform domain. *Exp. Sys. Appl.*, 42(21), 8184–8197.
7. Kasana, G. and Kasana, S. S. (2017). Reference based semi blind image watermarking scheme in wavelet domain. *Optik*, 142, 191–204.
8. Hu, H. T. and Hsu, L. Y. (2017). Collective blind image watermarking in DWT-DCT domain with adaptive embedding strength governed by quality metrics. *Multimed. Tools Appl.*, 76(5), 6575–659.
9. Thakkar, F. N. and Srivastava, V. K. (2017). A blind medical image watermarking: DWT-SVD based robust and secure approach for telemedicine applications. *Multimed. Tools Appl.*, 76(3), 3669–3697.
10. Makbol, N. M. and Khoo, B. E. (2013). Robust blind image watermarking scheme based on redundant discrete wavelet transform and singular value decomposition. *AEU-Inter. J. Elec. Comm.*, 67(2), 102–112.
11. Mishra, C. A., Sharma, A., and Bedi, P. (2014). Optimized gray-scale image watermarking using DWT–SVD and firefly algorithm. *Exp. Sys. Appl.*, 41(17), 7858–7867.
12. Makbol, N. M. and Khoo, B. E. (2014). A new robust and secure digital image watermarking scheme based on the integer wavelet transform and singular value decomposition. *Dig. Signal Proc.*, 33, 134–147.
13. Mehta, R., Rajpal, N., and Vishwakarma, V. P. (2016). LWT-QR decomposition based robust and efficient image watermarking scheme using Lagrangian SVR. *Multimed. Tools Appl.*, 75(7), 4129–4150.
14. Singh, D. and Singh, S. K. (2017). DWT-SVD and DCT based robust and blind watermarking scheme for copyright protection. *Multimed. Tools Appl.*, 76(11), 13001–13024.
15. Islam, G. M., Parmar, A. S., Kumar, A., and Laskar, R. H. (2017). SVM regression based robust image watermarking technique in joint DWT-DCT domain. *2017 Inter. Conf. Intell. Comput. Instrum. Control Technol. (ICICICT)*, 1406–1413.
16. Islam, M. and Laskar, R. H. (2018). Robust image watermarking technique using support vector regression for blind geometric distortion correction in lifting wavelet domain and singular value decomposition domain. J. Elec. Imag., 27(5), 053008.
17. Islam, M., Roy, A., and Laskar, R. H. (2020). SVM based robust image watermarking technique in LWT domain using different sub bands. *Neural Comput. Appl.*, 32, 1379–1403.

51 Blade shape optimization of H-Darrieus wind turbine using CFD analysis

Sourav Kanthal, Kazi Ganiur Rahman, Prodipta Samanta, Satyaki Birbanshi, Sagar Chemjong, Ashwini Routh, Anal Ranjan Sengupta[a], Thia Paul and Palash Biswas

Department of Mechanical Engineering, JIS College of Engineering, Kalyani, West Bengal, India

Abstract

At present time, vertical axis wind turbines (VWATs) are capable of showing improved performance in built environment condition than horizontal axis wind turbines (HAWTs). It has the capacity to contribute more significantly to the adoption of wind power. But due to complexity in VWATs aerodynamic design the progress has been hindered. If their performance and efficiency can be improved and maximized, their adoption and deployment will increase. Therefore, this study introduces a new optimized design of VAWT's blade profile to increase its aerodynamic efficiency. Adjoint method is an efficient way to perform blade shape optimization of VAWT's with great success. In this present investigation, optimization of the blade shape of NACA0018 & S1046 airfoils has been done using adjoint solver tool in ANSYS fluent at wind speed of 15 m/s. The aim of this optimization is to enhance the ratio of lift to drag of the airfoil by 20%. The results shows that the lift to drag ratios are significantly increased for the considered blade profiles for their optimized shape which can lead to enhanced performance of the H-Darrieus turbine.

Keywords: Vertical axis wind turbine, CFD, NACA0018, S1046, airfoil, optimization

Introduction

If sustainable energy sources are looked upon, wind energy is one of them. Among this small sized VAWTs are gaining more recognition because of high efficiency in power generation [1]. Wind energy can be categorized into three major types according to its geographical nature and power generation. Land-based wind energy, distributed wind energy, and offshore wind energy are some of them. For each type of energy, the most effective and highly potential methods of energy generation should be used to acquire a high output. The principle of working of wind turbine be about to turn the wind energy into kinetic energy. Power is generated when wind turns the blades of the turbine. Two basic types of wind turbine that are vertical axis wind turbine and horizontal axis wind turbine. The classification of wind turbine shown in Figure 51.1.

VAWTs and HAWTs are utilized to generate maximum power from wind. In comparison to HAWTs, VAWTs have several advantages. VAWT's are omnidirectional due to which they don't change the exposure to match the wind direction as HAWT's. This reduces the number of corridors. VAWT provides three major advantages that reduce the cost of wind energy, a low central gravity, less complexity of machine, and better flexibility to large sizes. Thus, VAWTs are largely effective and most effective than HAWTs for their high affair and durability for energy generation.

VAWT exploration has gained momentum over many decades through experimental research and numerical simulation. Still, certain crucial factors are impacting the performance. A huge amount of investigations are done with three-bladed H-Darrieus turbine as compared to two-bladed H-Darrieus turbine in the matter of performance enhancement to assess the effects on blade shape performance, depression, tip speed ratio (TSR), and Reynolds' number.

Zamani et al., studied about introducing an advance design for VAWT's blade shape to avoid the drawbacks, and they worked on a new J-shape blade profile of conventional NACA0015 which was investigated through 3D simulations [2]. It was noticed that at low speeds of air, the turbine operates faster. Gupta and Biswas performed simulation at low wind velocities to attain a high power coefficient (C_p) on a two-bladed airfoil type NACA0012 with the twisted blade of 30 and 10% of chord length from the trailing edge of blades [3]. Sengupta et al., noticed that at the solidity of 0.30, a small H-Darrieus VAWT having the NACA0015 blade profile showed a C_p value of 0.34 [4]. Su et al., performed optimization on a V-shaped blade, enhancing the power coefficient, and obtained an efficiency of about 24.1% [5]. Zhu et al., showed us that gurney flap can incredibly increases the C_p

[a]analranjan.sengupta@jiscollege.ac.in

DOI: 10.1201/9781003540199-51

value of VAWT with decreased rational velocity, and by taking the chord-length of height 0.75% and chord-length of width 0.12%, the improvement has reached up to 21.32% [6].

Ismail and Vijayavaraghavan inspected the effect of altered profile on the NACA0015 airfoil by using computational fluid dynamics (CFD) simulations and found 35% increment of average tangential force in a constant condition and 40% in the twitching condition with each rotation using optimized Gurney flap and an internal, semi-circle dimple [7]. Sobhani et al., used the blade cavity to enhance the optimal performance of 18% and the average efficiency of the turbine of 25% as differ to the referred NACA0021 [8]. Wang et al., performed optimization on VAWT, which has 18.3% power enhancement of TSR 2 and better operational capability at low TSR using CFD and the Taguchi method [9]. Gupta and Biswas performed a simulation of a twisted blade to show the variation of the Cp which increases in time of increased TSR up to the max. of 0.127 at TSR of 1.437 [10]. Leelakrishnan et al., analyzed the Darrieus wind turbine by varying TSR in the range of 0–3 m/s speed and observed that the TSR of 1.8 generates maximum power and will provide better performance [11]. According to Mannion et al., the tangential force is improved by 35% over standard S1046 for a TSR of 3.50 [12].

Rezaeiha et al., performed CFD analysis on NACA0018, by placing the smallest azimuthal increment which will yield the highest power coefficient [13]. Sengupta et al., worked on two unsymmetric blades S815 and EN0005 and one ceremonial symmetrical NACA0018 blade [14]. It was witnessed that the dynamic torque and power coefficient of S815 blade rotor is greater than the EN0005 and the NACA0018 blade rotors. The work conducted by Stratan et al., showcased the enhanced geometric designs for optimization of shape across various applications of CFD [15]. Chen optimized wind turbine airfoils using genetic algorithms and CFD simulation [16].

From all these earlier researches, it is realized that very few studies had been done on optimizing the blade shape using the adjoint solver tool in ANSYS fluent through CFD analysis.

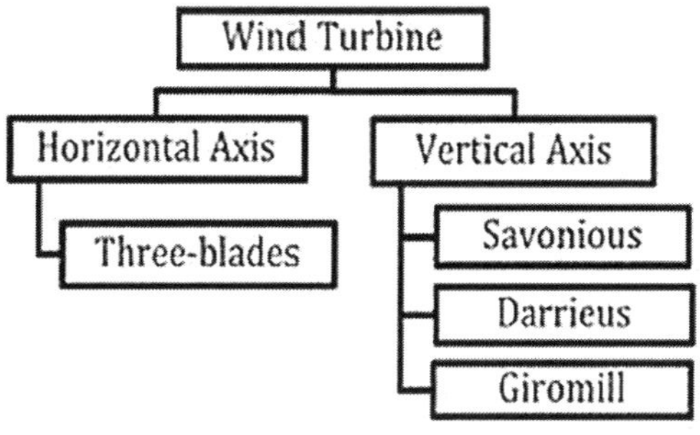

Figure 51.1 Wind turbine classification

Geometry modeling

For this current study, H-Darrieus rotors are primarily constructed using S1046 and NACA 0018 airfoil blade shapes. The profiles of these blades are portrayed in Figures 51.2 and 51.3 (airfoiltools.com). To mitigate the impacts of blockage, the computational domain had been carefully chosen so that the borders are placed quite a ways away from the blade. The

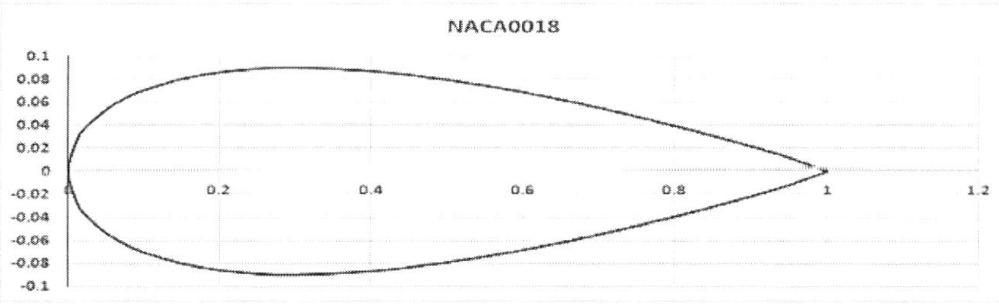

Figure 51.2 NACA0018 blade profile

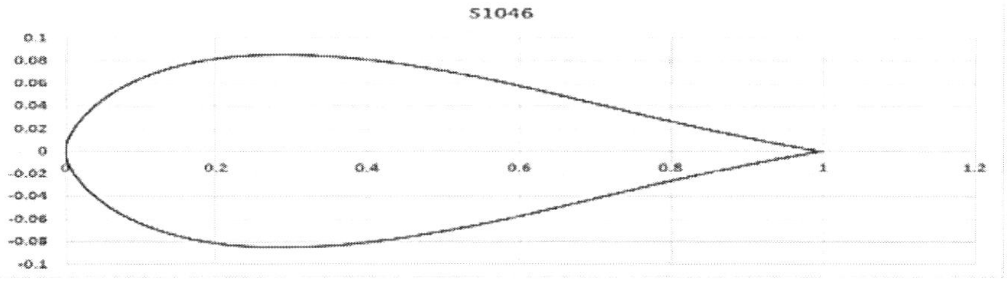

Figure 51.3 S1046 blade profile

chord length (c) is 1 m of the rotor blades. The lengths of the inlet and outlet domains are 4.5 times of the blades' chord length. Figure 51.4 shows that the right and left edges are subject to the velocity inlet and exit pressure conditions respectively, while the top and bottom edges are designated as symmetrical edges.

Meshing

In the computational domain, unstructured triangular meshing was executed for complex blade designs. A mesh size of 0.02 was chosen, and inflation layers were created in the viscous region with a total thickness of 0.08. A y+ value of 4 was selected, focusing on the viscous region. The grids close to the blade edges were denser than those farther away to capture the viscous sublayer accurately. Skewness was maintained below 0.5 for the domain geometry. In our case, the element size was taken 236304. Figure 51.5 below illustrates the meshing of the computational province as well as a close-up view of the mesh and inflation layer of the airfoil. In Figure 51.6, the mesh independency test chart has been given. Input parameters like wind velocity and air density remained consistent across simulations. The grid independence research identified the ideal compromise between simulation time and result accuracy.

Methodology

The adjoint method is a powerful mathematical technique that enables the efficient computation of gradients and sensitivities in complex simulations. By leveraging the adjoint equations and a backward solver, it avoids the need for direct differentiation of the governing equations, making it an invaluable tool in computational engineering and scientific research. The adjoint method utilizes an adjoint solver and gradient-based optimizer in ANSYS fluent to optimize turbine blade aerodynamics. Initial CFD analysis is executed with a baseline configuration, and objective functions such as lift, drag, and lift to drag ratio are defined for improvement. Mesh sensitivity data is obtained, and a mesh morpher is used to vary the airfoil geometry. To attain the targeted 20% raise (lift to drag ratio), the gradient-based optimizer performs up to 200 iterations.

This study focuses on implementing and comparing two distinct objective functions: optimizing the lift to drag ratio and increase the current situation.

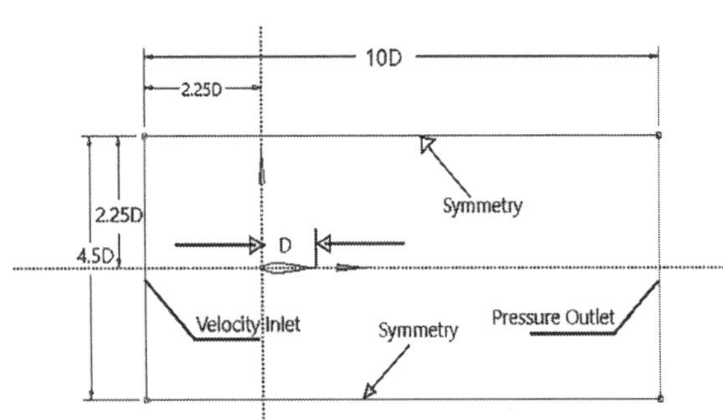

Figure 51.4 Computational domain

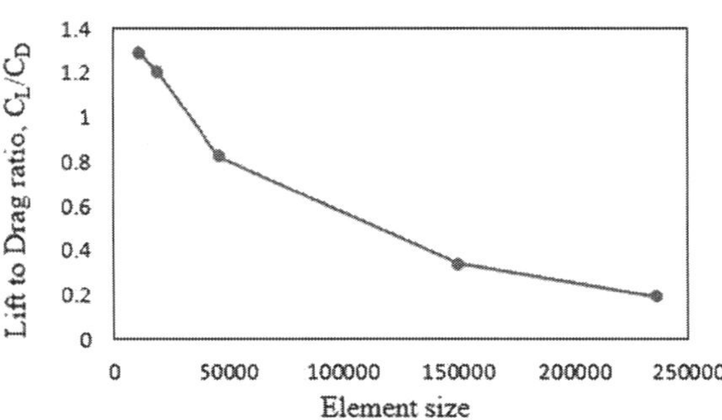

Figure 51.6 Mesh independency test

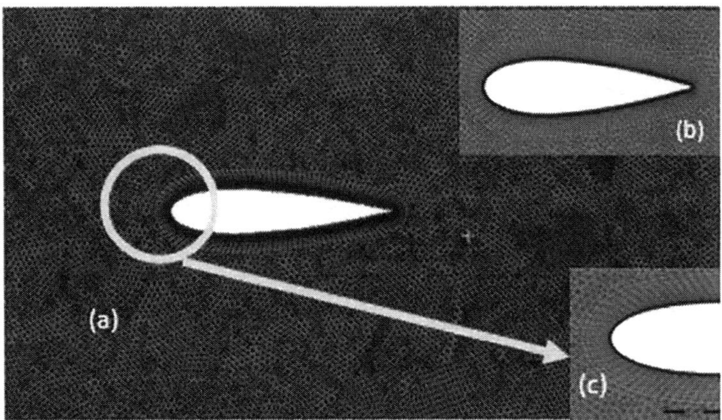

Figure 51.5 (a) Meshing of the full flow domain (b) close look of the airfoil mesh (c) close look of inflation layer

The objective of airfoil shape optimization is to maximize the selected objective function while considering the control points, represented by the letter (c). The formula can be written as [17]:

{*maximize* ((*c*), *c*) *Subject to* $R = (q(c), c)$ *equals to 0*}

where, we defined q as flow solution, c as inputs to the problem, ((*c*), *c*) as quantity of interest, and R(q(c);c) as residuals of the Navier stokes equation.

The *J* is objective function can be explained as Equation 1.

$$\delta J = \delta q\left(\frac{\partial J}{\partial q} - \varphi T\frac{\partial R}{\partial q}\right) + \delta c\left(\frac{\partial J}{\partial c} - \varphi T\frac{\partial R}{\partial c}\right) \quad (1)$$

The value of Langragian multiplier φT is carefully selected to simplify the given Equation 2 to the following form:

$$\frac{\partial J}{\partial q} = \varphi T\frac{\partial R}{\partial q} \quad (2)$$

The adjoint sensitivity equation is finally described as Equation 3.

$$dJ = \delta c\frac{\partial J}{\partial c} - \delta c\left(\varphi T\frac{\partial R}{\partial c}\right) \quad (3)$$

This work proposes a forced air (positive air pressure) solution to the problem.

Numerical simulation setup for blade shape optimization

This analysis shows the equations which were calculated using ANSYS fluent. An aerodynamic study was performed using a computational method that accounted for two-dimensional, incompressible flow, employing the Reynolds-averaged Navier-Stokes equations and the SST k-ω turbulence model. The flow is deemed to be near-wall in the range of y+ values between 0 and 5. This indicates that the cells are extremely close to the wall and that viscous effects are prevalent. The SST k-omega model is specifically intended for near wall region, moving seamlessly between the k-epsilon model (for low y+ values) and the k-omega model (for max y+ values). Wind velocities are maintained as 15 m/s. The bounding region around an airfoil geometry has been shown in Figure 51.7 when the adjoint solver optimization was being applied.

Results

Figures 51.8 and 51.9 show the comparison between NACA0018 and Opt_NACA0018. These diagrams reveal the differences in stagnation points, flow separation locations, and pressure distributions between the NACA0018 blade and the optimized blade. Opt_NACA0018 has a lower-side drift of the stagnation point and separation above the surface. The cambered bottom surface and sleek upper surface of Opt_NACA0018 create a pressure difference favoring lift. The pressure contour of NACA0018 is predictable and symmetrical, while Opt_NACA0018 exhibits chaotic

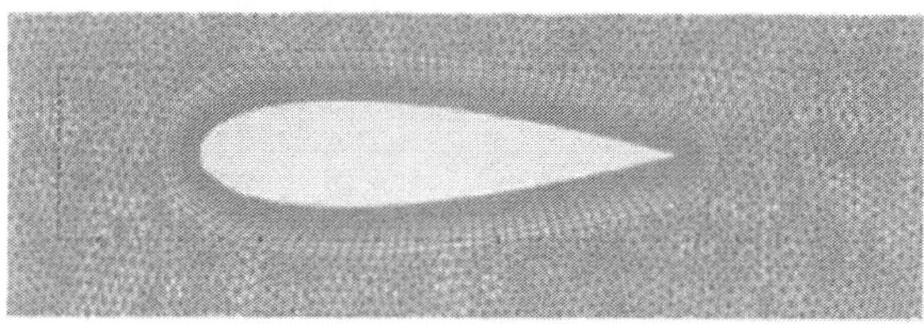

Figure 51.7 Bounding region around an airfoil geometry

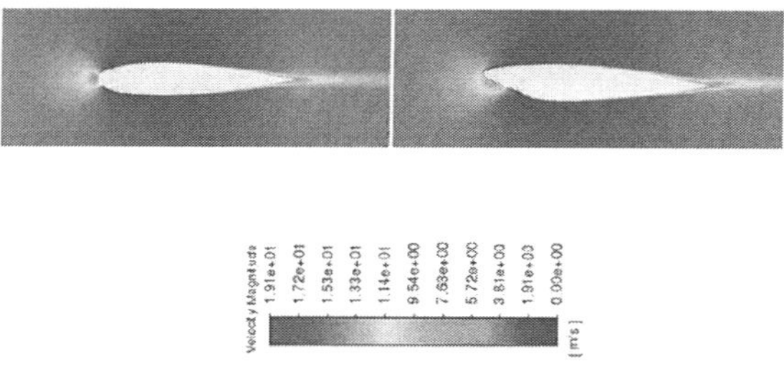

Figure 51.8 Velocity profile of NACA0018 and Opt_NACA0018

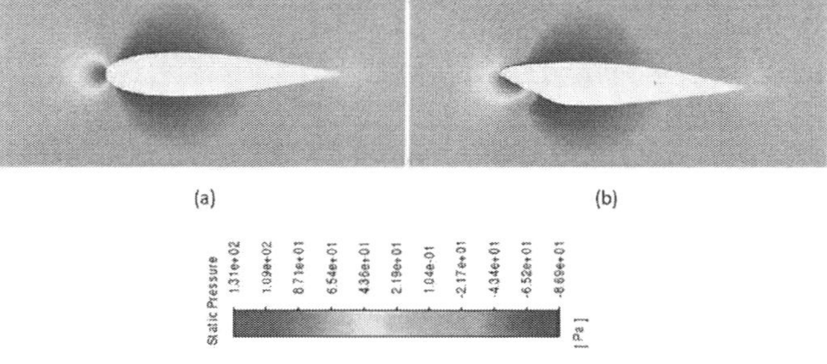

Figure 51.9 Pressure profile of NACA0018 and Opt_NACA0018

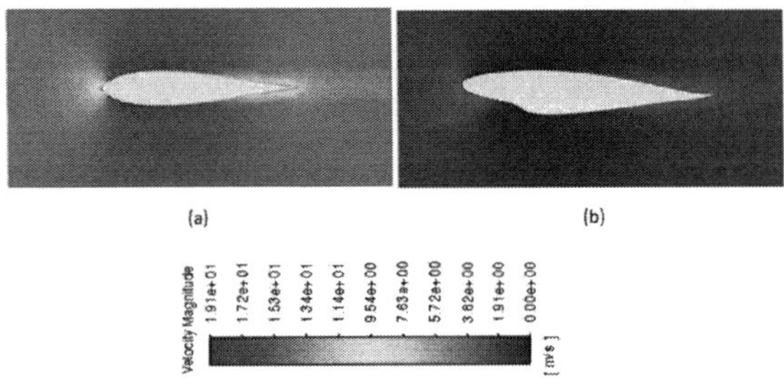

Figure 51.10 Velocity profile of S1046 and Opt_S1046

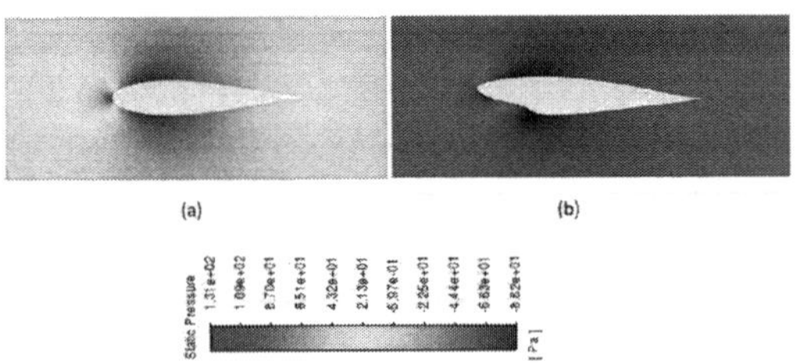

Figure 51.11 Pressure profile of S1046 and Opt_S1046

and uneven distribution with high tensile pressure on the bottom side and suction on the top layer. NACA0018's symmetrical pressure distribution results in inadequate lift, whereas the unsymmetrical Opt_NACA0018 generates higher lift due to stronger pressure on the leading lower edge surface even at 0° angle of attack. Opt_NACA0018 shows greater overall pressure differential than NACA0018.

Flow field contour comparisons of NACA0018 & Opt_NACA0018

Flow field contour comparisons of S1046 & Opt_S1046

Figures 51.51.10 and 11 compare the pressure and velocity contours of S1046 with optimized blade profile Opt_S1046. S1046's stagnation point has relocated to the bottom edge of the airfoil. Flow separation occurs on the upper surface of the cutting edge in Opt_S1046, whereas S1046 exhibits separation at the middle of cutting side. The streamlined higher surface and cambered lower surface of Opt_S1046 promote a pressure difference that favors lift. The pressure contour of S1046 shows even distribution on its upper and lower surfaces, while Opt_S1046 exhibits chaotic and uneven pressure distribution. The leading edge of Opt_S1046 experiences strong suction on its leading surface and significant tensile pressure on the bottom part, which contributes to lift creation. The symmetrical pressure distribution in S1046 results in inadequate

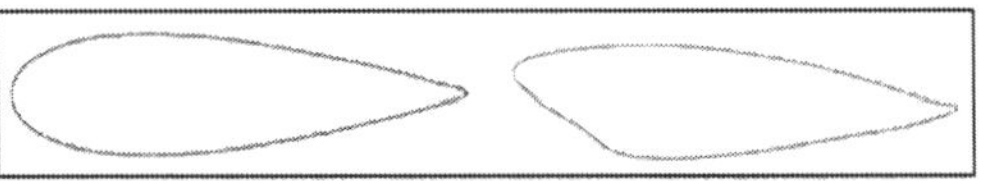

Figure 51.12 Blade shape comparison between NACA0018 & Opt_NACA0018

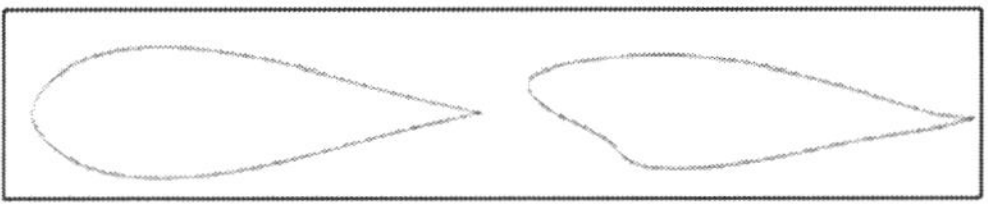

Figure 51.13 Blade shape comparison between S1046 & Opt_S1046

lift at 0° angle of attack, whereas the unsymmetrical shape of Opt_S1046 generates higher lift due to stronger pressure on the leading lower edge surface even at angle of attack. The variation in pressure is greater in Opt_S1046 as compared to S1046. Figures 51.12 and 51.13 show blade shape comparison among the considered conventional blade shapes along with their optimized profiles for better visualization.

Conclusions

In this present investigation, the blade shape optimization is done using the adjoint solver tool of ANSYS fluent software considering NACA0018 and S1046 blade profiles of H-Darrieus rotor. The primary findings of this present study are summarized below.

a. Adjoint method optimization significantly improved the lift to drag ratio by 99% and lift coefficient by 100%, of NACA0018, as well as the lift to drag ratio by 112% and lift coefficient by 113.4% of S1046. Negligible changes were observed in drag.
b. Pressure contour exhibited a higher shift in the stagnation point and uneven pressure distribution along the airfoil surfaces in the optimized geometries (Opt_NACA0018 and Opt_S1046). These findings demonstrate the adjoint method's effectiveness in enhancing the airfoil's aerodynamic performance.

In future, experimental investigation can be done using these optimized blade profiles. Again, CFD analysis in 3D may be performed to contrast the outcome with this present 2D study. Again, the aerodynamic features can be compared for velocity and pressure contours beyond zero-degree angle of attack.

References

1. Global Wind Energy Council (GWEC) (2022). Global Wind Report Annual Market Update. *Technical Report.* (Accessed on 15/02/2023). https://gwec.net/global-wind-report-2022/.
2. Zamani, M., Nazari, S., Moshizi, S. A., and Maghrebi, M. J. (2016). Three dimensional simulation of J-shaped Darrieus vertical axis wind turbine. *Energy*, 116, 1243–1255. https://doi.org/10.1016/j.energy.2016.10.031.
3. Biswas, A. and Gupta, R. (2014). Unsteady aerodynamics of a twist bladed H-Darrieus rotor in low Reynolds number flow. *J. Renew. Sustain. Energy*, 6(3), 033108. https://doi.org/10.1063/1.4878995.
4. Sengupta, A. R., Biswas, A., and Gupta, R. (2019). Comparison of low wind speed aerodynamics of unsymmetrical blade H-Darrieus rotors-blade camber and curvature signatures for performance improvement. *Renew. Energy*, 139, 1412–1427. https://doi.org/10.1016/j.renene.2019.03.054.
5. Su, J., Chen, Y., Han, Z., Zhou, D., Bao, Y., and Zhao, Y. (2020). Investigation of V-shaped blade for the performance improvement of vertical axis wind turbines. *Appl. Energy*, 260, 114326. https://doi.org/10.1016/j.apenergy.2019.114326.
6. Zhu, H., Hao, W., Li, C., Luo, S., Liu, Q., and Gao, C. (2021). Effect of geometric parameters of Gurney flap on performance enhancement of straight-bladed vertical axis wind turbine. *Renew. Energy*, 165, 464–480. https://doi.org/10.1016/j.renene.2020.11.027.
7. Ismail, Md. F. and Vijayaraghavan, K. (2015). The effects of aerofoil profile modification on a vertical axis wind turbine performance. *Energy*, 80, 20–31. https://doi.org/10.1016/j.energy.2014.11.034.
8. Sobhani, E., Ghaffari, M., and Maghrebi, M. J. (2017). Numerical investigation of dimple effects on darrieus vertical axis wind turbine. *Energy*, 133, 231–241. https://doi.org/10.1016/j.energy.2017.05.105.
9. Wang, Z., Wang, Y., and Zhuang, M. (2018). Improvement of the aerodynamic performance of vertical axis wind turbines with leading-edge serrations and helical blades using CFD and Taguchi method. Energy conversion and management 177, 107–121. https://doi.org/10.1016/j.enconman.2018.09.028.
10. Gupta, R. and Biswas, A. (2010). Computational fluid dynamics analysis of a twisted three-bladed H-Darrieus rotor. *J. Renew. Sustain. Energy*, 2(4), 043111, https://doi.org/10.1063/1.3483487.
11. Leelakrishnan, E., Sunil Kumar, M., David Selvaraj, S., Sundara Vignesh, N., and Abhesheka Raja, T. S. (2020). Numerical evaluation of optimum tip speed ratio for darrieus type vertical axis wind turbine. *Mater. Today Proc.*, 33, 4719–4722. https://doi.org/10.1016/j.matpr.2020.08.352.
12. Mannion, B., Leen, S. B., and Nash, S. (2018). A two and three-dimensional CFD investigation into performance prediction and wake characterisation of a vertical axis turbine. *J. Renew. Sustain. Energy*, 10(3), 034503. https://doi.org/10.1063/1.5017827.
13. Rezaeiha, A., Montazeri, H., and Blocken, B. (2018). Towards accurate CFD simulations of vertical axis wind turbines at different tip speed ratios and solidities: Guidelines for azimuthal increment, domain size and convergence. *Energy Conver. Manag.*, 156, 301–316. https://doi.org/10.1016/j.enconman.2017.11.026.
14. Sengupta, A. R., Biswas, A., and Gupta, R. (2016). Studies of some high solidity symmetrical and unsymmetrical blade H-Darrieus rotors with respect to starting characteristics, dynamic performances and flow physics in low wind streams. *Renew. Energy*, 93, 536–547. https://doi.org/10.1016/j.renene.2016.03.029.
15. van Stratan, G. S. M., Roy, S., and San, Y. K. (2023). Aerodynamic shape optimization of a NACA0018 airfoil using adjoint method and gradient-based optimizer. *MATEC Web Conf.*, 377, 01016. EDP Sciences, 2023. https://doi.org/10.1051/matecconf/202337701016.
16. Chen, X. (2014). Optimization of wind turbine airfoils/blades and wind farm layouts. Washington University in St. Louis, 2014. https://openscholarship.wustl.edu/cgi/viewcontent.cgi?article=2228&context=etd.http://airfoiltools.com/airfoil/details?airfoil=naca0018-il (Accessed on 14/05/2023). http://airfoiltools.com/airfoil/details?airfoil=s1046-il (Accessed on 14/05/2023).
17. Carrigan, T. J., Dennis, B. H., Han, Z. X., and Wang, B. P. (2012). Aerodynamic shape optimization of a vertical-axis wind turbine using differential evolution. *Inter. Scholar. Res. Notices*, Volume 2012, 1–12. Article ID 528418, https://doi.org/10.5402/2012/528418.

52 An investigation of blockchain technology and its application in the medical domain

Debadrita Roy[1,a], *Arnab Kundu*[2,b] *and Ripon Patgiri*[1,c]

[1]Department of Computer Science & Engineering, National Institute of Technology Silchar, Cachar, Silchar, Assam-788010, India

[2]Department of Electronics & Communication Engineering, National Institute of Technology Silchar, Cachar, Silchar, Assam-788010, India

Abstract

Blockchain technology has received much attention over recent decades in various medical fields due to its decentralization, transparency, and security properties. It uses a peer-to-peer (P2P) network for creating an uninterrupted, cumulative list of records termed blocks to form a digital ledger. This digital ledger records all transactions coordinated between all nodes within the same network. Blockchain can be applied for various resolutions, such as creating smart contracts, cryptocurrency transactions, recording intellectual property, managing medical records, and observing the supply chain. This article provides an overview of blockchain technology and its utilization in the healthcare sector, which includes a review of existing literature to outline potential future directions in this field. In addition, our paper explores the challenges to adopting blockchain in the medical environment in the future and overcomes to promise enhanced use of this technology.

Keywords: Blockchain, consensus protocol, distributed ledger, markle tree, radiology, smart contract

Introduction

The migration to electronic supervision of medical data has required physicians and their patients to create a practice of some innovative abbreviations such as EMRs (electronic medical records), EHRs (electronic health records), and PHRs (personal health records) [1]. These medical data cover information like patient names, medical practitioners' names, case studies of patients, various medical images, and reports processed by the physician in a documented format. It has been observed that the same patient's medical investigation reports are stored and monitored by different private hospitals. Such a data maintenance process is time-bound and contains various stages. Sometimes, patients must move from one hospital to another to collect the compact disc (CD). They must repeatedly even pay out of pocket for the collection of the disc [2]. Due to its confidentiality, storage, and security concerns, existing technologies for sharing radiographic images and patient data are based on central systems that are hectic to manage. Over recent years, data like medical record breaks within huge health data centers have made extra complications for all corporations looking for new medical image processing applications [3]. Results are derived from blockchain that offers safety against undesirable data exposure [4]. Blockchain technology was first familiarized in the market by Satoshi Nakamoto in 2008. A blockchain is a sequence of blocks in which every block is connected via crypto graphical functions. It is a distributed, immutable ledger that enables storing and recording transactions in a network. Blockchain has gained so much acceptance worldwide for its decentralized, immutable, transparent, and secure nature. It is decentralized. Decentralization means all the data within the blocks are distributed over the whole network. Whenever a new block wants to be included within the chain, it is verified by all the nodes within the network. Blockchain is immutable, i.e., nobody can tamper with the data. As blockchain is transparent, anybody can track the data. Important parameters like privacy and security of blockchain characterize encoding the data and storing it in the block using the hash value.

Literature review

We conducted a literature survey to find out the potential applications of Blockchain in medical domain.

Kiania et al., in their article, authors proposed conducting a systematic review of the current literature to evaluate how blockchain-based methods are being used to enhance privacy and security

[a]debadrita@gkciet.ac.in, [b]a.kunduwb@gmail.com, [c]ripon@cse.nits.ac.in

DOI: 10.1201/9781003540199-52

within electronic health systems [5]. Hang, et al., in their research paper, provided a meticulous classification system for blockchain technology in the context of clinical trials [6]. The findings demonstrated that blockchain has the capacity to encompass all stages of clinical trial investigations in a decentralized and transparent manner. The paper authored by Tagliafico, et al., outlines the utilization of blockchain in the medical field, specifically by radiologists, to enhance the value of radiological data [7]. Chen, et al., introduced the concept of utilizing medical data stored on the blockchain for practical applications such as remote diagnosis, treatment, and data mining [8]. Ploder, et al., in their research paper, discussed multiple risks associated with the implementation of blockchain technology in the healthcare sector and proposed strategies for mitigating these risks [9]. Researchers De Aguiar, et al., introduced a framework for overseeing medical data, administering medical records and images, and monitoring drug distribution while maintaining privacy and security through blockchain technology [10]. Additionally, the researchers explored various blockchain-based healthcare applications, including the integration of healthcare equipment monitoring with the Internet of Things (IoT). Zhuang, et al., outlined a blockchain based model designed to ensure the security and privacy of health records, allowing patients to maintain complete control over their own healthcare data [11]. In their research, Jabarulla and Lee devised a system aimed at verifying the security and integrity of patient health data without relying on a centralized organization [12]. Shen, et al., authors introduced a data-sharing plan in which they detailed how MedChain integrates blockchain technology, digest chain mechanisms, and peer-to-peer (P2P) networks to address efficiency challenges in the sharing of healthcare data [13]. Zhuang, et al., outlined a blockchain model that incorporates multiple trial-based contracts for overseeing clinical trials and organizing patient participation [14]. They established a central smart contract for automated subject matching and patient enrollment in trial management.

Background of blockchain technology

- In 1991, Research scientists Stuart Haber and W. Scott Stornetta wanted to implement a network by using cryptographic hash functions so that it never be tampered with.
- In 1992, Markle trees were introduced in the strategy to create a secure chain of blocks.
- Stefan Konst published the theory of cryptographically secured chains and implemented various ideas in 2000.
- Computer Scientist Hal Finney presented a method for digital cash in 2004, which was named "Reusable Proof of Work."
- In 2008, an anonymous person named Satoshi Nakamoto introduced the idea of blockchain by publishing a whitepaper. The paper was identified as "Bitcoin: A Peer-to-Peer E-Cash System" [4]. Nakamoto first executed blockchain as the public ledger for transactions using Bitcoin in January 2009.
- In 2014, blockchain was separated from digital currency, and Blockchain 2.0 came into the market

Organization of the article

The following section provides a concise tutorial on blockchain technology, explaining why it's referred to as a decentralized network. It delves into the fundamental principles of blockchain and explores various types of blockchain. Subsequently, it offers a brief examination of how blockchain is applied in the field of medical domain. Finally, the section summarizes key findings, addresses potential challenges associated with adopting blockchain in the medical sector, and proposes solutions.

Blockchain technology

Blockchain is one kind of distributed ledger. It records all transactions and helps in P2P transactions. It is open, accessible to all, and secure also. It is an increasing list of records, called blocks, linked using cryptography. Every block holds a cryptographic hash of the previous block, a timestamp, and transaction data.

In Figure 52.1, three blocks have been chained, i.e., the first block (genesis block) with index 0 and the rest of the two blocks index 1 and index 2. Every block keeps the timestamp, data, hash of the previous block, nounce, and hash it contains. Index are used for ordering. Timestamp shows date and time of transaction. The data is the data that will be stored within every node/block. When these blocks are chained, the data within the block will not be modified.

Why blockchain is called a decentralized network?

A blockchain enables the distribution of data across multiple users (nodes) located at various points. This not only results in data redundancy but also

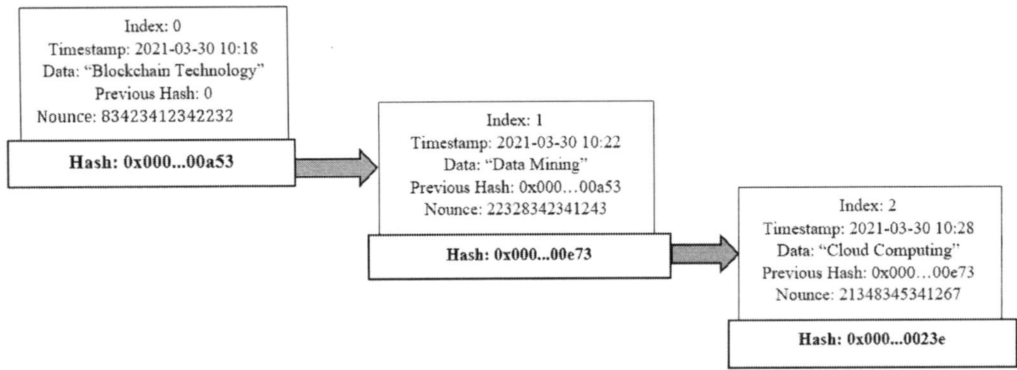

Figure 52.1 Concept of blockchain with three blocks

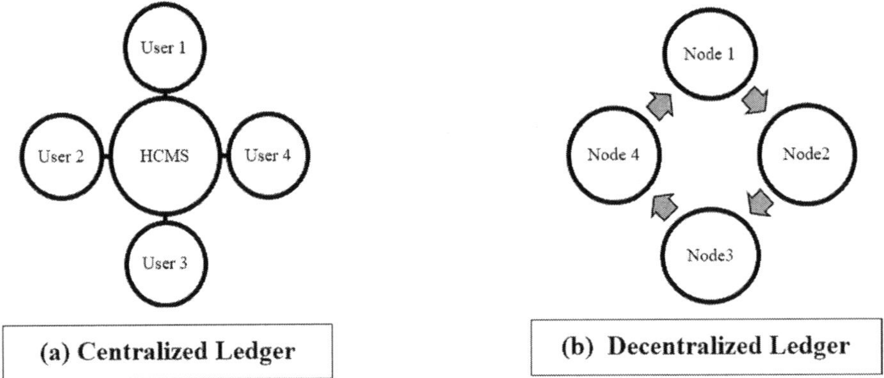

Figure 52.2 Centralized and decentralized ledger

ensures data consistency. Decentralization involves shifting decision-making authority and control from a centralized entity to a distributed system. By enabling multiple parties to conduct transactions in a distributed environment without the need for a central authority, blockchain addresses the issue of a single point of failure [15].

In Figure 52.2(a) Four users are present, but all records are stored and monitored by a health center monitoring organization (HCMO); (b) There is no central authority to monitor the data. Every user has some level of access to a ledger.

Key principles of blockchain

The fundamental principles of blockchain technology are described as follows:

- ***Immutability:*** Blockchain is immutable because as soon as data has been inserted within the chain, it cannot be changed (information is tamper-resistant) [16]. It is responsible for providing transparency and trust within the network.
- ***Consensus protocol:*** Consensus protocol is a set of logical instructions that perform blockchain operations. It allows users to coordinate in a distributed set so that all users can agree on a single source of truth even if one fails in the network.
- ***Decentralization:*** It is not issued by a central authority, and all the participants are created equal.
- ***Transparency:*** Each transaction is visible to anyone with access to the system.

Types of blockchain

Based on the usage, blockchain technology can be classified in three ways, i.e., public, private, and consortium. The basic differences are discussed in Table 52.1.

Utilizing blockchain in the medical system

Blockchain technology can provide many benefits in medical systems for its decentralized structure, which allows proper authentication, verification, security, and integrity. The decentralized blockchain network enables patients, physicians, and healthcare agencies to share data quickly and safely. The ledger technology helps medical researchers to find genetic code by simplifying the safe transmission of patient medical data and handling the drug supply

Table 52.1 The basic differences between different blockchain

Features	Public blockchain	Private blockchain	Consortium blockchain
Joining capability	It has entirely no access limitations. Anybody who have internet connection, become member of the network.	It is fully restrictive. One can only participate after getting invitation by network controller	This type of blockchain only permits working with group of members to read or mark data from or to the ledger [17]
Consensus protocol	Consensus process is permission less	Permission is needed	To carry out operation, permission is needed
Use of resources	Efficiency is low	Efficiency is high	Efficiency is high
Merits	It is complete transparent and trustable	Transaction speed is very high	More secured, efficient compared to public blockchain, best suited for organization
Demerits	It consumes a lot of energy, speed of transaction is very less	Less secure compared to public blockchain, trust gain is difficult, less transparent	Less transparent
Example	Bitcoin blockchain and the Ethereum blockchain	Corda, Hyperledger Fabric etc. are the examples of private block chain	IBM Food Trust is the example of consortium block chain

chain [18]. The various applications of blockchain in the medical domain are illustrated below.

Electronic health record (EHR)

An EHR, i.e., an electronic health record, is an automated system that is used to collect and monitor individual patients' sensitive health data. It stores patients' observation reports, test details, physician information, etc. EHR decreases the inefficiency, disorder, and data repetition difficulties linked with traditional paper-based medical data [19]. Blockchain has many benefits of decentralization, data transparency, confidentiality, and data security [20]. The process of medical (health) information storing and sharing among patients and health centers is discussed below.

Step 1: The doctor starts treatment and stores the patient's reports. The reports contain case history, medical test data, medication, etc., on their existing health system. The data fields assigned by the patient's public ID are despatched to the blockchain through APIs. Here, a transaction is made.

Step 2: Each transaction on the chain is certified and specified by a unique character (public key) that is stored on the blockchain itself.

Step 3: Medical professionals and healthcare providers generate a request via APIs and utilize the patient's public key to access the patient's encrypted data.

Step 4: Patients can approve the doctor or the healthcare providers by sharing their private key (i.e., password) to decrypt the data.

Managing electronic medical reports

Another blockchain application in the medical domain exists to handle EMR. The EMR-centered system provides immediate access to patient records in real-time, enhancing record accessibility, increasing accuracy, facilitating data sharing among various researchers, and ensuring the safety and security of patient data in comparison to the paper-based system [21]. With decentralization, immutability, reliability, robustness, smart contracts, security, and privacy, blockchain provides suitable and secure technology for storing patients' records.

Drug traceability

Blockchain has crucial applications in pharmacological supply chain management. Blockchain offers trustworthy data on how, when, and where data was created. The decentralization nature of blockchain virtually confirms full transparency in the drug delivery process. When a ledger for medicine is generated, it will be marked as the source. The ledger will maintain the data at every stage, such as where and when it has been handled and by whom, etc. until the consumer finally receives it.

Equipment tracking

The supply chain in the medical domain provides various medical and surgical laboratory equipment for doctors, nurses, and hospitals. In all phases, every transaction is stored and monitored within the blockchain. Dealers supply the raw materials to the industrial unit, companies create the products and send them to data centers, the data center delivers the products to the vendors, and vendors finally deliver the products to the hospital or individual user.

Clinical trial

Blockchain has essential applications in the clinical trials of drugs. Generally, most pharmaceutical companies try to develop 1–2 new drugs yearly, which takes high costs, and sometimes its success rate is nearly zero. Blockchain offers smart contract features that validate transparency and traceability over clinical trials. This distributed ledger could permit different patients to store their medical records by unique methods, making it visible to trial persons, who could then touch the patients if the data meet the requirements for the clinical trial. For example, Chainscript [22] is a proof-of-concept efficient process based on blockchain, which spontaneously verifies consent for updated protocols on a principal document for trial applicants.

Radiology and research

Radiology plays a critical role in the modern medical domain because it generates vital information for diagnosing and treating patients. There are numerous possible applications of blockchain in radiological clinical practice, such as maintaining patient records, image sharing, quantitative image analysis, etc. Once images are entered into the chain, they can't be modified. This could permit healthcare organizations to control access to their images and documents while allowing collaboration and image-sharing between various sectors.

Conclusions

Blockchain technology is experiencing continuous growth. When considering the possibility of its full-scale adoption in the future of the medical system in our country, several potential challenges warrant discussion. Firstly, there is a shortage of individuals in India with blockchain expertise. Secondly, hiring current blockchain experts can be costly. Thirdly, there are concerns about confidentiality, as within a blockchain network, all users have visibility into the entire system. Therefore, ensuring the security of the blocks within the chain is essential for adopting this technology. Fourthly, despite blockchain's inherent security, there have been instances of malicious actors creating unauthorized blocks to disrupt network performance in pursuit of financial gain. To address this, hospitals can implement a private blockchain to scrutinize all users. However, there is currently limited research on the impact of blockchain in the field of radiology. We intend to conduct a comprehensive literature review in the future to explore blockchain-based applications in medical imaging.

Acknowledgement

I want to convey my heartfelt appreciation to my mentor, Dr. Ripon Patgiri, an Assistant Professor at the NIT Silchar's Department of Computer Science and Engineering, for his invaluable guidance and steadfast support throughout the duration of this research project. I would also like to extend my gratitude to my co-author, Mr. Arnab Kundu, who is a Research Scholar (QIP-AICTE) in the Department of ECE at the NIT Silchar, for his valuable assistance and contributions.

Furthermore, I am deeply thankful to the Ghani Khan Choudhury Institute of Engineering & Technology in Malda, India, for providing me with the opportunity to undertake this research at the NIT Silchar in Assam-788010. Finally, I would like to express my profound gratitude to my family, including my daughter and parents, for their unwavering and constant encouragement throughout my research journey.

Appendix

Abbreviation with definition

Abbreviation	Definition
P2P	Peer-to-peer
EMR	Electronic medical record
EHR	Electronic health record
PHR	Personal health record
CD	Compact disc
IoT	Internet of Things

References

1. Heart, T., Ben-Assuli, O., and Shabtai, I. (2017). A review of PHR, EMR and EHR integration: A more personalized healthcare and public health policy. *Health Policy Technol.* 6(1), 20–25. https://doi.org/10.1016/j.hlpt.2016.08.002.

2. Morin, R. L. (2005). Outside images on CD: a management nightmare. *J. Am. Coll. Radiol.*, 2(11), 958. https://doi.org/ 10.1016/j.jacr.2005.08.006.
3. Seh, A.H., Zarour, M., Alenezi, M., Sarkar, A.K., Agrawal, A., Kumar, R. and Ahmad Khan, R., 2020, May. Healthcare data breaches: insights and implications. In Healthcare (Vol. 8, No. 2, p. 133). MDPI. https://doi.org/10.3390/healthcare8020133
4. Nakamoto, S. (2008) . Bitcoin: A peer-to-peer electronic cash system. *Decentral. Busin. Rev.* pp. 1-9
5. Kiania, K., Jameii, S. M., and Rahmani, A. M.. (2023). Blockchain-based privacy and security preserving in electronic health: a systematic review. *Multimed. Tools Appl.*, 1–27. https://doi.org/10.1007/s11042-023-14488-w.
6. Hang, L., Chen, C., Zhang, L., and Yang, J. (2022). Blockchain for applications of clinical trials: Taxonomy, challenges, and future directions. *IET Comm.*, 16(20), 2371–2393. https://doi.org/10.1049/cmu2.12488.
7. Tagliafico, A. S., Campi, C., Bianca, B., Bortolotto, C., Buccicardi, D., Francesca, C., Prost, R., Rengo, M., and Faggioni, L. (2022). Blockchain in radiology research and clinical practice: Current trends and future directions. *La Radiologia Medica*, 127(4), 391–397. https://doi.org/10.1007/s11547-022-01460-1.
8. Chen, Z., Xu, W., Wang, B., and Yu, H. (2021). A blockchain-based preserving and sharing system for medical data privacy. *Fut. Gen. Comp. Sys.*, 124, 338–350. https://doi.org/10.1016/j.future.2021.05.023.
9. Ploder, C., Spiess, T., Bernsteiner, R., Dilger, T., and Weichelt, R. (2021). A risk analysis on blockchain technology usage for electronic health records. *Cloud Comput. Data Sci.*, 20–35. https://doi.org/10.37256/ccds.222021777.
10. De Aguiar, Júlio, E., Faiçal, B. S., Krishnamachari, B., and Jó Ueyama. (2020). Survey of blockchain-based strategies for healthcare. *ACM Comput. Sur. (CSUR)*, 53(2), 1–27. https://doi.org/10.1145/3376915.
11. Zhuang, Y., Sheets, L. R., Chen, Y.-W., Shae, Z.-Y., Tsai, J. J. P., and Shyu, C.-R. (2020). A patient-centric health information exchange framework using blockchain technology. *IEEE J Biomed. Health Inform.*, 24(8), 2169–2176. https://doi.org/10.1109/JBHI.2020.2993072.
12. Jabarulla, M. Y. and Lee, H.-N. (2020). Blockchain-based distributed patient-centric image management system. *Appl. Sci.* 11(1), 196. https://doi.org/10.3390/app11010196.
13. Shen, B., Guo, J., and Yang, Y. (2019). MedChain: Efficient healthcare data sharing via blockchain. *Appl. Sci.*, 9(6), 1207. https://doi.org/10.3390/app9061207.
14. Zhuang, Y., Sheets, L. R., Shae, Z., Chen, Y. W., Tsai, J. J. and Shyu, C. R. (2019). Applying blockchain technology to enhance clinical trial recruitment. *AMIA Ann. Symp. Proc.*, 2019, 1276.
15. Agbo, C. C., Qusay, H. M., and Mikael Eklund, J. (2019). Blockchain technology in healthcare: a systematic review. *Healthcare*, 7(2), 56. https://doi.org/10.3390/healthcare7020056
16. Tasca, P. and Tessone, C. J. (2017). Taxonomy of blockchain technologies. Principles of identification and classification. *arXiv preprint arXiv:1708.04872*. pp. 1-43. https://doi.org/10.48550/arXiv.1708.04872
17. Hofmann, A., Gwinner, F., Windkelmann, A., and Janiesch, C. (2021). Security implications of consortium blockchains: the case of ethereum networks. *J. Intell. Prop. Info. Tech. Elec. Com. L.*, 12, 347. https://doi.org/doi/10.1145/2994581.
18. Haleem, A., Javaid, M., Singh, R. P., Suman, R., and Rab, S. (2021). Blockchain technology applications in healthcare: An overview. *Inter. J. Intell. Netw.*, 2, 130–139. https://doi.org/10.1016/j.ijin.2021.09.005.
19. Han, Y., Zhang, Y., and Vermund, S. H. (2022). Blockchain technology for electronic health records. *Inter. J. Environ. Res. Public Health*, 19(23), 15577. https://doi.org/10.3390/ijerph192315577.
20. Yaga, D., Mell, P., Roby, N., and Scarfone, K. (2019). Blockchain technology overview. *arXiv preprint arXiv:1906.11078.*
21. Liu, P. T. S. (2016). Medical record system using blockchain, big data and tokenization. *Inform. Comm. Sec. 18th Inter. Conf. ICICS 2016 Proc.*, 18, 254–261. https://doi.org/10.1007/978-3-319-50011-9_20.
22. Zhou, L., Wang, L., and Sun, Y. (2018). MIStore: A blockchain-based medical insurance storage system. *J. Med. Sys.*, 42(8), 149. https://doi.org/10.1007/s10916-018-0996-4.

Biographical notes

Debadrita Roy is currently affiliated with the Ghani Khan Choudhury Institute of Engineering & Technology in India, where she holds a faculty position in the Department of Computer Science and Engineering. She also works as a part-time research scholar in the Department of Computer Science and Engineering at NIT Silchar in India. She earned her Bachelor of Technology degree in Information Technology and her Master of Technology degree in Computer Science and Engineering from MAKAUT (formerly known as WBUT) in West Bengal, India. Her primary research interests revolve around blockchain technology and wireless communication. Furthermore, she maintains professional memberships with organizations such as the ACM, IEEE, International Association of Engineers, and ISRS.

Arnab Kundu is presently associated with NIT Silchar in India, where he is a research scholar participating in the AICTE-QIP program within the Department of Electronics and Communication Engineering. He completed his

Bachelor of Technology degree in Electronics and Communication Engineering and his Master of Technology degree in Mobile Communication and Network Technology at MAKAUT (formerly known as WBUT) in West Bengal, India. His primary research emphasis is on wireless communication and cognitive radio technology. Furthermore, Arnab Kundu is a member of various professional organizations, including the Society of EMC Engineers (India), IETE, International Association of Engineers, ISRS, ISTE, and IEEE.

Dr. Ripon Patgiri obtained his Bachelor's Degree from the Institution of Electronics and Telecommunication Engineers in New Delhi in 2009. He completed his M.Tech. degree at the Indian Institute of Technology Guwahati in 2012, and he earned his Doctor of Philosophy from the National Institute of Technology Silchar in 2019. Following the completion of his M.Tech. degree, he assumed the position of Assistant Professor in the Department of Computer Science & Engineering at the National Institute of Technology Silchar in 2013. Dr. Patgiri has authored numerous papers published in well-regarded journals, conferences, and books. His research interests encompass a range of topics, including distributed systems, file systems, Hadoop and MapReduce, big data, bloom filter, storage systems, and data-intensive computing. He holds senior membership status in IEEE and is a member of ACM and EAI. Additionally, he is a lifetime member of ACCS in India and holds an associate membership with IETE.

53 First-principles study of electronic structure of $Fe_{50}Rh_{50}$ alloy

Rakesh Das[1,a] *and S. K. Srivastava*[2]

[1]Department of Physics, Ghani Khan Choudhury Institute of Engineering and Technology, Malda-732141, India

[2]Department of Physics, Indian Institute of Technology Kharagpur, Kharagpur-721302, India

Abstract

FeRh possesses a CsCl crystalline structure and undergoes a phase transition from an antiferromagnetic (AF) to a ferromagnetic (FM) state when heated beyond room temperature. Understanding the electronic structure of FeRh is crucial for unraveling the underlying cause of this magnetostructural phase transition. Here, we present a methodical first-principles study of the electronic properties of the $Fe_{50}Rh_{50}$ system under density functional theory (DFT). Our analysis involved studying the spin-polarized (SP) total density of states (DOS) of this compound as well as the individual total DOS of Fe and Rh. Notably, close to Fermi energy level (E_F), the DOS primarily originates from the spin-up DOS of Fe in FeRh. Conversely, the spin-down DOS of FeRh is predominantly contributed by the total spin-down DOS of Rh. Importantly, both the spin-down and spin-up DOS equally contribute to the formation of the antiferromagnetic ground state in the system.

Keywords: Electronic structure, FeRh, density of states, phase transition

Introduction

Among the diverse class of materials, alloys have garnered significant interest owing to their unique properties derived from the combination of different elements. It is a promising alloy with potential applications in various fields, including magnetic storage devices and catalysis. The $Fe_{50}Rh_{50}$ (FeRh) system is an alloy that has garnered significant interest owing to its captivating electronic and magnetic properties. FeRh formed CsCl structure, is AF at room temperature. It undergoes a transition from AF to FM state by heating above room temperature T ≈ 350 K, observed by Fallot et al., over 8 years ago [1, 2]. Although a para(PM) – ferromagnetic(FM) phase transition occurs at $T_C \approx$ 650 K, the low-temperature transition of the FeRh system holds great promise for a wide range of technological applications, spanning from sensors and magnetic refrigerants to magnetic recording [1, 3–12]. Neutron diffraction measurements revealed that only the iron atom carry magnetic moment ($\mu_{Fe} \approx 3.2\mu_B$) in the AF state whereas both the atoms Fe ($\mu_{Fe} \approx 3.2\mu_B$) and Rh ($\mu_{Rh} \approx 0.9\mu_B$) carry moment in the FM phase [13]. The thermodynamically first-order magnetic phase transition from AF to FM is accompanied by a noteworthy 1% increase in the volume of a unit cell. This increase can be attributed to the remarkable giant volume magnetostrictive effect [14]. Additionally, there is a significant reduction in resistivity [15], leading to a striking room-temperature magneto-resistive effect. Moreover, a substantial change in magnetic entropy results in a remarkable giant magnetocaloric effect. It's crucial to emphasize that this distinctive transition heavily relies on the concentration of its components, and any deviation of more than 5% in stoichiometry can potentially disrupt it [1, 2]. Furthermore, both transition temperatures exhibit high sensitivity to minor changes in concentration. It is found that stain, pressure, chemical substitution, temperature, concentration, magnetic field, etc., are the controlling parameter for this phase transition across a wide temperature spectrum (100 K < T < 600 K) [3, 9, 14, 1–19, 21–32].

The root of the magnetostructural transition is still a topic of debate in both bulk and nano forms of $Fe_{50}Rh_{50}$, despite numerous efforts to understand its underlying mechanism. First-principles densities of states (DOS) calculations were performed for studying the transition but previous calculations were not consistent enough towards this goal. Here, we provide a comprehensive investigation of the electronic structure of $Fe_{50}Rh_{50}$ compounds using first-principles methods. By understanding the electronic properties of this system, we aim to gain insights into its electronic and magnetic behavior, which can further enhance its technological applications. In this approach, we have employed the generalized gradient approximation (GGA) within the framework of DFT. The study focuses on examining the SP total and partial DOS of this compound. The

[a]icon7117@gmail.com

DOI: 10.1201/9781003540199-53

investigation aims to understand the root cause of the phase transition through the band hybridization between Rh and Fe in the system.

Calculation details

In this work, we conducted DFT calculations employing the full-potential linearized augmented plane wave (FLAPW) method [33–35], utilizing the Wien2K code. The FeRh unit cell was constructed in the CsCl structure, adopting the iron lattice constant (2.87 Å) obtained from the experiment. To minimize atomic forces, the atomic coordinates were relaxed to a threshold of less than 1 mRy/au. Consequently, the relaxed lattice constant was determined to be 2.99 Å. We have embraced the GGA to account for electron exchange and correlation effects, following the work of Perdew et al. [36], The energy of separation in this method was taken as -0.6 Ry. The basis set was taken in line with Slater [37] to initiate the FLAPW method. In our computations, the wave function was constructed with a limited basis set, employing a cutoff parameter $K_{max}R^{min}_{MT} = 7.0$ for plane waves. Whereas for the spherical wave functions, maximum multipolarity of $l_{max} = 10$ was used. Additionally, $G_{max} = 12$ was taken to confine the density of charge for the Fourier expansion of the wave function. To sample the Brillouin zone effectively, a k-mesh was constructed with using 110 irreducible k-points.

Results

Figure 53.1 shows the SP total DOS of Fe in $Fe_{50}Rh_{50}$. E_F is taken at 0 eV and DOS is taken in the unit of per eV per atom. Given that calculations were performed at absolute zero temperature; it is anticipated that the compound would have exhibited an anti-ferromagnetic state. If we concentrate on the DOS of Fe, we can imagine that the Fe total DOS is mostly contributed from the DOS of Fe d state. Both spin down and spin up DOSs are plotted along the positive and negative y-axis. In case of Fe total DOS, we can observe that the center of gravity (CG) of the Fe up DOS is lying below E_F, whereas the CG of the DOS of Fe down is lying above the E_F. Consequent upon which, Fe moment is happened to be positive.

Figure 53.2 shows the SP total DOS of Rh in $Fe_{50}Rh_{50}$. Here also, E_F is taken at 0 eV. We can observe in this DOS that unlike the Fe DOS, the CG of down DOS has been lying below the E_F. Whereas, spin up DOS CG is still lying below E_F. But, the contribution from spin down DOS is almost equal. Because of which Rh total moment is expected to be zero in this case and it is in line with the experimental resu

Figure 53.3 shows the SP total DOS of FeRh. In the total density of states of FeRh, we must expect contributions from both Fe and Rh total DOS. At the ground state, FeRh is expected to be antiferromagnetic. Thus, spin up and spin down DOS should

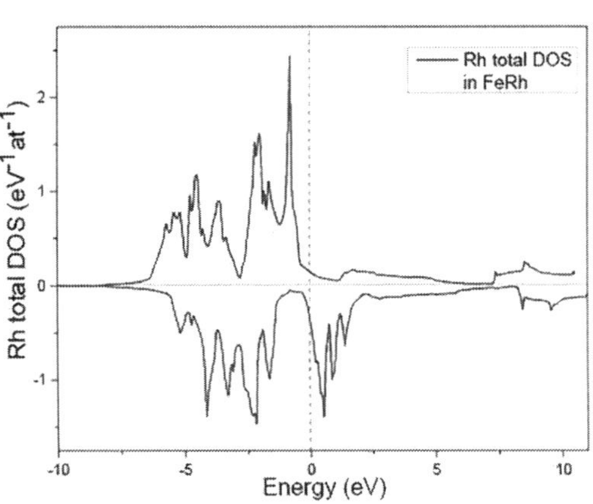

Figure 53.2 Spin-polarized total DOSs of Rh in $Fe_{50}Rh_{50}$. E_F is taken at 0 eV

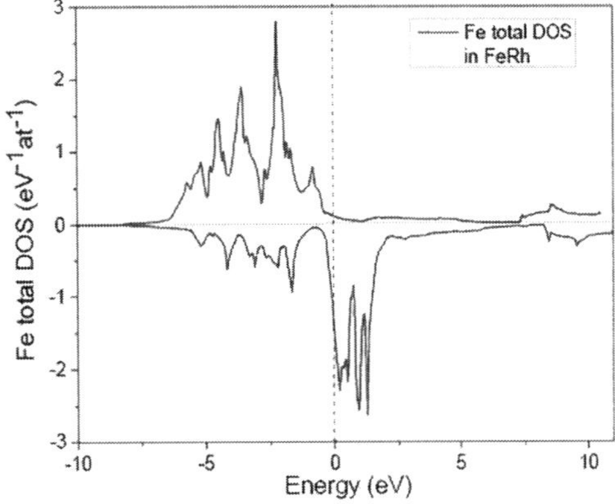

Figure 53.1 Spin-polarized total DOSs of Fe in $Fe_{50}Rh_{50}$. E_F is taken at 0 eV

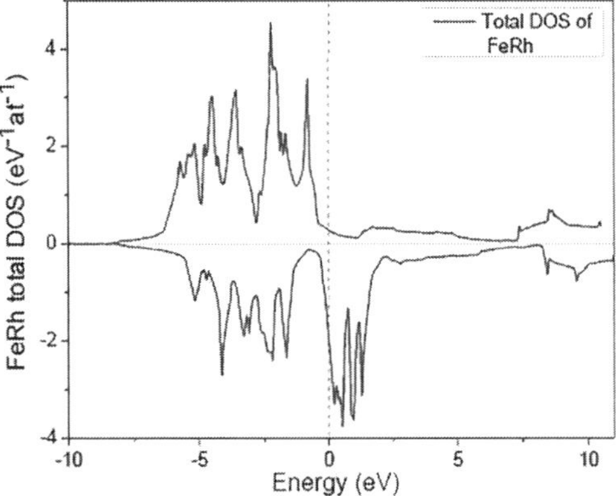

Figure 53.3 Spin-polarized total DOSs of $Fe_{50}Rh_{50}$. E_F is taken at 0 eV

contribute almost equally in the moment formation. If we observe closely the up and down DOS, we can see that spin up DOS is taken a shape similar to the Fe spin up DOS below E_F except in the region close to E_F. In the region close to E_F, the DOS is contributed from Rh spin up DOS. On the other hand, the total spin down DOS of FeRh is contributed mostly from Rh total spin down DOS. Although Fe spin down DOS is heavily contributed to the FeRh spin down DOS in the antibonding state, it will not contribute in the moment formation. In addition to the above, even the height of the DOS for the spin up case is slightly more than spin down DOS, the spin down DOS is widely spread with a higher contribution at the E_F. In this way, spin up and spin down DOS are contributing equally to the system for the formation of anti-ferromagnetic ground state.

Conclusions

In summary, we conducted more advanced and refined calculations to study the ground-state electronic properties of the $Fe_{50}Rh_{50}$ system. The computations were conducted utilizing the FLAPW technique, operating within the framework of DFT, and executed through the Wien2K code. The antiferromagnetic ground state of the system is thoroughly elucidated by analyzing the distribution of up and down density of states (DOS) for FeRh, Fe, and Rh individually. Specifically, we observed that the spin-up DOS in the final system is primarily contributed by Fe spin up DOS, while the spin-down DOS is predominantly influenced by Rh. Additionally, the moment formation is explained based on the DOS distribution. However, to gain a comprehensive understanding of the temperature-dependent magnetic transition in the compound for various technological applications, temperature-dependent DFT calculations are necessary. This work represents a significant step towards achieving that goal.

References

1. Fallot, M. and Hocart, R. (1939). On the appearance of ferromagnetism upon elevation of the temperature of iron and rhodium. *Rev. Sci.*, 77, 498–499.
2. Ibarra, M. R. and Algarabel, P. A. (1994). Giant volume magnetostriction in the FeRh alloy. *Phys. Rev. B*, 50(6), 4196–4199.
3. Vries, M. A. d., Loving, M., Mihai, A. P., Lewis, L. H., Heiman, D., and Marrows, C. H. (2013). Hall-effect characterization of the metamagnetic transition in FeRh. *New J. Phys.*, 15, 013008(1-15).
5. Barua, R., Jimenez-Villacorta, F., Shield, J., Heiman, D., and Lewis, L. (2013). Nanophase stability in a granular FeRh-Cu system. *J. Appl. Phys.*, 113, 17B523(1-3).
6. Barua, R., Jiang, X., Jimenez-Villacorta, F., Shield, J. E., Heiman, D., and Lewis, L.H. (2013). Predicting magnetostructural trends in FeRh-based ternary systems. *J. Appl. Phys.*, 113, 023910(1-6).
7. Loving, M., Vries, M. A. d., Jimenez-Villacorta, F., Graet, C. L., Liu, X., Fan, R., Langridge, S., Heiman, D., Marrows, C. H., and Lewis, L. H. (2012). Tailoring the FeRh magnetostructural response with Au diffusion. *J. Appl. Phys.*, 112, 043512-(1–8).
8. Manekar, M. and Roy, S. B. (2011). Very large refrigerant capacity at room temperature with reproducible magnetocaloric effect in $Fe_{0.975}Ni_{0.025}Rh$. *J. Phys. D: Appl. Phys.*, 44, 242001 (5pp).
9. Suzuki, I., Koike, T., Itoh, M., Taniyama, T., and Sato, T. (2009). Stability of ferromagnetic state of epitaxially grown ordered FeRh thin films. *J. Appl. Phys.*, 105, 07E501 (1–3).
10. Manekar, M. and Roy, S. B. (2008). Reproducible room temperature giant magnetocaloric effect in Fe–Rh. *J. Phys. D: Appl. Phys.*, 41, 192004 (4pp).
11. Jia, Z., Harrell, J. W., and Misra, R. D. K. (2008). Synthesis and magnetic properties of self-assembled FeRh nanoparticles. *Appl. Phys. Lett.*, 93, 022504 (1–3).
12. Maat, S., Thiele, J. U., and Fullerton, E. E. (2005). Temperature and field hysteresis of the antiferromagnetic-to-ferromagnetic phase transition in epitaxial FeRh films. *Phys. Rev. B*, 72, 214432 (1–10).
13. Kittel, C. (1960). Model of exchange-inversion magnetization. *Phys. Rev.*, 120, 335–342.
14. Moruzzi, V. L. and Marcus, P. M. (1992). Antiferromagnetic-ferromagnetic transition in FeRh. *Phys. Rev. B*, 46, 2864–2873.
15. Bergevin, F. d. and Muldawer, L. (1961). Crystallographic study of an iron-rhodium alloy. *C. R. Acad. Sci.*, 252, 1347–1349.
16. Kouvel, J. S. and Hartelius, C. C. (1962). Anomalous magnetic moments and transformations in the ordered alloy FeRh. *J. Appl. Phys.*, 33, 1343–1344.
17. Wayne, R. C. (1968). Pressure dependence of the magnetic transitions in Fe-Rh alloys. *Phys. Rev.*, 170, 523–527.
18. Zakharov, A. I., Kadomtseva, A. M., Levitin, R. Z., and Ponyatovskii, E. G. (1964). Magnetic and magnetoelastic properties of a metamagnetic iron-rhodium alloy. *Sov. Phys. JETP*, 19, 1348–1353.
19. Kamenev, K., Arnold, Z., Kamar′ad, J., and Baranov, N. V. (1997). Pressure induced giant volume magnetostriction in the $(Fe_{1-x}Ni_x)_{49}Rh_{51}$ alloys. *J. Appl. Phys.* 81, 5680–5682.
20. Hume-Rothery, W., Mabbott, G. W., and Evans, K. C. (1934). The freezing points, melting points, and solid solubility limits of the alloys of sliver and copper with the elements of the B sub-groups. *Philos. Trans. R. Soc. London, Ser. A* 233, 1–97.
21. Cullity, B. D. and Graham, C. D. (2011). *Introduction to Magnetic Materials.* John Wiley & Sons: Hoboken, NJ.

22. Kouvel, J. S. (1966). Unusual nature of the abrupt magnetic transition in FeRh and its pseudobinary variants. *J. Appl. Phys.*, 37, 1257–1258.
23. Yuasa, S., Otani, Y., Miyajima, H., and Sakuma, A. (1994). Magnetic properties of bec $FeRh_{1-x}M_x$ systems. *IEEE Trans. J. Magn. Jpn.* 9, 202–209.
24. Walter, P. H. L. (1964). Exchange inversion in ternary modifications of iron rhodium. *J. Appl. Phys.*, 35, 938–939.
25. Tu, P., Heeger, A. J., Kouvel J. S., and Comly, J. B. (1969). Mechanism for the FirstOrder magnetic transition in the FeRh system. *J. Appl. Phys.*, 40, 1368–1369.
26. Fogarassy, B., Kem´eny, T., P´al, L., and T´oth, J. (1972). Electronic specific heat of iron-rhodium and iron-rhodium-iridium alloys. *Phys. Rev. Lett.*, 29(5), 288–291.
27. Schinkel, C. J., Harthog, R., and Hochstenbach, F. H. A. M. J. (1974). On the magnetic and electrical properties of nearly equiatomic ordered FeRh alloys. *Phys. F.*, 4, 1412–1422.
28. Vinokurova, L. I., Vlasov, A. V., and Pardavi-Horvath M. (1976). Pressure effects on magnetic phase transitions in FeRh and FeRhIr alloys. *Phys. Status Solidi B*, 78(1), 353–357.
29. Miyajima, H. and Yuasa, S. (1992). Structural phase transition and magnetic properties of $FeRh_{1-x}Co_x$ alloys. *J. Magn. Magn. Mater.* 104–107, 2025–2026.
30. Makhlouf, A., Nakamura, T., and Shiga, M. (1994). Structure and magnetic properties of $FeAl_{1-x}Rh_x$ alloys. *J. Magn. Magn. Mater.* 135(3), 257–264.
31. Baranov, N. V. and Barabanova, E. A. (1995). Electrical resistivity and magnetic phase transitions in modified FeRh compounds. *J. Alloys Compd.* 219, 139–148.
32. Baranov, N. V., Zemlyanski, S. V., and Kamenev, K. (1998). In itinerant electron magnetism: Fluctuation effects. *Springer Netherlands* 55, 345–351.
33. Kreiner, K., Michor, H., Hilscher, G., Baranov, N. V., and Zemlyanski, S. V. (1998). Evolution of the electronic specific heat and magnetic order in $(Fe_{1-x}Ni_x)$ Rh. *J. Magn. Magn. Mater.*, 177–181, 581–582.
34. Hohenberg, P. and Kohn, W. (1964). Inhomogeneous electron gas. *Phys. Rev.* 136, 864–871.
35. Kohn, W. and Sham, L. (1965). Self-consistent equations including exchange and correlation effects. *J. Phys. Rev. A*, 140, 1133–1138.
36. Blaha, P., Schwarz, K., Madsen, G., Kvasnicka, D., and Luitz, J. (1999). WIEN2k an augmented plane wave + local orbitals programme for calculating crystal properties. *Karlheinz Schwarz, Techn. Universitat Wien, Austria ISBN 3-9501031-1-2.*
37. Perdew, J. P., Burke, K., and Ernzerhof, M. (1996). Generalized gradient approximation made simple. *Phys. Rev. Lett.*, 77, 3865–3868.
38. Eriksson, O., Nordström, L., Brooks, M. S. S., and Johansson, B. (1988). 4*f*-band magnetism in $CeFe_2$. *Phys. Rev. Lett.*, 60, 2523–2526.

54 Modeling and simulation of piezoelectric and pyroelectric hybrid energy harvesting through the bi-metallic beam

Umamaheshwaran T, Reshma Liyakath and Shaikh Faruque Ali[a]

Department of Applied Mechanics and Biomedical Engineering, IIT Madras, Chennai, India

Abstract

Energy harvesting from ambient sources has been in research from last few decades. Attention has moved from vibration-based sources to more perennial sources like thermal and light sources. Waste heat has drawn significant attention to be recovered and utilized as thermal energy harvesting in various applications. The pyroelectric phenomena, which rely on temperature variation with respect to time, can overcome the difficulty of maintaining a spatial temperature differential between hot and cold sides in thermoelectric devices. In addition, as all pyroelectric materials are piezoelectric, it makes a way to hybrid energy harvesting concept. A bi-metallic beam with the pyroelectric patch is modeled and analyzed in COMSOL Multiphysics®. Supply of airflow with an inlet temperature ranging from 293.15 K to 393.15 K is used to change the temperature of the bi-metallic beam. The thermal harvester is modeled as an Euler Bernoulli beam with the thin film pyroelectric patch covered with electrodes which are exposed to a laminar airflow. The piezoelectric and pyroelectric effects of this thermal harvester have been studied and compared for their power output.

Keywords: Energy harvesting, hybrid harvester, piezoelectricity, pyroelectricity

Introduction

Present day utilization of energy has risen exponentially without any increase in the availability of ambient sources. Search for novel alternate sources of energy and medium of transduction have gained a lot of attention in current scientific research. On the other hand, development in electronics has led to design of miniaturized sensors with reduction in their power requirement. As the power requirement for sensors reduced, innovations in energy generation (from ambient sources) have increased over two decades. Devices are designed to generate energy from vibrations that can power small sensors [1, 2].

Scavenging energy from the surrounding environment and converting it into usable electrical energy is known as energy harvesting. The target is low powered sensors. Energy harvesting systems offer autonomous and sustainable power generation potential by utilizing light, heat, vibration, motion, and radio frequency transmissions [2]. One such abundant source of energy is heat or thermal. Heat released is of various ranges in perspective of energy harvesting, for example, energy released from boilers, chimneys, engines are of very high intensity whereas heat energy released from human body is perennial but of relatively low intensity. Furthermore, in applications like computers/laptops/home appliances, where heat released is not desirable, heat/thermal energy scavenging can serve dual purpose of energy generation using thermoelectric and pyroelectric devices [3], which can also keep the host system relatively cooler. Pyroelectric devices use a temporal temperature gradient, whereas thermoelectric devices operate on the Seebeck effect, which uses a spatial temperature difference to generate energy. The difficulty of sustaining a spatial temperature gradient leads researcher to concentrate on pyroelectricity [14]. Materials that exhibit polarization in the absence of an electric field, called spontaneous polarization, produce the pyroelectric effect. Ferroelectric materials can reverse spontaneous polarization by applying an external electric field, whereas pyroelectric materials cannot be reversed. As both piezoelectric and pyroelectric materials come under the classification of ferroelectrics, all pyroelectrics are piezoelectrics [5]. So, pyroelectric materials can be used for hybrid energy harvesting. Cottrill et al., designed a thermal resonator that converts the spatial temperature difference into electrical power using highly porous and thermally conductive phase change composites [6]. Ravindran et al. presented a micro thermo-mechanic pyroelectric device using a bistable membrane reciprocating between hot and cold sides [7]. Hunter et al., developed a MEMS-based microstructure with a bimorph cantilever beam that operates by providing the heat source at the tip [11]. Lee et al. fabricated a thin film hybrid

[a]am21s023@smail.iitm.ac.in

DOI: 10.1201/9781003540199-54

stretchable nanogenerator by coupling the piezoelectric and pyroelectric effect which has applications in wireless sensors and wearable devices. These are just a few references that have shown the promise of pyroelectric energy generation [9].

This work deals with developing and analyzing a harvesting system that can harvest power from the waste heat produced by machinery, engines, and other related sources. A finite element model is built in COMSOL Multiphysics®. The model is used to explore the temperature variation rate in the pyroelectric patch attached to the bi-metallic beam. Initially the beam is base excited, and the maximum power output is observed with a resistance sweep. Then the beam is exposed to temperature variation for the analysis of pyroelectric effect for the attained optimal load resistance. Since the bi-metallic beam experiences strain the pyroelectric effect is compared with overall combined piezo and pyroelectric effect.

The basic principle of piezoelectric and pyroelectric effect, modeling and working mechanism of the thermal harvester were discussed in the following sections.

Pyroelectric and piezoelectric effect

Origin of pyroelectric effect can be traced back to ancient times by Greek philosopher Theophrastus described a stone called lyngourion that had the property of attracting straws and bits of wood. The fundamental scientific principles of particular material's unique behavior were not fully understood at that time [10]. Modern understanding of the pyroelectric effect is based on the concept of crystal symmetry and polarization. The structure of a material can be classified into 14 bravais lattices based on the geometric arrangements of atoms in seven possible crystal systems. The symmetry operation in these bravais lattices makes a combination of 32-point groups. Out of which, there are 21 non-centrosymmetric crystal structures. In these crystal structures, the center of gravity of positive and negative charges does not coincide, which gives rise to a dipole moment. Polarization is net dipole moment per unit volume [11]. The crystal lattice is deformed or displaced when an external mechanical stress or force is applied to a piezoelectric material. This changes the total dipole moment by disturbing the equilibrium of positive and negative charges within the material. As a result, the initially neutral material becomes polarized. An electric field develops inside the material due to this polarization which produces potential difference or voltage. This phenomenon, where mechanical energy is directly converted into electrical energy, is called the piezoelectric effect. The coupled constitutive equations for the stress-charge form are given as

$$\sigma = cES - e^{\sigma}E \tag{1}$$

$$D = eS + \varepsilon_s E \tag{2}$$

where σ is stress, S is strain, E is electric field, D is electric displacement, c_E is elasticity matrix at a constant electric field E, e is coupling matrix (at a constant σ), and ε_s is permittivity matrix (at a constant strain, S).

Figure 54.1 explains in brief the working principal of the pyroelectric effect. This is also explained in the following steps.

a) Pyroelectricity is the result of spontaneous polarization, which makes the free charges build up on the surface of materials.
b) At steady state temperature, spontaneous polarization remains constant, and there is no current flow.
c) As an increase in temperature, decreases spontaneous polarization, the surface charges redistribute themselves to cause a current flow.
d) During cooling phase, spontaneous polarization increases, causing current flow in the reverse direction [12].

The pyroelectric current produced during this phenomenon is given as

$$i_p = p\,A\,\frac{dT}{dt} \tag{3}$$

where i_p is pyroelectric current, p is pyroelectric coefficient, A is the cross-sectional area, T is temperature, and t is time [13].

Device modeling and operation

Materials and geometry

For the simulation study presented in this work a bi-metallic cantilever beam with PVDF pyroelectric

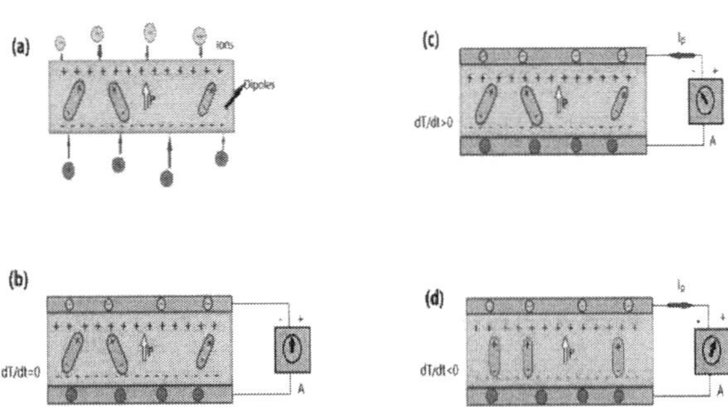

Figure 54.1 Mechanism of pyroelectric effect

material as the top layer is considered. 2D model of the beam is shown in Figure 54.2, and its dimensions and properties are listed in Table 54.1.

The pyroelectric material is chosen as polyvinylidene difluoride (PVDF) for its high level of biocompatibility and mechanical flexibility. PVDF is very well suited for simultaneous piezoelectric and pyroelectric energy harvesting. The pyroelectric material is covered with two electrodes to accumulate the charges generated during temperature variation. The silver electrodes were used because they are chemically stable, especially when air exposure. The bending phenomenon by a beam can be done using two different materials with a significant difference in coefficient of thermal expansion (CTE). Aluminum is taken in the upper part of the beam where CTE is higher than the lower part of the beam made up of steel AISI 4340. The pyroelectric patch is placed on the beam with a fixed left end. When a beam gets heated, the aluminum expands more than the steel, which makes the beam bend downwards. The beam setup is surrounded by the air channel of dimension 0.5 m × 0.1 m for temperature variation [8].

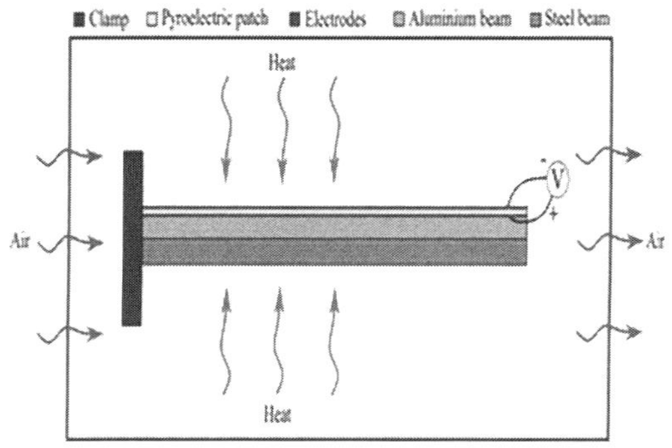

Figure 54.2 A bi-metallic beam with pyroelectric patch

Meshing of domains

The cantilever beam and the pyroelectric patch have been meshed using 2500 quadrilateral elements and an element ratio of 1 with a linear growth rate and symmetrical distribution. The air domain is meshed with quadrilateral and triangular mesh of elements sizes 580 and 3742, respectively.

Piezo-pyro analysis of PVDF patch

The piezoelectricity physics is chosen in COMSOL for the structure eigenfrequency and frequency domain study. The beam is base excited in y direction with an acceleration of 1 g. The bottom layer of the electrode surface attached to the beam is taken as the terminal, and the top electrode surface as ground. The circuit is connected with 1 kΩ of resistance, and the maximum electric power output and maximum load resistance for its natural frequency is evaluated.

Next, the beam is placed in an air environment with the fluid-solid interaction physics chosen for time-dependent study. No heat input is provided for a minute. The heat source of 80 W is applied after a minute on the beam with a uniform distribution. The heat is turned off after five minutes, allowing the beam to cool naturally without airflow. Following that, the procedure is repeated by cooling with an inlet airflow of 0.5 m/s at room temperature (293.15 K). The temperature of the air entering the channel rises to 393.15 K after 25 minutes. The pyroelectric patch captures the energy as there is temperature fluctuation in the beam. The deflection of the bi-metallic beam produces strain that causes the piezoelectric effect.

54. Results and discussion

Eigenfrequency and frequency domain study

The eigenfrequency of the bi-metallic cantilever beam with the pyroelectric patch is 157.91

Table 54.1 Specifications and properties of the device

Description	Aluminum	Steel	Silver	PVDF
Length × Height (mm × mm)	100 × 1	100 × 1	100 × 0.01	100 × 0.08
Density (Kg/m3)	2700	7850	10500	1780
Young's modulus (GPa)	70	205	83	2.17
Poisson's ratio	0.33	0.28	0.37	0.33
Thermal conductivity (W/mK)	283	44.5	429	0.13
Coeff. thermal expansion (/K)	23×10^{-6}	12.3×10^{-6}	18.9×10^{-6}	13×10^{-5}
Specific heat capacity (J/kgK)	900	475	235	1400
Pyroelectric coefficient (μC/m2K)	-	-	-	{0,0, -27}

Hz. Figure 54.3 shows that the power generated nearer to the beam natural frequency and its value increases with the rise in load resistance gets saturated at the optimal value and starts decreasing as shown in the Figure 54.4. The electric power of 1.42 mW is harvested at 159 Hz, which is equivalent to its natural frequency, and the maximum power of 24.38 mW is harvested at the load resistance of 33,884 Ω.

Time-dependent study

Initially, the temperature of the entire system is maintained at 293.15 K. The heat input is applied after 1 minute, which makes the device reach the temperature of 326.39 K in the next 5 minutes. Once the heat is cut-off for 5 minutes, the temperature comes down to 325.9 K due to self-cooling. Again, the heat is applied so that the devices reach 338.25 K. Now, the air at room temperature with an inlet velocity of 0.5 m/s flows into the channel, which cools down the device to 332.52 K because of forced convection. When the temperature of airflow increases, it again gets heated up to 427.26 K. Figures 54.5 and 54.6 show the temperature and velocity distribution of the beam during airflow

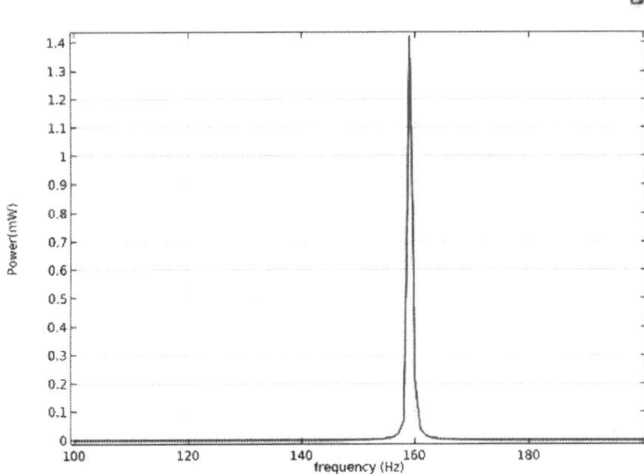

Figure 54.3 Electric power vs, frequency

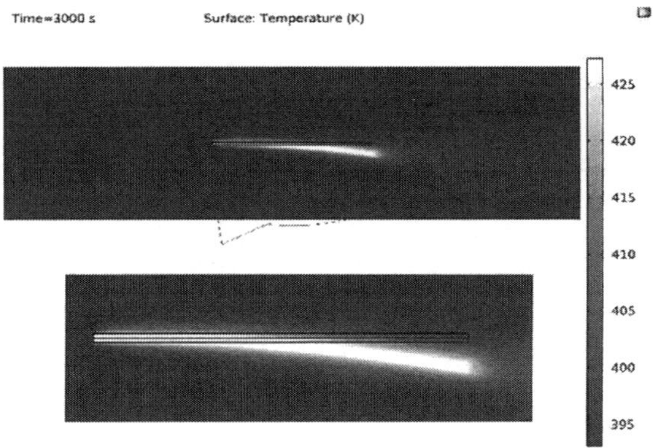

Figure 54.5 Temperature distribution at 3000s

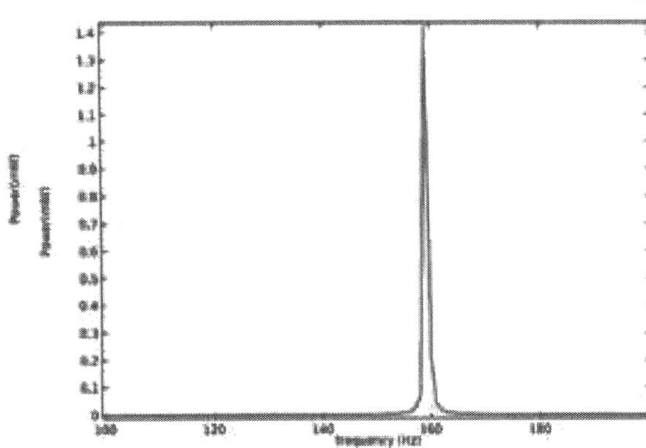

Figure 54.4 Electric power vs. resistance

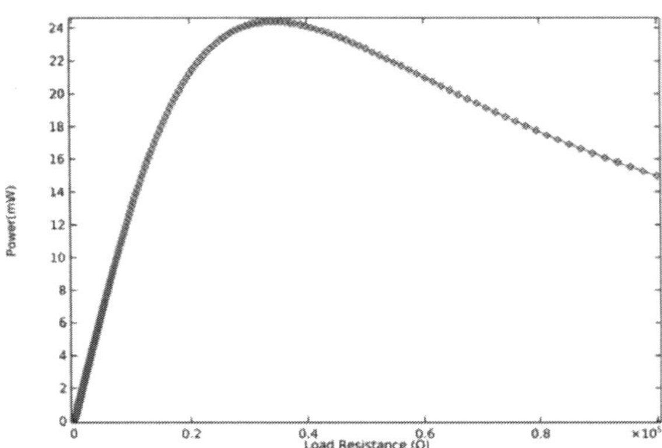

Figure 54.6 Velocity distribution at 3000s

Table 54.2 Electrical output of pyroelectric patch at heating and cooling

Process		Heating of beam	Cooling by air	Cooling by air
Voltage (mV)	Pyro	3.88	-1.49	10.6
	Piezo+Pyro	4.86	-1.84	13.23
Current (μA)	Pyro	0.11	-0.04	0.31
	Piezo+Pyro	0.14	-0.05	0.39
Power (nW)	Pyro	0.45	0.07	3.32
	Piezo+Pyro	0.69	0.10	5.17

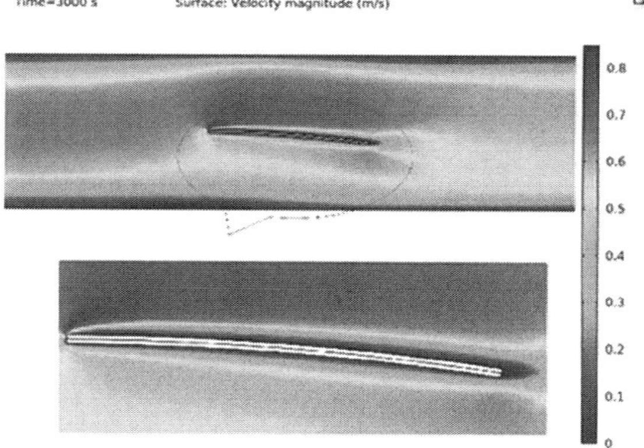

Figure 54.7 Temperature vs. time

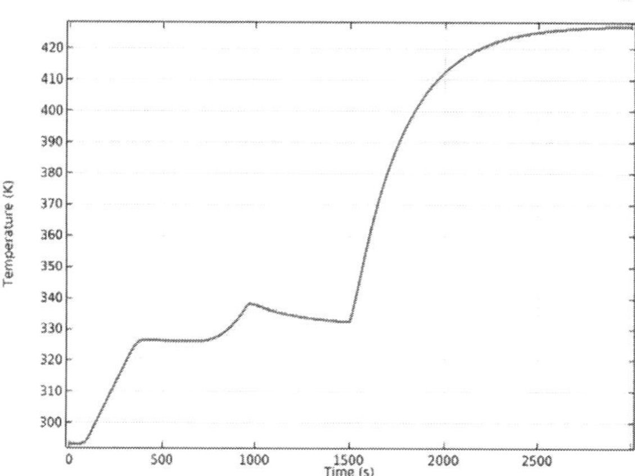

Figure 54.8 Voltage vs Time

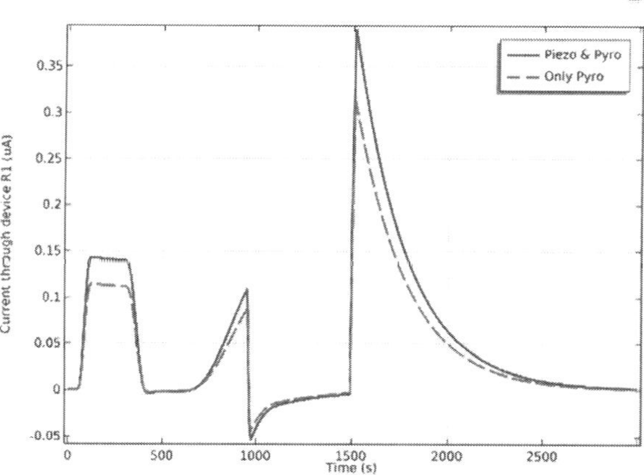

Figure 54.9 Current vs Time

where the bending phenomenon of the beam can be observed.

The pyroelectric patch is connected with a resistance of 33,884 Ω shows the electric output of both pyroelectric and piezoelectric effect during different processes is listed in the Table 54.2. The

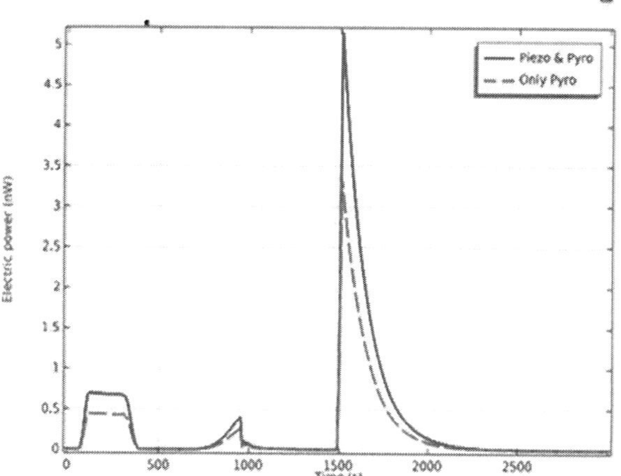

Figure 54.10 Power vs. time

temperature, voltage, current, power varies with respect to time is plotted in the Figures 54.7–54.10. The maximum electrical output is attained during the passage of varying temperature inlet air, and its output voltage, current, power value obtained as 13.23 mV, 0.39 μA and 5.17 nW, respectively.

Conclusions

The pyroelectric patch produces piezoelectric effect in addition to pyroelectric effect because of the usage of bi-metallic beam where strain generates during the deflection of the beam caused by temperature fluctuations. This kind of hybrid energy harvesting can be used in applications like inlet pipe of heat engines where the exhaust gas of the engines can be recirculated for heat exchange in inlet pipes to preheat the air, which reduces carbon emissions by complete combustion of fuels. Further work needs to be done for the suitable design of a bi-metallic beam that can be placed in inlet pipes of heat engines and different functions of inlet temperature.

References

1. Williams, C. B. and Yates, R. B. (1996). Analysis of a micro-electric generator for microsystems. *Sens. Actuat. A Phy.*, 52, 1–3, 8–11. https://doi.org/10.1016/0924-4247(96)80118-X.
2. Anton, S. R. and Sodano, H. A. (2007). A review of power harvesting using piezoelectric materials (2003–2006). *Smart Mater. Struct.*, 16, 3. DOI: 10.1088/0964-1726/16/3/R01.
3. Sultana, M. M., Alam, T. R., Middya, D. M. (2018). A pyroelectric generator as a self- powered temperature sensor for sustainable thermal energy harvesting from waste heat and human body heat. *Appl. Energy*, 221, 299–307. https://doi.org/10.1016/j.apenergy.2018.04.003.

4. Bowen, C. R., Taylor, J., LeBoulbar, E., Zabek, D., Chauhan, A., and Vaish, R. (2014). Pyroelectric material and devices for energy harvesting. *Energy Environ. Sci.*, 7, 3836–3856. https://doi.org/10.1039/C4EE01759E.
5. Liang, L., Kang, X., Sang, Y., and Liu, H. (2016). One-dimensional ferroelectric nanostructures: Synthesis, properties, and applications. *Adv. Sci.*, 3(7), 1500358. https://doi.org/10.1002/advs.201500358.
6. Cottrill, A. L., Zhang, G., Liu, A. T., Bakytbekov, A., Silmore, K. S., Koman, V. B., Shamim, A., and Strano, M. S. (2019). Persistent energy harvesting in the harsh desert environment using a thermal resonance device: Design, testing, and analysis. *Appl. Energy*, 235, 1514–1523. https://doi.org/10.1016/j.apenergy.2018.11.045.
7. Ravindran, S., Huesgen, T., Kroener, M., and Woias, P. (2011). A self-sustaining micro thermomechanic-pyroelectric generator. *Appl. Phy. Lett.*, 99, 104102. https://doi.org/10.1063/1.3633350.
8. Hunter, S. R., Lavrik, N. V., Bannuru, T., Mostafa, S., Rajic, S., and Datskos, P. G. (2011). Development of MEMS based pyroelectric thermal energy harvesters Development of mems based pyroelectric thermal energy harvesters. *Energy Harves. Storage Mater. Dev. Appl. II*, 8035, 199–210. https://doi.org/10.1117/12.882125.
9. Lee, J. H., Lee, K. Y., Gupta, M. K., Kim, T. Y., Lee, D. Y., Oh, J., Ryu, C., Yoo, W. J., Kang, C. Y., Yoon, S. J., Yoo, J. B., Kim, S. W. (2014). Highly stretchable piezoelectric-pyroelectric hybrid nanogenerator. *Adv. Mater.*, 26(5), 765–769. https://doi.org/10.1002/adma.201303570.
10. Lang, S. B. (2005). Pyroelectricity: from ancient curiosity to modern imaging. *Phy. Today*, 58(8), 31–36. https://doi.org/10.1063/1.2062916.
11. McKinstry, S. T. (2008). Crystal chemistry of piezoelectric materials. *Piezoelec. Acous. Mater. Trans. Appl.*, 39–56. Springer US. https://doi.org/10.1007/978-0-387-76540-2_3.
12. Kishore, R. A. and Priya, S. (2018). A review on low-grade thermal energy harvesting: Materials, methods and devices. *Materials*, 11(8), 1433. https://doi.org/10.3390/ma11081433.
13. Whatmore, R. W. (1986). Pyroelectric devices and materials. *Reports Prog. Phy.*, 49, 1335. https://doi.org/10.1088/0034-4885/49/12/002.

55 The concept behind auction-based game theoretic approach for target channel selection in the cognitive radio network

Arnab Kundu[1,a], *Debadrita Roy*[2,b] *and Wasim Arif*[1,c]

[1]Department of Electronics & Communication Engineering, National Institute of Technology Silchar, Cachar, Silchar, Assam-788010, India

[2]Department of Computer Science & Engineering, National Institute of Technology Silchar, Cachar, Silchar, Assam-788010, India

Abstract

The game theory becomes smart to investigate the performances of wireless communication systems by the capability to model autonomous decision-makers and their actions potentially affecting the actions of other decision-makers. Wireless systems have moved toward the booming sector by assuming game theoretic approaches to wireless networks for controlling power, the adaptation of rate, and accessing the channel. It applies dynamic game theory-like means to organize cognitive radio networks (CRN), wireless local area networks, and long term evolution systems toward 5G (fifth generation) onwards. The collaboration between intelligent and rational decision-makers has been described in game theory (GT), and it is one kind of mathematical approach that explains the phenomenon of conflict. The cognitive user (CU) or secondary user (SU) belongs to that cooperative communication and may improve its activity in a licensed spectrum by dynamically choosing a vacant spectrum. Therefore, it is the technology to deal with the scarcity problem in the radio spectrum. In our study, we propose and discuss the consequences and determinations of GT and its applications in CRNs. The GT approach allows us to assume that each player knows other players' payoffs. However, it is only sometimes valid for all types of games. Auction-based games, where the other player's payoffs have yet to be discovered. So that there will be some uncertainty regarding other players' payoff, such a game model comes under the Bayesian game model. According to current research findings, cognitive radio and cooperative communication are two vibrant network technologies. Therefore, incorporating these two latest technologies is a novel solution to the spectrum scarcity problem.

Keywords: Auction theory, channel payoff, cognitive radio network, cooperative channel selection, game theory, handoff

Introduction

A clearly defined constraint is the radio spectrum [1], and the allocation of spectrum for wireless applications remains fixed. Yet, recent studies have revealed that a significant portion of the assigned spectrum is being utilized haphazardly, resulting in the improper use of frequency resources [2]. The cognitive radio initially interacts with the adjacent radio conditions and senses the spectrum holes to acclimate their operational constraints optimally. So that unlicensed users can get access to the blank part of the spectrum without interrupting the ongoing communication of the primary user (PU), hence causing dynamic spectrum access possible [3]. The CR networks' basic idea is to accommodate SU to evacuate the channel once the PU is identified. Therefore, pausing the ongoing communication and migrating to a new target channel for restarting is known as spectrum handoff [4], affecting longer switching delays and total service time (TST) for SU.

On the other hand, cooperative communication is a nascent technology with significant potential to enhance data transfer rates between two wireless devices, all without the need for infrastructure assistance [5]. Naturally, integrating these two hot technologies is expected to solve the spectrum scarcity and capacity limitation problems. In this study, we mainly focus on the procedure where a CR can find a suitable license band to transmit data so that its end-to-end throughput may increase and end-to-end delay may reduce.

In applied mathematics, the game theory (GT) is a branch that furnishes models and mechanisms for investigating circumstances where considerable rational agents may relate with each other to acquire their objectives. Again, the GT has been frequently used for distributed resource management algorithms in communications engineering. It may examine system operations for decentralized and self-organizing networks. It also simplifies the behavior of players/competitors in a game.

[a]a.kunduwb@gmail.com, [b]debadrita@gkciet.ac.in, [c]arif@ece.nits.ac.in;

DOI: 10.1201/9781003540199-55

Therefore, we made the more realistic assumption of incorporation of GT in target channel selection. The auction-based GT approach is one of the branches if they need help to achieve an expected quality of service (QoS) on their own [6]. Auctions provide an incentive. And to share their possessions centered on financial encouragement among more robust users. Therefore, the auction-theoretic approach vides a subset of the game theory proposed by auction-theoretic algorithms [7]. The application of game theory approaches in CRN and 5G-6G (sixth generation) is essential. It is also necessary to define the impact of the same on several wireless communication technologies [8].

The key contributions of this work are accentuated as: next section portrays the basic components of game theory following auction as a non-cooperative game theoretic approach. We made some assumptions for our study in section tiled assumptions. Section system model and analysis, presents the overview of the system model and analysis including proposed algorithm. The performance evaluation has been performed in the next section following simulation results, and in last section conclusion has been drawn.

Basic components of game theory

A game comprises competitive engagements among multiple participants, striving to optimize adherence to rules or protocols. In wireless communication, competitiveness emerges from individuals, users, or devices vying for resources like bandwidth, spectrum, or radio channels. Thus, the game's objective revolves around maximizing efficient bandwidth usage while minimizing interference through adherence to regulations governing registered devices [9]. A game includes a set of players/competitors consisting of a set of strategies. It also includes a set of correlated utility functions. n no. of SUs belong to a CRN is given by a 3-tuple $G = <N, S, U>$ to form a game. For a specific game G, the $N = \{n_1, n_2, n_3, ..., n_n\}$ is a finite set of secondary or cognitive users. $S = \{S_1, S_2, S_3, ..., S_n\}$, is the strategy of handoff of the cognitive user i can be selected and depicted by S_i $(i = 1, 2, ..., n)$. The corresponding payoff function of CU is $U = \{u_1, u_2, u_3, ..., u_n\}$, where CU i is defined by u_i $(i = 1, 2, ..., n)$, and u_i is the value of utility for individual CU received at the action end.

A player's strategy defines a comprehensive plan of action in all probable circumstances in the game. The players act selfishly to maximize their outcomes according to their priorities. A strategy is selected to frame the payoff functions and enable the node i, a strategy s_i is selected, representing the best response. The other $(n - 1)$ CUs will choose the other strategies. If the node i chooses a strategy s_i then all the other CUs in the game will determine the specific strategy s_{-i}. The strategy profile of a game is defined by $s = \{s_i, s_{-i}\}$. Sometimes it is called a strategy combination. Different selections of strategies of different combinations produce a unlike strategy profile like $s = \{s_1, s_2, s_3, ..., s_n\}$, which desires to be placed for NE. Then a game's steady state condition is described. It involves two or more players, and the other players' equilibrium strategies are considered. It is also known to each player. But there are no players who have gain anything by changing only their strategy unilaterally [10].

The player consists of $N = \{n_1, n_2, n_3, ..., n_n\}$ belongs to Bayesian games;

$$(x_i) = (BR)_i(x_{-i}) \tag{1}$$

where $(BR)i$ in Equation 1 represents best response of player i and (x_{-i}) is the chosen strategy profile. The NE is the intersection of best responses. It is a self-enforcing agreement that does not need any other external agency to enforce. Therefore, for N number of players' game i.e., $x^*_1, x^*_2, x^*_3, ..., x^*_n$. The Nash equilibrium for every player i is

$$U_i(x_i^*, x_{-i}^*) \geq U_i(x_i, x_{-i}^*) \tag{2}$$

The Equation 2 implies that for player i, the action x^*_i yields a higher payoff than any alternative action x_i when considering the actions of all other players. Here x^*_i represents the optimal response to the actions of all other players x^*_{-i} and this condition applies to every player and every possible action i.

Auction a non-cooperative game theory

It outlines the approach for placing bids to acquire the item and optimize earnings, the concept of strategic interaction. Auctions can be modeled as games of incomplete information [11, 12]. A game γ consists of certain parameters like a) a set of players $= N$, b) for each player i, $i \in N$, and a nonempty auction set is λ_i and c) payoff function δ. Therefore, the auction for N set of players in a strategic form, is defined by Equation 3

$$\gamma = [N, \lambda_i, \delta_i]_{i \in N} \tag{3}$$

Bids are submitted secretly in a sealed-bid auction. It is only known to the seller. The awards go to the bidder who contains the highest proposal and pays his proposal in a first-price sealed-bid auction, whereas those who spend the second-highest amount among all bids placed may get a second-price auction award. In both formats, losers (bidders) don't have to pay anything. In this study, we accept only the standard auction mechanism belonging to the auction-theoretic approach in the subsequent manners: (i) the highest bidder consistently gets the award of the sold object(s), (ii) consider all bidders are unknown entities, and (iii) the highest bidder gets the object by random allotment, and ties are cracked. In the CRN model, a group of users with weak capacity always tries to pay uneven shares of "money/proposal amount" for higher grades of QoS. At the same time, a group of potential supporters is always ready to receive financial dividends for temporarily allocating their resources and helping such users as predicted [13].

Assumptions

When the primary user (PU) reverts to its designated channel, the secondary user (SU) is required to vacate that channel and shift its ongoing transmission to an alternative one. The SU should detect that the new channel is unoccupied either before or after the transition. However, difficulties arise when the desired channel is not accessible to the SU. In this scenario, the SU must wait in a queue along with other SUs, as they all intend to use the same channel for resuming their ongoing transmissions. Consequently, various hypotheses have been proposed to address this predicament.

A single potential PU (seller) exists per group; If the target channel is accessible, the SU requests a transmission link like a single-hop connection.

If all competitive waiting SUs start their transmission simultaneously, then all of them may interfere with each other because the distance between them is small.

Interrupted and primary users need to be made aware of the exact number of neighboring SUs and their channel conditions. For simplicity, we have considered a game that is played between two SUs but may extend up to *N* no. of users.

System model and analysis

Since three special queues for each wireless channel are: (i) High priority queue (HPQ) for PUs with priority high, (ii) Low priority queue (LPQ) dedicated to the newly arrived SUs, and (iii) Interrupted user queue (IUQ), which is another queue for interrupted users. The new appearance of CUs has lower priority than the IUs. Nonetheless, PUs is given precedence over IUs in terms of minimizing handoff delay and reducing TST. CUs anticipate the sequence of target channels by analyzing the traffic intensity of PUs within the network. As $\rho_{PU} = \lambda_{PU}/m_{PU}$ = PUs' channel utilization factor. A channel payoff vector $V_i = V_1, V_2, \ldots, V_c$ for each secondary user *i*. Once a SU is interrupted, the value of payoff for the available channels is being calculated based on the following Equation 4.

$$V_p(t) = \beta P_i^p(t) + \frac{(1-\beta)}{W_2^p(t)}\beta \quad \in (0,1); \quad p \in (1,\ldots,N) \quad (4)$$

Consequently, a channel is chosen as the target channel for a handoff when its associated V_p value surpasses other channels, with β as a coefficient to adapt to different channel conditions. Thus, the selection process for the target channel is determined by identifying the highest payoff value, as described in Equation 5 [14].

$$P_{channel} = \frac{V_{channel}}{\sum_{i=1}^{C} V_i} \quad (5)$$

where, $P_{channel}$ is channel selection probability and $V_{channel}$ represents a specific channel payoff value for among C no. of channels.

Figure 55.1 shows the auction-based communication control structure. During the contention process where the bidding starts, the potential PU is being contacted by *N* no. of SUs of a group, and each of them is seeking the benefits of switching to a new channel. Then PU, as per its potentiality, computes the number of SU it can provide assistance concurrently. If (a) the potential PU receives cooperation requests ≥2 and (b) the QoS of its own is already satisfied during the same, then the auction contention occurs. The contention procedure during T_{ds} initiates in the sensing phase, followed by the bidding and transmission phase during T_{db} and T_{dt}, respectively. It is considering T_{dt} = data delivery time or service time. Once the bidding stage during T_{db} is over, the cooperative transmissions start. Otherwise, PU is unable to process any cooperative requests received during T_{dt}, assuming $T_{ds} \approx T_{db} < T_{dt}$. Single-stage auction-based CU selection needs to define the bidder's private information, which is defined by random variable S_i, representing the bargaining object's value to user *i*. Where the variable S_i is independently and identically distributed over ($-\infty$ to $+\infty$).

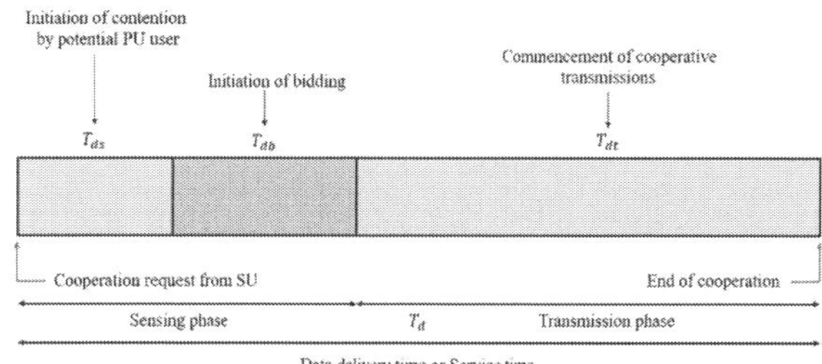

Figure 55.1 Single stage auction-based communication control structure

Again consider $S\alpha$ is the number of identified bidders during the auction, specifically in T_{db}, and $S\alpha < N$. The total number of SUs is always more significant than the number of identified bidders. It affects and improves the arrival in channel λ_i, cs during the cooperation request; otherwise, it moves into another channel. Moreover, a SU proceeds for bidding if it consists of positive private values x_i. The arrival of a new SU during the period T_{db} will not vary due to the lack of information on α, which is required to place a bid. Therefore, based on the number of cooperation requests which is received by potential PU, the bidder i evaluates the number of competitors, which is to be $S\alpha_i$. However, the exact number of bidders is uncertain, which is contestant uncertainty to the potential PU, nothing more than that.

Proposed algorithm

The algorithm for auction theoretic approach has been proposed in the following Table 55.1.

Performance evaluation

Simulation parameters

The simulation of the suggested algorithm is defined in section named simulation results. It is performed in a MATLAB environment. Due to the varying number of SUs, the channel will be utilized a number of times. Therefore, channel utilization is measured for different numbers of SUs, including PU. More accurately, we have considered four schemes for 5, 10, 15, and 20 PUs. We also assumed 10,000 seconds as the simulation time for each individual scheme. Due to the continuous variation of PUs in the licensed channel, SU may get the opportunity to access the same. However, the number of iterations refers to the algorithm running so that the best-licensed channel gets allocated to the SUs.

Simulation results

Figure 55.2 shows the result of an alteration of utilization due to varying no. of SUs. Randomly selected channels get allocated to the SUs. An effective increase of 28.7% has been observed in the utilization with SU against allocated channel utilization without SU [15]. Whereas considering the proposed algorithm, an improvement of 11.09% can be observed in utilization after auction-based communication control, and overall it can be measured as 39.79%. However, interference may reduce among channels in low-dense network structures. It may also increase due to a high-dense network, and overall system utilization may decrease.

An alteration in utilization for different iterations (3 PUs and 2 SUs) has been shown in Figure 55.3. As the network becomes dense, the inclusion of interference among channels may increase. The overall increase in utilization is 35.6%. However, a 6.9% increase in utilization has been observed by the involvement of SUs after considering the auction-based communication control structure as per

Table 55.1 Algorithm for auction

Algorithm: Auction()

Step 1: Define an array of available channels *AvailableChannel*[]
Step 2: Define an array for storing the interference value of each channel *ChannelInterference*[]
Step 3: Define the number of secondary users *SU*
Step 4: Define an array for storing the utility of every available channel *ChannelUtility*[]
Step 5: **if** (*SU*==(*N* > =2)) **then** [where, N is defined as total number of users]
print "Allocate the channel for bidding"
else
print "Allocation failed and exit"
Step 6: $T_{ds} \approx T_{db} < T_{dt}$ [where, *T* is defined as total amount of time, T_{ds} = initiation of the contention process, T_{db}= initiation of bidding, T_{dt} = initiation of transmission]
Step 7: Identified bidders during $T_{db} = S\alpha$ & $S\alpha <= N$
Step 8: **while** (End of *AvailableChannel*[]) **do**
Step 9: **if** (*ChannelInterference*[]==0)
print "Allocate the channel for bidding"
Step 10: **else**
ChannelUtility[]=*ChannelInterference*[] * *SU*[]
Step 11: **end if**
Step 12: **end while**
Step 13: Sort the *ChannelUtility*[]
Step 14: Next channel will be allocated as per increasing order of channel utility
Step 15: **end**

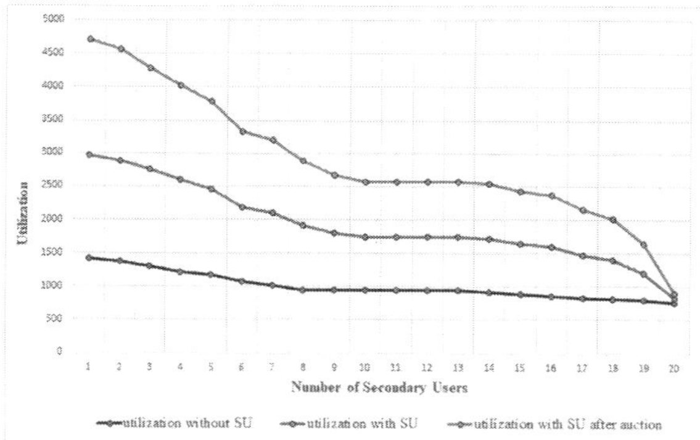

Figure 55.2 Alteration of utilization due to varying no. of SUs

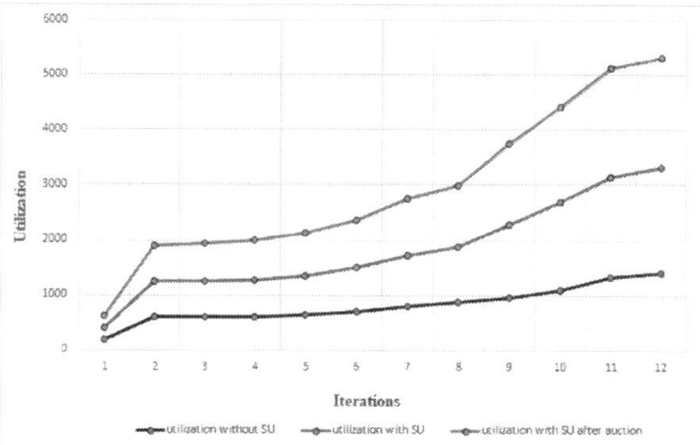

Figure 55.3 Alteration in utilization for different iterations (3 primary users and 2 secondary users)

proposed system model. The only channel with a minimum amount of interference is allocated.

Conclusions

Once the PU is noticed, SU has to leave the channel as per the basic properties of CRN. The process is known as spectrum handoff, which includes halting the continuous transmission and swapping to a fresh target channel for recommencing its communication. It affects longer switching delays and the total service time required by the SU. So incorporating the game theory approaches in CRN and 5G–6G communication systems is essential to determine the uninterrupted wireless communication technologies. This study contributes significantly by providing an auction-based game-theoretic investigation in CRN and 5G communication. The auction theoretic analysis shows that the number of users, interference, bidding time, and transmission time are significant for the overall network utilization. This scheme defines the total number of SUs as consistently more effective than the number of identified bidders, and it affects and improves the arrival in the channel during the bidding process. The proposed communication control structure also may be considered for post-handoff target channel selection in the cognitive radio network. In the future, we have targeted to include additional parameters like waiting time, average waiting time, overall system utilization, and inactive probability. We will also compare the system performance with another stable system network.

Acknowledgement

I want to sincerely thank Dr. Wasim Arif, Associate Professor of ECE at the NIT Silchar, for his invaluable guidance and unwavering support throughout this research endeavor. I would also like to thank my co-author, Ms. Debadrita Roy, Trainer in CSE at Ghani Khan Choudhury Institute of Engineering & Technology, Malda, for her valuable assistance and contributions. Additionally, I am deeply grateful to the AICTE, New Delhi, India, for providing me with the opportunity to conduct this research under the QIP scheme at the NIT Silchar, Assam-788010. Finally, I wish to express my profound gratitude to my family, including my daughter and parents, for their steadfast and constant encouragement in my research pursuits.

Appendix

Abbreviation with definition

Abbreviation	Definition
CRN	Cognitive radio networks
5G	Fifth generation
GT	Game theory
CU	Cognitive user
SU	Secondary user
PU	Primary user
TST	Total service time
QoS	Quality of service
6G	Sixth generation
NE	Nash equilibrium
HPQ	High priority queue
LPQ	Low priority queue
IUQ	Interrupted user queue

References

1. Liang, Y.-C., Chen, K.-C., Li, G. Y., and Mahonen, P. (2011). Cognitive radio networking and communications: An overview. *IEEE Trans. Veh. Technol.* 60(7), 3386–3407. doi:10.1109/TVT.2011.2158673.
2. Hoque, S., Arif, W., Baishya, S., Singh, M., and Singh, R. (2014). Analysis of spectrum handoff under diverse mobile traffic distribution model

in cognitive radio. *2014 Inter. Conf. Dev. Cir. Comm. (ICDCCom)*, 1–6. doi: 10.1109/ICDCCom.2014.7024696.

3. Gupta, N., Dhurandher, S. K., Woungang, I., and Rodrigues, J. J. P. C. (2017). Game theoretic analysis of post handoff target channel sharing in cognitive radio networks. *2017 IEEE Glob. Comm. Conf., IEEE*, 1–6. doi: 10.1109/GLOCOM.2017.8254068.
4. Akyildiz, I. F., Lee, W.-Y., and Chowdhury, K. R. (2009). CRAHNs: Cognitive radio ad hoc networks. *AD Hoc Netw.*, 7(5), 810–836. doi: 10.1016/j.adhoc.2009.01.001.
5. Laneman, J. N., Tse, D. N. C., and Wornell, G. W. (2004). Cooperative diversity in wireless networks: Efficient protocols and outage behavior. *IEEE Trans. Inform. Theory*, 50(12), 3062–3080. doi: 10.1109/TIT.2004.838089.
6. Mukherjee, A. and Kwon, H. M. (2010). General auction-theoretic strategies for distributed partner selection in cooperative wireless networks. *IEEE Trans. Comm.*, 58(10), 2903–2915. doi: 10.1109/TCOMM.2010.082010.080248.
7. Maillé, P. (2004). Auctioning for downlink transmission power in CDMA cellular systems. *Proc. 7th ACM Inter. Symp. Model. Anal. Simul. Wire. Mobile Sys.*, 293–296. doi: 10.1145/1023663.1023714.
8. Isnawati, A. F. (2022). A survey of game theoretical approach in cognitive radio network and 5G-6G communications. *J. Comm.*, 17(10), 830–843. doi:10.12720/jcm.17.10.830-843.
9. Fang, F., Liu, S., Basak, A., Zhu, Q., Kiekintveld, C. D., and Kamhoua, C. A. (2021). Introduction to game theory. *Game Theory Mac. Learn. Cyber Sec.*, 21–46. doi: 10.1002/9781119723950.ch2.
10. Shi, H.-Y., Wang, W.-L., Kwok, N.-M., and Chen, S.-Y. (2012). Game theory for wireless sensor networks: A survey. *Sensors*, 12(7), 9055–9097. doi: 10.3390/s120709055.
11. Krishna, V. (2002). *Auction theory*, Academic press: San Diego, USA.
12. Myerson, R. B. (1991). *Game theory: analysis of conflict*. Harvard university press.
13. Sendonaris, A., Erkip, E., and Aazhang, B. (2003). User cooperation diversity. Part I. System description. *IEEE Trans. Comm.* 51(11), 1927–1938. doi: 10.1109/TCOMM.2003.818096.
14. Kahvand, M., Soleimani, M. T., and Dabiranzohouri, M. (2013). Channel selection in cognitive radio networks: A new dynamic approach. 2013 *IEEE 11th Malaysia Inter. Conf. Comm. (MICC)*, 407–411. doi:10.1109/MICC.2013.6805863.
15. Shrivastav, V., Dhurandher, S. K., Woungang, I., Kumar, V., and Rodrigues, J. J. P. C. (2016). Game theory-based channel allocation in cognitive radio networks. *2016 IEEE Global Comm. Conf. (GLOBECOM)*, 1–5. doi: 10.1109/GLOCOM.2016.7841855.

Biographical notes

Arnab Kundu is currently affiliated with the NIT Silchar in India, where he serves as a research scholar participating in the AICTE-QIP program within the Department of ECE. He earned his Bachelor of Technology degree in Electronics and Communication Engineering and his Master of Technology degree in Mobile Communication and Network Technology from MAKAUT (formerly known as WBUT) in West Bengal, India. His primary research focus revolves around wireless communication and cognitive radio technology. Additionally, Arnab Kundu holds professional memberships with organizations such as the Society of EMC Engineers (India), IETE, International Association of Engineers, ISRS, ISTE, and IEEE.

Debadrita Roy is presently associated with the Ghani Khan Choudhury Institute of Engineering & Technology in India, holding a faculty position in the Department of CSE. Additionally, she serves as a part-time research scholar in the CSE Department of NIT Silchar, India. She completed his Bachelor of Technology degree in IT and his Master of Technology degree in Computer Science and Engineering at MAKAUT (formerly known as WBUT) in West Bengal, India. Her primary research interests encompass Blockchain Technology and wireless communication. Additionally, she holds professional memberships with organizations such as the ACM, IEEE, International Association of Engineers, and ISRS.

Dr. Wasim Arif is currently affiliated with the NIT Silchar in India, where he holds the position of Associate Professor within the Department of ECE. He completed his B.E. in ECE at Burdwan University in W.B., India, achieving the top rank in his class. He also earned a M.E. in Telecommunication Engineering from Jadavpur University in India and a Ph.D. in Electronics and Telecommunication Engineering from NIT Silchar in India. His research primarily centers around wireless communication, cognitive radio, and the next generation of wireless technology and its applications. He is a Senior member of IEEE and IEEE Communications Society.

56 Safeguarding data in the digital world: Exploring cryptography for enhanced security and confidentiality

Ankur[1], *Abinaya S.*[2], *Awani Bhushan*[1,a], *Ashna Verma*[3] *and RajaRam Kumar*[4]

[1]School of Mechanical Engineering, Vellore Institute of Technology, Chennai, Tamil Nadu, India

[2]School of Computer Science and Engineering, Vellore Institute of Technology, Chennai, Tamil Nadu, India

[3]School of Electronics and Communication Engineering, Vellore Institute of Technology, Chennai, Tamil Nadu, India

[4]Department of Electrical Engineering, GhaniKhan Choudhury Institute of Engineering & Technology, India

Abstract

Data security presents an ongoing challenge for both inventors. To address the colorful attacks employed byhackers, the need for further robust security technologies is imperative. This exploration paper introduces a new approach aimed at secure textbook transmission, achieved through the development of a low-complexity array data structure security algorithm. The algorithm's strength lies in its application of multiple complicating variable factors, making it significantly more grueling for bushwhackers to recover the original communication. The primary idea of this paper is to advance an encryption algorithm that leverages the Fibonacci matrix metamorphosis to improve data security. Through this process, data is converted into ciphertext guaranteeing increased sequestration, data authenticity, and data integrity. By incorporating vectors into the algorithm,several benefits are achieved, such as upgraded effectiveness, dynamic resizing capabilities, andstreamlined error running. ThedesignandmodelingoftheencryptionalgorithmencompassthegenerationofFibonaccimatricesandthe encryption process employing XOR operations. Likewise, the algorithm's Fibonacci format and theintegrationofvarying gruelingfactorsoffer rigidityto acclimate thealgorithm'ssecurity grounded onspecific circumstances..

Keywords: Encryption algorithm, fibonacci matrix metamorphosis, data security, sequestration, data authenticity, data integrity, effectiveness, scalability, error running, complexity analysis

1. Introduction

In the digital period of present days, data security has surfaced as a pivotal concern due to the rapid-fire advancement of internet technology and the wide transfer of textual and multimedia data over the web. The vulnerability of data during transfer, owing to insecure communication channels, has given rise to unauthorized access and breaches. To attack these security challenges, colorful ways, including cryptography, have been espoused [1–3].

Cryptography involves securing data by encrypting it with an encryption key, which can either be symmetric, known to both the sender and receiver or asymmetric, differing on both ends [4]. These cryptographic styles play a vital part in ensuring data security and confidentiality, securing sensitive information from unauthorized access. Both the public and private sectors have employed a range of techniques to cover data from unauthorized users, with cryptography being a prominent technology extensively employed to achieve high levels of security and confidentiality. By encrypting data, its contents become hidden, making it challenging for individualities to pierce or understand the information. Also, the use of Fibonacci matrix metamorphoses adds a redundant subcaste of security, as implicit interferers may not indeed be apprehensive of the actuality of hidden data [5].

Employing encryption algorithms that incorporate Fibonacci matrix metamorphoses enables associations and individuals to bolster the security of their data and communications, shielding sensitive information from unauthorized access and mitigating pitfalls in the digital realm [6]. The idea of this exploration is to develop an encryption algorithm that focuses on reducing unauthorized access and data corruption throughout the data lifecycle, while also enhancing sequestration, data authenticity, and data integrity during transmission.

Moreover, this paper addresses the limitations of existing encryption systems, such as computational complexity, key management issues, potential vulnerabilities, implementation bugs, and encryption overhead. The proposed model offers improvements in efficiency, dynamic resizing, ease

[a]awani.bhushan@vit.ac.in

DOI: 10.1201/9781003540199-56

of implementation, scalability, and error handling, making it a promising approach for advancing encryption and decryption algorithms [7].

The primary contributions of this research paper are as follows:

1. Development of a practical algorithm with reduced complexity.
2. Proposal and development of a two-layered line data encryption and decryption scheme using a symmetric key cryptography approach.

2. Background

The encryption process described in the method uses a cryptographic algorithm along with a specific key denoted as "e = (E, D)."The main goal is to convert the original data, known as plaintext (M), into encrypted data called ciphertext (C) [8]. The process involves two main steps: encryption (E) and decryption (D).

The encryption process (E) takes both the key (K) and the plaintext message (M) as input and produces the ciphertext (C) using the formula: C = E (K, M). Here, C represents the encryption of the message M using the key K.

On the other hand, the decryption process (D) takes the same key (K) and the ciphertext (C) as inputs and retrieves the original plaintext message (M) using the formula: M = D(K, C).

To achieve encryption, the method converts each character in the plaintext message (M) into its corresponding ASCII binary representation. For example, the character "H"is converted to "01001000", "E"to "01000101", "L"to "01001100", and "O"to "01001111". Then, each bit of the ASCII binary representation is combined using the XOR operation with the corresponding bit in the key (K).

For instance, let's encrypt the word "HELLO"using the key "KEY":

Plaintext (M): H E L L O
ASCII Binary: 01001000 01000101 01001100 01001100 01001111
Key (K): K E Y
ASCII Binary: 01001011 01000101 01011001
XOR Result: 00000011 00000000 00010101 00010100 00000010

The resulting XOR values are organized and stored in a matrix, as shown above. This matrix is then transmitted to the receiver. Since the key is symmetric and known to both the sender and receiver, this process is commonly referred to as XOR encryption [9-10].

It's important to note that while XOR encryption is a basic encryption technique, it might not provide strong security on its own, especially against sophisticated attacks. For more robust security, modern encryption algorithms often employ complex mathematical operations and use longer, randomly generated keys.

3. Proposed scheme

The proposed research paper introduces an advanced encryption scheme that builds upon the conventional XORmethod. This new scheme incorporates a fresh subcaste of defense by converting the final XOR-used textbook into a Fibonacci series before transmission.

To achieve this, a dynamic array is employed as the data structure. This improvement introduces a new encryption and decryption function, denoted as d = (C), where the binary cipher text (C) is transformed into its decimal representation (d). Subsequently, the decimal value (d) is utilized to generate the Fibonacci series (F).

The integration of this new function contributes to a more robust and secures encryption process.

Algorithm 1 generates the Fibonacci matrix
Input: n (number)
// n ε[0, ∞] are positive values
Output: a row of Fibonacci series for each number.

```
functiongenerateFibonacciMatrix(num):
matrix = empty vector of vectors
fib1 = 0
fib2 = 1
whilenum> 0:
row = empty vector
while fib2 <= num:
row.push_back(fib2)
fib1, fib2 = fib2, fib1 + fib2
num -= row.back()
matrix.push_back(row)
return matrix
End Function
```

Algorithm 2: generates cipher text by using Algorithm 1
Input: plaintext, key
// length of plaintext and key must be the same
Output:matrix containing each row as Fibonacci series

```
function encrypt(plaintext, key):
binaryMatrix = empty vector of vectors
firstRow = empty vector
firstXORResult = ASCII value of plaintext[0] XOR ASCII value of key[0]
firstRow.push_back(firstXORResult)
for j = 1 to length of key - 1:
row = empty vector
```

```
xorResult = ASCII value of plaintext[j] XOR ASCII value of key[j]
row.push_back(xorResult)
binaryMatrix.push_back(row)
binaryMatrix.insert(binaryMatrix.begin(), firstRow)
seriesMatrix = empty vector of vectors
for each row in binaryMatrix:
rowSeries = empty vector
for each num in row:
fibonacciMatrix=generateFibonacciMatrix(num)
for each fibNum in fibonacciMatrix:
rowSeries.push_back(fibNum)
seriesMatrix.push_back(rowSeries)
returnseriesMatrix
End Function
```

4. Performance

The presented algorithm for encryption aims to outperform the traditional XOR algorithm in terms of speed. Although it has been assessed for complexity, a more in-depth discussion is necessary to completely understand its capabilities [11]. An important point taking further explanation is the implicit violation of an abecedarian property caused by the use of a circular queue. A thorough explanation of this violation is essential to grasp the counteraccusations and downsides of this approach, particularly regarding the algorithm's security and implicit vulnerabilities. Despite the absence of analogous approaches being in the literature, the developed algorithm claims to achieve the lowest possible complexity for the given encryption conception. Likewise, it asserts superior performance compared to traditional cryptographic algorithms. To strengthen these claims, empirical substantiation and relative studies should be included in the research paper. Conducting experiments demonstrating the algorithm's speed and effectiveness under various scenarios would be largely salutary.

The encryption function's overall time complexity is stated as O(NM + NK), with N representing the size of the plain text, M representing the size of the key, and K indicating a constant time factor. Additionally, the presence of a matrix formed during transmission contributes to a final complexity of O(N * M).

In Figure 56.1, the flowchart depicts the process of generating a matrix that encapsulates a cipher text based on the Fibonacci series. This is achieved by executing the functions "generate FibonacciMatrix(num)"and "encrypt(plaintext, key)", resulting in the production of a series matrix. Subsequently, this series matrix is intended tobetransmitted.

To further enhance the research paper, a comprehensive analysis, and evaluation of the proposed algorithm's complexity are essential. This analysis should consider factors similar to space complexity, computational outflow, and scalability. Understanding the algorithm's resource conditions and its performance under different input sizes is pivotal for assessing its practicality and effectiveness in real-world applications. To validate the claims made in the research paper, concrete exemplifications or experimental results demonstrating the algorithm's performance advancements should be provided. A comparison of its speed and effectiveness with other well-established encryption algorithms under colorful scripts will offer strong evidence of its advantages and support the claims of improved speed.

5. Conclusions

A novel end-to-end security algorithm based on a unique data structure is proposed in this research. The algorithm leverages arrays as the abecedarian data structure and employs the Fibonacci sequence to generate the cipher text. The encryption process provides a high level of security, but the decryption process poses challenges due to the utilization of variable factors. One of the crucial features of the proposed algorithm is its flexible size-tunable mechanism. The sender and receiver must agree on the array size, keyword letter/symbol, and the representation of Fibonacci numbers before establishing a connection. This agreement is pivotal to ensure successful encryption and decryption of the data. In addition to its connection to text messages, our algorithm aims to extend its usage to other types of data, including images, voice, and video. By incorporating the proposed algorithm, we anticipate enhanced security and confidentiality for a wide range of data formats. Furthermore, the proposed algorithm boasts faster encryption and decryption processes compared to traditional algorithms. This advantage is attributed to the low complexity of the encryption and decryption operations, resulting in improved efficiency and reduced computational overhead.

6. References

1. Hassaballah, M., Hameed, M.A.,Awad, A.I., and Muhammad, K. (2021).A novel image steganography method for industrial Internet of Things se-

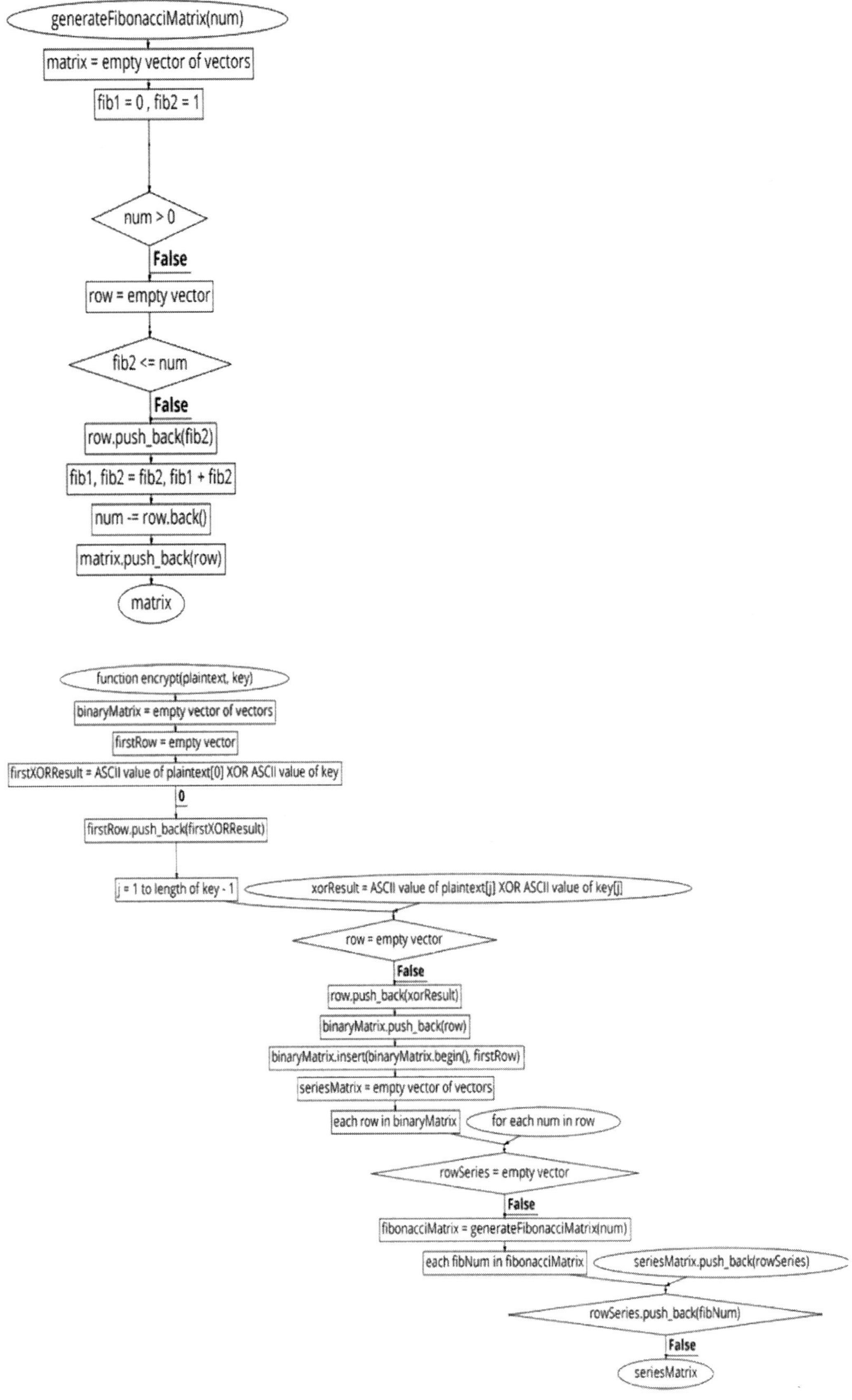

Figure 56.1 The illustration of the process of generating a matrix that encapsulates a cipher text based on the Fibonacci series

curity.*IEEE Trans.Indus. Informat.*, 17(11),7743–7751.doi: 10.1109/TII.2021.3053595.

2. Ahvanooey, T. M., Li, Q., Hou, J., Mazraeh, H.D. and Zhang, J.(2018).AITSteg: An innovative text steganography technique for hidden transmission of text message via social media. *IEEE Acc.*, 6,65981–65995,doi: 10.1109/ACCESS.2018.2866063.
3. Baagyere, E.Y.,Agbedemnab,P.A.N., Qin, Z.,Daabo, M.I.,and Qin,Z.(2020).A multi-layered data encryption and decryption scheme based on genetic algorithm and residual numbers. *IEEE Acc.*, 8, 100438–100447.doi: 10.1109/ACCESS.2020.2997838.
4. Merkle, R.C. and Hellman,M.E. (1981). On the security of multiple encryption.*Comm. ACM*, 24(7), 465–467.
5. Kasapbasi, M.C.(2019).A new chaotic image steganography technique based on Huffman compression of Turkish texts and fractal encryption with post-quantum security. *IEEE Acc.*, 7, 148495–148510,doi: 10.1109/ACCESS.2019.2946807.
6. Fatima, A. andKinani, E.H.E.(2012).An elliptic curve cryptography based on matrix scrambling method. *National Days Netw. Sec.Sys.*,31–35.doi: 10.1109/JNS2.2012.6249236.
7. Liu, J., Ke,Y., Lei,Y., Zhang, Z., Li,J., Luo,P., Zhang,M.,and Yang, X. (2020).Recent advances of image steganography with generative adversarial networks. *IEEE Acc.*, 8, 60575–60597.doi: 10.1109/ACCESS.2020.2983175.
8. Phull, S. and Som, S. (2016)Symmetric cryptography using multiple access circular queues (MACQ).*Proc. Second Inter. Conf. Inform. Comm. Technol.Compet. Strat.*, 107, 1–6. DOI: 10.1145/2905055.2905166.
9. Suli, W., Zhang, Y., and Jing,X. (2008).A novel encryption algorithm based on shifting and exchanging rule of bi-column bi-row circular queue. *Inter. Conf.Comp. Sci.Softw.Engg.*, 841–844.doi: 10.1109/CSSE.2008.987.
10. Kahate, A. (2017). Cryptography and Network Security.Tata McGraw-Hill Education, 3rd Edition. pp.59-61.
11. Albu-Rghaif, A., Jassim, A., Abboud, A.J.(2018).A data structure encryption algorithm based on circular queue to enhance data security. *1st Inter. Sci. Conf.Engg.Sci. 3rd Sci. Conf.Engg.Sci. (ISCES)*, 24–29.doi: 10.1109/ISCES.2018.8340522.

57 Performance analysis of energy efficient WSN using blockchain model for IOT application

Monika Malik[a], *Gayatri Sakya*[b], *Ruchi*, *Suyash Gupta*[c] and *Rohit Sambharkar*

Department of Electronics and Communication Engineering, JSS Academy of Technical Education Noida, Noida, India

Abstract

Wireless sensor networks (WSNs) play a key role in the Internet of Sensor Things (IoST). IoST helps collect data from environments and is used in energy trading, monitoring, smart grids, and more. Connect to the Internet and automate your surveillance system without third-party involvement. An IoST network consists of sensor nodes that perform environmental monitoring. Wireless sensor networks (WSN) and the Internet of Things (IoT) have gained popularity in recent years as the underlying infrastructure for connected devices and sensors in a variety of sectors. The data generated by these sensors are used in smart cities, agriculture, transportation systems, healthcare systems, toll collection systems, automatic identification of road data, automatic identification of vehicle license plates, and more. WSN and IoST are two sides of a coin since WSN are deployed with tiny battery operated sensors when we are implementing real time applications with IoST security is at most concern. In this proposed paper the security aspect is carried out by blockchain mechanism. The main problems and challenges of WSN are effectively reduced by using the LEACH routing protocol which follows the proficient cluster head (CH) selection scheme.

Keywords: WSN, IoST, blockchain, Python

Introduction

In recent years, IoT has started to play an important role in our daily life, and it has transformed our perception and ability to change the environment around us. IoT is defined as the connection of physical devices that enable the collection of data and the exchange of that information. The global Internet of Things standards initiative has identified IoT as the infrastructure of the Information Society. IoT enables devices to be sensed and remotely controlled by established systems, creating opportunities to directly connect the universe to computerized systems for improved performance. Since IoT is combined with other technologies such as WSN and blockchain, this technology can be seen in the more general class, which includes inventions such as smart grid, smart city, and smart home, as shown in Figure 57.1 which represents many real time applications handled by IoST. Since the key requirement of IoST is the sensing of any event by the sensor nodes, this drain off the battery life. A lot of research is going on to apply optimization techniques to enhance the network lifetime. This paper presents one such routing protocol which shows the improvement of network lifetime.

Existing system review

In some clustering methods, resource-rich nodes are pre-determined as cluster heads (CHs). The problem with this method is that most WSNs are homogeneous and resource constrained. Therefore, in some cases, this procedure will not work. Also, even if a resource-rich node is found in a heterogeneous network and can be selected as CH, if it is CH for a long time, the power of the node will deplete rapidly, causing node death. Moreover, if the CH is fixed, the dynamics of the mobile nodes and the network can lead to unbalanced clusters in terms of the number of members or amount of data transmitted, resulting in unbalanced network load and resource consumption. The application of blockchain at different base station (BS) is discussed in ref [1]. Collapses [2] Taheri et al. proposed clustering method has three stages. In the initial stage, the knowledge about the neighborhood is modified and the crisp fuzzy output is calculated. Each node sensor is placed at some point in the later phase until the delay time to listen for CH messages. If it fails, impersonate a temporary CH while simultaneously placing the message within the boundaries of the cluster. In the next iteration, it will be the last CH and communicate when it has the lowest cost among the temporary CHs near. Researchers

[a]monikamalik@jssaten.ac.in, [b]gayatrisakya@jssaten.ac.in, [c]guptasuyash04@gmail.com

DOI: 10.1201/9781003540199-57

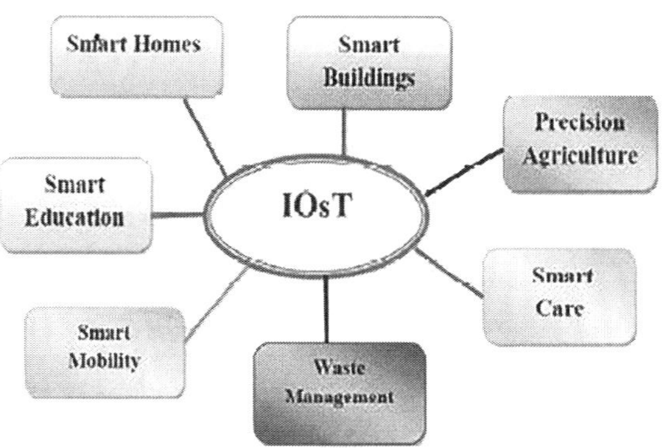

Figure 57.1 IoST-based smart city real time applications

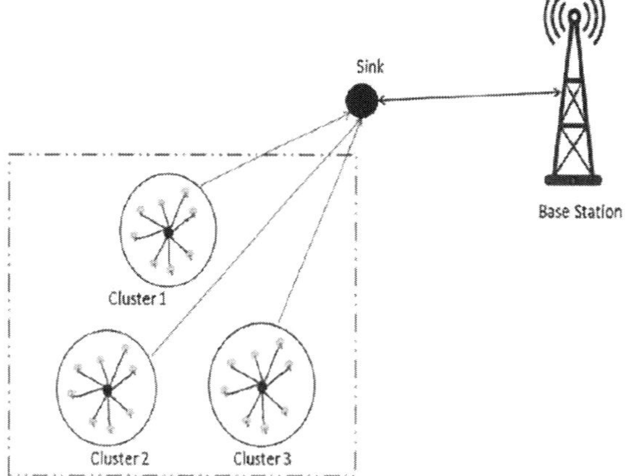

Figure 57.2 IoT-based clustering scheme

have proposed a ZigBee protocol for tracking agricultural environments. A wireless sensor node is used to collect live information on the farm and transmit that information to a base station (BS) via his ZigBee protocol [2]. Authors in ref [3] give the polling approach for better throughput. The network efficiency is improved by applying linear coding [4]. Discuss the usability of LoRa for IoT application [5]. The authors propose an energy-efficient clustering-based multistage routing protocol that evenly distributes CHs and minimizes unnecessary energy consumption [6]. The protocol uses forwarder nodes to minimize routing distance and multiple power amplification levels to conserve energy. The proposed protocol outperforms existing routing protocols concerning of energy efficiency, throughput, and network lifetime, as demonstrated through MATLAB simulation. Further research directions include extending the proposed solutions to improve energy efficiency in the MAC layer and making the routing protocol application-specific. Additionally, the paper suggests modeling and implementing quality of service (QoS) in WSNs. The paper proposes a mathematical method for secure data storage in a dynamic manner through distributed node cooperation [6]. It also introduces an ownership transition feature and dynamic storage system architecture for blockchain-based WSN to reduce storage overhead for each node without compromising data integrity. The paper also suggests a novel ownership determination secure data (ODSD) framework for real-time applications that use incentives in digital currency to collect data through storage nodes. The proposed ODSD distributed storage Blockchain-IoT architecture creates a feasible solution to reduce projected capacity for every node without affecting overall information reliability [7]. The paper proposes a blockchain and IoT-based network to secure data and reduce computational costs [8]. Only authorized nodes can participate, and data is sent to CHs for processing. A DDR-LEACH protocol is suggested to select CHs based on specific criteria and replace those with low energy levels. Data is stored in IPFS, and blockchain incentivizes long-term storage. The network uses AES 128-bit encryption for secure service provisioning and incurs low transaction costs. Simulation results compare the performance of DDR-LEACH and LEACH in terms of energy consumption, network lifetime, and throughput. Authors evaluated the proposed smart contract's effectiveness against attacks and suggest the use of machine learning for detecting malicious nodes [7]. Authors discuss about WSN for real time applications in ref [9]. Ala et al., used MATLAB to try to compare LEACH and LEACH-C protocols [10]. Figure 57.2 shows the clustering used with IoT. The first node death occurred earlier in LEACH than in LEACH-C. This is an important factor that supports the purpose of LEACH-C, which is designed to increase network life. Another measure they used was energy dissipation and according to its results LEACH-C showed less energy dissipation than LEACH.

Proposed model design

As shown in Figure 57.3, a solution for secure IoT using energy efficient WSN and block-chain scheme is proposed.

Initialization phase

To conduct the investigation, we selected an area of 100×100 $m^{2,}$ divided this area into clusters, and randomly placed the IoST nodes in different clusters as shown in Figure 57.4. The nodes in a cluster are shared across clusters, so each cluster absorbs all nodes and does not communicate with each other within the same cluster, but most effectively communicates with its CHs to talk about

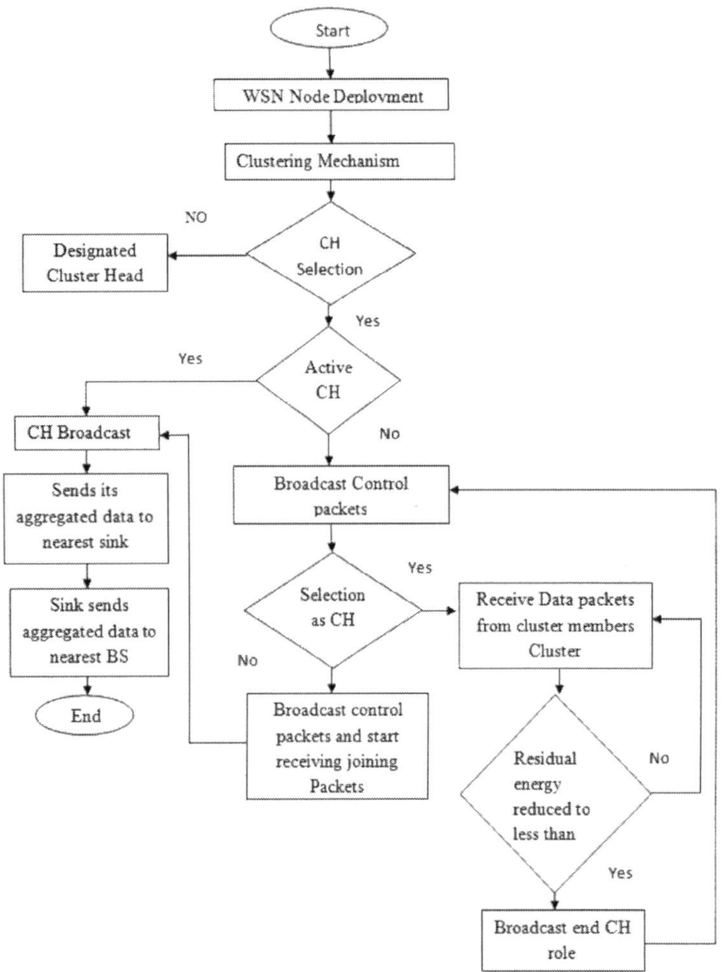

Figure 57.3 Flowchart for CHs election in the LEACH protocol

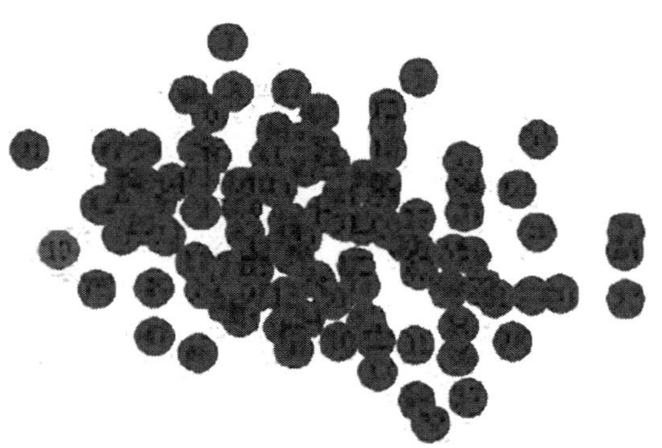

Figure 57.4 Nodes deployment in the field of 100*100

sinks. Figure 57.4 shows the deployment of 100 nodes. Send information to the base station. A system model creates a node (destination and source node) by using a network and we can transfer the data to source to destination as shown in Figures 57.3 and 57.4 of deployed sensor nodes in mentioned area.

Mechanism of cluster formation

The clustering process starts after the IoT nodes are deployed in the cluster farm, each cluster has G. Each cluster has similar or different node shapes, depending on the actual requirements. A CH is elected per cluster and all nodes forward information to their respective CH.

Routing phase

For efficient routing, we first adopted a new IoST clustering protocol and then integrated IoST with blockchain to achieve better results. A three-stage data transmission technology is introduced. In the first step, member node collects information to share with each cluster his head, in the second step the CH forwards the information to the server, and lastly the data is send to Base station Suppose the sink is outside the farm and the nodes deployed in the cluster farm know exactly where the sink is and cannot change it. CH is chosen based on its high energy and short Euclidean distance to the sink and GH. Take the Euclidean distance between any two

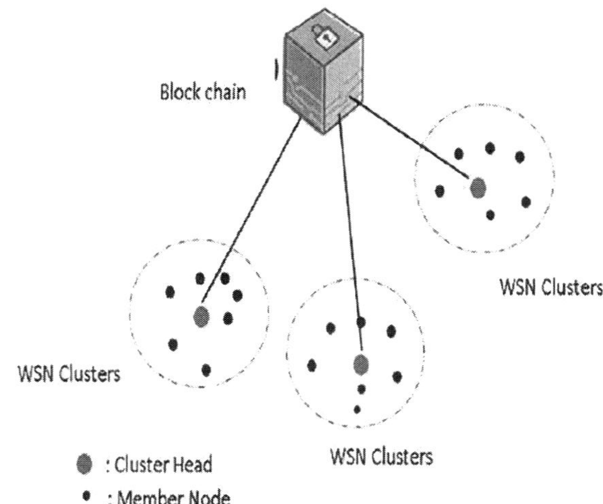

Figure 57.5 IoST and blockchain integration

Table 57.1 Performance parameters of WSN

Parameters	Output
First node died at round	87
Tenth node died at round	990
Packet drop rate:	8.33
Energy	0.26

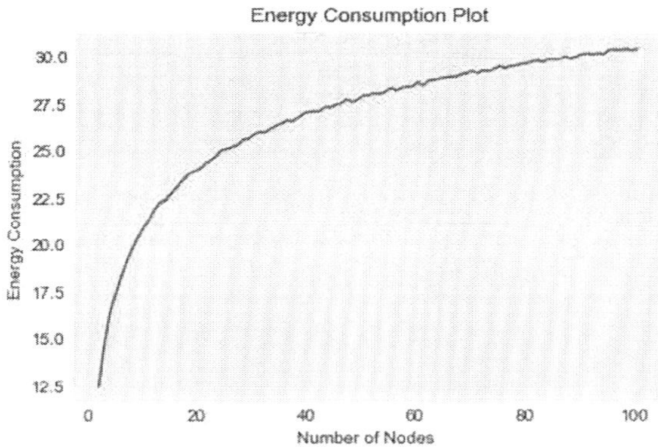

Figure 57.6 IoST-LEACH energy consumption

nodes a and b in the two dimensions with the closest Euclidean distance given in Equation (1)

$$(a. b) = \sqrt{((x_2-x_1)^2+(y_2-y_1)^2)} \qquad (1)$$

Here a, b represent the node position in the deployed area and x_1, x_2, y_1, y_2 are the vector position of sensor nodes.

Integration of blockchain and IoT

Blockchain smart contracts have the ability to make routing protocols more secure by removing redundancies from the aggregated data collected from

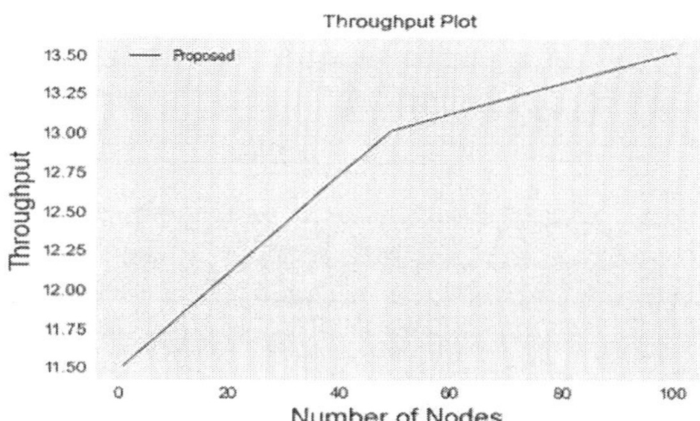

Figure 57.7 IoST – LEACH throughput

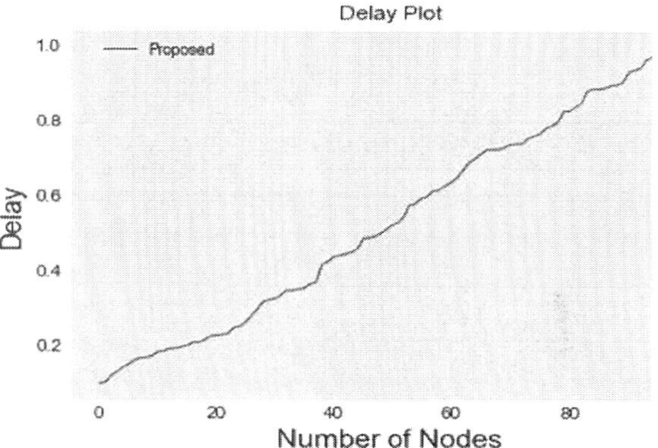

Figure 57.8 IoST – LEACH delay

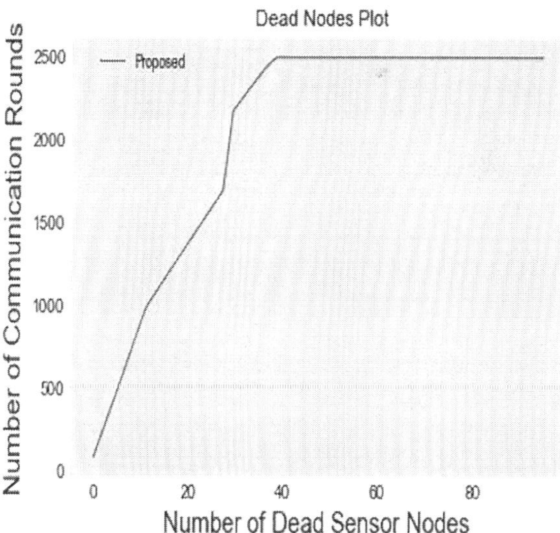

Figure 57.9 IoST – LEACH dead nodes

IoT nodes and blocking attacks on IoT networks-grid., resulting in lower energy consumption and leading to longevity.

This research approach focuses on the use of smart blockchain smart contracts that run autonomously. Thousands of collectively distributed mining nodes implement smart contract functionality and code and mutually agree on the final result.

One thing that needs to be mentioned here is that the blockchain network is made up of mining nodes as depicted in Figure 57.5.

Results

We select input data using Panda's package and perform pre-processing on the collected data In this step It can handle the missing data, can perform label encoding and drop unwanted columns. Some performance metric are analyzed such as, accuracy, precision, recall, F1-score, confusion matrix, ROC curve, energy consumption, bandwidth, throughput, delay, packet delivery ratio, network traffic. Table 57.1 depicts WSN parameters. To test the effectiveness of the system in handling anonymous attacks, an attack was conducted. The Python-based Spyder platform was used to simulate the proposed methodology, and this section presents the results of the network transactions. The results include the throughput, number of alive nodes, and number of dead nodes for each round of wireless communication transmission. Various parameters performance for using IoST-LEACH clustering model with blockchain is given in various Figures below. Figure 57.6 shows energy consumption for IoT-LEACH, Figure 57.7 depicts throughput, Figures 57.8 and 57.9 shows delay and number of dead nodes, respectively.

Conclusions

IoT is one of the newest technologies, and its combination with blockchain opens up new avenues. The IoT nodes have the ability to capture the environmental characteristics of the cluster and send the captured information to the base station via the sink, thus offloading the CH. Blockchain further eliminates redundant data, it prevents IoT nodes from consuming a lot of power during transmission. In this study, we used a blockchain model to propose energy efficient WSN for his IoT application scheme; simulation results show that the proposed scheme has a longer network life, lower power consumption, and superior performance than his LEACH in agriculture. We have shown that it has a high throughput. Future research will develop intelligent models based on IoT and blockchain for clustered environmental monitoring and information.

References

1. Aljumaie, S. G. and Alhakami, W. (2022). A secure LEACH-PRO protocol based on blockchain. *Sensors,* 22(21), 8431. https://doi.org/10.3390/s22218431.
2. Taheri, H., Neamatollahi, P., Younis, O. M., Naghibzadeh, S., and Yaghmaee, M. H. (2012). An energy-aware distributed clustering protocol in wireless sensor networks using fuzzy logic. *Ad Hoc Networks*, 10(7), 1469–1481. https://doi.org/10.1016/j.adhoc.2012.04.004.
3. Zhang, Z. (2008). Energy-efficient multihop polling in clusters of two-layered heterogeneous sensor networks. *IEEE Trans. Comput.*, 231–245.
4. Aloÿs, A., Yi, J., Clausen, T., and Mark, W. (2016). A study of LoRa: Long range & low power networks for the internet of things. *Sensors*, 16(9), 1466.
5. Saleh, N. A. and Ayoub, M. (2021). Energy-efficient and blockchain enabled model for Internet of Things (IoT) in smart cities. *Comp. Mater. Contin.*, 66(3), 25092524. https://doi.org/10.32604/cmc.2021.014180.
6. Hussain, S. A., Ahmed, S., and Nawaz, A. (2020). Blockchain with IoT, an emergent routing scheme for smart agriculture. *Inter. J. Adv. Comp. Sci. Appl* 11(4), 2020. http://dx.doi.org/10.14569/IJACSA.2020.0110457
7. Ibrahim, O. K. and Abdulsahib, M. G. (2023). Optimized dynamic storage of data in Iot based on blockchain for wireless sensors networks. *Peer-to-Peer Network. Appl.*
8. Amjad, S., Abbas, S., Abubaker, Z., Mohammed, H., Jahid, A. A., and Javaid, N. (2022). Blockchain based authentication and cluster head selection using DDR-LEACH in Internet of sensor Things. *Sensors,* 22(5), 1972. https://doi.org/10.3390/s22051972.
9. Malik, M., Sakya, G., and Joshi, A. (2021). Enhancement of network life time in the LEACH protocol for real time applications. Thirteenth Inter. Conf. Contemp.Comput,Pages426–431 https://doi.org/10.1145/3474124.3474188
10. Al-Shaikh, A., Khattab, H., and Saleh, A. F. (2018). Performance comparison of LEACH and LEACH-C protocols in wireless sensor networks. *J. ICT Res. Appl* 12(3), 219-236, https://doi.org/10.5614/itbj.ict.res.appl.2018.12.3.2.

58 Design and comparison of a H-infinity-based speed controller with a conventional PID controller for a brushless DC motor

Arindom Dehingia[1,a], Chiranjeeb Kumar Das[1,b] and Mrinal Buragohain[2,c]

[1]Department of Instrumentation and Control Engineering, Jorhat Engineering College, Jorhat, India

[2]Department of Electrical Engineering, Jorhat Engineering College, Jorhat, India

Abstract

Purpose of this work is to regulate the speed of a brushless direct current (BLDC) motor under varying external loads and thus maintaining a desired speed. For this, two different control approaches are adopted. A H-infinity controller and a traditional PID controller is designed in MATLAB/SIMULINK. System identification toolbox in MATLAB is the tool for figuring the transfer function of BLDC motor. The speed of the BLDC motor in both the cases are examined under variations in external load as well as changes in reference speed. It is found that the H-infinity controller is an effective replacement for the conventional PID controller.

Keywords: Brushless dc motor, H-infinity control, pid control, robust toolbox

Introduction

Traditional DC motors, which utilize brushes for commutation and brushless DC (BLDC) motors, which use permanent magnets and electronic commutation circuits, are two different types of DC motors [1]. Brushless DC motors regulate the current flow to the motor windings using a series of electronic switches as opposed to brushed DC motors, which transmit power to the rotor using brushes and a commutator. When it comes to convert electricity into mechanical power, brushless motors outperform brushed motors. The convenience is primarily due to the fact that the frequency of electricity switching rather than voltage, determines the brushless motor's speed. In comparison to brushed DC motors, brushless motors provide a number of benefits, including greater torque per weight, greater torque per watt (improved efficiency), increased dependability, lower noise and longer lifetime [2]. But the BLDC motors, in order to generate torque requires position data. Different devices such as hall sensors, position sensors and resolvers are required to collect the exact position data. This increases the machine's weight, cost and rotor inertia [3].

As a standard way to govern the speed and positioning of BLDC motor drives, many researchers have adopted PI or PID controllers [4, 5]. Although use of PI or PID controller is easy, but they sometimes give inaccurate results as they do not have the required robustness to both internal and external perturbations. Therefore, to successfully overcome those inaccuracy, researchers are working on various robust speed control methods such as sliding mode control (SMC), backstepping control, model predictive control (MPC), H-infinity control, etc. [6]. In ref [7], H-infinity control approach is used to regulate a permanent magnet synchronous motor (PMSM's) speed. Whereas in ref [8] it is used to regulate a DC servo motor's speed. This article shows how to regulate the speed of a brushless DC motor using H-infinity control approach.

Figure 58.1 displays the block diagram of the recommended system. The H-infinity controller controls the motor's speed by controlling the DC voltage input to the 3-phase voltage source inverter through a voltage regulator. Whereas, the hall effect signal from the motor is used to generate the gate pulse for the VSI through a commutation logic. An error signal is produced when the feedback signal which is the actual present speed of the motor is compared with the user defined reference speed and this error signal acts as the controller's input.

Modeling of proposed system

The mathematical model of BLDC motor in transfer function format is determined with the help of system identification toolbox available in MATLAB. The first step is to collect the input and output data of the simulation in MATLAB workspace by performing simulation on open loop model of BLDC motor. Then we import the data from workspace to the system identification toolbox, which is opened by using the MATLAB command *"ident"* and

[a]dehingiaarindom96@gmail.com, [b]chiranjeebkumardas07@gmail.com, [c]mrinalburagohain@gmail.com

DOI: 10.1201/9781003540199-58

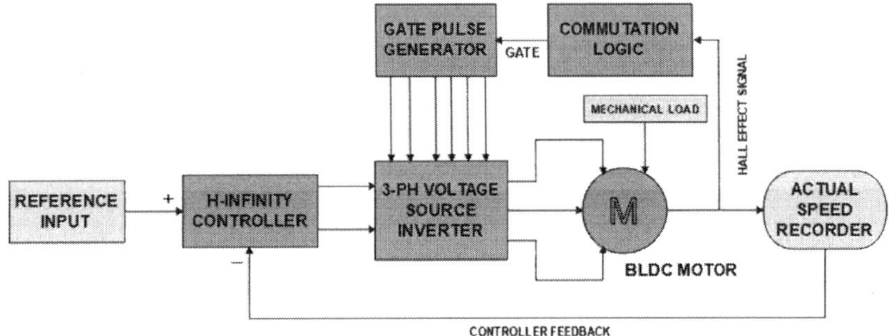

Figure 58.1 Block diagram demonstrating speed control of BLDC motor

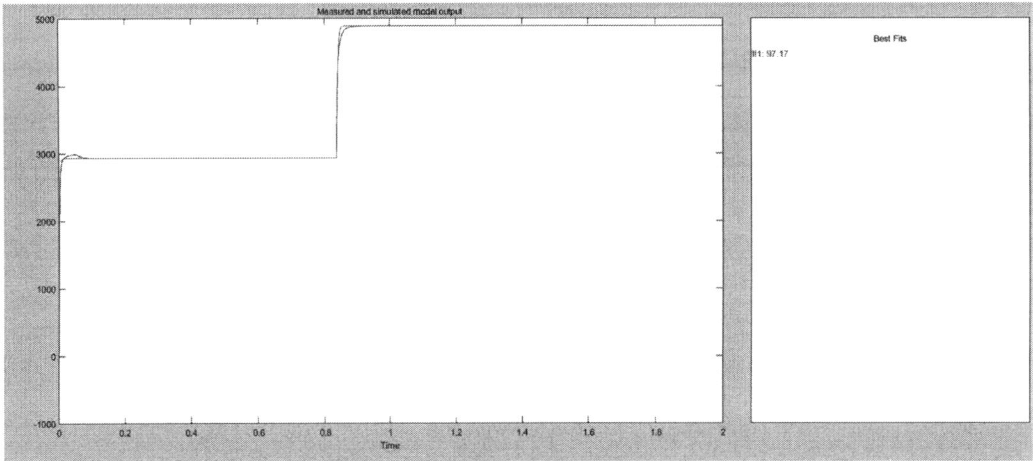

Figure 58.2 Fitness of transfer function of BLDC motor

estimate the transfer function of the BLDC motor. System identification is done with linear parametric model (ARX). Fitness of ARX model is found to be 97.17%. Figure 58.2 displays the fitness in percentage of the transfer function. The transfer function after simulation is found as shown in Equation 1.

$$G(s) = \frac{(4.8746 \times 10^5 s) + (1.203 \times 10^8)}{s^2 + (3.677 \times 10^5 s) + (8.995 \times 10^7)} \quad (1)$$

H-infinity control

In a practical plant model different uncertainties may arise due to several factors like uncertainties while modeling the plant, uncertainties identified from noisy input-output data, saturation of the actuator, deteriorating hardware over a long span of time, deliberate model simplification, reducing the controller's order deliberately, etc. A control system is termed as robust control system if it is insensitive to model uncertainties. In control theory, H_∞ method is used to produce controllers that can eliminate these unwanted perturbations and give guaranteed performance. H_∞ norm of a function is its highest singular value over the Hardy-space. For a SISO system, it is the maximum magnitude of frequency response [8].

The basic block diagram of a generalized plant is shown in Figure 58.3. "P" is the plant model, "K" is the H-infinity controller gain. "w" is the exogenous input to the plant "P" and it consist of the disturbance as well as the reference signal. "u" represents control input to plant "P". "z" represents the error signal to be minimized. "y" represents measurable variable use for controlling the overall system. Getting a reliable output "z" that is independent of "w" is prime goal of controller design.

The block diagram illustrating how the proposed H-infinity controller would be used as a speed controller is displayed in Figure 58.4. The BLDC motor is shown as plant "G" and its speed is to be maintained. Exogenous input "w" and the control signal "u" are two inputs to plant. "y" and "z" represents calculated speed and robust output, respectively, and are the two outputs of the plant. In order to generate a controlled output or reference torque "u", the calculated speed "y" is compared with

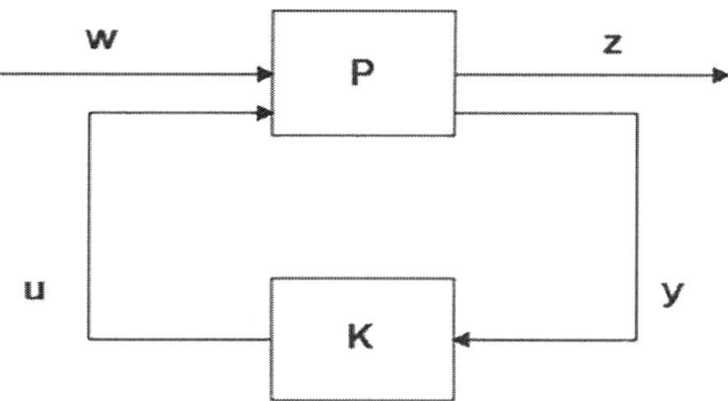

Figure 58.3 Basic block diagram of a generalized plant

Source: 10.11591/ijpeds.v12.i3.pp1379-1389

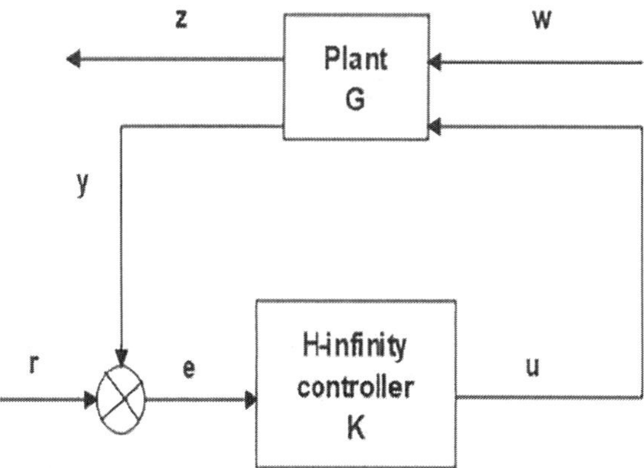

Figure 58.4 Block diagram for a brushless DC motor that incorporates H-infinity approach

reference speed "r". The error "e" generated acts as input to H-infinity controller "K".

The aim is to realize a controller gain "K" that stabilizes the closed loop system and satisfies the Equation 2.

$$\|T_{\infty}\| \triangleq \begin{bmatrix} W1S \\ W2R \\ W3T \end{bmatrix} \leq \gamma \tag{2}$$

Here $\|T_{\infty}\|$ is weighted mixed sensitivity function. "S" is sensitivity function and it is given by $= \frac{1}{1+GK}$. It refers to transfer matrix from "w" to "z". Input sensitivity function is shown as "R" and it refers to transfer matrix from "w" to "u". "T" is called complementary sensitivity function and is given by $T = 1 - S$ [9].

H-infinity control problem can be resolved by augmenting the plant "G" with suitable weighting functions. Weight "W1" for error signal "e", weight "W2" for control signal "u" and weight "W3" for output signal "y". The augmented plant "P" is found with MATLAB command *"augw"*. The

Table 58.1 Ratings and parameters of brushless dc motor

S. No.	Parameters	Values
1	Rating of motor	1.1 HP
2	Stator phase resistance, (Rs)	0.7 Ω
3	Stator phase inductance, (Ls)	2.7e-3 *H*
4	Flux linkage	0.1194 *V. s*
5	Back EMF flat area	120°
6	Inertia, J	0.0027 kgm^2
7	Viscous damping, F	0.0004924 N.m. s
8	Pole pairs	4
9	No. of phases	3

command in MATLAB is written as P=augw (G, W2, W2, W3). For developing a H-infinity controller "K" for augmented plant "P", MATLAB function *"hinfsyn"* is used. The command in MATLAB is written as K=hinfsyn(P).

The weighting functions used in this simulation are found by using trial and error method. The value of the H-infinity norm (□) is found to be 0.0752 which is less than 1, which is admissible. The transfer function of the H-infinity controller gain "K" is found as shown in Equation 3.

$$K = \frac{(9.036s^2) + (3.322 \times 10^6 s) + (8.127 \times 10^8)}{s^3 + (3.677 \times 10^5 s^2) + (9 \times 10^7 s) + (9 \times 10^5)} \tag{3}$$

Simulation

The ratings and parameters of BLDC motor utilized in Simulink model are shown in Table 58.1 below:

The Simulink model for regulating speed of BLDC motor using H-infinity controller in MATTLAB/SIMULINK is shown in Figure 58.5. Blocks developed in Simulink are simulated in MATLAB R2021a version. The simulation type is discrete and the sample time is 50e-7. Simulation is performed for 10 seconds.

Results and discussion

The reference speed is determined first at 100 rpm, followed by 500 and 300 rpm at (0–3), (3–7), (7–10) seconds, respectively. The DC voltage is regulated as shown below in Figure 58.6. At first, the open loop speed control is performed and as shown in Figure 58.7, the actual speed does not track the reference speed. Actual speed is shown by yellow line and

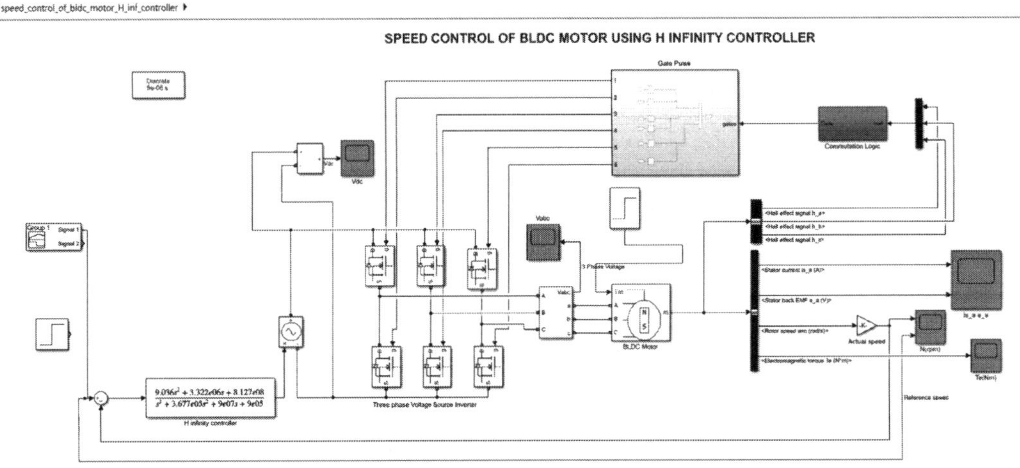

Figure 58.5 Simulink model for regulating speed of brushless DC motor

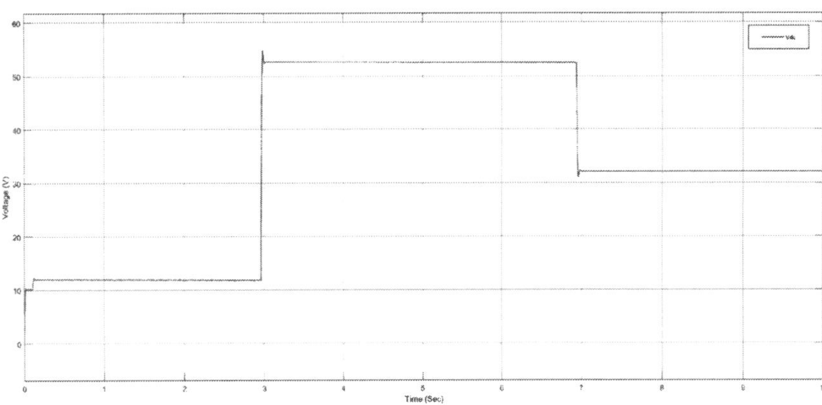

Figure 58.6 The regulated DC voltage

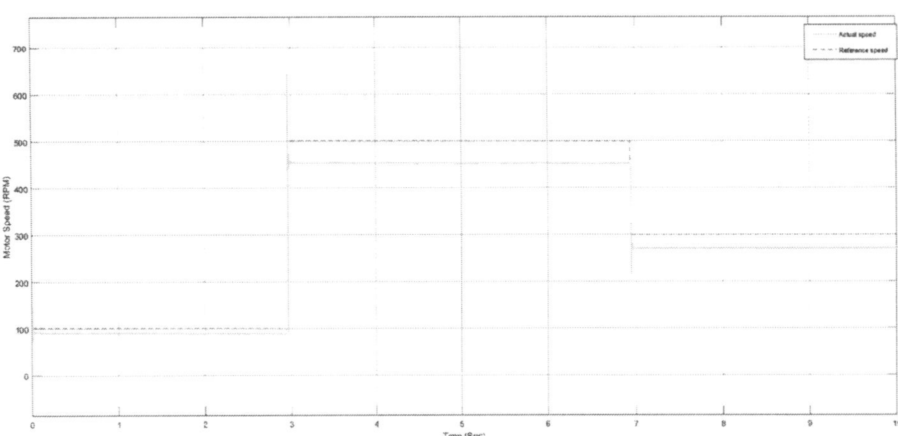

Figure 58.7 Open loop speed control of brushless DC motor

reference speed is shown by blue line. Moreover, the overshoot is found to be almost 20%.

With an H-infinity controller, closed loop speed control is carried out. The actual rpm after regulation is recorded and it is observed to be tracking the user defined reference rpm as shown in Figure 58.8. This proves the success of the proposed controller.

In order to check the controller's performance in external load disturbance, a mechanical load of 10 Nm at 5 seconds of simulation is being loaded while keeping the reference rpm constant at 500 rpm. The

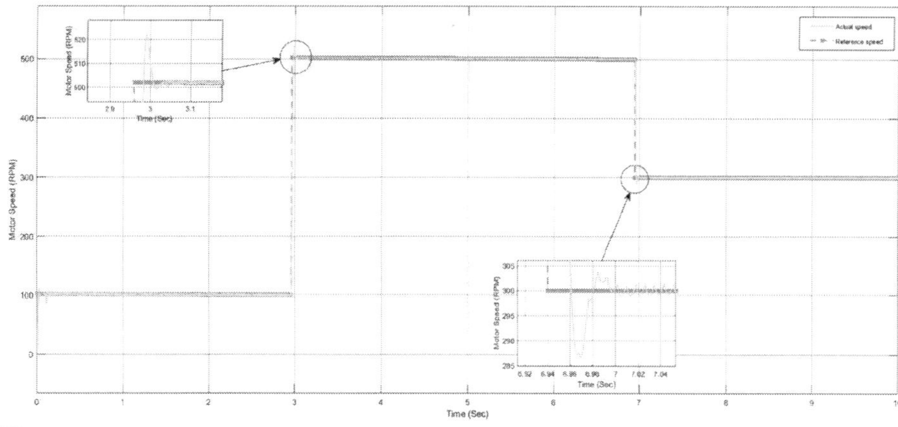

Figure 58.8 Actual speed tracks reference speed using H-infinity controller

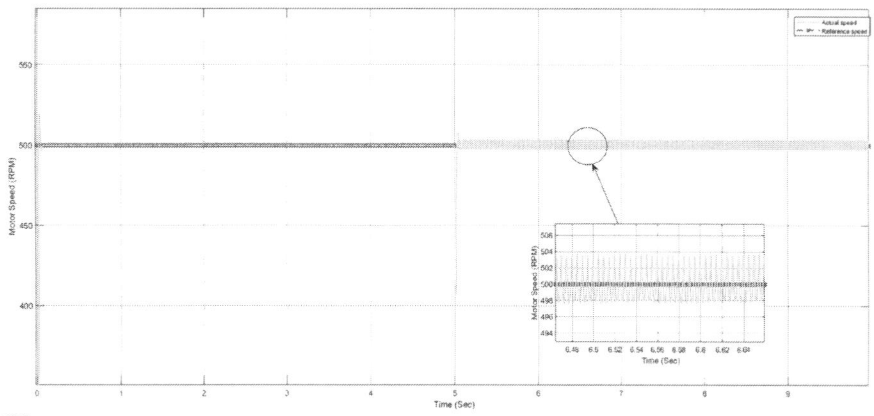

Figure 58.9 Speed tracking under varying load condition using H-infinity controller

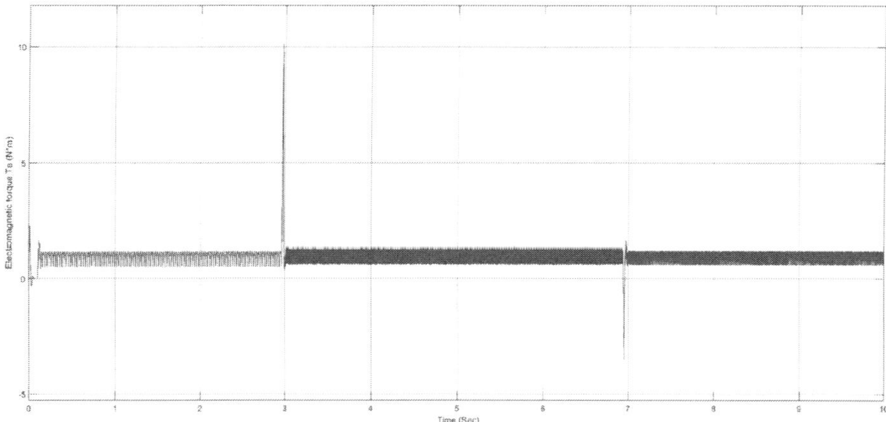

Figure 58.10 Electromagnetic torque

controller as shown in Figure 58.9 shows a commendable job in keeping the speed of the motor constant at 500 rpm.

The BLDC motor's electromagnetic torque, stator current and back emf are shown in Figure 58.10 and Figure 58.11, respectively.

In order to compare the proposed controller with a conventional controller, a PID controller is designed and is placed in the system replacing the H-infinity controller. PID controller parameters are $K_P = 0.01035, K_I = 12.777, K_D = 12.564$. While keeping the conditions same simulation is

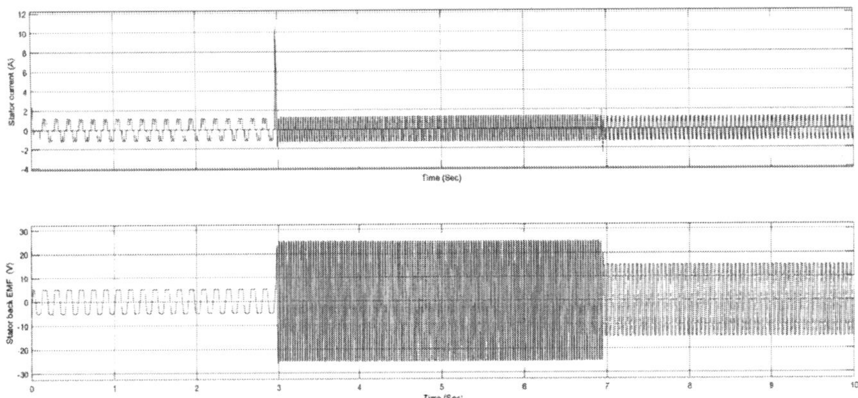

Figure 58.11 Stator current and Back emf

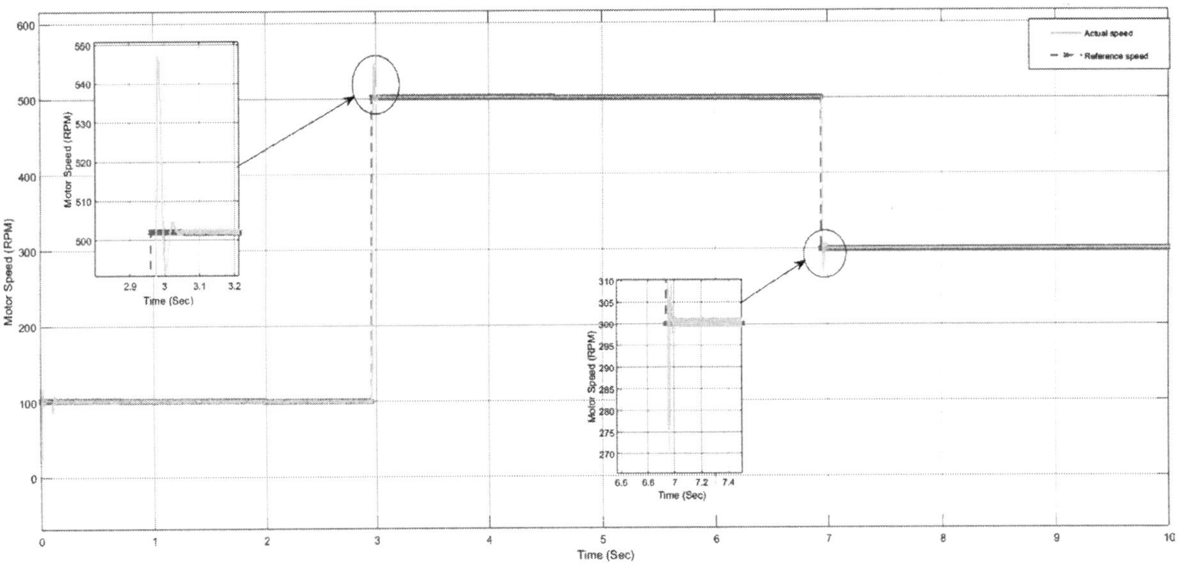

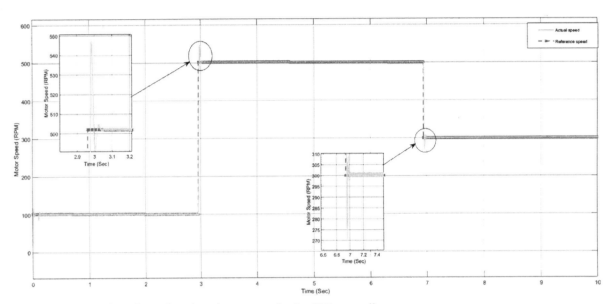

Figure 58.12 Actual speed tracks reference speed using PID controller

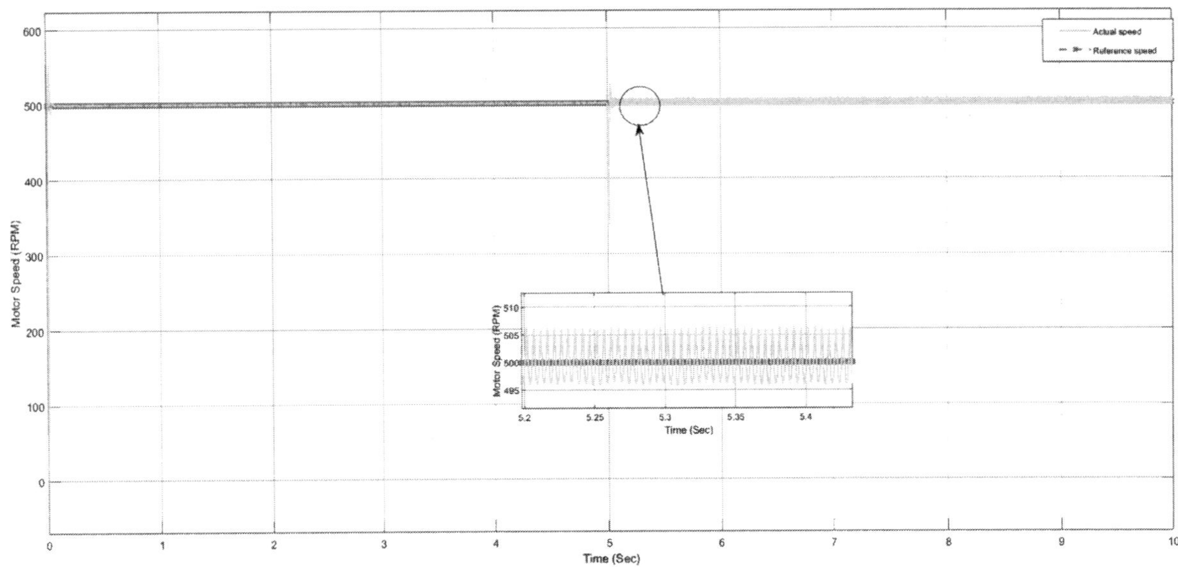

Figure 58.13 Speed tracking under varying load conditions using PID controller

Table 58.2 Comparison of performance parameters

Parameters	H-infinity controller	PID controller
Rise time (second)	2.9617	2.9614
Settling time (second)	6.9745	9.9984
Overshoot	73.9334	83.2642
Peak	54.6979	57.8469
Peak time (second)	2.9891	2.9826

performed and recorded. Figure 58.12 shows the tracking of the reference speed by the actual speed and Figure 58.13 shows the maintenance of constant speed at changing load, respectively while using a PID controller.

58.Performance metrics for the PID and H-infinity controllers are compared as shown in Table 58.2. The parameters of the controllers are obtained by using the "*stepinfo*" command in MATLAB.

Conclusions

Here, simulation is carried out for regulating speed of a brushless DC motor with the help of H-infinity controller in MATLAB/SIMULINK. System identification toolbox, available in MATLAB is used for obtaining the transfer function of the brushless DC motor. H-infinity synthesis is carried out and a controller of order third is obtained. To certain the effectiveness of H-infinity controller, simulation is carried out in the same model but with a PID controller and comparison is done between the two methods. While comparing both the controllers, it is found that the overshoot of the actual speed is more in the case of PID controller as compared to the H-infinity controller. The increase in overshoot is almost 12% approximately while using a PID controller. The settling time is also found to increase by almost 43% while using a PID controller instead of H-infinity controller. Similarly, when changing the mechanical load, at 5 s while keeping the speed constant at 500 rpm the H-infinity controller shows a better performance than a traditional PID controller. From the results of the simulation, it is seen that H-infinity controller has a better reference tracking and disturbance rejection than the PID controller. So, it can be concluded that the H-infinity controller is a better option than the traditional PID controller.

Acknowledgement

The authors gratefully acknowledge the staff and authority of Electrical Engineering department for their invaluable guidance and support in the research.

References

1. Walekar, V. R. and Murkute, S. V. (2018). Speed control of bldc motor using pi & fuzzy approach: A comparative study. *2018 Inter. Conf. Inform. Comm. Engg. Technol. (ICICET)*, 1–4. https://doi.org/10.1109/ICICET.2018.8533723.
2. Gupta, J. B. (2013). Theory and Performance of Electrical Machines. 15th ed. New Delhi: S.K. Kataria & Sons.
3. Manda, P. and Veeramalla, S. K. (2021). Brushless DC motor modeling and simulation in the MATLAB/SIMULINK software environment. *J. homepage*, 64(1–4), 27–33. http://iieta. org/journals/ama_b. https://doi.org/10.18280/ama_b.641-404.
4. Mahmud, Md., Motakabber, S. M. A., Zahirul Alam, A. H. M., and Nordin, A. N. (2020). Control BLDC motor speed using PID controller. *Inter. J. Adv. Comp. Sci. Appl.*, 11(3). https://doi.org/10.1109/CSPA55076.2022.9782030.
5. Hameed, H. S. (2018). Brushless DC motor controller design using MATLAB applications. *2018 1st Inter. Sci. Conf. Engg. Sci.-3rd Sci. Conf. Engg. Sci. (ISCES)*, 44–49. https://doi.org/10.1109/ISCES.2018.8340526.
6. Krishnan, T. V. D., Krishnan, C. M. C., and Panduranga Vittal, K. (2017). Design of robust H-infinity speed controller for high performance BLDC servo drive. *2017 Inter. Conf. Smart Grids Power Adv. Control Engg, (ICSPACE)*, 37–42. https://doi.org/10.1109/ICSPACE.2017.8343402.
7. Zhetpissov, Y., Kaibaldiyev, A., and Do, T. D. (2019). Robust H-infinity speed control of permanent magnet synchronous motor without load torque observer. *2019 IEEE Vehicle Power Propul. Conf. (VPPC)*, 1–4. https://doi.org/10.1109/VPPC46532.2019.8952432.
8. Dey, N., Mondal, U., and Mondal, D. (2016). Design of a H-infinity robust controller for a DC servo motor system. *2016 Inter. Conf. Intell. Control Power Instrum. (ICICPI)*, 27–31. https://doi.org/10.1109/ICICPI.2016.7859667.
9. Zhou, K. and Doyle, J. C. (1998). Essentials of robust control. Upper Saddle River, New Jersey: Prentice Hall.

59 Cost-effective retrofit electric vehicle solutions empowered by IoT dashboard integration

Kalyan Dusarlapudi[1,a], *K. Sudhakar*[2], *J. S. V. Gopala Krishna*[3], *Kandi Sai Teja*[4] *and Naredla Venkata Sai*[4]

[1]Department of Electrical and Electronics Engineering, Koneru Lakshmaiah Education Foundation, Vaddeswaram, Vijayawada, Andhra Pradesh, India

[2]Assistant Professor, Department of AI&DS, Lakireddy Bali Reddy College of Engineering, Mylavaram, Andhra Pradesh, India

[3]Associate Professor, Department of CSE, Sir CRR Engineering College, Eluru, Andhra Pradesh, India

[4]UG Student, Department of Electrical and Electronics Engineering, Koneru Lakshmaiah Education Foundation, Vaddeswaram, Vijayawada, Andhra Pradesh, India

Abstract

This paper presents a review of the current latest advancements in retrofit electric vehicle technology. It examines advantages and disadvantages of retrofitting existing vehicles with electric components, and the potential for such conversions to reduce emissions and increase fuel efficiency. The paper also considers the regulatory and economic implications of retrofitting and discusses the potential for retrofits to be an attractive alternative to purchasing a new electric vehicle. Results obtained explain how Retrofit model Electric vehicles are cost efficient way to produce two-wheelers. India holds the title of the world's largest manufacturer of two-wheeler and two-wheeler industry growth has been increased with new technologies in the market. Due to problems caused on the environment by fuel powered vehicles, auto industries turned towards alternative that is electric powered vehicles. This paper explains about different components of two-wheeler electric vehicles like battery, brushless direct contact (BLDC) motor, motor controller, DC-DC converter, charger, Internet of Things (IoT) dashboard, sensors like fingerprint sensor, voltage sensor, temperature sensor. This paper refers to how using IoT technology in electric vehicles makes vehicles even more efficient. This paper presents an IoT dashboard for electric vehicles that enables users to monitor their vehicle's performance and receive real-time status about the vehicle's parameters. The dashboard utilizes a variety of sensors to collect data of the vehicle's parameters, such as battery voltage, battery temperature, and vehicle position. This data is then presented in a visually appealing dashboard that allows users to monitor their vehicle's performance quickly and easily. Results also explains how an IoT dashboard with cloud interface provides user flexible way to access electric vehicle parameters.

Keywords: Retrofit model, electric vehicle, IoT-Internet of Things, cloud interface, IoT dashboard, fingerprint unlock, voltage sensor, current sensor, temperature sensor

Introduction

Electric vehicles were first introduced at the end of 1800s and from then their popularity has been increasing with use of emerging technologies and reduction in costs (especially the batteries). Due to problems caused by the internal combustion engine vehicles on the environment, there needed an alternative to power the vehicle, then electric vehicles came into light. The motivation for this work is to explore cost-effective solutions for retrofitting existing vehicles into electric vehicles. Retrofitting provides a sustainable alternative to conventional fuel powered vehicles, contributing to environmental friendliness by reducing emissions and promoting sustainability. The paper aims to examine the advantages and challenges of retrofit electric vehicle technology, with a focus on the use of IoT dashboard integration to enhance the efficiency and performance of these vehicles. Particularly from 1960 to present, many attempts were made to produce efficient practical electric vehicles and will continue to make attempts. Electric vehicles basically consist of electric motor, motor controller, battery, (direct current) DC-DC converter, charger. Electric motors used are DC motors or (brushless direct current) BLDC motors because DC power source is directly used from the batteries. Batteries used are lead acid type, lithium-ion type and many more [1, 2]. Mostly lithium-ion batteries are because of their efficiency and less maintenance. DC-DC converter powers the low-voltage circuit like headlight, horn. Electric vehicles are turning into promising ways to sustainable management of fossil fuels, improving

[a]kalyandusarlapudi@gmail.com

DOI: 10.1201/9781003540199-59

environmental conditions like air quality. Electric vehicles do not emit greenhouse gases and reduce global warming.

An accelerometer is a sensor designed to gauge the acceleration forces affecting an object, aiding in determining its spatial orientation and tracking its motion. Acceleration, a vector quantity, denotes the rate of change of an object's velocity, with velocity being the object's displacement divided by the change in time. When integrated into an electric vehicle, an accelerometer sensor can detect the vehicle's orientation, distinguishing between its upright position and a potential fall or tilt [4].

Internet of Things (IoT) technology is interconnection of devices to communicate that is to send and receive data between the devices through internet. IoT enables IoT devices to control other IoT enabled devices. IoT devices are interfaced to cloud to store and perform analytics in the cloud platform [3]. So, the data can be accessed from anywhere from any device as IoT and cloud platforms have wide network connectivity. IoT used in electric vehicles can provide user the flexibility of accessing vehicle parameters like battery temperature, state of charge (SoC), vehicle position in real-time.

Security is one of the main features of an electric vehicle. Secured vehicles are less prone to theft. So, to increase the security of the vehicles, current technologies can be used like unlocking the vehicle using owner's biometrics. Biometrics registration is not limited to only the vehicle owner, but also other persons can also be registered. The owner has the flexibility to allow others to use the vehicle. Registering different people's biometrics helps different people to unlock the vehicle.

Literature review

Numerous papers were published exploring the work in this field of electric vehicles. There were many papers exploring the work on the applications of IoT in electric vehicles [3]. IoT is used in electric vehicles for easier maintenance and management of charging the vehicle which provides possibility of charging at charging stations by providing real-time information of charging stations. For destination and route information, estimation of travel distance, battery current state of charge and for switching between electric and internal combustion engine, fuel in hybrid electric vehicles. Accelerometer is used for estimation of requirements for hybrid electric vehicles [5].

But not many of the works are concentrated as follows. An IoT dashboard intended to provide the user vehicle parameters as well as upload the parameters data into the cloud [6]. The data in cloud is updated in real-time. So, users can access the information in real-time anywhere. Accelerometer provides the information of the vehicle position and is uploaded to the cloud. Fingerprint sensor increases the security of the vehicle by only allowing the registered users to unlock the vehicle problem [17].

Methodology

Various types of existing electric vehicles and technologies used in them have been studied. Additionally, the extension of the use of IoT in electric vehicles has been identified. Using IoT to provide real-time monitoring of vehicle parameters to the users [16]. A detailed overview of an electric vehicle and its parts was studied and using this a retrofit model was designed and modified. Electric motor was selected based on the rating requirements for modeling the electric vehicle. Battery pack was designed by calculating the power requirement for the motor. Sensors were selected based on the required characteristics. The controller board used was chosen with Wi-Fi compatibility. Cloud platform was chosen with less data uploading time. This paper is organized into different sections as follows: Section 2 describes the literature review. Section 3 describes the methodology. Section 4 describes the block diagram of electric vehicle. Section 5 describes flow chat of IoT dashboard. Section 6 describes the modeling and experimental setup. Section 7 describes the cost analysis of retrofit model electric vehicle. In section 8 result and analysis are described, conclusion in section 9 followed by future scope in section 10.

3.1 Block diagram

Figure 59.1 describes the parts and components of an electric vehicle in detail. An electric vehicle consists of different blocks where each block has its own functionality even though each block is connected to the other for their operation.

Fingerprint sensor provides better security, and it can be used in an electric vehicle to unlock the vehicle. Fingerprint sensor identifies the biometrics of a registered person. This data can be used to unlock the vehicle. A number of people can register their biometrics and by verifying their biometrics vehicle can be unlocked.

IoT dashboard (see Figure 59.2) is same as the normal dashboard of a vehicle which displays the voltage, current, battery temperature and vehicle position. The special feature of IoT dashboard is to display all the parameters present on physical dashboard on a real-time dashboard with the help of cloud. Data is uploaded and updated every second on the cloud platform. From the cloud, data can

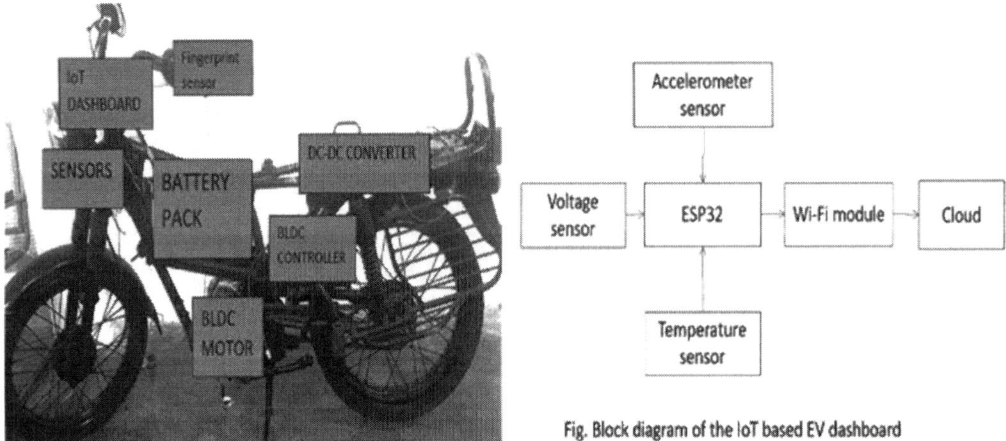

Figure 59.1 Electric vehicle block diagram

Figure 59.2 IoT dashboard

be shared to a mobile application and authorized people can access the application [14].

By accessing the Google firebase real-time database, Real-time monitoring is possible. IoT technology allows for real-time monitoring in an electric vehicle by integrating sensors and communication technologies throughout the vehicle. These sensors collect data on various parameters such as battery level, charging status, which is then transmitted to the dashboard. Figure 59.3 shows the real-time database where the parameters are updated in real time. The dashboard provides the driver with a complete overview of the vehicle's status and can be accessed remotely, giving the driver access to valuable information such as the charging status of the vehicle [4].

Real-time monitoring plays a critical role in electric vehicle ownership, empowering drivers to make informed decisions and enhance the efficiency of their driving experience. For example, if the driver sees that the battery is running low, they can quickly find a charging station and avoid running out of power while on the road. Additionally, real-time monitoring helps to increase the overall safety of electric vehicle ownership by alerting the driver to any potential issues, such as a high battery temperature that could affect the vehicle's performance [7].

The use of IoT in electric vehicles also helps to improve the overall user experience by providing drivers with access to valuable information about their vehicle, such as its status and driving behavior. This information can be used to optimize driving behavior, reduce costs, and improve the overall efficiency of electric vehicle ownership.

In Figure 59.2, the parameters that are displayed can be seen. Sensors are implemented to measure the battery voltage, battery temperature, and position of the vehicle. For that voltage sensor, current

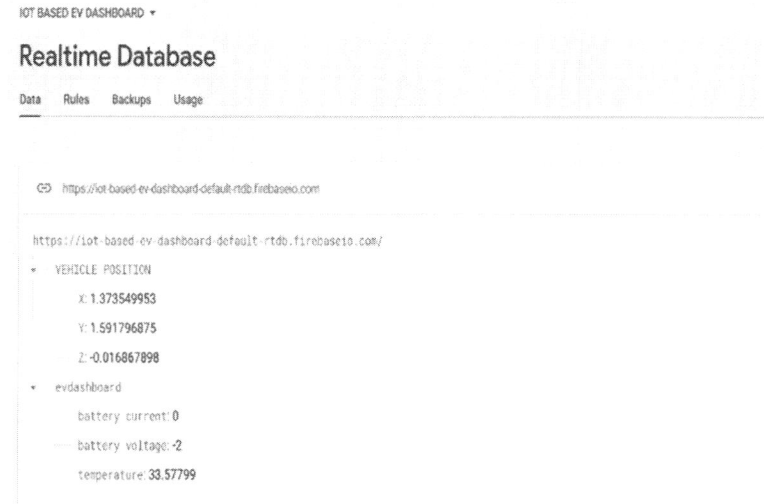

Figure 59.3 Data uploading in cloud real-time database

sensor, thermistor, accelerometer is used respectively for measuring the above parameters.

Accelerometer X, Y, Z provides the information of the vehicles position. This can be helpful for detecting vehicles position in a parking area or even can detect whether vehicle had an accident while travelling.

Battery pack is the combination cells connected in series or parallel or combination of series and parallel connection to obtain 48 volts as output voltage. Lead acid batteries are used for the battery pack. Four lead acid batteries each of 12 volts are connected in series to obtain 48 volts output. These are connected in series with BLDC controller to supply power to the (brushless direct current) BLDC motor of 48 volts and DC-DC converter [11].

BLDC motor uses direct current and do not have brushes. This motor is used for high torque and high efficiency. The special feature of the BLDC motor is its commutation. The commutation is done electronically so there will not be any brushes and brush losses will be zero. Due to this efficiency of BLDC motor is very high. The vehicle requires high torque to get on to the slopes so for this reason BLDC motor is used since it produces high torque. The BLDC motor can produce variable speeds and variable torques. The BLDC motor can be controlled by the help of BLDC controller.

BLDC (Brushless direct current) controller is used for controlling the BLDC motor. The BLDC controller has 3 half effect sensors for 3-phases of the BLDC motor which determines the position of the rotor. The stator windings of the motor are powered through the transistor switches in the controller depending on how much rotation is required. The BLDC controller will be controlling the rotation of the motor in clockwise and anti-clockwise direction. It can also provide regenerative braking. Controller controls the speed of the motor by changing the voltage supply and also produces the torque required to drive the electric vehicle.

The retrofit model is a process of converting a traditional petroleum-powered two-wheeled vehicle into an electric vehicle. This involves taking an existing two-wheeled vehicle that is damaged, particularly its engine, and modifying it to make it an electric vehicle. The engine is removed and replaced with a brushless DC motor, which is an electric motor that runs on direct current and is controlled by an electronic controller. The fuel tank is also replaced with a battery hub that stores the electricity needed to power the vehicle. The brake system is connected to the brushless DC motor controller to provide control and stopping power to the vehicle.

By converting a conventional petroleum two-wheeled vehicle into an electric vehicle, the retrofit model provides an environmentally friendly and cost-effective solution for those who still want to use their existing vehicle but with the benefits of electric power. The conversion process is relatively simple, and the resulting electric vehicle is easier to maintain and operate than a traditional gasoline-powered vehicle. It also provides a greener alternative to petroleum-powered vehicles, reducing emissions and improving air quality [9].

The conversion of a conventional petroleum two-wheeled vehicle into an electric vehicle involves several steps. The first step is to remove the engine from the vehicle. Then, an iron plate is attached to the bottom of the vehicle through welding. This iron plate serves as the base for the brushless DC motor, which is then attached to the iron plate.

Next, the fuel tank of the vehicle is dismantled and replaced with a battery hub in the form of a hollow cuboid. This battery hub is attached to the vehicle and serves as the storage for the batteries. The batteries are placed inside the battery hub and connected to the BLDC motor controller. This allows the motor to be powered by the batteries and controlled by the controller.

Finally, any unused parts of the original petroleum-powered vehicle are removed, such as the silencer. This helps in reducing the weight of the vehicle and also improves the overall efficiency.

3.2 Flowchart

Figure 59.4 describes the flow of the proposed model. When the vehicle starts, sensors and controller board are powered up. Sensors starts reading the vehicle parameters. ESP32 connects to the internet via portable modem or personal hotspot. Voltage sensor reads the battery voltage, Thermistor reads the temperature of the battery pack, accelerometer reads the X, Y, Z values. Each sensor data is sent to the ESP32.ESP32 connects to the cloud platform, here Google Firebase. The data is uploaded to the Google firebase real-time database. Each time data sent to ESP32, the data is updated in the real-time database. The data in real-time database can be monitored by login into the Google firebase account and even a mobile application can be developed to monitor the data.

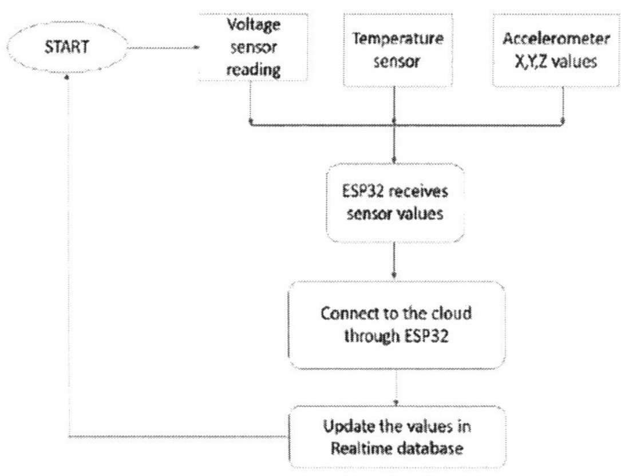

Figure 59.4 IoT dashboard flow chart

3.3 Experimental setup and modeling

The electric vehicle is a retrofit based model so the setup mainly consists of battery pack, electric motor, motor controller, battery charger, DC-DC converter for low voltage circuit. IoT dashboard setup consists of sensors, microcontroller board and a display.

The voltage sensor measures the DC voltage of battery in range of 50 volts. The current sensor measures the DC current through batteries by indirect sensing method. Thermistor is used to measure the temperature of the battery. to avoid overheating of battery. The accelerator will determine the positioning of the vehicle, i.e., in X, Y, Z axes to determine the posture of the vehicle. Speed sensor will determine the speed of motor in the form of rpm. DC-DC converter is employed to convert the voltage from 48 volts to 12 Volts powering the lights, horns, and indicators of the vehicle.

The IoT dashboard main feature is to display all the monitoring parameters on the display as well as upload them into cloud. The parameters can also be seen in a smartphone with the help of cloud platforms and mobile applications. Cloud will allow the data to be accessible in the mobile application to the authorized users.

Table 59.1 Pros and cons of retrofit electric vehicle

S. No.	Pros of retro fitting	Cons of retro fitting
1	Retrofit electric vehicle conversions are typically much cheaper than purchasing a new vehicle. This makes them a great option for those who want to get into electric vehicles but don't have the budget for a new model	The cost of maintaining an electric vehicle is typically higher than a fuel-based vehicle due to the complexity of the electric components
2	Customization is being done as per the requirements	Complexity increases in conversion because it requires additional calculations to install components
3	Retrofit electric vehicles helps reduce scraps wastage	Repair works may occur suddenly and need specialized operator to repair the vehicle
4	Retrofit electric vehicle conversions help reduce air pollution and greenhouse gas emissions	Some components of the conversion may be incompatible with the existing vehicle, which could lead to safety issues

Motor calculation

1) *Motor specifications*
    - Voltage (V) = 48 v
    - Power (P) = 800 w
2) *Power equation*
    - Power (P) = Current (I) x Voltage (V)
    - I = P ÷ V = 800 ÷ 48 = 16.6666667 A.
3) *Torque of the motor (T)*
    - T = (P×60) ÷ (2×π×N)
    - N = speed of BLDC motor in rpm
    - T = (800×60) ÷ (2×3.14×450)
    - T = 16.98 Nm
4) *Torque in kw*
    - P (kw) = 0.105 x T (Nm) x N (rpm) / 1000
    - P (kw) = 0.105 x 16.98 x 450 / 1000
    - P (kw) = 0.80 kw

The motor used for the vehicle is a BLDC - brushless DC motor of 48 volts and power rating of 800 w. This motor provides torque of 16.98 Nm i.e., 0.80 kw.

3.5 Battery calculation

From the motor calculation,

- Wattage W = 800 Watts
- Voltage V = 48 Volts

So, watt.hr = 800 x 1 = 800 w.hr

It is recommended to utilize 80% of the full battery capacity, leaving 20% remaining, for optimal battery performance and longevity. So,

- 800 w.hr x 1.20 = 960 w.hr
- Considering the air resistance and other forces, the watt.hr is 1300w.hr

Therefore, current (Ah) = 1300 w.hr / 48v = 28Ah

- *Battery specifications*
    - Voltage rating = 48 V
    - Wattage of battery = 1300 w.hr

Battery pack is 48 volts. It consists of four 12 volts, 28 Ah batteries connected in series. So, the total voltage is 48 volts and Ampere-hour rating is same i.e., 28 Ah as in series current rating remains same.

3.6 Battery charger selection

- Considering, to charge the battery in 5 hours.
- Based on the above consideration,

Charger wattage = 1300 w hours /5 hours = 260 Watts

- Charger current rating = 260 / 48 volts

Current rating of charger = 5.41 A = 6 A

As per the above calculations, to charge the 48 volts, 28 Ah battery in 5 hours the required charger rating is 48 volts, 6 A charger.

3.7 Cost analysis

- Cost of the IoT dashboard
    - Including LCD display, Sensors, Controller board = ₹ 1,500.00
- Cost of the battery pack
    - Each battery of 12 volts, 28 Ah = ₹ 3,000
    - Total 4 batteries = 3,000 x 4 = ₹ 12,000.00
- Cost of BLDC motor & Controller = ₹ 13,000.00
- Cost of chassis = ₹ 10,000.00
- Total cost of retro fit electric two-wheeler = ₹ 37,000.00
- cost of cheapest petroleum two-wheeler in the market = ₹ 59,104.00
- Cost of electric two-wheeler in the market = ₹ 63,660.00

Compared to both petroleum and electric two wheelers, retro fit electric two-wheeler is cheaper.

4. Results and analysis

The retrofit electric vehicle, equipped with an IoT dashboard, was successfully developed and tested. The analysis includes the evaluation of the vehicle's performance, the effectiveness of the IoT dashboard, and the cost-effectiveness of the retrofit model. The vehicle's performance during testing was satisfactory, with a maximum speed of 30 kmph and a range of 30–40 km on a single full charge. The BLDC motor, rated at 48 volts and 800 watts, provided a torque of 16.98 Nm, ensuring smooth acceleration and deceleration. The battery pack, consisting of four 12 volts, 28 Ah lead-acid batteries connected in series, delivered a total output voltage of 48 volts and charged fully in 5 hours using 48 volts, 6 A charger. The IoT dashboard, integrated with various sensors, provided real-time monitoring of critical vehicle parameters. The sensors included a voltage sensor, current sensor, thermistor, accelerometer, and speed sensor. The IoT dashboard displayed essential parameters such as battery voltage, battery temperature, and vehicle position on the vehicle's physical dashboard. Additionally, the data was uploaded in real-time to the Google Firebase real-time database, allowing users to access the data from anywhere using the cloud platform or a mobile application. As outlined in Table 59.1, the pros and cons of retrofitting electric vehicles offer valuable insights into the advantages and challenges associated with such conversions. The cost analysis showed that the retrofit electric vehicle was

Figure 59.5 Final embedded electric vehicle model

a cost-effective alternative compared to both conventional petroleum two-wheelers and new electric two-wheelers available in the market.

As of September 20, 2021, a total of 2,930 vehicles that are 15 years or older have been dismantled. This marks a significant rise of 91% compared to the same period last year. From the cost analysis section as detailed, retrofit electric vehicles are cost effective way to produce electric vehicles. Analysis also explains that using IoT dashboard provides the user flexibility to access the vehicle parameter from the comfort of their home [5].

The proposed system described in the paper can indeed be implemented in practical electric vehicle systems. The system serves as an extension to the existing electric vehicle setup, enhancing its monitoring capabilities and ultimately providing users with a better riding experience.

By integrating IoT technology with an electric vehicle, the proposed module allows for real-time monitoring of various parameters such as battery voltage, temperature, and vehicle position. This enables users to stay informed about their vehicle's performance and make informed decisions while riding.

Furthermore, the system can be designed to be compact and integrated onto a printed circuit board (PCB), which can then be concealed with 3D print to ensure precision and seamless integration into the vehicle's design. This ensures that the monitoring system is robust and reliable in various operating conditions (Figure 59.5).

5. Conclusions

Electric vehicles are the future of the automotive industry. They offer several advantages for both environmental and economic reasons. They are efficient, quiet, and cleaner than traditional vehicles and can reduce air pollution and greenhouse gas emissions. Additionally, electric vehicles can save money and time for consumers, as they require less maintenance and fuel costs. Fuel vehicles usage results in more air pollution. To control this, electric vehicles are used. Electric vehicles use electric power from the batteries to run the electric motor and thus the electric vehicle. So, electric vehicles emits zero greenhouse gases and thus does not participate in air pollution and global warming. Retrofit model electric vehicle saves money spent on the vehicle because the chassis used is of a fuel engine vehicle.

6. Future scope

The potential for electric vehicles is immense. As technology advances and the cost of batteries and electric motors decreases, the viability of electric vehicles increases. Further research on IoT dashboard can be used to track the location of the vehicle in real time, providing an enhanced level of security. This can be used to prevent theft and provide an early warning of any potential issues. Additionally, Speed sensor can also be interfaced to cloud to provide real-time speed of the vehicle. Finally, the dashboard can be used to monitor the performance of the charging station, allowing drivers to ensure that their vehicle is properly charging.

Acknowledgement

We would like to acknowledge the Electrical and Electronics Engineering Department of Koneru Lakshmaiah University (Deemed to be University) for the support of this project. We thank our team for their challenging work and dedication to developing a comprehensive understanding of electric vehicles.

References

1. Bade, R., Ramana, B. V., and Dusarlapudi, K. (2019). A six servo six DOF position controlled platform for sorting application. *Inter. J. Engg. Adv. Technol.*, 8(4), 1215–1219.
2. Praveen, V. S., Sankar, M. S., Jyothi, B., and Dusarlapudi, K. (2022). Solar-powered in situ IoT monitoring for EV battery charging mechanism. *Lec. Notes Mec. Engg.*, 51–60.
3. Dusarlapudi, K., Raju, K. N., Charishma, N., and Vandana, A. (2022). The modelling and analysis of novel 2-axis MEMS accelerometer for energy harvesting applications. *IEEE Inter. Conf. Distrib. Comput. Elec. Circuits Elec., ICDCECE*, 2022.
4. Dusarlapudi, K. and Narasimha Raju, K. (2020). Embedded prototype of 3 DOF parallel manipulator for endoscope application using 3 axis MEMS accelerometer. *J. Adv. Res. Dynam. Control Sys.*, 12(2), 1225–1235.

5. Kataboina, S. K., Reddy, M. J., and Dusarlapudi, K. (2020). Multi-functional electrical vehicle for agricultural applications. *J. Adv. Res. Dynam. Control Sys.*, 12(2),101–110.
6. Chandrika, K., V. S. P., Annepu, C. R., Dusarlapudi, K., Sreelatha, E., and Tiruvuri, C. S. (2020). Smart approach of harvesting rainwater and monitoring using IoT. *J. Adv. Res. Dynam. Control Sys.*, 12(2), 101–110.
7. Chundhu, S., Dusarlapudi, K., Narayanam, V. S. K., and Narasimha Raju, K. (2022). Novel programmable solar based SIMO converter for SMPS applications with IoT infrastructure. *Lec. Notes Elec. Engg.*, 828, 1011–1023.
8. Cannon, B. L., Hoburg, J. F., Stancil, D. D., et al. (2009). Magnetic resonant coupling as a potential means for wireless power transfer to multiple small receivers. *IEEE Trans. Power Elec.*, 24(7), 1819–1825.
9. Yugendra Rao, K. N. (2015). Dynamic modeling and calculation of self and mutual inductance between a pair of coils for wireless power transfer applications using ANSYS Maxwell. *Inter. Adv. Res. J. Sci., Engg. Technol.*, 2(10),page 7-9.
10. Gurjar,D.S., Nguyen, Ha H., and Tuan, H.D.. Wireless information and power transfer for IoT applications in overlay cognitive radio networks, *IEEE Internet of Things Journal, vol. 6, no. 2, pp. 3257-3270, April 2019, DOI 10.1109/JIOT.2018.2882207.*
11. Takumi, N. (2016). Design procedure for wireless power transfer system with inductive coupling-coil optimizations using PSO.2016 IEEE International Symposium on Circuits and Systems (ISCAS) (2016): pp 646-649.
12. Young-Sik, S. (2012). Wireless power transfer by inductive coupling for implantable batteryless stimulators.2012 IEEE/MTT-S International Microwave Symposium Digest, Montreal, QC, Canada, 2012, pp. 1-3, doi: 10.1109/MWSYM.2012.6259403.
13. Ricardo, M. (2013). Modeling inductive coupling for wireless power transfer to integrated circuits.2013 IEEE Wireless Power Transfer (WPT), Perugia, Italy, 2013, pp. 198-201, doi: 10.1109/WPT.2013.6556917.
14. Narayana, M. V., Dusarlapudi, K., Uday Kiran, K., and Sakthi Kumar, B. (2017). IoT based real time neonate monitoring system using Arduino. *J. Adv. Res. Dynam. Control Sys.*, 9(Special issue 14), 1764–1772.
15. Dusarlapudi, K. and Narasimha Raju, K. (2020). Embedded prototype of 3 DOF parallel manipulator for endoscope application using 3 axis MEMS accelerometer. *J. Adv. Res. Dynam. Control Sys.*, 2(2), 1225–1235.
16. Debata, S., Mantoliya, R., Sahithi, V., and Kolluru, V. R. (2018). Implementation of IoT based smart street light intensity control system using IR and LDR sensors. *Inter. J. Engg. Technol. (UAE)*, 7, 316–319.
17. Geetha Pratyusha, M., Misra, Y., and Anil Kumar, M. (2018). IoT based reconfigurable smart city architecture. *Inter. J. Engg. Technol. (UAE)*, 7(2), 175–178.

60 House price prediction using ensembled machine learning model

Jeevan Mallik[1,a], Mayuri Kundu[1,b], Binduswetha Pasuluri[2,c] and Ashwinkumar U. M.[1,d]

[1]School of Computer Science and Engineering, REVA University, Bengaluru, India

[2]Department of Electronics and Communication Engineering, Vardhaman College of Engineering, Hyderabad, India

Abstract

This paper compares the efficacy of two well-known methods in machine learning (ML), decision tree, and linear regression, to forecast property prices. This study's dataset covers numerous residential property parameters such as the price, floors and square area, among others. The algorithms performance are assessed through frequently used measures such as R^2 score and mean scored error (MSE). The results reveal that both decision tree and linear regression are successful at forecasting housing prices, with decision tree regression marginally outperforming linear regression in accuracy. Overall, this demonstration research demonstrates the potential of ML algorithms for house pricing and provides insights into their comparative performance in this sectors.

Keywords: House price prediction, machine learning, linear regression, decision tree regression

Introduction

Accurately predicting home prices is a difficulty that the real estate business faces simply due to a variety of economic factors that might create swings in the housing market. To address this, we conducted research that employed two machine learning (ML) algorithms: decision tree and linear regression. We collected data from various sources and applied feature selection techniques to decide the most important variables in predicting house prices. Our findings show that both algorithms are effective in predicting house prices, with decision tree regression performing slightly better in terms of accuracy [1]. To analyze the performance of the models, we utilized commonly used measures such as R^2 score and mean squared error (MSE). Furthermore, we discovered that the number of bedrooms and square area are the most important determinants in determining property values [2].

Related work

Shelter is a basic human need, and owning a house is a significant financial investment. Unfortunately, many people make mistakes when dealing with properties. Agreeing to a deal without understanding the actual value of a property can led to financial consequences. Our goal is to create a prototype that can benefit the real estate industry. One powerful tool we use to analyze data is ML. ML models have been developed across various domains due to their flexibility, including for predicting housing prices. Past research has implemented ML algorithms on housing datasets to make predictions of several types. The research purpose is to lead the development of a feasible housing price projection model utilizing a ML approach [3].

Machine learning

This technology is a field of research concerned with the development of algorithms and mathematical analysis that computer systems may employ to do activities that do not require explicit programming. The main goal of ML is to ease machines in learning from data. The ML method entails developing a machine learning approach on a big data set in order to detect patterns and correlations in the data [4]. The ML model then applies these patterns to new, unseen data to make accurate predictions or classifications. The goal of machine learning is to create intelligent machines that can learn from data, adapt to new situations, and make decisions or predictions with minimal human intervention. Researchers have conducted many studies on how to make machines learn without explicit programming [5].

Supervised learning in machine learning

Supervised ML involves feeding labeled data into a ML model. The model is trained using input and output data with known values, allowing it to predict future outputs accordingly [6].

[a]jeevanmallik47@gmail.com, [b]kundu.mayuri@gmail.com, [c]bswetha.pasuluri@gmail.com, [d]ashwinkumar.um@reva.edu.in

DOI: 10.1201/9781003540199-60

Unsupervised learning in ML
Unsupervised learning is a ML method that does not require explicit direction for the model. This approach mainly deals with unlabeled data. Some examples of unsupervised learning algorithms include anomaly detection, clustering, and neural networks [7].

Prediction in ML
Prediction is the result of using an algorithm that has been trained on past data and then applied to new data to estimate the probability of a specific outcome. We can generate very accurate predictions based on previous data using machine learning algorithms. ML techniques for prediction include random forest, regression models and many more [8].

Proposed system

The proposed system employs a new set of parameters and a different technique to predict land prices and compensation for property settlement. Mathematical relationships expressed with precise numbers can offer additional clarity and understanding of many aspects of daily life. In this case, decision tree and linear regression techniques are used to describe the connection between independent and dependent predictor variables [9].

The regression model that we are using is a popular approach for modeling the dependent and independent variables' linear relationship. It is a simple yet effective technique applicable in many fields. By employing linear regression, the proposed system can predict future house prices, which can help programmers determine the price of a house, and customers decide the optimal time to purchase a property [10].

The decision tree method, on the other hand, is a structured representation that recursively splits the data depending on the most essential attributes. This approach is often used to address classification and regression issues in ML and data mining. Utilizing the decision tree algorithm, the proposed system can provide an accurate and efficient prediction for land prices and compensation for property settlement [11] (Figure 60.1).

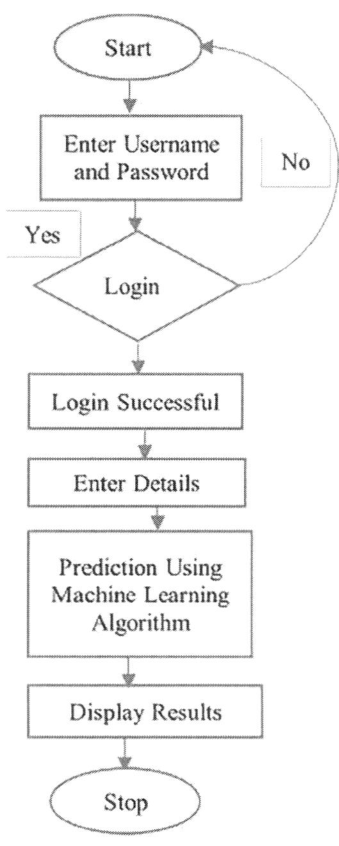

Figure 60.1 System – block diagram

Linear regression model
This model is a popular supervised learning in ML technique that involves estimating the value of a variable (dependent) (Y) based on a given variable (independent) (X). The goal of this model is to establish a linear connection between two variables and then use that relationship to predict Y given new values of X. The following equation may be used to express the correlation between different variables:

$$Y = m^*X + b$$

- Dependent variable - Y.
- Independent variable - X.
- "m" approximates the slope.
- "b" denotes the estimated intercept (Figure 60.2).

To implement linear regression, the initial stage is to preprocess the data by scaling it using standard scaler from the scikit-learn library. The model is then fitted to the training dataset and predictions are made on the test data. The model's performance is assessed using attributes such as R^2 score, MSE, root mean squared error (RMSE).

Decision tree regression
A decision tree is a flexible tool that may be used for categorization as well as prediction. The structure of a decision tree resembles a tree, with internal nodes representing tests on specific attributes and branches indicating the test outcomes. Considering

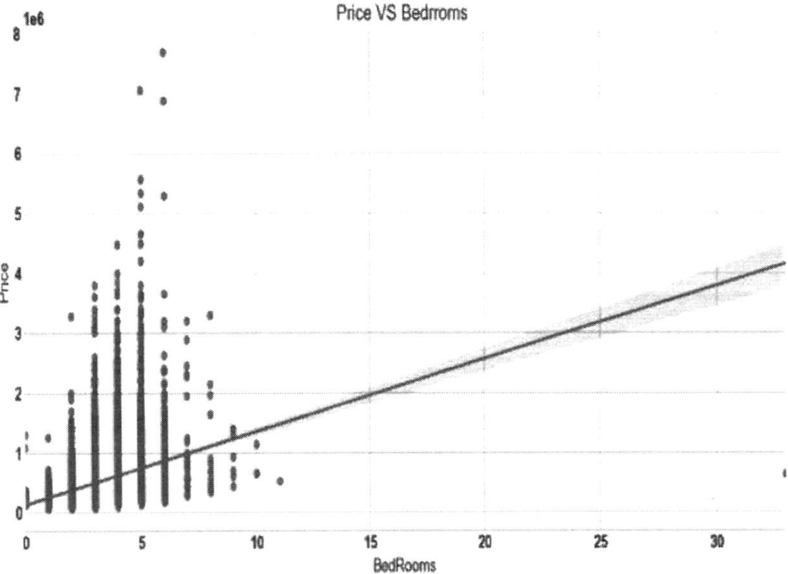

Figure 60.2 Linear regression

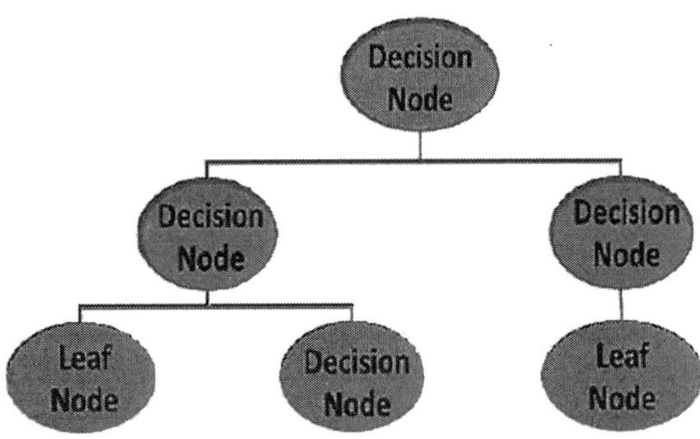

Figure 60.3 Decision tree regression

the tree – like structure from the roots to the leaf nodes, new instances may be simply categorized after establishing the decision tree. One of the benefits of using decision trees for classification is that it does not require much computation. Moreover, decision trees can handle both continuous and categorical types of attributes (Figure 60.3).

To apply decision tree regression, the initial step is to preprocess the data by scaling it using standard scaler from the scikit-learn library. After fitting the model to the training dataset, predictions are generated on the test data. The model's performance is assessed using metrics such as MSE, RMSE and R^2 score.

Dataset

The King's country dataset is well-known in the machine learning community, and it is frequently used to create forecast models for property prices. The dataset consists of 21,613 samples, each containing 21 features such as floors, bedrooms, condition, etc. The target variable in this dataset is house's price.

To evaluate the performance of a predictive model constructed using a dataset, it is usual to divide data in to the 70/30 training set and test them. The model is being trained using training data and its performance is evaluated using testing data. This strategy helps to prevent the model from overfitting to the training data and ensures that it can effectively generalize to new, previously unseen data.

Result and analysis

In this part of experiment, comparison of the accuracy levels of the linear regression and decision tree regression techniques for the house price prediction scenario and showcase the best solution.

We construct a scikit-learn for evaluation purposes, which allows us to compare the models using the following information (Figure 60.4).

As demonstrated in Table 60.1, house prices can be predicted with reasonable accuracy using both models. Finally, we compare both linear regression (Figure 60.5) and decision tree regression (Figure 60.6) in Table 60.160.. The best model performance is decision tree regression with 0.75 (75%) R-squared (R^2 score) value and minimum RMSE values (Figure 60.7).

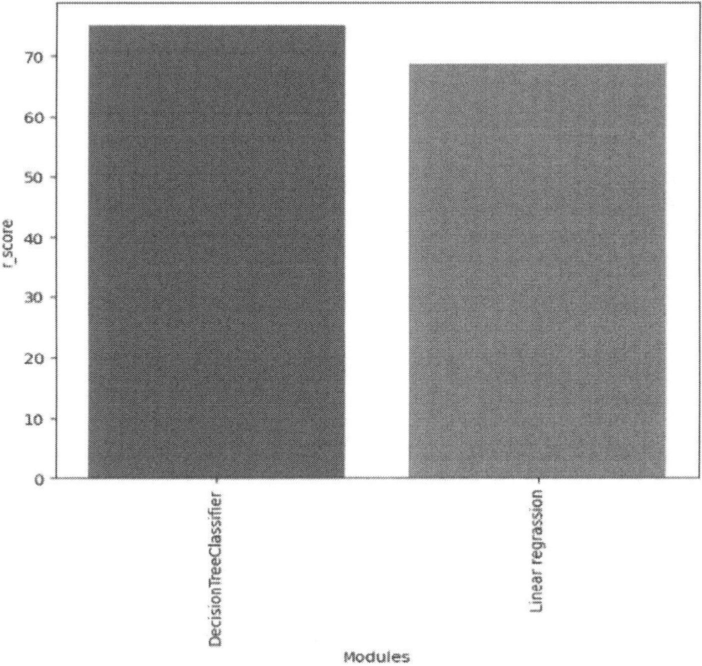

Figure 60.4 R2_score (accuracy)

Table 60.1 Evaluation of decision tree and linear regression

Models	R-squared (R^2)	RMSE
Linear regression	0.68	207035.94
Decision tree	0.75	185405.33

```
from sklearn.tree import DecisionTreeRegressor

de = DecisionTreeRegressor(random_state=0)

de.fit(X_train,y_train)

▾        DecisionTreeRegressor
DecisionTreeRegressor(random_state=0)

de_pred = de.predict(X_test)

de_r = r2_score(y_test,de_pred)

de_r

0.7496201523182228

score = mean_squared_error(de_pred,y_test)
score

34375137938.74082

score1 = np.sqrt(score)
score1

185405.33417013875
```

Figure 60.5 Model performance (decision tree)

```
regressor_LR = LinearRegression()

regressor_LR.fit(X_train,y_train)

▾ LinearRegression
LinearRegression()

y_pred_lin = regressor_LR.predict(X_test)
print (y_pred_lin)

[ 384885.98850091 1522091.68631228  541596.44539847 ...  320303.71125752
  224536.90644659  140903.44896077]

from sklearn.metrics import r2_score
lrr = r2_score(y_test,y_pred_lin)
lrr

0.6877902899299526

#This is MSE for LR
from sklearn.metrics import mean_squared_error
score = mean_squared_error(y_pred_lin,y_test)
score

42863880415.45748

# rmse
score1 = np.sqrt(score)
score1

207035.9399125125
```

Figure 60.6 Model performance (linear regression)

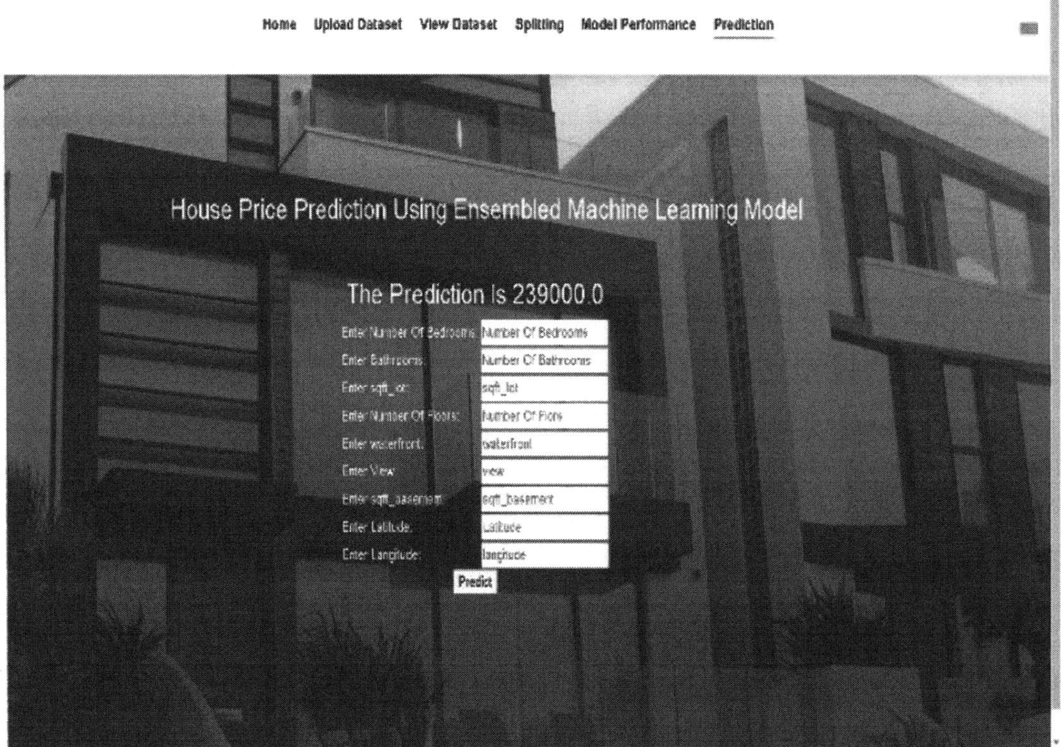

Figure 60.7 Interface

Conclusions

The study was able to apply linear regression and decision tree regression algorithms to predict house prices using a dataset of house features. The findings showed that both algorithms were effective in predicting house prices, but decision tree regression performed better in terms of prediction accuracy. Feature selection and engineering were also found to be crucial factors in improving the models' performance. The study's contributions to the field of machine learning for Real estate provide valuable insights into the suitability of various algorithms for this task.

Acknowledgement

The authors gratefully acknowledge REVA University for infrastructural support to carry on the work.

References

1. Adetunji, A. B., Akande, O. N., Ajala, F. A., Oyewo, O., Akande, Y. F., and Oluwadara, G. (2022). House price prediction using random forest machine learning technique. *Proc. Comp. Sci.*, 199, 806–813.
2. Matey, V., Chauhan, N., Mahale, A., Bhistannavar, V., and Shitole, A. (2022). Real estate price prediction using supervised learning. *2022 IEEE Inter. Conf.*, 1–5.
3. Kundu, M. and Nagendra Kumar, D. J. (2021). Route optimization of unmanned aerial vehicle by using reinforcement learning. *J. Phy. Conf. Ser.*, 1921(1), 1–7.
4. Manasa, J., Gupta, R., and Narahari, N. S. (2020). Machine learning based predicting house prices using regression techniques. *2020 2nd Inter. Conf. Innov. Mec. Indus. Appl. (ICIMIA)*, 624–630.
5. Bhagat, N., Mohokar, A., and Mane, S. (2016). House price forecasting using data mining. *Inter. J. Comp. Appl.*, 152(2), 23–26.
6. Kundu, M., Pasuluri, B., and Sarkar, A. (2023). Vehicle with learning capabilities: A study on advancement in urban intelligent transport systems. *2023 Third Inter. Conf. Adv. Elec. Comput. Comm. Sustain. Technol. (ICAECT)*, 58–67.
7. Lu, S., Li, Z., Qin, Z., Yang, X., and Goh, R. S. M. (2017). A hy-brid regression technique for house prices prediction. *2017 IEEE Inter. Conf. Indus. Engg. Engg. Manag. (IEEM)*, 319–323.
8. Jain, M., Rajput, H., Garg, N., and Chawla, P. (2020). Prediction of house pricing using machine learning with Python. *2020 Inter. Conf. Elec. Sustain. Comm. Sys. (ICESC)*, 570–574.
9. Durganjali, P. and Pujitha, M. V. (2019). House resale price prediction using classification algorithms. *2019 Inter. Conf. Smart Struct. Sys. (ICSSS)*, 1–4.
10. Islam, M. and Laskar, R. H. (2018). Robust image watermarking technique using support vector regression for blind geometric distortion correction in lifting wavelet transform and singular value decomposition domain. *J. Elec. Imag.*, 27(5), 053008–053008.
11. Islam, M., et al. (2017) . SVM regression based robust image watermarking technique in joint DWT-DCT domain. *2017 Inter. Conf. Intell. Comput. Instrum. Control Technol. (ICICICT)*, 1426–1433.

61 An ensemble of filter methods for disease classification enhancement

Sayantan Dass[1,a], *Sukhen Das Mandal*[2,b], *Sujoy Mistry*[1,c] *and Pradyut Sarkar*[1,d]

[1]Department of Computer Science and Engineering, Maulana Abul Kalam Azad University of Technology, Kolkata, West Bengal, India

[2]Department of Computer Science and Engineering, Ghani Khan Choudhury Institute of Engineering & Technology, Malda, West Bengal, India

Abstract

In the realm of medical diagnostics, disease classification is of paramount importance for early diagnosis and effective treatment planning. Gene expression data is correlated with disease states which is useful for detection of the diseases. However, the increasing size and complexity of the expression data pose a challenge for automatic diagnosis of disease condition using computing devices. To overcome this, feature selection methods offer a solution by eliminating irrelevant information from the dataset. Yet, relying solely on one filter-based method may lead to the inadvertent removal of critical features. This is where our ensemble-based approach, i.e., incorporating information gain, reliefF, and chi-square filters, proves its worth. By aggregating their rankings or scores, the ensemble selects the most relevant features for classification. Experimental results on colon and leukaemia cancer datasets demonstrate high classification accuracies of 94.7% and 97.2%, respectively. This ensemble-based methodology improves disease classification and enhances early detection in the medical field, providing a valuable tool for accurate diagnosis and effective treatment planning.

Keywords: Chi-square, classification, disease diagnosis, ensemble model, microarray data, information gain, reliefF

Introduction

Disease classification is a fundamental task in the field of medical diagnostics, playing a vital role in early detection and effective treatment planning [1]. Change of gene expression has been seen in disease condition and used as datasets for classify diseases conditions for diagnosis. As the volume and complexity of this type disease datasets continue to grow, there is an increasing demand for automated methods that can accurately classify diseases using computing devices [2]. However, this task poses a significant challenge due to the sheer amount of data and the presence of irrelevant information within the datasets. To address this challenge, feature selection methods have been developed to identify and retain the most relevant features for disease classification [3]. These methods aim to diminish the dimensionality of the dataset by eliminating irrelevant or redundant information, thereby improving the efficiency and accuracy of the classification process. It may also be useful to identify the important genes involved in disease development and may be useful for biomarker for disease diagnosis. Among the various approaches to feature selection, filter-based methods have gained significant attention due to their simplicity and effectiveness [4].

However, relying solely on one filter-based technique for feature selection can have limitations. The inherent biases and assumptions of a single method may lead to the inadvertent removal of critical features, resulting in suboptimal classification performance [5]. To overcome this limitation, an ensemble-based approach that combines multiple filter methods can provide a more robust and reliable solution.

In this study, we propose an ensemble of filter methods to enhance disease classification accuracy. Our ensemble incorporates three widely used filter techniques: information gain, reliefF, and chi-square filters. By leveraging the strengths of each method and aggregating their rankings or scores, the ensemble aims to identify the most significant features to ensure precise disease classification.

The effectiveness of our ensemble-based approach is evaluated through experimental analysis using colon and leukaemia cancer datasets/microarray gene expression data. These datasets represent two distinct types of cancer, offering a diverse range of disease characteristics and complexities. The classification accuracies achieved on these datasets serve as a measure of the effectiveness of our ensemble method in improving disease classification performance.

[a]dasssayantan@gmail.com, [b]sukhen@gkciet.ac.in, [c]sujoy.mistry@makautwb.ac.in, [d]pradyut.sarkar@makautwb.ac.in

DOI: 10.1201/9781003540199-61

In the following sections, we will provide an overview of related work, discuss the methodology employed in our study, and analyze the obtained results. Lastly, we will explore the implications of our findings and emphasize potential directions for future research.

Related work

Several studies have focused on feature selection methods for disease classification in the medical domain, aiming to improve accuracy and efficiency in automated diagnosis. Our study utilized gene expression data from two distinct cancer types: colon cancer and leukaemia. The colon cancer dataset, obtained from [6] 6, consisted of 62 cases with 2000 genes. Among these cases, 40 corresponded to abnormal tumor biopsies, while 22 represented normal samples. The leukaemia dataset, obtained from Alon et al., 7, comprised of 72 cases with 7129 genes, including 25 cases of acute myeloblastic leukaemia (AML) and 47 cases of acute lymphoblastic leukaemia (ALL) [7].

In the field of colon cancer and leukaemia diagnosis, various studies have investigated feature selection methods to improve accuracy. A two-stage approach with Kruskal-Wallis (KWs) test and correlation-based feature selection (CFS) achieved 80.28% accuracy [8]. Distributed feature selection obtained 83.87% accuracy with selected key attributes [9]. Use of fuzzy decision trees gave 80.28% accuracy [10]. Combination of fast correlation based feature selection (FCBFS) and optimized support vector machines (SVM) showed 93.55% accuracy [11], whereas use of genetic programming (GP), genetic algorithm (GA), and information gain (IG) showed 85.48% accuracy [12]. Modified analytic hierarchy process (MAHP) coupled with a probabilistic neural network (PNN) gained 88.89% accuracy [13].

Furthermore, numerous studies have explored feature selection methods for enhancing accuracy in diagnosing various other diseases using machine learning techniques. For instance, a combination of KW test and Bonferroni correction (BC) was utilized to select significant genes, followed by employing the CFS method to assess gene relationships to remove duplicate genes, leading to the identification of potential gene subsets [14]. Another hybrid model was developed by mixing reliefF with grey Wolf optimization (GWO) procedures, significantly improving predictive capabilities for early and accurate cancer diagnosis [15].

These studies have demonstrated the efficacy of different feature selection methods, and our work extends the research by proposing an ensemble approach to further enhance classification accuracy.

Methodology

The methodology employed in this study involves the use of an ensemble-based approach that combines three filter methods: information gain, reliefF, and chi-square filters. This ensemble aims to enhance the accuracy of disease classification by selecting the most relevant features from the datasets. Figure 61.1 presents a detailed flow-diagram illustrating the proposed methodology.

Dataset pre-processing

In the initial step of our methodology, we preprocess the disease dataset using Min-Max normalization. This normalization technique scales the feature values to a range between 0 and 1, ensuring that all features contribute equally to the subsequent feature selection process. Min-Max normalization is applied to eliminate potential biases caused by differences in the scales of the features.

Ensemble of filter methods

Our methodology incorporates an ensemble-based approach that combines three distinct filter methods: information gain (IG), reliefF, and chi-squared (χ^2). Each filter method assesses the significance of features using distinct criteria, enhancing the robustness and reliability of the ensemble

Information gain (IG)

Information gain quantifies the level of information provided by a feature regarding the class labels. It measures the reduction in entropy or uncertainty about the class labels when considering a particular feature. Higher information gain values indicate features that contribute more to distinguishing different disease classes.

ReliefF

ReliefF is a distance-based feature selection method that assesses the significance of features by considering their capability to discriminate between instances of different classes. It estimates feature weights based on the differences in feature values among instances with the same and different class labels. ReliefF focuses on identifying features that are most informative for classification.

Chi-square (χ^2)

Chi-square is a statistical measure used for assessing the independence between categorical variables. In our methodology, chi-square evaluates the association between every feature & the

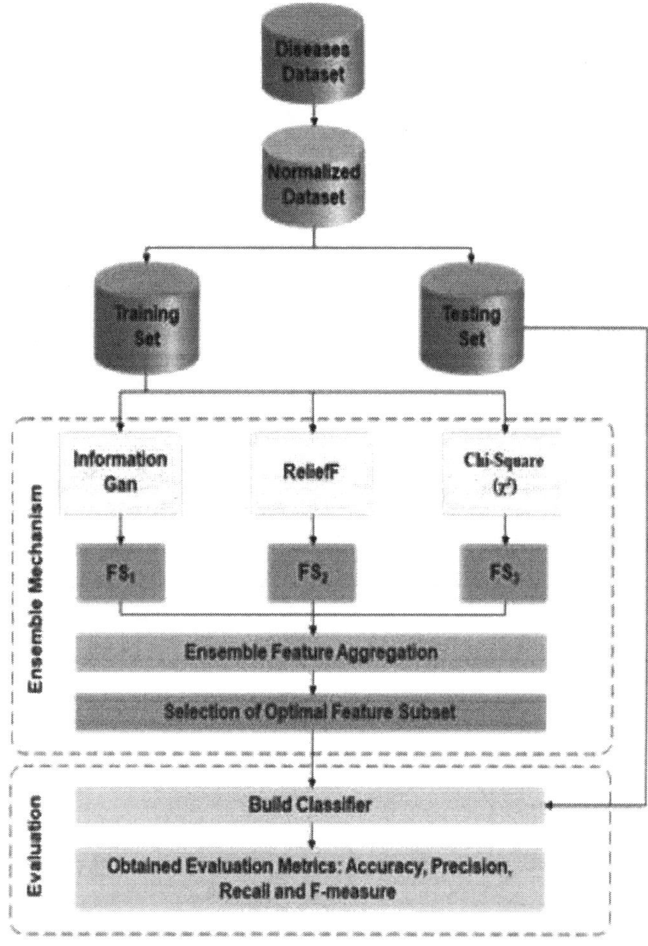

Figure 61.1 Flow diagram of proposed ensemble model

target class by comparing observed and expected frequencies of feature-value and class-label combinations. Higher chi-square scores indicate stronger associations between features and disease classes.

To form the ensemble, the rankings or scores obtained from each filter method are aggregated. In our methodology, features that are selected by a minimum of two filter methods are considered for inclusion in the final feature set. This aggregation approach ensures that only the features deemed relevant by multiple filter methods are included, enhancing the robustness and reliability of the selected feature set.

Feature selection and classification

After obtaining the aggregated rankings or scores from the ensemble, we select the top-ranked features for disease classification. The numeral of features to be selected can be determined based on a thorough systematic evaluation of different feature subset sizes.

For classification, we employ two popular algorithms: support vector machines (SVM) and k-nearest neighbors (k-NN) [16]. SVM constructs an optimal hyperplane to separate instances of different classes, while k-NN allots class labels by considering on the majority vote of the k-NN.

Result

The ensemble-based methodology's performance is assessed by measuring the classification accuracy of both SVM and k-NN models. To compare the performance of the ensemble-based approach with individual filter methods, comparative experiments are carried out. Table 61.1 demonstrates that the ensemble-based methodology outperforms individual methods and achieves improved classification accuracy. The selected features contribute significantly to disease classification, leading to better discriminatory power and generalization ability. Table 61.1 presents a comprehensive comparison between the existing methods and the

Table 61.1 Comparative performance of the proposed method and individual feature selection methods across disease datasets

Dataset	Classifier	Method	#Feature Selected	Performance Metrics			
				Accuracy	Precision	Recall	F-measure
Colon Cancer	SVM	IG	30	0.846	0.875	0.875	0.875
		ReliefF	5	0.923	1	0.875	0.933
		χ^2 Test	10	0.923	1	0.875	0.933
		Proposed	**2**	**0.941**	**1**	**0.9**	**0.947**
	k-NN	IG	30	0.769	0.857	0.75	0.799
		ReliefF	10	0.912	1	0.867	0.913
		χ^2 Test	15	0.905	1	0.825	0.905
		Proposed	**6**	**0.947**	**1**	**0.9**	**0.947**
Leukemia	SVM	IG	15	0.867	0.6	1	0.749
		ReliefF	10	0.933	0.75	1	0.857
		χ^2 Test	10	0.933	0.75	1	0.857
		Proposed	**4**	**0.955**	**0.8**	**1**	**0.889**
	k-NN	IG	15	0.8	0.5	1	0.666
		ReliefF	10	0.867	0.6	1	0.749
		χ^2 Test	15	0.933	0.75	1	0.857
		Proposed	**7**	**0.972**	**0.9**	**1**	**0.947**

Table 61.2 Performance comparison against different methods across disease datasets

Publication	Method	Dataset	# Attributes	Acuuracy(%)
Ludwig et al. (2018)	FDT	Colon	6	0.8028
Gao et al. (2017)	FCBFS+SVM	Colon	14	0.935
Salem et al. (2017)	IG+GA+GP	Colon	60	0.8548
Nguyen et al. (2015)	MAPH+PNN	Colon	5	0.8889
Dass et al. (2021)	KWs Test+CFS	Colon	23	0.909
Proposed	IG+ReliefF+ χ^2 Test	Colon	6	**0.947**
Ludwig et al. (2018)	FDT	Leukemia	4	0.875
Salem et al. (2017)	IG+GA+GP	Leukemia	3	0.9706
Nguyen et al. (2015)	MAPH+PNN	Leukemia	5	0.9278
Proposed	IG+ReliefF+ χ^2 Test	Leukemia	7	**0.972**

proposed model, highlighting the key details and differences.

Table 61.2 presents the comparison of classification accuracy between our proposed model, which utilizes an SVM classifier, and the previous research for leukaemia and colon cancer datasets. The comparison table reveals that our model surpasses the previous approaches, achieving an accuracy of 97.2% for the leukaemia dataset and 94.7% for the colon cancer dataset.

Conclusions

In summary, our ensemble-based feature selection methodology for disease classification has demonstrated significant improvements compared to individual feature selection methods. By combining information gain (IG), reliefF, and chi-square (χ^2) filters, the ensemble approach enhances the robustness and reliability of feature selection, resulting in more accurate and reliable disease classification models.

Conducting rigorous experimentation and performance analysis on benchmark disease datasets validates the methodology's effectiveness. The ensemble-based method consistently excels over individual methods, revealing its superiority in classification accuracy, F1-score, recall, and precision. The methodology's capability to select the most relevant features enhances the discriminatory power and generalization ability of the models. This has important implications for disease classification, facilitating accurate diagnosis and treatment decision-making.

In future, experimentation is needed to investigate the role of selected features in disease development and also future research directions could explore additional filter methods or classification algorithms to further improve the methodology's performance. The methodology may also be extended to handle high-dimensional datasets or incorporate domain-specific knowledge, contributing to advancements in disease classification and improving patient outcomes.

Acknowledgement

The authors express their gratitude to the students, staff, and administration of the Computer Science & Engineering department for their collaborative support in the research.

References

1. Ahsan, M. M., Luna, S. A,. and Siddique, Z. (2022). Machine-learning-based disease diagnosis: A comprehensive review. *Healthcare*, 10(3), 541. https://doi.org/10.3390/healthcare10030541.
2. Ghosh, T., Mitra, S., and Acharyya, S. (2021). Pathway marker identification using gene expression data analysis: A particle swarm optimisation approach. *Lec. Notes Netw. Sys.*, 127–136. https://doi.org/10.1007/978-981-16-4435-1_14.
3. Jain, D. and Singh, V. (2018). Feature selection and classification systems for chronic disease prediction: A review. *Egypt. Informat. J.*, 19(3), 179–189. https://doi.org/10.1016/j.eij.2018.03.002.
4. Remeseiro, B. and Bolon-Canedo, V. (2019). A review of feature selection methods in medical applications. *Comp. Biol. Med.*, 112, 103375. https://doi.org/10.1016/j.compbiomed.2019.103375.
5. Rajab, M. and Wang, D. (2020). Practical challenges and recommendations of filter methods for feature selection. *J. Inform. Knowl. Manag.*, 19(01), 2040019. https://doi.org/10.1142/s0219649220400195.
6. Golub, T. R., Slonim, K. D., Tamayo, P., Huard, C., Gaasenbeek, M., Mesirov, P. J., Coller, H., et al. (1999). Molecular classification of cancer: Class discovery and class prediction by gene expression monitoring. *Science*, 286(5439), 531–537. https://doi.org/10.1126/science.286.5439.531.
7. Alon, U., Barkai, N., Notterman, A. D., Gish, K., Ybarra, S., Mack, D., and Levine, J. A. (1999). Broad patterns of gene expression revealed by clustering analysis of tumor and normal colon tissues probed by oligonucleotide arrays. *Proc. National Acad. Sci.*, 96(12), 6745–6750. https://doi.org/10.1073/pnas.96.12.6745.
8. Dass, S., Mistry, S., Sarkar, P. and Paik, P. (2022). An optimize gene selection approach for cancer classification using hybrid feature selection methods. Advanced Network Technologies and Intelligent Computing. 751–764. https://doi.org/10.1007/978-3-030-96040-7_56.
9. Potharaju, S. P. and Sreedevi, M. (2019). Distributed feature selection (DFS) strategy for microarray gene expression data to improve the classification performance. *Clin. Epidemiol. Global Health*, 7(2), 171–176. https://doi.org/10.1016/j.cegh.2018.04.001.
10. Ludwig, S. A., Picek, S., and Jakobovic, D. (2017). Classification of cancer data: Analyzing gene expression data using a fuzzy decision tree algorithm. *Inter. Ser. Manag. Sci./Oper. Res.*, 327–347. https://doi.org/10.1007/978-3-319-65455-3_13.
11. Gao, L., Ye, M., and Wu, C. (2017). Cancer classification based on support vector machine optimized by particle swarm optimization and artificial bee colony. *Molecules*, 22(12), 2086. https://doi.org/10.3390/molecules22122086.
12. Salem, H., Attiya, G., and El-Fishawy, N. (2017). Classification of human cancer diseases by gene expression profiles. *Appl. Soft Comput.*, 50, 124–134. https://doi.org/10.1016/j.asoc.2016.11.026.
13. Nguyen, T., Khosravi, A., Creighton, D., and Nahavandi. (2015). A novel aggregate gene selection method for microarray data classification. *Patt. Recogn. Lett.*, 60–61, 16–23. https://doi.org/10.1016/j.patrec.2015.03.018.
14. Dass, S., Mistry, S., Sarkar, P., Barik, S., and Dahal, K. (2023). A proficient two stage model for identification of promising gene subset and accurate cancer classification. *Inter. J. Inform. Technol.*, 15(3), 1555–1568. https://doi.org/10.1007/s41870-023-01181-2.
15. Dass, S., Mistry, S., and Sarkar, P. (2023). Identification of promising biomarkers in cancer diagnosis using a hybrid model combining reliefF and grey Wolf optimization. *Lec. Notes Netw. Sys.*, 311–321. https://doi.org/10.1007/978-981-99-2322-9_23.
16. Manna, S. and Mistry, S. (2023). Presaging cancer stage classification by extracting influential features from breast/lung/prostate cancer clinical datasets based on TNM model. Third Congress on Intelligent Systems, Lecture Notes in Networks and Systems. 187–203. https://doi.org/10.1007/978-981-19-9225-4_15.

62 Classification of text and non-text components in offline unconstrained handwritten documents using generative adversarial network

Saikh Risat, Bhaskar Sarkar and Showmik Bhowmik[a]

Department of Computer Science and Engineering, Machine Learning and Systems Biology Research Lab, Ghani Khan Choudhury Institute of Engineering and Technology, Malda, India

Abstract

Detecting and distinguishing text from non-text elements in handwritten scripts or documents that are not constrained by any specific structure is crucial in the advancement of optical character recognition (OCR) systems. To deal with this issue, researchers predominantly rely on extracting manually crafted features that capture the texture details for classifying components as either text or non-text. However, when faced with noise, these feature descriptors tend to perform poorly. To address this limitation, we use a variant of generative adversarial networks called Pix2Pix for image segmentation [1] here. For evaluation, an in-house dataset comprising 200 handwritten pages was created. This model achieves an accuracy of 89.044% on this dataset.

Keywords: Text, non-text, GAN, Pix2Pix, handwritten documents, text non-text separation

Introduction

Documents hold significant importance in our society, as they serve as crucial means for sharing and storing information. However, safeguarding physical copies of documents containing vital information poses challenges. These valuable documents are susceptible to loss due to various factors such as natural disasters, mishandling, and the effects of time and aging [2]. Merely digitizing these documents is insufficient. To ensure efficient management and interpretation of digitized documents, optical character recognition (OCR) becomes essential. When dealing with real-world data, a significant challenge arises as OCR systems can only handle textual content. The existence of non-textual elements can potentially mislead the outcome of an OCR engine. Hence, it becomes crucial to detect and identify non-textual components at an early stage [3] to ensure accurate OCR processing. The task of detecting non-textual elements from text in documents has many challenges to address. This may be due to the variations in background color, component height, width, and orientation within the document. The task becomes much more difficult when dealing with unconstrained handwritten documents. Components within unconstrained handwritten scripts exhibit significant variations in shape, size, and orientation, of the components unlike printed documents. This disparity arises from the diverse writing styles of individuals, variations in paper and ink quality, among other factors. Furthermore, in handwritten documents, it is common for text to be intertwined with non-textual elements, which can further hinder the accurate recognition process. To tackle this challenge, the current study employs a modified version of the generative adversarial network (GAN) model known as Pix2Pix. This model is utilized to discern and distinguish the text elements from the non-textual components within a handwritten document.

Related work

Separation of textual and non-textual elements in unconstrained handwritten scripts got comparatively less focus than the printed ones. However, few works to address this problem are introduced later in the literature. For example, in ref [4], authors use rotational invariant local binary pattern (RILBP) based feature to characterize the extracted connected components at the feature space and then use a multi-layer perceptron (MLP) to identify these separately as textual or non-textual component. In ref [5], authors conduct experiments to showcase the usefulness of various local binary pattern based feature descriptors in identifying textual and non-textual components in handwritten documents. Authors in ref [6], apply feature selection method to isolate most pertinent regions of an image to classify this as text or non-text. In ref [7], authors use a shape based approach in two stages to classify the extracted components of scientific handwritten

[a]showmik@gkciet.ac.in

DOI: 10.1201/9781003540199-62

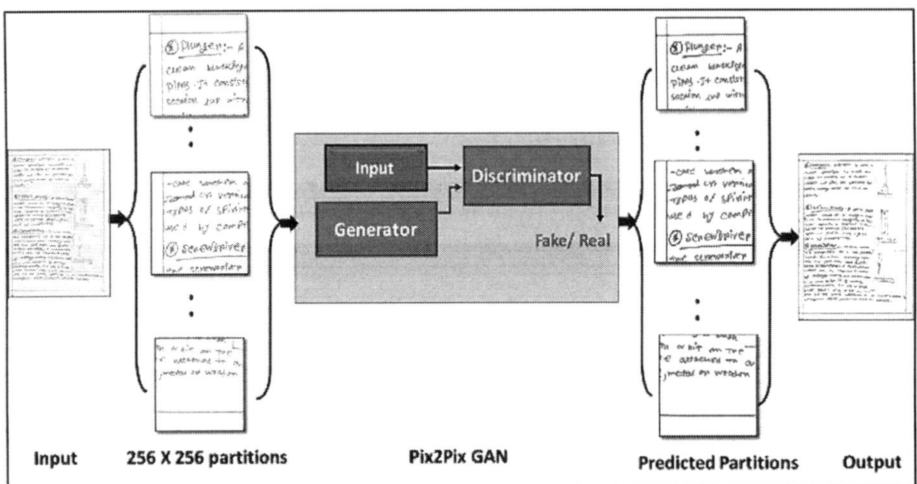

Figure 62.1 Illustration of entire workflow

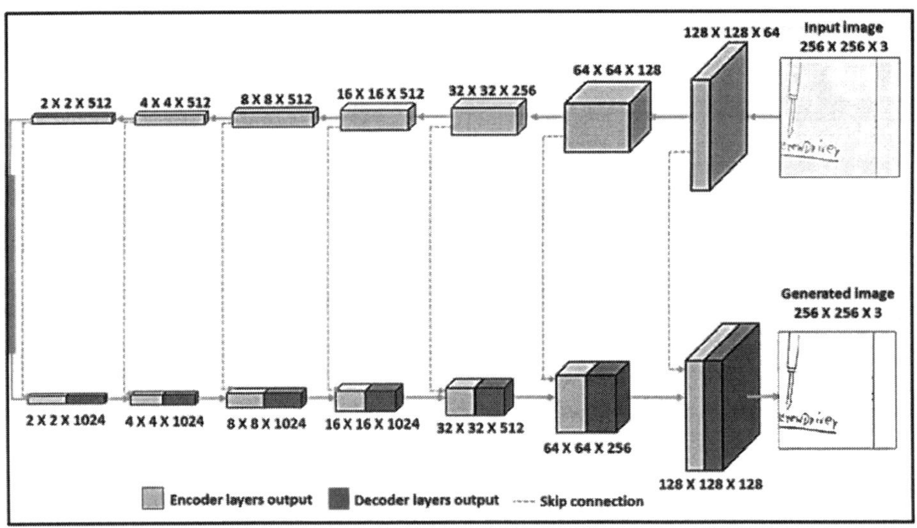

Figure 62.2 Generator network

documents. For separation of textual and non-textual elements in handwritten scripts, various LBP descriptors are mostly considered. Despite their popularity, these descriptors are very sensitive to noise. Considering this fact, the current study employs a modified version of the GAN model known as Pix2Pix to detect textual and non-textual components separately in handwritten documents.

Pix2Pix model follows an image-to-image translation approach that involves converting an input image from one domain to another. Pix2Pix, was proposed by Isola et al. [1], developed based on conditional generative adversarial networks (cGANs). It has shown remarkable performance in various image translation tasks, including semantic segmentation, image colorization, and edge-to-photo-synthesis. Pix2Pix operates by learning a mapping from the input domain (grayscale images in this case) to the output domain (color images). Recently, Bhowmik et. al. use a modified version of this model to segment the touching text non-text components [8]. Considering its performance in various image processing related problems, in the present scope of the work, this model is considered for detecting and distinguishing textual and non-textual elements in handwritten documents.

Present work

As said earlier, here, we follow the image-to-image translation approach to identify the textual and non-textual components separately in handwritten images. To do that we use the pix2pix model that is a type of conditional GAN, or cGAN, where the generation of the output image is conditional on an input image. The pix2pix network architecture comprises two major parts, the generator and the discriminator. The generator transforms the input image and generates the predicted output image. The discriminator measures the similarity of the input image to an unknown image (either a target image from the dataset or an output image from the generator) and tries to guess if this was produced by the generator.

In this work, we initially partition the input image into eight sub-images and then perform the rest of the processing (see Figure 62.1). The architecture of the generator and discriminator networks are also described in this section.

Generator architecture

The generator model considered here uses U-net architecture and it consists of an encoder (downsampler) and decoder (upsampler) networks. Generator network takes a 256×256×3 image as input and passes it through a convolutional layer. The Encoder network has seven convolutional blocks. Each convolutional block has a convolutional layer, followed by a LeakyRelu activation function (with a slope of 0.2 in the paper). Each convolutional block also has a batch normalization layer except the first convolutional layer. The decoder network has seven transpose convolutional blocks. Each upsampling convolutional block has an upsampling layer, followed by a convolutional layer, a batch normalization layer, and a ReLU activation function.

Discriminator architecture

The discriminator model used here is a PatchGAN consisting of a number of convolution blocks. This PatchGAN architecture takes an N×N part of the image and tries to find whether it is real or fake. Each block of the discriminator contains a convolution layer, batch norm layer, and LeakyReLU. This discriminator receives two inputs: the input image and target Image (which discriminator should classify as real) and the input image and generated Image (which they should classify as fake) (Figure 62.2).

Loss function

The loss function considered here is as follows,

$$L\_TNT = arg \llbracket max \rrbracket \_GN \llbracket min \rrbracket \_DS [L\_T (GN, DS) + \gamma L\_1 (GN)] \quad (1)$$

where,

$$L\_T (GN, DS) = E\_((IN, OT)) [logDS(IN, OT)] + E\_((IN, N)) [\log(1 - DS(IN, GN(IN, N)))] \quad (2)$$

$$L\_1 (GN) = E\_((IN, OT, N)) [\|OT - GN(IN, N)\|] \quad (3)$$

Here, GN and DN represent generator and discriminator respectively. IN, OT, and N represent the input image, mask image and noise, respectively (Figure 62.3).

Experimental results

For the evaluation of the proposed method, an in-house dataset of 200 pages is prepared. For that purpose, handwritten class notes are collected which are mostly written by the school level students. Initially, all the collected pages are scanned in 300 dpi using a flat board scanner and stored as a bmp file. The sample wise ground truth images are generated manually. From 200 pages total 150 pages are used for training and the remaining 50 pages are used for testing. As a performance metric, in the present experiment *Text detection accuracy* (TDA), *Non-text detection accuracy* (NTDA), and *Overall accuracy* (OA) are computed using the Equation 4, 5 and 6, respectively.

$$TDA = (Total\ accurately\ detcted\ text\ pixels) / (Total\ text\ pixels) \quad (4)$$

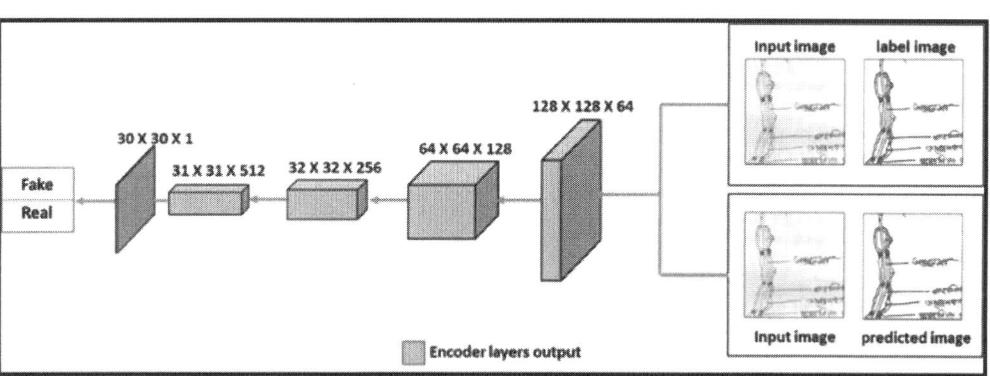

Figure 62.3 Discriminator architecture

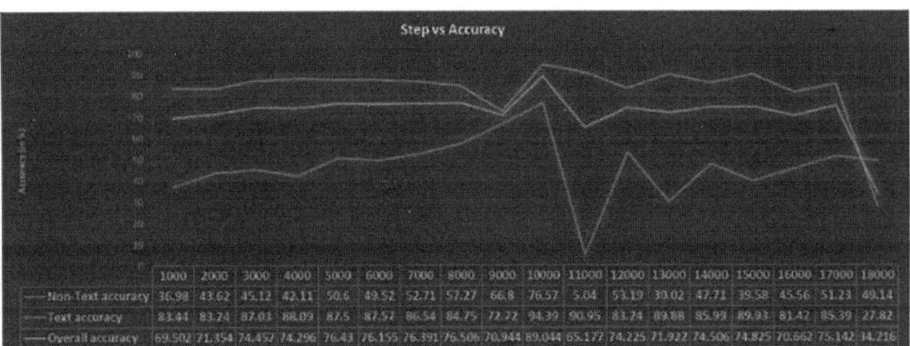

Figure 62.4 Effect of number of steps parameter on OA

Table 62.1 Performance comparison with a recent method

Method	TDA (in%)	NTDA (in%)	OA (in %)
Ghosh et. al. [6]	88.5	70.7	83.65
Present work	**94.39**	**76.57**	**89.004**

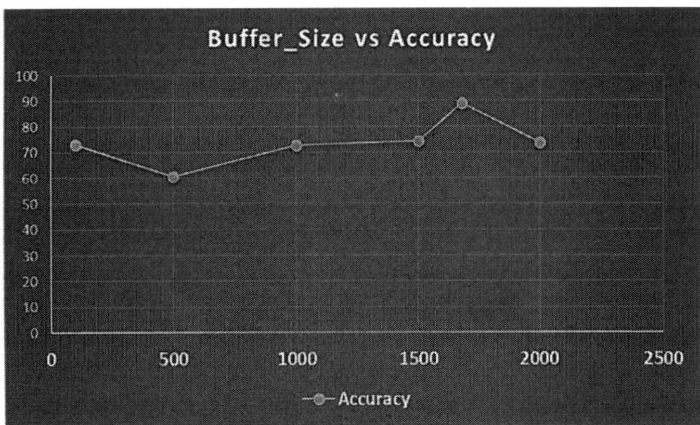

Figure 62.5 Effect of buffer_size parameter on OA

$$NTDA = (Total\ accurately\ detcted\ non - text\ pixels)\ /\ (Total\ non - text\ pixels) \quad (5)$$

$$OA = (0.7 \times TDA) + (0.3 \times NTDA) \quad (6)$$

From Equation 5, it can be noticed that, in the present scope of the work, the text detection accuracy is given more weightage than non-text. The reason for that is here our aim is to detect more text pixels. During this experiment, this method achieves 94.39% – TDA, 76.57% – NTDA and 89.044 % – OA.

Experimental parameter

During the experiment, we mostly tune two parameters, number of iteration or steps and buffer_size. The purpose of the buffer size is to shuffle the training data. It represents the number of samples from the training set that will be randomly sampled and stored in a buffer. During training, the model processes data in batches. The buffer size determines the number of elements that are preloaded into the buffer for efficient shuffling. In the present experiment the number of steps is experimentally set to 10,000, and buffer_size is set to 1700. The change in OA over the number of steps is shown in Figure 62.4 and same for buffer_size is provided in Figure 62.5.

From Figure 62.4, it can be seen that at 10,000 steps this method obtains the best result. Even after 10,000 steps a significant number of ups and downs are there but none of these reach the best accuracy. A similar effect can be seen in Figure 62.5, which shows for the buffer_size 1700, this method obtains best accuracy.

Performance comparison

In this section, we compare the result obtained by the present method with a recent method [6] and it is noticed that the present method obtained a better result (see Table 62.1).

Conclusions

Separating textual and non-textual components in offline unconstrained handwritten documents is an essential and difficult task in the pipeline of an efficient OCR system. OCR systems primarily focus on processing text and thus can experience degraded performance in the presence of non-textual elements. Despite its significance, Surprisingly, only a small number of works have focused on text non-text separation, specifically for handwritten documents. In contrast, a significant number of methods can be found for addressing this problem in printed documents [9]. Existing approaches predominantly rely on feature engineering-based methods, which tend to struggle when confronted with noisy data. In order to tackle this challenge, the current study employs a deep learning model based on GAN called Pix2Pix. This model is utilized to detect both textual and non-textual components within unconstrained handwritten documents. Compared to traditional handcrafted feature models, GAN-based approaches often demonstrate superior performance.

To assess the effectiveness of the developed model, a proprietary dataset of 200 pages was generated. This dataset comprises 150 images for training and 50 images for testing. Notably, our present method achieved an overall accuracy of 89.044% when tested on this in-house dataset. In future, authors have a plan to address more complex issues in text non-text separation like touching components and additionally enlarge this dataset to include more realistic samples.

References

1. Isola, P., Zhu, J.-Y., Zhou, T., and Efros, A. A. (2017). Image-to-image translation with conditional adversarial networks. *Proc. IEEE Conf. Comp. Vision Patt. Recogn.*, 1125–1134.
2. Bhowmik, S. (2023). Document Layout Analysis, 1st ed. Springer Singapore. 1–86.
3. Bhowmik, R. S., Nasipuri, M., and Doermann, D. (2018). Text and non-text separation in offline document images: A survey. *Int. J. Doc. Anal. Recogn.*, 21(1–2), 1–20. doi: https://doi.org/10.1007/s10032-018-0296-z.
4. Bhowmik, S., Kundu, S., and Sarkar, R. (2020). BINYAS: A complex document layout analysis system. *Multimed. Tools Appl.*, 8471–8504. doi: https://doi.org/10.1007/s11042-020-09832-3.
5. Ghosh, S., Hassan, S. K., Khan, A. H., Manna, A., Bhowmik, S., and Sarkar, R. (2022). Application of texture-based features for text non-text classification in printed document images with novel feature selection algorithm. *Soft Comput.*, 26(2), 891–909.
6. Ghosh, M., Ghosh, K. K., Bhowmik, S., and Sarkar, R. (2020). Coalition game based feature selection for text non-text separation in handwritten documents using LBP based features. *Multimed. Tools Appl.*, 1–21. doi: https://doi.org/10.1007/s11042-020-09844-z.
7. Bhowmik, S., Kundu, S., De, B. K., Sarkar, R., and Nasipuri, M. (2019). A two-stage approach for text and non-text separation from handwritten scientific document images. *Inform. Technol. Appl. Math.*, 41–51.
8. Mondal, R., Bhowmik, S., and Sarkar, R. (2020). tsegGAN: A generative adversarial network for segmenting touching nontext components from text ones in handwriting. *IEEE Trans. Instrum. Meas.*, 70, 1–10. doi: 10.1109/TIM.2020.3038277.
9. Binmakhashen, G. M. and Mahmoud, S. A. (2019). Document layout analysis: A comprehensive survey. *ACM Comput. Surv.*, 52(6), 1–36.

63 Mapping the scholarly landscape of Microsoft Azure: A bibliometric analysis and research trends

Rajesh T. Nakhate[1,a], *Parul Dubey*[1,b], *Prof Priti Kakde*[1,c], *Swati Shamkuwar*[1,d], *Swati Tiwari*[1,e] *and Rahul Bambodkar*[2,f]

[1]Department of I.T, G H Raisoni College of Engineering, Nagpur, India

[2]Department of CSE, JD College of Engineering & Management, Nagpur, India

Abstract

In this study, we conduct a bibliometric evaluation of Microsoft Azure, a prominent cloud platform. We surveyed the Azure cloud academic ecosystem by evaluating publishing patterns, author networks, and canonical works using the Web of Science database and VOSviewer. The study showed that the number of papers related to Azure cloud research has been rising steadily over time. It was possible to get insight into the dynamic and collaborative character of the research community by identifying key research fields and partnerships. VOSviewer was used to conduct a visual analysis of bibliographic data, which helped researchers spot patterns and connections in the published literature. This bibliometric study provides important insight into Azure cloud research and may be used as a reference by future scholars and policymakers.

Keywords: Bibliometric analysis, azure cloud, web of science, publication trends, co-occurrence analysis, bibliographic coupling

Introduction

Cloud computing is a game-changing innovation that is altering the way businesses store, access, and analyze their data. Microsoft Azure has risen rapidly in popularity to become one of the top cloud computing platforms. With Microsoft's Azure, organizations can use Microsoft-managed data centers to create, deploy, and manage their own apps and services. Increased interest in Azure calls for a review of the academic literature on Azure clouds, including an examination of current and future research trends, impacts, and directions [3].

In this study, we use the Web of Science (WOS) database to conduct a bibliometric analysis of Microsoft Azure. The World of Science (WOS) is an authoritative library of scientific publications covering a broad range of topics. We want to better understand the Azure cloud research environment by illuminating significant study fields, prominent publications, and emerging trends by making use of the wealth of information accessible in WOS.

We used a wide range of methods and processes in our bibliometric study. Bibliographic coupling, citation network analysis, author cooperation analysis, and other similar methods are all examples. By using these techniques, we are able to pinpoint developing trends in Azure cloud research, assess citation patterns and prominent publications, and get insight into the academic community's collaboration networks.

This study will shed light on the state of Microsoft Azure research and give a panoramic perspective of the field as a whole. Understanding the important study topics and the influence of Azure cloud in academia in a systematic way would be useful for researchers, practitioners, and decision-makers. Furthermore, this analysis will aid in determining where further study and growth is needed in this dynamic subject.

In conclusion, using the Web of Science database, this study attempts to add to the current body of knowledge by doing a thorough bibliometric analysis of Microsoft Azure. This study will provide academics, practitioners, and decision-makers with invaluable insights on the current state of the field as well as the past, present, and future of Azure cloud research.

Literature review

Azure is Microsoft's umbrella brand for its many cloud-computing offerings. It includes a wide variety of services that are essential to cloud computing and continue to expand [1].

Cloud computing is the outcome of the use of computer technology and includes the creation and integration of several disciplines. The data is from

[a]rajesh.nakhate@raisoni.net, [b]priti.kakde@raisoni.net, [c]dubeyparul29@gmail.com, [d]swati.shamkuwar@raisoni.net, [e]swati.tiwari@raisoni.net, [f]Rahul.bmbodkar@gmail.com

DOI: 10.1201/9781003540199-63

the article ref [11] came from the Web of Science core collection database. Exploring the shift, distribution of research forces, critical topics, research hot spots, and international collaboration in cloud computing via data mining and quantitative analysis using Excel, Bibexcel, VOSviewer, and CiteSpace. The findings demonstrate an increasing body of literature on cloud computing across a wide range of disciplines (most notably Computer Science, Engineering, and Telecommunications). China is at the forefront of international collaboration, and it has robust cooperative links with the main nations; the Chinese Academy of Sciences is heavily invested in cloud computing research, and its publications are years ahead of those from other institutions. The research mostly focuses on topics including cloud computing, big data, security, mobile devices, and virtualization. Mobile cloud computing, big data, fog computing, secure storage, access control, server consolidation, etc. are some of the most active research areas.

Another study in ref [5] focuses on three industry leaders and provides an in-depth analysis of their services and pricing structures. Furthermore, this paper provides a critical analysis of the price and service approach used by various cloud service providers as a means of differentiating themselves in the market. Cloud service providers seek to maximize profits while also meeting the requirements of both their businesses and their clients. Providers of infrastructure, software, secure data storage, and automated testing are all available on the cloud. Each cloud service provider's strengths and weaknesses are summed up in the last section of the study.

Especially in the scientific and practical domains, bibliometric approaches or "analysis" have become well-established as scientific specializations and form an essential element of research assessment methodology. These techniques are becoming more common in the scientific community and in the global rankings of universities and other academic organizations. Enough research has been conducted, and the ensuing body of literature allows for an in-depth examination of the bibliometric approach within the framework of that approach. The bibliometric literature in research [4], which was retrieved from Web of Science, is separated into two portions using a manner analogous to the approach of others.

Methodology

In this study, we used a methodical procedure to retrieve and examine Microsoft Azure-related scientific publications from the Web of Science (WOS) database. The research process used in this article to examine the Azure cloud is shown in Figure 63.1. The procedure consisted of the following actions:

- Data extraction: We searched the WOS database for publications about Azure cloud using appropriate keywords.
- Dataset curation: Each item was handpicked to ensure it had substantial coverage on Azure cloud.
- Bibliometric analysis: Analysis of publishing trends, author cooperation, citation networks, and theme analysis were among the methods used.

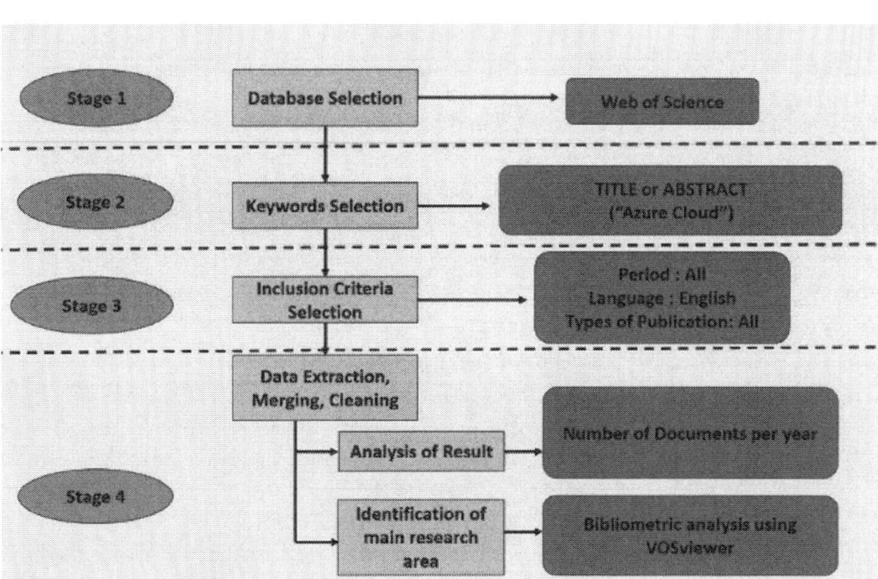

Figure 63.1 The study's approach to the Azure cloud

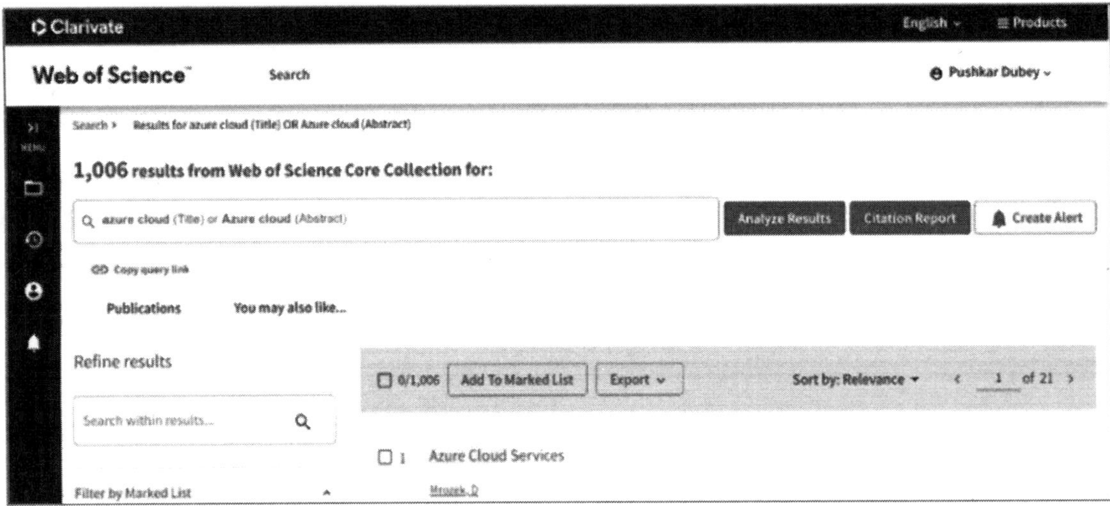

Figure 63.2 WOS database information extraction

- Interpretation: The results were analyzed to learn more about the present and future of Azure cloud research as well as its influence and trajectory.

This technique enabled a thorough evaluation of the Microsoft Azure academic ecosystem, yielding useful information for moving forward with decisions and studies.

Dataset description

Articles from the academic Web of Science (WOS) database were used to compile the dataset used in this study. Articles that included the term "Azure cloud" in their titles or abstracts were prioritized for inclusion in the collection. The search included papers published in any year, yielding a total of 1006.

The dataset guaranteed relevance to the study issue by restricting the search to papers with "Azure cloud" in the title or abstract, allowing for a thorough review of the academic literature relevant to Azure cloud. We used this data to perform our bibliometric study and learn more about the historical trajectory, current state, and potential future of Azure-related research. Figure 63.2 depicts information retrieved from the WOS database.

Publication trends

Analysis of publishing patterns suggests a rising tide of work on Microsoft's Azure cloud over time. Starting in 2009 with only two publications, by 2018 the number has risen to a staggering 120. This rising pattern reflects growing curiosity and use of Microsoft's Azure cloud among academics and professionals. The following years, 2019 through 2022, had a consistent number of papers published, ranging from 108 to 116. However, in 2023, there was a decline in the number of papers published, with just 30 being documented. The end of the survey period might explain this drop. Figure 63.3 displays the rising or falling rate of Azure cloud studies published in WOS. This rising tide of articles is a testament to Azure cloud's growing importance in the area of cloud computing, and it shows that researchers are continuing to devote significant time and energy to studying it.

Bibliometric analysis

The intellectual environment and trends in a certain study subject can only be fully understood by bibliometric analysis. VOSviewer is a popular programme for bibliometric evaluation. VOSviewer is a useful programme for locating important research topics, prominent publications, author partnerships, and developing trends via visual study of bibliographic data [8].

VOSviewer makes use of methods like co-citation analysis and co-authorship analysis to visualize the connections and trends found in the academic literature [9]. By analyzing the frequency with which two or more papers are mentioned by one another, co-citation analysis draws attention to significant research clusters and influential publications. The investigation of co-authorship patterns provides insight into research networks and inter-institutional partnerships.

Co-occurrence analysis is a bibliometric method for measuring the closeness and degree of associations between various phrases or keywords in a corpus of academic work. Researchers may learn

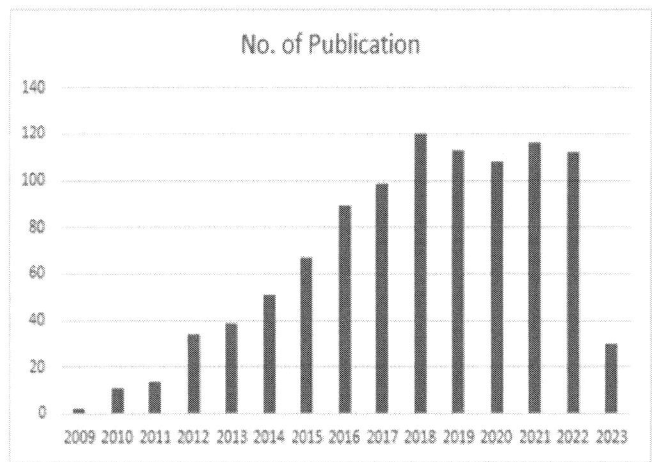

Figure 63.3 Azure cloud publication patterns in the WOS

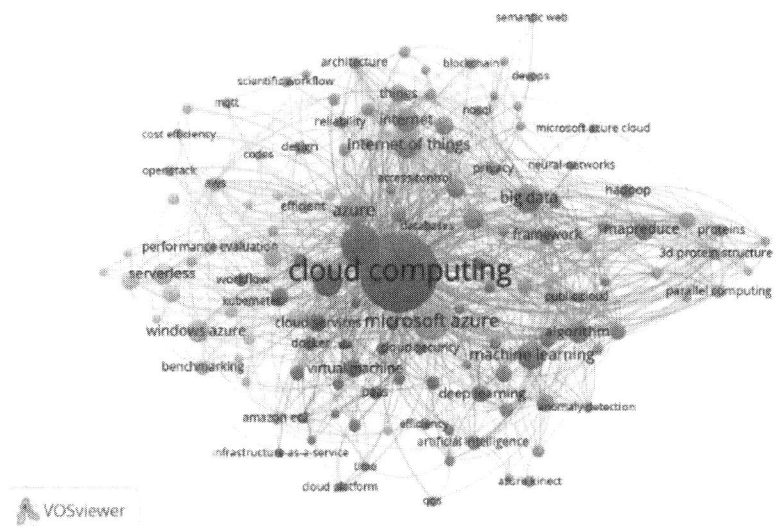

Figure 63.4 Analysis of all keywords in Azure cloud data for co-occurrence by WOS

more about the underlying structure of a study area by looking at how often related phrases appear together. Co-occurrence analysis provides a comprehensive overview of the knowledge landscape by highlighting key ideas, new trends, and the relationships across various fields of study. Figure 63.4 shows the analysis of all keywords in Azure cloud data for co-occurrence by WOS.

When doing a bibliographic study based on country, researchers take a close look at how various nations have contributed to the body of knowledge in a certain area of study. Researchers may learn about the global distribution of research output and partnerships by looking at author affiliations and the countries to which they belong [2]. Finding research partners, chances for cooperation, and regional research trends are all made easier with the information gleaned from this comparison of nations' research capabilities and skills.

By examining the journals, conference proceedings, and other venues in which articles appear, bibliographic analysis based on sources examines the scholarly output in a certain field. By analyzing the most prominent publications in the area, this study highlights the most important channels for sharing findings [6]. Publishing frequency, citation impact, and the dispersion of research across multiple publication venues are only some of the insights provided. By evaluating the available data, researchers may get a handle on the current publishing climate and make educated choices about where to submit their own research.

Researchers may see the relationships between papers, authors, and institutions by building interactive maps using VOSviewer. The structure and dynamics of a study area may be better understood with the use of these maps. Important publications, authors, or institutions are represented on the map

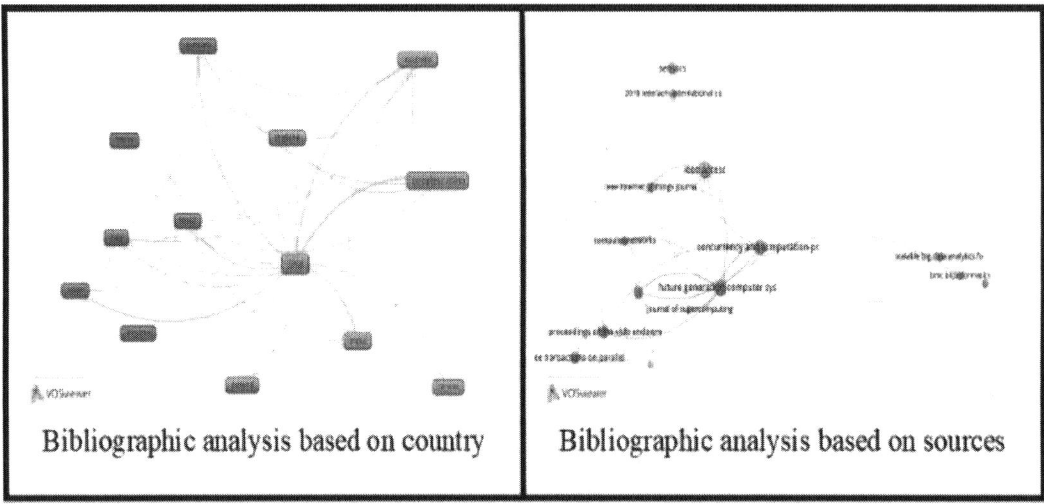

Figure 63.5 Country and source-specific bibliographic analysis

by larger, more central nodes. Color coding and node clustering make it easier to see connections between ideas and spot patterns in the data. Figure 63.5 shows the country and source-specific bibliographic analysis co-citation analysis and co-authorship analysis are only two of the methods that VOSviewer use to create eye-catching visualizations of connections and trends in the academic literature. Co-citation analysis identifies major research clusters and influential publications by highlighting works that are often cited together. Co-authorship analysis reveals relationships between writers and the institutions they are affiliated with.

Researchers may use VOSviewer to create dynamic diagrams of the interconnections of works of literature, authors, and institutions. These visual representations provide important insights about the composition and evolution of a subject of research. The size and proximity of nodes on the map indicate the significance of publications, authors, or institutions within the network. The use of color and the clustering of nodes help to classify topics and spot patterns in the data.

Researchers may tailor the visualizations in VOSviewer to their own needs and interests thanks to its flexible configuration options. In order to focus on certain time periods or keywords, researchers may utilize filters and a variety of layout methods.

Through bibliometric analysis with VOSviewer, scholars may get a comprehensive picture of the research ecosystem around a topic like Azure cloud. They are able to spot seminal works, eminent researchers, and critical collaborations, providing invaluable context for research, collaboration, and decision-making in the future.

Conclusions

The scientific environment of Azure cloud was illuminated by the bibliometric study performed using VOSviewer. According to the data, the number of articles using Azure cloud in academic research has been steadily rising over time. It was shown that the research community is both active and collaborative by identifying key study fields, prominent publications, and author collaborations. Using VOSviewer, researchers were able to visually investigate the bibliographic data and unearth hidden trends, patterns, and connections. These results help fill in the picture of Azure cloud research and provide the groundwork for future studies and policy decisions. The bibliometric study, when combined with VOSviewer, proved to be a potent tool for understanding the Azure cloud research environment and has broad applicability.

Acknowledgement

The authors gratefully acknowledge the students, staff, and authority of G H Raisoni College of Engineering for their cooperation in the research.

References

1. Copeland, M., Soh, J., Puca, A., Manning, M., and Gollob, D. (2015). Microsoft Azure and cloud computing. *Microsoft Azure*, 3–26. https://doi.org/10.1007/978-1-4842-1043-7_1.
2. Donthu, N., Kumar, S., Mukherjee, D., Pandey, N., and Lim, W. M. (2021). How to conduct a bibliometric analysis: An overview and guidelines. *J. Busin. Res.*, 133, 285–296. https://doi.org/10.1016/j.jbusres.2021.04.070.

3. Dubey, P., Dubey, P., and Sahu, K. K. (2023). A guide to cloud platform with an investigation of different cloud service providers. *Big Data, Cloud Computing and IoT*, 37–51. https://doi.org/10.1201/9781003298335-3.
4. Ellegaard, O. and Wallin, J. A. (2015). The bibliometric analysis of scholarly production: How great is the impact? *Scientometrics*, 105(3), 1809–1831. https://doi.org/10.1007/s11192-015-1645-z.
5. Gupta, B., Mittal, P., and Mufti, T. (2021). A review on Amazon Web Service (AWS), Microsoft Azure & Google Cloud Platform (GCP) Services. *Proc. 2nd Inter. Conf. ICT Digital Smart Sustain. Dev. ICIDSSD 2020*, 27–28. https://doi.org/10.4108/eai.27-2-2020.2303255.
6. Moral-Muñoz, J. A., Herrera-Viedma, E., Santisteban-Espejo, A., and Cobo, M. J. (2020). Software tools for conducting bibliometric analysis in science: An up-to-date review. *El Profesional De La Información*, 29(1). https://doi.org/10.3145/epi.2020.ene.03.
7. Morelli, D. A. and Ignacio, P. S. D. A. (2021). Assessment of researches and case studies on cloud manufacturing: A bibliometric analysis. *Inter. J. Adv. Manufac. Technol.*, 117(3–4), 691–705. https://doi.org/10.1007/s00170-021-07782-0.
8. Subject and Course Guides: Bibliometric Analysis and Visualization: VOSViewer. (2023). https://researchguides.uic.edu/bibliometrics/vosviewer.
9. VOSviewer - Visualizing scientific landscapes. (2023). VOSviewer. https://www.vosviewer.com//.
10. Wankhede, P., Talati, M., and Chinchamalatpure, R. (2020). Comparative study of cloud platforms -Microsoft Azure, Google Cloud Platform and Amazon EC2. *J. Res. Engg. Appl. Sci.*, 05(02), 60–64. https://doi.org/10.46565/jreas.2020.v05i02.004.
11. Yu, J., Yang, Z., Zhu, S., Xu, B., Li, S., and Zhang, M. (2018). A bibliometric analysis of cloud computing technology research. *2018 IEEE 3rd Adv. Inform. Technol. Elec. Autom. Control Conf. (IAEAC)*. 2353–2358. https://doi.org/10.1109/iaeac.2018.8577750.

64 Solar PV-based DC-AC-DC converter for power supply

Jeyashree Y.[1,a], *Sukhi Y.*[2,b], *Anita S.*[2], *Kavitha P.*[2], *Fayaz Ahamed A.*[2] *and Supriya S.*[2]

[1]Department of Electrical and Electronics Engineering, SRM Institute of Science and Technology, Kattankulathur, Tamil Nadu, India

[2]Department of Electrical and Electronics Engineering, R.M.K. Engineering College, Kavaraipettai, Tamil Nadu, India

Abstract

In this paper, the design of resonant converter having low output power with high switching frequency and a high step-down voltage gain is discussed. In this paper analysis is done for different types of inverters and rectifiers. The estimates of complexity and efficiency are used to choose the type of class E rectifier and inverter and are implemented with three distinct power stages.

Keywords: Resonant converter, resonant inverter, DC-DC power converter, SMPS

Introduction

In case of power converter circuits, it is necessary to increase the power density which reduces the cost and size of the converter [1]. SMPS have a high value of the power density to improve the voltage gain which reduces the size of passive elements in the converter. These components have the capacity to store electric energy. Switching frequency is the key factor to decide the size of the converter. The size depends on the value of the passive components [2, 3]. It is possible to lower the size of the SMPS with an increase in switching frequency. SMPS topologies use converter circuits which has power loss. These converters produce more losses with hard switching and the power is dissipated during each cycle of switching [4, 5]. This effect of power loss also depends on the frequency range. The power loss is not considered much for switching frequency ranging between 50 and 400 kHz and it dissipates more power during switching with very high frequency operation. This leads to the requirement of extreme cooling of MOSFET and reduces its efficiency. This led to the evolution of resonant converters. This has led to the development of the resonant concept in the power supply circuits. New topologies have been developed in order to enhance the frequency and to prevent switching losses [6, 7]. The research has been made on dc/dc converters using RF amplifiers and rectifiers as shown in Figures 64.1 and 64.2. It is possible to obtain zero losses during switching with the implementation of resonant elements in the switches. When there is zero voltage across the MOSFET with current flowing, it turns on with zero losses. Theoretically, when the switching occurs, immediately and at an accurate time, switching loss should be eliminated [8]. This is not practically achievable. As a result, switching frequency controls the loss occurring in the circuit. Thus, increasing the value of frequency, certain challenges are posed. So, in the case of passive elements, there is a reduction in size with an increase in frequency. This assumption is generally held, except for certain cases. When the value of the frequency reaches megahertz, for most of the core materials the losses in the magnetic core rise quickly and become highly unacceptable.

Air core and integrated inductors on PCB make up a viable solution due to the possibility of manufacturing inductors in small physical sizes at these frequencies and the primary losses could be prevented.

It is possible to have less expense with high switching frequency because high frequency operation of the circuit will result in small size capacitors. High switching frequency facilitates EMI standards compliance since switching harmonics are easily removed by simple, inexpensive filters. Fast transient responses that are in great demand in some applications. The application of faster and reliable closed loop operation will result in effective and beneficial converter circuits. Since the converter will achieve a steady state at high switching frequencies in few microseconds, it is possible to employ a series of small size converters and to turn it on and off whenever required. The implementation of converter connection in series manner, the

[a]jeyashry@srmist.edu.in, [b]ysi.eee@rmkec.ac.in

DOI: 10.1201/9781003540199-64

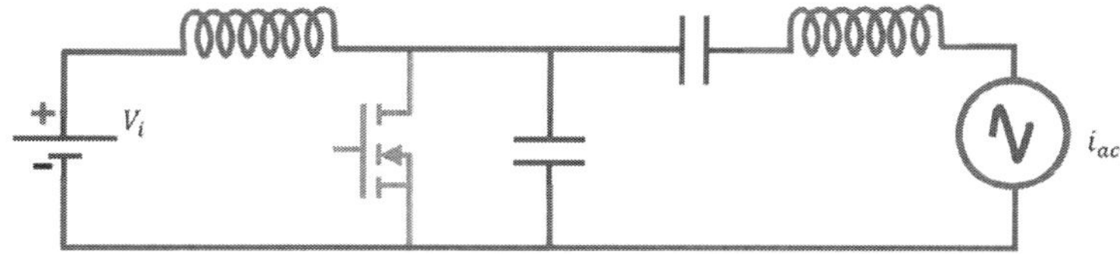

Figure 64.1 DC-AC converter circuit

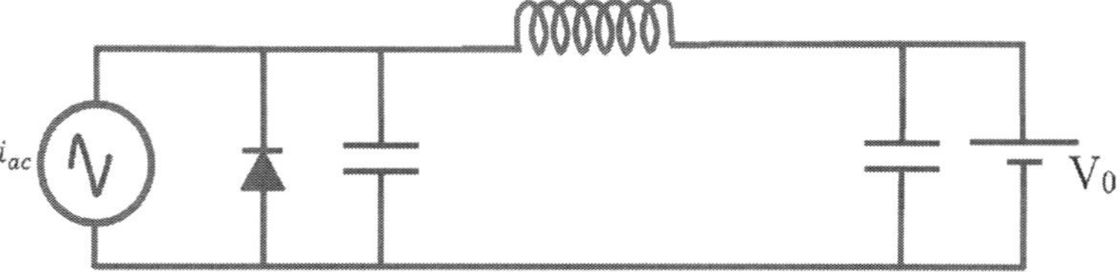

Figure 64.2 AC-DC converter circuit

load is divided among each converter which greatly simplifies the design. Autotransformer is added at the output to raise the load impedance. The implementation of parallel load with an array of converters having series of inputs will result in low output voltage. Class E inverter is the most commonly used inverter; however, other topologies do exist. For complete DC/DC converters the efficiency drops by 10%. Converters have limited gain and the highest gain is a step up of 2 and a step down of 6.6 times. A large reduction in gain is mainly due to the different ways of making load impedance appear larger, and it is not solely produced by the converter and gives the relationship in Equation 1.

$$V_{in}^2 \times f_s/P_{out} \propto \frac{1}{\eta} \tag{1}$$

This equation demonstrates that it is not easy to maintain high efficiency for low output power conditions under high value of supply voltage and high value of switching frequency. The input voltage determines the amount of electrical energy stored in the output capacitance of the switching device for each switching cycle and f_s determines how frequently this must occur each second. To make sure ZVS is proportionate to the sum of these values, energy must be circulated through the converter. Therefore, the relationship in Equation 1 is validated to attain high efficiency when the circulating current is high relative to the output power. This paper is developed to discuss low output power with low load voltages. This needs a serious discussion about a significant drop in voltage. Designing the rectifier often comes first, followed by designing the inverter for the input and load that are provided. However, it is also feasible to develop the inverter to work with a specific load. This approach frequently reduces the possible efficiency, size, and cost while unnecessarily increasing the converter complexity.

Resonant converter circuits

The process of converting alternating current to direct current involves the use of a rectifier. In this context, the junction capacitance of the diode can be compared with the output capacitance of the switching device. It is essential to ensure a smooth transition to prevent the diode from losing energy. This is achieved by ensuring that the discharge of charges in the capacitor has to take place before the turn on process of the diode. Already it is mentioned that achieving high efficiency becomes challenging for low capacity loads for high frequency switching conditions at high voltages. In order to overcome this problem, it is better to set the switching frequency to a particular frequency. When the required results are obtained at this frequency, then an incremental change in frequency can be done to make the converter size small. In this study, two approaches are considered to achieve the required results. The class E rectifier is widely used. The class DE inverter is the inverse of class DE rectifier. These methods provide ways to enhance the efficiency and performance of the conversion process.

Class E converter

One of the simple resonant converters, the class E converter that comprises one inductor, one diode and two capacitors as illustrated in Figure 64.3. These passive components added in the converter are capable of converting alternating current to DC current. When the high value of capacitance is of high value, it may result in a constant output voltage. Additionally, when the diode is ideal, it does not provide any reverse current and junction

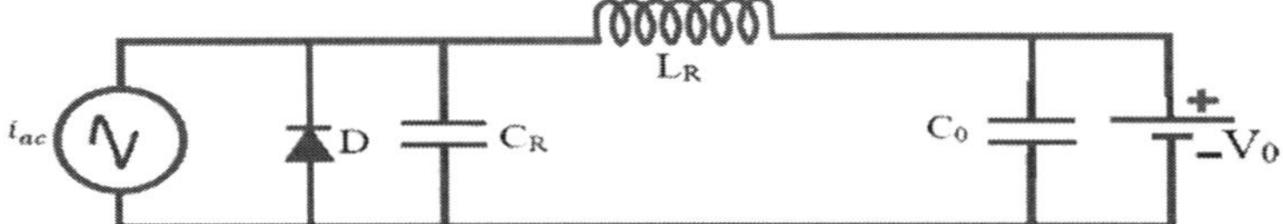

Figure 64.3 Class E rectifier configuration

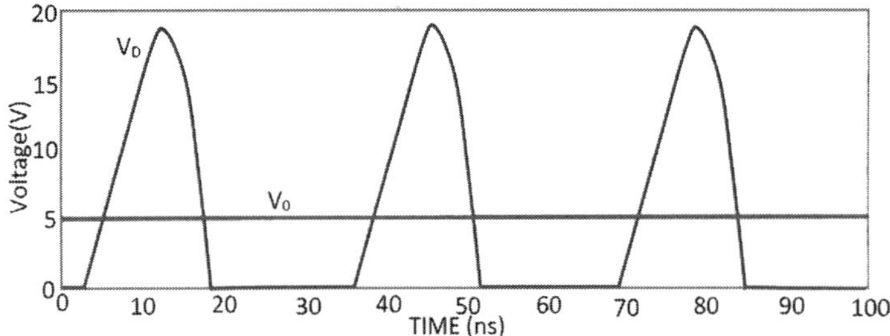

Figure 64.4 Waveforms of Class E configuration

capacitance with zero forward voltage drops. The rectifier load acts as resistor at resonance. At the resonant frequency, inductive reactance is equal to capacitive reactance. At this frequency, the design of the components is simple due the simplified formulae. Based on the inductive and capacitive components, the duty ratio is determined. If the value of duty ratio is less than the forward voltage drop is reduced. The reduced duty ratio results in less conduction period which results in reduced conduction losses.

The presence of a low forward voltage drop diode indeed imposes a limitation on the maximum efficiency. This limitation produces a decrease in efficiency of 10% during the converter configuration changes from inverter to converter. When the duty ratio is selected as 50%, then the maximum diode voltage is more than the diode voltage by 3.6 times.

$$C_R = \frac{1}{f_s \times R_L \times 2 \times \pi^2} = 67.5\text{pF} \tag{2}$$

This leaves a small margin below the maximum voltage rating of the diode. To calculate the value of CR, the following steps are followed using Equation 2.Using the above value, the inductance is determined below using Equation 3.

$$L_R = \frac{1}{4 \times \pi^2 \times f_s^2 \times R_L} = 417 \text{ nH} \tag{3}$$

The direct current flowing through the inductor is 0.2 A, which corresponds to the output current of 1 W and 5 V. Similarly, the maximum alternating current through is 120 mA as shown in Figure 64.4. The inductor has a DC resistance estimated as 25 mΩ and an AC resistance of 330 mΩ. The core size and the number of windings play a vital role in the calculation of inductance value. An inductor having 8 turns with 0.4 mm diameter wire and 6 mm core diameter is required for this value of inductance. When the losses are calculated for this inductor, it is found to be 2.4 mW due to alternating current and 1 mW due to DC current. These losses account for 0.34% of the output power. It is noted that large air core inductor has high resistance. On the other hand, the ceramic capacitors have lesser value of equivalent series resistance (ESR) of 200 mΩ [3]. Since these capacitors carry less value of current as compared with inductor, their losses are not considered as significant. The parasitic capacitance due to the charges stored in junction of the diode is added with the resonant circuit capacitance. According to the datasheet, the diode has a parasitic capacitance of 65 pF at a reverse voltage at 5 V.

Class DE converter

The class DE rectifier differs from the class E rectifier by having an additional diode and eliminating the use of inductors. However, the physical size and cost of both rectifiers are similar. Using the schematic shown in Figure 64.5 and the rectified output in Figure 64.6, the output capacitance is divided into two parts. For the purpose of analysis, both capacitances are assumed to be infinite, resulting in a pure DC output voltage. Since the diodes are directly connected in parallel with the output, the voltage developed in the diode becomes equal to the load circuit voltage, denoted as V_{OUT}. In the class DE rectifier configuration, the duty ratio of the diodes is chosen to be very from 0 to 0.5. However, a higher duty cycle makes both of the diodes to conduct simultaneously. From reference [6], Equations

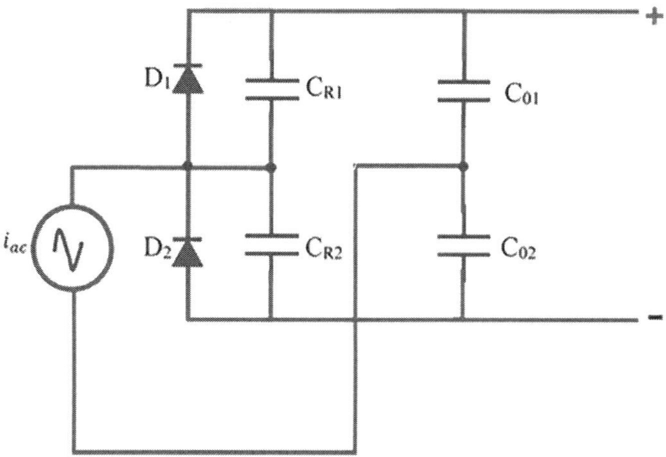

Figure 64.5 Configuration of class DE rectifier

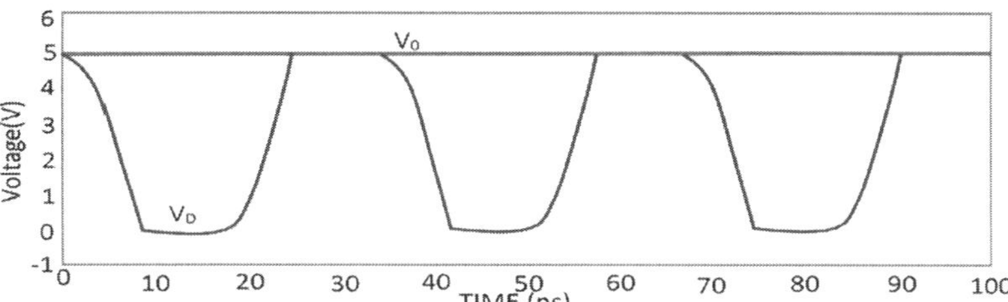

Figure 64.6 Waveforms of class DE rectifier configuration

4–6 provides relevant information. By isolating I_{imis} in Equation 4 and φ in Equation 6, the obtained results can be substituted into Equation 5. This allows for the isolation of resonant capacitor in order to determine the required capacitances for achieving a desired diode duty cycle, as described in Equation 7. With the duty ratio of diode as 25%, the capacitance is obtained as 667 pF.

$$V_0 = \frac{I_{imax} \times R_L}{\pi + \omega C_R R_L} \tag{4}$$

$$cos\emptyset = 1 - \frac{2\omega C_R V_0}{I_{im}} \tag{5}$$

$$D_D = \frac{\pi - \varphi}{2\pi} \tag{6}$$

$$C_R = \frac{\pi(1-\cos(\pi - 2\pi D_D)}{1+\cos(\pi - 2\pi D_D} \tag{7}$$

As previously mentioned, the diodes in the class DE rectifier are directly connected to the load, and it is able to provide a path for the DC current to flow to the output side load. Consequently, the dc value of current flowing through each of the diode will be equal to the output current, denoted as I_{OUT} in Figure 64.6. This leads to twice the diode losses compared to the class E rectifier, as both diodes carry the same output current. The higher diode losses in the class DE rectifier result in a total loss of more than 150 mW. Even if multiple diodes are placed in parallel for the implementation of drop in voltage in the forward biasing to some extent, the losses produced due to the flow of forward current exceeds 100 mW. When represented in percentage, these losses are found to be more than 10% of the power delivered to the load.

Selection of rectifier

After analyzing the two rectifiers, it is concluded that the class E rectifier can be considered as better choice than class DE. Although the difference of the size and cost of these rectifiers are not much considerable, the class DE rectifier relatively subjected to more switching losses. However, it should be noted that for high voltage applications, the class DE rectifier might be more suitable as it allows for lower voltage across the diodes, enabling the use of smaller diodes.

Whenever low output voltage required, the class E rectifier has the features to be one of the useful converters as per the application requirement.

With the possible design of the class E rectifier, it is found that the losses are above the acceptable range. To address this, one possible approach is to

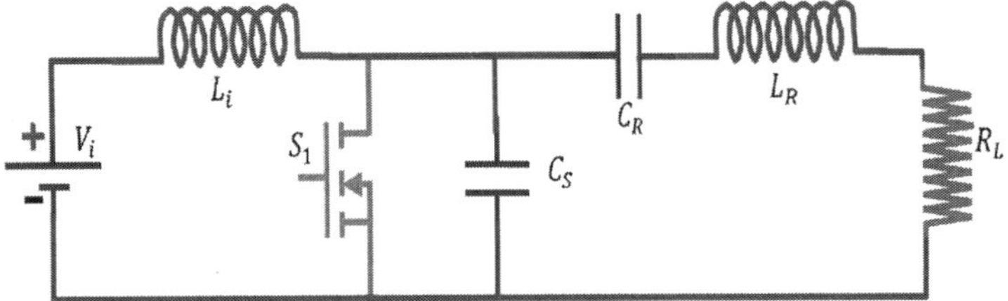

Figure 64.7 Class E inverter

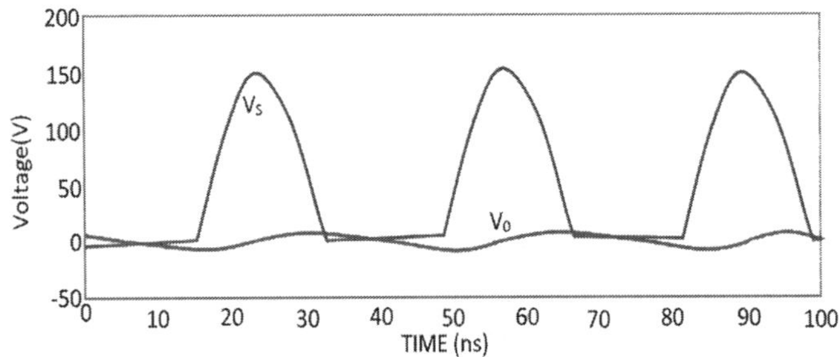

Figure 64.8 Waveforms of class E inverter

use a synchronous rectifier. By employing an additional MOSFET, the forward voltage losses can be eliminated. However, this would require appropriate gate drive and control for the added MOSFET.

Topology of resonant inverter

The resonance concept is used in resonant inverter to minimize switching losses. The circuits are designed to operate in either ZVS or ZCS in the inverter configuration. In some cases, both the switching strategy is employed simultaneously to reduce the losses. ZVS eliminates losses caused by parasitic capacitances, while ZCS eliminates losses due to parasitic inductance. In power applications involving MOSFETs and diodes, capacitors in the circuit contribute to the dominant losses during operating conditions. Therefore, ZVS is often the primary criterion for achieving high efficiency. However, in certain situations, it is easy to attain both ZVS and zero derivative switching (ZDS). The time at which the switches are operated plays a vital role for the amount of switching losses. During these switching, losses are very less because the voltage across the MOSFET momentarily reaches zero, minimizing the energy stored in the capacitor. This helps to reduce switching losses. When only ZVS can be achieved, precise timing becomes crucial. The MOSFET must be turned on precisely when the voltage across it reaches zero. If the switching occurs slightly too early, energy stored in the capacitor can cause switching losses. On the other hand, if the switching of the devices takes place in a delayed manner, there will be reduction in voltage below zero level, leading to conduction losses through the body diode.

Class E resonant inverter

The class E resonant inverter configuration and its waveform are shown in Figures 64.7 and 64.8, respectively. Two inductors, two capacitors, and one MOSFET form this circuit. This circuit offers a direct input current when the device is operating optimally. At low duty cycle, the circuit is inductive, and the inverter is built with both ZVS and ZDS switching. When the input voltage is 50 V with switching frequency of 50 Hz for an output of 1 W, the output capacitance required is 2.1 pF and an impedance of 1.44 kΩ.

Under special condition of fS, and P0, operation becomes optimal. For this condition, MOSFET switches having capacitance in the output with a value of around 2.1 pF is not available. The output capacitances that is at least ten times larger. The reactance is more than the rectifier input impedance. Consequently, ZVS and ZDS switching are not possible. But ZVS can be achieved with better control during transition which results in high efficiency. In this situation, the elements must be specifically selected to guarantee that ZVS is achieved to operate the converters in a sub-nominal situation [4].

Conclusions

The design of resonant elements of the resonant converter is carried out to develop the class E inverter. The different configurations of class E inverters are developed and the component values are calculated. As shown in the configuration, it does not have any switch in the high voltage side which reduces its complexity and improves the performance of the converter. The theoretical design of the DE inverter is found to be better due to its high voltage side gate drive. An efficient high voltage side gate drive design is crucial for maximizing the performance and reliability of power electronics systems, ensuring minimal power losses and achieving high efficiency in various applications. Different high voltage side gate drive techniques and specialized integrated circuits are able to address these challenges effectively. In class E topology, a combination of switching elements and passive elements is used to create a resonant circuit that allows for near-ideal switching behavior. The key principle behind class E topology is that the switching element operates as an ideal switch, meaning it has zero voltage and zero current when turned on or off.

Acknowledgement

The authors gratefully acknowledge the students, staff, and the management of R.M.K. Engineering College for their support in the research.

References

1. Junaid, K. A. M., Sukhi, Y., Jeyashree, Y., Jenifer, A., and A. Fayaz Ahamed. (2022). PV based electric vehicle battery charger using resonant converter. *Renew. Ener. Foc.*, 36, pp: 24-32. 10.1016/j.ref.2022.05.005.
2. Madsen, M., Knott, A., and Andersen, M. A. E. (2014). Low power very high frequency switch-mode power supply with 50 V input and 5 V output. *IEEE Trans. Pow. Electr.*, 29(12), pp:6569-6580. https://ieeexplore.ieee.org/document/6739180.
3. Lin, B. R. (2018). Soft switching resonant converter with duty-cycle control in DC micro-grid system. *Int. J. Electr.*, 105(1), pp:137-152. https://doi.org/10.1080/00207217.2017.1355020.
4. Sukhi Y., Jeyashree, Y., Perarasi, M., and Sarojini, B. (2017). Standalone PV-fed LED street lighting using resonant converter. *Elec. Power Compon. Sys.*, 45(5), pp: 548-559. . https://doi.org/10.1080/15325008.2016.1271063.
5. Jeyashree, Y. and Sukhi, Y. (2015). LCC resonant converter with power factor correction for power supply units. *J. Chine. Inst. Engg.*, 38, 843-854. 10.1080/02533839.2015.1037995.
6. Sugali, H. and Sathyan, S. (2023). Design and analysis of isolated high step-up Y-source DC/DC resonant converter for photovoltaic applications. *Ener. Sourc. Part A Recov. Utiliz. Environ. Effects*, 45(1), pp: 1604–1623. https://doi.org/10.1080/15567036.2023.2179698.
7. Sukhi, Y. and Jeyashree, Y. (2010). Analysis and implementation of series –parallel resonant converter for regulated power supply. *J. Inst. Eng. India*, 90, pp: 8–22.
8. Junaid, K. A. M., Sukhi, Y., Anjum, N., Jeyashree, Y., Fayaz, A. A., Debbarma, S., Chaudhary, G., Priyadarshini, S., Shylashree, N., Garg, S., Kumar, M., and Nath, V. (2023). PV-based DC-DC buck-boost converter for LED driver. *e-Prim. Ad. Elect. Eng. Electr. Ener.*, 5, pp: 100271. https://doi.org/10.1016/j.prime.2023.100271.

65 Zeta converter optimized with fractional order PID

Sumit Kumar[1,a], *Shimi S. L.*[2,b], *Lini Mathew*[3,c] *and Prakash Dwivedi*[4,d]

[1]Department of Electrical Engineering, NIT, Uttrakhand, India

[2]Punjab Engineering College, Chandigarh, India

[3]NITTTR Chandigarh, India

[4]NIT, Uttrakhand, India

Abstract

This study aims to control a DC-DC zeta converter feeded from PV-system or any other source using a fractional-order PID (FOPID) controller. Due to time varying V-I characteristic of the source, it necessitates a power electronic interface for achieving a desired and constant voltage level. To optimize system performance, the suggested controller parameters are adjusted utilizing the particle swarm optimization (PSO) algorithm. The responses of both the fractional order PID and conventional PID controllers are evaluated beneath varying power constraints by manipulating values for the load resistor as well as variable input conditions. The simulation's outcomes are examined based on distortion of its output response and for error constrained point of view its integral of time weighted squared error (ITSE). The results show that in terms of performance, FOPID controller outperforms the traditional PID controller.

Keywords: DC-DC converter, zeta converter, dPSO optimization, PID controller and FOPID controller

I Introduction

The global production of pollution is primarily attributed to non-renewable energy sources including coal, oil, and natural gas, among others. One significant limitation of these sources is their inability to regenerate after use. This drawback has paved the way for the introduction and adoption of renewable energy sources like wind, solar, and hydro power. One notable advantage of using renewable energy for electricity production is the low maintenance costs associated with these sources. By adjusting the main switches' duty cycles inside of the circuits, DC-to-DC converters are electrical circuits created to convert the direct current (DC) from one voltage level to another. These converters find extensive application in regulated switched-mode DC power supplies and DC motor drive systems. However, due to the operation of the switching devices, the DC/DC converters inherently exhibit nonlinear characteristics due to their fast switching properties [1]. These traits place restrictions on the closed-loop system's efficiency and make operating of DC-DC converters extremely challenging and difficult. The main objective of DC-DC converters is to maintain a stable output voltage regardless of fluctuations in the input voltage, load, or model being used. Typically, DC-DC converters transform the input voltage received from sources, such as a photovoltaic cell, into the desired level of voltage which is varying as per load requirement. The required output voltage is depends upon specific application [2]. In battery-operated portable devices, the converter plays a crucial role in converting the battery's input voltage into a suitable output voltage for electronic loads. With the variation in battery charge level, its voltage can span a wide range, potentially dropping below the load voltage at low charge levels [3]. D. W. Hart in ref [4] reported that the converter needs to be able to function in both boost and buck modes. To supply a consistent load voltage spanning the whole battery voltage range. The converters from Buck-boost, Cuk, SEPIC, and zeta satisfy this operational need. However, the basic forms of Buck-boost converter and Cuk converters produce output voltages with reversed polarity compared to the input voltage. Although the inclusion of an isolation transformer can solve this problem, it would result in larger and more expensive converters. However, as fourth-order DC-DC converters, the SEPIC and zeta converters can operate in both step-up and step-down modes, without facing polarity reversal issues.

As a result, they are highly attractive for the mentioned application [5]. Although the zeta converter boasts impressive features, it has not received as much attention as the Cuk and Sepic converters [6]. The zeta converters offer several key advantages: Firstly, they exhibit minimal output ripples, simplifying output regulation. Secondly, due to their constant output current, designers can optimize load performance by utilizing smaller capacitors. Most features of the zeta converter's operation are

[a]sumitkumar@nituk.ac.in, [b]shimisl@pec.edu.in, [c]linimathew@yahoo.com, [d]prakashdwivedi@nituk.ac.in

DOI: 10.1201/9781003540199-65

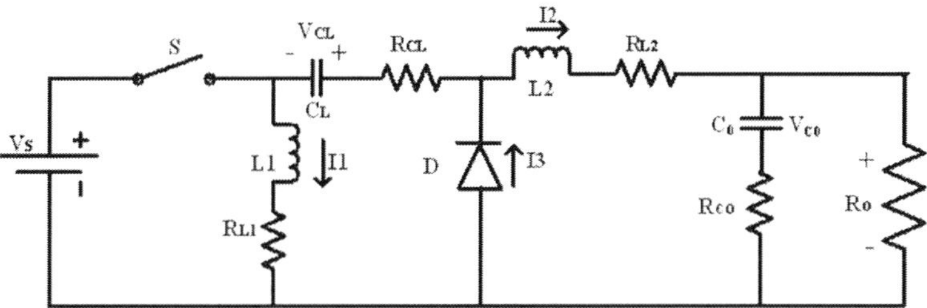

Figure 65.1 Zeta converter

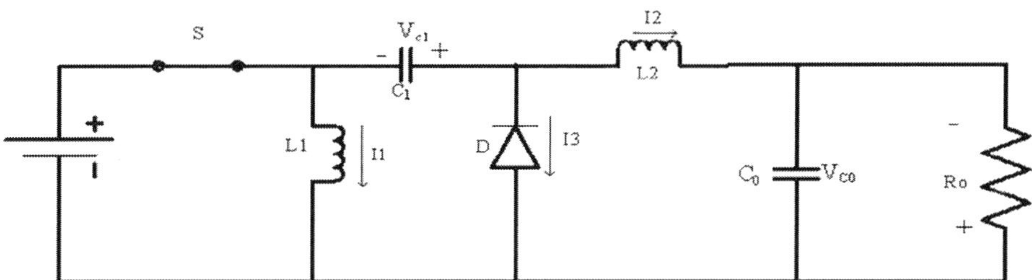

Figure 65.2 Mode-I

comparable to those of the Buck-boost converter, with the exception that its output polarity always coincides with its input polarity, (which is commonly used). The zeta converter is a fourth-order nonlinear system made up of a number of parts, including an operational power switch, a semiconductor diode, two inductors, and two capacitors. A PWM feedback control loop is frequently used to control the converter's output voltage. Create a linear model of the converter to make the feedback control design simpler or to perform a stability analysis [7]. To get a desired or constant output voltage from zeta converter a suitable controller is required. For this purpose PID and FOPID controllers can be used. In this particular study, we will discuss the performance analysis of PID and FOPID controllers tuned using particle swarn optimization (PSO) technique. PID controllers, known as proportional-integral derivative controllers, have gained widespread usage in controlling the speed and position of various applications. The Ziegler-Nichols approach is a commonly known technique for tuning of PID controllers but introducing significant overshoots. Consequently, in order to improve the effectiveness of conventional PID parameter tuning methods, alternative intelligent techniques such as genetic algorithms (GA) and particle swarm optimization (PSO) have been suggested in ref [8] Sahin in ref [9] stated that over the years, classical PID controllers have been widely employed in industry and literature due to their ease of implementation and robustness. However, researchers have recently shown interest in the FOPID controller in order to govern dynamic systems. In comparison of traditional PID controller, the fractional order PID controller is recognized to provide more precise system responses. When handling higher-order systems, the fractional order PID (FOPID) controller outperforms the standard PID controller in terms of flexibility and dependability and its ISO damping property, which enhances the closed-loop response's robustness against gain variations. To address the parameter adjustment challenges associated with the FOPID controller, For improving its settings, the particle swarm optimization (PSO) approach is used [10]. This research begins by investigating the zeta converter and subsequently focuses on the development of a robust controller using the FOPID controller to effectively tackle the identified issue.

This paper focuses on the design of FOPID optimized zeta converter. The controller's parameters are fine-tuned using the PSO algorithm to achieve the optimal system response. Paper is structured in following manner: Section II provides an explanation of the zeta converter's concise mathematical modeling. Section III describes the proposed controller design process. Implementation of PID controller its result and FOPID controllers and its subsequently result analysis is presented in Section IV. Observations are mentioned in Section V.

II. Zeta converter

A. Experimental framework

Kazimierczuk introduced the Zeta converter in 1980, which is a composition of two converters one is Buck-Boost, other is Buck converter. This configuration is illustrated in Figure 65.1. In the Figure

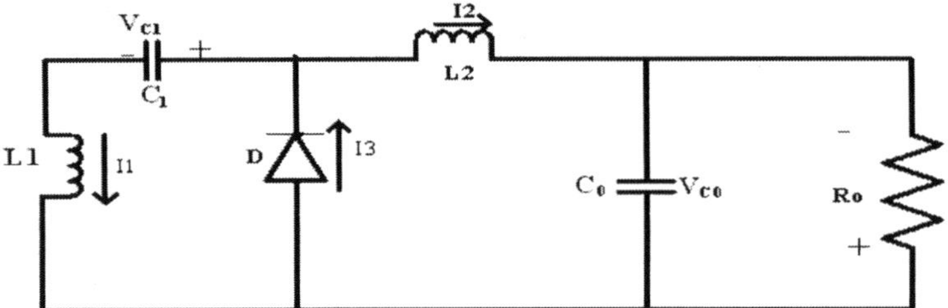

Figure 65.3 Mode-II

Table 65.1 Design requirement of zeta converter

S. No.	Parameters description	Notation	Experimental value	Unit
01	Source voltage	V_S	15	V
02	Output voltage	V_O	30	V
03	Source current	I_S	1.066	A
04	Inductor L_1 current	I_{l1}	0.833	A
05	Inductor L_2 current	I_{l2}	0.833	A
06	Output current	I_o	1	A
07	Duty ratio	D/d	0.4	-
08	Switching frequency	F_s	50	KHz
09	L_1,L_2 ripples	ΔI_{l1}	0.0083	A
10	Voltage ripple	ΔV_o	0.5	V
11	Inductor	L_1,L_2	28.91	mH
12	Capacitor	C	4	µF
13	Load resistance	R	50	Ω

65.1, R_0 represents the load resistance, while RC_L, RC0 represent the comparable resistance in series of capacitors and RL_1 and RL_2 equivalent resistance of inductors. In practical scenarios, the resistance values of capacitors and inductors are usually insignificant compared to the load resistance R0 due to their relative small magnitude.

B. Governing equations

The zeta converter functions in two modes: Mode-I when switch is in ON state and mode-II when switch is in off state. Depending on the duty ratio (D) value, the converter can either decrease or increase the supplied voltage. For both operational modes, mathematical modeling of the zeta converter is undertaken to generate an average state space model. This allows for a comprehensive understanding of its behavior. The state variables chosen are I1, I2, VCL and VCO.

Mode-ON

Figure 65.2 shows zeta converter switch on mode. L_1 stores energy first, and then C_1, L_2, and C_0 provide power to the load. Applying KCL and KVL in given circuit differential Equation (1)–(4) describes the dynamics of the system.

$$\frac{di_{L1}}{dt} = \frac{1}{L_1} V_S \tag{1}$$

$$\frac{di_{L2}}{dt} = \frac{1}{L_2} V_S + \frac{1}{L_2} V_{C_l} - \frac{1}{L_2} V_{C_0} \tag{2}$$

$$\frac{dV_{C_l}}{dt} = -\frac{1}{C_2} i_{L_2} \tag{3}$$

$$\frac{dV_{C_0}}{dt} = \frac{1}{C_0} i_{L_2} - \frac{1}{Rco} V_{C_0} \tag{4}$$

State space model of zeta converter for mode-1can be written as in (5), (6) using Equation (1)–(4).

$$\begin{bmatrix} \frac{di_{L_1}}{dt} \\ \frac{di_{L_2}}{dt} \\ \frac{dV_{C_l}}{dt} \\ \frac{dV_{C_0}}{dt} \end{bmatrix} = \begin{bmatrix} 0 & 0 & 0 & 0 \\ 0 & 0 & 1/L_2 & -1/L_2 \\ 0 & -1/C_2 & 0 & 0 \\ 0 & 1/C_0 & 0 & -1/Rco \end{bmatrix} \begin{bmatrix} iL_1 \\ iL_2 \\ VC_l \\ VC_0 \end{bmatrix} + \begin{bmatrix} 1/L_1 \\ 1/L_2 \\ 0 \\ 0 \end{bmatrix} V_s \tag{5}$$

Table 65.2 Designing inductor, core and wire parameters

S. No.	Parameter description	Notation	Desired values
01	Winding resistivity	ρ	1.427*10^{-6}
02	I_{Coil}	I_{max}	2
03	$R_{winding}$	R	0.65
04	Flux $_{Max\ Density}$	B_{max}	0.30
05	Fill factor	K_u	0.5
06	Cross-sectional area of core	AC	2.30
07	Core winding area	WA	2.56
08	Average length/turn	MLT	10.0

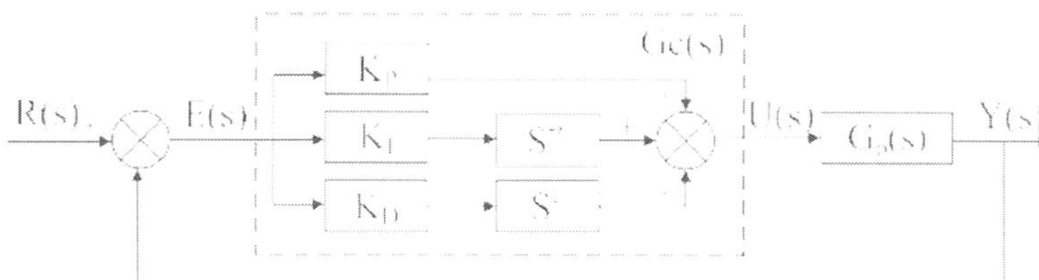

Figure 65.4 FOPID block-diagram

$$[V_0]=[0 \quad 0 \quad 0 \quad 1]\begin{bmatrix} iL_1 \\ iL_2 \\ VC_l \\ VC_0 \end{bmatrix} \tag{6}$$

Mode-OFF

Figure 65.3 shows the circuit diagram for the zeta converter's off state. While the load is receiving electricity from C_0, C_L and L_2 will store energy through L_1. We have gotten Equations (7)–(10) by using KVL and KCL in Figure 65.3.

$$\frac{di_{L1}}{dt} = -V_{C_l}\frac{1}{L_1} \tag{7}$$

$$\frac{di_{L2}}{dt} = -\frac{1}{L_2}V_{C_0} \tag{8}$$

$$\frac{dV_{C_l}}{dt} = \frac{1}{C_1}iL_1 \tag{9}$$

$$\frac{dV_{C_0}}{dt} = \frac{1}{C_0}iL_2 - \frac{1}{Rco}V_{C_0} \tag{10}$$

From Equation (7)–(10) state space model of zeta converter for mode 2 is as follows:

$$\begin{bmatrix} \frac{di_{L_1}}{dt} \\ \frac{di_{L_2}}{dt} \\ \frac{dV_{C_l}}{dt} \\ \frac{dV_{C_0}}{dt} \end{bmatrix} = \begin{bmatrix} 0 & 0 & -1/L_1 & 0 \\ 0 & 0 & 0 & -1/L_2 \\ 1/C_1 & 0 & 0 & 0 \\ 0 & 1/C_0 & 0 & -1/RC_0 \end{bmatrix}\begin{bmatrix} iL_1 \\ iL_2 \\ VC_l \\ VC_0 \end{bmatrix} + \begin{bmatrix} 0 \\ 0 \\ 0 \\ 0 \end{bmatrix} Vs \tag{11}$$

$$[V_0]=[0 \quad 0 \quad 0 \quad 1]\begin{bmatrix} iL_1 \\ iL_2 \\ VC_l \\ VC_0 \end{bmatrix} \tag{12}$$

The zeta converter's final state space is obtained by averaging the state space equations mentioned above (5), (6), (11) and (12) are necessary. Equations (13)–(15) provide the average matrices A, B, and C are.

$$A=(D)A_1+(1-D)A_2 \tag{13}$$

$$B=(D)B_1+(1-D)B_2 \tag{14}$$

$$C-(D)C_1+(1-D)C_2 \tag{15}$$

Where the ON and OFF modes' associated system matrices are A_1, A_2, respectively. Similarly B_1, B_2 are the input matrices and C_1, C_2 are the output matrices. Final state space equation is presented in (16) and (17).

$$X = AX + BVs \tag{16}$$

$$V_0 = CX \tag{17}$$

The table below lists the calculated values for design specifications, as well as the values of the capacitor and inductor.

By placing above values in state space model the transfer function is derived as:

$$G(s) = \frac{Vo(S)}{Vi(S)} = \frac{82.5(1 + \frac{s}{17.85} * 10^6)(1 - \frac{s}{1238.15})}{(1 + \frac{s}{2377.45})(1 + \frac{s}{2624.49})} \quad (18)$$

III Controller design

A. Classical PID controller

Because of its simplicity and ability to provide satisfactory results across a wide range of processes, PID controller is frequently employed in industries for closed loop control. PID controller offers a cost benefit ratio that is challenging to attain with alternative controllers. Industry data reveals that an overwhelming majority of regulatory controllers, accounting for the simplicity of their processes for designing and the effectiveness they deliver to system performance were the primary motivations for PID controllers' continued popularity [11]. Because of those characteristics, PID controllers have been able to seamlessly shift from mechanical and pneumatic networks to microprocessors, electronics tubes, the transistors, circuits with integrated circuits, and other technologies. The lasting appeal of PID controllers can be credited to their straightforward design procedures and the remarkable performance improvements they offer to systems [12]. The parallel form T/F is given by,

$$G(s) = KP + (KI)1/S + KD\,S \quad (19)$$

The transfer function equation for PID controller in its ideal form is given by,

$$G(S) = KP\,(1 + S/T1 + TD\,S) \quad (20)$$

In given system, the parameters are defined as follows: Kp represents the proportional gain. Ki represents the integral gain. Kd represents the derivative gain. Ti represents the integral time constant. Td represents the derivative time constant. Using an all-pass gain factor, the proportional term nurtures a control action that is directly proportional to the error signal. By introducing low-frequency compensation via an integrator, the integral term tackles steady-state faults.

B. The fractional order control system

In spite of the widespread use of the traditional PID controller in industry, applications of fractional-order calculus in control theory have been made possible by advances in fractional calculus. One notable application is the fractional-order PID controller, which has garnered significant recognition in both academic research and commercial use. The fractional order PID controller represents an development over the conventional integer-order PID controller and has demonstrated superior performance in many cases [12]. A comparative analysis between the classical PID controller and the fractional-order PID controller optimized through PSO has been conducted. A development of conventional integer-order calculus is fractional calculus that dates back to the early days of calculus' development. Despite its historical origins, fractional calculus remained relatively overlooked from a research standpoint until the last century. In recent times, scholars have increasingly focused on exploring the applications of fractional calculus in a number of fields, like control systems, and speech signal processing, and process modeling [13]. In comparison to the standard PID controller, FOPID controllers offer greater flexibility. While the standard PID controller has three controllable parameters (P, I, and D), FOPID introduces two additional parameters (λ and α) resulting in a total of five variables (P, I, D, λ and α). The inclusion of the two more parameters (λ for integration and α for derivative) adds

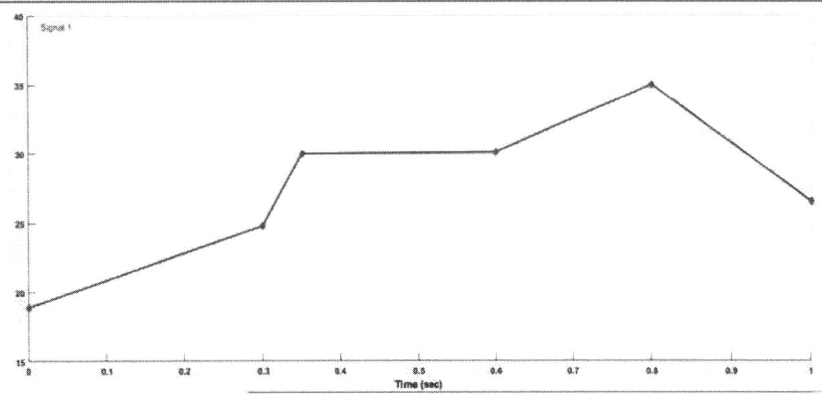

Figure 65.5 Variable input voltage waveform

Table 65.3 ITSE iteration

K_P	K_I	K_D	λ	α	ITSE
10	3	0.10	0.80	0.65	1.136
8.7	3.32	0.08	0.84	0.65	1.112
8	4.5	0.06	0.90	0.7	1.118

complexity to the tuning process of the FOPID controller. Its transfer function is given by

$$S=K_P+K_I S^{-\lambda}+K_D S^{\alpha}$$

where λ and α is greater than zero. The time domain equation of FOPID is given by

$$U\ (t) = K_P\ e(t) + K_I\ D^{-\lambda}e(t) + KD\ D^{\alpha}e(t)$$

Genetic algorithm approach is used for determining the optimal settings of the five parameters in fractional controllers to obtain optimized values for the parameters (P, I, D, λ, and α). By employing this method, we can achieve an optimized output from the FOPID controller (Figure 65.4).

C. PSO-based controller parameter optimization

The computational search algorithm known as the particle swarm optimization method, proposed by Kennedy and Eberhart, offers an iterative approach to problem optimization. This algorithm emulates the behaviors of a bird flock utilizing mathematical formulas for the velocity and position of particles [14]. Each particle in the population maintains a memory that stores its previous best position, referred to as Pbest (representing candidate solutions or local minima), along with its corresponding fitness value. Additionally, the particle with the minimum fitness value is referred to as Gbest (representing the global minima).

The mathematical formulas for the velocity and position of the particles are provided below, respectively in equation. In these formulas, the variable "i" represents the particle number, "d" denotes the dimension, "c1" and "c2" correspond to the acceleration constants of the velocity, "w" represents the inertia weight, and "r1" and "r2" are uniformly random numbers.

$$Vid = wVid + c1 \times r1 \times (Pid - Xid) + c2 \times r2 \times (Gid - Xid). \quad (23)$$

$$Xid = Xid + Vid \quad (24)$$

The algorithm concludes once the stopping criteria are satisfied. In this research, the integral of time weighted squared error (ITSE) criterion, a frequently utilized for performance measure, serves as the fitness function. The objective of the PSO algorithm is to determine the controller parameters that minimize the ITSE criteria.

IV Implementation of controllers and subsequent analysis of the obtained results

A. Implementation of PID controller on zeta converter topology

It is required to get desired and constant output from zeta converter as per requirement of load. As output of zeta converter is varying w.r.t input a controller in closed loop is required to provide a constant output irrespective of input variation [15]. In this section we will apply PID controller in closed loop for controlling output voltage. Different tuning techniques are available for tuning of PID controller. Traditional methods for tuning PID controllers, such as the Ziegler-Nichols (ZN) method, often result in significant overshoot. To overcome this limitation, modern heuristic approaches such as GA and PSO have been utilized to augment the capabilities of conventional techniques. These advanced approaches are intended to boost the effectiveness and performance of PID controller tuning. The PSO

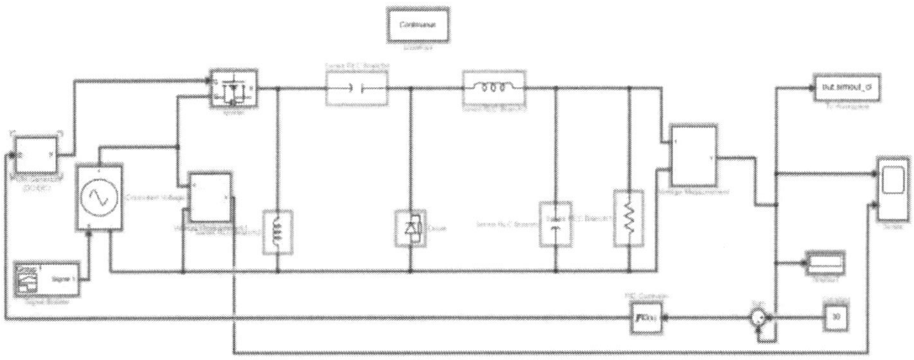

Figure 65.6 MAT lab Simulink model

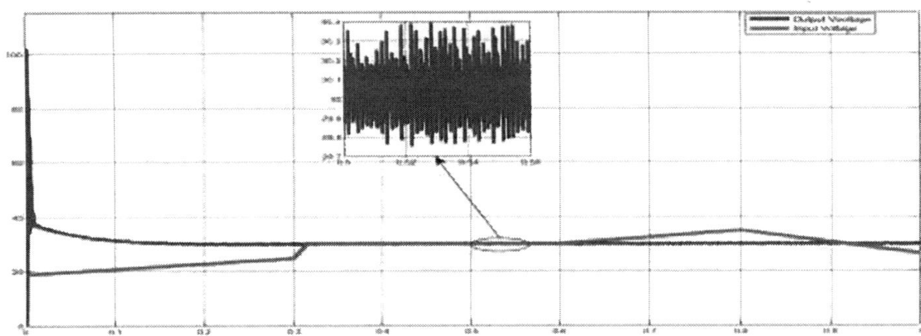

Figure 65.7 Input and output response with PID controller

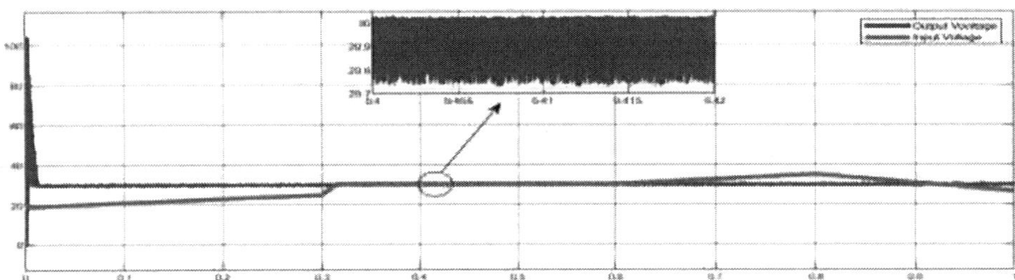

Figure 65.8 Input and output response with FOPID controller

Table 65.4 PSO optimization result

Controller	K_P	K_I	K_D	λ	α
PID	10.7	4.48	0.2	-	-
FO-PID	8.7	3.32	0.08	0.84	0.65

algorithm is utilized for offline tuning of PID parameters within the search space, which is expressed by a matrix, an initial swarm of particles evolves in the PSO process. With values ranging from 0 to 100, each particle represents a potential solution for the PID parameters. In this 3-dimensional problem, position and velocity are represented by matrices with dimensions of 3xSwarm size, where a swarm size of 40 is considered sufficient. By finding an optimal set of PID controller parameters, the system response can be significantly improved, leading to enhanced performance and minimized errors [16].

B. Implementation of FOPID controller on zeta converter topology

When compared to conventional PID controllers, fractional order PID (FOPID) controllers offer increased flexibility and reliability in controlling higher-level systems. FOPID controllers demonstrate ISO-damping property, which enhances the roughens against variations in closed loop system's gain. To address parameter adjustment challenges in FOPID controllers, The FOPID controller gets optimized via the particle swarm optimization (PSO) algorithm [17]. Key performance factors such as peak overshoot, rise time, and settling time are minimized using the PSO technique. The outcomes of the simulation verify the suggested methodology and exhibit the efficacy of the FOPID controller concerning control performance and resilience in contrast to PID controllers. There are several criteria for judging performance in this domain, including, in order to guarantee the efficient tuning of controllers in the time domain and appraise their performance.

$$IAE = \int_0^{\infty} e(t)dt$$

Integral Time Absolute Error.

$$ITAE = \int_0^{\infty} t^2 te(t)dt$$

Integral of Square Error

$$ISE = \int_0^{\infty} t^2 e(t)dt \ te^2$$

Integral Time Square Error

$$ITSE = \int_0^{\infty} te^2(t)dt$$

C. Simulation findings

Simulation results used to validate the FOPID and PSO-optimized PID controllers for the zeta converter is represented under findings.

To assess system performance, the entire system is modeled in the Matlab/Simulink platform and tested for variable input voltage Figure 65.5. Design and system configuration in the Matlab/Simulink platform are depicted in Figure 65.6.

ITSE iteration for different values of controllable variables (P, I, D, λ and α) is mentioned in Table 65.3.

The best parameters for PSO optimized FOPID was selected for minimum value of ITSE given in Table 65.4.

Figure 65.7 shows the closed loop input and output voltage using PSO optimized PID controller. In this case, the incoming voltage changes from 19 to 25 V for the first 0.0–0.3 s, then is increased to 30 V from 0.35 to 0.36 s and remains constant at 30 V for the next 0.6 s, before being increased once more to 35 V and then made to decrease to 26.5 V for the final 0.8–1 s. Here we use PSO optimized PID controller for getting constant output voltage. The required voltage of system is 30 V while the actual system output varies from 29.70 to 30.40 V. The distorted resultant voltage is shown in zoomed section.

Figure 65.8 depicts the input and output voltage waveform of PSO optimized FOPID controller. The input voltage is remain same as Figure 65.6. The corresponding output voltage of system varies from 29.75 to 30 Volt with less distortion.

V. Conclusions

Different type of controllers and optimization techniques has been discussed in literature. Both the controllers, classic PID and FOPID is tested for variable input voltage. The corresponding output voltage waveforms depicts that output voltage is having very less distortion and better system response with FOPID controller instead of classic PID controller.

References

1. Babu, P. R., Prasath, S. R., and Kiruthika, R. (2015). Simulation and performance analysis of CCM zeta converter with PID controller. *Inter. Conf. Circuit Power Comput. Technol. [ICCPCT].*
2. Chandan, B., Dwivedi, P., and Bose, S. (2019). Closed loop control of SEPIC DC-DC converter using loop shaping control technique. *2019 IEEE 10th Control Sys. Grad. Res. Colloq. (ICSGRC 2019).*
3. Vuthchhay, E. and Bunlaksananusorn, C. (2010). Modeling and control of a zeta converter. *Inter. Power Elec. Conf.*
4. Hooshmandi, K., Bayat, F., Jahed-Motlagh, M. R., and Jalali, A. (2018). Robust sampled-data control of non-linear lpv systems: time-dependent functional approach. *IET Control Theory Appl.*
5. Erickson, R. W. and Maksimovi, D. (2001). Fundamentals of power electronics. 2nd edition. Kluwer Academic Publishers.
6. Tan, S.-C., Lai, Y.-M., and Tse, C.-K. (2011). Sliding mode control of switching power converters: techniques and implementation. CRC press.
7. Admane, A. and Naidu, H. (2018). Analysis and design of zeta converter. *Inter. J. Innov. Res. Multidis. Field.*
8. Solihin, M. I., Tack, L. F., and Kean, M. L. (2011). Tuning of PID controller using particle swarm optimization (PSO). *Inter. Conf. Adv. Sci. Engg. Inform. Technol.*
9. Erdinc, S., Mustafa, S., Ayas, T., Turkey, I., and Hakki, A. (1990). Design of pi controllers for dc-to-dc power supplies via extended linearization. *Inter. J. Control*, 51(3), 601–620.
10. Jain, R. V., Aware, M. V., and Junghare, A. S. (2016). Tuning of fractional order PID controller using particle swarm optimization technique for DC motor speed control. *1st IEEE Inter. Conf. Power Elec. Intell. Control Ener. Sys.*
11. Plaza, D., De Keyser, R., and Bonilla, J. (2008). Model predictive and sliding mode control of a boost converter. In power electronics, electrical drives, automation and motion. *SPEEDAM Inter. Symp.*, 37–42.
12. Qi, Z., Yang, H., Li, H., and Li, J. (2019). Research on fractional order PID controller in servo control system. *IEEE Conf. Ener. Internet Ener. Sys. Integ.*
13. Rao, G. K., Subramanya, M. V., and Sathyaprakash, K. (2014). Study on PID controller design and performance based on tuning techniques. *Inter. Conf. Control Instrum. Comm. Comput. Technol.*
14. Komurcugil, H. (2012). Adaptive terminal sliding-mode control strategy for dc–dc buck converters. *ISA Trans.*
15. Bonilla, J., De Keyser, R., and Plaza, D. (2007). Nonlinear predictive control of a dc-dc converter: A nepsac approach. *Control Conf. (ECC).*
16. Olalla, C., Leyva, R., El Aroudi, A., and Queinnec, I. (2009). Robust lqr control for pwm converters: an lmi approach. *IEEE Trans. Indus. Elec.*, 56(7), 2548–2558.
17. Chander, S., Agarwal, P., and Gupta, I. Design modelling and simulation of DC-DC converter.

66 Social welfare in the smart grid scenario with an optimal schedule of plugin vehicle charging

Susmita Dhar Mukherjee[1,a], *Rituparna Mukherjee*[1,b], *Sandip Chanda*[2,c] *and Abhinandan De*[3,d]

[1]Department of Electrical Engineering, Swami Vivekananda University, Barrackpore, West Bengal, India

[2]Department of Electrical Engineering, Ghani Khan Choudhury, Institute of Technology, West Bengal, India

[3]Department of Electrical Engineering, Indian Institute of Engineering, Science & Technology, West Bengal, India

Abstract

In a scenario with a smart grid, the addition of demand response may show to be a fantastic input for solving the "optimal power flow (OPF)" problem and achieving operational excellence. The price-demand properties of plugin vehicles (PEV) can help the OPF programmes to provide fruitful results in the grid optimization. In order to support operational standards in power system networks with the inclusion of PEV demand response in the smart grid space, this paper offers a methodology. The unique price elastic properties of PEVs are backed by a methodology and objective function that are presented in this research in order to reduce the cost of electricity generation while providing maximum load catering. In order to guarantee the bare minimum dispatch demanded by customers and PEVs even in the worst-case scenarios for the system, the methodology includes a "revolutionary load curtailment strategy". When this methodology is used effectively with PEVs, the total advantage of all market participants in the power sector or social welfare may be guaranteed, the operation will be in a stable state, and the amount of electricity wasted will be kept to a minimum.

Keywords: PEV, IEEE 30 bus system, battery

Introduction

A smart grid is a reorganized electrical system that combines the behaviors of providers and consumers through modern information and communications technology in order to boost the "efficiency, dependability, economics, and sustainability" of the production and distribution of power. In order to be reliable and cost-effective, local grids had to be networked in the 20th century. The number of power plants increased between the 1970s and 1990s as a result of rising demand. Some areas experienced poor power quality, including blackouts and brownouts (bad voltage profile), due to the inability of the electricity supply, especially during peak hours, to keep up with this demand. Since the beginning of the twenty-first century, the potential for utilizing electronic communication technology to lower the cost of the electrical power generation, transmission, and distribution grid and improve power system performance while operating within limits has increased. Peak power prices were equitably spread among all participants in the power market thanks to the deployment of these new technologies, which successfully averaged them out. Engineers are concentrating on the efficient utilization of sizable amounts of renewable energy as a result of the tremendous and growing environmental issues brought on by conventional fossil-fired power plants. Due to their high cost of production and erratic nature, wind and solar energy were not compatible with existing grid control systems, necessitating the development of more sophisticated ones. The traditional grid was turned into the smart grid—smarter in the sense that it emphasizes social welfare, which is a merited benefit for all participants broadly—power producers and consumers. This was done in response to the necessity for dependable ways of controlling bidirectional electronic communication equipment.

The smart grid was developed by combining the power system and the information network and integrating renewable energy sources, demand side management, and standard operating procedures. The smart grid is also designed from the ground up to be self-healing in order to boost dependability and respond to intentional sabotage or natural calamities [1]. An efficient information network deployment boosted by a smart grid may seem to be an invaluable instrument for managing the system's operational status and maximizing system performance to provide a successful result [2, 3]. In an effort to maximize the use of these additional resources, researchers have lately proposed local approaches and solution algorithms. The plug-in

[a]susmitadm@svu.ac.in, [b]rituparnamukherjee@svu.ac.in, [c]sandipee978@gmail.com, [d]abhinandan.de@gmail.com

DOI: 10.1201/9781003540199-66

electric vehicles (PEVs) are one of those promising resources that have recently emerged in the field of automotive technology, with enormous potential for regulating the functioning of the future grid with the addition of a clean environment. PEVs are so well-liked because they maximize the usage of the power system's off-peak hours for battery charging and supplying the vehicles at times of high electrical demand. But PEV's advantages go well beyond this; it has also been discovered to be able to supply gird during times of peak loading. Although in ref [4] it approaches are highly relied upon by power system planners, ref [5] provided a unique work in which operating conditions, such as loss and voltage profiles, were optimized with a coordination methodology for plug-in vehicle charging. In a smart grid, decentralized demand side control of plug-in vehicles has been shown in refs [6, 7]. While charging PEVs sufficiently during peak loads, the method efficiently flattens loads, but it was unable to anticipate the total PEV demand and its timing. A prediction-based strategy for charging PEVs was shown in ref. [8], where the best charging schedule for PEVs was created using dynamic price information. The demand response characteristics were successfully utilized in the technique, however, the designed algorithm's impact on operating conditions was not made clear. Although ref [9] developed an OPF to produce a PEV charging plan that takes operating factors into consideration, the method was unable to lower generation costs by maximizing load catering. Battery storage advantages were employed in ref [10] to optimize the generation and loading schedules of microgrids, but it was unable to ensure a uniform level of operational conditions. All of these methods recommended utilizing demand response effectively to improve PEV and smart grid performance. While in refs [12–14] examined the challenges of incorporating demand response in the grid's distribution, authors in ref [11] proposed an architecture for the future transmission grid. This study demonstrates it amply. To improve the operational state of a power system that uses a certain renewable energy source, authors in refs [15] and [16] suggested optimization approaches. Ref [17] showed a novel algorithm for controlling system parameters when there are several intermittent sources. The energy hub concept was used in the model to optimize system functioning, but its implementation would require effective infrastructure, which may not be accessible. In addition, the model does not take into account problems like line flow control and load curtailment. Ref [18] worked with integrating demand response with variable energy sources. The proposed multi-objective algorithm reduced costs and curtailed but was unable to maximize load catering for the system operator. For the best use of erratic or renewable energy sources, charged batteries can be deployed effectively. Another benefit of PEVs in the smart grid is seen here. The fundamental goal of the smart grid is to promote social welfare, or the benefit of all market players. According to a literature review, PEVs can be designed to function with demand response and intermittent renewable energy sources. They can also be used to standardize operating settings for the system's voltage and loss profile Additionally, peak load flattening, which lowers the dynamics of electricity price, is effective with it. The study discussed in this chapter aims to increase social welfare through PEVs. The objectives in refs [19] and [20] incorporated social welfare as a goal but placed less focus on maintaining system operation or the load shedding mechanism. In the smart grid, load shedding or curtailment techniques are a problem for reliability. While fresh solutions are provided in refs [23] and [24], the present load curtailment tactics are described in refs [21] and [22]. The majority of these solutions are optimal load curtailment (OLC) programmes, which under typical operating conditions cannot assure a client of a reliable supply. In order to segregate stations based on their willingness to pay for the best curtailment of charging schedules, power systems with PEV charging stations need to use a particular load curtailment strategy. It is clear from the aforementioned survey that a standard parametric operational condition is necessary to keep power costs as low as possible while also ensuring ideal load catering and maximizing PEV utilization. A pricing model that takes demand response, generation characteristics, price sensitivity of voltage profile, and line length into account is necessary for the algorithm to work well. The choice of a stochastic optimization method like particle swarm optimization (PSO) was influenced by the convex character of the solution algorithm with nonlinear working surface. A typical OPF [25] has been used to compare the answer that was achieved. The simulations were run in the IEEE 30 bus system, and the results were fairly encouraging.

Plug-in electric vehicles

The most cutting-edge and environmentally benign alternative to merely conventional fuel-based vehicles is emerging as plug-in hybrid electric vehicles. The electric power storage system that is integrated into these kinds of cars to store electrical power from power outlets is referred to as a "plug-in" system. More of these kinds of cars are being developed as a result of the scarcity of fossil fuels and the

Table 66.1 Generator characteristics of IEEE 30 bus system

Gen bus no	A	b	c	P_{max}	P_{min}
1	0.00375	2	0	50	200
2	0.0175	1.75	0	20	80
5	0.0625	1	0	15	50
8	0.0083	3.25	0	10	35
11	0.025	3	0	10	30
13	0.025	3	0	12	40

resulting price increase. It is referred to as a hybrid when it has both a battery storage system for an electric motor and an internal combustion engine (ICE) that runs on gasoline. But in contrast to commercially available hybrid electric vehicles (HEVs) that use hefty nickel-metal hydride (NiMH) batteries, plug-in electric vehicles (PEVs) use lighter lithium-ion (Li- ion) batteries that can be charged from any household outlet.[28]

In terms of autos, vehicles of the PEV-20, PEV-40, and PEV- 60 types are found to operate effectively. Here, the numbers 20, 40, and 60 represent the electrical range of the cars, i.e., how far their completely charged batteries can take them, 20, 40, and 60 miles, respectively [29].

No matter how effective a PEV is, after a capacity run, it needs a power source to recharge. Home outlets may be used for charging, or charging stations may be set up in designated areas. Therefore, the grid and PEVs need to be able to communicate in both directions so that the grid can inform the PEV of its limitations, generation capabilities, load catering priorities, and the price at which it can afford to sell power to the PEVs, and the PEVs can express their needs for time of charging and price at which they can purchase power from the grid. It is important to note in this context that PHEV vehicles can run on gasoline because they are hybrid vehicles Therefore, if the CO_2 emission considerations are disregarded, the price of power is a crucial grid component that will determine whether PEVs can be charged straight from the grid [30].

Social welfare

Social welfare is a wide word that describes the general well- being of power market players. The supplier and the consumer are the two possible groups of participants in a deregulated power system. They however, offer contradictory objective of participation. The supplier wants to sell maximum of its produced power at maximum profit whereas the consumer wants to purchase power at the lowest possible rate. Their dynamic response is also opposite with respect to price of electricity. When prices are higher, the producer increases production and the customer decreases demand. Simply a compromise between these two qualities, price equilibrium. By overall welfare an operating condition is meant where the participants with different characteristics negotiate and compromise to benefit each other. In social welfare theory, a participant would rather think of the benefit of other participants to cause social benefit thereby causing its own benefit.

1.1 Social welfare optimization problem

With regular operating conditions, it maximizes load catering while concurrently minimizing expense. If the market clearing price falls, the consumer is encouraged to use more energy, which benefits the suppliers by driving them to sell more energy. The cost optimization objective is though popular but fails to restrict the misuse of electricity and cannot associate the benefit of supplier with the consumers. In the following section an objective function has been formulated where the operational standard constraint generation cost has been subtracted from the consumer cost benefit function. With regular operating conditions, it maximizes load catering while concurrently minimizing expense. If the market clearing price falls, the consumer is encouraged to use more energy, which benefits the suppliers by driving them to sell more energy.

1.2 Description of the methodology

The description of the methodology developed in this work has been depicted in Figure 66.1.

Illustrative case study

A modified IEEE 30 bus scenario was used for the case study, and the inputs to OPF were generator characteristics as well as price-responsive demand characteristics or demand response. The appendix contains a single line diagram, busdata, and line data.. In the example study, generator characteristics and price-responsive demand characteristics, or demand response, were inputs to OPF using a modified IEEE 30 bus scenario. One line The appendix contains the schematic, the busdata, and the linedata. The off-peak PEV has a price-responsive demand response, and its ability to reduce demand while maintaining operational standards. The price at which a load of the continuous charging type can be charged relies on the market price of gasoline, which was implied in the preceding section. The shiftable electrical load and PEV both have characteristics that allow them to switch to a new time

Table 66.2 Consumer cost benefit function

Bus no	Type of load	a	b	c	Pmin	Pmax
2	PEV off-peak	-0.2	95	0	21.7	31.7
3	PEV shifted	-0.3	87.2	0	2.4	5
4	Shiftable electrical load	-0.5	85.5	0	7.6	10
5	PEV continuous charging	0	90	0	110	110
7	Shiftable electrical load	-0.15	70	0	22.8	40
8	Shiftable electrical load	-0.2	90	0	30	45
10	PEV continuous charging	0	70	0	10	10
12	Shiftable electrical load	-0.2	50	0	11.2	15
14	Shiftable electrical load	-0.3	80	0	6.2	12
15	PEV continuous charging	-0.3	60	0	10	10
16	PEV Off peak	-0.3	70	0	3.5	10
17	PEV continuous charging	-0.1	65	0	15	15
18	Shiftable electrical load	-0.2	90	0	3.2	5
19	Shiftable electrical load	-0.5	68	0	9.5	12
20	PEV off peak	-0.5	90	0	2.2	5
21	PEV continuous charging	-0.4	95	0	20	20
23	Shiftable electrical load	-0.2	60	0	3.2	6
24	Shiftable electrical load	-0.4	95	0	8.7	10
26	Shiftable electrical load	-0.6	90	0	3.5	5
29	PEV continuous charging	-0.4	95	0	4	4
30	Shiftable electrical load	-0.4	200	0	10.6	15

Table 66.3 Forecasted percentage availability of intermittent sources

Hour	Gen 1	Gen 2	Gen 3	Gen 4	Gen 5	Gen 6
h-1	100%	100%	100%	100%	100%	100%
h-2	100%	100%	60%	90%	100%	50%
h-3	100%	100%	40%	90%	100%	30%
h-4	100%	100%	20%	90%	100%	10%

period when the price of fuel or electricity changes (Table 66.1).

1.3 Base case

The sensitivity matrix A between these two states is obtained by linear approximation. If all the Eigen values of matrix A, have zero real part then the solution becomes asymptotically stable against small signal disturbances. To justify this equilibrium point Lyapunov's concept has been followed that is $\dot{x} = 0$

As declared earlier state variables are the generation of the six generators, operational status that is minimum bus voltage, maximum line flow, total transmission loss, load curtailment and active and reactive power loads. The optimizer is trained to increase the loading up to maximum possible level. Hence the convergence of optimizer would suggest even with increase of loading the state variables are efficiently manages loads of PEVs to cater maximum demand at minimum cost in the most optimal way maintaining all operating constraints. not changing or in other words $x(n+1) = x(n)$ or $x = 0$. Thus, base case refers to a particular operating condition at which the state variable does not change and constitute to offer a particular price of electricity.

1.4 Performance of developed methodology with intermittent energy sources

Promotion of distributed generation with renewable energy sources as stated earlier is one of the

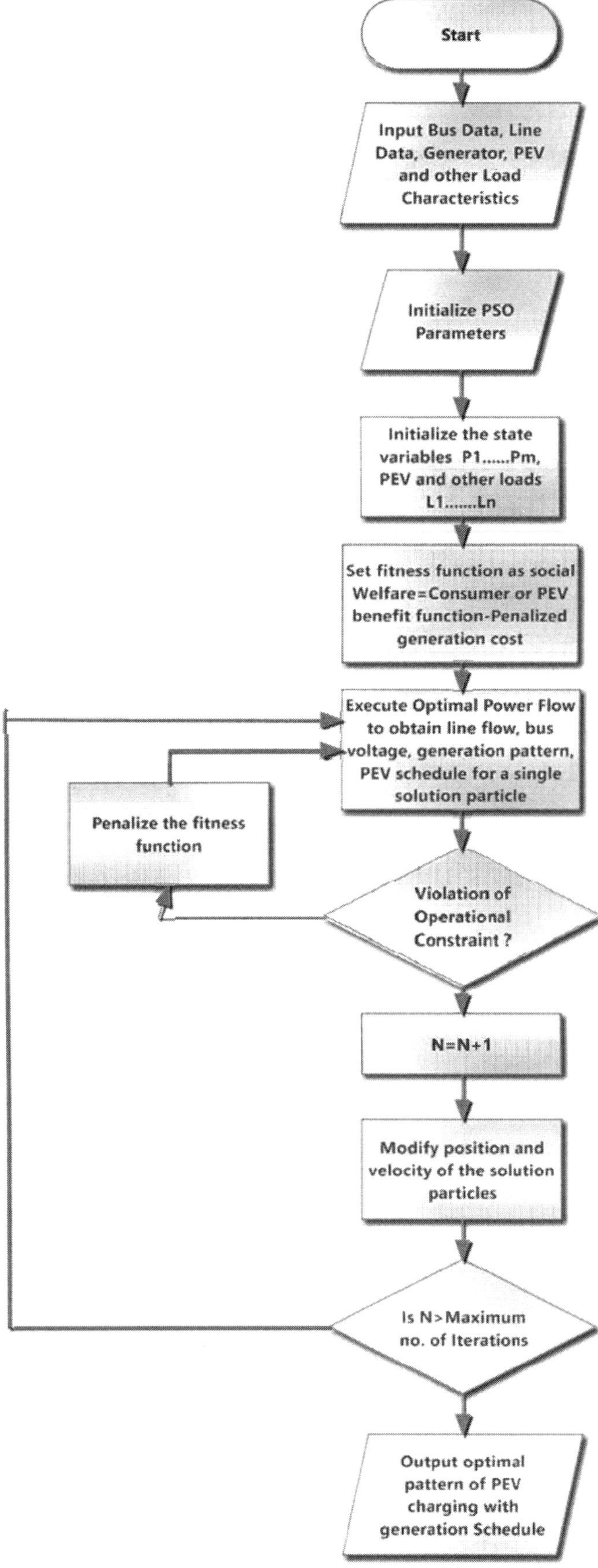

Figure 66.1 Flowchart of the developed methodology

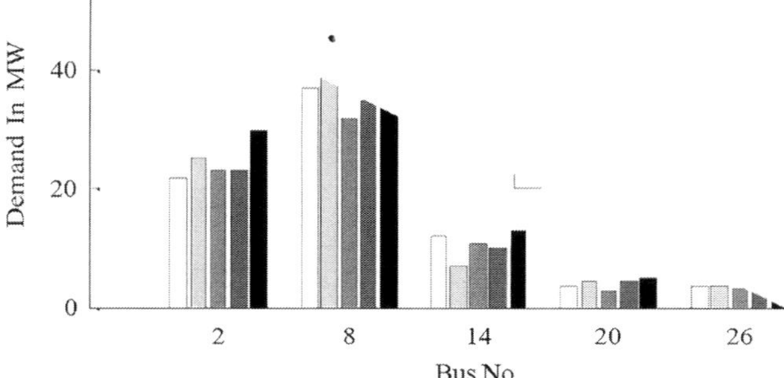

Figure 66.2 Peak shaving of demand during intermittency of generation

Table 66.4 Performance comparison

Observation	MCP ($/M Wh)	Traditional optimization			Proposed optimization		
		Line Loss (MW)	Max Line Flow (MW)	Min Bus Voltage (p.u.)	Line Loss (MW)	Max Line Flo (MW)	Min Bus Volt age (p.u.)
h-1	3.33	12.	124	0.97	6.7	71.	0.97
		06	.85	40	5	91	63
h-2	3.12	14.	140	0.97		129	0.97
		20	.80	41	13.		01
					26		
h-3	3.08	13.	142	0.97		105	0.97
		27	.96	46	12.	.13	67
					50		
h-4	2.75	18.	121	0.97		110	0.99
		14	.28	39	15.	.09	00
					80		

non-negotiable objectives of the present and future grids. However, their intermittent generation profile requires extensive load management for power market equilibrium. With the intermittency of the generators as shown in Table 66.3, the performance of the developed algorithm has been observed to be remarkably superior in sustaining the operating conditions within safe limit (Table 66.3), owing to the fact that the developed algorithm not only manages generation for an optimal cost (as in case of traditional method) but also

The efficiency of the proposed algorithm to self heal under the alterations of operating conditions particularly during peak periods, when the available generation is less than requested demand of PEVs is depicted in Figure 66.2, where the peak shaving attribute of the proposed algorithm has been demonstrated at selected buses.

Conclusions

The accessibility to smart metering infrastructure has brought about radical changes in power grid operation and simultaneously has thrown newer operational challenges for electric power system operators, where a departure from traditional operations planning, scheduling, and dispatch practices needs to be altered to take into account tribulations like voltage instability, line congestion, line loss intensification and payment cost minimization. In this pursuit this paper presents a new state space based pricing model and a methodology to illustrate optimal and efficient operations of smart grid with the presence of plug in electric vehicles (PEVs). The model effectively identifies the state variables of MCP and regulates the same utilizing PSO to reach prolific solutions negotiating with generator characteristics, demand response of shiftable loads, PEVs both continuous and off peak charging type , voltage stability, and congestion, curtailment and line loss limits. As the market participants are aware of pricing signals, the proposed model will enable them to regulate their consumption, production or curtailment limits to cause an overall social welfare. The use of PEVs is constricted by the price of gasoline in the market. The methodology adopted takes that into account also to achieve optimum utilization of their characteristics. Simulation results convincingly demonstrate that the combined operations of the proposed model and the methodology with PEVs mitigate network constraints while catering higher demand levels and reducing the energy costs even under worst possible states of the system operation. In comparison with the conventional system operation which is based on only GENCOs scheduling between maximum and minimum possible generation, the proposed algorithm incorporates consumer or PEV scheduling with a smart communication facility where not only GENCOs but also PEV characteristics can participate to change their stand for their own and social benefit.

References

1. Gang, L., Debraj, D., et al. (2010). Smart grid lab: A laboratory-based smart grid test bed. 1st IEEE Inter. Conf. Smart Grid Comm., 143–148.
2. Rahimi, F. and Ipakchi, A. (2010). Demand response as a market resource under the smart grid paradigm. *IEEE Trans. Smart Grid*, 1(1), 82–88. doi: 10.1109/TSG.2010.2045906.
3. Ipakchi, A. (2011). Demand side and distributed resource management — A transactive solution. *2011 IEEE Power Ener. Soc. Gen. Meeting*, 1–8. doi: 10.1109/PES.2011.6039272.
4. Deilami, S., Masoum, A. S., Moses, P. S., and Masoum, M. A. S. (2011). Real-time coordination of plug-in electric vehicle charging in smart grids to minimize power losses and improve voltage profile. *IEEE Trans. Smart Grid*, 2(3), 456-467. doi: 10.1109/TSG.2011.2159816.
5. Erol-Kantarc, M., et al. (2010). Decentralized demand side management of Plug-in hybrid vehicles in a smart grid. *Proc. First Inter. Workshop Agent Technol. Ener. Sys.* pp. 720-728.
6. Erol-Kantarci, M. and Hussein, T. M. (2010). Prediction- based charging of PHEVs from the smart grid with dynamic pricing. *IEEE Local Comp. Netw. Conf.*, 1032–1039, doi: 10.1109/LCN.2010.5735676.
7. E.Sortomme, M.A. El-Sharkawi, (2009), Optimal Power Flow for a System of Microgrids with Controllable Loads and Battery Storage, Power Systems Conference and Exposition, 2009. PSCE '09.
8. F. Li et al., "Smart Transmission Grid: Vision and Framework," in IEEE Transactions on Smart Grid vol. 1, no. 2, pp. 168-177, Sept. 2010, doi: 10.1109/TSG.2010.2053726.
9. A. -H. Mohsenian-Rad, V. W. S. Wong, J. Jatskevich, R. Schober and A. Leon-Garcia, "Autonomous Demand-Side Management Based on Game-Theoretic Energy Consumption Scheduling for the Future Smart Grid," in IEEE Transactions on Smart Grid, vol. 1, no. 3, pp. 320- 331, Dec. 2010, doi: 10.1109/TSG.2010.2089069.
10. Y. Riffonneau, S. Bacha, F. Barruel and S. Ploix, "Optimal Power Flow Management for Grid Connected PV Systems With Batteries," in IEEE Transactions on Sustainable Energy, vol. 2, no. 3, pp. 309-320,July 2011, doi: 10.1109/TSTE.2011.2114901.
11. M. Geidl and G. Andersson, "Optimal Power Flow of Multiple Energy Carriers," in IEEE Transactions on Power Systems, vol. 22, no. 1, pp. 145-155, Feb. 2007, doi: 10.1109/TPWRS.2006.888988.
12. A. Kiani and A. Annaswamy, "The effect of a smart meter on congestion and stability in a power market," 49th IEEE Conference on Decision and Control (CDC), Atlanta, GA, USA, 2010, pp. 194- 199, doi: 10.1109/CDC.2010.5717141.
13. Y. Gu and J. McCalley, "Market-based transmission expansion planning," 2011 IEEE/PES Power Systems Conference and Exposition, Phoenix, AZ, USA, 2011, pp. 1-9, doi: 10.1109/PSCE.2011.5772507.
14. Garng. M. Huang, Nirmal-Kumar C Nair, An OPF based Algorithm to Evaluate Load Curtailment Incorporating Voltage Stability Margin Criterion , Conference Proc. of NAPS 2001. pp. 1547-1552
15. F. Kjetil, Joint State-Space Model for Electricity Spot and Futures Prices, Norwegian Computing Centre, pp- 16-18https://www.risk.net/sites/default/files/import_unmanaged/risk.net/data/eprm/pdf/november/technical.pdf
16. C. Florin, G. Mevludin, Interior-point based algorithms for the solution of optimal power flow problems, Electric Power Systems Research, Elsevier, 508–517
17. H. M. A. Zeeshan, M. Naeem and F. Nisar, "Advanced techniques for control of smart grids," 2017 International Smart Cities Conference (ISC2), Wuxi, 2017, pp. 1-5, doi: 10.1109/ISC2.2017.8090814.
18. B. Terry, Energy Efficiency, Demand Response and PHEVs and the Smart Grid: pulling it all together , Summer Seminar. http://mydocs.epri.com/docs/summerSeminar11/2010_summerSeminar.pdf
19. R. Couillet, S. M. Perlaza, H. Tembine and M. Debbah, "Electrical Vehicles in the Smart Grid: A Mean Field Game Analysis," in IEEE Journal on Selected Areas in Communications, vol. 30, no. 6, pp. 1086-1096, July 2012, doi: 10.1109/JSAC.2012.120707.
20. C. Gerkensmeyer, Technical Challenges of Plug In Hybrid Electric Vehicles and impacts to the US Power System: Distribution System Analysis, Prepared for the U.S. Department of Energy under ContracDE-AC05-76RL01830 pp-19165
21. Majid O, Buygi,Gerd B, et al (2004), Market-Based Transmission Expansion Planning, IEEE Transactions on Power Systems, pp. 1-9
22. Thelma S. P. Fernandes, et al , Load Shedding Strategies Using Optimal Load Flow With Relaxation of Restrictions, IEEE Transactions on Power Systems, 23. pp- 712 - 718
23. Garng. M, Nirmal-K, An OPF based Algorithm to Evaluate Load Curtailment Incorporating Voltage

Stability Margin Criterion , www.pserc.wisc.edu pp-1-16

24. Kjetil F, et al, (2002), Joint State-Space Model for Electricity Spot and Futures Prices, Norwegian Computing Centre, pp 269–281
25. Florin C, Mevludin G, et al (2007) Interior-point based algorithms for the solution of optimal power flow problems, Electric Power Systems Research, Elsevier, pp-508–517
26. David S, Laurent S, (2010)Advanced Power System Operations with Smart Grid Technologies, 2010 IEEE PES Panel Session, pp. 353-358
27. Ho C, Joung-H, 92010) Development of Smart Controller with Demand Response for AMI connection, International Conference on Control, Automation and Systems 2010 Korea, pp- 752-755
28. Terry B, (2010) Energy Efficiency, Demand Response and PHEVs and the Smart Grid: pulling it all together , Summer Seminar.pp. 1-4
29. Romain C, Samir M, et al, (2012) Electrical Vehicles in the Smart Grid: A Mean Field Game Analysis, IEEE Journal on Selected Areas in Communications 30(6) 1086-1096 pp. 1086-1096
30. C. Gerkensmeyer, et al, (2010), Technical Challenges of Plug In Hybrid Electric Vehicles and impacts to the US Power System: Distribution System Analysis, Prepared for the U.S. Department of Energy under Contract DE-AC05-76RL01830 pp- 1119-1126

67 Evaluation of dynamic characteristics of a pilot operated pressure relief valve: Mathematical modeling and simulation

Nitesh Mondal[1,a], *Md Abbas S.K.*[2,b] *and Rana Saha*[2,c]

[1]Department of Mechanical Engineering, Ghani Khan Choudhury Institute of Engineering and Technology, Malda, India

[2]Department of Mechanical Engineering, Jadavpur University, Kolkata, India

Abstract

In modern days fluid power plays an essential role in various power applications. Pressure control valves are used in fluid power circuits to regulate the dynamics of an actuator and control the reference pressure levels at which the whole system works to reduce the chance of an accident and protect the system from excess pressure. Mainly the reference pressure or the pre-set pressure is controlled by the spring force. In a hydraulic or electrohydraulic circuit, a pressure relief valve unlocks and bypasses the pressurized fluid when the system pressure exceeds its reference setting. The valve then locks again when pressure falls. In this work, a pilot-operated pressure relief valve (PRV) has been modeled mathematically and studied the performance of the PRV through MatLab/Simulink environment.

Keywords: Fluid power, pre-set pressure, pressure relief valve, pilot operated

Introduction

Pressure relief valve (PRV) or relief valve (RV) is a safety valve that is used in a hydraulic circuit to prevent any type of accident due to excess pressure in the circuit. The maximum pressure of a hydraulic circuit is sometimes controlled by the PRV. In general, pre-set spring force is used as a control force of a pressure relief valve. PRV allow the flow through an auxiliary line when pressure goes beyond the pre-set or reference set pressure of the hydraulic circuit. In general, two types of relief valve are available in the commercial market one is direct operated another one is pilot operated. The life of the hydraulic circuit, safety, performance and economic impact directly or indirectly depend on the relief valve. Handroos and Vilenas [1] developed a mathematical model of a pressure controlled valve and studied the behavior of internal elements. An alternate method also describes for the development of single stage pressure control valve. Morselli et al. [2] presented a research concept on the importance of lumped parameter dynamics along with the wave equations in the piping system and discussed about the damping effect against the wave propagation of pressure. Kappi and Ellman [3] modeled a proportional mobile valve and discussed about the usability of the model in the hydraulic system. A pressure relief valve with a proportional solenoid-controlled pilot stage was designed through simulation and compared the performance with the experimental result by Maiti et al. [4] and found that the overall dynamic action had been shown to be controlled by the solenoid characteristic involving force to given voltage on the system. A numerical study considering the various pressure-flow characteristics has been simulated through the Bondgraph technique and identified some crucial factors that have a significant role in the transient response of the system [5]. Osterland and Weber [6] represent a state forward model of a pressure relief valve and formulated the stationary and dynamic behavior of the valve with analytical solution of static pressure flow character. The intrinsic relation between the gradient of static pressure-flow character and stability has been proved mathematically.

Hubballi and Sundor [7] studied the dynamics characteristic of a compound pressure relief valve theoretically through a Matlab-Simulink environment, especially during the transient discharge flow. Moon et al. [8] studied the dynamic characteristics of a spool-type pressure relief valve and minimized the oscillation incorporating orifices in the inner pilot line and tank line of the system. Ghosh et al. [9] developed a special type of spool valve for high-pressure application trough finite element approach. This work mainly focuses on the mathematical modeling of a pressure relief valve with pilot line through simulation environment and studies the pressure and flow character of each and every section of the valve.

[a]niteshju@yahoo.com, [b]skmdabbas@gmail.com, [c]rana.saha@jadavpuruniversity.in

DOI: 10.1201/9781003540199-67

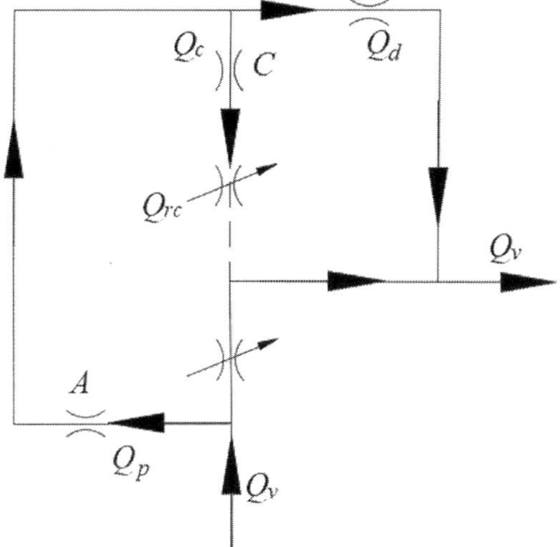

Figure 67.1 Schematic diagram of the proportional solenoid PRV

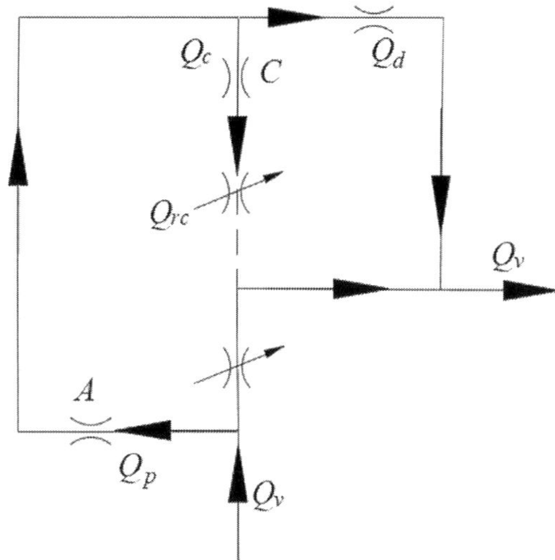

Figure 67.2 Representation of flow through different orifices

Description of the valve

The configuration of the valve can be assumed from the schematic diagram as shown in Figure 67.1. The pilot spool of the valve is driven by a proportional solenoid linear actuator which is fixed at the end of the pilot stage. All the arrows in the diagram show the flow direction inside the valve. Orifice A is the main orifice which creates the connection between the main line and pilot stage. Orifice C has a significant role as damper. Any variation in pressure at any part of the system, for any set voltage of the solenoid, is detected by the pilot spool and the essential change of the pilot spool activates the main spool to re-establish the valve pressure setting. The flow network of the valve is presented in the Figure 67.2.

Mathematical modeling of the PRV

A mathematical model has been developed to describe the PRV. To understand the complete function of pilot operated PRV needs a comprehensive knowledge of fluid flow through orifice, the relation between flow and force both at the pilot and the main stage, and also the valve geometry. To analysis, the dynamic performance of the valve, the momentum principle along with continuity equation have been incorporated considering the following cases

1. The reservoir or tank pressure of the system is atmospheric pressure.
2. Leakage flow around the main stage poppet is negligible.
3. The effective bulk modulus of the fluid inside the valve chamber is constant.
4. Neglect the cavitation effect and air entrapment in the valve
5. All kind of friction effects are ignored.

Fundamental relations of pressure and flow

The pressure dynamics of the main stage valve has been evaluated using the fundamental relationship among the pressure, flow rate or discharge and the bulk modulus of the working fluid as

$$\left(Q_v - Q_p - Q_{rv} - A_s\frac{dy_s}{dt}\right)\left(\frac{\beta}{V_1}\right) = \frac{dP_s}{dt} \tag{1}$$

Now the flow through the orifice A is defined as

$$Q_p = C_d\frac{\pi}{4}d_a^2\sqrt{\frac{2(P_s-P_c)}{\rho}} \tag{2}$$

And the flow through the variable orifice which is created due to the movement of the poppet can be presented as [4]

$$Q_{rv} = C_q\pi d_s y_s\, sin\varphi_s\sqrt{\frac{2(P_s-P_0)}{\rho}} \tag{3}$$

The flow continuity has been illustrated as

$$Q_v = Q_p + Q_{rv} = Q_d + Q_c + Q_{rv} \tag{4}$$

Reynolds number typically plays a great role in the flow equation at the pilot stage and the coefficient of discharge is a function of it. Flow through the pilot stage is formulated as

$$Q_d = C_p y_p^2(P_c - P_0) \tag{5}$$

Where Cp, the flow coefficient of discharge for the pilot stage, is expressed with a constant (=0.066 in the range 0.2<Re<15) and other parameters as [4, 10]

$$C_p = \frac{2\varepsilon^2 \pi d_p (sin\varphi_p)^2}{\rho v} \tag{6}$$

Considering the laminar flow, the flow through the orifice C, has been modeled as

$$Q_c = \frac{\pi d_c^4}{128\mu L_c}(P_v - P_c) \tag{7}$$

Now, the pressure inside the poppet chamber has been evaluated with a well-known mathematical relation

$$\left(Q_c - A_s \frac{dy_s}{dt}\right)\left(\frac{\beta}{V_a}\right) = \frac{dP_v}{dt} \tag{8}$$

There are several pressures drop across the different orifice associated in the valve. So, the pressure between the main line and the pilot stage has been controlled by the equation of compressibility as

$$\left(Q_p - Q_d - A_p \frac{dy_p}{dt} - Q_c\right)\left(\frac{\beta}{V_2}\right) = \frac{dP_c}{dt} \tag{9}$$

Force dynamic of the valve

There several forces acting on the system such as pressure force, spring force, viscous force and flow force. Now, the dynamics of the main stage and the pilot stage are illustrated as

$$A_s(P_s - P_v) - K_s(y_s + y_{s0}) - b_s\dot{y}_s - F_s = M_s\ddot{y}_s \tag{10}$$

$$A_p(P_c - P_0) - K_p(y_p + y_{p0}) - b_p\dot{y}_p - F_p = M_p\ddot{y}_p \tag{11}$$

Where F_s and F_p are the flow forces [11–13] acting on the main spool and the pilot spool at the dynamic phase. Those forces are represented as

$$F_s = C_d C_v \pi d_s y_s \, sin(2\varphi_s) P_s \tag{12}$$

$$F_p = C_d C_v \pi d_p y_p \, sin(2\varphi_p)(P_c - P_0) - 8C_d^2 \pi y_p^2 (sin\varphi_p)^2 (P_c - P_0) \tag{13}$$

Results and discussion

Figure 67.3 represents the flow dynamics through the various orifices. Initially, the main stage flow overshoot is less than the flow through orifice A and pilot stage flow because the spring inside the main spool cup and the flow through the orifice C to the spool cup act as a damping element. Flow through the orifice C has less overshoot. The variation of flow through the pilot stage flows the variation of flow through the orifice A as they are connected in a series. The steady state of the system in terms of flow rate is achieved within 30 ms.

Interestingly, the nature of the pressure dynamics of the system for various position is almost same. Although initially the overshoot is very high, the dynamics achieved the steady state with the flow dynamics as shown in Figure 67.4. The size of the valve is small so the flection of pressure throughout the valve is all most same. The pressure at all ends of the valve leads the quick response characteristic of the valve although there is huge difference in the flow and displacement dynamics of the various parts of the system.

Figure 67.5. represent the spool displacement of the valve. The main spool dynamics is very smooth than the pilot stage. The effect of upstream pressure

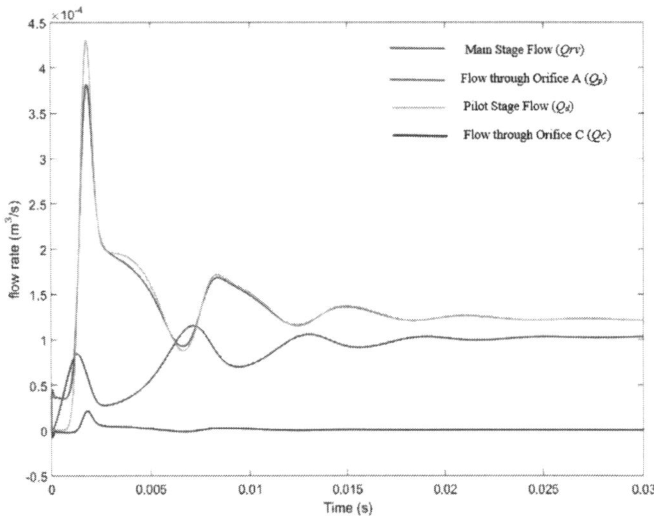

Figure 67.3 Flow dynamics through different orifices

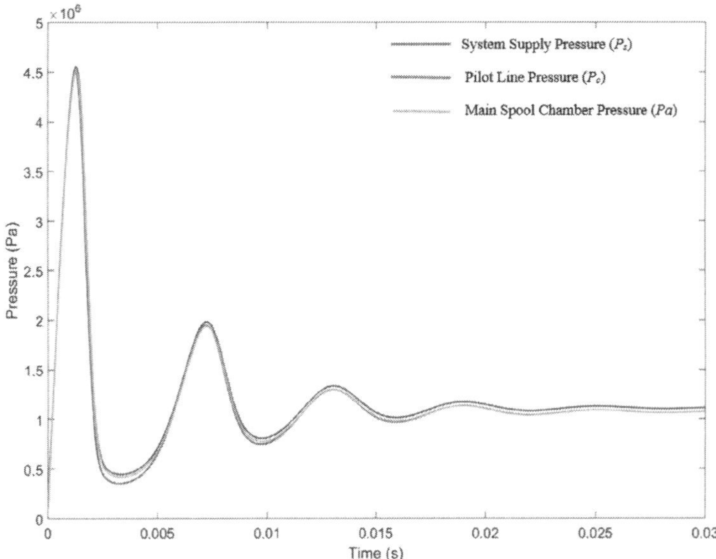

Figure 67.4 Pressure dynamics of the system

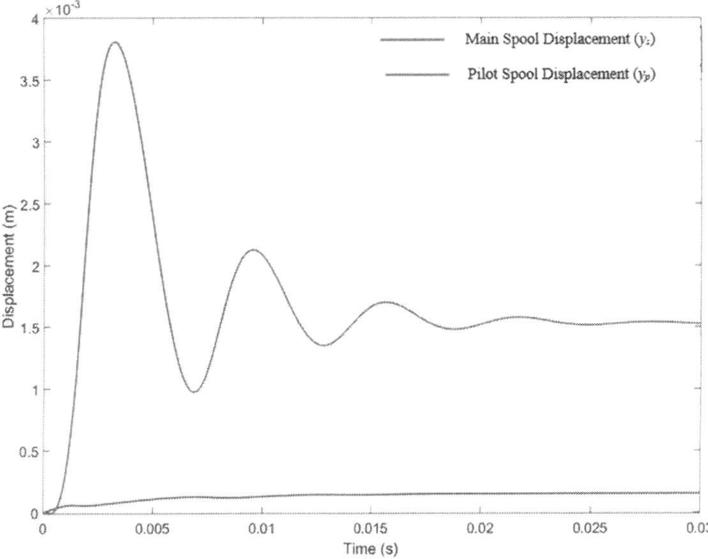

Figure 67.5 Spool displacement of the system

dynamics of the pilot stage has a great influence on the pilot spool dynamics. It is clear from the pilot stage spool displacement and the pressure dynamics that the pilot stage spool displacement and the pressure dynamics of the system has a strong relationship.

Conclusions

Mathematical model of the pilot stage pressure relief valve has been established through MatLab/Simulink considering the effect of orifice flow and fluid compressibility. The dynamic response or characteristics of the valve is established by the flow, pressure and spool displacement. It is clear from this work that the pressure is controlled by the flow dynamics or vice versa. The pilot stage pressure and the displacement of the pilot spool have a great influence throughout the valve dynamics. According to the study of spool dynamics, it may conclude that the pressure inside the main stage cup and spring together has a great damping characteristic.

Acknowledgement

The authors gratefully acknowledge the students, staff, and authority of Mechanical System

Laboratory, Mechanical Engineering Department, Jadavpur University.

References

1. Handroos, H. M. and Vilenius, M. J. (1990). The utilization of experimental data in modelling hydraulic single stage pressure control valves. *J. Dyn. Sys. Meas. Control*, 482–488. doi: https://doi.org/10.1115/1.2896168.
2. Morselli, S., Gessi, S., Marani, P., Martelli, M., De Hieronymis, and Rozzi, C. M. (2019). Dynamics of pilot operated pressure relief valves subjected to fast hydraulic transient. *AIP Conf. Proc.*, 2191(1), 020116: 1-8. https://doi.org/10.1063/1.5138849.
3. Käppi, T. and Ellman, A. (1999). Modelling and simulation of proportional mobile valves. *Proc. JFPS Inter. Symp. Fluid Power*, 1999(4), 531–536. chrome-extension://efaidnbmnnnibpcajpcglclefindmkaj/https://www.jstage.jst.go.jp/article/isfp1989/1999/4/1999_4_531/_pdf/-char/en.
4. Maiti, R., Saha, R., and Watton, J. (2002). The static and dynamic characteristics of a pressure relief valve with a proportional solenoid-controlled pilot stage. *Proc. Instit. Mec. Eng. Part I J. Sys. Control Engg.*, 216(2), 143–156. doi: https://doi.org/10.1243/0959651021541.
5. Dasgupta, K. and Karmakar, R. (2002). Dynamic analysis of pilot operated pressure relief valve. *Simul. Model. Prac. Theory*, 10(1–2), 35–49. doi: https://doi.org/ 10.1016/S1569-190X(02)00061-8.
6. Osterland, S. and Weber, J. (2019). Analytical analysis of single-stage pressure relief valves. *Inter. J. Hydromec.*, 2(1), 32–53. doi: https://doi.org/ 10.1504/IJHM.2019.098951.
7. Hubballi, B. and Sondur, V. (2016). Dynamical analysis of compound pressure relief valve. Proc. 2016 Inter. Conf. Hydraul. Pneumat., 24–35.
8. Moon, K. H. and Huh, J. Y. (2018). An analysis of the dynamic characteristics of a spool type pressure control valve. *J. Drive Control*, 15(4), 61–66. doi: https://doi.org/ 10.7839/ksfc.2018.15.4.061.
9. Ghosh, A., Gupta, A., and Mondal, N. (2023). Design and design investigations of a flow control spool valve. *Inter. J. Interac. Design Manufac. (IJIDeM)*, 17(1), 115–124. doi: https://doi.org/ 10.1007/s12008-022-01135-1.
10. Shin, Y. C. (1991). Static and dynamic characteristics of a two-stage pilot relief valve. *J. Dyn. Sys. Meas. Control*, 280–288. doi: https://doi.org/ 10.1115/1.2896376.
11. Mondal, N. and Datta, B. (2019). Performance evaluation and frequency response analysis of a two stage two spool electrohydraulic servovalve with a linearized model. *Engg. Trans.*, 67(3), 411–427. doi: https://doi.org/ 10.24423/EngTrans.945.20190509.
12. Mondal, N. and Datta, B. N. (2013). A study on electro hydraulic servovalve controlled by a two spool valve. *Inter. J. Emerg. Technol. Adv. Engg.*, 23(3), 479–484. https://www.semanticscholar.org/paper/A-STUDY-ON-ELECTRO-HYDRAULIC-SERVOVALVE-CONTROLLED-Mondal-Datta/81bbaab49c545230ac0a8d4e4482fd7fbe883ff1.
13. Manring, N. D. and Zhang, S. (2012). Pressure transient flow forces for hydraulic spool valves. *J. Dyn. Sys. Meas. Control*, 034501. doi: https://doi.org/ 10.1115/1.4005506.

68 Sub-transient torque estimation for discrimination of non-linear IPMSM and BLDC state model

Tapash Kr Das[a], Debdeepta Ghosh, Didhiti Dey and Surajit Chattopadhyay

Department of Electrical Engineering, Ghani Khan Choudhury Institute of Engineering & Technology, Malda, West Bengal, India

Abstract

This paper presents an analytical comparison of the electro-mechanical state behavior of brushless direct current (BLDCs) motors and interior permanent magnet synchronous motors (IPMSMs) used in vehicular applications. This has been achieved by developing their nonlinear state model and analyzing sub-transient behavior. To perform the comparison investigation, involving the installation of BLDC motors and IPMSM in several electric cars, a comprehensive Simulink model has been created. After simulation, different torque parameters like estimated, measured, request, and reference torque, are determined. Their sub-transient natures in the time domain are compared. Distinct time domain features are extracted from the comparative study. The study may be helpful in designing a starter model for EVs.

Keywords: BLDC motor, electric vehicle, Fast Fourier transform (FFT), interior permanent magnet synchronous motor (IPMSM)

1. Introduction

To sustain the journey towards a carbon-free environment, dependency on electric vehicles (EV) has become a first and easy choice for professionals. Researchers are trying their best to design better mathematical models and prototypes that will provide better performance for different new models of EVs. Different motor actuators are also being tested in EV applications. Among them, interior permanent magnet synchronous motors (IPMSM) and brushless direct current motors (BLDCM) have drawn lots of attention from researchers due to their flexible and easy-to-design electro-mechanical torque behavior. A detailed comparative study has been carried out to study the Sub-transient based performance study of EV. Zhao et al. has introduced a fault-tolerant scheme for the breakdown of one certain bridge arm [1]. Husain et al. in 2021 has submitted an electric drive technology, trends, challenges, and opportunities for future electric vehicles [2]. Tripathi et al. in the year 2018, have proposed an optimal pulse-width modulation [3]. Rathore et al. has introduced an isolated multilevel bidirectional DC-DC converter [4]. Recently in 2023 February Sadabadi et al. presented a decoupled dq current control of grid-tied packed E-cell inventors in V2G technology [5]. Mutoh et al. proposed a suitable method in the year 2008 for eco vehicles to control surges [6]. Du et al. have introduced cascaded H-bridge multi-level boost inverters for electric vehicle (EV) and hybrid electric vehicle (HEV) applications without using inductors [7]. Guha et al. proposed the impact of dead time on inverter input current, DC- Link dynamics, and light load instability in rectifier-inverter-fed induction motor drives in 2018 [8]. Akin et al. in the year 2009 proposed DSP-based sensor less electric motor fault-diagnosis tools [9]. Haskew et al. has proposed the examination of the impact of torque ripple induction [10]. Yang et al. in 2015 announced a new current ripple prediction method [11]. Singh et al. introduced a method of direct electric vehicle-to-vehicle (V2V) power transfer (2022) [12]. In the year 2020, Ranjith et al. invented a fast onboard integrated battery charger [13]. Zhu et al. introduced a paper on electrical machines and drives for electric, hybrid, and fuel cells of vehicles in 2007 [14]. Savaresi et al. proposed a design of an automatic motion inverter [15]. Janabi et al. invented a switched capacitor voltage boost converter for electric and hybrid electric vehicle drives [16]. Das et al. has proposed a sub-harmonic data-driven adaptive line to ground fault detection in solar-fed microgrids in the year [17]. They also introduced an FFT-based classification of solar photo voltaic microgrid systems in 2019 [18]. In the year 2021, they invented an energy-efficient cooling scheme [19]. They have also proposed sub-harmonics-based string fault assessment in solar PV arrays [20].

After a detailed review study of the above articles, the authors have noticed that very few studies

[a]tapash@gkciet.ac.in

DOI: 10.1201/9781003540199-68

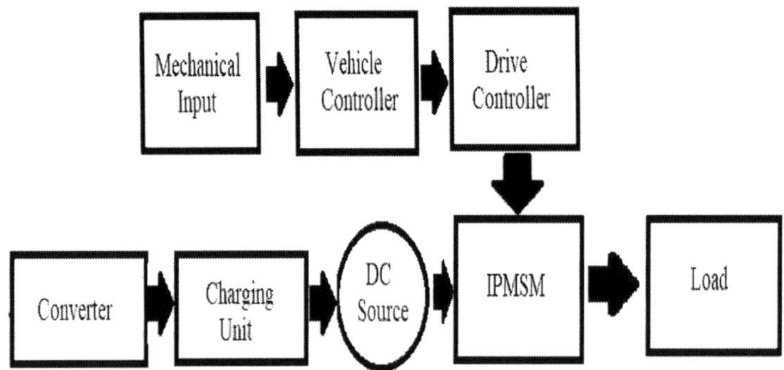

Figure 68.1 The model architecture of IPMSM-based EV

Table 68.1 Detailed specification of components used in IPMSM-based EV model

Vehicle controller	**Acceleration pedal characteristics**	**1:1**
	Brake torque	0.02
Drive controller	Current controller anti-windup gain	1
	Outer loop control magnetic flux linkage	0.04 Wb
Motor	Rated current	180 A
	Rated torque	72Nm
Converter	Forward voltage	0.8V
	Threshold voltage	0.5 V
DC supply	Range	80V,50A
Charging unit	Regulation point	14.4 V

have been carried out, which has motivated the authors to work on the sub-transient torque estimation for discrimination of non-linear IPMSM and BLDC state model. The objective of this paper is to survey the IPMSM-based and BLDC-based EV models and analyze the nature of the sub-transient torque performed while used as part of EVs. In this study, firstly layout of the architecture has been modeled in Section 2. Further, the features have been extracted from sub-transient limit torque and electro-mechanical correlations in Sections 3 and Section 4. Thereafter, sub-transient torque estimation and discriminative feature comparison of both IPMSM & BLDCM-based EV are carried out in Section 5 followed by a conclusion in Section 6.

2. Modeling

Firstly, IPMSM and BLDCM-based EV architecture has been considered. Thereafter modeling of EV layouts has been designed as shown in Figures 68.1 and 68.2. Moreover, the detailed specification of both EVs has been extracted and presented in Tables 68.1 and 68.2. The interior permanent magnet synchronous motor (IPMSM) is a highly efficient and compact electric motor. Figure 68.1 shows the detailed model architecture of IPMSM-based EV. It shows how the mechanical input is fed into the vehicle controller. Further signals have been disbursed to the next stage i.e., the drive controller. With the help of a converter and charging unit, the DC voltage source has been charged. Finally, the IPMSM-based drive has been driven using the signal received from the drive controller to run the load under various conditions accordingly.

The detailed specification of components used in IPMSM-based EV model has been demonstrated in Table 68.1. The specifications of the vehicle controller, drive controller, motor, converter, DC supply, and charging unit have been shown in the Table 68.1.

The brushless DC motor (BLDCM) is a type of electric motor that is known for its reliability and efficiency. Figure 68.2 shows the detailed model architecture of IPMSM-based EV. It shows how the mechanical input is converted into a vehicle controller and then into a drive controller. With the help of a converter, the charging unit is charged and it supplies voltage to the DC source. The DC source is connected to the BLDCM driven by the controller. Then it is used to take load and work accordingly.

The detailed specification of components used in BLDCM-based EV model has been demonstrated in Table 68.2. The specifications of the vehicle

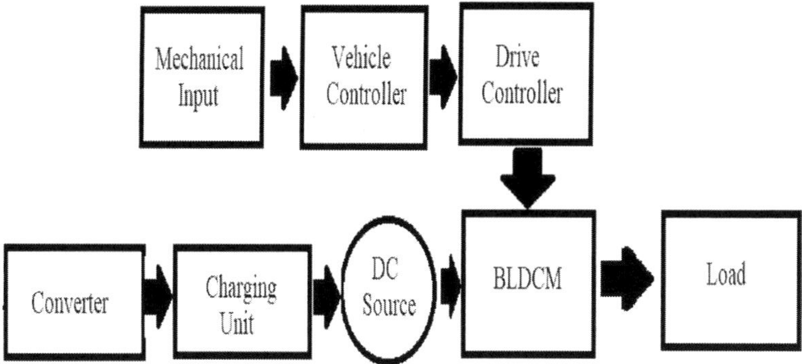

Figure 68.2 The model architecture of BLDCM-based EV

Table 68.2 Detailed specification of components used in BLDCM-based EV model

Vehicle controller	Acceleration pedal characteristics	1:1
	Brake torque	0.02
Drive controller	Current controller anti-windup gain	1
	Outer loop control magnetic flux linkage	0.04 Wb
Motor	Rotor-induced back emf	9.6 V
	Stator axis inductance	0.0002 H
Converter	On state resistance	0.001 ohm
	Off state conductance	0.0067 V
DC supply	Range	80V,50A
Charging unit	Regulation point	14.4 V
	Boost duration	2 Hrs

controller, drive controller, motor, converter, DC supply, and the charging unit have been shown in Table 68.2.

3. Feature extracted from sub-transient limit torque

The sub-transient torque limit features have been extracted for IPMSM and BLDCM to capture the limit torque nature of both the motors which are shown in Figures 68.3 and 68.4. Figure 68.3(a) shows the torque vs. time characteristic of limit torque. In this limit torque, torque is increasing very slowly as subjected to time. From 0.05 s torque is constant as increasing the time. At 0.6 s torque decreases to 0.65 s and again torque will be constant up to 1 s. Thus, the graph first increases to 200 Nm and becomes stable up to 0.04 s and sharply decreases to 140 Nm and continues.

Figure 68.3(b) shows the nature of the curve of torque concerning the time nature of limit torque in BLDC. The torque is initially 0 at time zero, suggesting that no applied torque is present and that the motor is not running. The torque rapidly rises from zero to a greater value over time. Thus, the graph first increases to 200 Nm and becomes stable up to 0.60 s and sharply decreases to 140 Nm and continues. The motor has likely established a stable working condition during this phase of constant torque.

4. Sub-transient torque estimation

The sub-transient torque of IPMSM and BLDCM-based EVs is estimated for the comparative study. During the comparative study, estimated torque, reference torque, measured torque, and request torque have been closely monitored and analyzed. Sub-transient torque estimation refers to the estimation of torque in an interior permanent magnet synchronous motor (IPMSM) during the sub-transient period. The sub-transient period is the initial period when the motor is subjected to a sudden change in the electrical or mechanical parameters, such as a sudden change in the applied voltage or load torque. The block estimates torque using known machine parameters and the measured phase current vectors in the dq_0 reference frame. In this nature, the torque will reduce with increasing time. From 0 s to 0.10 s torque is constant and then torque will reduce up to 0.20 s, again torque will be constant up to 0.65 s. Figure 68.4 (a) shows the

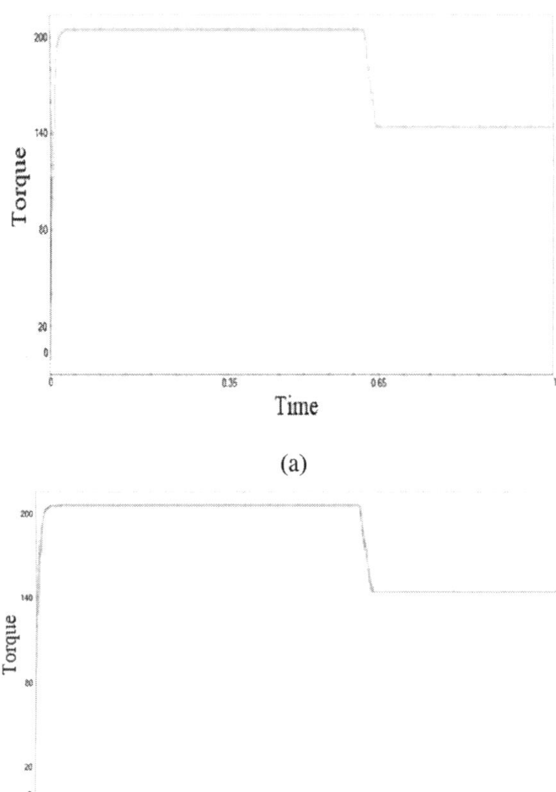

Figure 68.3 Torque vs. time nature of (a) limit torque in IPMSM (b) limit torque in BLDC

nature of the curve between torque concerning the time nature of estimated torque in IPMSM. In the figure, the torque sets to a higher value at 0 s up to 0.10 s then it sharply decreases up to -0.0004 at 0.20 s. Again, this state runs stable up to 0.60 s and again decreases to -0.0009 at 0.66 s and the stable state continues. Thus, the graph shows decreasing in nature concerning time.

Figure 68.4(b) represents the time vs. torque characteristic of reference torque in an IPMSM. This is a square type of graph, starting from 0 graph is constant up to 0.11 s, then from 0.11 s graph fully increases up to 100 Nm torque, again it is constant up to 0.23 s, then it highly increasing the torque up to 150 Nm, again it is constant up to 0.37 s. Again, torque is increasing the peak point of the graph at 200 Nm. Then it is highly decreases the torque up to 40 Nm. By doing this order graph will decrease the torque negative of the graph at -140 Nm in 1 s.

Figure 68.4(c) depicts the time vs. torque characteristic of the measured torque in the IPMSM may be seen in this figure. Up until 0.10 s, torques starting at 0 is constant. Then it starts to decrease in the same sequence as the graph's negative side, reaching -0.00020 Nm in 0.20 s. After that, the torque will remain steady for 0.65 s. Up until 0.65 s, torque will once again fall in the same sequence until constant. Request torque determines the amount of engine output torque that will be required quickly by the machine to run; it also refers to the desired torque. Figure 68.4(d) represents the time vs. torque characteristic of reference torque. This is a square type of graph, starting from 0 s, torque height in the graphs are constant up to 0.11 s. Then from 0.11 s graphic fully increases up to 100 Nm torque, again it is constant up to 0.23 s, then it highly increases the torque up to 150 Nm, again it is constant up to 0.37 s again torque increasing the peak point of the graph at 200 Nm. Then it is highly decreasing the torque up to 40 Nm. By doing this order graph will decrease the torque negative of the graph at -160 Nm in 1 s. Thus, the curve depicts the decreasing nature of torque with time after increasing to a certain limit. The torque characteristics of the electric motor with time or speed are depicted in an electric vehicle (EV) inverter's output torque diagram. The two axes on the torque output diagram commonly reflect time or speed on the x-axis and torque values on the y-axis.

Figure 68.4 (e) shows the torque nature concerning the time nature of request torque in BLDC. The torque is initially 0 at time zero, DC motor has high starting torque. At 0.11 s the torque increased up to 3 Nm and the state continues. After some time, torque will reduce suggesting that no applied torque is present. The motor applies an increasing torque throughout this period to overcome the inertia and begin motion. The torque achieves a constant amount after the initial quick increase and stabilizes for a while. Again, the torque decreases sharply at 0.56 s and continues. At the beginning of the graph stays steady, when the duration increases the curve shows that from 0.10 s to 0.60 s the torque fluctuates and then suddenly increases and fluctuates. Figure 68.5 shows the nature of torque concerning the time nature of estimated torque in BLDC starting from 0 Nm up to 0.11 s, the torque shows its transient nature. This continues up to 0.62 s and the torque comes to its sub-transient state. Thus, the Simulink plot shows the rapid state of transformation of toque as the time progresses.

The torque vs. time characteristic is depicted in Figure 68.4(f) with the x and y-axes standing for time and torque, respectively. The torque is transitory when the time increases and the graph initially exhibit a steady state. From the initial point torque is constant up to 0.10 s, after that torque is suddenly reduced. At 0.10 s the torque remains 0 Nm, then it decreases straight at -10 Nm. Again, this pre-result

(a) (b)

(c) (d)

(e) (f)

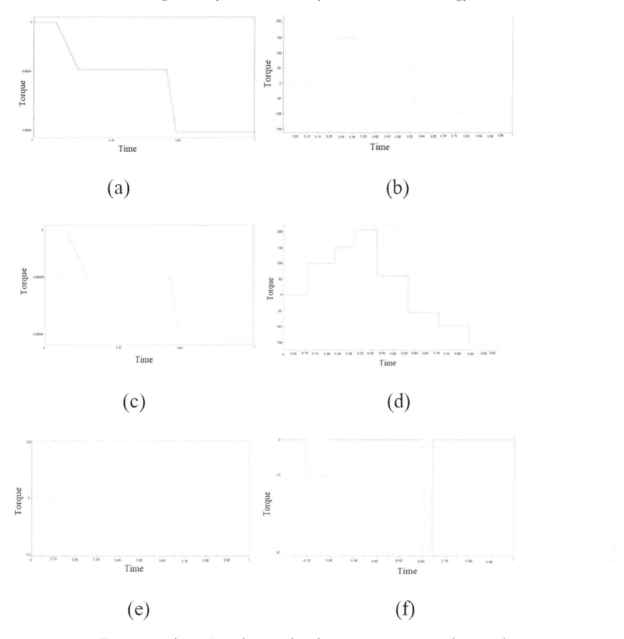

Figure 68.4 Time nature of (a) estimated torque (b) reference torque (c) measured torque (d) request torque in IPMSM and (e) request torque (f) measured torque in BLDC

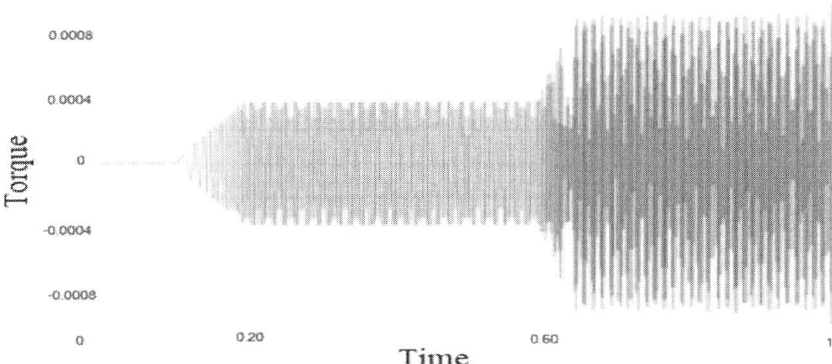

Figure 68.5 Time nature of estimated torque in BLDC

Table 68.3 Torque comparison of IPMSM and BLDCM used in EV

S. No.	Parameters	IPMSM	BLDC
1	Time nature of estimated torque (Figures 4(a) and 5)	It decreases step by step	It is oscillatory in nature with increasing amplitude
2	Time nature of measured torque (Figures 4(c) and (f))	It decreases step by step	It shows step nature with drooping and interruption; deviated from estimated nature
3	The magnitude of request torque (Figures 4(d) and (e))	It step-wise increases and then decreases	After a low constant initial value, it shows a step jump and attains constant magnitude
4	Reference torque (Figure 4(b))	It step-wise increases and then decreases	-

torque continued up to 0.20 s and it increases again to 0 Nm torque. But now the torque continues to stay at 0 for a longer time and decreases sharply up to -32 Nm next at 0.60 s. It rises again and the plot continues. The graph of torque concerning time is inversely related to time growth and reduction at beginning conditions, time zero, and the highest torque.

5. Discriminative feature comparison

Table 68.3 shows the discriminative feature comparison between BLDCM and IPMSM. The features made to compare are estimated torque analysis, measured torque analysis, request torque analysis, and reference torque analysis.

6. Conclusions

In this paper, a comparative study between them is shown and discussed, enabling the analytical study between IPMSM and BLDC. For both the motors torque is studied to show the variation in sub-transient limit torque. From the study it is shown that the torque is initially zero at starting, suggesting that no applied torque is present or the motor is not running. Then the torque is increasing very slowly as subjected to time. From 0.05 s, torque is constant as increasing the time. At 0.6 s torque is decreasing to 0.65 s and again torque will constant for up to 1 s to study sub-transient torque estimation, estimated torque and reference torque of IPMSM is designed, and thus measured torque is estimated. The comparisons are made regarding time and torque and a brief study is executed. According to this research, IPMSMs have less mechanical stress and more mechanical stability, which improves the efficiency of the system as a whole. The present concept for both motors has a vast field of study and research with upgrading the algorithms, power density, efficiency, and cost-effectiveness, for the advancement of the growth of electric vehicle technology. Studying the features from this paper, it can be analyzed that IPMSM has high-performance analysis over BLDCM due to more efficiency, more stability, and higher output torque. IPMSM-based EV model has better performance than BLDC-based EV model.

Acknowledgement

The authors are thankful to the Head of the Department of Electrical Engineering, GKCIET for providing adequate laboratory facilities and also to the Honorable Director, GKCIET for encouraging and providing overall support.

References

1. Zhao, W., Chen, Z., Xu, D., Ji, J., and Zhao, P. (2018). Unity power factor fault-tolerant control of linear permanent-magnet vernier motor fed by a floating bridge multilevel inverter with switch fault. *IEEE Trans. Indus. Elec.*, 65(11), 9113–9123. doi: 10.1109/TIE.2018.2793224.
2. Husain. (2021). Electric drive technology trends, challenges, and opportunities for future electric vehicles. *IEEE*, 109(6), 1039–1059. doi: 10.1109/JPROC.2020.3046112.
3. Tripathi and Narayanan, G. (2018). Torque ripple minimization in neutral-point-clamped three-level inverter fed induction motor drives operated at low-switching-frequency. *IEEE Trans. Indus. Appl.*, 54(3), 2370–2380. doi: 10.1109/TIA.2018.2804325.
4. Rathore, V., Rajashekara, K., Nayak, P., and Ray, A. (2022). A high-gain multilevel dc–dc converter for interfacing electric vehicle battery and inverter. *IEEE Trans. Indus. Appl.*, 58(5), 6506–6518. doi: 10.1109/TIA.2022.3185183.
5. Sadabadi, M. S., Sharifzadeh, M., Mehrasa, M., Karimi, H., and Al-Haddad, K. (2023). Decoupled dq current control of grid-tied packed E-cell inverters in vehicle-to-grid technologies. *IEEE Trans.*

Indus. Elec., 70(2), 1356–1366. doi: 10.1109/TIE.2022.3156160.

6. Mutoh, N. and Kanesaki, M. (2008). A suitable method for ecovehicles to control surge voltage occurring at motor terminals connected to PWM inverters and to control induced EMI noise. *IEEE Trans. Vehicul. Technol.*, 57(4), 2089–2098. doi: 10.1109/TVT.2007.912174.
7. Du, Z., Ozpineci, B., Tolbert, L. M., and Chiasson, J. N. (2009). DC–AC cascaded H-bridge multilevel boost inverter with no inductors for electric/hybrid electric vehicle applications. *IEEE Trans. Indus. Appl.*, 45(3), 963–970. doi: 10.1109/TIA.2009.2018978.
8. Guha and Narayanan, G. (2018). Impact of dead time on inverter input current, DC-link dynamics, and light-load instability in rectifier-inverter-fed induction motor drives. *IEEE Trans. Indus. Appl.*, 54(2), 1414–1424. doi: 10.1109/TIA.2017.2768524.
9. Akin, S. B. O., Toliyat, H. A., and Rayner, M. (2009). DSP-based sensorless electric motor fault-diagnosis tools for electric and hybrid electric vehicle powertrain applications. *IEEE Trans. Vehicul. Technol.*, 58(6), 2679–2688. doi: 10.1109/TVT.2009.2012430.
10. Haskew, T. A., Schinstock, D. E., and Waldrep, E. M. (1999). Two-phase on drive operation in a permanent magnet synchronous machine electromechanical actuator. *IEEE Trans. Ener. Conver.*, 14(2), 153–158. doi 10.1109/60.766970.
11. Yang, F., Taylor, A. R., Bai, H., Cheng, B., and Khan, A. A. (2015). Using d–q transformation to vary the switching frequency for interior permanent magnet synchronous motor drive systems. *IEEE Trans. Trans. Elec.*, 1(3), 277–286. doi: 10.1109/TTE.2015.2443788.
12. U. B. S., Khadkikar, V., Zeineldin, H. H., Singh, S., Otrok, H., and Mizouni, R. (2022). Direct electric vehicle to vehicle (V2V) power transfer using on-board drivetrain and motor windings. *Trans. Indus. Elec.*, 69(11), 10765–10775. doi: 101109/TIE.2021.3121707.
13. Ranjith, S., Vidya, V., and Kaarthik, R. S. (2020). An integrated EV battery charger with retrofit capability. *IEEE Trans. Trans. Elec.*, 6(3), 985–994. doi: 10.1109/TTE.2020.2980147.
14. Zhu, Z. Q. and Howe, D. (2007). Electrical machines and drives for electric, hybrid, and fuel cell vehicles. *IEEE*, 95(4), 746–765. doi: 10.1109/JPROC.2006.892482.
15. Savaresi, S. M., Tanelli, M., Taroni, F. L., Previdi, F., and Bittanti, S. (2006). Analysis and design of an automatic motion inverter. *IEEE/ASME Trans. Mec.*, 11(3), 346–357. doi: 10.1109/TMECH.2006.875552.
16. Janabi and Wang, B. (2020). Switched-capacitor voltage boost converter for electric and hybrid electric vehicle drives. *IEEE Trans. Power Elec.*, 35(6), 5615–5624. doi: 10.1109/TPEL.2019.2949574.
17. Das, T. K., Chattopadhyay, S., and Das, A. (2020). Sub-harmonics data-driven adaptive line to ground fault detection in solar fed microgrids. *Michael Faraday IET International Summit 2020 (MFIIS 2020)*, 155–163. doi: 10.1049/icp.2021.1136.
18. Das, T. K., Banik, A., Chattopadhyay, S., and Das, A. (2019). FFT-based classification of solar photo voltaic microgrid system. *2019 Second Inter. Conf. Adv. Comput. Comm. Paradigms (ICACCP)*, 1–5. https://doi.org/10.1109/ICACCP.2019.8882995.
19. Das, T. K., Banik, A., Chattopadhyay, S., and Das, A. (2021). Energy-efficient cooling scheme of power transformer: An innovative approach using solar and waste heat energy technology. Ghosh, S. K., Ghosh, K., Das, S., Dan, P. K., Kundu, A. (eds). *Adv. Thermal Engg. Manufac. Prod. Mana. ICTEMA 2020. Lec. Notes Mec. Engg.*, Springer: Singapore. https://doi.org/10.1007/978-981-16- 2347-9_17.
20. Das, T. K., Banik, A., Chattopadhyay, S., and Das, A. (2019). Sub-harmonics based string fault assessment in solar PV arrays. Chattopadhyay, S., Roy, T., Sengupta, S., Berger-Vachon, C. (eds). *Model. Simul. Sci. Technol. Engg. Math. Adv. Intel. Sys. Comput.*, 749. https://doi.org/10.1007/978-3-319-74808-5_25.

69 A survey on measuring various security policies for voice assistant applications

Geetha Rani E.[1,a], *Bhuvaneshwari P.*[2,b], *Anusha D.*[3,c], *Ch. Suguna Latha*[4,d], *Tejas S.*[5,e], *Sunil Gowda S.*[5,f], *Shashank S.*[5,g] *and Suman A.*[5,h]

[1]Assistant Professor, Department of CSE, Alliance University, Bengaluru, India

[2]Associate Professor, Department of CSE, Manipal Institute of Technology Bengaluru, Manipal Academy of Higher Education, Manipal, India

[3]Associate Professor, Department of CSE, SRKIT, AP, India

[4]Assistant Professor, Department of CSE, NRIIT, AP, India

[5]Scholar, Department of CSE, MVJCE, Bengaluru, India

Abstract

Researchers have also investigated the use of machine learning and natural language processing techniques to improve the efficacy and precision of voice assistants while upholding high security standards. These techniques involve listening for anomalies in speech patterns that might indicate fraudulent or unauthorized access. Maintaining user confidence and ensuring the security of sensitive data requires the development of a safe framework for voice assistants. The strong foundation provided by the current work may be expanded upon in future research in this field, and ongoing efforts will continue to protect voice assistants improving. The introduction to speaking assistants has fundamentally altered the way we utilize technology. However, with the increasing use of voice assistants' usage in sensitive fields such as finance and medical, security concerns have emerged. To address this issue, Scholars have been diligently striving to develop secure voice assistant frameworks. Prior studies in this area focused on several security-related issues, such as permission, data privacy, encryption, and authentication. Several researchers have recommended using multi-factor authentication to ensure that the voice assistant can only be used by authorized users. Other people have developed safe methods for storing and sending voice data.

Keywords: Privacy and security, voice assistants, intelligent personal assistants, security and privacy

Introduction

Smart cars use speech gratitude to improve driving experience and provide hands-free access to the vehicle's functions so the driver can focus on the road. The possibility that kids could gain access to the keys or keys is one of the main concerns regarding voice-activated cars. Generally, speaking the assisted wake word will allow you to start the car without any problems. It helps with a lot of things, including starting cars [1]. PVAs are useful, and the rate at which they are deployed is rapidly increasing. For example, 21% of Americans own at least one smart speaker, and 81% of Americans own a smartphone. Because of this, users are very likely to always be in the vicinity of at least one PVA. Users might not be able to deactivate them, be unaware of them, or have any control over how they act. Due to the devices' ability to monitor and understand speech, people are worried about their privacy. Users may be interested in knowing answers to questions such as "How do I manage which PVAs are hearing what I'm saying?" How can I hear conversations that have been recorded? And how can I express that I require seclusion? [2]. The most popular applications that make use of virtual user interfaces, or VUIs, include voice-activated navigation apps like Google Maps and Apple Maps, and well-known assistants like Apple's Siri. Some apps need access to calendars, geolocation, microphone, contacts, and storage, among other sensitive permissions, to function properly. When users ask questions, they can then respond using the internal speakers. Sensitive personal information is often contained in these VUI responses. The innovative and helpful threat known as stealthy IMU uses motion sensors without authorization to gather personal data that is shielded by consent from the VUI responses. These permissions are particularly given to the VUI by the user and are closely related to user privacy [3]. Therefore, the primary motivation for our study is to advance the concepts

[a]geetha.edupuganti@alliance.edu.in, [b]bhuvaneshwari.p@manipal.edu.in[2], [c]d.anushasrk@gmail.com, [d]sugunachitturi@gmail.com, [e]1mj19cs177@mvjce.edu, [f]1mj19cs170@mvjce.edu.in, [g]1mj19cs167@mvjce.edu.in, [h]1mj19cs191@mvjce.edu.in

DOI: 10.1201/9781003540199-69

that enable voice assistant-driven systems that oversee authorization and information collection have informed agreement [4]. The SPA connects integrated home appliances and Internet-based remote services. For example, the user can check the traffic and weather, listen to podcasts, shop online, have audio and video chats, and control other gadgets like temperature sensors and smart lights. In contrast to traditional voice interaction systems, SPA manages the gathered user commands differently. The user's question is answered in the best possible way. Using recent advances in natural language processing, when a user gives voice instructions, SPA "understands" them and interprets the message. With a traditional system, the user needs to give a clear command. that follows a strict structure and is pronounced correctly to initiate a function. Typically, a voice assistant continuously samples audio from its microphone while listening to keywords like "Alexa," "Siri," or "Google". The primary technique for locating keywords is "Keyword Spotting" (KWS), which involves connecting a voice assistant that is on to a contact regulator. When the VAs identifies the wake keyword, they begin streaming the audio recording that will be used for voice command analysis [7]. To be proactive and offer contextual recommendations, the assistant needs to continuously record conversations. This poses an obvious risk to privacy and will only exacerbate the many concerns that consumers already have regarding smart speakers. One way to limit what assistants can hear is to use permissions, like those that mobile devices are used to restrict an application's ability to access personal resources like cameras or locations. Voice assistants that are currently in use already have permissions: For example, Alexa shows third-party add-ons called "skills" when it retrieves usernames, email addresses, or addresses [8]. The way that voice-based VA consent is currently implemented has several serious issues and disregards informed consent guidelines [9].

Literature survey

PVA usage has increased dramatically over the last 10 years; 80% of Americans now own a smartphone, and over 21% of them own a smart speaker. Personal voice assistants are devices designed to recognize speech input from the user and identify whether the user can perform the required action in response to a command. PVAs usually combine software and hardware, like speakers and microphones, to facilitate recording, listening, analyzing, and acting. A PVA typically looks for a wake word nonstop [1]. Recognition from the sensor is a very under explored topic. However, stealthy can overcome the challenge because of two important insights. To begin with, there aren't many voices produced by machines. Second, compared to real human voices, the sound characteristics of VUIs turned out to be more preset [2]. There have already been several ethical concerns with voice assistants identified. Most of these concerns center on interpersonal relationships and privacy. There is consensus that people's impressions of how voice assistants operate, including their comprehension of confidentiality constraints, are vague and/or incorrect. Despite their reservations, people use VAs because they believe in external safeguards, particularly privacy laws. Bad actors may watch SPAs specifically as they become more prevalent in our daily lives. Furthermore, because these features of user-device interaction enable malicious users, they increase the attack surface [4]. The KWS task must locate a predetermined set of keywords in an audio stream. The unapproved audio stream is usually captured by the VA's microphone or microphones. Lastly, audio preprocessing and KWS classification are carried out by the VA. The functionality of KWS is crucial to the user experience of VA. A close to ideal true-positive rate is necessary for the device's functionality and responsiveness. On the other hand, a KWS error compromises the virtual assistant's integrity as well as the privacy of its users. When the VA is activated by an unauthorized command, it is called mis activation [6]. The idea of privacy is vague, impersonal, and full of ethical conundrums. Because of its inherent complexity and openness, the concept of privacy is frequently contested. A public conversation concerning IPAs "listening in" and taking actions without permission in homes has been triggered by several issues in recent years. Due to a malfunctioning touch button that has been removed from all Google Mini devices, many people were shocked to learn that data was being continuously recorded and sent to Google servers by a Google Home Mini [7]. Customers are using virtual assistants (VAs) more often despite privacy concerns. However, research on the factors influencing consumers' perceptions of VAs and their post-adoption behavior is currently scarce (for instance, Pal et al., 2020; Ashfaq et al., 2021; Kowalczuk, 2018; Moriuchi, 2019). For example, Ashfaq et al. (2021) investigated the effects of functional, hedonic, economic, and social values on attitudes and intentions to continue using smart speakers. They found that attitudes toward the use of virtual assistants (VAs) are not explained by any other connections, aside from the one between attitudes and social values [8]. Dynamic analysis tracks an application's actions while it is in use and collects evidence that confidential data is being removed from the device. In ref [9] Similarly,

399 new capabilities requesting permissions were added between 2020 and 2021; Amazon expects developers to disclose their archiving in the privacy statement. It's intriguing to note that the number of abilities with full traceability has increased relative to those with partial or broken traceability, according to the data. Specifically, 518 (52%) and 256 (64%) capabilities are among the in 2020, new talent will be added with complete traceability. This skill gathers device address and location services. Nevertheless, the links to the privacy statement direct users to a nonexistent webpage, which undermines the traceability of the skill. While things are generally improving, privacy issues are still present in many recently submitted skills. Therefore, we propose that there is still room for improvement in the screening process [10]. Buildings runtime permission requests may be a component of any one of the many helper architectures, which could benefit from the studies reported in this paper. To be clear, the study allows cross-checking to a certain extent. Lastly, we now describe a particular architecture that forms the basis for our study's application of permission. If you are unsure about the veracity of the information provided for design decisions and assumptions, it is a good idea to obtain a second opinion. We observe that these options only represent one possible set in a large design space [11]. The voice forward consent process has several major issues, which we have outlined here after reviewing the relevant literature and regulations. the lengthening of the permissions process; the challenge of verbally conveying the required amount of information; the lack of distinction between the operating system (OS) of Alexa and voice from external skills, 2 all violate the generally accepted guidelines for informed consent, even though they do it in different ways [12].

Related work

When dealing with audio data, machine learning algorithms cannot handle raw audio representations. Rather, the audio characteristics that were taken out of the audio recordings are utilized by the models. Features such as pitch, loudness, and timbre can be used to classify aspects of audio that are audible to humans. We refer to these qualities as perceptual traits. Other features that can only be represented mathematically and statistically are known as physical features [14]. Power Spy recommends locating the user by utilizing the change in power consumption caused by cell phone modems. With just one navigation voice, locate the location and even identify any sensitive areas [15]. The first step of MSS employs a standard threshold determined empirically to determine the vibration of interest to exclude scenarios in which the smartphone is either stationary or suddenly moving. In the second stage, 500 Hz is the resampled signal frequency. and the magnitude values are then normalized. Since very few MSS make it past the initial stage, resampling in this case results in a tolerable processing burden. Subsequently, a discrete Fourier transform (DFT) is employed to eliminate the segments that lack significant high-frequency components. The signal buffer connected to an entire voice sequence must finally be divided [16]. First cycle results: treaty and careful consideration of the function of permits for points of pact, the legal necessity of authorized acts and the best training, joint controllerships, conflict resolution, and unresolved issues, giving consent decisions more nuance, content for consent dialogues, and opportunities created by speech—Round two outcomes [17]. Sometimes, "virtual assistants" are used to refer to personal assistants. Many different families of electronic devices, such as wearables, desktop computers, mobile phones, and speakers, support personal assistants. SPAs are linked to the cloud services of the businesses that offer these external components for access [18]. The default was caused by Facebook, Apple, Google, and Amazon.

Enrollment in response to public criticism over users the option to opt-in or opt-out of the manual transcription of spoken instructions. Google recently released a keyword sensitivity setting that users can adjust to give them more control over the trade-off between privacy and utility. These initiatives, however, ignore the privacy issues brought up by VA miss activations and do not live up to users' expectations for hands-free interactions [19]. The chosen groups describe in detail the experiences and beliefs that underpin a range of privacy concerns and demonstrate the connections between these and affordances. Interestingly, many respondents prioritize physical and immediate hazards while taking a pragmatic approach to hypothetical and long-term concerns. Regarding the latter, the results of the focus groups show that our Dutch respondents and earlier research share some similarities. A US-based study claims that realistic viewpoints regarding how challenging platforms should regularly preserve and review recordings from every device to ease worries about being listened in on and being recordable (Lau et al., 2018) [7]. The quartet declarations the privacy cynicism was measured using a scale from a Choi et al. study (2018). The instrument used to gauge attitudes toward VAs consisted of five claims drawn from the study carried out by Moriuchi et al. There is evidence to support the existence of this association

Table 69.1 A summary of the contrasts between different parameters

Reference papers	[6]	[2]	[3]	[8]	[12]
Space	Large space	Medium space	Less space	Less space	Medium space
Accuracy	Medium	High	High	High	Medium
Sensitivity	Medium	Medium	Low	Medium	High
Specificity	-	High	-	High	Medium

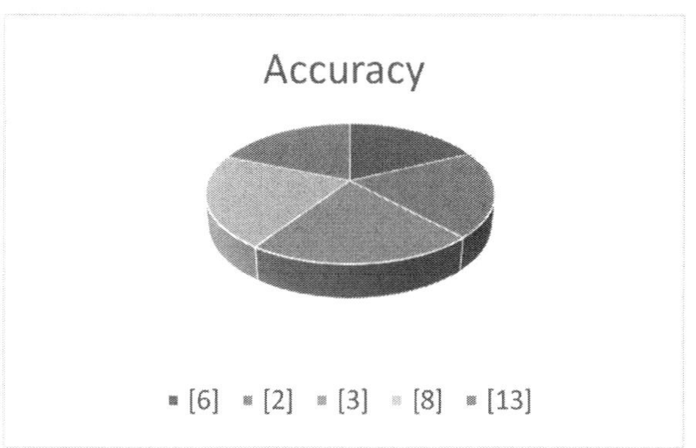

Figure 69.1 Accuracy parameters graphically represented across various sources

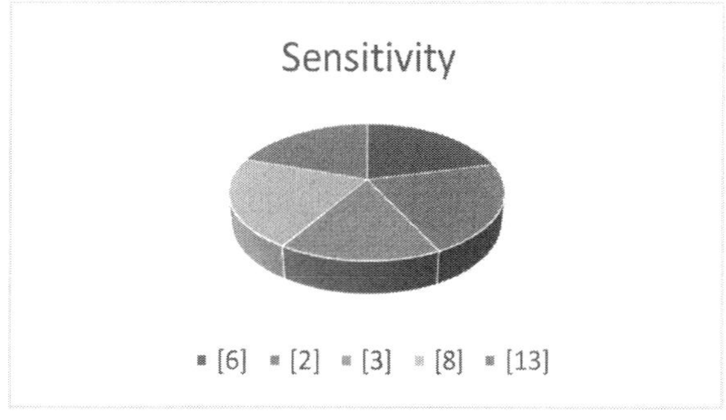

Figure 69.2 Sensitivity parameter visualization using graphics from various studies

in this scenario as well, where consumers' skepticism about privacy may cause their concerns to be minimized even when they are genuine. As a result, Vas's growing levels of trust are preceded by privacy cynicism [20]. We describe the steps we took to collect and look through decrypted network traffic to check for any possible exposure of PMI. We finally discussed in detail how we handled tracking behaviors that were found through code and network analysis in Section 3.4. We limited the apps we included in our research to those that could be accessed through both UK stores because these apps need to comply with the general data protection regulation [21]. This problem of obtaining permissions while using insufficient, our prior research serves as additional motivation for privacy procedures. This study expands on our earlier findings by revealing how traceability has evolved over time and what factors have contributed to these changes. Research on There has also been research on the traceability of privacy in other fields, such as mobile phone applications, social media collectors, and online networks or nodes [22]. In order to find meeting minutes, Ambient Spotlight automatically searched through files. Research by Carrascal et al. examined the process of extracting important data from phone call transcripts [23]. The consent procedure has a sense of urgency that has never been seen before thanks to the placement of VFC, Alexa's rotation approach, and the communication itself. If, on the other hand, Alexa's timer lasts for 8 seconds and then prompts the user again, people will learn to respond within that window of time to avoid the device pursuing them. Customers' consent is transferred by VFC to the voice boundary when they allow the sharing of their data because they cannot withdraw their consent via speech without being informed or shown how to do so (or even that they can do so). Alexa OS oversees voice-forward consent delivery. Speech generated by the Alexa OS cannot be distinguished from other speech in any way [12].

Results and discussions

Figure 69.1 shows a pie chart illustrating the overall review and the distribution of accuracy scores among the various references. Every area of the pie chart, which represents a single data source, is labeled with the name of the reference or project and the accuracy percentage attained. The medium accuracy ref [6] is represented by the blue section. The highly accurate ref [2] is represented by the yellow section. The highly accurate reference [3] is represented by the grey section. The overall review, which was highly accurate, is represented by the orange section [8]. The reference [24] is represented by the light blue section. The distribution of sensitivity scores across various references and overall reviews is displayed in this pie chart on Figure. 69.2.

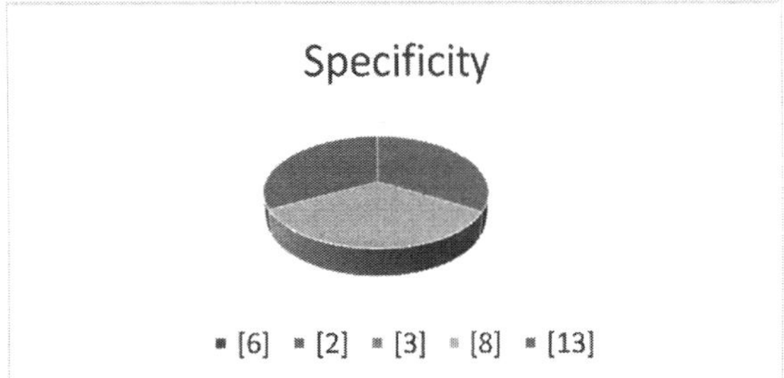

Figure 69.3 Specificity for various scenarios represented graphically, along with how it affects the sensitivity parameter.

The name of the reference or project and the level of sensitivity attained are labeled on each section of the pie chart, which represents a single data source. The blue section is the reference [6], which achieved medium sensitivity. Reference [2], which is represented by the yellow section, achieved medium sensitivity. The grey section illustrates reference [3] low sensitivity. The reference [8] review, which attained medium sensitivity, is represented by the orange section. The reference [12] is represented by the light blue section. Figure 69.3 shows a pie chart with the specific score distribution for the three references and the review. Each section of the pie chart, which represents a single data source, is labeled with the name of the reference or project and the level of specificity attained. Table 69.1 illustrates that the reference with the highest specificity, reference [2], is shown in yellow. Reference [8], which achieved a high degree of specificity is represented by the orange section. High specificity is represented by the light blue section of [13].

Conclusions and future scope

Multiple levels of protection, including safe authentication procedures, encryption of data while it's both in motion and at rest, as well as routine risk assessment and system audits, should be part of a voice assistant security architecture. When creating a secure voice assistant, it is imperative to consider the security risks associated speech recognition technologies, including unauthorized microphone access and voice spoofing. Robust security measures, like multi-factor authentication, user verification, and access restrictions, can be implemented to mitigate these risks and protect user data. In the future, voice assistants that detect wake words using machine learning (ML) will be much more prevalent in our daily lives. Thanks to machine learning, these voice-activated devices will be far more adept at comprehending user commands and accurately identifying their wake word. If a wake-word can be consistently identified, then such interactions will be far more realistic and intuitive. Voice assistants with machine learning-based wake-word detection can be used in an interesting domain: virtual personal assistants. These assistants could be designed to perform a range of tasks. As voice assistant technology develops, It may be possible for those with impairments or restricted mobility to use it to operate numerous devices throughout their house that integrate ML and wake-word detection appears to be very promising, and we can expect to see the release of many more innovative applications in the years to come.

References

1. Konrad, K., Shuba, A., Binns, R., Van Kleek, M., and Shadbolt, N. (2022). Are iPhones really better for privacy? A comparative study of iOS and android apps. DOI: 10.2478/popets-2022-0033.
2. Suarez-Tangil, G. IMDEA Networks Institute Madrid, SpainJide Edu King's College London London, UK, Xavier Ferrer-Aran King's College London London, UK, Jose Such King's College London London, Measuring Alexa Skill Privacy Practices across Three Years© 2022 Copyright held by the owner/author(s). Publication rights licensed to ACM. ACM ISBN 978-1-4503-9096-5/22/04.
3. Malkin, N. and Wagner, D. (2022). University of California, Berkeley; Serge Egelman, University of California, Berkeley & International Computer Science Institute Runtime Permissions for Privacy in Proactive Intelligent Assistants USENIX Symposium on Usable Privacy and Security (SOUPS).
4. Vyshali Rao, K. P., Srividya, R., and Dhanalakshmi, M. (2023). Reliable informational data and secured deviation notification over networks using iot. *2023 Inter. Conf. Recent Trends Elec. Comm. (ICRTEC)*, 1–6.
5. Dhanalakshmi, M. and Basu, A. Energy efficient task provisioning networks using meta heuristic technique. *Inter. J. Appl. Engg. Res. (IJAER)*, 201
6. Shukla, S. (2023). Unlocking the power of data: An introduction to data analysis in healthcare. *Inter. J. Comp. Sci. Engg.*, 11(3), 1–9.
7. Shukla, S. (2022). Developing pragmatic data pipelines using Apache airflow on Google Cloud Platform. *Inter. J. Comp. Sci. Engg.*, 10(8), 1–8.
8. Shukla, S. (). Exploring the power of Apache Kafka: A comprehensive study of use cases suggest topics to cover. *Inter. J. Latest Engg. Manag. Res. (IJLEMR)*, 8, 71–78. doi: 10.56581/IJLEMR.8.3.71-78.
9. Shukla, S. (2023). Enhancing healthcare insights, exploring diverse use-cases with K-means clustering. *Inter. J. Manag. IT Engg.*, 13, 60–68.
10. Shukla, S. (2023). Real-time monitoring and predictive analytics in healthcare: Harnessing the power of data streaming. *Inter. J. Comp. Appl.*, 185, 32–37. doi: 10.5120/ijca2023922738.

11. Shukla, S. (2023). Streamlining integration testing with test containers: Addressing limitations and best practices for implementation. *Inter. J. Latest Engg. Manag. Res. (IJLEMR)*, 9, 19–26. doi: 10.56581/IJLEMR.8.3.19-26.
12. Shukla, S. (2019). Examining Cassandra constraints: Pragmatic eyes. *Inter. J. Manag. IT Engg.*, 9(3), 267–287.
13. Shukla, S. (2019). Data visualization with Python pragmatic eyes. *Inter. J. Comp. Trends Technol.*, 67(2), 12–16.
14. Shukla, S. (2019). Debugging microservices with Python. *IIOAB J.*, 10, 32–37.
15. Bhondge, S. K., Bhoyar, D. B., and Mohod, S. (2016). Strategy for power consumption management at base transceiver station. *2016 World Conf. Futur. Trends Res. Innov. Soc. Welfare (Startup Conclave)*, 1–4. doi: 10.1109/STARTUP.2016.7583988.
16. Raut, P. W. and Bhoyar, D. B. (2009). Design FPGA based implementation of MIMO decoding in a 3G / 4G wireless receiver. *2009 Second Inter. Conf. Emerg. Trends Engg. Technol.*, 1167–1172. doi: 10.1109/ICETET.2009.7.
17. Nanare, K. H. S., Bhoyar, D. B., and Balam war, S. V. (2021). Remote satellite image analysis for deforestation in Yavatmal district, Maharashtra, India. *2021 3rd Inter. Conf. Sig. Proc. Comm. (ICPSC)*, 684–688. doi: 10.1109/ICSPC51351.2021.9451744.
18. P.W Raut, D. Bhoyar, K.D Kulat, R. Deshmukh ,(2020)"Design and Implementation of MIMO-OFDM Receiver Section for Wireless Communication" International Journal of Advanced Research in Engineering and Technology (IJARET), IAEME Publication vol 11, issue 12 pages 44-54
19. Apeksha Yadav, Sandeep Kakde, Atish Khobragade, Dinesh Bhoyar, and Shailesh Kamble, (2018)" LDPC Decoder's Error Performance over AWGN Channel using Min-Sum Algorithm, International Journal of Pure and Applied Mathematics Volume 118 No. 20 2018, 3875-3879 ISSN: 1314-3395
20. R. Bondare, D. B. Bhoyar, C. G. Dethe and M. M. Mushrif, (2010)"Design of high frequency phase locked loop," *2010 INTERNATIONAL CONFERENCE ON COMMUNICATION CONTROL AND COMPUTING TECHNOLOGIES*, Nagercoil, India, 2010, pp. 586-591, doi: 10.1109/ICCCCT.2010.5670772.
21. N. A. Pande and D. Bhoyar, (2022) "A comprehensive review of Lung nodule identification using an effective Computer-Aided Diagnosis (CAD) System," 2022 4th International Conference on Smart Systems and Inventive Technology (ICSSIT), Tirunelveli, India, 2022, pp. 1254-1257, doi: 10.1109/ICSSIT53264.2022.9716327.
22. Dinesh B. Bhoyar, ; Chandrashekhar G. Dethe, ; Milind M. Mushrif, (2014) "Modified Step Size LLMS Channel Estimation Method for MIMO-OFDM System", Recent Advances in Communications and Networking Technology (Formerly Recent Patents on Telecommunication) (Discontinued), Volume 3, Number 2, 2014, pp. 98-105(8) Publisher: Bentham Science Publishers
23. K. Kumar and A. Dwivedi, "Big Data Issues and Challenges in 21 st Century," International Journal on Emerging Technologies (Special Issue NCETST-2017), vol. 8, no. 1, pp. 72–77, 2017.
24. Dwivedi, R. P. Pant, S. Pandey and K. Kumar, "Internet of Things' (IoT's) Impact on Decision Oriented Applications of Big Data Sentiment Analysis," 2018 3rd International Conference on Internet of Things: Smart Innovation and Usages (IoT-SIU), Bhimtal, India, 2018, pp. 1-10, doi: 10.1109/IoT-SIU.2018.8519922.

70 Production of biodiesel from waste chicken fat for sustainable vehicular application

Santosh Kumar Dash[1,a], *Dharmeswar Dash*[1], *Pritam Kumar Das*[2] *and Debabrata Barik*[3]

[1]Ghani Khan Choudhury Institute of Engineering and Technology, Malda, West Bengal-732141, India

[2]Aditya Engineering College, Surampalem, Andhra Pradesh-533437, India

[3]Karpagam Academy of Higher Education, Coimbatore, Tamil Nadu-641021, India

Abstract

Energy crisis is real and the consequence of using conventional energy sources has already been realized in the form of environmental pollution and depletion of fossil fuel reserves. Biodiesel is one of the most sought alternative fuel technology that tries to supplement or replace petroleum based fuel. Though there are many literature available in the context of tree seed oil based biodiesel, less work has been done on waste fat based biodiesel. This research work deals with the production of biodiesel from waste chicken fat through the transesterification process using homogeneous catalyst. Sodium hydroxide (NaOH) as a homogeneous catalyst and methanol (CH_3OH) as an alcohol has been used in the transesterification reaction. Reaction temperature 60°C, reaction time 1.5 hour and speed of reaction 400 rpm has been maintained in the esterification process. In the esterification process, molar ratio of methanol to fat oil is kept as 6:1 with acid catalyst sulfuric acid amount 1 wt.% of waste chicken fat oil. In the transesterification reaction, molar ratio of methanol to oil 6:1, NaOH catalyst 1.5 wt.%, reaction temperature 60°C, reaction time 2 hour and agitation speed 500 rpm kept as optimal parameters. Finally, the biodiesel yield obtained to be 73.4%. All the important properties of the prepared biodiesel found to be comparable to that of petroleum diesel fuel.

Keywords: Biodiesel, chicken fat, diesel fuel, waste, pollution

1 Introduction

It is already realized that conventional fossil fuel causes environmental pollution. Though their demand is ever increasing, they are depleting at a fast pace [1]. This necessitates that more attention should be focused on the renewable resources such as biofuel, solar energy, wind energy, etc. Both biodiesel and ethanol can be synthesized from bio-origin feedstock which is renewable resource. Biodiesel is carbon neutral as the oil seed bearing plant has the ability to sequester atmospheric CO_2, which neutralizes the CO_2 emitted when the biodiesel combusted in the diesel engine. Sincere efforts have been made by scientists around the world to strategically reduce the consumption of diesel fuel and replace them with biofuel [2, 3].

Biodiesel is basically a mono-alkyl ester prepared by chemically reacting vegetable or animal fats with short chain alcohol such as methanol, ethanol, etc. Neat oil use is not recommended due to its high viscosity (10–60 mm^2/s), which causes detrimental effects on the behavior of the combustion engine [4]. Many researchers have tried using neat oil albeit limited success [5, 6]. High viscosity and density of animal fats causes engine seizure and choking of injector nozzle. From last two decades it is highly recommended to use treated fuels to lower viscosity and density. There are many studies that report animal fats have high viscosity and density, which needs two stage transesterification to reduce viscosity and density [7, 8]. In the transesterification process, the vegetable oil converted step wise from triglyceride to diglyceride and then from diglyceride to monoglyceride. Finally, ester and glycerol obtained as product. The glycerol has medicinal value and is a value-added byproduct in the biodiesel preparation process. Feedstock scarcity mainly creating problems in the large-scale production of biodiesel. This is the reason why animal fats must be exploited to prepare biodiesel. Biodiesel has immense scope to increase rural economy, improve environmental stability, align order of the eco system [9].

The major objective of biodiesel is to replace petroleum diesel or supplement diesel fuel. The derived fuel should meet the standard fuel characteristics as per ASTM limit or other country-wise set standards. Basically, three main parameters govern the popularization of biodiesel.

- Biofuel feedstock availability
- Policy
- Economy

[a]santosh@gkciet.ac.in

DOI: 10.1201/9781003540199-70

2 Literature review

There are many literatures available on the synthesis of animal fat and vegetable based triglyceride for biodiesel production [10, 11].

Awad et al. [12] have investigated animal fat residue based biodiesel for use in diesel engine. They observed drastic reduction of unburned hydrocarbon when fueled with biodiesel. They also found the PM emission also reduced significantly at low load operation. On the penalty side, they found the power dropped by 9% for the biodiesel.

Chakraborty et al. [13] made an excellent review on the slaughterhouse animal fats and poultry animal fats considering parametric sensitivity. They encouraged exploiting animal waste fat-based biodiesel production citing their low cost and easy availability. They cautioned that the cold filter plug point (CFPP) and cloud point for poultry based biodiesel has high values. However, by blending with diesel fuel the net effective CFPP can be normalized.

Shi et al. [14] developed catalytic composite membrane and sodium methoxide to prepare biofuel from poultry fats. They observed that the esterification conversion greatly affected by methanol/oil mass ratio. They also found the reaction time, temperature and catalyst amount also plays vital role on the conversion.

From the review it was learned that limited works have been explored on study of biofuel potential of waste chicken fat. More study on this is recommended and this work is the result of the research gap established.

3 Materials and methodology

The waste chicken fats were collected from local butchery located in Malda, West Bengal, India. It was manually washed in the tap water and subsequently defeathered and stored in the kitchen. Waste fats and skin fats were collected and washed again thoroughly after which it was cut into small pieces. Then in a cook pot the cut pieces were placed and cooked in a very low flame and maintained at a temperature of 110°C for 1 hour. The fats got melted and subsequently collected from the top layer of the cooking pot. The oil collected was poured in a beaker and were allowed to cool at room temperature. After which it was filtered off to separate any possible impurities. To avoid any gum formation of the oil or possible damage, without any delay it was proceeded for biodiesel production. The biodiesel synthesis was carried out in a laboratory setup. The setup includes beakers, flasks, condenser, magnetic stirrer with hot plate device and chemicals such as methanol, sodium hydroxide (NaOH), sulfuric acid, etc. A three necked flask is used for the production of biodiesel via esterification and transesterification process.

4 Production of chicken fat biodiesel

4.1 Acid catalyzed esterification

High free fatty acid (14.2%) of the raw fat resulted failed transesterification reaction and formation of unwanted soaps. The FFA needs to be reduced to below 2% at least in order to proceed for the base catalyzed transesterification process. To lower the FFA content sulfuric acid catalyzed esterification process was followed. Basically, in esterification an alcohol (e.g., methanol) and an acid (e.g., FFA) is reacted in the presence of catalyst (usually sulfuric acid) [15]. To perform esterification of chicken fat oil, 500 ml oil was poured in a beaker and the temperature of chicken fat oil was raised to 40°C. Molar ratio 6:1 methanol to oil was added in preheated oil. Then slowly sulfuric acid 1 wt.% of oil is dropped in the three necked flask containing reaction mixture of methanol and chicken fat oil. The reaction mixture was agitated at 400 rpm stirring rate at 60°C temperature and stirred for 90 minutes at atmospheric pressure and temperature of 32°C. The mixture was allowed to cool in the setup. Then three necked flask is removed from the setup and the reacted mixture is poured into a separating funnel, where different layers of liquid visible confirmed the successful reaction. The top layer being unreacted methanol, the middle layer is triglyceride methyl ester and the lower layer contains moisture. The middle layer is separated and the stored in beaker. The FFA value reduced significantly in the esterification process and found to be as low as 1.2%. As the FFA is lowered to safe level, it was proceeded for NaOH catalyzed transesterification reaction process.

4.2 Base catalyzed transesterification

Transesterification is carried out by reacting an alcohol (e.g., methanol) with plant seed based oil or animal fat in the presence of a base catalyst such as NaOH. To perform esterification of chicken fat oil, 450 ml oil was poured in a beaker and the temperature of chicken fat oil was raised to 40°C. Methanol to oil molar ratio has been kept 6:1. Then slowly NaOH pellet powder 1.5 wt.% of oil is dropped in the three necked flask containing reaction mixture of methanol and chicken fat oil. The reaction mixture was agitated at 500 rpm stirring rate at 60°C temperature and stirred for 120 minutes at atmospheric pressure and temperature of 32°C. The mixture was allowed to cool in the setup. Then three necked flask is removed from the setup and the reacted mixture is poured into a separating funnel, where different layers of liquid visible confirmed the successful reaction. The top layer being unreacted methanol, the middle layer is biodiesel and the lower layer contains glycerol. The middle layer is separated and the stored in beaker. The acid value reduced significantly in the transesterification process and found to be as low as 0.11 kg KOH/g. The density and viscosity reduced significantly in the transesterification process. The crude biodiesel needs purification before practical application and storing. The crude biodiesel contains traces of moisture, soup as a result of base catalyst and traces of glycerol. The impurities in the crude biodiesel removed by water washing by adding warm distilled water to the crude biodiesel in the separating funnel. This causes the biodiesel to float at the upper layer and the heavy water with impurities along with soap to settle down at the lower layer. In order to remove soap completely, water washing steps have been repeated two more times. The clear water indicated the biodiesel is free of impurities. In the process of water washing, there might be chances that the biodiesel has hints of moisture in it. To remove the moisture, it was heated on the hot plate by regulating temperature not exceeding 100°C. The photo of biodiesel is presented in Figure 70.1.

Figure 70.1 Chicken fat biodiesel

Few important physic-chemical parameters have been checked to confirm that the obtained biodiesel is of standard quality (Table 70.1). The density of the biodiesel obtained to be 0.876 g/cc and the viscosity of the biodiesel found to be 3.91. Though the density and viscosity is comparatively higher than petroleum diesel, still it is within the ASTM limits. The higher flash point has advantage of limiting storing safety issue compared to diesel fuel.

5 Conclusions

This paper discusses the production of biodiesel from waste chicken fat collected from local slaughterhouse. Both acid and base catalyzed process adopted for preparing biodiesel as the free fatty acid was higher. The process parameters were not optimized in this study but the experiments were carefully conducted. The esterification was conducted at reaction temperature 60°C, reaction time 1.5 hour and speed of reaction 400 rpm. In the esterification process, molar ratio of methanol to fat oil is kept as 6:1 with acid catalyst sulfuric acid amount 1 wt.% of waste chicken fat oil. In the transesterification reaction, molar ratio of methanol to oil 6:1, NaOH catalyst 1.5 wt.%, reaction temperature 60°C, reaction time 2 hour and agitation speed 500 rpm kept. Finally, the biodiesel yield obtained to be 73.4%.

6 References

1. Shahid, E. M. and Jamal, Y. (2011). Production of biodiesel: A technical review. *Renew. Sustain. Energy Rev.*, 15(9), 4732–4745. doi: 10.1016/j.rser.2011.07.079.
2. Atadashi, I. M., Aroua, M. K., Abdul Aziz, A. R., and Sulaiman, N. M. N. (2012). Production of biodiesel using high free fatty acid feedstocks. *Renew. Sustain. Energy Rev.*, 16(5), 3275–3285. doi: 10.1016/j.rser.2012.02.063.
3. Dash, S. K. and Lingfa, P. (2017). A review on production of biodiesel using catalyzed transesterification. *AIP Conf. Proc.*, 1859, 020100.
4. Dash, S. K., Lingfa, P., and Chavan, S. B. (2018). An experimental investigation on the application poten-

Table 70.1 Properties of waste chicken fat biodiesel

S. No.	Properties	Unit	Chicken fat biodiesel	Diesel	ASTM method
1.	Color	-	Pale yellow	Pale yellow	-
2.	Density	g/cc	0.876	0.834	D1298
3.	Kinematic viscosity	cSt@40 °C	3.91	2.73	D4809
4.	Calorific value	MJ/kg	41.29	-	D6751
5.	Flash point	°C	174	68	D93
6.	Pour point	°C	2.1	-7.5	D97

tial of heterogeneous catalyzed Nahar biodiesel and its diesel blends as diesel engine fuels. *Ener. Sources Part A Recover. Util. Environ. Eff.*, 40(24), 2923–2932. doi: 10.1080/15567036.2018.1514433.

5. Agarwal, A. K. and Rajamanoharan, K. (2009). Experimental investigations of performance and emissions of Karanja oil and its blends in a single cylinder agricultural diesel engine. *Appl. Ener.*, 86(1), 106–112. doi: 10.1016/j.apenergy.2008.04.008.
6. Agarwal, D., Kumar, L., and Agarwal, A. K. (2008). Performance evaluation of a vegetable oil fuelled compression ignition engine. *Renew. Ener.*, 33(6), 1147–1156. doi: 10.1016/j.renene.2007.06.017.
7. Barik, D. and Vijayaraghavan, R. (2020). Effects of waste chicken fat derived biodiesel on the performance and emission characteristics of a compression ignition engine. *Int. J. Ambient Ener.*, 41(1), 88–97. doi: 10.1080/01430750.2018.1451370.
8. Altikriti, E. T., Fadhil, A. B., and Dheyab, M. M. (2015). Two-step base catalyzed transesterification of chicken fat: Optimization of parameters. *Ener. Sources Part A Recover. Util. Environ. Eff.*, 37(17), 1861–1866. doi: 10.1080/15567036.2012.654442.
9. Dash, S. K., Lingfa, P., Das, P. K., Saravanan, A., Dash, D., and Bharaprasad, B. (2023). Effect of injection pressure adjustment towards performance, emission and combustion analysis of optimal nahar methyl ester diesel blend powered agricultural diesel engine. *Energy*, 263, 125831. doi: https://doi.org/10.1016/j.energy.2022.125831.
10. Atabani, A. E., et al. (2013). Non-edible vegetable oils: A critical evaluation of oil extraction, fatty acid compositions, biodiesel production, characteristics, engine performance and emissions production. *Renew. Sustain. Ener. Rev.*, 18, 211–245. doi: 10.1016/j.rser.2012.10.013.
11. Srivastava, A. and Prasad, R. (2000). Triglycerides-based diesel fuels. *Renew. Sustain. Ener. Rev.*, 4(2), 111–133. doi: 10.1016/S1364-0321(99)00013-1.
12. Awad, S., Loubar, K., and Tazerout, M. (2014). Experimental investigation on the combustion, performance and pollutant emissions of biodiesel from animal fat residues on a direct injection diesel engine. *Energy*, 69, 826–836. doi: 10.1016/j.energy.2014.03.078.
13. Chakraborty, R., Gupta, A. K., and Chowdhury, R. (2014). Conversion of slaughterhouse and poultry farm animal fats and wastes to biodiesel: Parametric sensitivity and fuel quality assessment," *Renew. Sustain. Ener. Rev.*, 29, 120–134. doi: 10.1016/j.rser.2013.08.082.
14. Shi, W. et al. (2013). Biodiesel production from waste chicken fat with low free fatty acids by an integrated catalytic process of composite membrane and sodium methoxide. *Bioresour. Technol.*, 139, 316–322. doi: 10.1016/j.biortech.2013.04.040.
15. Papu, N. H., Lingfa, P., and Dash, S. K. (2021). An experimental investigation on the combustion characteristics of a direct injection diesel engine fuelled with an algal biodiesel and its diesel blends. *Clean Technol. Environ. Policy*, no. 0123456789. doi: 10.1007/s10098-021-02058-3.

71 Design, manufacturing and testing of parabolic solar heating/cooking system

Dharmeswar Dash[a], *Santosh Kumar Dash*[b], *Amit Nath*[c], *Rintu Prasad Gupta*[d] *and Samim Ali*[e]

Department of Mechanical Engineering, GKCIET, Malda, West Bengal-732141, India

Abstract

This paper discussed a parabolic solar heating system where solar energy receiver is made of a copper tube to absorb the heat energy from the sun's rays after a parabolic disc has focused them. The heat from the solar energy receiver is transferred to the cooking platform by using a heat-transfer carrying fluid called glycerol, where the food is to be cooked or heated. On the cooking platform, food is heated/cooked by absorbing heat from the carrying fluid. Experiments were conducted to heat water over the cooking platform, and the temperature variations of the carrying fluid, cooking platform, and water were documented. It has been discovered that the heating temperature of water at 01.00–01.35 pm reaches its maximum temperature, followed by the time from 10.00 am to 10.35 am and 03.00–03.35 pm.

Keywords: Parabolic solar heating system, parabolic disc concentrator, glycerol, pump with motor, glass wool

1. Introduction

Cooking is the way humans prepare the food they need to survive and which accounts for a significant portion of the energy used in many nations. About 37–53% of the total energy consumed in affluent nations like the United States is used for cooking. In underdeveloped nations in Africa, Asia, and South America, cooking consumes the majority of household energy. With extremely high rates of greenhouse gas emissions and deforestation, almost 80% of sub-Saharan Africa still cooks food using firewood, fossil fuels, biomass, and electricity [1]. This shows the need for substitute cooking techniques that are safe for the environment and use on clean energy sources. Only approximately 14% of the world's renewable energy resources are now exploited, but due to coordinated international efforts to reduce greenhouse gas emissions that cause global warming, renewable energy utilization is likely to rise soon.

Approximately 5,000 trillion kWh of energy are produced on land in India each year, with the majority of the country receiving between 4 and 7 Kwh per square meter every day, according to the Ministry of New and Renewable Resources. Solar energy is the safest energy source available in terms of energy security because it is widely accessible. Theoretically, a small portion of the total incident solar energy could supply all of a nation's electricity demands if it were successfully captured. And it is found that in Asian countries the solar energy received is the highest from other temperature countries because the duration of sunshine is high in a year [2]. Therefore, as a nation, we have great advantages in using solar energy for cooking and power generation. Many researchers from different countries are researching renewable energy to reduce our dependence on non-renewable sources of energy. The massive use of these resources (that is, fossil fuels) produces greenhouse gases, which contribute significantly to climate change. On the other hand, sustainable, free, and renewable forms of energy like solar and wind energy do not release greenhouse gases [3].

The heating/cooking component is the main emphasis of this investigation. On the market, there are numerous types of solar cookers with a wide range of functionalities [4], including box-style solar cookers [5], parabolic dish cookers [6], which are among the most popular types of parabolic solar cookers, and indirect solar cookers, which operate similarly to solar geysers [7]. In this project, we use a parabolic solar dish for one experiment that is cooking. The surface finish, like a mirror, concentrates the solar energy into one focal point [8], where a rectangular copper box is installed inside the copper box. There is a round-shaped copper tube along with copper dust. Inside the copper tube, a fluid called glycerol flows.

[a]dharmeswar@gkciet.ac.in, [b]santosh@gkciet.ac.in, [c]amitnath860@gmail.com, [d]4rintugupta1@gmail.com, [e]5sksamimali441@gmail.com

DOI: 10.1201/9781003540199-71

2. Components and manufacturing of parabolic solar heating/cooking system

The important components of this solar heating/cooking are a parabolic solar dish, solar energy receiver, storage tank, cooking platform, and gear pump with a motor. The assembled of components is shown in Figure 71.1. A solar receiver is positioned at the parabolic dish's focal point. Within this solar receiver, a curved copper pipe is carefully positioned. Copper powder is used to seal the space between the receiver base plate and the curved copper pipe for maximizing the copper pipe's capacity to absorb heat and improving efficiency. Except for the receiver and cooking platform, the remaining exposed sections are coated with glass wool to reduce heat loss. This copper pipe serves as a conduit for the carrying fluid. Copper pipe connects to a circular curved cooking platform, providing a pathway for carrying fluid. From there, the carrying fluid flows into a storage tank dedicated to its containment. To ensure a continuous flow, the storage tank is connected to a gear pump that is powered by an induction motor. The gear pump effectively propels the carrying fluid back to the solar receiver, completing the cycle of carrying fluid circulation.

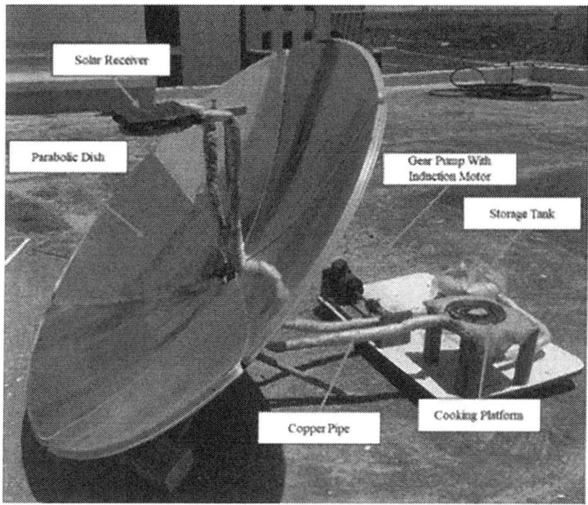

Figure 71.1 Manufacturing of parabolic solar heating/cooking system

3. Experimental procedure and performance test

3.1 Experimental procedure

Solar energy receiver that absorbs the heat coming from the parabolic dish when the incident rays are focused on the receiver's base. The receiver is a rectangular copper box, and a curved copper pipe is placed in it. This pipe is filled with carrying fluid (glycerol). To increase its efficiency, the receiver is filled with copper powder, which is distributed evenly. The copper powder inside the receiver is for maximum heat absorption to heat the fluid flowing through the pipe and to trap the heat inside the receiver. To pump the fluid, a gear pump is used which is connected to an induction motor with the help of jaw coupling, the motor is powered by electricity. When the rays are incident on the parabolic dish, it heats the receiver box and heats the glycerol that is running through the copper pipe. Then the hot glycerol is transferred to the cooking platform. The cooking platform is also made of copper pipe, and its shape is like a circular coil. When the fluid is passed through it, it will heat the cooking platform and will be able to heat/cook food. The circulation of the fluid is maintained throughout the pipe so that carrying fluid picks up heat from the receiver at each time and transfers it to the cooking platform.

3.2 Performance test

The performance of this parabolic solar heating system was tested at GKCIET, Malda, West - Bengal, INDIA -732141. In this experiment, 0.5 liters of water is used for testing. The water is kept in the steel bowl and placed on the cooking platform to observe a rise in temperature w.r.t. the time during the operation. Performance test of a parabolic solar system has been conducted in three-time zone that

Table 71.1 Performance test from 10.00 am to 10.35 am

S. No.	Time (min)	Fluid temp.()	Cooking platform temp. ()	Water temp. ()
1	5	68.2	40.6	32
2	10	69.8	64.2	35.5
3	15	72.2	69.2	38.8
4	20	74.6	71.8	40.2
5	25	78.1	74.6	41.5
6	30	81.1	78.2	41.9
7	35	83.3	80.6	42.8

Table 71.2 Performance test from 01.00 pm to 01.35 pm

S. No.	Time (min)	Fluid temp.()	Cooking platform temp. ()	Water temp. ()
1	5	70.4	49	35.1
2	10	78.7	70.3	37.5
3	15	86.4	77.6	41.3
4	20	91.5	77.9	42.4
5	25	91.9	85.2	43.2
6	30	92.1	86.6	43.8
7	35	92.1	87.6	44.8

Table 71.3 Performance test from 03.00 pm to 3.35 pm

S. No.	Time (min)	Fluid temp.()	Cooking platform temp. ()	Water temp. ()
1	5	63.2	38.5	31.4
2	10	66.4	62.5	32.5
3	15	69.4	67.5	34.9
4	20	72.5	69.6	36.7
5	25	75.6	72.5	39.4
6	30	78.4	74.4	40.7
7	35	79.8	77.3	41.4

Table 71.4 Performance test fluid temperature after 5 and 35 mins of operations

Performance test	Fluid temperature ()	
	Test after 5 min	Test after 35 min
From 10.00 am to 10.35 am	68.2	83.6
From 1.00 pm to 1.35 pm	70.4	92.1
Performance test from 3.00 pm to 3.30 pm	63.2	79.8

is 10.00–10.35 am, 01.00–01.35 pm, and 03.00–03.35 pm.

3.2.1 Performance test 10.00 am – 10.35 am

Before starting the test operation, the atmospheric temperature was measured as 28 at 10.00 am. The test value of carrying fluid, and cooking platform, water temperature from 10.00 am to 10.35 am as shown in Table 71.1.

3.2.2 Performance test from 01.00 pm to 1.35 pm

The performance test was carried out on the same day afternoon from 01.00 pm to 01.35 pm and the cooking platform temperature was noted as 39. The testing value of fluid, and cooking platform, water temperature from 01.00 pm to 01.35 pm as shown in Table 71.2.

3.3.3 Performance test from 03.00 pm to 3.35 pm

The performance test was carried out on the same day afternoon from 3.00 pm to 3.35 pm and the surrounding temperature was noted as 32. The testing value of fluid, and cooking platform, water temperature from 3.00 pm to 3.35 pm as shown in Table 71.3.

4. Result and discussion

4.1. Observation and analysis of performance test of solar heating/cooking system

Tables 71.4–71.6 show the temperature of fluid, cooking platform, and water during the time from 10.00 am to 10.35 am, 01.00–01.35 pm, and 03.00–03.35 pm, respectively over a period of 35 mins. Comparison of carrying fluid, cooking platform,

Table 71.5 Performance test of cooking platform temperature after 5 and 35 mins of operations

Performance test	Cooking platform temperature ()	
	Test after 5 min	Test after 35 min
From 10.00 am to 10.35 am	40.6	80.6
From 1.00 pm to 1.35 pm	49	87.6
From 3.00 pm to 3.30 pm	38.5	77.3

Table 71.6 Performance test of water temperature after 5 and 35 mins of operations

Performance test	Water temperature ()	
	Test after 5 min	Test after 35 min
From 10.00 am to 10.35 am	32	42.8
From 1.00 pm to 1.35 pm	35.1	44.8
From 3.00 pm to 3.30 pm	31	41.4

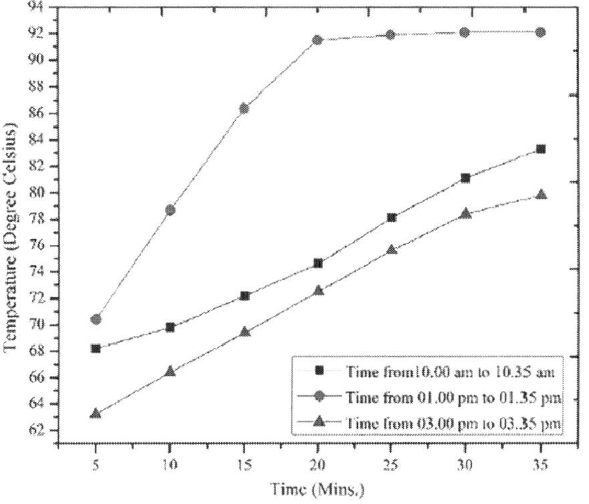

Figure 71.2 Comparison of carrying fluid temperature in different time schedules

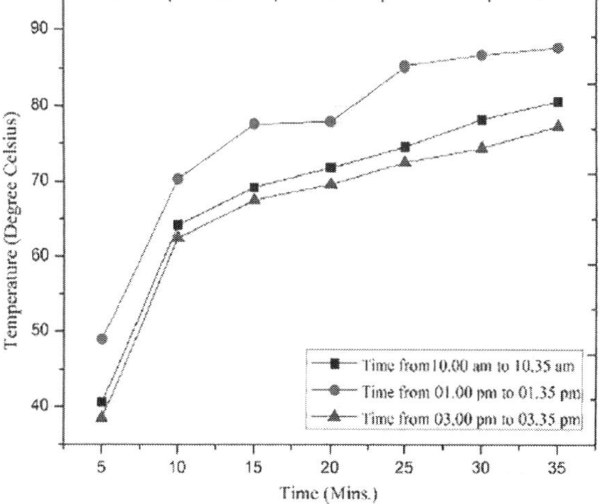

Figure 71.3 Comparison of cooking platform temperature in different time schedules

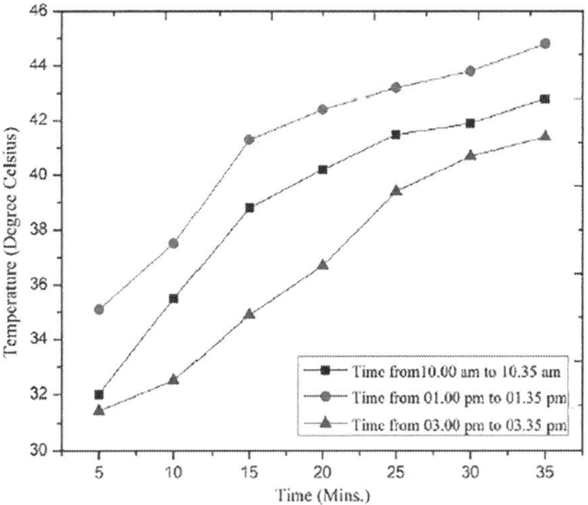

Figure 71.4 Comparison of water temperature in different time schedules

and water temperatures w.r.t. time is shown in Figures 71.2–71.4, respectively. These observations suggest that the heating process is effective in raising the temperatures of the carrier fluid, cooking platform, and water. The cooking platform receives heat from the carrier fluid, which is then transferred to the water to heat it. The results indicate that the temperature of carrying fluid is high out of measurement in all temperatures as it receives heat directly from the parabolic disc due to radiation. However, heat transfer from the carrying fluid to the cooking platform is decreased as heat loss takes place during the heat transfer. Again heat is transferred from the cooking platform to the container and finally heat transferred from the container to the water for heating/cooking purposes. Here it is observed that the temperature of heated water is less as compared to the temperature of carrying fluid as many mode heats are transferred from parabolic disc to water and each time heat loss takes place. The heat loss also depends upon the atmosphere condition and surface smoothness of the cooking platform and container. Overall, the data shows that in all three tests, the carrying fluid, cooking platform, and the cooking/heating process successfully raised water temperatures. Additionally, it has been discovered that afternoons are the times when water temperatures are at their highest, followed by morning and evening.

5. Conclusions

This section presents the results of tests conducted at different time intervals during the cooking/heating process. The tests were conducted at 10.00–10.35 am, 01.00–01.35 pm, and 03.00–03.35 pm, and the temperature variation of the carrying fluid, cooking. platform, and water were recorded. The outcome of the results are as follows:

1. It was noted that the cooking platform temperature grew consistently during the course of the test from 10.00 am to 10.35 am, peaking at 80.6°C after 35 min. A similar pattern could be seen in the fluid temperature, which increased from 68.2°C to 83.3°C in 35 min. On the other hand, the water's temperature rose from 32.0°C to 42.8°C by the end of the experiment. During the test period, the rates of temperature change for all three components were noticeably faster.
2. The cooking platform temperature climbed from 49°C to 87.6°C throughout the course of 35 min during the test period from 1:00 pm to 1:00:35 pm. Both the fluid temperature and the water temperature exhibited an upward trend, rising from 35.1°C to 44.8°C for the water and from 70.4°C to 92.1°C for the carrying fluid.
3. During the 03.00–03.35 pm test time, it is found that the cooking platform temperature remained relatively stable between 60°C and 70°C throughout the experiment. The fluid temperature increased gradually from 63.2°C to 79.8°C, over a period of 35 min, whereas the water temperature steadily increased and then maintained a constant value of around 40°C.
4. The data show that the temperatures of carrying fluid, cooking platform, and water were successfully increased by the cooking/heating process in each of the three experiments. Additionally, it was found that the maximum temperature of water was achieved during the time 01.00 pm–01.35 pm followed by 10.00 am–10.35 am and 03.00 pm–03.35 pm.

Acknowledgements

Authors are grateful to the GKCIET, Malda, West Bengal for providing the necessary facilities to carry out this work successfully.

References

1. Lentswe, K., Mawire, A., Owusu, P., and Shobo, A. (2021). A review of parabolic solar cookers with thermal energy storage. 7(10), 1–6. DOI: https://doi.org/10.1016/j.heliyon.2021.e08226.
2. Kannan, N. and Vakeesan, D. (2016). Solar energy for future world - A reviews. *Renew. Sustain. Ener. Rev.*, 62, 1092–1105. DOI: https://doi.org/10.1016/j.rser.2016.05.022.
3. Zhang, Q., Stewart, S. W., Jeffrey, R. S. B., and Witmer, L. T. (2012). Geyser pump solar water heater system modelling design optimization. *41st Am. Solar Ener. Soc. Meeting Proc. 41st Am. Solar Ener. Soc. Meeting*. 1–8.
4. Cuce, E. and Cuce, P. M. (2013). A comprehensive review on solar cookers. *Appl. Ener.*, 102, 1399–1421. DOI: https://doi.org/10.1016/j.apenergy.2012.09.002.
5. Harmim, A., Merzouk, M., Boukar, M., and Amar, M. (2013). Design and experimental testing of an innovative building – integrated box type solar cooker. *Solar Energy*, 98, 422–433. DOI: https://doi.org/10.1016/j.solener.2013.09.019.
6. Joyee, E. B. and Mizanur Rahman, A. N. M. (2014). Design and construction of a parabolic dish solar cooker. *Inter. Conf. Mec. Indus. Ener. Engg.* 1–5.
7. Hebbar, G., Hegde, S., Sanketh, B., Sanith, L. R., and Udupa, R. (2021). Design of solar cooker using evacuated tube solar collector with phase change material. *Mater. Today Proc.* 46 (7), 2888–2893.
8. Masum Ahmed, S. M., Rahmmatullah, Md., Amim, Al., Ahammed, S., Ahmed, F., Saleque, A. M., and Abdur Rahman, Md. (2020). Design, construction and testing of parabolic solar cooker for rural households and refugee camp. *Solar Energy*, 205, 230–240. DOI: https://doi.org/10.1016/j.solener.2020.05.007.

72 Modeling of DC-DC converters using basic converters

Mohamed Junaid K. A.[1,a], *Sukhi Y.*[2,b], *Jeyashree Y.*[3,c] *and Priscilla Whitin*[4,d]

[1]Department of Electronics and Communication Engineering, R.M.K. Engineering College, Thiruvallur, Tamil Nadu, India

[2]Department of Electrcal and Electronics, R.M.K. Engineering College, Thiruvallur, Tamil Nadu, India

[3]Department of Electrical and Electronics Engineering, SRM Institute of Science and Technology, Kattankulathur, Tamil Nadu, India

[4]Department of Electrical and Electronics Engineering, Veltech Rangarajan Dr.Sagunthala R&D Institute of Science and Technology, Avadi, Tamil Nadu, India

Abstract

In this paper, the DC/DC converter is modeled which uses pulse width modulation principle along with basic converter configurations. There are general power electronic converters operating with PWM like Sepic, Buck, Cuk, Boost, Zeta, Buck-boost and these converters are discussed. The buck and boost operation of the converters are dealt using basic converter units. The response of these converters is studied in continuous conduction mode for different duty cycles using basic converters and the comparison of output voltage for these configurations is discussed with the graph.

Keywords: Pulse width modulation (PWM), DC-DC converter, basic converter units (BCUs), modeling

Introduction

The dynamic behavior of the converter is important to understand the performance characteristics of DC converters using pulse width modulation (PWM). The process of modeling a converter and the design of controllers is a fundamental task in power electronics. To achieve this, the state-space analysis is employed. This technique operates at frequencies less than switching frequency, allowing for an effective representation of the converter dynamics. The method involves the determination of average power flows over a switching period and formulation of state and output equations to describe behavior of the converter [1]. This approach treats each converter independently and enables separate analysis for each converter. Various approaches have been proposed to simplify the derivation of circuit models. This involves replacing the switching components with pseudo circuit components [2, 3]. This method takes advantage of recognizing the presence of PWM switches in each converter, streamlining the modeling process. Cuk utilized the standard switching technique to systematically model the converter. This technique aids in efficiently representing the complex dynamics of converters and provides insights into their behavior [4]. In order to reduce the complexity of matrix inversions required in the modeling process, the averaged converter model is used [5]. This enhances the computational efficiency of the analysis. Furthermore, researchers proposed a method to incorporate parasitic components into PWM converter models [6]. This approach, based on the principle of energy conservation, ensures more accurate representation of real-world converter behavior. In an effort to enhance the accuracy of PWM DC/DC converter models, an innovative averaging approach was introduced in ref [7]. This approach yields averaged models that explicitly account for the switching frequency, contributing to improved model fidelity.

The research in power electronics is explored various techniques to simplify and improve the process of deriving small-signal models for converters and designing effective controllers.

These methodologies take advantage of averaging, parameterization, and advanced mathematical techniques to accurately capture the behavior of converters in a wide range of applications. By building on the foundation of these research efforts, engineers and scientists try to advance in power electronics and drive innovations in energy conversion and control systems.

Essentially, the works in ref [4–7] focused on converters similar to those discussed in ref [1]. These research efforts contribute to the development of efficient modeling techniques for PWM converters, enabling a deeper understanding of their dynamic behavior and facilitating the design of effective control strategies. An innovative approach is used to model PWM dc/dc converters without transformers with a concept known as basic converter units

[a]kam.ece@rmkec.ac.in, [b]ysi.eee@rmkec.ac.in, [c]jeyashry@srmist.edu.in, [d]priscilla150682@gmail.com

DOI: 10.1201/9781003540199-72

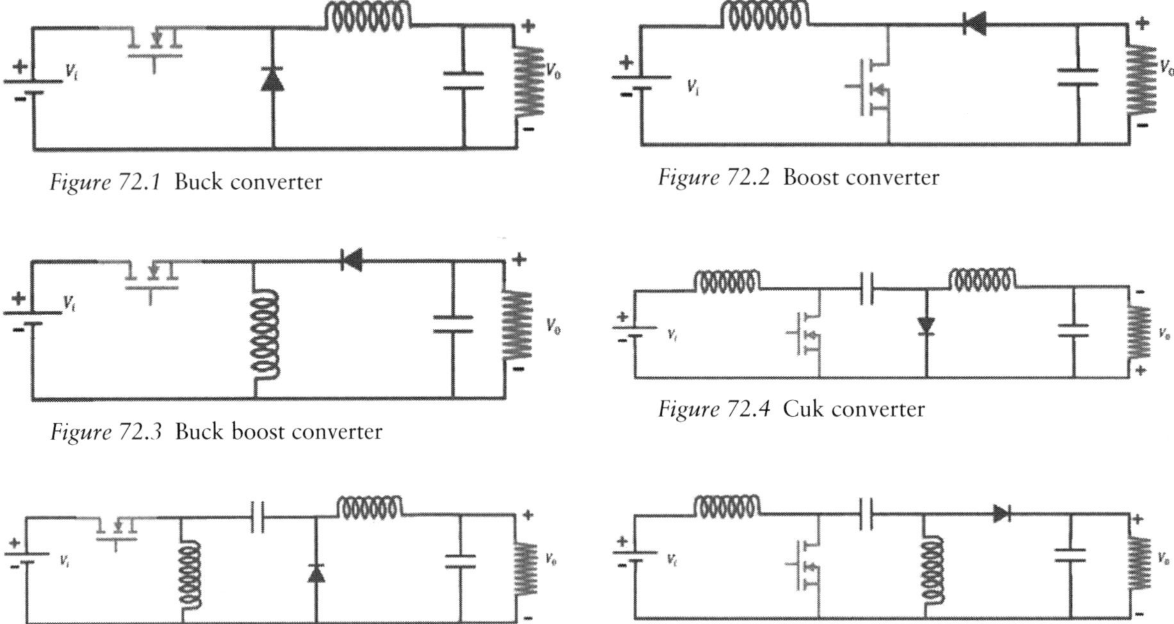

Figure 72.1 Buck converter

Figure 72.2 Boost converter

Figure 72.3 Buck boost converter

Figure 72.4 Cuk converter

Figure 72.5 Zeta converter

Figure 72.6 Sepic converter

(BCUs). The conventional buck converter used in various applications is given in Figure 72.1, while PWM step-up converter configuration is depicted in Figure 72.2. Additionally, the circuit for buck boost as shown in Figure 72.3, Cuk converter is given in Figure 72.4, followed by the diagram of the Zeta topology as shown in Figure 72.5, and PWM SEPIC topology is given in Figure 72.6.

These converters are designed to operate as buck converter and/or boost converters, using the passive components in the circuit. In this proposed framework, the converters are classified into two BCU types: buck and boost converters. Furthermore, the entire array of PWM converters is working in either buck mode or boost mode.

The significance of this methodology lies in its relevance not only to model transformerless converters, which is the primary focus, but also its general applicability. This generality stems from the fact that many isolated converter circuits do not possess transformer.

The paper delves into the concept of reconfiguring PWM DC/DC converters, exploring the feasibility and implications of such reconfigurations. By proposing this novel approach and demonstrating its versatility across various converter types, the paper contributes to the advancement of transformer-less power conversion methodologies. This research is expected to have broader implications for designing efficient and adaptable power electronics systems in diverse applications where transformer-less converters are essential. The general process of modeling converters using BCUs is discussed. The paper concludes with a brief summary.

A converter unit typically refers to a basic building block in power electronics used to transform or convert electrical energy from one form to another. These units are often combined in various configurations to create more complex power converters that serve specific purposes, such as voltage regulation, power conversion, and energy distribution.

PWM converter reconfiguration

In the case of buck converters, applying the principle of the inductor volt second balance principle for the circuit shown in Figure 72.1 with CCM operation, a consistent relationship emerges between the input and output parameters. This relationship yields a gain of $\frac{V_O}{V_i=D}$, where D is the duty cycle of the switch.

For ease of understanding and reference, the circuit of the buck converter along with its input-output connection is visually shown in Figure 72.7. For the buck-boost topology operating under CCM, the voltage gain is expressed as $\frac{V_0}{V_i} = \frac{D}{1-D}$. This derivation is derived by considering that the buck converter feedback gain is one as shown in Figure 72.8. The circuit diagram to illustrate this concept is presented in Figure 72.9.

Through appropriate reconfiguration of the components, the circuit in Figure 72.9 takes the form of the buck-boost converter, as given in Figure 72.10. Expanding on this idea using the feedback concept,

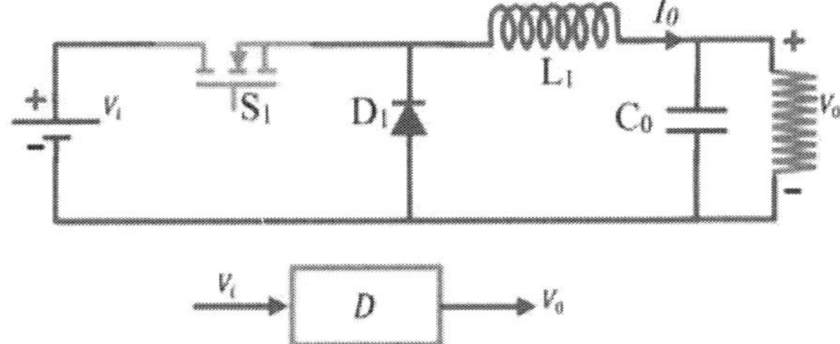

Figure 72.7 Buck converter with transfer function

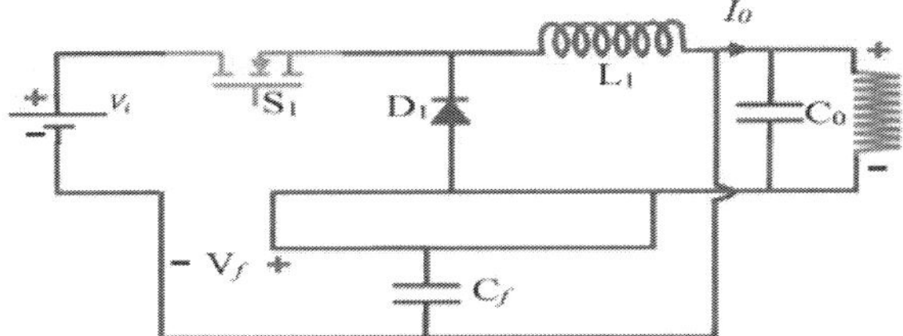

Figure 72.9 Buck boost converter

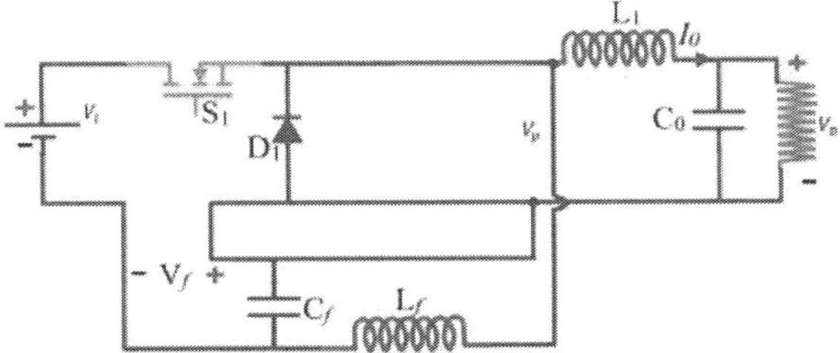

Figure 72.11 Boost Converter with feedback

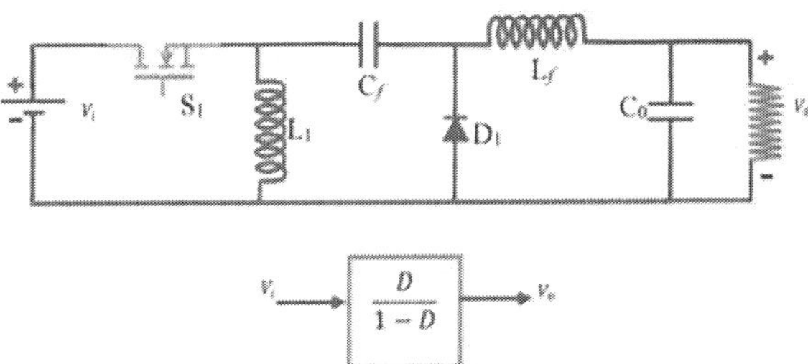

Figure 72.13 Zeta converter with transfer function

an alternative form of a PWM converter is derived, as demonstrated in Figure 72.6.

Given that the voltage V_p exhibits pulsations and an inductor filter is judiciously introduced into the feedback path. This inclusion serves to mitigate voltage ripples, yielding a smoothed voltage V_f. The voltage gain which is usually determined for the different topology is effectively obtained from the block diagram shown in Figure 72.11.In this depiction, D_p conceptually represents the ratio of , while "TF1" and "TF2" denote the transfer functions of specific low-pass filters. As a result, the overall configuration depicted in Figure 72.12 is a concise representation of the circuit behavior.

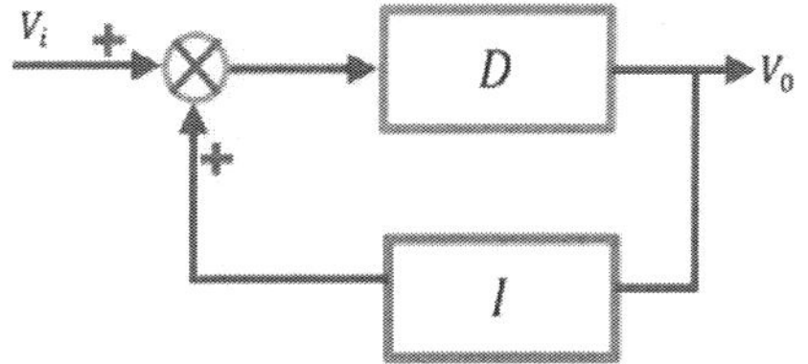

Figure 72.8 Converter closed loop transfer function

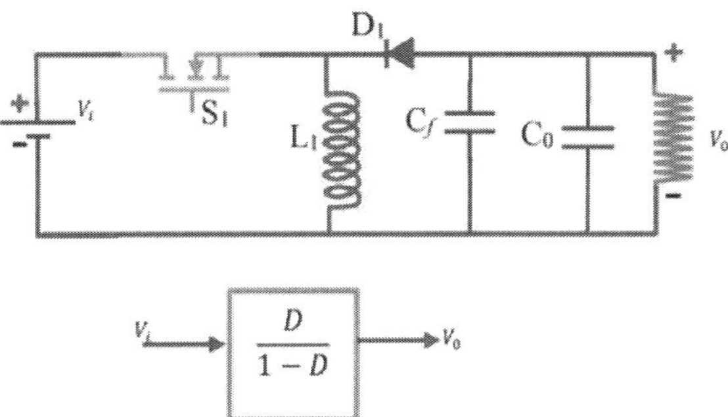

Figure 72.10 Buck boost converter with transfer function

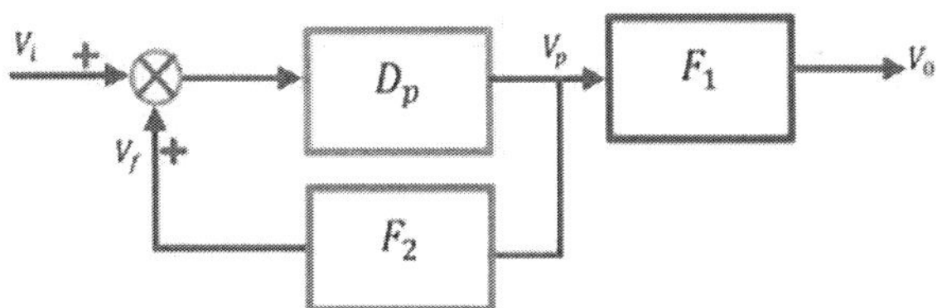

Figure 72.12 Buck boost converter with transfer function block

By employing these concepts and building upon the various configurations discussed, this approach provides a comprehensive framework for understanding and analyzing different PWM converter topologies.

This analysis holds particular significance for designing efficient and effective power conversion systems with minimized ripples and optimized voltage transfer characteristics by the Equation 1.

$$\mathrm{TF1}=\frac{V_0}{V_i}=\frac{D_X TF_1}{1-D_X TF_2} \tag{1}$$

From the buck converter, it can be noted that TF2 can be written in terms of TF1 as given in Equation 2.

$$TF2 = D_X TF_1 = D \tag{2}$$

Considering that , the Eq.1 can be written as given in Equation 3.

$$\frac{V_0}{V_i}=\frac{D}{1-D} \tag{3}$$

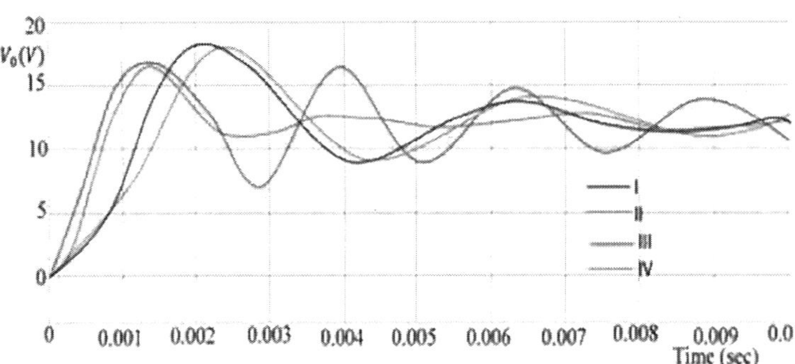

Figure 72.14 Output voltage of different topology: (I) boost converter, (II) buck–boost, (III) Zeta, (IV) Cuk for D = 0.3

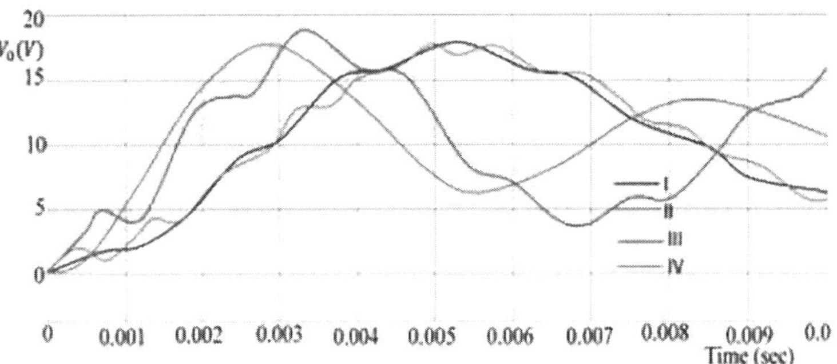

Figure 72.15 Output voltage of different topology: (I) boost converter, (II) buck–boost, (III) Zeta, (IV) Cuk for D = 0.7

The preceding conversation highlights that through a strategic rearrangement of circuits and expressions illustrated in Figures 72.11 and 72.12, the Zeta converter configuration and its corresponding input-output correlation is showed in Figure 72.13. The Zeta converter emerges from a buck BCU, in conjunction with linear components. This categorizes them within the buck converter family. Pulse width modulation DC/DC converters are widely used in power electronics to efficiently convert voltage levels for various applications. These converters are often constructed using basic converter units in combination to achieve the desired regulation. Some of the common basic converter units used in PWM DC/DC converters is given here. Buck converter consists of a switch in series with an inductor and a diode. The control of duty cycle (on-time vs. off-time) can regulate output voltage. The boost converter also includes a switch, an inductor, and a diode. By controlling the duty cycle (on-time vs. off-time), the output voltage is regulated. A buck-boost converter can increase or decrease the input voltage to provide a positive or negative voltage. Cuk converter provides non-inverting voltage conversion with good efficiency. By combining these basic converter units and appropriately controlling their switching operations using PWM techniques, design of complex PWM DC/DC converters is made that meet specific voltage and current regulation requirements for a wide range of applications.

Small-signal modeling

The process of modeling using small signal for the converters can be systematically depicted here. First, the h or g parameters for the BCUs as well as the parameters for the feedback network are derived. These parameters capture the linearized relationship between the input and output variables of the BCUs. Second sum them up in accordance with the converter configuration. This summation allows the aggregation of the effects of individual components and subsystems to yield the overall transfer function of the converter. After summing up the parameters, the small signal model is obtained. This will provide insights into the dynamic behavior in response to small changes or perturbations around their steady-state operating points. The utilization of h or g parameters facilitates the creation of compact and easily interpretable small-signal models. These parameters represent the linearized relationships between input and output variables in a straightforward manner. By carefully deriving the parameters for the BCUs and feedback network and then combining them appropriately, the converter small-signal dynamics is made. The specific details and equations involved will depend on the converter topologies, the chosen modeling technique, and the parameterization method. Additionally, the derived parameters are accurate and consistent, contributing to the reliability of the subsequent small-signal models. The small-signal transfer function of a buck converter can be derived using control theory techniques. The small-signal model allows to analyze the dynamic behavior of the converter around its steady-state operating point. One common way to represent the small-signal transfer function is using control loop equations or Laplace-transformed transfer functions. For a simplified buck converter, assuming a voltage-mode control scheme, the small-signal transfer function can be derived. This transfer function is typically represented in terms of duty cycle (D), output voltage ripple, and control loop parameters. Please note that the derivation of these equations can be involved and might depend on the specific assumptions and parameters of the buck converter. The basic representation of the small-signal transfer function is given in Equation 4.

$$G(s) = \frac{V_0(s)}{V_i(s)} = \frac{D}{1-D} \quad (4)$$

Relationship among converters

By applying the proposed approach, PWM converters can be systematically classified into distinct families. This classification enables a structured understanding of converter behavior, and it raises the intriguing possibility that there could be relationships among the dynamics of different converters within these families. However, as of our current understanding, delving into these relationships can be challenging due to the complexity inherent in the general expressions of the models.

The classification of PWM converters into families provides a structured framework for grouping converters based on shared characteristics, topologies, or operational principles. This categorization aids in simplifying the analysis and comparison of different converters, as converters within the same family may exhibit similar behaviors and responses to perturbations. Despite the potential for these families to offer insights into relationships among converter dynamics, the intricate nature of the general model expressions poses a significant challenge.

The complexity arises from the interaction of various components, feedback loops, and nonlinearities present in the converters. Investigating relationships between different converter dynamics would require a comprehensive understanding of the underlying mathematical formulations and the ability to identify patterns among them.

Researchers in the field recognize the importance of exploring such relationships as they could lead to valuable insights into converter design, control strategies, and system optimization. However, due to the intricate nature of these systems, thorough investigations into these relationships will necessitate advanced mathematical analysis, simulation studies, and potentially empirical validations. The proposed approach enables the classification of PWM converters into families, uncovering relationships among their dynamics remains a complex challenge due to the intricate nature of the general model expressions. Recent research in this direction holds the potential to uncover valuable insights that could contribute to advancements in power electronics and converter design.

In order to determine the relationship among the converters in the buck topology, the following specifications are taken $P_0 = 12\ W$, $f_s = 50\ kHz$, $L = 0.1\ mH$, $C = 470\ \mu F$, $V_0 < 0.01\ V_0$. In case of = 0.3 , $V_i = 24\ V$, $V_0 = 12\ V$, $R_0 = 1.2\ \Omega$, and for $D = 0.7$, $V_i = 12\ V$, $V_0 = 28\ V$, $R_0 = 6.0\ \Omega$

In Figure 72.14, the step response is shown in the plot for comparison, where $L_1 = L_f = L_x = 0.1\ mH$. For a scenario where D=0.7, the output power and component parameters remain consistent with the situation where D=0.3. However, this leads to distinct Q (quality factor) values for these two operational conditions. When D=0.7, the system exhibits a higher Q compared to the case with D=0.3. As a consequence, there is a more pronounced oscillation that endures for an extended duration. The responses stemming from a step change in duty ratio, transitioning from 0 to 0.7, are depicted in Figure 72.15 for various converter configurations.

Conclusions

The proposed approach is proper reconfiguration of general converter representations within the buck and boost families. This reconfiguration enables systematic and efficient modeling of these converters when operated in continuous conduction mode. The proposed approach can also be applied to model converters operated in the discontinuous conduction mode. A notable feature is its ability to classify PWM converters into families. This classification allows for the exploration of potential relationships among converter dynamics within a family. However, in the paper, the investigation of such relationships is done with an example. The ability to classify PWM converters into families provides a framework for organized study, efficient design, and targeted research and development in the field of power electronics. This enhances the understanding of the converter to have optimum performance. It contributes to the advancement and standardization of PWM converter technology, benefiting both academia and industry.

Acknowledgement

The authors gratefully acknowledge the students, staff, and the management of R.M.K. Engineering College for their support in the research.

References

1. Bindi, M., Corti, F., Grasso, F., Manetti, S., Piccirilli, M. C., and Reatti, A. (2023) Failure prevention in DC–DC converters: theoretical approach and experimental application on a Zeta converter. *IEEE Trans. Indust. Elect.*, 70(1), pp: 930-939. 10.1109/TIE.2022.3153827.
2. Chen, Y. Y., Chang, Y. C., and Wei, C. L. (2022). Mixed-ripple adaptive on-time controlled non-inverting buck-boost DC-DC converter with adaptive-window-based mode selector. *IEEE Trans. Circ. Sys. Exp. Bri.*, 69(4), pp: 2196-2200. https://ieeexplore.ieee.org/document/9664803.
3. Junaid, K. A. M., Sukhi, Y., Jeyashree, Y., Jenifer, A., and Fayaz Ahamed, A. (2022). PV based electric vehicle battery charger using resonant converter. *Renew. Ener. Foc.*, 36, pp: 24-32. 10.1016/j.ref.2022.05.005.
4. Jeyashree, Y. and Sukhi, Y. (2015). LCC resonant converter with power factor correction for power supply units. *J. Chine. Inst. Engg.*, 38, 843-854. 10.1080/02533839.2015.1037995.
5. Tasi-Fu Wu, T. F. and Chen, Y. K. (1998). Modeling PWM DC/DC converters out of basic converter units. *IEEE Trans. Pow. Elect.*, 13(5), pp: 870-881 . 10.1109/63.712294.
6. Khodabandeh, M., Afshari, E., and Amirabadi, M. (2019). A family of Cuk, Zeta, and SEPIC based soft-switching DC–DC converters. *IEEE Trans. Pow. Elec.*, 34(10), pp: 9503-9519 . 10.1109/TPEL.2019.2891563.
7. Junaid, K. A. M., Sukhi, Y., Anjum, N., Jeyashree, Y., Fayaz, A. A., Debbarma, S., Chaudhary, G., Priyadarshini, S., Shylashree, N., Garg, S., Kumar, M., and Nath, V. (2023). PV-based DC-DC buck-boost converter for LED driver. *e-Prim. Ad. Elect. Eng. Electr. Ener.*, 5, pp: 100271. https://doi.org/10.1016/j.prime.2023.100271.

73 Unmanned surface vehicle for remote water top pollution monitoring

Surajit Chottapadhyay[a], Bhaskar Roy[b], Mrinmoy Nayek[c] and Suvajit Ghosh[d]

Department of Electrical Engineering, Ghani Khan Choudhury Institute of Engineering & Technology, Malda, West Bengal, India

Abstract

This paper aims to develop an unmanned surface vehicle (USV) to monitor pollution at the water surface. Pollution has become an important concern for the sustainable development and growth of the country. Researchers are facing challenges to deal with different kinds of pollutions that are coming out as a result of multi-dimensional activities. It may occur in different areas like soil, air, sound, water, etc. This work focuses on monitoring pollution that may occur on top of the water surfaces and at the top layer of water. This has been achieved in three major steps: (a) designing an USV to move around the water surface. (b) Different image-capturing devices will be installed to capture the image of different objects and (c) analyzing them by advanced image processing tool to detect pollution.

Keywords: Electric vehicle, mini USV, remote control water pollution

1. Introduction

Along with many technological advancements, the world is facing many challenges that create obstacles to the ecosystem of all living beings. Along with many social and economic challenges, pollution is an important threat associated with this. There are many areas where pollution is important. Air, sound, and water are three major areas where pollution has many direct and indirect effects on humans and other living beings. Scientists and engineers should come forward to give proper direction on how to face and overcome different kinds of pollution. To deal with pollution, the first job is to detect and analyze pollution occurring in a system.

Many methodical advancements have been observed in the last few years. Mendoza-Chok et al. presented a method which is based on the concept of a hybrid system controller for an unmanned surface vehicle. It improves the quality of the data and minimizes human exposure in hazardous areas. The critical aspects such as requirement, risk-management, functional classification, design variability, etc. were considered during the development of this project. It proposed the main component for surface vehicle, the data computing system which runs the robotics operation with the deliberative control architecture, and another computer that runs the state-machines with the help of hierarchical control architecture. For different maneuvers, manual and automatic modes are suitable and show a better performance which is achieved by the proposed hybrid control architecture [1]. Weiser et al. proposed a performance analysis based on the unmanned vehicle of fixed wing which is capable of aerial operation and underwater operation with low-energy loitering capabilities. It combines the persistence of aircraft with the capability of diving. The literature describes the concept of vehicles including their execution, main components and key subsystems such as the combination of motor and propeller, and hydrodynamic control surfaces. Results of tests were presented which may help the authors to characterize and illustrate each state of operations of surface surveillance, water egress, water ingress, etc. [2]. Madeo et al. introduced a low-cost architecture of unmanned surveillance vehicle which includes crucial parameters about the quality of water of river, pond, lake, etc. The vehicle has been provided with sensors and data capturing units to measure physical/chemical parameters to monitor the operations. Various data analytic tools were provided to run the vehicle automatically along our suitable and desired paths. These data can be used to estimate the oxygen concentration with the help of an easy computing model in the water sources along with the biophysical inputs. The aim is to develop a complete monitoring eco system of collecting, storing and analyzing data [3].

Zhang et al. introduced unmanned vehicles for surveillance and autonomous navigation. The paper is mainly focused on the building of a USV based

[a]surajitchattopadhyay@gmail.com, [b]bhaskarroy28@gmail.com, [c]mrinmoybdn20@gmail.com, [d]suvajitghosh177@gmail.com

DOI: 10.1201/9781003540199-73

perception network using multiple sensors. The information fusion of LIDAR and vision based detection of objects on the water and how to deal with obstacles have been taken into consideration [4]. Shan et al. introduced a scheme to process the 3-D LiDAR data directly for the achievement of accurate and stable navigable region for sailing. An approach of deep learning is adapted for the segmentation of objects on rivers. After that, an object tracking model which is based on the Kalman filter (KF) is deployed for accurate tracking of river bank objects. Finally, a region which is properly navigable is modeled based on Wave Frontier detection (WFD) [5].

Zhou et al. proposed a network of waterway segmentation which is collision-free using deep learning to obtain a pixel-level classification result. The accuracy of segmentation is improved and the details of refined water lines are achieved with the help of an asymmetric encoder-decoder model. Integration and augmentation of data are performed to learn the waterway features [6].

Liu et al. developed a protocol that uses pragmatic distribution for passing water channels smoothly for multiple unmanned surveillance vehicles. They proposed a predictive system for trajectory tracking for path tracking with the help of two leader USVs. They have also designed a scanning-formation controller with an estimator of trajectory state for the followers for passing through the channel efficiently [7]. Wang et al. introduced an autonomous pilot framework with the facility of waypoints of an unmanned surveillance vehicle in congested water area. Elite-duplication GA has been incorporated in ref [8]. Klinger et al. conducted an open loop maneuvering test to characterize the dynamics of the USV. It has been observed during the variable mass and drag test that it has been controlled efficiently with the model reference of adaptive backstepping speed and heading controller [9]. Vasilijević et al. introduced a multi-robot combination for monitoring of the underwater ocean environment. A systematic description of the design and the cooperative robotic model implementation has been introduced in the literature. For environment monitoring a human-on-the-loop approach is used in the robotic system. The proposed scheme provides real data of pollution with less human effort and time [10]. Ring back et al. has proposed a control technique considering the possibility of collision while tracking an underwater target. With the help of these protocols, the ASVs form a desired formation in order to track the moving target [11]. Bovcon et al. proposed a novel monitoring model consisting using deep encoder-decoder scheme. It has been observed that with this method false positive detection has been reduced significantly and it has also increased the true positive detection rates [12]. Along with these developments, many advancements have been observed in the field of signal analysis and performance improvement of different types of unmanned vehicles [13–16]. Feature extraction based methods have been found very effective in discrimination of different system parameters and system conditions [17–21].

However, comparatively, little work has been found that deals with water surface pollution monitoring from the remote end. This has motivated authors to focus on this area and come out with a proposal for an unmanned surface vehicle for water top pollution navigation. This will be achieved in three major steps: (a) Unmanned surface vehicle (USV) will be built to move around the water surface. (b) Different image-capturing devices will be installed to capture the image of different objects and (c) Images will be analyzed and compared by some advanced image processing tool to detect pollution. The USV unit will be tested in the tap water bed and in moving water having different wave movements. The proposed method is expected to have the outcome of monitoring top water pollution. After the introduction, design and modeling

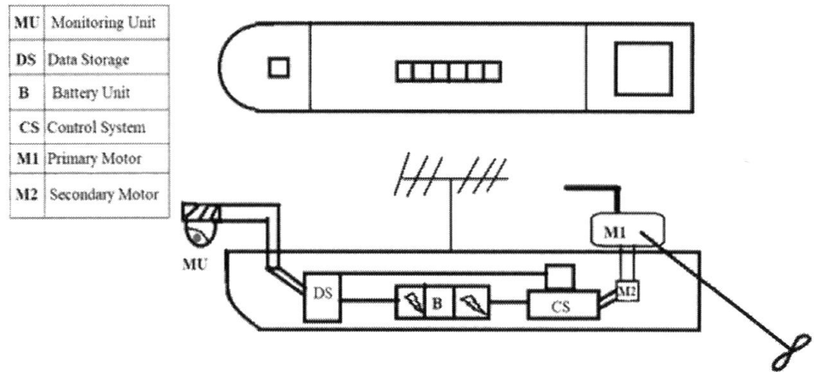

Figure 73.1 Top and side view of unmanned surface vehicle

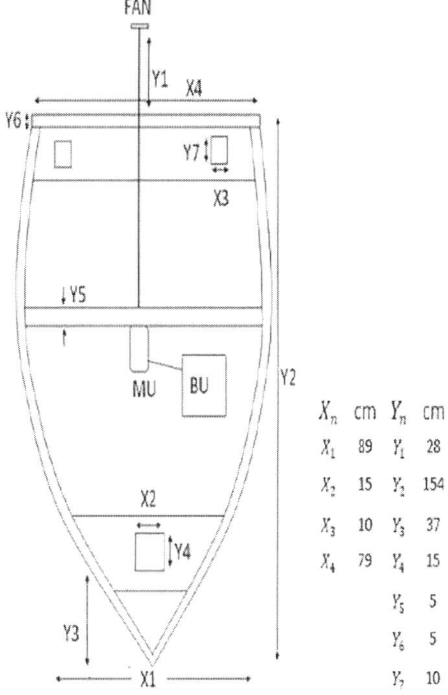

Figure 73.2 Top view of the vehicle

Table 73.1 Dimensions of the Mini-USV

Xn	cm	Yn	cm
X_1	89	Y_1	28
X_2	15	Y_2	154
X_3	10	Y_3	37
X_4	79	Y_4	15
-	-	Y_5	5
-	-	Y_6	5
-	-	Y_7	10

have been presented in Section 2. Then case studies along with data capture are presented in Section 3. A conclusion is made in Section 4.

Design of a water surface vehicle

A small USV will be developed for monitoring purposes. The top view and side view of proposed and designed mini USV are shown in Figure 73.1. The tentative dimension of the USV has been planned as length of 4 feet, width of 2.5 feet, and depth of 1.5 feet. It will consist of different units such as a battery unit, data storage unit, control system unit, monitoring unit, and signaling unit.

Figure 73.3 Designed USV

The USV was made unmanned with the help of a battery, motor, and remote-control unit. Velocity and direction of movement can be controlled from the remote end. A monitoring unit has been installed with the USV. It will be a waterproofed camera that is connected to capture images and store them in the memory unit. The physical location of different subunits has been shown in Figure 73.2. Top view dimensions have been provided in Table 73.1. Complete designed USV has been shown in Figure 73.3. The USV has been operated in closed water (pond and swimming pool) with different wave conditions and then on moving water (river) having different wave conditions. The specifications of different items have been presented in Table 73.2.

Table 73.2 The specifications of different items

S. No.	Item name	Specifications
1	Motor	Type-PMDC, 12 V, 4 Ampere
2	Batter	12 V, lead acid
3	Camera	16 megapixel
4	Transmitter/receiver Tower	Long type – 2.5 ft, 01 no, Short type- 1 ft, 02 nos
5	Controller	Range – 200 meters, AC/DC switch control option

3. Discrimination of surface pollution

Data have been captured in image form and have been stored in memory units; however, they may also be stored in remote server locations through cloud communication. Figure 73.4 shows few sample captured images of water surface in field trial. Images have been analyzed by different image processing tools and machine learning tools. Attempts has been taken to discriminate surface particles causing pollution and features will be extracted for pollution monitoring that may occur on surface water and top water layer. Based on the observation, an algorithm has been framed to detect surface pollution as shown in Figure 73.5 and has been tested by different water surface environments.

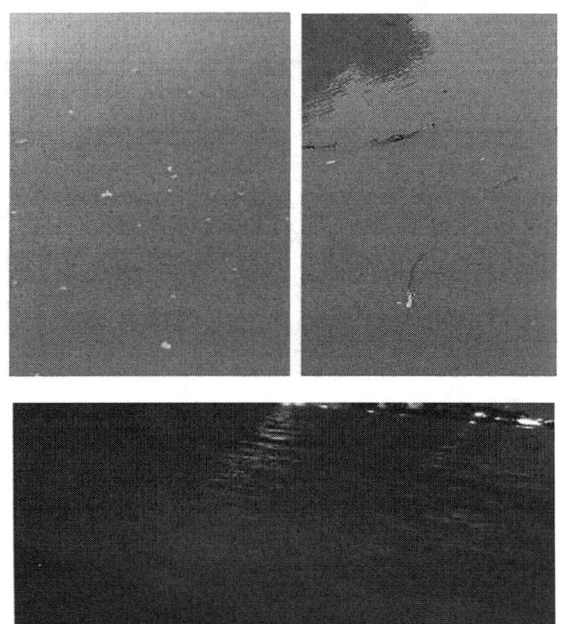

Figure 73.4 Few sample captured images of water surface in field trial

Skewness and Kurtosis of wavelet decomposition based detailed are determined and compared at different condition. Skewness of detailed (SD) has been found best. Its values in different decomposition levels have been determined. Their mean has been shown in bar chart comparison as presented in Figure 73.6. The study has made three level discrimination on pollution levels. It shows well discrimination among different samples of pure, medium and worst category.

4 Discussion

This work developed a scheme for monitoring pollution that may occur on top of the water surfaces and at the top layer of water. This has been by developing an USV that can move around the water surface. Different image-capturing devices have been installed to capture the image of different objects. The captured images have been analyzed and compared by some advanced image processing tools to discriminate pure and polluted water. The monitoring system has been tested in tapped water bed

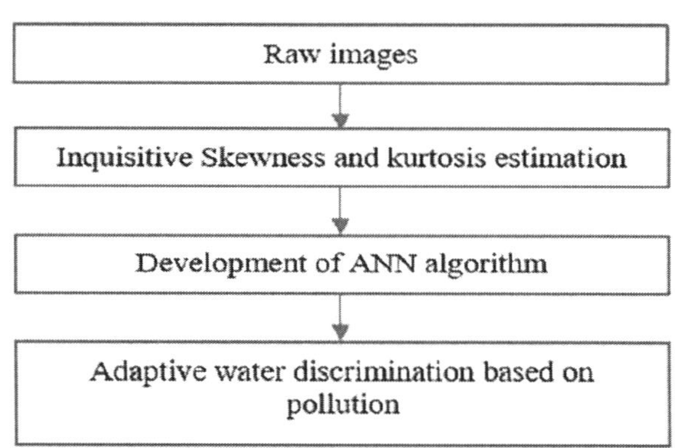

Figure 73.5 Working topology for case discrimination

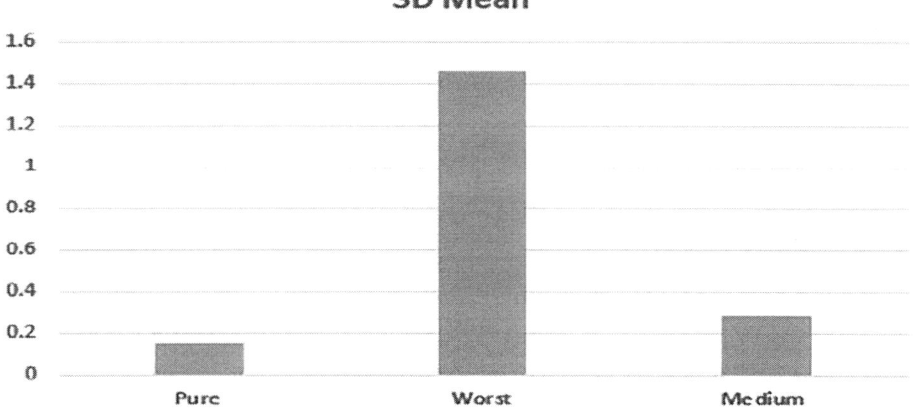

Figure 73.6 Learning outcomes: comparison of mean of SD for different pollution levels

(pond) at different wave conditions. The hardware monitoring unit has been fitted with the tower of mounted on USV. The proposed method has shown the satisfactory outcome for discrimination of pollution levels.

4. Conclusions

This work focuses on monitoring pollution that may occur on top of the water surfaces and at the top layer of water. This will be achieved in three major steps: (a) Mini unmanned surface vehicle has been built to move around the water surface. (b) Different image-capturing devices has been installed to capture the image of different objects and (c) Images has been analyzed and compared by some advanced image processing tool to detect pollution. The harder unit will be tested in tapped water bed and in moving water having different wave movements. The proposed method shows expected outcome of monitoring top water pollution. Also, the USV unit can have the opportunity to be utilized in different other future applications like underwater monitoring, obstacle monitoring, collection of water samples from remote locations, etc. Thus, the research findings / developed equipment may be used in the following areas: for water surface monitoring, underwater (top layer) monitoring, water surface pollution monitoring, underwater monitoring, obstacle monitoring, etc.

Acknowledgement

This project of USV for pollution monitoring has been funded by "The Institution of Engineers (India), 8 Gokhale Road, Kolkata 700020 under R&D Grant-in-Aid scheme". It has been implemented and tested in GKCIET, Malda. Authors acknowledge and thank them for all their cooperation.

References

1. Mendoza-Chok, J., Luque, J. C. C., Salas-Cueva, N. F., Yanyachi, D., and Yanyachi, P. R. (2022). Hybrid control architecture of an unmanned surface vehicle used for water quality monitoring. *IEEE Acc.*, 10, 112789–112798. doi: 10.1109/ACCESS.2022.3216563.
2. Weisler, W., Stewart, W., Anderson, M. B., Peters, K. J., Gopalarathnam, A., and Bryant, M. (2018). Testing and characterization of a fixed wing cross-domain unmanned vehicle operating in aerial and underwater environments. *IEEE J. Oceanic Engg.*, 43(4), 969–982. doi: 10.1109/JOE.2017.2742798.
3. Madeo, D., Pozzebon, A., Mocenni, C., and Bertoni, D. (2020). A low-cost unmanned surface vehicle for pervasive water quality monitoring. *IEEE Trans. Instrumen. Meas.*, 69(4), 1433–1444. doi: 10.1109/TIM.2019.2963515.
4. Zhang, W., Jiang, F., Yang, C. -F., Wang, Z. -P., and Zhao, T. -J. (2021). Research on unmanned surface vehicles environment perception based on the fusion of vision and Lidar. *IEEE Acc.*, 9, 63107–63121. doi: 10.1109/ACCESS.2021.3057863.
5. Shan, Y., Yao, X., Lin, H., Zou, X., and Huang, K. (2021). Lidar-based stable navigable region detection for unmanned surface vehicles. *IEEE Trans. Instrum. Meas.*, 70, 1–13. doi: 10.1109/TIM.2021.3056643.
6. Zhou, R., et al. (2022). Collision-free waterway segmentation for inland unmanned surface vehicles. *IEEE Trans. Instrum. Meas.*, 71, 1–16. doi: 10.1109/TIM.2022.3165803.
7. Liu, B., Zhang, H.-T., Meng, H., Fu, D., and Su, H. (2022). Scanning-chain formation control for multiple unmanned surface vessels to pass through water channels. *IEEE Trans. Cybernet.*, 52(3), 1850–1861. doi: 10.1109/TCYB.2020.2997833.
8. Wang, N., Zhang, Y., Ahn, C. K., and Xu, Q. (2022). Autonomous pilot of unmanned surface vehicles: Bridging path planning and tracking. *IEEE Trans. Veh. Technol.*, 71(3), 2358–2374. doi: 10.1109/TVT.2021.3136670.
9. Klinger, W. B., Bertaska, I. R., von Ellenrieder, K. D., and Dhanak, M. R. (). Control of an unmanned surface vehicle with uncertain displacement and drag. *IEEE J. Oceanic Engg.*, 42(2), 458–476. doi: 10.1109/JOE.2016.2571158.
10. Vasilijević, A., Nađ, Đ., Mandić, F., Mišković, N., and Vukić, Z. (2017). Coordinated navigation of surface and underwater marine robotic vehicles for ocean sampling and environmental monitoring. *IEEE/ASME Trans. Mechatron.*, 22(3), 1174–1184. doi: 10.1109/TMECH.2017.2684423.
11. Ringbäck, R., Wei, J., Erstorp, E. S., Kuttenkeuler, J., Johansen, T. A., and Johansson, K. H. (2021). Multi-agent formation tracking for autonomous surface vehicles. *IEEE Trans. Control Sys. Technol.*, 29(6), 2287–2298. doi: 10.1109/TCST.2020.3035476.
12. Bovcon, B. and Kristan, M. (2022). WaSR—A water segmentation and refinement maritime obstacle detection network. *IEEE Trans. Cybernet.*, 52(12), 12661–12674. doi: 10.1109/TCYB.2021.3085856.
13. Chattopadhyay, S., Du, B.-X., Dang, Z. M., and Chen, G. (2021). Nano-materials for engineering application. *IET Nanodielectr.*, 4, 81–83. https://doi.org/10.1049/nde2.12028.
14. Kar Ray, D., Roy, T., and Chattopadhyay, S. (2021). Switching transient-based state of Ampere-hour prediction of lithium-ion, nickel-cadmium, nickel-metal-hydride and lead acid batteries used in vehicles. *IET Nanodielectr.*, 4(3), 121–129. https://doi.org/10.1049/nde2.12017.
15. Das, T. K., Chattopadhyay, S., and Das, A. (2021). String fault detection in solar photo voltaic arrays. *IETE J. Res.* 69(5), pp. 2670–2682. DOI: 10.1080/03772063.2021.1905081.

16. Kar Ray, D., Roy, T., and Chattopadhyay, S. (2020). Single and diagonal double thrust failure assessment of quad-copter. *Measurement*, Volume 156, 2020, https://doi.org/10.1016/j.measurement.2020.107591.
17. Chattopadhyay, S. (2022). Nanogrids and picogrids and their integration with electric vehicles. *IET, London*. pp. 1-367, 2022. ISBN: 978-1-83953-482-9.
18. Karmakar, S., Chattopadhyay, S., Mitra, M., and Sengupta, S. (2016). Induction motor fault diagnosis approach through current signature analysis. Springer: Singapore. ISBN: ISBN 978-981-10-0624-1.
19. Chattopadhyay, S., Mitra, M., and Sengupta, S. (2011). Electric Power Quality. pp. 1-177, Springer: Netharland. ISBN: 978-94-007-0635-4.
20. Midya, M., Ganguly, P., Datta, T., and Chattopadhyay, S. (2023). ICA feature extraction-based fault identification of vehicular starter motor. *IEEE Sens. Lett.* vol. 7, no. 2, pp. 1-4, Feb. 2023, DOI: 10.1109/LSENS.2023.3242814.
21. Chattopadhyay, S., Karmakar, S., Mitra, M., and Sengupta, S. (2012). Assessment of crawling of an induction motor by stator current Concordia analysis. *IET Electron. Lett.*, 48(14), 841–842. DOI: 10.1049/el.2011.4008.

74 Voltage stability constraint optimal power flow-based estimation of IMO pay for specific position generators

Dipu Sarkar[a] *and Rupali Brahmachary*

Department of Electrical and Electronic Engineering National Institute of Technology Nagaland, Nagaland, India

Abstract

Electric-energy retail market is opening up to more opportunities around the world. The transition compelled the energy sector to establish new independent market operator (IMO). IMO's function is to control markets used for energy scheduling and reserve operations, and to obtain additional ancillary services through contracts as well. Two different case studies for the different location of generators are carried out in the present article to find the IMO pays, L-index stability, and bid loss. The proposed approach has been verified using IEEE 14-bus system. The results show that the generator positions can influence IMO pay, total losses, and bid losses.

Keywords: Bid loss, energy scheduling, independent market operator, L-index stability, optimal power flow

Introduction

In present market scenario, restructuring of generation and operation costs are done, but restructuring of transmission line is not taken into consideration, which leads to the congestion of the transmission line. The priority of a transmission system operator is the proper maintenance of the reliability of that area's entire system. The assumption of system operator not only includes the current flowing through system equipment and line but also measuring the post contingency limit. The market where each supplier is required to fix bidding price for each unit of generation is said to have used dynamic programming-based approach [1]. Paper [2] refers to the AC optimal power flow model of a system having energy reference bus independent of locational marginal price decomposition thus presenting a way to get over the various disadvantage of depending on the conventional systems. The basic concepts of locational pricing strategies in the electricity markets is suggested in ref [3]. The genetic algorithm approach is used to model the cases which does not have accurate knowledge of the rival's bidding strategy and those cases which are unsymmetrical [4]. Concept of the impact of losses on locational marginal pricing checking the Karush-Kuhn-Tucker (KKT) condition is given in paper [5]. Paper [6] has used the approach of ant colony optimization meta-heuristic method for the optimal power flow solution. The main intention is the lessening of the overall fuel cost of generation. Use of Lévy flight spider monkey optimization (LFSMO) algorithm [7] and testing is done by considering various important functions. The proposed algorithm improves the usability of spider monkey optimization algorithm. The power flow solution using rotary power flow controller, is explained in ref [8] it helps in injecting series voltage in transmitting line and also helps in series compensation of the line. It also plays a role in phase shifting. Paper [9, 10] mainly focuses on congestion management as a must necessary task done by the system operator to keep the transmission system within operating limits.

A simple and effective two step approach model is used for optimizing the location of FACTS devices in paper [11–13]. Wind farms installations at the optimal sites for mitigating congestion is suggested in paper [14]. The optimal location is searched by bus sensitivity factor and wind availability factor. The differential advancement for solving nonlinear optimization problem is carried out based on a population based heuristic algorithm. Various factors which lead to the development of distribution system operator and the corresponding flexibility associated with the transmission system operator when resources are connected in the power grid is discussed in ref [15]. Paper [16] discusses a meta-heuristic algorithm for congestion management by real time power rescheduling of the generators, in pool-based electricity market. The paper also accounts for the various security constrains including the voltage at the load bus and loading parameter of the line. Different strategies for operating battery energy storage system explained in ref [17], to do secondary balancing power in the energy market. Ref [18] suggests using a hybrid robust stochastic approach to get ideal contribution and bidding

[a]dipusarkar5@rediffmail.com

DOI: 10.1201/9781003540199-74

approaches for the big manufacturing consumers, considering the various uncertainties. Bidding strategy with the addition of wind energy to maximize the turnover by the use of probability distribution factor is incorporated in ref [19]. Voltage stability of the of distribution network is improved in refs [20–23] with the help of graph theory and Kruskal's maximal spanning tree algorithm. A day ahead market with maximum profit that can be gained from thermal power producing unit is discussed in ref [24], which delivers the information required in determining the optimal plans of day ahead market. A new virtual synchronous generator strategy is used along with DG units in ref [25] to improve the dynamic performance and stability of the system and to study the improvement of the developed new method in comparison to the conventional VSG control method. An equivalent of grouped generators of an external area is replaced by a fourth order model synchronous generator with a common exciter and low-level teamwork hybrid algorithm along with cuckoo search and the chaotic salp swarm algorithm is used to obtain the system dynamics during post fault condition as explained in ref [26]. Aneke et al. have proposed modal analysis to determine the voltage stability for Nigeria 44-bus grid network [27]. The proposed method also helps to find out the weakest buses in the system. The above said network has been simulated in PSAT environment and it has been shown that the Yola bus is the weakest bus based on the participating factors. Authors Pandya et al. have introduced the moth-flame optimizer (MFO) to solve the OPF including wind turbines and FACTS devices [28, 29]. The proposed method has been verified with IEEE 30-bus test system. It has been shown that multi-objective can solve the OPF problems including various constraints.

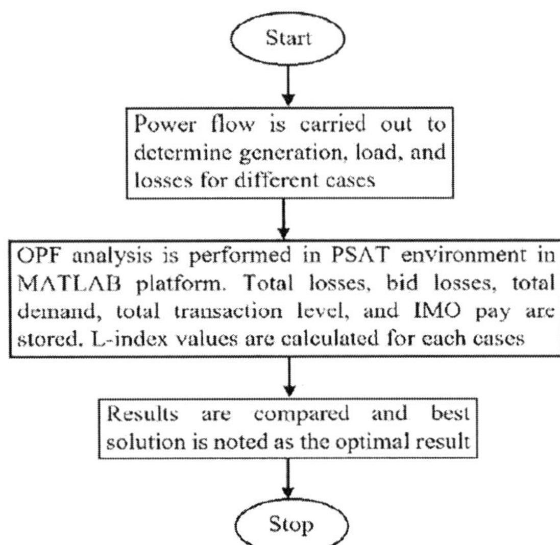

Figure 74.1 Flow chart of the proposed approach

Through this work a novel analysis method has been discussed so that voltage stability constraint optimal power flow-based estimation of IMO pay for specific position generators is achieved by positioning generator in specific buses and calculating the L-Index value. The rest of the paper is arranged in the following sequence: the next section discusses about the theory of the optimal power flow analysis, then discussion on the basics of L-index stability is done. Thereafter, the test system is described and the results obtained after the optimal power flow analysis and the L-index stability calculation is also discussed and lastly the conclusion is discussed.

Optimal power flow analysis

The optimal power flow analysis is carried out so as the supply power to the consumers at the minimum cost possible, which will also help in providing profit to the suppliers.

The power flow equations are:

$$\min \textstyle\sum_{i=0}^{n} f_i(P_{G,i}Q_{G,i}) \tag{1}$$

$$P_{ij} = |V_i||V_j|\left(G_{ij}\cos\theta_{ij} + B_{ij}\sin\theta_{ij}\right) \tag{2}$$

$$Q_{ij} = |V_i||V_j|\left(G_{ij}\sin\theta_{ij} + B_{ij}\cos\theta_{ij}\right) \tag{3}$$

The generator cost function for AC optimal power flow is represented in Equation (1). The real and reactive power generated are given by Equations (2) and (3). "V" gives the voltage magnitude at each bus. "G" gives the conductance value and "B" gives the susceptance value of the system. represents the voltage angle between the adjacent buses.

The optimal power flow equations are:

$$CF = \min \textstyle\sum_{i=0}^{n} C_{i,1}P_i + C_{i,2}P_i^2 \tag{4}$$

Subjected to:

$$0 \leq P_i \leq P_{i,max} \tag{5}$$

$$P_i - D_i - \min \textstyle\sum_{j=0}^{n} X F_{ij} = 0 \tag{6}$$

$$x_{12}F_{12} + x_{23}F_{23} + x_{13}F_{13} = 0 \tag{7}$$

Equation (4) gives the cost function (CF) for DC optimal power flow, gives the marginal cost for each generator. gives the real power produced and gives the real power demanded is the power flow between nodes. "x" denotes the reactance between the corresponding nodes buses.

L-index stability

For reduced network,

$$L=4[(x_{eq}P - r_{eq}Q)^2 + x_{eq}Q + r_{req}P] \quad (8)$$

where,
r_{eq} = equivalent resistance for line
x_{eq} = equivalent reactance for line
P = total real power in the distribution network
Q = total reactive power in the distribution network

$$r_{eq} = \frac{\sum P_{loss}}{\{(P+\sum P_{loss})^2+(Q+\sum Q_{loss})^2\}} \quad (9)$$

$$x_{eq} = \frac{\sum Q_{loss}}{\{(P+\sum P_{loss})^2+(Q+\sum Q_{loss})^2\}} \quad (10)$$

Thus, the range of the stability index (L) is in between 0 and 1, when the value is close to 0 the system is indicated as a stable system and when the value is close to 1 then there is the chance of voltage instability. The proposed approach has been depicted in Figure 74.1.

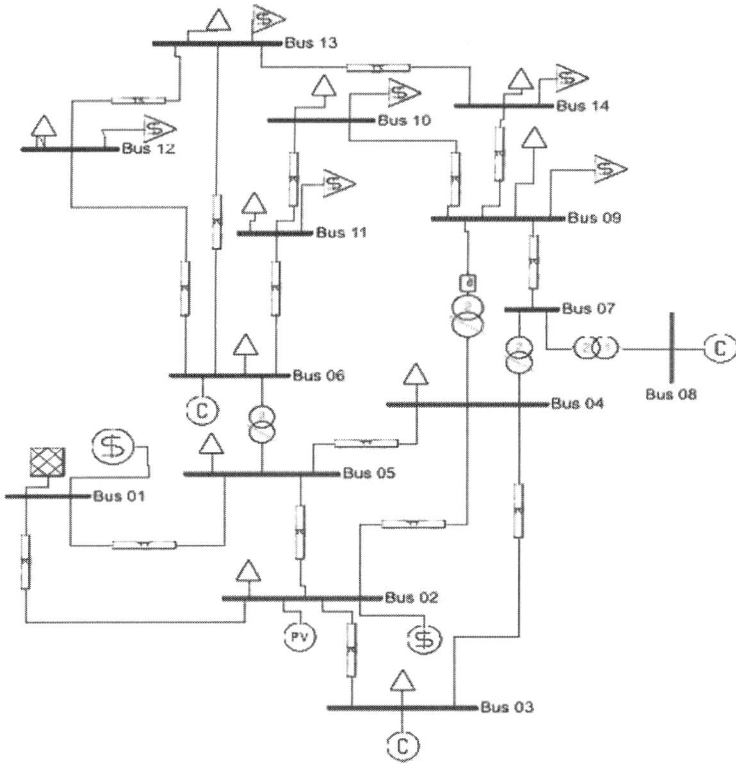

Figure 74.2 Both the generating units are connected

Test system description

The base power rating is 100 MVA and the base frequency being 50 Hz. The system is divided is two cases and each case is further divided into three parts. The first case consists of bus 1 using fixed generating unit and bus 2 having another generating unit i.e., generator 2. In case two the generating unit's positions are interchanged i.e., bus 2 having the fixed generating unit and bus 1 is connected with generator 2. Part one refers to the use of both the generators as shown in Figure 74.2. Part two only uses the fixed generating unit to supply the required demand of the system and here the generator 2 is disconnected from the system (from Figure 74.1). Lastly, part three uses only the generator 2 to supply the demand, and here fixed generating unit is disconnected from the system (from Figure 74.1). In each cases the above-mentioned parts are applied and the results are simulated in power system analysis toolbox (PSAT), version 2.1.9 in MATLAB platform. The power flow and optimal power flow results are discussed in the next section.

Result and discussion

The entire result simulation is divided into two cases and three parts. The cases indicate the generator position and the parts indicates the connection and disconnection of the generators.

Table 74.1 discusses the power flow result comparison. It is seen that when both the generators are connected i.e., Case I, Part 1 and Case II, Part 1 the power flow generation comes out as 392.05 MW and 205.54 MVAr and 380.67 MW and 166.69 MVAr, respectively.

The result comes similar to when only the generator 2 is connected to the system i.e., Case I, Part 3 and Case II, Part 3 which are again 392.05 MW and 205.54 MVAr and 380.67 MW and 166.69 MVAr, respectively. It is seen that interchanging the generator's position has reduced total generation of real power as is seen in Table 74.1. Again, when only the fixed generating unit is kept connected and the generator 2 is disconnected from the network i.e., Case I, Part 2 and Case II, Part 2 it is seen that the generated power is 397.20 MW and 224.77 MVAr and 381.17 MW and 169.92 MVAr, respectively, which is more than that of the other cases discussed above. The result shows that when only fixed generating unit is used as the power supplying unit and the generating unit is kept in bus 1 the power generated is the highest which is equal to 397.20 MW and 224.77 MVAr. If losses are considered for each part of Case I and Case II, it is seen that losses are most when only fixed generating units are supplying power i.e., 34.603 MW and 110.81 MVAr and 18.573 MW and 53.963 MVAr, respectively. Loss is highest when fixed generating unit is connected in bus 1 as seen from Table 74.1.

Table 74.2 discusses the comparison of optimal power flow results and the L-index values for different cases and parts. It also extends various bid losses and total transaction level (in MW). However, for comparison purpose only the mod-values of bid losses are considered. The positive and the negative bid losses represent the excess supply and system deficiency (in MW), respectively. From the simulated results, an outlook of the IMO pay for specific location of the generators, i.e., the most economic process of supplying the required energy to the system is achieved.

It is seen that when the fixed generating unit is used in bus 1 and generator 2 at bus 2 i.e., Case I and Part 1 the losses are equal to 30.213 MW, bid losses are 0.761 MW and IMO pay is 371.3261$/h. When the generator 2 is disconnected from the system and the entire system is getting supply from the fixed generating bus the total loss, bid loss and IMO pay are the highest, i.e., 32.794 MW, 1.809

Table 74.1 Comparison of different case studies w.r.t. generation, load, and losses (in MW)

	Case – I			Case – II		
	Part - 1	Part - 2	Part - 3	Part - 1	Part - 2	Part - 3
Total generation						
Real power	392.05	397.20	392.05	380.67	381.17	380.67
Reactive power	205.54	224.77	205.54	166.69	167.92	166.69
Total load						
Real power	362.60	362.60	362.60	362.60	362.60	362.60
Reactive power	113.96	113.96	113.96	113.96	113.96	113.96
Total losses						
Real power	29.452	34.603	29.452	18.069	18.573	18.069
Reactive power	91.576	110.81	91.576	52.725	53.963	52.725

Table 74.2 Optimal power flow result comparison

	Case – I			Case – II		
	Part - 1	Part - 2	Part - 3	Part - 1	Part- 2	Part - 3
Total losses [MW]	30.213	32.794	25.993	19.707	20.566	17.372
Bid losses [MW]	0.761	-1.809	-1.459	1.638	-1.007	-0.697
Total demand [MW]	44.2392	21.809	28.459	43.3615	26.0068	20.6972
Total transaction level [MW]	406.839	384.409	391.059	405.961	388.606	383.297
IMO pay [$/h]	371.326	444.348	305.200	231.793	240.100	230.773
L-index values	0.5072	0.5757	0.5072	0.3297	0.3368	0.3297

MW, and 444.3482 \$/h, respectively. Again, when only generation 2 is connected and supplying the entire power to the system the total losses, bid loss, and IMO pay are 25.993 MW, 1.459 MW, and 305.2007 \$/h, respectively. It is seen that lowest IMO pay availed only when the fixed generating unit is not connected to the system to supply power. When the fixed generating unit is used in bus 2 and generator 2 in bus 1, the positions of the generating units are interchanged, i.e., Case II and Part 1, the losses are equal to 19.707 MW, bid losses are 1.638 MW and IMO pay is 231.7934 \$/h. When the generator 2 is disconnected from the system and the entire system is getting supply from the fixed generating unit in bus 2, the losses, bid loss and IMO pay are 20.566 MW, 1.007 MW and 240.1002 \$/h, respectively. Again, when only generation 2 is connected and supplying the entire power to the system the losses, bid loss and IMO pay are the lowest, i.e., 17.372 MW, 0.697 MW and 230.7733 \$/h, respectively.

The comparison of the L-index values of the different cases and part, from the simulated result, gives an outlook of the L-index values for specific location of the generators so as to obtain the most economic process of supplying the required energy to the system, without hindering the stability. It is seen that when the fixed generating unit is used in bus 1 and generator 2 in bus 2 i.e., Case I and Part 1 the value is equal to 0.5072. When the generator 2 is disconnected from the system the value becomes 0.5757 and when only generation 2 is connected and supplying the entire power to the system the L-index value is 0.5072. Again, when the fixed generating unit is used in bus 2 and generator 2 in bus 1 i.e., Case II and Part 1 the value is equal to 0.3297. When the generator 2 is disconnected from the system the value becomes 0.3368 and when only generation 2 is connected and supplying the entire power to the system the L-index value is 0.3297. It is seen that comparative lower L-index value is recorded when the fixed generating unit is connected with bus 2 and generator 2 is connected with bus 1, in the system to supply power.

Conclusions

The analyzed results verify that the method used for this novel approach for voltage stability constraint optimal power flow gives the desired improvement in the result. It is seen from the above discussion that the lowest IMO pay, lowest total losses and lowest bid losses can be availed only when the fixed generating unit is not connected to the system to supply power and the generator 2, which is connected to bus 1, is used as the only source of power supply. The second-best case is where generator 2 is connected in bus 1 and fixed generating unit is connected in bus 2. The estimated result of IMO pay, losses, power flow, and L-index values has improved significantly for specific position generators.

Acknowledgement

The authors gratefully acknowledge the EEE department of NIT Nagaland for their cooperation in the research.

References

1. Singhal, P. K. and Sharma, R. N. (2011). Dynamic programming approach for solving power generating unit commitment problem. *ICCCT*, 298–303.
2. Cheng, X. and Overbye, T. J. (2006). An energy reference bus independent LMP decomposition algorithm. IEEE Trans. Power Sys., 21(3), 1041–1049.
3. Fu, Y. and Li, Z. (2006). Different models and properties on LMP calculations. *2006 IEEE Power Engg. Soc. Gen. Meet.*, 1–11, doi: 10.1109/PES.2006.1709536.
4. Prabavathi, M. and Gnanadass, R. (2018). Electric power bidding model for practical utility system. *Alexandria Engg. J.*, 57(1), 277–286.
5. Yang, Z., Bose, A., Zhong, H., Zhang, N., Lin, J., Xia, Q., Kang, C. (2017). LMP revisited, A linear model for the loss-embedded LMP. *IEEE Trans. Power Sys.*, 32(5), 4080-4090.
6. Allaoua, B. and Laoufi, A. (2009). Optimal power flow solution using ant manners for electrical network. *Adv. Elec. Comp. Engg.*, 9(1), 34–40.
7. Sharma, A., Sharma, H., Bhargava, A., Sharma, N., and Bansal, J. C. (2016). Optimal power flow analysis using Lévy flight spider monkey optimisation algorithm. *Inter. J. Artif. Intell. Soft Comput.*, 5(4), 320–352.
8. Haddadi, A. M. and Kazemi A. (2011). *Optimal power flow control by rotary power flow controller. Adv. Elec. Comp. Engg.*, 11(2), 79–86.
9. Kumar, A., Srivastava, S. C., and Singh, S. N. (2005). Congestion management in competitive power market: A bibliographical survey. *Elec. Power Sys. Res.*, 76(1–3), 153–164.
10. Sarkar, D., Brahmachary, R., and Barma, S. D. (2021). Transient stability controlling and assessment of a congested power system in a deregulated environment. *Intell. Elec. Sys. Step Towards Smarter Earth*, 151.
11. Brahmachary, R. and Sarkar, D. (2021). Small signal stability assessment in presence of SSSC for a power system under fault disturbance. *Inter. J. Smart Grid Sustain. Ener. Technol.*, 4(1), 136–139.
12. Sarkar, D., Kumari, K., and Brahmachary, R. (2021). Impact of DGs in competitive deregulated environment for congestion management. *Recent Adv. Power Sys. Sel. Proc. EPREC 2020*, Springer: Singapore, 467–475.

13. Singh, S. N. and David, A. K. (2001). Optimal location of FACTS devices for congestion management. *Elec. Power Sys. Res.*, 58(2), 71–79.
14. Suganthi, S. T., Devaraj, D., Ramar, K., and Thilagar, S. H. (2018). An improved differential evolution algorithm for congestion management in the presence of wind turbine generators. *Renew. Sustain. Ener. Rev.*, 81, 635–642.
15. Hadush, S. Y. and Meeus, L. (2018). *DSO-TSO cooperation issues and solutions for distribution grid congestion management. Ener. Policy*, 120, 610–621.
16. Verma, S., Saha, S., and Mukherjee, V. (2017). A novel symbiotic organisms search algorithm for congestion management in deregulated environment. *J. Exper. Theoret. Artif. Intell.*, 29(1), 59–79.
17. Olk, C., Sauer, D. U., and Merten, M. (2019). Bidding strategy for a battery storage in the German secondary balancing power market. *J. Ener. Storage*, 21, 787–800.
18. Abedinia, O., Zareinejad, M., Doranehgard, M. H., Fathi, G., and Ghadimi, N. (2019). Optimal offering and bidding strategies of renewable energy based large consumer using a novel hybrid robust-stochastic approach. *J. Cleaner Prod.*, 215, 878–889.
19. Singh, S. and Fozdar, M. (2019). Optimal bidding strategy with the inclusion of wind power supplier in an emerging power market. *IET Gener. Trans. Distrib.*, 13(10), 1914–1922.
20. Sarkar, D., Goswami, S., De, A., Chanda, C. K., and Mukhopadhyay, K. (2011). Improvement of voltage stability margin in a reconfigured radial power network using graph theory. *Can. J. Electr. Electron. Engg.*, 2(9), 454–462.
21. Sarkar, D., De, A., Chanda, C. K., and Goswami, S. (2015). Kruskal's maximal spanning tree algorithm for optimizing distribution network topology to improve voltage stability. *Elec. Power Compon. Sys.*, 43(17), 1921–1930.
22. Gunturi, S. K., et al. (2022). A combined graph theory–machine learning strategy for planning optimal radial topology of distribution networks. *Elec. Power Compon. Sys.*, 49(13–14), 1158–1168.
23. Sarkar, D. and Gunturi, S. K. (2021). Machine learning enabled steady-state security predictor as deployed for distribution feeder reconfiguration. *J. Elec. Engg. Technol.*, 16, 1197–1206.
24. Nojavan, S., Zare, K., and Feyzi, M. R. (2019). Optimal bidding strategy of generation station in power market using information gap decision theory (IGDT). *Elec. Power Sys. Res.*, 96, 56–63.
25. Daili, Y. and Harrag, A. (2021). Improved decoupling virtual synchronous generator control strategy. *Rev. Roum. Sci. Techn. Élec. et Énerg.*, 66(1), 153–160.
26. Benmiloud, O. and Arif, S. (2021). Hybrid method for determination of power systems dynamic equivalents based on measurements. *Rev. Roum. Sci. Techn. Élec. et Énerg.*, 66(2), 97–10.
27. Aneke, E. N., Ibekwe, B. E., Iyidobi, J. C., and Okafor, E. N. C. (2021). Voltage stability evaluation in the Nigeria 44 bus grid network using modal analysis. *J. Engg. Res. Reports*, 20(11), 80–89.
28. Pandya, S., Jangir, P., and Trivedi, I. N. (2021). Multi-objective moth flame optimizer: A fundamental visions for wind power integrated optimal power flow with FACTS devices. *Smart Sci.*, 1–24.
29. Ataeizadeh, Md., Kargar, A., and Abazari, S. (2020). Optimal location of flexible alternating current transmission systems devices to achieve multiple goals in power systems. *Rev. Roum. Sci. Techn. Élec. et Éner.*, 65, 27–34.

75 Price determination of EOQ in cobweb dynamics of exponential demand and exponential supply

Prem Prakash Mishra[a], *Deepraj Dutta*[b] *and Limainla Kichu*[c]

Department of Science and Humanities (Mathematics), National Institute of Technology Nagaland, India

Abstract

In nonlinear cobweb dynamics, the sequence of seasonal prices, the price of goods of daily use as a function of continuous time and the stochastic prices are being computed. We also establish the relation of economic order quantity (EOQ) of an item with dynamic prices under exponential demand and exponential supply.

Keywords: Price, cobweb dynamics, economic order quantity, demand, supply

Introduction

In the cutting-edge competition of demand of commodity and its supply, there is a challenging situation to determine the economic order quantity (EOQ). Since demand and supply are dependent on the price of a commodity but there is a significant difference between the quantity of supply and demand at a particular time. Hence the quantity of supply at time t does not match the quantity of demand at time t. This mismatching makes chaos in the market by which consumers suffer a lot for essential commodities. We deal with the non-linear demand and supply functions of price. Several authors have studied the determination of price of EOQ in cobweb sense. Hommes focused on the role of non linear dynamics, learning, and heterogeneity within the cobweb framework [5]. He finds out the effect of price fluctuations by observations of individual and aggregate behavior from laboratory experiments and fits on cobweb theory. Askar has explained the dynamic characteristics of an economic cobweb model in which producers adopt a gradient-based mechanism [2]. He has calculated the equilibrium price point and its local stability conditions are discussed through analytical and numerical investigations. The memory factor is introduced in the 2D model which, can be destabilized through chaotic behaviors and investigates the influence of the speed of adjustment parameter on the stability of equilibrium price. Brianzoni et al. consider the dynamics of a stochastic cobweb model with linear demand and a backward bending supply curve [3]. They studied the Markov chain in the case of discrete and the Markov process in the case of continuous time. They found for discrete values of prices the corresponding Markov chain is asymptotically stable.

Nadarajah and Lyu illustrated new multivariate log-normal distributions and their application to an insurance data set [8]. Mishra and Mishra provide an overview of the significant time lag between the production and consumption of inventory items and its effect on fuzzy price-dependent demand and supply [6]. They have also done a comparative study between fuzzy and crisp model (Q, R) model. Sen et al. have presented a case study of the Noakhali district on a linear cobweb model [9]. They found that the fluctuation of vegetable prices is normal in a market economy. The price of the commodity and its fluctuation has a significant influence on farmers and consumers. But reasonable and stable price has an irreplaceable effect on the safe running of the vegetable market. Salahshour et al. study about dynamic differential-based cobweb model using a non-singular kernel fractional derivative. A comprehensive analysis has been done to study the asymptotic behaviors of solutions. This model demonstrates the effectiveness of the proposed cobweb model in different times interval.

Sajjad and Simin (2022) investigated the impact of time delay in flight schedules which is referred to the Bullwhip effect (BWE) in production systems. The time delay increases demand fluctuations over the time in supply chain. The effect of the arrival time of aircraft (landing time) to the airport, as well as the delayed flights (depart time with delay) to the next destination on the demand variance, were examined. They show that demand fluctuations increased significantly by delaying flights in closing hours. Anokye et al. introduces a novel approach to solving a delay differential cobweb model, utilizing the Lambert W-function with real model parameters [1]. The novelty of this model is that the authors not only have considered the price elasticity ratios but also the crucial time-delay parameter

[a]maths.prem79@gmail.com, [b]deepraj.dutta299@gmail.com, [c]aienmkichu365@gmail.com

DOI: 10.1201/9781003540199-75

which offers a more comprehensive understanding of stability in systems with time delay. Daniel et al. (2013) extended the classical cobweb theorems with production lags and price forecasting. Price forecasting based on a longer period has a stabilizing effect on prices. Longer production lags do not necessarily lead to unstable prices, very long lags lead to cycles of constant amplitude. Price forecasting has a stabilizing effect where random shocks are also considered.

We focus here on the determination of the price at any time *t*. If the price changes seasonally in sequential form the EOQ is determined based on the current seasonal price. But if the price changes with the deterministic rate continuously, the EOQ for such type of commodity depends on the continuous price at the current time. The other situations, where the price changes randomly with time are termed price processes. Since the stock price is a non-negative process, it is well suited to a lognormal distribution. The expected price of a commodity has been computed to supply the commodity for the fulfillment of future demand and the EOQ is determined in terms of the stochastic price process.

2 Assumptions and notations

- i All items are identical
- ii Instantaneous supply to fulfill the demand at a time *t* by the supply
- iii If supply and demand are not equal at time *t*, then a case of shortage or surplus occurs
- iv Similarity between the inventory and cobweb problems are shown below
- v The demand curve in the cobweb model is proportional to the ordering cost curve of the inventory model
- vi The supply curve in the cobweb model is proportional to the carrying cost. The total cost function is a convex function.

Following notations are used in the development of this model

d_0: Initial demand at time $t = 0$.

λ: value of the parameter as the coefficient of price P_t in best fit of demand on the exponential form.

P_t: Price of a unit item at time *t*.

s_0: Initial stock at time $t = 0$.

μ: value of the parameter as the coefficient of price P_{t-1} in best fit of supply on the exponential form.

γ: Adjustment factor which measures the rate of change in price at time *t* when demand is more than one unit of supply at time *t*.

α: Ratio between the price at time *t* and the price at the time t-τ.

3 Mathematical model

Economic order quantity in nonlinear demand and supply in cobweb dynamics

We consider the demand function of the item as best fit on the exponential form. The corresponding supply function of items is also best fit in exponential form. The cobweb model is a best-suited model for agricultural products.

If the demand and supply are the nonlinear function

$d(P_t) = d_0 e^{\lambda P_t}, \lambda \geq 0$ and $q_s(P_{t-\tau}) = s_0 e^{-\mu P_{t-\tau}}, \mu \geq 0$.

Then the equilibrium point of the Cobweb is $d(P_t) - q(P_{t-\tau}) = 0$

Case I*:* Let price be a discrete random sequence. Hence, Let $\tau = 1$, we get $d_0 e^{\lambda P_t} = s_0 e^{-\mu P_{t-1}}$

$$\frac{s_0}{d_0} = e^{\lambda P_t + \mu P_{t-1}};$$

Taking log both side $\lambda P_t + \mu P_{t-1} = log\left(\frac{s_0}{d_0}\right)$;

$$P_t = -\left(\frac{\mu}{\lambda}\right) P_{t-1} + \frac{1}{\lambda} log\left(\frac{s_0}{d_0}\right);$$

$$P_1 = -\left(\frac{\mu}{\lambda}\right) P_0 + \frac{1}{\lambda} log\left(\frac{s_0}{d_0}\right);$$

$$P_2 = -\left(\frac{\mu}{\lambda}\right)\left\{-\left(\frac{\mu}{\lambda}\right) P_0 + \frac{1}{\lambda} log\left(\frac{s_0}{d_0}\right)\right\} + \frac{1}{\lambda} log\left(\frac{s_0}{d_0}\right);$$

$$P_2 = \left(\frac{\mu}{\lambda}\right)^2 P_0 - \left(\frac{\mu}{\lambda}\right)\frac{1}{\lambda} log\left(\frac{s_0}{d_0}\right) + \frac{1}{\lambda} log\left(\frac{s_0}{d_0}\right)$$

Similarly, using mathematical Induction we get

$$P_n = (-1)^n \left(\frac{\mu}{\lambda}\right)^n P_0 + \frac{1}{\lambda} log\left(\frac{s_0}{d_0}\right)\left\{(-1)^{n-1}\left(\frac{\mu}{\lambda}\right)^{n-1} + (-1)^{n-2}\left(\frac{\mu}{\lambda}\right)^{n-2} + \cdots (-1)\left(\frac{\mu}{\lambda}\right) + 1\right\}$$

$P_n = (-1)^n \left(\frac{\mu}{\lambda}\right)^n P_0 + \frac{1}{\lambda} log\left(\frac{s_0}{d_0}\right)\frac{\left\{1-(-1)^n\left(\frac{\mu}{\lambda}\right)^n\right\}}{\left\{1+\left(\frac{\mu}{\lambda}\right)\right\}}$; where $s_0 > d_0$ and $\frac{\mu}{\lambda} < 1$ (1)

Hence, the price during an even cycle (off-session of production) is given as below

$$\boldsymbol{P_{2m} = \left(\frac{\mu}{\lambda}\right)^{2m} P_0 + \frac{1}{\lambda} log\left(\frac{s_0}{d_0}\right)\frac{\left\{1-\left(\frac{\mu}{\lambda}\right)^{2m}\right\}}{\left\{1+\left(\frac{\mu}{\lambda}\right)\right\}}} \quad (2)$$

And the price during an odd cycle (on-session of production) is given as below

$$P_{2m+1} = \frac{1}{\lambda} log\left(\frac{s_0}{d_0}\right)\frac{\left\{1+\left(\frac{\mu}{\lambda}\right)^{2m+1}\right\}}{\left\{1+\left(\frac{\mu}{\lambda}\right)\right\}} - \left(\frac{\mu}{\lambda}\right)^{2m+1} P_0 \quad (3)$$

EOQ in off-season of agriculture product:

$$\Rightarrow \; q(P_{2m}) = \sqrt{\left(\frac{2c_o d_0 e^{-\lambda P_{2m}}}{c_h}\right)} \quad (4)$$

When $n = 2m + 1$ (For odd season): $P_{2m+1} =$

$$-\left(\frac{\mu}{\lambda}\right)^{2m+1} P_0 + \frac{1}{\lambda} log\left(\frac{s_0}{d_0}\right)\frac{\left\{1+\left(\frac{\mu}{\lambda}\right)^{2m+1}\right\}}{\left\{1+\left(\frac{\mu}{\lambda}\right)\right\}}$$

EOQ within the on-season of agriculture product:

$$\Rightarrow q(P_{2m+1}) = \sqrt{\left(\frac{2c_o d_0 e^{-\lambda P_{2m+1}}}{c_h}\right)} \quad (5)$$

Case II: When price continuous random process

The rate of change of price is proportional to the difference between demand and supply at any time t in the cobweb model

Hence,

$$\frac{dP_t}{dt} = \gamma\{d(P_t) - q(P_{t-\tau})\} = \gamma\{d_0 e^{\lambda P_t} - s_0 e^{-\mu P_{t-\tau}}\},$$

where γ is taken as an adjustment factor.

When the price rate is in a decreasing trend then

$$\frac{dP_t}{dt} = \gamma\{q(P_{t-\tau}) - d(P_t)\} = \gamma\{ s_0 e^{-\mu P_{t-\tau}} - d_0 e^{\lambda P_t}\}.$$

$$e^{-\lambda P_t}\frac{dP_t}{dt} - \gamma s_0 e^{-\lambda P_t - \mu P_{t-\tau}} = -\gamma d_0$$

$$\Rightarrow \quad \frac{de^{-\lambda P_t}}{dt} + \lambda\gamma S_0 e^{-\mu P_{t-\tau}},\; e^{-\lambda P_t} = \lambda\gamma d_0$$

Let $e^{-\lambda x(t)} = y(t)$: $\frac{dy(t)}{dt} + \gamma\lambda s_0 e^{-\mu P_{t-\tau}} y(t) = \lambda\gamma d_0$

$-\lambda P_t + \int \lambda s_0 e^{-\mu P_{t-\tau}} dt = log\lambda\gamma d_0 + log$ $\left\{\int e^{\gamma\lambda s_0 \int e^{-\mu P_{t-\tau}dt}}\right\}$;

Let $P_{t-\tau} = P$ price which is known and constant then

$$P_t = \frac{1}{\lambda}[\; -log\lambda\gamma d_0 - \gamma t\lambda s_0 e^{-\mu P} + \lambda s_0 e^{-t\mu P} + log\gamma\lambda s_0 - \mu P] \quad (6)$$

$$P_t = \frac{1}{\lambda}\left[\; \lambda s_0 e^{-t\mu P} + log\frac{s_0}{d_0} - \mu P - \gamma t\lambda s_0 e^{-\mu P}\right]$$

If $\gamma > 0$ (price rate is in a decreasing trend) then

$$P_t = \frac{1}{\lambda}\left[\; \lambda s_0 e^{-t\mu P} + log\frac{s_0}{d_0} - \mu P - \gamma t\lambda s_0 e^{-\mu P}\right] \quad (7)$$

And If (price rate is in an increasing trend) then

$$P_t = \frac{1}{\lambda}\left[\; \lambda s_0 e^{-t\mu P} + log\frac{s_0}{d_0} - \mu P + \gamma t\lambda s_0 e^{-\mu P}\right] \quad (8)$$

Economic order quantity in decreasing and increasing trend of price rate

Minimize the total cost when the total cost is given as below:

Total Cost at time n, i.e. $C(P_t) = c_h\frac{q(P_t)}{2} + c_o\left(\frac{d(P_t)}{q(P_t)}\right)$;

When $t \geq 0$

To minimize the total cost:

$$\frac{dC(P_t)}{dq(P_t)} = \frac{c_h}{2} - c_o\left(\frac{d(P_t)}{(q(P_t))^2}\right) = 0$$

When the price rate is in a decreasing trend

$$q(P_t) = \sqrt{\left(\frac{2c_o d_0 e^{-\lambda P_t}}{c_h}\right)}$$

where

$$P_t = \frac{1}{\lambda}\left[\; \lambda s_0 e^{-t\mu P} + log\frac{s_0}{d_0} - \mu P - \gamma t\lambda s_0 e^{-\mu P}\right] \quad (9)$$

When the price rate is in an increasing trend

$$q(P_t) = \sqrt{\left(\frac{2c_o d_0 e^{-\lambda P_t}}{c_h}\right)}$$

where

$$P_t = \frac{1}{\lambda}\left[\; \lambda s_0 e^{-t\mu P} + log\frac{s_0}{d_0} - \mu P + \gamma t\lambda s_0 e^{-\mu P}\right] \quad (10)$$

Case III: The lognormal model of stock price

Suppose denotes the stock price at time t. It is a random quantity at time *t*. Let the probability distribution of be continuous and lognormal distribution process because . Hence lognormal is best-suited model for the stock random price model.

We let an accumulation factor $\alpha(t-\tau, t) = \frac{P_t}{P_{t-\tau}}$;

$\alpha(t-\tau, t) < 1$ implies stock price goes down and

$\alpha(t-\tau, t) > 1$ implies stock price goes up.

$\alpha(t-\tau, t) - 1 = \frac{P_t - P_{t-\tau}}{P_{t-\tau}}$ is known as a proportionate change in the stock price from time $t - \tau$ upto time *t*.

We are interested in modeling the probability distribution accumulation factor $\alpha\,(t-\tau, t)$.

$P[\alpha(t-\tau, t) < x]$ for any given x.

It is perhaps reasonable to assume that probability distribution α $(t-\tau, t)$ depends on τ but not t. Also, assume that it is independent of the current stock price or on the past history of stock price. Making these assumptions, we can let n be a positive integer and $i = 0,1,2,3 \dots n$, set $t_i = (t-\tau) + \frac{i}{n}\tau$;

$t_0 = t - \tau$, $t_n = t$ then $\frac{P_t}{P_{t-\tau}} = \frac{P(t_n)}{P(t_0)} = \left(\frac{P(t_1)}{P(t_0)} \cdot \frac{P(t_2)}{P(t_1)} \cdot \frac{P(t_3)}{P(t_2)} \cdots \frac{P(t_{n-1})}{P(t_n)}\right)$

So that

$\alpha(t-\tau, t) = \alpha(t_0, t_1)\alpha(t_1, t_2)\alpha(t_2, t_3) \dots \alpha(t_{n-1}, t_n)$;
Taking log both side, we get

$$log\alpha(t-\tau, t) = log\alpha(t_0, t_1) + log\alpha(t_1, t_2) + log\alpha(t_2, t_3) \dots + log\alpha(t_{n-1}, t_n);$$

$$L(t-\tau, t) = L(t_0, t_1) + L(t_1, t_2) + L(t_2, t_3) + \cdots + L(t_{n-1}, t_n)$$

Our assumption is the random variables $L(t_{i-1}, t_i)$ are independent and identically distributed $1 \le i \le n$.

Taking n is very large, we use the central limit theorem to deduce the $L(t - \tau, t)$ is normally distributed

$$L(0, t+\tau) = L(0, t) + L(t, t+\tau);$$

Mean of the processes $m(t+\tau) = m(t) + m(\tau)$
And variance $Var(t+\tau) = Var(t) + Var(\tau)$..
From the above one can deduce that
$m(t) = ct$; where c is constant, $Var(t) = \sigma^2 t$, where σ^2 is nonnegative constant.
So, that

$$log\left(\frac{P_t}{P_{t-\tau}}\right) = log\alpha(t-\tau, t) = L(t-\tau, t) \sim N(c\tau, \sigma^2\tau);$$
$$\frac{L(t-\tau, t) - c\tau}{\sigma\sqrt{\tau}} = Z$$

$$L(t-\tau, t) = Z\sigma\sqrt{\tau} + c\tau \Rightarrow log\left(\frac{P_t}{P_{t-\tau}}\right) = Z\sigma\sqrt{\tau} + c\tau$$

Hence, $\left(\frac{P_t}{P_{t-\tau}}\right) = e^{(Z\sigma\sqrt{\tau} + c\tau)}$

Let $P_{t-\tau} = P =$ *constant and known value.*

Then $P_t = P\, e^{(Z\sigma\sqrt{\tau} + c\tau)} \Rightarrow P_t = Pe^{c\tau}\, e^{(Z\sigma\sqrt{\tau})}$;

The expected value of the price at time t is

$$E[P_t] = Pe^{c\tau}E\left[e^{(Z\sigma\sqrt{\tau})}\right];$$

$$[P_t] = Pe^{c\tau}.e^{\frac{\sigma^2\tau}{2}} \Rightarrow E[P_t] = Pe^{\left(c+\frac{\sigma^2}{2}\right)\tau}$$

This expected price is suitable expected price for deciding the supply to fulfill the demand at time t.

It means that we may able to approximately fulfill the demand at time t with this expected price $E[P_t]$ in place of $P_{t-\tau}$ price. Hence, $d(P_t) \cong q(E[P_t])$

$$\Rightarrow d_0e^{\lambda P_t} \cong s_0e^{-\mu Pe^{\left(c+\frac{\sigma^2}{2}\right)\tau}} \Rightarrow d(P_t) \cong q(E[P_t]):$$

$$e^{\lambda P_t + \mu Pe^{\left(c+\frac{\sigma^2}{2}\right)\tau}} = \frac{s_0}{d_0}$$

$$\Rightarrow \lambda P_t + \mu Pe^{\left(c+\frac{\sigma^2}{2}\right)\tau} = log\left(\frac{s_0}{d_0}\right);$$

$$P_t = \frac{1}{\lambda}\left[log\left(\frac{s_0}{d_0}\right) - \mu Pe^{\left(c+\frac{\sigma^2}{2}\right)\tau}\right] \quad (11)$$

Let we say that the stock price process is a geometric Brownian Motion (or lognormal process) with an expected rate of return (or infinitesimal drift) ξ and volatility σ. Note ξ is related to c by

$$\xi = \left(c + \frac{\sigma^2}{2}\right) \Rightarrow c = \xi - \frac{\sigma^2}{2}$$

3.1 Definition: Let $\sigma > 0$ and let ξ be a constant. The lognormal process (or geometric Brownian motion) with expected rate of return ξ and volatility σ, is a family of random variate P_t, $t \ge 0$ with proposition $\forall\, t-\tau \ge 0,\ \tau \ge 0$, then $log\left(\frac{P_t}{P_{t-\tau}}\right) \sim N\left[\left(\xi - \frac{\sigma^2}{2}\right)\tau,\ \sigma^2\tau\right]$.

3.2 Economic order quantity(EOQ) under the stochastic price

Price taken as stochastic random process $P(P, t) = P_t$ then EOQ is obtained by

$$q(P_t) = \sqrt{\left(\frac{2c_0d_0e^{-\lambda P_t}}{c_h}\right)} \quad (12)$$

Where $P_t = \frac{1}{\lambda}\left[log\left(\frac{s_0}{d_0}\right) - \mu Pe^{\left(c+\frac{\sigma^2}{2}\right)\tau}\right]$

4 Results and discussions

4.1 For discrete case

$m = 1$, $s_0 = 1000$, $d_0 = 950$, $P_0 = 100$, $c_0 = 10$, $c_n = 5$, μ=0.0001 (for Table 75.1) , λ =1, (for Table 75.2)

Table 75.1 λ vs. price (P_{2m}, P_{2m+1}) and quantity (q_{2m}, q_{2m+1})

λ	P_{2m}	q_{2m}	P_{2m+1}	q_{2m+1}
0.200000	0.256363	60.083897	0.256313	60.084198
0.240000	0.213651	60.083794	0.213616	60.084042
0.280000	0.183138	60.083717	0.183112	60.083935
0.320000	0.160251	60.083664	0.160232	60.083851
0.360000	0.142450	60.083622	0.142434	60.083786

Table 75.2 μ vs. price (P_{2m}, P_{2m+1}) and quantity (q_{2m}, q_{2m+1})

λ	P_{2m}	q_{2m}	P_{2m+1}	q_{2m+1}
0.000010	0.051293	60.083290	0.051293	60.083290
0.000028	0.051292	60.083317	0.051292	60.083321
0.000046	0.051291	60.083340	0.051291	60.083351
0.000064	0.051290	60.083363	0.051290	60.083385
0.000091	0.051289	60.083389	0.051288	60.083439

4.2 For continuous case

P = 100, t = 10, Υ = 0.01, μ = 0.0025 (For Table 75.3), λ = 0.1 (For Table 75.4)

Table 75.3 λ vs. price (P_t) and quantity (q_t)

λ	P_t	q_t
0.050000	0.230937	61.289268
0.051000	0.308862	61.160538
0.052000	0.383789	61.032085
0.053000	0.455888	60.903896
0.054500	0.559076	60.712120

Table 75.4 μ vs. price (P_t) and quantity (q_t)

μ	P_t	q_t
0.002520	0.728178	59.440334
0.002522	0.610379	59.791431
0.002523	0.492783	60.143993
0.002524	0.434089	60.320740
0.002526	0.258236	60.853401

4.3 For stochastic case

ξ = 1, τ = 0.8, μ = 0.0001(for Table 75.5), λ = 0.1 (For Table 75.6)

Table 75.5 λ vs. expected price ($\tilde{P}_t$) and expected quantity ($\tilde{q}_t$)

λ	$\tilde{P}_t$	$\tilde{q}_t$
0.054500	0.532806	60.755596
0.153500	0.189172	60.755596
0.252500	0.115002	60.755596
0.351500	0.082611	0.082611
0.450500	0.064457	60.755596

Table 75.6 μ vs. expected price ($\tilde{P}_t$) and expected quantity ($\tilde{q}_t$)

μ	$\tilde{P}_t$	$\tilde{q}_t$
0.000118	0.512933	60.083275
0.000136	0.512933	60.083275
0.000145	0.512933	60.083275
0.000163	0.512933	60.083275
0.000181	0.000181	60.083275

4.4 Price vs. quantity graph for exponential demand and supply

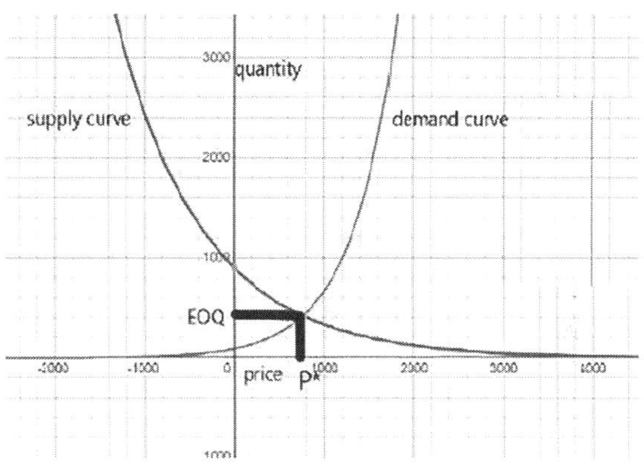

Figure 75.1 EOQ and optimal price

4.5 Description about table and figure

In the Table 75.1, we can observe that if price coefficient λ increases (λ) then the on and off season price and decrease (↓), respectively and corresponding EOQ also decreases (↓).

From Table 75.2, if μ the coefficient of price increases then the price , decrease but the corresponding EOQ increases in both cases.

From Table 75.3, we conclude that if λ increases then continuous price increases and decreases.

From Table 75.4, μ increases then decreases rapidly and increases linearly.

From Table 75.5, if λ increases the is rapidly decreases but remains in steady nature.

From Table 75.6, if μ increases then and remain in steady state situation.

In the Figure 75.1, we conclude that the exponential demand and exponential supply curves intersect at least one point, corresponding to this points there exist optimal price and EOQ for the nonlinear cobweb model.

Conclusions

The mathematical formulae have been derived to find the value of balanced price of seasonal agricultural products as well as its associated economic order quantity. In case II, we have obtained the EOQ for continuous price signals. And case III is dedicated to that type of products which price signal is stochastic in nature at any time. We have determined the EOQ which depends on the stochastic price under cobweb phenomenon of exponential demand and supply. The future scope of this paper is in stock market as well as markets for tomato, onion, and other vegetables and material market. The results of case III have scope to apply in the share market to know the price of share. Future scope of research is to do the case study on empirical data set and check the validity of this model in real life.

References

1. Anokye, M., Barnes, B., Ohene Boateng, F., Adom-Konadu, A., and Amoah-Mensah, J. (2023). Price dynamics of a delay differential cobweb model. *Dis. Dynam. Nat. Soc.* 1–11.
2. Askar, S. S. (2022). Nonlinear dynamics and multistability in a cobweb model. *Dis. Dynam. Nat. Soc.* 1–12.
3. Brianzoni, S., Mammana, C., Michetti, E., and Zirilli, F. (2008). A stochastic cobweb dynamical model. *Dis. Dynam. Nat. Soc.* 1–19.
4. Dufresne, D. and Vázquez-Abad, F. (2013). Cobweb theorems with production lags and price forecasting. *Economics*, 7(1), 20130023.
5. Hommes, C. (2018). Carl's nonlinear cobweb. *J. Econ. Dynam. Control*, 91, 7–20.
6. Mishra, S. S. and Mishra, P. P. (2011). A (Q, R) model for fuzzified deterioration under cobweb phenomenon and permissible delay in payment. *Comp. Math. Appl.*, 61(4), 921–932.
7. Khiavi, A. S. and Dastghiri, S. S. (2022). The impact of time and time delay on the bullwhip effect in supply chains. *J. Quality Engg. Prod. Optimiz.*, 7(2), 1–13.
8. Nadarajah, S. and Lyu, J. (2022). New bivariate and multivariate log-normal distributions models for insurance data. *Results Appl. Math.*, 14, 100246.
9. Sen, B. R., Rahman, M. M., and Hossain, M. I. (2018). Cobweb Model for the stabilization of vegetable prices in some selected villages of Noakhali district. *Inter. J. Sci. Busin.*, 2(4), 707–717.
10. Soheil, S., Ahmadian, A., and Allahviranloo, T. (2021). A new fractional dynamic cobweb model based on non-singular kernel derivatives. *Chaos, Solitons Frac.*, 145, 110755.

76 Study of gate on source-based tubular TFET for biosensing application

Avtar Singh[1], *Dereje Tekilu*[1,a], *Gangiregula Subarao*[1] *and Manash Chanda*[2]

[1]Department of ECE, SOEEC, ASTU, Adama, Ethiopia

[2]Department of ECE, MSIT, KOLKATA, India

Abstract

This work includes, a nano regime TFET made up of silicon tube, based biosensor is studied and proposed for the detection of biomolecules. In this FET based biosensor the micro or nano organism can immobilize at the surface above the source and the gate, interact on the upper wall of the biosensor as well as on the inner part of the tube of the silicon nanotube transistor. The gate electrode is extended on the source and also extended on the little part of drain. Utilizing the TCAD 3D tool, simulations are performed. A nano gap is formed on the whole volume of the silicon nano tube inside as well as the outside of the tube except some part of the drain. The ON current is taken as the sensitivity parameters. For sensing the microorganism dielectric modulation-based technique is utilized. It has been found that incorporating tubular tunnel FET and gate engineering improves the device's TFET performance and qualifies it for low power and more practical applications.

Keywords: Tunnel FET, silicon nano tube FET, tubular-based biosensor, drain current sensitivity parameter, gate on source

1 Introduction

Currently, research advances in FET-based mechanisms for the physical access of biomolecules possess desirable qualities like as downsizing, acquire complete efficiency, excellent receptivity, and label free detection [1, 2]. The fundamental function of the biosensor is an alteration for electrical conductivity of the substance in question, that continues as a monitoring mechanism due to a variation in conductivity, and due to the presence of biomolecules [3, 4]. The biosensors have been widely utilized in certain applications such as disease examination, toity contact, health diagnostics, and so on. FETs based ion sensing FETs operate via a mechanism that cause variations in electrical parameters like as driving capability, conductivity, ambipolar current and threshold voltage. The fundamental idea behind the dielectric modulated FET is that variations in the efficacious interaction across the gate-area with the channel owing toward potential changes in dielectric constant of molecules in cells [3]. Because of their numerous benefits, several nanogap implanted field effect transistors (FETs)-based bio sensors are currently investigated as effective label-free sensors. In comparison with ELISA-based devices (enzyme-linked immunosorbent assay technology), FET-based devices are less expensive, smaller, and better able to identify biological species [5]. However, due to kT/q restrictions, the FET biosensors also have some issues with the highest attainable sensitivity and require a longer time to recognize biological molecules. However, the following limitations are brought on by scaling in FET devices subthreshold swing (SS) >60 mV/dec [6], short-channel impacts (SCEs), drain induced barrier lowering (DIBL), weak I_{ON}/I_{OFF} ratio, and higher power results in more leakage [6].

Tunnel field effect transistor (TFET) inspired biosensors are a kind of FET type biosensor. TFETs are a good alternative for solving the scaling challenges which conventional metal oxide semiconductor FETs (MOSFETs) confront. Although, TFETs possess ambipolar conduction, which is that means that drain current moves across both directions of the gate voltage that is present [7]. The ambipolar characteristic of TFETs limits their utility as switching devices. As a result, many approaches have been developed. Therefore, a number of methods have been suggested to manage this undesirable ambipolar current, among which one of them depends on projecting the gate onto the drain, as described in ref [8]. The tunneling barrier thickness across the channel-drain interface broadens for negative gate voltages as a result of the aforementioned overlap of the gate-on-drain, which lowers the ambipolar nature and make the device suitable for the low power applications. The

[a]avtar.ju@gmail.com

DOI: 10.1201/9781003540199-76

dielectric constant beneath the overlapping gate impacts the decrease in ambipolar current as well. When the overlapping capacitance coupling grows and the overlaid gate-on-drain's dielectric constant rises, the ambipolar current is reduced [9]. The overlapping drain area is thereafter more depleted, resulting in a wider tunneling barrier region at the drain-channel intersection. Biomolecules that are immobilized in the nanogap produced by the overlaying gate over drain region can be detected using a label-free sensing approach based on the correlation amongst the dielectric constant and the ambipolar current [10].

In this paper we have utilized the silicon based tubular structure for the immobilizing of the biomolecules [11]. To increase the area of interaction between surface and biomolecules here we have utilized the gate overlapping over the source region due to which the surface to volume ratio is increased in the tubular structure as compared to the conventional gate all around structures. More surface is available to come in contact with the biomolecules, therefore the sensitivity is also better in these types of structures. Drain current is utilized as the sensing parameter.

Section 2 through some light on the device description and simulation parameter, in Section 3 we presented the detailed discussion on the results. Finally, Section 4 concluded the paper.

2 Device descriptions and simulation setup

Figure 76.1 depicts the silicon nanotube-based tunnel FET biosensor's horizontal cross-section. The channel's length is 40 nm, and the tube's width is assumed to be 10 nm. The length of source and the drain are taken as 30 nm each. The source part's doping is considered to be of the p type and value is 1×10^{20} cm^{-3} and of the drain part it is taken as $5\times10^{18}cm^{-3}$. The channel is region is taken just as intrinsic. Hafnium oxide is taken as the gate dielectric with 1 nm thickness. The hafnium oxide is taken on the channel region as well as 10 nm extended towards the drain region as shown in figure The source region is isolated with the biomolecules using the thin layer of air (k=1). For better immobilization of the biomolecules two different metal i.e., aluminum and tungsten are taken as a gate electrode.

Here the gate is overlapping over source region and the cavity region is also increased towards the source region as shown in the figure. Some part of the drain region is also acquired to increase the surface to volume ratio.

The enzymes, bacteria, and other organisms are identified at the inner and outer gate utilizing the dielectric fluctuation, resulting in a change in the sensitivity parameter's value. The drain current is being used in this study as a sensitivity factor. The parameter specifications listed in Table 76.1 are used to approximate the device. By comparing the findings with the databases or previously validated results, it is feasible to determine which categories of biomolecules have come into interaction with a sensor. In this biosensor, the biomolecules can immobilize on the outer and as well as inner surface of gate and source region. The 10 nm of drain region (from 0.022 μm to 0.032 μm) is also available for the biomolecules to interact with the surface. A greater number of biomolecules may be detectable inside or outside the tube due to a higher surface to volume ratio. The sensitivity parameter is the drain current. The TCAD oriented 3D device simulator is employed to simulate the operation of the device, and several simulated physics algorithms are included.

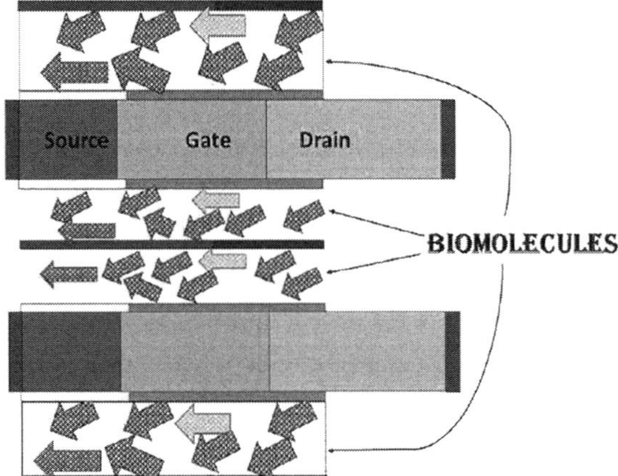

Figure 76.1 Horizontal cross-section of silicon nanotube tunnel FET-based biosensor

Table 76.1 Parameter specifications for devices

Parameter	Value
Channel length	40 nm
Tube diameter	10 nm
Source/drain length	30 nm
Oxide thickness	1 nm
Source doping	1×10^{20} P type
Drain doping	1×10^{18} $/cm^3$
Gate oxide material	HfO_2
Dielectric constant (k) modulation range	K=1 to K=9

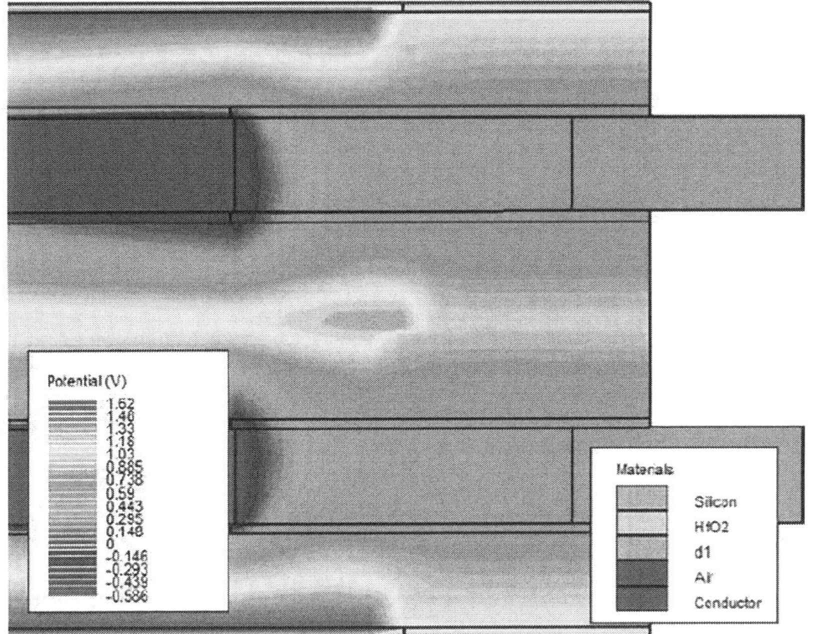

Figure 76.2 Potential contour of silicon nanotube tunnel FET biosensor

In addition to the Shockley Read Hall Recombination Model (SRH), CONMOB, and FLDMOB, we applied the KANE model for band-to-band tunneling process. Regarding numerical iteration, Newton trap model is utilized.

3 Result and discussion

In this section the silicon nanotube-based biosensors results are going to be discussed. For all the simulation the drain current is taken as 1 V and both aluminum and tungsten are employed as a gate electrode. The are two types of techniques in biosensors i.e., gating and dielectric modulated. In this work we have utilized the dielectric modulated type biosensor to realize the different type of biomolecules. To realize the enzymes, proteins and other biomolecules, values ranging from 1 to 9 for the dielectric constant are considered. Figure 76.2 shows the potential contour of tubular-based biosensor at k=9. As we observed potential is maximum at the gate electrodes but as we are moving the silicon region the potential is changing and when we immobilize with different dielectric constant values than we are getting the sufficient difference so we can able to discriminate between the different types of biomolecules.

Figure 76.3 shows transfer characteristics of tubular based biosensor. It is observed from the shown figure that on increasing the value of k, drain current is increasing and there is a vast change in the I_{ON} current as well as subthreshold swing. On increasing the k value from 3 to 7 the

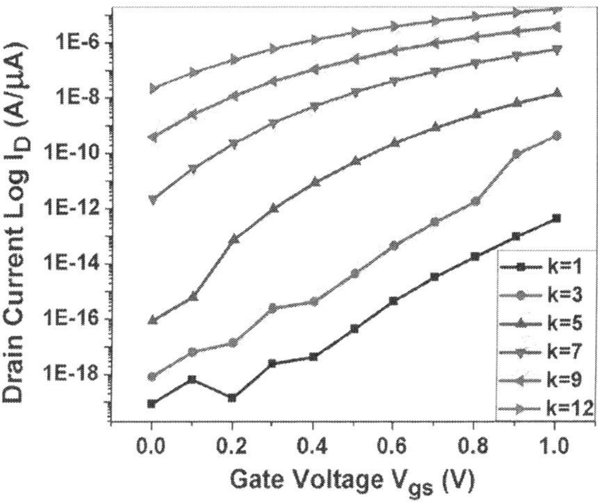

Figure 76.3 Transfer characteristic of silicon nanotube tunnel FET-based biosensor with different k values

drain current is changing about 2-folds, which is comparatively good among the other sensitivity parameter.

Figure 76.4 (a) depicts the impact of k values varying from 1 to 9 on vertical electric field along with x-axis and Figure 4(b) presents the effect of different k value on the electron mobility in x direction.

It is observed from the Figure 76.4 (a) that on escalating the k value the vertical electric field is actually increasing in -ve direction and the electron mobility is also decreasing on increasing the value of k. Actually, electron mobility has to increase but

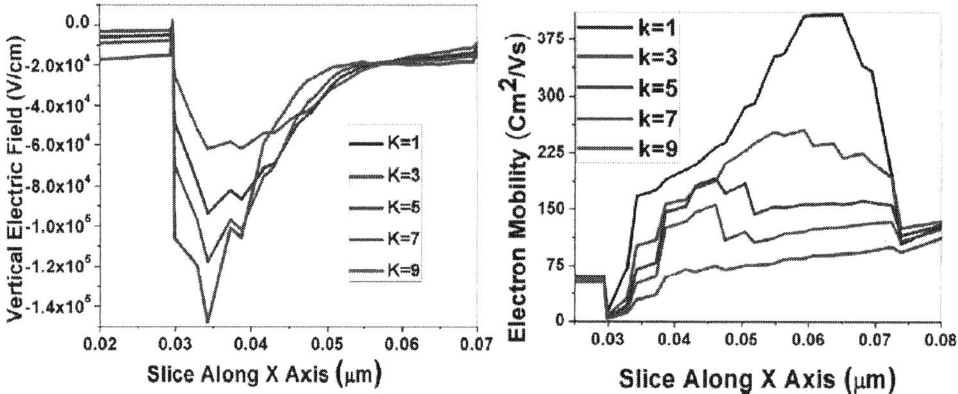

Figure 76.4 (a) Vertical electric field (b) Electron mobility potential contour of silicon nanotube tunnel FET biosensor

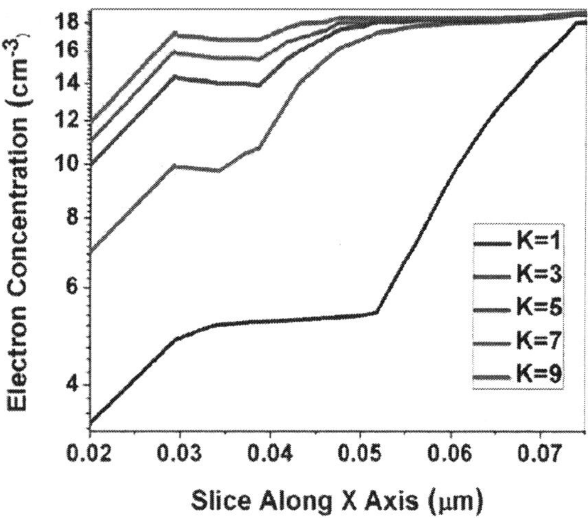

Figure 76.5 Impact of K value on electron concentration along x-axis

the figure shows the electron mobility, which is perpendicular to the structure.

The electron concentration is also one of the major parameters while working on FET based sensors. The impact of k value on the electron concentration inside the silicon region is shown in Figure 76.5. The electron concentration is slight increasing at the larger value of k but at the smaller value of k the variation in the electron concentration at the source channel region is very high.

Finally, to check the sensitivity of biosensor, calculation of sensitivity parameter is very important. The drain current was used as a sensitivity parameter in this work. The sensitivity is determined by following formula and the found values are written in the Table 76.2.

$$S_I = \frac{I_D(bio) - I_D(air)}{I_D(air)} \quad (1)$$

Table 76.2 Sensitivity values

K=1	0.00E+00
K=3	9.99E+02
K=5	3.29E+04
K=7	1.36E+06
K=9	8.79E+06
K=12	3.92E+07

As the values of the k increasing the sensitivity is also increasing, for example k=3 the sensitivity in 10^2 but when the value is increasing to k=9 the sensitivity is in 10^6. The main benefit of this structure is that more biomolecules can immobilize on the surface since there is increased surface area relative to volume, that is, both within and outside the tube.

4 Conclusions

In the current paper, we presented a silicon-based tubular tunnel FET providing extremely sensitive biosensors. Here we have utilized the dielectric modulation to detect the biomolecules. ON current (I_{ON}) or drain current is used as the sensitivity parameter. It was also found through vast simulations of a 3D device utilizing the TCAD tool that every time biomolecules connect to surfaces of the tubular FET, the intensity of the drain current varies, causing the tunneling barrier width at the channel drain interface to fluctuate. The sensitivity of the proposed FET based biosensor is also more because of the overlapping gate on the source region. Sensitivity is measured for k= 1 to k=9 and it has been found on increasing the value of the k the sensitivity is also increased with noted amount. In the future, this form of biosensor might be used for low power DM FET-based biosensors.

References

1. Bergveld, P. (2003). Thirty years of ISFETOLOGY. *Sens. Actuat. B Chem.*, 88(1), 1–20. doi: 10.1016/S0925-4005(02)00301-5.
2. Guan, W., Duan, X., and Reed, M. A. (2014). Biosensors and bioelectronics highly specific and sensitive non-enzymatic determination of uric acid in serum and urine by extended gate fi eld effect transistor sensors. *Biosens. Bioelectron.*, 51, 225–231. doi: 10.1016/j.bios.2013.07.061.
3. Venkatesh, P., Nigam, K., Pandey, S., Sharma, D., and Kondekar, P. N. (2017). A dielectrically modulated electrically doped tunnel FET for application of label free biosensor. *Superlattices Microstruct.*, 109, 470–479. doi: 10.1016/j.spmi.2017.05.035.
4. Singh, A., Chaudhury, S., Chanda, M., and Sarkar, C. K. (2020). Split gated silicon nanotube FET for biosensing applications. *IET Circuits Devices Syst.* 14(8), 1289–1294. doi: 10.1049/iet-cds.2020.0208.
5. Pandey, C. K., Dash, D., and Chaudhury, S. (2019). Approach to suppress ambipolar conduction in Tunnel FET using dielectric pocket. *Micro Nano Lett.*, 14(1), 86–90. doi: 10.1049/mnl.2018.5276.
6. Kao, K. H., Verhulst, A. S., Vandenberghe, W. G., and De Meyer, K. (2013). Counterdoped pocket thickness optimization of gate-on-source-only tunnel FETs. *IEEE Trans. Electron Devices*, 60(1), 6–12. doi: 10.1109/TED.2012.2227115.
7. Singh, A., Pandey, C. K., Chaudhury, S., and Sarkar, C. K. (2020). Tuning of threshold voltage in silicon nano-tube FET using halo doping and its impact on analog/RF performances. *Silicon*. 3871–3877. doi: 10.1007/s12633-020-00698-6.
8. Dash, S., Jena, B., Kumari, P., and Mishra, G. P. (2016). An analytical nanowire Tunnel FET (NW-TFET) model with high-k dielectric to improve the electrostatic performance. *2015 IEEE Power, Commun. Inf. Technol. Conf. PCITC 2015 - Proc.*, 5, 447–451. doi: 10.1109/PCITC.2015.7438207.
9. Chakraverty, M. (2014). A Compact Model of Silicon-Based Nanowire Field Effect Transistor for Circuit Simulation and Design. 1–202. Available: http://arxiv.org/abs/1407.2358.
10. Singh, A., Chaudhary, S., Sharma, S. M., and Sarkar, C. K. (2020). Improved drive capability of silicon nano tube tunnel FET using halo implantation. *Silicon*, 12(11), 2555–2561. doi: 10.1007/s12633-019-00350-y.
11. Singh, A., Shifaw, A. F., Tekilu, D., and Chanda, M. (2021). Silicon nanotube tunnel FET as a label free biosensor. *IJNEAM Unimap*, 14(3), 229–236.

77 Modeling and performance analysis of an electric vehicle using Matlab Simulink

Goutam Kumar Ghorai[a], *Rohit Nayak*[b], *Nairita Das*[c] *and Priyanshi Gupta*[d]

Department of Electrical Engineering, Ghani Khan Choudhury Institute of Engineering & Technology, West Bengal, India

Abstract

This research explores the modeling and performance analysis of electric vehicle (EVs) using MATLAB simulink. The study commences by constructing a dynamic model grounded in the mathematical representation of the electric vehicle, thereby capturing the intricate interplay between its key components. Torque-speed data is extracted from this dynamic model, serving as a fundamental input for subsequent analyses. Subsequently, a simulation model is developed to encompass the motor drive system integrated with the EV's dynamic response, considering a passive load scheme. This integration allows for a comprehensive evaluation of the vehicle's performance under varying conditions and loads. The primary objective of this investigation is to access the efficiency of electric vehicles through the lens of Simulink simulations. By considering both the motor drive system and vehicle dynamics within a holistic framework, we gain valuable insight into the performance characteristics of electric vehicles. The outcomes of this research paper contribute to a deeper understanding of EV behavior and efficiency, offering crucial insights for optimizing their design and operation.

Keywords: Aerodynamic drag coefficient, electric vehicle, road gradient, rolling resistance coefficient

1. Introduction

The global automotive landscape is undergoing a profound transformation, driven by the imperative to reduce emissions and curb the environmental impact on transformation. At the forefront of this transformation are electric vehicles (EVs), promising zero-emission, energy-efficient alternatives to traditional internal combustion engine vehicles [1]. To harness the full potential of electric mobility, a comprehensive understanding of the intricate interplay between vehicle dynamics and the electric drive train is crucial [2, 3]. One of the critical factor affecting the performance of EVs is their energy efficiency, which directly impacts their driving range, charging frequency, and overall user experience [4]. In this context, three key parameters come to the forefront. This research paper aims to delve into the intricate relationship between these factors and electric vehicle performance. Aerodynamic drag is a fundamental force opposing the motion of a vehicle through the air. It is influenced by the vehicle's shape, size, and design features, wind velocity, on the other hand, plays a pivotal role in determining the intensity of aerodynamic drag, with higher wind speed resulting in increased resistance [5]. Rolling resistance, which arises from the friction between the tires and the road surface, also has a substantial impact on an EV's energy consumption [6]. In this paper, we shall examine the existing body of research on aerodynamic drag, wind velocity, and rolling resistance in the context of electric vehicles. Additionally, we shall present our own empirical data and analysis, contributing to a deeper understanding of how these factors influence EV performance [7, 8]. This research paper is dedicated to exploring the realms of modeling and performance analysis in the context of electric vehicles, with a specific focus on the integration of MATLAB Simulink [9, 10]. Electric vehicle modeling and performance analysis have evolved into pivotal domains of study, critical for optimizing efficiency, range, and overall drivability [11]. However, the integration of the vehicle dynamics with the electric drive train adds an additional layer of complexity and this investigation. In realism to this context, this paper embarks on a journey to delve into the following fundamental aspects.

2. Vehicle dynamics modeling

Tractive force is the force produced by the power plant and transferred to the wheels through a transmission system in order to make the vehicle move. It is the sum of aerodynamic force, rolling resistance force, gradient force, and force due to linear acceleration.

[a]goutamghorai79@gmail.com, [b]rohitnayak500@gmail.com, [c]nairita94316@gmail.com, [d]priyanshigupta541@gmail.com

DOI: 10.1201/9781003540199-77

a) Aerodynamic force and power

Aerodynamics drag is a force that the vehicle encounters while moving at a certain speed v through the air. Aerodynamic force F_{ad} and aerodynamic power P_{ad} shown in Equations 1 and 2, respectively.

$$F_{ad} = \frac{1}{2}\rho c_d A\,(v \pm v_{air})^2 \tag{1}$$

$$P_{ad} = F_{ad} \times v \tag{2}$$

where ρ is air density, Cd is drag coefficient, A is frontal area of the vehicle, v, and v_{air} are the velocity of the vehicle and air velocity.

b) Rolling resistance and rolling resistance power

When wheel is rolling on the road, the rolling resistance is acting between tire and road. The front portion of tire encounter higher pressure than the back portion, as a result pressure center of wheel shift from its center to the high-pressure region, and due to depressed of tire the radius of wheel effectively decreased. The rolling resistance force F_{rr} depends on weight of the vehicle, tire surface and pressure, road roughness, presence or absence of liquid on the road. If mg is the weight of the vehicle, μ_{rr} be the rolling resistance coefficient, and vehicle moving with a path of slope θ degree then rolling resistance and rolling resistance power P_{rr} outlined in Equations 3 and 4.

$$F_{rr} = \mu_{rr} mg \cos\theta \tag{3}$$

$$P_{rr} = F_{rr} \times v \tag{4}$$

c) Gradient force and power

The vehicle climbs upward on inclined road then motor delivers positive force and when it is moving downward the force is negative. The gradient force F_{te} is counterbalance by the tractive effort of the motor. An extra gradient power P_{te} is required to overcome this gradient force. The gradient force and gradient power shown in Equations 5 and 6, respectively.

$$F_{te} = mg\ \sin\theta \tag{5}$$

$$P_{te} = F_{te} \times v \tag{6}$$

d) Accelerating force

Linear acceleration force F_{la} changes the linear velocity of the vehicle and angular acceleration force F_{aa} generates the torque which provides spin in the rotating part of the drive train. According to Newton's law of motion mathematically written in Equations 7 and 8

$$F_{la} = m\frac{dv}{dt} \tag{7}$$

$$F_{aa} = \frac{J}{r^2}\frac{dv}{dt} \tag{8}$$

Therefore, total tractive effort written as Equation 9.

$$F_{TE} = mg\sin\theta + \mu_{rr} mg\cos(\theta) + \frac{1}{2}\rho c_d A(v + v_{air})^2 + m\frac{dv}{dt} + \frac{J}{r^2}\frac{dv}{dt} \tag{9}$$

3. Vehicle performance

Vehicle performance depends on external parameters like wind velocity, air drag coefficient, rolling resistance of the road, and gradient of the road especially in hill regions.

a) Effect of wind velocity on power demand

Aerodynamic power requirement increases with the increasing wind velocity. A vehicle running from zero to 100 km/hr with a body mass of 1500 kg, frontal area of 2.2 m^2, air density 1.25 kg/m^3, and air drag coefficient of 0.25 for wind velocity of 10 km/hr, 20 km/hr and 30 km/hr is showing the aerodynamic power demand with the increasing vehicle speed. The aerodynamic power demand with increasing velocity is shown in Figure 77.1 and the corresponding data is presented in Table 77.1

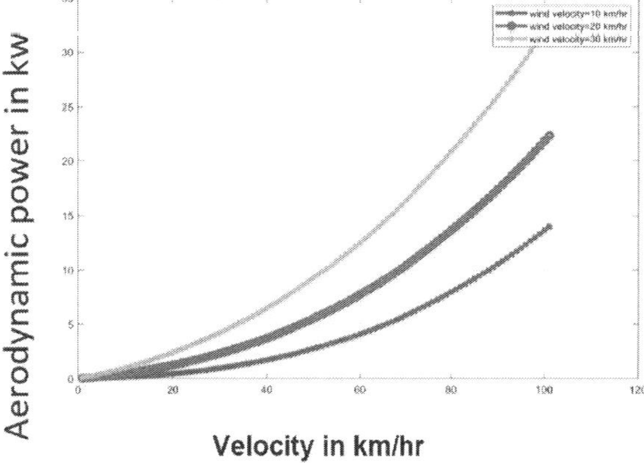

Figure 77.1 Aerodynamic power versus speed curve obtained from simulation

Table 77.1 Aerodynamic power versus speed data obtain from Figure 77.1

Speed in Km/h	Aerodynamic power in KW when wind velocity 30 km/hr	Aerodynamic power in KW when wind velocity 20 km/hr	Aerodynamic power in KW when wind velocity 10 km/hr
20	2.414	1.247	0.462
40	6.455	3.697	1.702
60	12.48	7.703	4.074
80	20.83	13.62	7.931
100	31.75	21.8	13.63

Table 77.2 Aerodynamic drag power versus speed data obtain from Figure 72.2

Speed in Km/hr	Aerodynamic drag power in KW when drag coefficient 0.35	Aerodynamic drag power in KW when drag coefficient 0.30	Aerodynamic drag power in KW when drag coefficient 0.25
20	0.082	0.082	0.082
40	0.660	0.565	0.560
60	2.228	1.91	1.591
80	5.281	4.527	3.772
100	10.31	8.841	7.368

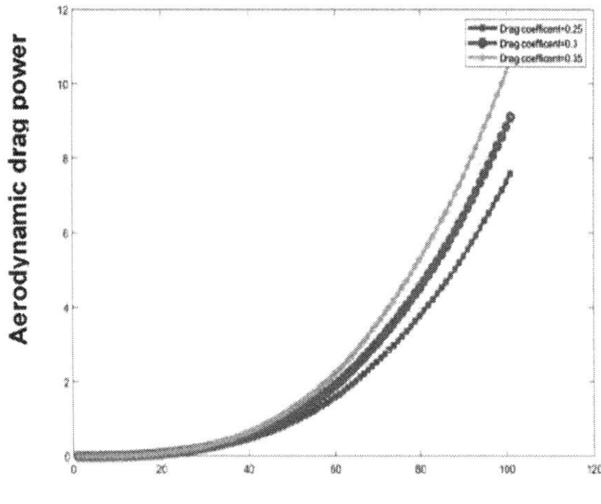

Figure 77.2 Aerodynamic drag power versus speed curve obtained from simulation

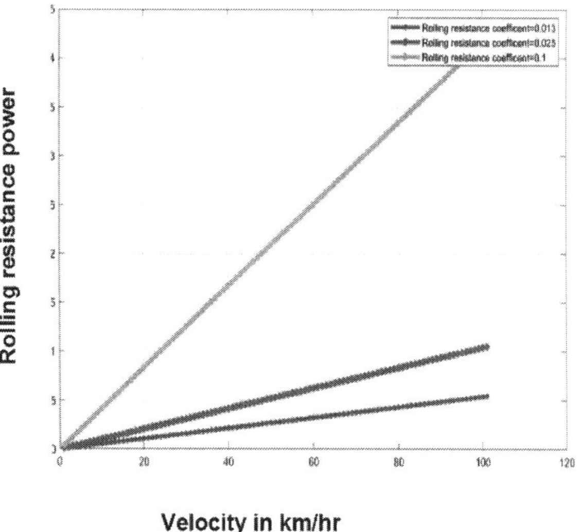

Figure 77.3 Rolling resistance power versus speed curve obtained from simulation

b) Effect of air drag coefficient on air drag power demand

Air drag coefficient depends on the shape of the vehicle. Air drag coefficient of open type of vehicle generally considered 0.5–0.7 whereas for a K-shaped vehicle it is consider 0.2. A vehicle running from zero to 100 km/hr with a body mass of 1500 kg, frontal area of 2.2 m^2, air density 1.25 kg/m^3 and air drag coefficient of 0.25, 0.3, and 0.35 with zero wind velocity is showing aerodynamic drug power with the increasing vehicle speed. The aerodynamic power with increasing velocity is shown in Figure 77.2 and the corresponding data is shown in Table 77.2.

c) Effect of rolling resistance coefficient on rolling resistance power demand

Rolling resistance is an important factor for the vehicle performance calculation. Rolling resistance coefficient for concrete road 0.013,

Table 77.3 Rolling resistance power versus speed data obtained from Figure 77.3

Speed in Km/hr	Rolling resistance power in KW when rolling resistance coefficient 0.1	Rolling resistance power in KW when rolling resistance coefficient 0.025	Rolling resistance power in KW when rolling resistance coefficient 0.013
20	0.833	0.2083	0.1083
40	1.677	0.4167	0.2167
60	2.5	0.625	0.325
80	3.333	0.833	0.433
100	4.166	1.042	0.5417

Table 77.4 Gradient power versus speed data obtained from Figure 77.4

Speed in Km/hr	Gradient power in kw for road inclination 20 degree	Gradient power in kw for road inclination 10 degree	Gradient power in kw for road inclination 0 degree
20	0.833	0.2083	0.1083
40	1.667	0.4167	0.2167
60	2.5	0.625	0.325
80	3.333	0.833	0.433
100	4.166	1.042	0.5417

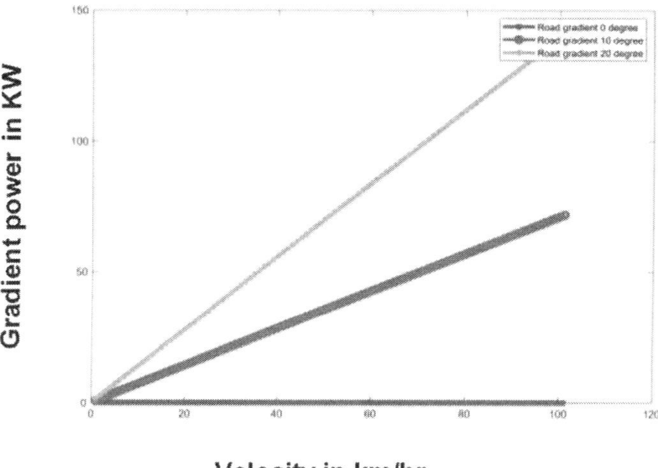

Figure 77.4 Gradient power versus road gradient curve obtained from simulation

gravel road 0.02, Tar road 0.025, fields 0.1–0.35 and minimum when wheels on rail. The power requirement with the vehicle velocity considering rolling resistance coefficients 0.013, 0.25, and 0.1 is shown in Figure 77.3 and the corresponding data is presented in Table 77.3 for the same vehicle.

d) Effect of road gradient on gradient power demand

Road inclination especially in hill road demands extra tractive effort to overcome the gradient force (F_{te}= mg sinθ). A vehicle running from zero to 100 km/hr with a body mass of 1500 kg, frontal area of 2.2 m^2, air density 1.25 kg/m^3, and air drag coefficient of 0.25 for road gradients 0 degrees, 10 degrees, and 20 degrees is shows the gradient power (P_{te}) demand with the increasing vehicle speed in Figure 77.4 and corresponding data is shown in Table 77.4.

4. Simulation model of vehicle dynamics based on mathematical model

Following the formulation of the Simulink model by integrating the mathematical expressions

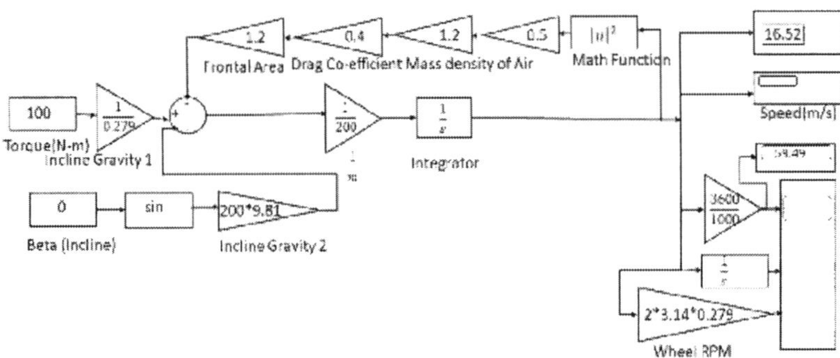

Figure 77.5 Simulation architecture

Table 77.5 Vehicles parameters

Parameter	Vehicle No-1	Vehicle No-2	VehicleNo-3	Vehicle No-4	Vehicle No-5
Mass(kg)	200	500	1335	1400	1800
Drag coefficient	0.4	0.3	0.35	0.38	0.31
Radius of wheel in meter	0.279	0.29	0.3	0.3	0.39
Air density(kg/m^3)	1.2	1.35	1.127	1.2	1.36
Frontal area in m^2	1.2	1.6	2.19	2.31	2.4

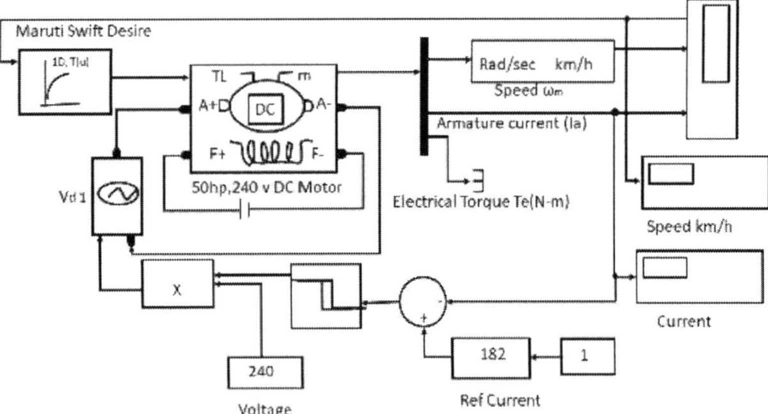

Figure 77.6 Simulation model of dc motor with vehicle dynamics

outlined in Equations 10 and 11, thereafter comprehensive representation of the electric drivetrain has been developed in the MATLAB Simulink environment. This model incorporates various components, including the DC motor, electronic components, control elements, and measuring instruments. The integration of these components facilitates the simulation of the interaction between vehicle dynamics and the electric propulsion system in a holistic manner, as depicted in Figure 77.5.

$$F_d = \frac{1}{2}\rho c_d A V_x^2 \sin V_X \tag{10}$$

$$F_x = m\frac{dV_x}{dt} + F_d + mg\sin\beta \tag{11}$$

Where m = mass, Cd = Air drag coefficient, ρ = air density, β = angle, A = frontal area, V_x = velocity of the vehicle.

5. Vehicle data of different five vehicles and simulation result

The simulation process used different parameters of five numbers of vehicles shown in Table 77.5. The torque versus speed data and efficiency of each vehicle obtained from the simulation are presented in Tables 77.6 and 77.7.

6. Result

Using the simulation model (Figure 77.6), we derived the efficiency values for various vehicles, and these results are documented in Table 77.7.

Table 77.6 Simulation result of torque versus speed data obtain from Figure 77.5

Vehicle 1		Vehicle 2		Vehicle 3		Vehicle 4		Vehicle5	
Torque in (N-m)	*Speed (km/hr)*	*Torque in (N-m)*	*Speed (km/hr)*	*Torque in (N-m)*	*Speed (km/hr)*	*Torque in (N-m)*	*Speed (km/hr)*	*Torque in (N-m)*	*Speed (km/hr)*
10	9.6	10	10.31	10	15.36	10	17.36	10	40.03
30	25.77	30	27.81	30	38.56	30	43.75	30	69.56
50	38.99	50	40.13	50	55.87	50	63.53	50	89.8
70	50.17	70	52.17	70	69.76	70	79.42	70	106.3
90	61.2	90	64.26	90	82.14	90	93.94	90	120.23
100	67.59	100	69.81	100	86.49	100	117.14	100	127.13

Table 77.7 Efficiency of different vehicles

Vehicle	Vehicle 1	Vehicle 2	Vehicle 3	Vehicle 4	Vehicle 5
Efficiency	59.16	84.53	78.4	77.44	73.63

7. Conclusions

The study has been done to understand the power requirement and efficiency variation due to external factors like wind velocity, air drag resistance, rolling resistance, and gradient of the road with increasing vehicle speed. The result shows the power requirement increases with increasing speed for each case. The simulation model ultimately identifies the vehicle type and the strategic adjustment of the DC motor rating. The result exhibits the potential for enhancing electrical vehicle efficiency by modifying the DC motor rating. In the present scenario, industry is moving toward an environmentally conscious future, and understanding and optimizing these factors will be pivotal in shaping the next generation of high-efficiency electrical vehicles.

8. References

1. Mamo, T., Gopal, R., Yoseph, B., Tamirat, S., and Seifu, Y. (2023). Numerical study on low speed electric car for public transportation to predict its aerodynamic performance and stability. *Engg. Technol. J.*, 8(5), 2183–2190.
2. Sharmila, B., Srinivasan, K., Devasena, D., Suresh, M., Hitesh Panchal, R., Ashok Kumar, R., Meena Kumari, Kishor Kumar, S., and Ronak Kumar, R. S. (2022). Modeling and performance analysis of electric vehicle. *Inter. J. Ambient Ener.*, 43(1), 5034–5040. DOI: 10.1080/01430750.2021.1932587.
3. Yu, H. and Huang, M. (2008). Potential energy analysis and limit cycle control for dynamics stability of in-wheel driving electric vehicle. *2008 IEEE Veh. Power Propul. Conf.*, 1–5.
4. Rahman, K. M. and Ehsani, M. (1996). Performance analysis of electric motor drives for electric and hybrid electric vehicle applications. *Power Elec. Trans.*, 49–56.
5. Skuza, A. and Jurecki, R. S. (). Analysis of factors affecting the energy consumption of an EV vehicle – A literature study. DOI 10.1088/1757-899X/1247/1/012001.
6. Muratori, M., Moran, M. J., Serra, E., and Rizzoni, G. (2013). Highly-resolved modeling of personal transportation energy consumption in the United-States. *Energy*, 58, 168–177.
7. CarFolio. (2021). BMW i4 e-Drive 40 specifications technical data available at www.carfolio.com/bmw-i4-edrive40-728508 (accessed May, 2023).
8. Tata Motors. (2022). Tata Nexon EV specifications available at www.nexonev.tatamotors.com/features/(accessed June, 2023).
9. Vempalli, S. K., Ramprabhakar, J., Shankar, S., and Prabhakar, G. (2018). Electric vehicle designing, modeling and simulation. *2018 4th Inter. Conf. Converg. Technol. (I2CT)*, 1–6.
10. Math Works. Explore the electric vehicle reference application, available at www.mathworks.com/help/autoblks/ug/exploretheelectrical.
11. Sinoquet, D., Rousseau, G., and Milhau, Y. (2011). Design optimization and optimal control for hybrid vehicles. *Optimiz. Engg.*, 12, 199–213.

78 Modelling and optimization of Si/GaP quantum dot solar cells using Silvaco TCAD

Pratik De Sarkar[a], *Subhajit Kar and Nirban Chakraborty*

Renewable Energy Laboratory, Institute of Engineering & Management Kolkata, University of Engineering & Management Kolkata, India

Abstract

Achieving higher conversion efficiencies across the entire solar spectrum is a pivotal focus within the contemporary solar cell industry. To enhance these efficiencies, quantum dots (QDs) have become integral components in modern solar cell technology. Keeping in view both commercial viability and financial considerations, we have developed a silicon solar cell featuring multiple layers of gallium phosphide (GaP) quantum dots. In contrast to conventional silicon solar cells, this innovative design yields an impressive approximately 20% conversion efficiency for specific wavelengths of light. Moreover, this efficiency can be further tailored to encompass a range of frequencies by adjusting the orientation of the quantum dots.

Keywords: QD, solar cell, efficiency, open circuit voltage, short circuit current

1. Introduction

Because technology is developing so quickly, today's cutting-edge devices become obsolete tomorrow. The important element of any device nowadays, given the development of current technology, is effective energy utilization. Which implies that an energy source should be created so that it has high conversion efficiency [1]. The sun is a huge source of energy if we can harness it effectively, thus this is a crucial consideration when designing our solar cells. If we look at the construction of contemporary products, the solar cells that are now on the market are enormous. So every solar cell's structure needs be scaled down to the nanometer range in order to adapt to contemporary technology [2]. Modern solar cells are produced as thin sheets as a result, which are extremely flexible and may be adapted to any modern gadget.

Even though these thin film solar cells are compatible with contemporary gadgets, their design prevents many solar light wavelengths from being absorbed, or if they are, they cannot contribute to the production of solar energy. As a result, conversion efficiency declines [3]. After extensive investigation, it has been shown that the conversion efficiency of solar cells that contain QDs may reach very high levels. A significant number of photons may be integrated into the QD's interstitial band transition to produce free electrons. The choice of materials for QDs, however, is the fundamental obstacle. It has been observed that some alloys may be used to create extremely effective QDs, however those alloys are not cost-effective enough to satisfy commercial demands [4].

The research team at NREL has noted a significant slowdown in the hot electron cooling process in InP QDs [5]. An Auger process is one method for QDs that breaks the phonon bottleneck that is projected to limit carrier cooling in QDs and hence enabling rapid cooling. Here, a hot electron can thermalize a hole by using the Auger process to transfer its excess kinetic energy to the hole, which can subsequently cool fast due to its larger effective mass and closer-spaced quantized states. But if the hole is taken out of the QD core by a quick hole trap at the surface, the Auger process will be halted and a phonon bottleneck will develop, which would result in sluggish electron cooling. This effect was initially demonstrated for CdSe QDs [6, 7], and it has since been demonstrated for InP QDs, where a rapid hole trapping species (Na biphenyl) was discovered to reduce the cooling time of the electrons to roughly 7 ps [5]. This should be contrasted with the 0:3 ps electron cooling time for passivated InP QDs, which are characterized by the absence of a hole trap and, as a result, by the presence of holes that can engage in an Auger process with the electrons.

After surveying various aspects this solar cell structure is modeled with Si (Silicon) and GaP (Gallium-Phosphate). These materials are very economic to design any solar cell which can acceptable in society.

[a]pratik.desarkar@iem.edu.in

DOI: 10.1201/9781003540199-78

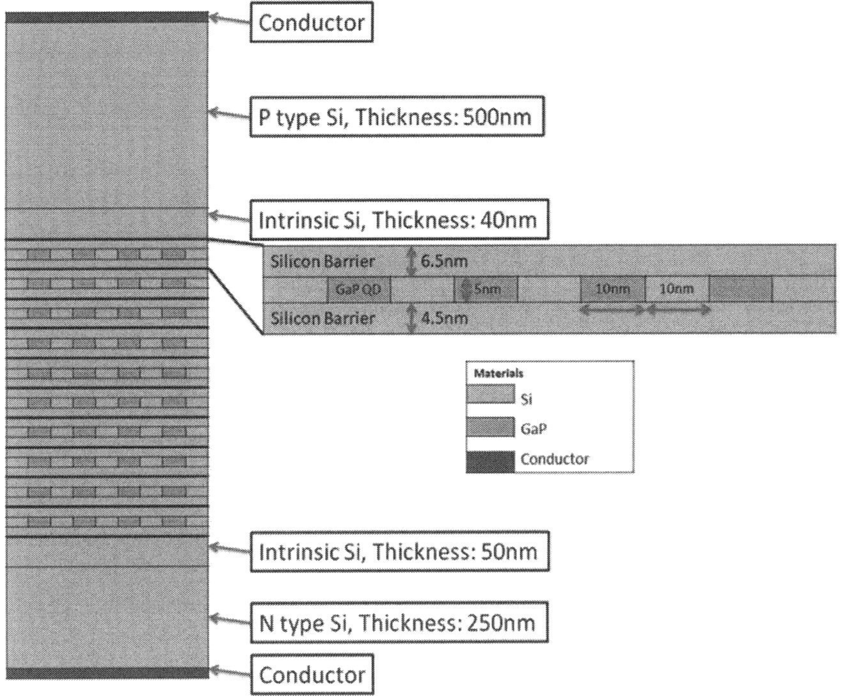

Figure 78.1 Front view of 10 layer QD solar cell

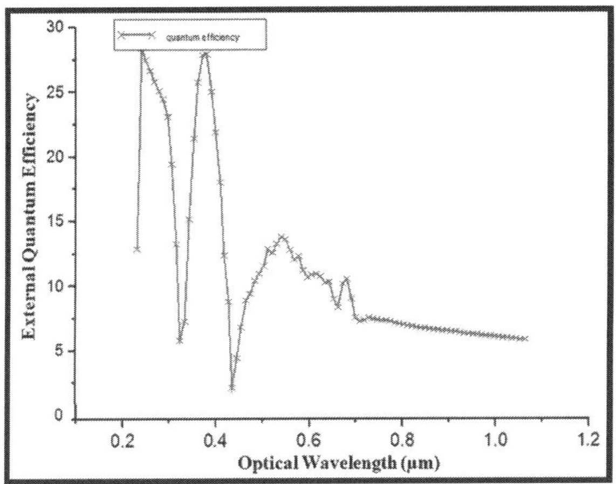

Figure 78.2 External quantum efficiency vs. optical wavelength for 10 layer QD structure

2. Modeling and simulation

A straightforward PiN solar cell structure has been used as a reference for the modeling portion [8, 9], and QDs layers have been inserted into the intrinsic layer.

As shown in the Figure 78.1 the top and bottom most layers are conductors. The second layer from the top is P type Si layer having uniform doping concentration of 2e18/cm^3 [10]. Then there are 10 layers of QDs sandwiched between two intrinsic layer of silicon having thickness of 40 nm and 50 nm, respectively [11]. Second layer from the bottom is the N type Si layer having uniform doping concentration of 1e17/cm^3 [12–16].

All the 10 QDs layers are having same structure. As shown in the figure the GaP QDs are incorporated into the intrinsic layer of silicon. This QD layer is sandwiched between two layers of intrinsic silicon. By this design the 3D confinement is achieved in the QDs. All the QDs are having dimension of 5 × 10 nm and spaced 10 nm apart. In each layer there are 4 dots. This structure is simulated using Silvaco TCAD. This structure is illuminated under AM 1.5.

3. Results and discussions

The graphic below shows how Silvaco created the quantum efficiency vs. optical wave length relationship. Two peaks can be seen in Figure 78.2; one is located at 325 nm, while the other is located around 450 nm. This shows that the QDs in this solar cell have been adjusted to produce increased efficiency at those particular wavelengths. Now, as illustrated in Figure 78.3, the height of the peaks similarly decreases when the number of QD layers is raised to 20. It implies that as the quantity of QDs layers increases, so does efficiency.

Due to the same size and doping concentration of the QDs, the peaks do not change when the number of QD layers is increased. The peaks in the QDs will move to a different wavelength when the dimension and doping concentration are changed.

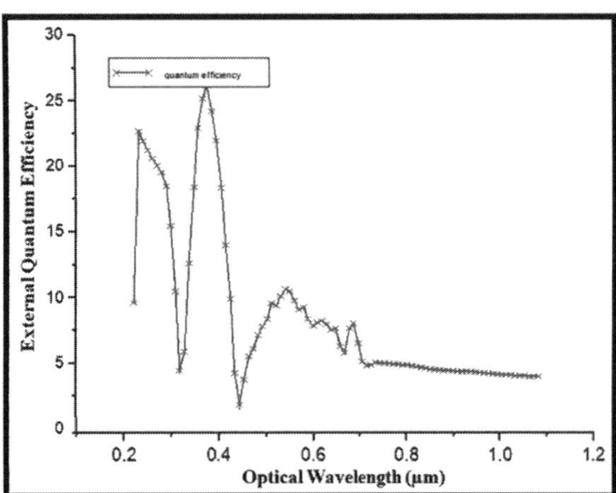

Figure 78.3 External quantum efficiency vs. optical wavelength for 20 layer QD structure

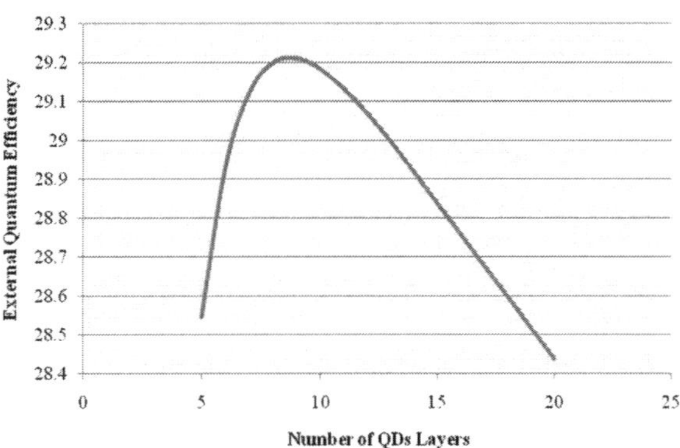

Figure 78.5 External quantum efficiency vs. number of layers

The efficiency of this solar cell will then increase for some other light wavelengths. The J-V curve of the solar cell changes as the number of QD layers is altered. The various color graphs are for the various numbers of QD layers, as illustrated in Figure 78.4.

Figure 78.5 makes it abundantly evident that the maximum efficiency of 29.21 may be attained for this specific design provided the number of QD layers is kept to a maximum of 10. A comparative study has been carried out with some state of the art solar cells and it has been shown in the Table 78.1. The proposed design has been proven to be best in case efficiency output.

4. Conclusions

This research paper has developed the modeling and simulation of Si/GaP quantum dot solar cells utilizing the powerful Silvaco TCAD software. Through meticulous investigation and analysis, we have explored the potential of quantum dots as key elements in enhancing the efficiency of solar cell technology. Our findings demonstrate that the integration of multiple layers of GaP quantum dots into silicon solar cells offers a remarkable boost in conversion efficiency, particularly for specific wavelengths of light. Moreover, the flexibility to fine-tune these cells for various frequencies by adjusting quantum dot orientations opens up exciting possibilities for tailoring solar cell performance to diverse environmental conditions. This research underscores the critical role of simulation and modeling tools like Silvaco TCAD in the advancement of photovoltaic technology. It has illuminated a path toward more efficient and adaptable solar cells, addressing the pressing need for sustainable energy solutions in our rapidly changing world.

As the field of semiconductor devices and solar cells continues to evolve, the insights and methodologies presented in this study pave the way for further exploration, experimentation, and innovation. Through ongoing research and development efforts, we can look forward to a future where Si/GaP quantum dot solar cells, as modeled and simulated in this study, contribute significantly to a cleaner and more energy-efficient world.

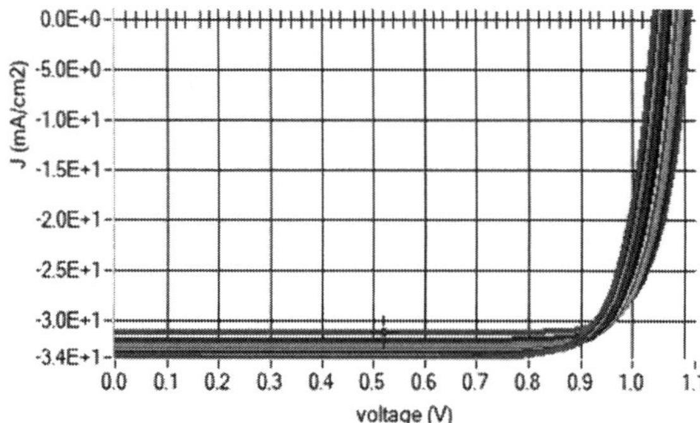

Figure 78.4 Variation of J-V curve for different number of layers

Table 78.1 Comparative study with some state of the art design

Solar Cells	Efficiency
Reference cell-1 [13]	24%
Reference cell-2 [14]	25.53%
Reference cell-3 [15]	13.61%
Reference cell-4 [16]	25.6%
The proposed design	**29.21%**

5. References

1. Aissat, A., Chenini, L., Nacer, S., and Vilcot, J. P. (2022). Modeling and simulation of GaAsPN/GaP quantum dot structure for solar cell in intermedi-

ate band solar cell applications. *Inter. J. Ener. Res.*, 46(8), 10133–10142.
2. Mukherjee, I., Somay, S., and Pandey, S. K. (2022). Comprehensive device modeling and performance analysis of quantum dot-Perovskite solar cells. *J. Elec. Mat.*, 51(4), 1524–1532.
3. Rodhuan, M. B., Abdul-Kahar, R., and Ameruddin, A. S. (2021). Simulation on optical absorption for amorphous silicon thin film solar cell with CdSe/ZnS quantum dots. *Proc. 7th Inter. Conf. Appl. Sci. Math. Sciemathic 2021*, 81–93. Singapore: Springer Nature Singapore.
4. Ghoshal, M. N. and Gaidhane, V. H. (2022). Analysis of hybrid tandem solar cell using neural network. *Mater. Today Proc.*, 49, 2052–2057.
5. Lao, X., Bao, Y., and Xu, S. Impact of excitation energy on the excitonic luminescence of cesium lead bromide perovskite nanosheets. *Optics Lett.*, 45(14), 3881–3884.
6. Diroll, B. T., Chen, M., Coropceanu, I., Williams, K. R., Talapin, D. V., Guyot-Sionnest, P., and Schaller, R. D. (2019). Polarized near-infrared intersubband absorptions in CdSe colloidal quantum wells. *Nat. Comm.*, 10(1), 4511.
7. Pavlenko, V. and Beloussov, I. (2023). Amplified spontaneous emission from CdSe/CdS/CdZnS quantum dot films. *J. Lumines.*, 257, 119643.
8. Bailey, C. G., Forbes, D. V., Polly, S. J., Bittner, Z. S., Dai, Y., Mackos, C., Raffaelle, R. P., and Hubbard, S. M. (2012). Open-circuit voltage improvement of InAs/GaAs quantum-dot solar cells using reduced InAs coverage. *IEEE J. Photovolt.*, 2(3), 269–275.
9. Bailey, C. G. (2012). pp 14-47, Optical and mechanical characterization of InAs/GaAs quantum dot solar cells. *Rochester Institute of Technology*.
10. Queisser, H. J. (1961). Slip patterns on boron-doped silicon surfaces. *J. Appl. Phy.*, 32(9), 1776–1780.
11. Green, M. A. (2006). pp 95–109, Third generation photovoltaics.
12. Kolodinski, S., Werner, J. H., Wittchen, T., and Queisser, H. J. (1993). Quantum efficiencies exceeding unity due to impact ionization in silicon solar cells. *Appl. Phy. Lett.*, 63(17), 2405–2407.
13. Wang, Y., Zhang, S.-T., Li, L., Yang, X., Lu, L., and Li, D. (2023). Dopant-free passivating contacts for crystalline silicon solar cells: Progress and prospects. *EcoMat, 5*(2), e12292.
14. Liu, Z., Lin, Z., Liu, W., Yang, L., Lin, N., Liao, M., Yu, X., Yang, Z., Zeng, Y., and Ye, J. (2023). Regionalizing nitrogen doping of polysilicon films enabling high-efficiency tunnel oxide passivating contact silicon solar cells. *Small*, 2304348.
15. Wang, X., Xu, L., Ge, S., Foong, S. Y., Liew, R. K., Chong, W. W. F., Verma, M., et al. (2023). Biomass-based carbon quantum dots for polycrystalline silicon solar cells with enhanced photovoltaic performance. *Energy*, 274, 127354.
16. Khokhar, M. Q., Hussain, S. Q., Kim, Y., Dhungel, S. K., and Yi, J. (2023). A novel passivating contact approach for enhanced performance of crystalline silicon solar cells. *Mater. Sci. Semicond. Proc.*, 155, 107231.

79 Performance assessment of PMSG-based small-scale stand-alone wind turbine system for battery charging in rural application

Raja Ram Kumar[a], *Rajib Mondal*[b], *Narayanasetti Gayatri*[c], *Aftab Ansary*[d], *Gyanvi Sharma*[e], *Priyanka Pal*[f] *and Arpita Roy*[g]

Department of Electrical Engineering, Ghani Khan Choudhury Institute of Engineering & Technology, Malda, India

Abstract

This paper presents a performance assessment of a permanent magnet synchronous generator (PMSG)-based small-scale stand-alone wind turbine system (PMSG-BSSSAWTS) for battery charging in rural applications. The proposed system mainly consists of a wind turbine, PMSG, rectifier, direct current (DC) to DC converter, battery pack, and charge controller. Among these components, the generator is the main component to generate the electricity. For this purpose, PMSG is chosen because of its numerous advantages like high efficiency, reliability, low maintenance, and no rotor winding for excitation over conventional generators. For the performance analysis of the proposed battery charging scheme, a model has been created in the MATLAB/Simulink environment. Different characteristics like percentage (%) SOC vs. time, line voltage vs. wind speed, rectifier voltage vs. time, and discharging battery voltage vs. time are carried out from the simulation. Faster battery charging by the proposed system and limited reach of electricity reinforced the need of the proposed system in rural applications.

Keywords: Battery, DC to DC converter, PMSG, rectifier, wind turbine

1. Introduction

In recent decades, the global reliance on non-renewable energy sources, such as coal, gasoline, and crude oil, has grown significantly to meet rising energy demands. However, these resources are finite, posing concerns about future availability and cost volatility. Furthermore, their utilization is linked to environmental degradation, with emissions of pollutants that contribute to air and water contamination, harming ecosystems, and public health. Notably, the combustion of these fuels releases substantial greenhouse gases, primarily carbon dioxide, exacerbating global warming and climate change [1]. Transitioning to sustainable, renewable energy alternatives is critical for mitigating these issues and ensuring a cleaner, more sustainable energy future [2]. Renewable energy sources like solar energy, biomass energy, wind energy, tidal energy, geothermal energy, and hydro energy. Among all the renewable energy sources, wind energy is more reliable for generating electricity. Wind energy offers numerous advantages; it is renewable and abundant, environmentally friendly, cost-effective, and available day and night [3]. A wind farm consists of several components that work together to generate electricity The primary elements of the wind farm are a wind turbine, tower, foundation, substation, power grid connection, etc. Among all those components, the generator is the most important component. There are various types of generators used over time, but doubly fed induction generator (DFIG) and permanent magnet synchronous generator (PMSG) are widely used. PMSG has many advantages. There is no need for external excitation for operation. It is energy efficient and more reliable. The implementation of PMSG eliminated the need for a gearbox, which was previously employed to boost the turbine speed [4]. The efficiency and cost-effectiveness of wind energy further solidify its position as a viable alternative. Unlike non-renewable resources, wind energy can be harnessed day and night, providing a consistent and reliable source of power. Wind farms, comprised of various components such as wind turbines, towers, foundations, substations, and power grid connections, work together to harness and convert wind energy into electricity. Within the realm of wind energy technology, the choice of generator plays a pivotal role. The DFIG and PMSG are widely used in wind farms. The PMSG has gained popularity due to its numerous advantages. Notably, it eliminates the need for external excitation during operation,

[a]rajaram@gkciet.ac.in, [b]rajibmondal2307@gmail.com, [c]gayatrinarayanasetti2003@gmail.com, [d]aftabansary2001@gmail.com, [e]gyanvisharma29102@gmail.com, [f]palp5331@gmail.com, [g]sanjitaray50@gmail.com

DOI: 10.1201/9781003540199-79

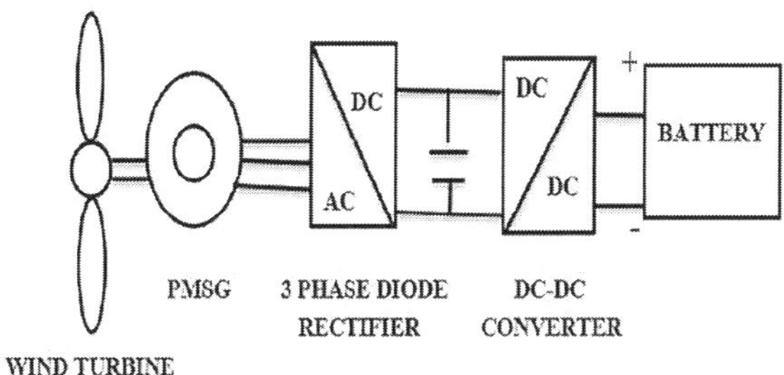

Figure 79.1 Wind turbine system for battery charging

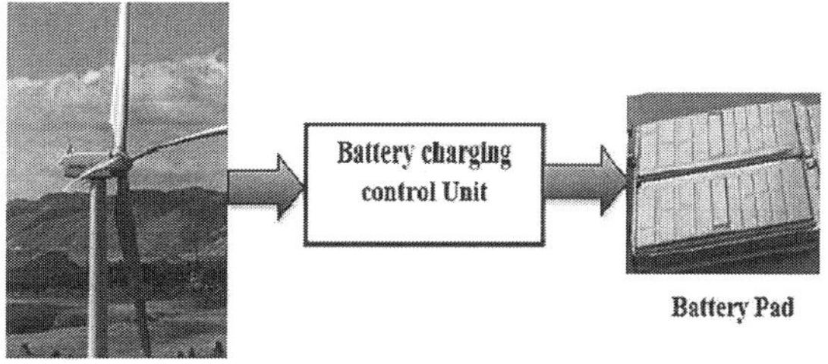

Figure 79.2 Wind turbine-based battery charging scheme

Table 79.1 Battery details

Parameters	Rating
Battery-type	Lithium-ion
Rated-capacity	100Ah
Nominal-voltage	48V
Battery-rating	4.8 KWh
Cutoff-voltage	36 V
Full charge-voltage	55.87 V

Table 79.2 Rating of generator

Parameters	Values
Power-kW	2
Voltage-volts	440
Phase	3
Efficiency-%	92
Frequency-Hz	50
Speed-rpm	1500

enhancing energy efficiency and reliability. One significant advancement associated with the use of PMSG is the elimination of the gearbox, previously employed to boost turbine speed. This innovation has far-reaching implications, reducing maintenance requirements and minimizing losses compared to conventional machines. As a result, the PMSG has become a preferred choice in wind farm applications. AS a result, utilizing PMSG reduces maintenance requirements and significantly decreases losses compared to conventional machines used in the past. Among all these advantages of PMSG, it more popular in wind farms. The main goal of this paper is to assess the performance of PMSG-based small-scale standalone wind turbine system for battery charging in rural applications. The proposed system mainly consists of a wind turbine, PMSG, rectifier, DC to-DC converter, battery pack and charge controller. For the performance analysis of the proposed battery charging scheme, a model has been created in the MATLAB/Simulink environment. Different characteristics like %SOC vs. time, line voltage vs. wind speed, rectifier voltage vs. time, and discharging battery voltage vs. time are carried out from the simulation. The paper's structure is as follows: Section 2 introduces the model and battery specifics, Section 3 discusses performance characteristics, and Section 4 offers concluding remarks and a summary.

2. Model description

Figure 79.1 illustrates the wind turbine system designed for rural area applications. The main components of the system are a wind turbine, PMSG, rectifier, DC to DC converter and battery. For generating electricity, PMSG has been chosen because of its simple, robust construction, and cheaper in cost. When the wind appears on the blade of a wind turbine, then the turbine rotates the directly coupled rotor of PMSG. Due to this three-phase electricity is generated. The generated alternating current (AC) voltage then further rectified by AC to DC converter. This DC voltage can be utilized to charge the battery by adjusting the voltage level using DC to DC converter.

In Figure 79.2, it shows that the components working together seamlessly to facilitate the energy conversion and storage process. The wind turbine, a central element of the system, captures the kinetic energy present in the wind, initiating the rotation of its blades. This rotational motion is then translated into electrical power by the generator, marked by its specifications outlined in Table 79.2. The battery, a critical component for energy storage, is detailed in Table 79.1. Here, you'll find specifics about its capacity, voltage, and other essential characteristics. Understanding the battery's capabilities is crucial as it directly influences the system's ability to store and supply energy as needed.

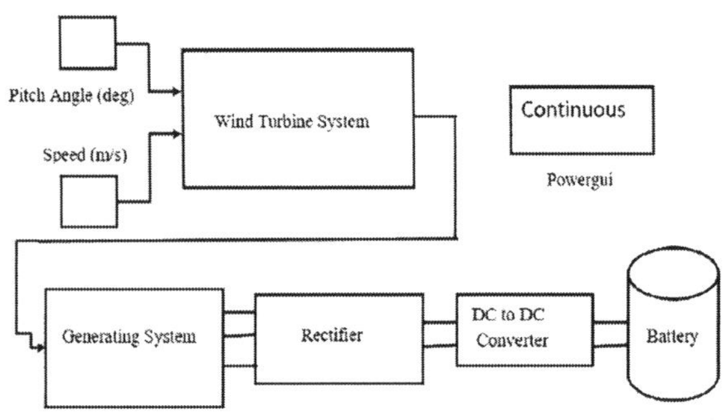

Figure 79.3 Circuit diagram of wind turbine based charging system

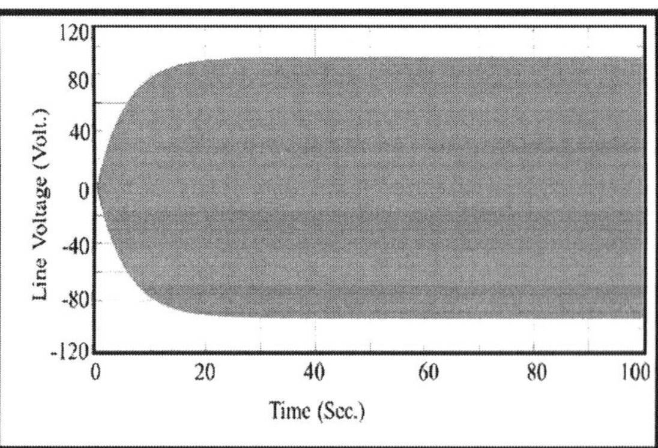

Figure 79.4 Line voltage vs. time

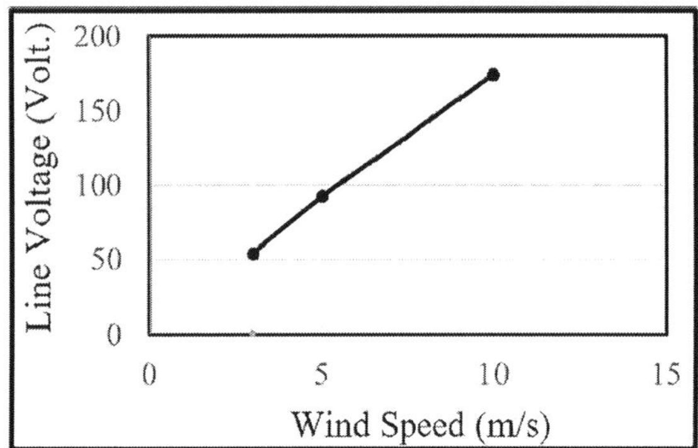

Figure 79.5 Line voltage vs. wind speed.

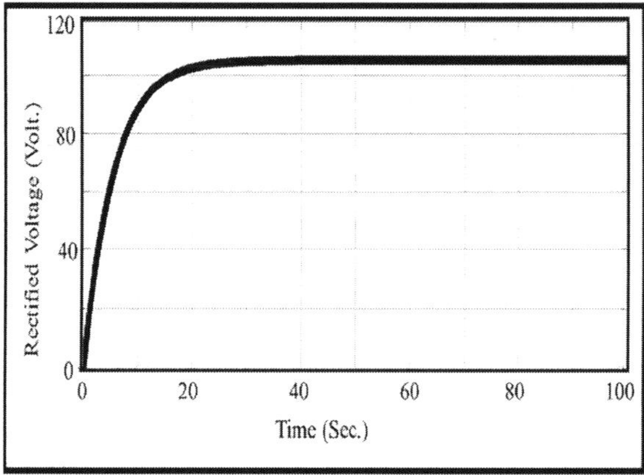

Figure 79.6 Rectified voltage vs. time

3. Simulated model and performance analysis

A proposed PMSG-BSSSAWTS for battery charging is modeled in MATLAB/Simulink environment as shown in Figure 79.3. The model is simulated for the performance analysis like line voltage vs. time, line voltage vs. wind speed, rectified voltage vs. time, charging voltage vs. time, %SOC vs. time, and discharging battery voltage vs. time of the battery charging.

Figure 79.4 shows that line voltage vs. time at the wind speed 5 m/s and pitch angle 15°. Line voltage increases as the wind- speed increases. When wind-speed is 5 m/s, the line voltage is 92.6 volts. Line voltage increases rapidly for up to 10 seconds after that it will attain the maximum value 92.6 volt.

In Figure 79.5, the relationship between line voltage and wind speed is evident. As wind speed rises, so does the line voltage, revealing a notable correlation. At 3 m/s, the line voltage registers at 54.1 volts, a figure that escalates to 92.6 volts at 5 m/s. The trend continues with a significant increase to 174 volts at 10 m/s. This graphical representation underscores the direct impact of wind speed on the system's electrical output, offering valuable insights into the dynamic nature of the relationship between these two crucial parameters.

In Figure 79.6, the graph illustrates the relationship between rectified voltage and time. Notably, the rectified voltage experiences a rapid increase within the initial 10 seconds. It reaches its pinnacle at 20 seconds, achieving a maximum value of 105.3 volts. This peak coincides with a wind speed of 5 m/s and a pitch angle of 15 degrees. The characteristic curve provides crucial insights into the temporal evolution of rectified voltage, highlighting key milestones in its ascent and offering valuable data for assessing the system's efficiency and response to varying conditions.

In Figure 79.7, the graph depicts the relationship between wind speed and the charging time of the battery. The characteristic reveals an inverse correlation: as wind speed decreases, the charging time increases. At 10 m/s, the battery charges in 1.15 hours, while at 3 m/s, it extends to 8.77 hours. This insight into the charging dynamics under different wind speeds is crucial for understanding the system's performance variability and optimizing its efficiency across diverse environmental conditions.

Figure 79.8 illustrates the percentage state of charge (%SOC) over time, assuming an initial battery charge of 80% at the simulation's outset. Notably, the %SOC exhibits a swift increase, reaching 90% within 30 seconds and achieving a full

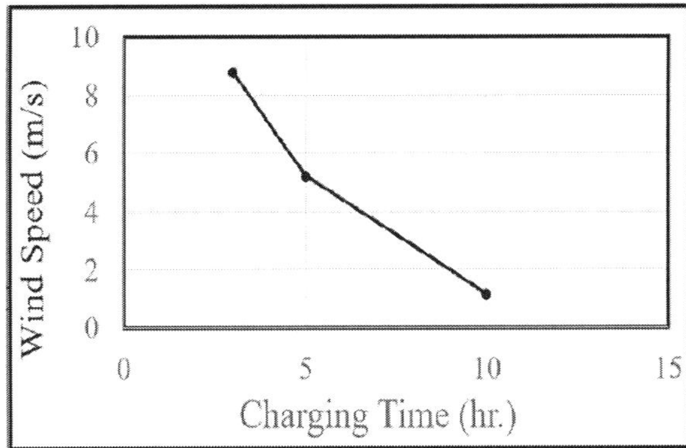

Figure 79.7 Wind speed vs. charging time of battery

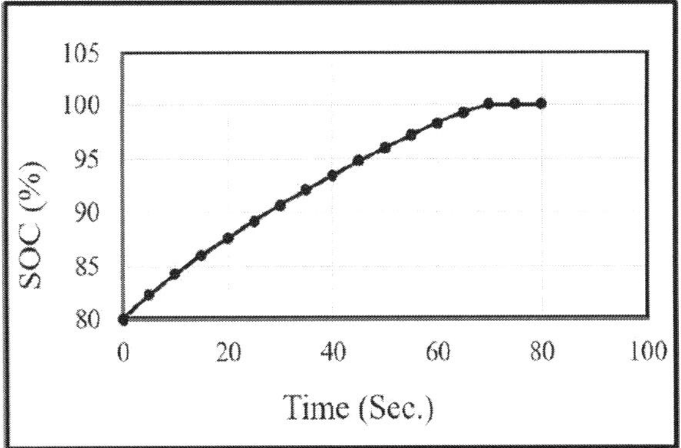

Figure 79.8 % SOC vs. time of wind-based charging of battery

100% charge within 80 seconds. This depiction of charging progression offers a concise yet insightful overview, highlighting the battery's rapid response and efficient utilization of wind-generated power during the simulated timeframe.

In Figure 79.9, the graph illustrates the discharging voltage vs. time of the battery, considering varying load currents. Examining the characteristics reveals a significant impact: at an 8A load, the battery discharges over 12.2 hours, while at 35 A, the discharge time dramatically reduces to 2.5 hours. This underscores a crucial observation – as load current increases, the battery experiences a rapid depletion. The clear relationship between load current and discharge time provides essential insights for managing and optimizing the battery's performance under different usage scenarios, aiding in the development of strategies to enhance overall system efficiency and longevity.

4. Conclusions

This paper proposes a PMSG-BSSSAWTS for a battery charging system designed for rural applications. For this purpose, a PMSG is chosen over a

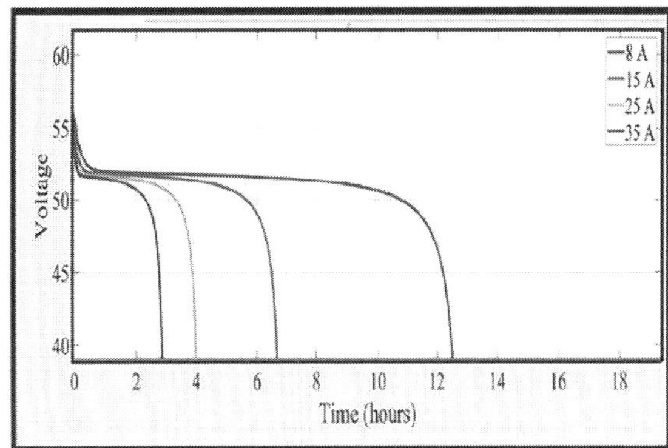

Figure 79.9 Discharging battery voltage vs. time

conventional generator because of its several advantages, such as high efficiency, eco-friendliness, and no requirement for rotor winding. To evaluate the system's performance, the proposed model is developed within the Simulink/MATLAB environment. The performance evaluation shows that the line voltage developed with wind speed increases rapidly, reaching 92.6 volts at a wind speed of 5 m/sec. Similarly, the rectified voltage reaches 105.3 volts at 5 m/sec. The charging time also reduces with the increase in wind speed, while battery discharging time increases at reduced load current. Based on these performances, it can be inferred that the suggested battery charging system is better suited for rural applications.

References

1. Paulin-Bessette, T., Ninad, N., and Lopes, L. A. C. (2023). PV and wind utilization analysis for a Canadian small arctic PV-wind-diesel hybrid microgrid. *2023 IEEE 14th Inter. Conf. Power Elec. Drive Sys. (PEDS)*, 1–6. 10.1109/PEDS57185.2023.10246606.
2. Kumar, R. R., et al. (2023). Performance analysis of dual stator six-phase embedded-pole permanent magnet synchronous motor for electric vehicle application. *IET Elec. Sys. Trans.*, 1–13. https://doi.org/10.1049/els2.12063.
3. Tumminelli, F., Scozzaro, M., Nevelson, C., Iones, A., Trapanese, M., and Franzitta, V. (2023). Simulations of an off-shore WEC: Analysis of the potential energy producibility in the Mediterranean Sea. OCEANS 2023 - Limerick, Limerick, Ireland, 1–6. 10.1109/OCEANSLimerick52467.2023.10244280.
4. Mahmoud, M. M., Ratib, M. K., J. R. I, Swaminathan, J., Aly, M. M., and Abdel-Rahim, A.-M. M. (2021). Application of grey Wolf optimization for PMSG-based WECS under different operating conditions: Performance assessment. *2021 Innov. Power Adv. Comput. Technol. (i-PACT)*, 1–7, 10.1109/i-PACT52855.2021.9696881.

80 PWM generation and regulation of DC to DC converter using fuzzy logic controller

Radak Blange[a] *and A. K. Singh*

Department of Electrical Engineering, North Eastern Regional Institute of Science & Technology (NERIST), Nirjuli. Arunachal Pradesh, India

Abstract

DC–DC converter plays prominent roles today in generating smooth switching of the electrical and electronic circuit switches. The ON-OFF switching of the gates of the converters is electronically triggered by generation of pulse width modulation (PWM) either through controllers such as conventional proportional-integral-derivative (PID) controller or fuzzy logic controller using suitable algorithms. In this study a dc/dc buck/boost converter has been used for regulation of output-voltage using fuzzy logic controller. With the appropriate fuzzy logic based algorithms the required PWM and its duty cycle have been smoothly and successfully generated with the help of which the required output voltage of the converter circuit for both the buck and boost mode have been achieved. The scheme has the vast scope of implementing using DSP/PIC microcontroller future for the smooth turning ON-OFF with high switching frequency which will be suitable regulation of solar PV output voltage.

Keywords: Fuzzy Logic Controllers, pulse-width modulation, Proportional-Integral-Derivatives

I. Introduction

With the use of suitable controlling technique, dc/dc converter has high potential of being used as switching regulator to regulate the desired DC output-voltage by use of pulse-width-modulation to control the way of switching to metal oxide field effect transistor (MOSFET) of Buck/Boost circuit. By altering the duty-cycle (D) of the MOSFET with appropriate switching-frequency, the output-voltage (V_0) across the load can be increased or decreased depending upon our requirement [1. 2]. Using a particular reference voltage, the output voltage has been compared thereby making the DC/DC buck-boost converter closed loop as it has high performance efficient than that of open loop. With the feedback of output voltage, the uncertainties and disturbances occurred in power semiconductors leading to fluctuating output is regulated by controlling the PWM. Such type of scheme can be easily implementing using hardware of DSP processor for the proper ON-OFF of MOSFET with high switching frequency for the regulation of solar PV output voltage. There are various algorithms to generated PWM to control the converter and these techniques include PID control. However, the PID controller is linear one due to which it finds difficult for controlling the non-linear signals occurs in the DC-DC converter [3, 4]. Fuzzy-logic-controller (FLC) being a nonlinearity in controlling is most commonly applied techniques for nonlinear applications particularly in photovoltaic (PV) controlling and is much better than PID controller. Fuzzy logic controllers do not require precise information about the system but it works based on approximation in modeling. FLC is flexible and robust one as more and more new fuzzy rules can be supplemented /added onto it. Hence fuzzy logic control provides smooth and enhanced switching pulse control which in turns controls the output of the converter which is more robust than other controls techniques. [5–10]. The fuzzy logic controller with appropriate fuzzy rules coupled with suitable method for diffuzification is nonlinear but easy and faster that too without going through details of the system parameters as that of other methods involves which includes PID controller associating with complicated circuit parameters of system like photovoltaic cell etc.

In this paper, Fuzzy logic controller with control rules of fuzzy based on Mamdani type with center of gravity for defuzification have been used for achieving the required PWM so that duty cycle of for keeping ON-OFF to MOSFET is easily made for the smooth regulation of the voltage-output of DC converter for a value of reference-voltage (V_{ref}). The DC-DC buck boost converter having a static loads has been provided in section 2 and 5 Membership-functions (MF) and a fuzzy-rule Table have been discussed in the Section 3 [7, 8]. Hence unlike linearity in conventional controller finding difficult for handling non-linear signals,

[a]radakblange@gmail.com

DOI: 10.1201/9781003540199-80

Fuzzy-logic-controller (FLC) due to its nonlinearity in nature is preferably accepted applied techniques in controlling the nonlinear uses and have bright scope of doing further researches in fuzzy world. For convenient this paper has been organized into five sections ahead namely Section 1–5, respectively.

II. DC–DC Buck-Boost converter

The DC-DC converter-model having the parameters as provided in Table 80.1 have been used for the purpose of simulation in Sim-Power-System-tool box of the Matlab library. The converter has input-voltage-source (V_{in}) a MOSFET-M, inductor-L, diode-D_1 a capacitor-C for filtering and a load-resistance (R_0). When M is turned On, the input/voltage (V_{in}) linked to L. The D being in its reverse biased mode does not conduct. Hence the voltage drop across the element L is equal to input voltage.

$$V_{in} = V_L \tag{1}$$

Table 80.1 Details of the converter

Buck-Boost converter	
Input power	55 W
Output power	52.3 W
Min. input voltage	12 V
Max input voltage	14 V
Output voltage	12 V
Output current	4.35 A
Load resistance	2.75 Ω
Switching frequency	10 khz;
Diode volt drop	0.6 V
Min inductor (L) at D=0.5	34.5 μH
Capacitor (C)	145 μF,
Initial voltage of capacitor	-2 V

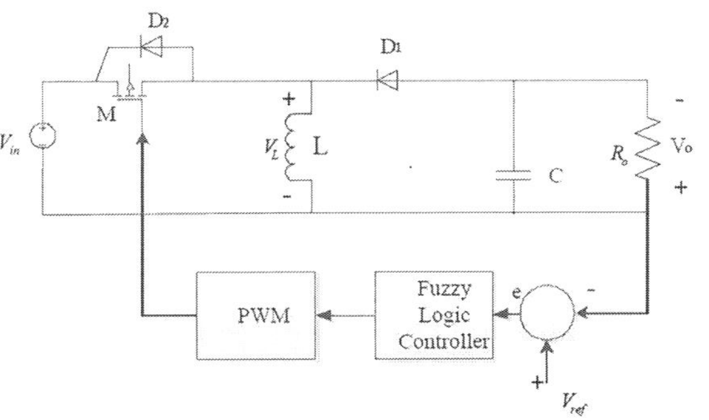

Figure 80.1 DC-DC buck-boost converter with static load

Thereafter, the current flowing the inductor (L) starts increasing and hence the capacitor (C) produces output voltage (V_0) across the load (R_0). The moment switch (M) is turned-OFF, Then the diode (D) is in forward-biased and the D_1 start conducting and thereby provides a complete circuit for the flow of current through the inductor (L) [11–18]. Then voltage (V_1) across the inductor(L) is equal to output voltage.

$$V_L = V_o \tag{2}$$

The change in the inductor (L) current should be zero over one-switching complete cycle. Hence the voltage-second balance equation as

$$V_{in}D + V_o(1-D) = 0 \tag{3}$$

D being the duty ratio expressed as $D = \frac{T_{on}}{T_s}$, T_{on} being ON-STATE-time of M and T_s is the switching-time-period. Using Equation (3), we got

$$V_o = -\frac{D}{1-D}V_{in} \tag{4}$$

Hence, the Equations 1 and 2 indicates the energy being transferred from L to R_0 via C. Figure 80.1 shows the closed loop DC/DC converter with fuzzy logic block. The V_0 and V_{ref} are compared for the generation of error-signal. Now comparing these signals with a given saw tooth signal, required PWM which drives the switching of MOSFET. For the purpose of simulations, the required parameters of DC/DC converter being used were taken in Table 80.1 and converter is shown in Figure 80.1.

III. Fuzzy logic controller

The FLC have been used for PWM generation in concurrent with the tracking of corresponding response in a given reference value of voltage. The indicative block-diagram (BD) for FLC with the converter is as provided at Figure 80.2 [10].

The desired output-voltage of buck/boost converter is achieved by controlling the duty cycle using fuzzy algorithms. Here error means the difference of reference-voltage (V_{ref}) and the out-put voltage (V_0). If is V_{ref} and y(k) is *Vo* values then the error of voltage signal is expressed as follows:

$$e(k)=r(k)-y(k) \tag{5}$$

The change in error of voltage signal (Δe) is also given by

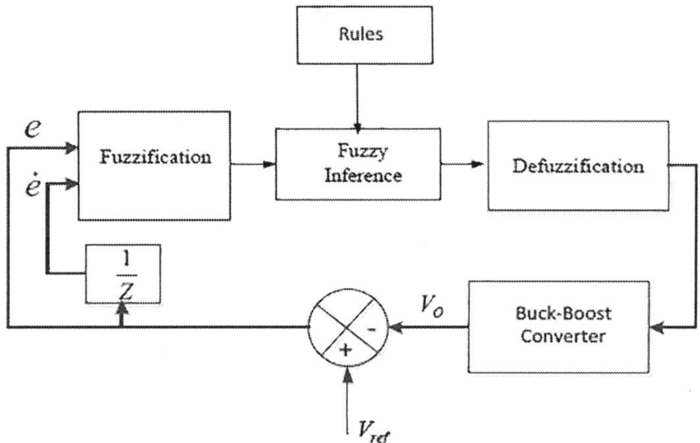

Figure 80.2 BD of fuzzy logic controller

Table 80.2 Fuzzy rules-Mamdani fuzzy rules with center of gravity method for defuzzification

Δe \ e	PB	PS	Z	NS	NB
NB	Z	NL	NL	NH	NH
NS	PL	Z	NL	NL	NH
Z	PL	PL	Z	NL	NL
PS	PH	PL	PL	Z	NL
PB	PH	PH	PL	PL	Z

$$\Delta e = e(k) - e(k-1) \tag{6}$$

In this fuzzy techniques, first the error and the change in error are made compatible for generation of fuzzy sets in the fuzzifier. Duty cycle signals output and resultant fuzzy sets is then converted into the corresponding CRISP values having highest degree of membership function (MF) being 1.0 in the fuzzy sub-sets. The input membership functions being NB-Negative Big, NS-Negative Small, Z means Zero, PS – Positive Small, PB – Positive Big and the output membership functions being NH means negative high, NL means negative low, Z means zero, PL means positive low and PH positive High respectively. The MF of Input and Output fuzzy-variables are provided in Figures 80.3– 80. 80.5, respectively.

The FLC flowchart and control rules of fuzzy algorithms provided in Figure 80.6 and Table 80.2 of Mamdani fuzzy rules with adoption of centroid method for defuzzification is adopted to determine the output (duty-cycle). The generated PWM are provided in Figures 80.7– 80. 80.9.

IV. Results and discussions

(a) For Buck-mode: Taking *Vref* = 13 V and *Vin* = 14 V the output-voltage (V_0) is calculated. Rise time (T_r), peaktime (T_p), peak overshoot (M_p) and settling time (τ_s) the simulation-result of output, the output-voltage (V_0) of the converter using FLC shown in Figure 80.10.

Results: T_r= 0.0011s, T_p= 0.00175 s, M_p = 0.1V, τ_s= 0.005 s.

Discussions: Hence the output voltage have been successfully tracked as per reference voltage of 13 volt and thereby buck mode of the converter is regulated.

(**b**) For Boost-mode: Taking *Vref* = 29 V and *Vin* = 14 V the output-voltage (V_0) is calculated. The

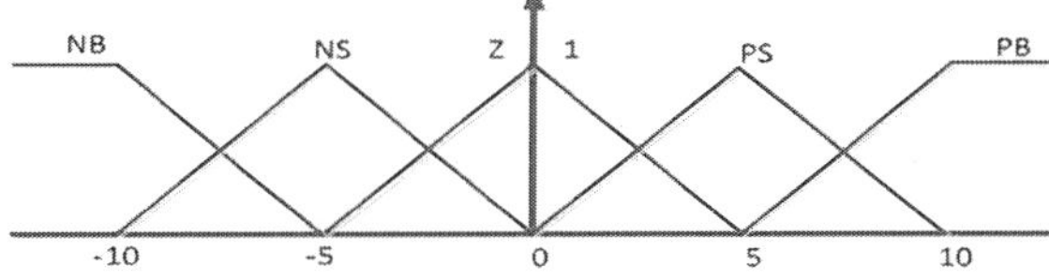

Figure 80.3 MF of error

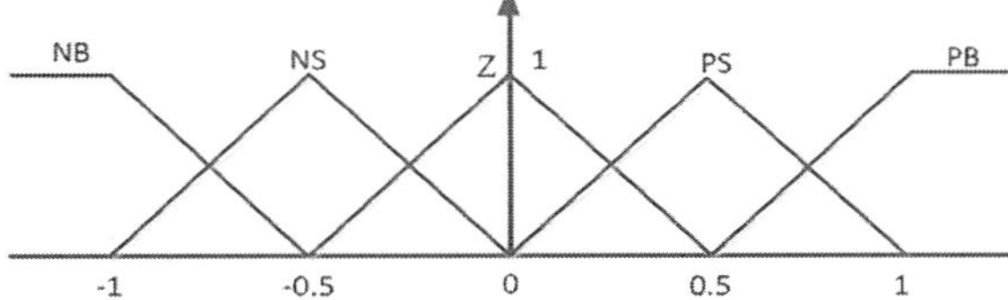

Figure 80.4 MF of change in error

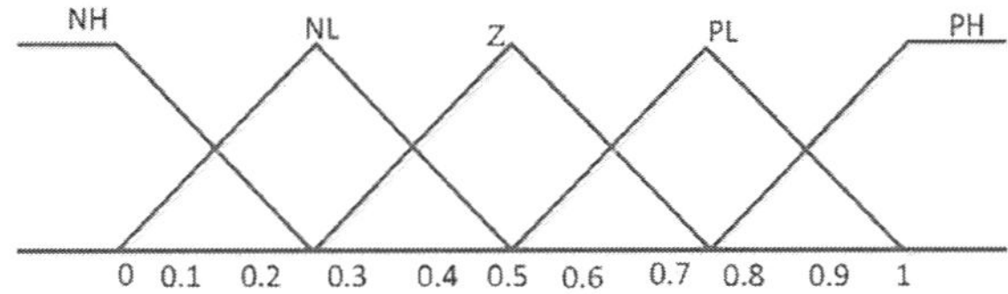

Figure 80.5 Membership function (MF) of output-duty cycle

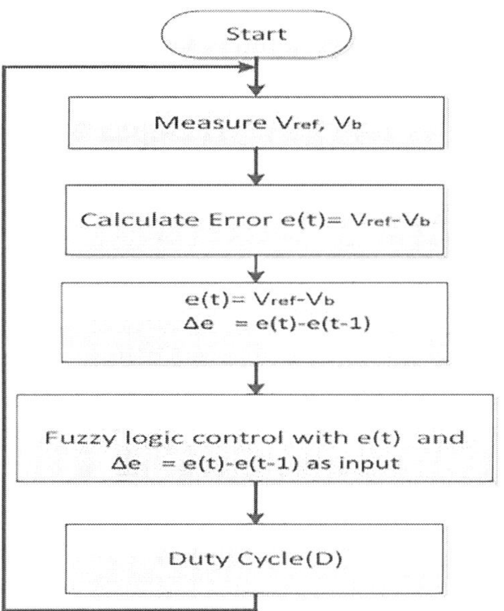

Figure 80.6 FLC algorithm flow chart

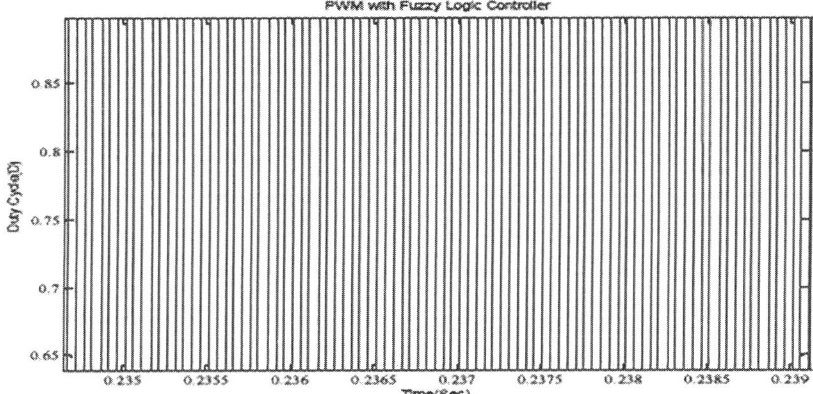

Figure 80.7 PWM output FLC

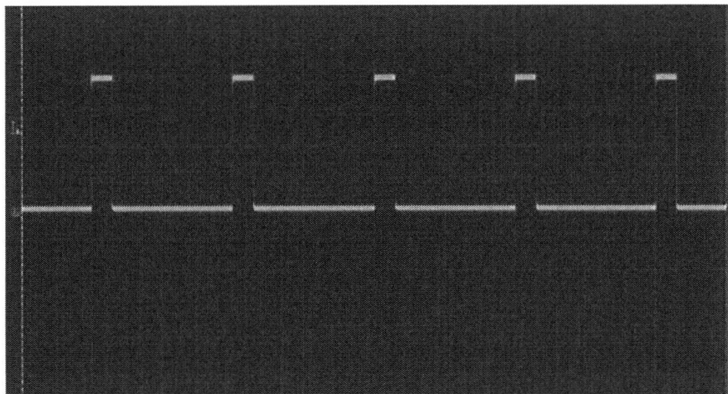

Figure 80.8 PWM output with fuzzy logic controller at 14.8% duty cycle

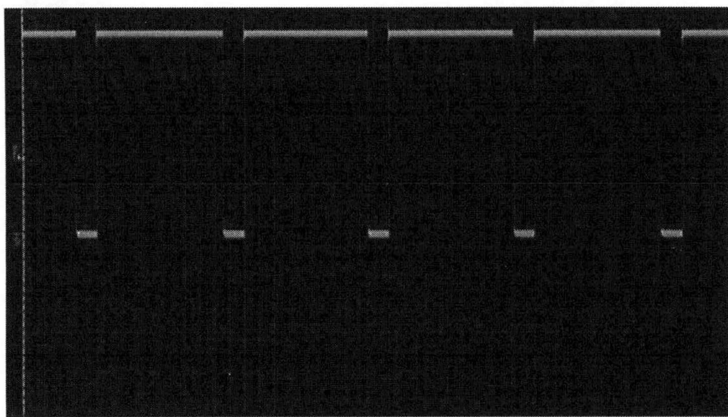

Figure 80.9 PWM output with fuzzy logic controller at 85.8% duty cycle

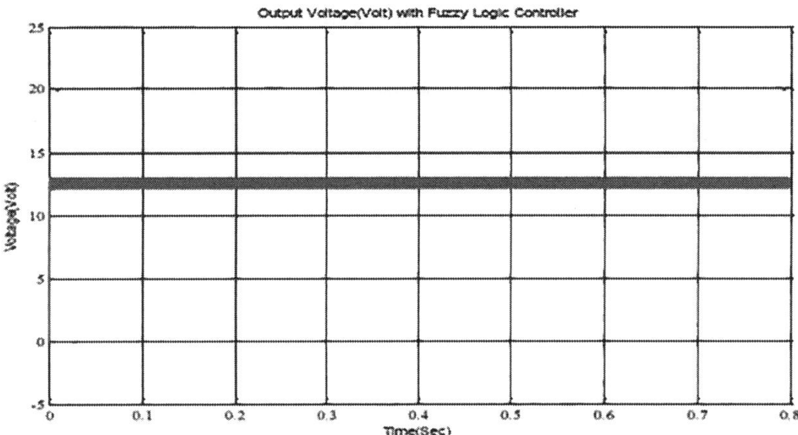

Figure 80.10 Output-voltage with Vref =12 V

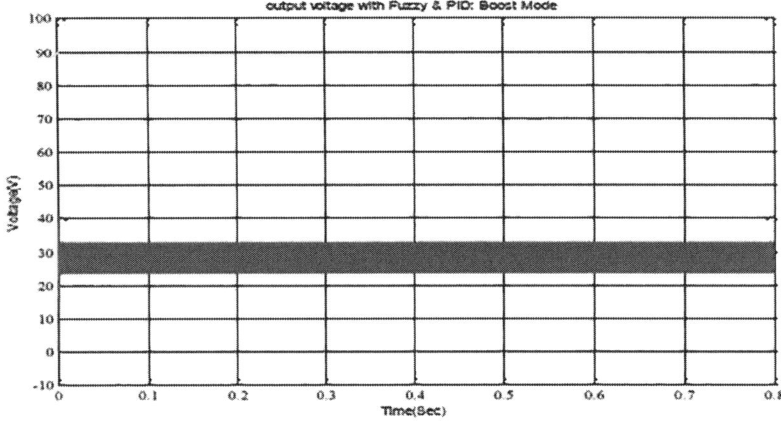

Figure 80.11 Output-voltage with Vref =29 V

simulation-result of voltage output of the converter is provided at Figure 80.11.

The result: T_r=0.0018s, T_p=0.00205s, M_p= 10.5V, τ_s = 0.0032s.

Discussion: Hence the output voltage have been successfully tracked as per reference voltage of 29 volt and thereby output voltage for boost mode of the converter is regulated (Figure 80.12).

V. Conclusions

In this study, Buck-Boost DC/DC Converter has been used for the generation of PWM and the regulation of output voltage thereof. A fuzzy logic controller based algorithm have been applied to achieve PWM and duty cycle for proper switching pulse of gating of the DC/DC Buck/Boost converter with a resistor. The modeling and simulations were carried out on Sim-Power-System and Mamdani type fuzzy based algorithm was implemented on Fuzzy-Tool-Box on Matlab/Simulink platform. With the appropriate fuzzy logic based algorithms, smooth PWM, duty cycle for different duty cycle of the getting signal and the output voltage have been successfully achieved. From the results it is observed that there is smooth generation of PWM with which the response of output voltage of the converter is tracked successfully as per given reference voltage with faster settling time having less overshooting and ripple. Hence the scheme with fuzzy logic controller technique is easy, faster and can be very useful for regulation of output voltage in an application where the input voltage changes besides having the vast scope of implementing on DSP to regulate output voltage solar PV. At the same time FLC is very quick and useful for system of nonlinear without going the complicated system parameters complicated circuit parameters due to which the fuzzy technique has been adopted.

Acknowledgement

We gratefully acknowledge the students, staff, and authority of electrical engineering department for their cooperation in the research.

References

1. Ceelho, R. F., Concer, F., and Martins, D. C. (2009). A study of the basic of DC-DC converters applied in maximum power point tracking power. *Electronics*, 673–678.
2. Chihchiang, H. and Shen, C. (1998). Study of maximum power tracking techniques and control of DC/DC converters for photovoltaic power system. *Power Elec. Spec. Conf.*, 1, 86.
3. Stokes, J. and Sohie, G. R. (n.d.). Implementation of PID ontrollers on the Motorola DSP56000/DSP56001. 1–84.
4. Coelho, R. F., Concer, F. M., and Martins, D. C. (2010). Analytical and experimental analysis of DC-DC converters in photovoltaic maximum Power Point Tracking applications. *IECON 2010 - 36th Ann. Conf. IEEE Indus. Elec. Soc.*, 2778–2783.
5. Altas, H. and Sharaf, A. M. (2007). A generalized direct approach for designing fuzzy logic controllers in Matlab/Simulink GUI Environment. *Inter. J. Inform. Technol. Intell. Comput.*, 1(4), 1–27.
6. Radak, B., Mahanta, C., and Gogoi A. K. (2019). DSP based photovoltaic hardware development for controlling voltage and current. *IEEE 3rd Inter. Conf. Recent Dev. Control Autom. Power Engg.* 1–104.
7. Radak, B., Mahanta, C., and Gogoi A. K. (2015). MPPT of solar photovoltaic cell using perturb & observe and fuzzy logic controller algorithm for Buck-Boost DC-DC converter. *Inter. Conf. Ener. Power Environ. Towards Sutain. Growth (ICEPE)*. 87–92.
8. Wu, Y., Zhang, B., Lu, J., Du, K. L. (2011). Fuzzy logic and neuro-fuzzy system: A systematic introduction. *Int. J. Artif. Intell. Expert Sys. (IJA)*, 47–80.
9. Sharaf, M. and Sahin, M. E. (2011). A novel photovoltaic PV-powered battery charging scheme for electric vehicles. *ICEAS IEEE Conf., India, 2011*. 176–192.
10. Reshmi, R. and Babu, S. (2013). Design and control of DC-DC converter using hybrid fuzzy PI controller. *Inter. J. Res. Engg. Technol.*, 1(3), 73–78.
11. Mohan, N., Undulant, T. M., and Robbins, W. P. (1995). Power Electronics, John Wiley. 1–820.
12. Kuo, B. C. (1995). Automatic control systems. New Jersey: Prentice-Hall. 1–189.
13. Masoum, M. A., Dehbonei, H., Fuchs, E. F. (2002). Theoretical and experimental analyses of photovoltaic systems with voltage and current-based maximum power-point tracking. *IEEE Trans. Ener. Conver.*, 17(4), 514–522.
14. Hansen, D., Sørenson, P., Hansen, L. H., and Bindner, H. (2000). Models for a stand-alone PV system. Risø National Laboratory, Roskilde. 231–245.
15. Nevruzov, V. (2010). Computer controlled solar house for measurement Rize city solar energy potential. *NURER, Ankara*. 404–414.
16. Vijayalakshmil, S., Arthika, E., Shanmuga Priya, G. (2015). Modeling and simulation of interleaved BuckBoost converter with PID controller. *IEEE Sponsored 9th Inter. Conf. Intell. Sys. Control (ISCO)*. 231–243.
17. Kurokawa, F. and Ishibashi, T. (2009). Dynamic characteristics of digitally controlled buck-boost DC-DC converter. *Power Elec. Drive Sys.*, 300–303.
18. Chao, P. C.-P., Wei-Dar, C. and Chih-Kuo, C. (2012). Maximum power tracking of a generic photovoltaic system via a fuzzy controller and a two-stage DC–DC converter. *Microsys. Technol.*, 18, 1267–1281.

Printed in the United States
by Baker & Taylor Publisher Services